Magnetic Susceptibility of Superconductors and Other Spin Systems

Magnetic Susceptibility of Superconductors and Other Spin Systems

Edited by

Robert A. Hein

University of Washington
Seattle, Washington

Thomas L. Francavilla

U.S. Naval Research Laboratory
Washington, D.C.

and

Donald H. Liebenberg

Office of Naval Research
Arlington, Virgnia

Springer Science+Business Media, LLC

Library of Congress Cataloging-in-Publication Data

Magnetic susceptibility of superconductors and other spin systems
edited by Robert A. Hein, Thomas L. Francavilla, and Donald H.
Liebenberg.
p. cm.
"Proceedings of the Office of Naval Research Workshop on Magnetic
Susceptibility of Superconductors and Other Spin Systems, held May
20-23, 1991, in Coolfont, West Virginia"--T.p. verso.
Includes bibliographical references and index.
ISBN 978-1-4899-2381-3
1. Superconductors--Magnetic properties--Congresses. 2. Magnetic
susceptibility--Measurement--Congresses. I. Hein, Robert A.
II. Francavilla, Thomas L. III. Liebenberg, D. H. IV. United
States. Office of Naval Research. V. Office of Naval Research
Workshop on Magnetic Susceptibility of Superconductors and Other
Spin Systems (1991 : Coolfont, W. Va.)
QC611.97.M34M34 1991
537.6'236--dc20 92-12566
CIP

Proceedings of the Office of Naval Research Workshop on
Magnetic Susceptibility of Superconductors and Other Spin Systems,
held May 20–23, 1991, in Coolfont, Berkeley Springs, West Virginia

ISBN 978-1-4899-2381-3 ISBN 978-1-4899-2379-0 (eBook)
DOI 10.1007/978-1-4899-2379-0

Preface

The workshop entitled Magnetic Susceptibility of Superconductors and other Spin Systems (S^4) was held at Coolfont Resort and Health Spa, located near Berkley Springs West Virginia on May 20-23, 1991. There were over sixty attendees, approximately half from the United States, the remainder representing over twelve different countries. The international character of the workshop may be gleaned form the attendee list, included in this volume.

The intent of the workshop was to bring together those experimentalists and theoreticians whose efforts have resulted in significant recent contributions to the development and use of the ac susceptibility technique as well as to the interpretation of data obtained from these measurements. Many spirited discussions occurred during and after the presentations. These are reflected in the manuscripts contained in these proceedings. Although camera ready manuscripts were required from all participants at registration, all manuscripts were revised and reflect the lively exchanges that followed each presentation. The small size of the workshop allowed the participants a high degree of flexibility. Consequently when a controversial topic such as "the irreversibility line" emerged, a special session was organized on the spot. At the suggestion of Ron Goldfarb, participants were invited to contribute a one page summary containing their thoughts on the topic. These stand alone contributions were retyped and included as submitted, with only minor editorial changes.

These proceedings are intended for those experienced scientists new to the field and graduate students just beginning their research. We have all at one time or another experienced frustration in trying to follow the detailed arguments in a paper. Frequently terms are not defined, and crucial steps are omitted which are familiar to the author but not to the uninitiated experienced scientist or graduate student beginning his or her research. Thus in our initial contacts with the invited speakers, it was emphasized that it was our perception that many articles appearing in the scientific literature lack sufficient detail, be it experimental or theoretical, to allow the working scientist to readily evaluate the relative merit or correctness of a given experimental result or theoretical model. Therefore we stressed that manuscripts to be published in the workshop proceedings should contain sufficient experimental and theoretical detail so as to overcome our perceived shortcomings of the current literature. Quoting from a letter sent to the attendees by the organizing committee: Referring the readers to unpublished works for details is "streng Verboten", i.e. not allowed. Also in place of the "it is easily shown that - ", theorists were urged to take a few paragraphs and show us just how easy it is to do whatever it is that "is easily shown". These same principles applied to the contributed articles. In general everyone cooperated and the results of their efforts are contained in these proceedings.

Because of the emphasis on basics and fundamentals, some repetition was unavoidable. We believe that this will provide access to alternative derivations and discussions that should aid in the communication, and as a result most contributions can "stand alone" thus avoiding much cross referencing.

Without the financial support of the Office of Naval Research and DARPA, as well as the help and cooperation of the University of Washington and the Naval Research Laboratory, this workshop would not have been possible. We also would like to acknowledge at this time, the support of the Coolfont Staff. Finally we would like to thank the conferees for their patience and cooperation that made the organization, running of the workshop, and the preparation of this book a pleasant and enjoyable venture.

Robert A. Hein
Thomas L. Francavilla
Donald H. Liebenberg

CONTENTS

SECTION 3 - BULK AND SINGLE CRYSTAL SAMPLES

SECTION 4 - THIN FILMS

PROLOGUE

R. A. Hein

University of Washington
Seattle, WA

INTRODUCTION

The discovery, in 1987, of superconductivity in the $YBa_2Cu_3O_{7-\delta}$ system with transition temperatures in the vicinity of 90K, has led to the wide spread use of ac induction techniques to characterize the superconducting parameters of this new class of high T_c superconductors. These workshop proceedings document, in considerable detail, the experimental techniques and the theoretical models and concepts that have evolved during the past four years to account for the magnetic and electrical properties of these granular superconductors. To appreciate some points of controversy discussed by the participants in the workshop, one should be aware of a few historical facts associated with ac magnetic susceptibility measurements.

For 22 years between the discovery of superconductivity and the experiment of Meissner and Ochsenfeld[1], who used a small flip coil and a ballistic galvanometer or flux meter to probe the magnetic field around a superconducting sphere in the presence of an external dc magnetic field, the magnetic properties of superconductors were regarded as uninteresting. It was obvious that these properties follow from Maxwell's equations in the limit of infinite electrical conductivity, hence why waste precious liquid helium to check the obvious? The discovery by Meissner and Ochsenfeld that superconductors possess unique magnetic properties, not derivable from the infinite electrical conductivity aspect, led to the development of ac and dc mutual induction techniques to search for new superconductors and to study the magnetic properties of known superconductors.

Kurti and Simon[2] used the dc ballistic inductance method to discover superconductivity in Cd and Zn with T_c values of 0.54K and 0.87K respectively. Shoenberg[3] used an ac self inductance bridge in his study of the intermediate state in Sn (T_c = 3.72K) and Daunt[4] used an ac mutual inductance technique to investigate ac shielding effects. In the 50 years which have preceded the discovery of high T_c superconductors, there has been much discussion about the interpretation of magnetic susceptibility data obtained by induction type measurements. A few selected highlights will serve to give the reader an appreciation for difficulties encountered in the interpretation of the data.

BULK VERSUS NONBULK RESPONSE

It appears that until the 1960s, the dc ballistic mutual inductance technique was preferred over ac techniques because of complications associated with eddy current losses in the intermediate state[5] and because of the relative simple experimental requirements; one needs only a battery, coil system and a ballistic galvanometer. In addition, one can readily measure the total magnetic moment of the sample by simply moving the sample from the center of one secondary to the center of the other in the presence of a dc magnetic field as the resulting deflection or "throw" of the ballistic galvanometer is proportional to twice the sample's magnetic moment, M. Thus in a given experiment one can measure M in the presence of a dc magnetic field as well as $\Delta M/\Delta H$; where ΔM is the change in M caused by the application of the incremental dc magnetic field, ΔH. Early popularity of the dc ballistic inductance measurements also stemmed from the belief that these responses reflect bulk properties of the sample. Consequently, one could use magnetic measurements of the critical magnetic field curve of bulk samples to derive thermodynamic quantities, i.e. the specific heat jump, $\Delta C(T=T_c)$, the electronic specific heat coefficient, γ, etc. Such calculations were routinely done even though R. P. Hudson[6] had shown that the superconducting "bulk" transition in PbTe does not occur when the sample was in the form of a powder. This experiment clearly demonstrated that small amounts of superconducting impurities, in this case free Pb, can give rise to magnetic responses that mimic a "bulk" response.

In 1949, Daunt and Heer[7] used the dc ballistic method to measure the critical magnetic field curve of Zn chips imbedded in a paramagnetic salt pill used to produced the required low temperatures. They reported the observation of an "excessive paramagnetism" as the sample warmed in the presence of an applied dc magnetic field. This feature was attributed to the formation of multiply connected superconducting regions by the Zn chips. This interpretation was cast into doubt by the data of Steele and Hein[8] on Cd. These workers at the Naval Research Laboratory, Washington, D.C. were new to the field and were part time graduate students. Their results, shown in Figure 1 were obtained on a chemically pure, annealed cylinder of Cd, hence the mutiply-connected region argument of Daunt and Heer seemed inappropriate. Clearly what one is seeing is the response of a reversible intermediate state. Shoenberg had shown, in 1937, that the positive, i.e. paramagnetic, dM/dh of the intermediate state is readily observed in the behavior of the real part, $\chi'_H(T)$, of the complex ac magnetic susceptibility. Thus Steele[9] pointed out that the "excessive" paramagnetism observed by Daunt and Heer need not imply multiply-connected regions. How this comes about as a result of the Meissner-Ochsenfeld effect is shown with the aid of Figure 2. Figure 2a consists of a series of isothermal magnetization curves appropriate for a spherical sample exhibiting the Meissner-Ochsenfeld effect. In this case the magnetization curves are thermodynamically reversible and M(T) attains its maximum diamagnetic moment of $(-1/4\pi)H_c(T)$ when the applied dc magnetic field, H^{dc} is equal to $(1-N)H_c(T)$. Here N is the demagnetization factor which equals 1/3 in the case of a sphere and $H_c(T)$ is the thermodynamical critical magnetic field. There are two basic types of susceptibility measurements employed in the search for, and study of, superconductors: (a) the incremental dc or ac measuring field is the only magnetic field acting on the sample and (b) the measuring field is a superposition on an applied dc magnetic field. In the latter case, this discussion will be restricted to the situation where the measuring field and H^{dc} are collinear in direction and $H^{dc}>\Delta H$.

Case (a). This is the one usually employed in the search for new superconductors where one observes $\chi_0(T) = (\Delta M/\Delta H)_{H^{dc}=0}$ as a function of temperature: here ΔH is the incremental dc field applied in the dc ballistic technique. In the ac techniques one is concerned with (dM/dh) where h is the ac magnetic field used in the measurement, sometimes referred to as the excitation field. If the sample is cooled in zero applied magnetic field and $\chi_0(T)$ is measured as the sample warms from $T<<T_c$, the initial value of $\chi_0(T)$ is $[1/(1-N)](-1/4\pi)$ which for a sphere will be $-3/8\pi$, see Figure 2b. As the sample warms, $\chi_0(T)$ is a constant until the temperature T_{in} is attained at which temperature the perfect shielding property of the superconducting state breaks down as the sample begins to enter the intermediate state. T_{in} is a function of the measuring field and sample's demagnetization factor. It is given by $\Delta H = (1-N)H_c(T_{in})$. With further increase in T, $\chi_0(T)$ decreases in magnitude and attains the value appropriate to the normally conducting state of the material, usually zero, at a temperature T_{com} which is a function of ΔH alone and is given by $\Delta H = H_c(T_{com})$. In this case the transition width $T_{com}-T_{in}$ is governed by

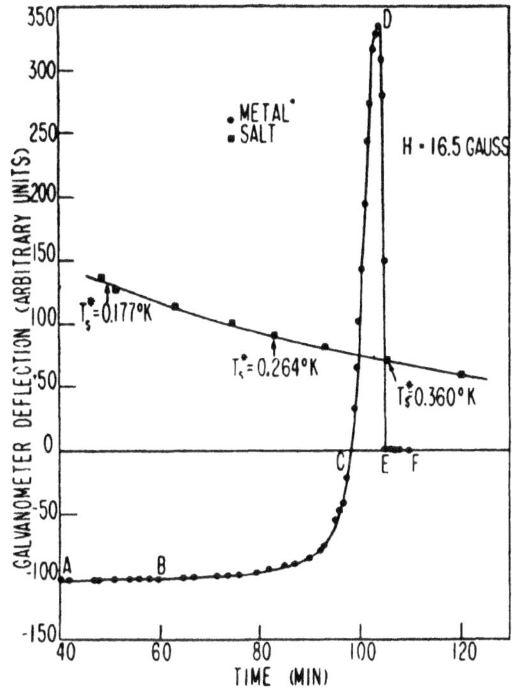

Figure 1. The incremental magnetic susceptibility $\chi_H(T) = (\Delta M/\Delta H)_{Hdc} \neq 0$ of a cylindrical shaped sample of cadmium (Cd) as measured by a dc ballistic mutual inductance technique. The galvanometer deflection, proportional to $\chi_H(T)$, has been set equal to zero for the sample in its normally conducting state. For these measurements H^{dc} = 16.5 Oersteds (Gauss) and ΔH was approximately 1.0 Oersteds. The magnetic susceptibility of the paramagnetic salt, included in the figure, was used to determine the temperature of the sample as the system warmed from the low temperatures produced by the adiabatic demagnetization of the paramagnetic salt. The region AB denotes the full diamagnetic shielding state of the superconducting Cd sample while EF denotes its normal state. CE denotes the region of "excessive paramagnetism". (taken from reference 8)

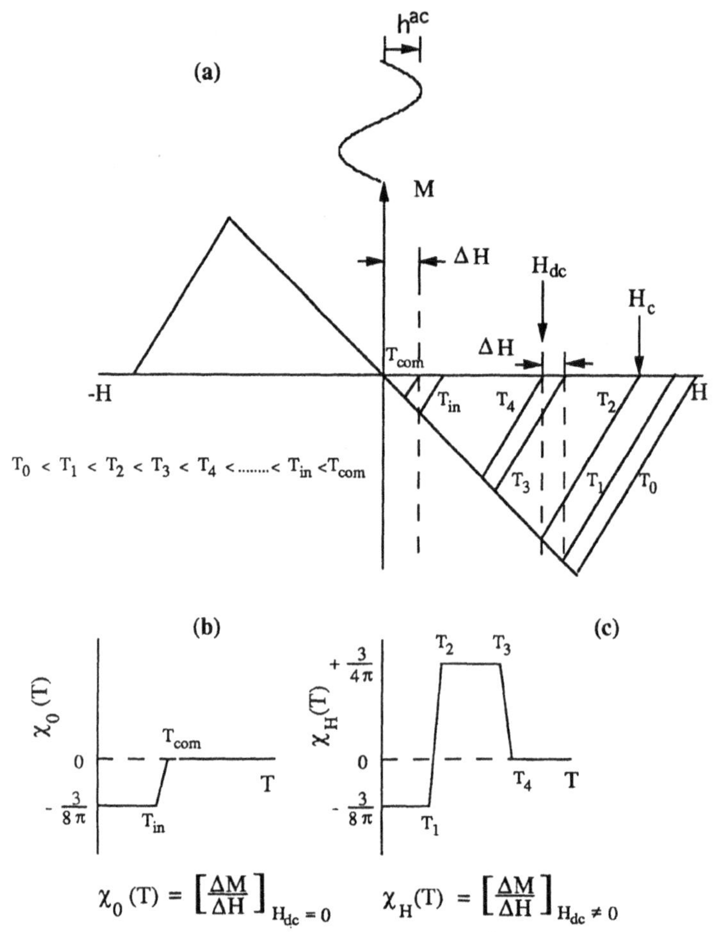

Figure 2. (a) Isothermal magnetization curves, for selected temperatures, modeled for a superconducting sphere, of dimensions large compared to the superconducting penetration depth, which exhibits the Meissner-Ochsenfeld effect. Included in this figure are two fiducial points, or values for the dc magnetic field, H^{dc}, i.e. $H^{dc} = 0$ and $H^{dc} \neq 0$ as well as schematic representations of the magnitudes of ΔH and h^{ac}. (b) Model response for $\chi_0(T)$ for a given ΔH or h^{ac} as a function of temperature. (c) Model response for $\chi_H(T)$ for a given ΔH as a function of temperature showing the differential paramagnetism of the reversible intermediate state. A similar response will be observed with h^{ac} but the definitions of T_1----T_4 will be slightly different due to the ac nature of dh.

N, ΔH and the initial slope of the critical magnetic field curve, $\{dH_c(T)/dT\}_{T=T_c}$. Note that $\chi_0(T)$ is always diamagnetic.

Case (b), here one has, see Figure 2c, a dc magnetic field applied to the sample such that $H^{dc} + \Delta H < (1-N)H_c(T_0)$ where $T_0 << T_c$ is the temperature at which the measurement of $\chi_H(T)$ is initiated. Once again $\chi_H(T)$ starts out at $-3/8\pi$ and remains constant at this value until a temperature T_1 is reached at which temperature $H^{dc} + \Delta H = (1-N)H_c(T_1)$. As the sample warms $\chi_H(T)$ becomes less diamagnetic, passes through zero, and takes on positive values reaching a maximum positive value of $(1/N)(1/4\pi)$ at $T = T_2$ which is given by $H^{dc} = (1-N)H_c(T_2)$ and the maximum positive value, $+3/8\pi$, in the case of a sphere is just twice the magnitude of the full diamagnetic shielding value. Upon further warming $\chi_H(T)$ remains constant at this positive value until a temperature T_3 is attained at which $\chi_H(T)$ starts to decrease. T_3 is given by $H^{dc}+\Delta H = H_c(T_3)$. Note that while $\chi_H(T)$ defined as $M(T)/H$ is always diamagnetic, the incremental or differential susceptibility is positive for T in the interval T_2 to T_3. With further increase in T, $\chi_H(T)$ continues to decrease and attains the value appropriate for the normally conducting state at T_4, where T_4 is given by $H^{dc}=H_c(T_4)$. Note that T_4 is not a function of N, and hence the point at which $\chi_H(T) = 0$ is a measure of $H_c(T)$. Note also that the relative magnitudes of the positive and negative constant levels is a measure of N. Any deviations from this result is an indication of nonideality in the magnetic response of the sample caused by either time effects or magnetic losses.

Figure 3a displays data[10] on an annealed 1.2 cm diameter Sn sphere. The M vs. H and $\Delta M/\Delta H$ vs. H data were obtained by the dc mutual inductance technique whereas the dM/dh vs. H data were obtained with the same coil system using the ac mutual inductance method. The agreement with the model results is viewed as excellent. Deviations from model predictions are presumed to be related to time effects and eddy current losses. Figure 3b displays data obtained on a machined 1.2 cm diameter sphere of Ta. The M vs. H data show considerable hysteresis and no differential paramagnetic effect (DPE) is evident in the ac data, i.e. dM/dh vs. H.

The dc data show a modified DPE in that the first application of ΔH after an increase in H^{dc} results in a positive value for $\Delta M/\Delta H$. The subsequent removal of ΔH and all subsequent applications and removals result in the full diamagnetic value for $\Delta M/\Delta H$. Generally speaking most published reports of $\Delta M/\Delta H$ obtained by the dc technique are based on an average of several on and off readings - thus the "initial" $\Delta M/\Delta H$ value is ignored or lost in the averaging procedure as well as in any ac measurement. For a detailed discussion of how these data relate to the minor hysteresis loop associated with ΔH and the effects which "sweeping" the H^{dc} field has on the ac data, the reader is referred to reference 10.

The above explanation of the "excessive paramagnetism" of Daunt and Heer as being a manifestation of the DPE clearly rules out their "multiply - connected" explanation. The DPE can only be observed if the sample is exhibiting a reversible magnetization curve, i.e. one which has magnetic hysteresis that is small compared to ΔH or dh which are usually in the 10^{-1} to 10^{-3} Oe range.

This explanation raises the question of why earlier workers using the dc mutual inductance technique did not observe a DPE. In fact, Kurti asked Hein this very question. In Kurti's case it was because the incremental field change, $-\Delta H$ to $+\Delta H$, in his technique was larger than the intermediate state interval of his Cd and Zn samples. In the case of

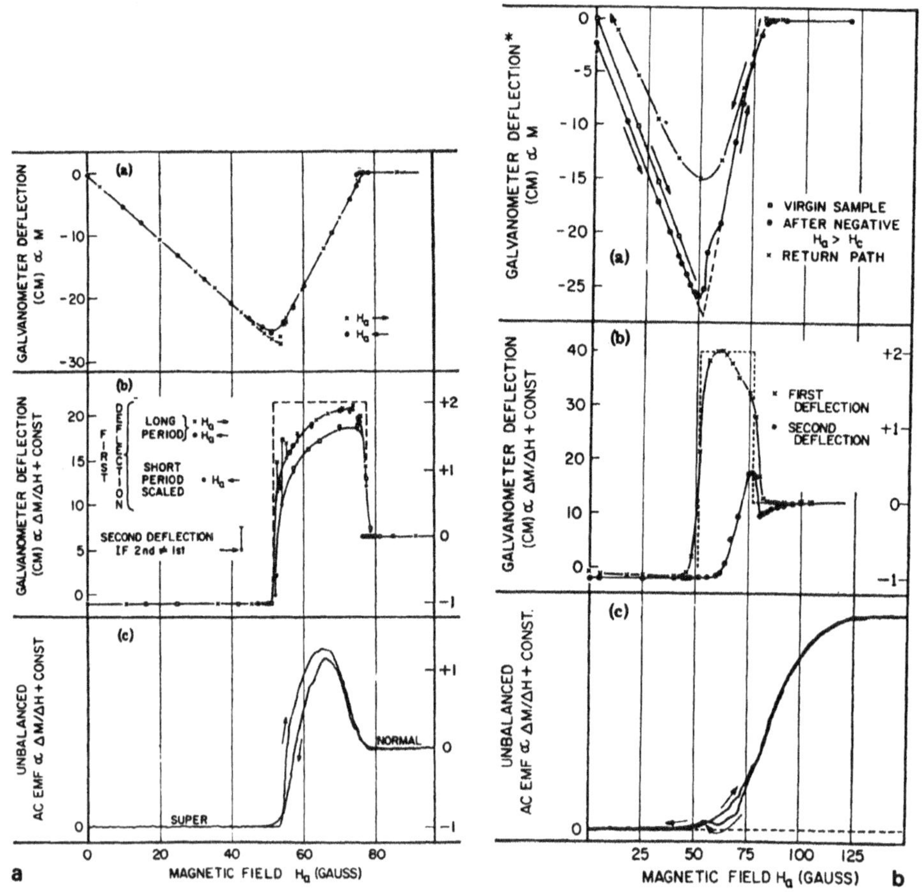

Figure 3. (a) Measured isothermal magnetization curve, M vs. H, of a nominal 1.3 cm diameter machined and annealed sphere of tin along with the corresponding magnetic responses as measured by the dc and ac mutual inductance methods. The temperature of the sample was 3.17 K and the magnitudes of ΔH and dh were both 1.8 Oersteds. The frequency of the ac measuring field was 30 cps. In the dc ballistic technique, galvanometers of two different time constants were employed. Note that in some cases the application, removal and reapplication of ΔH resulted in slightly different values for $\chi_H(3.17K)$. (taken from reference 10) (b) Same type of data as in (a) for a machined but unannealed 1.3 cm diameter sphere of tantalum. Note the large discrepancy in the responses to the first and second application of ΔH. (taken from reference 10)

other workers one presumes that it was a lack of chemical and/or physical purity in the samples, i.e. the magnetization curves were too hysteretic.

The DPE, if of the correct magnitude for the sample shape, allows one to definitely conclude that the sample is exhibiting "bulk" superconductivity. Note that to see this effect the incremental measuring field ΔH must be small compared to the range of field over which the intermediate state exists. The DPE has been observed in most "soft" superconductors. Serin et al.[11] in their study of the isotope effect in Hg used the ac mutual inductance method and observed a DPE. They took the peak in $\chi'_H(T)$ as a measure of

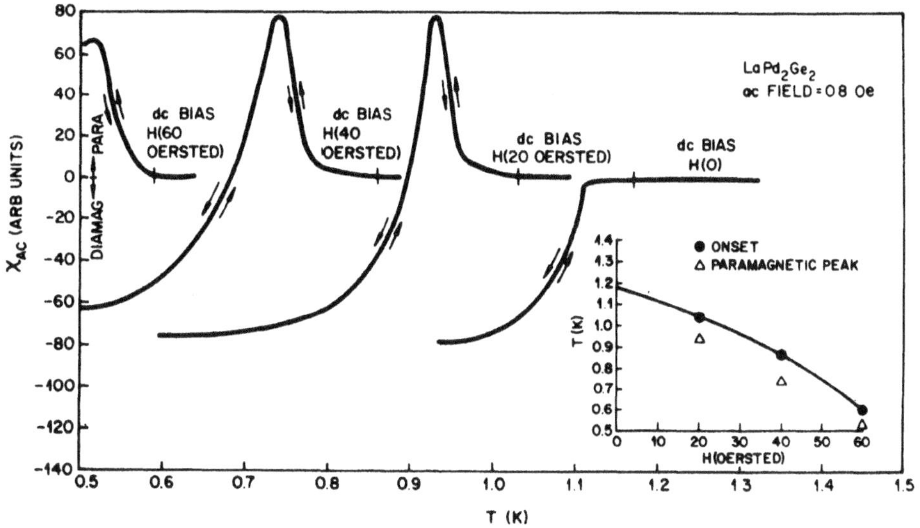

Figure 4. $\chi_H(T)$ of the superconducting alloy $LaPd_2Ge_2$ for four values of H_{dc}. Note the case for $H_{dc} = 0$ was denoted as $\chi_0(T)$ in figure 2b. (taken from reference 13)

$H_c(T)$. This is an incorrect assignment; but, if the demagnetization factor is small the error involved is small and clearly of no consequence in the comparative type study utilized to observe the isotope shift.

It was quite a surprise when Smith et al.[12] reported the observation of a DPE in the compound $AuGa_2$ when the material was subjected to pressures in excess of 15 kbar. Since then, Hull et al.[13] have also reported that the ternary intermetallic alloy $LaPd_2Ge_2$ also exhibits the DPE, see Figure 4. To the author's knowledge these are the only two "alloys" for which a DPE has been reported in ac magnetic susceptibility data.

If under correct experimental conditions a DPE is not observed. than one must render the sample into the form of a powder in order to rule out the effects of trace amounts of superconducting impurities. Conversely if one knows he is dealing with a bulk superconductor, the DPE can be used as a measure of the samples "effective" demagnetization factor. Clearly $\chi_0(T)$ data on bulk samples can not be used as a definitive test for "bulk superconductivity", for such data do not reflect the "Meissner Ochsenfeld" effect.

FILAMENTARY VESRSUS SURFACE SUPERCONDUCTIVITY

Maxwell and Strongin's work[14] on alloys gave rise to a renewed interest in the behavior of the imaginary, or loss, component of the ac magnetic susceptibility. Whereas Shoenberg's pioneering work showed that an extra loss peak in $\chi''_H(T)$ is due to increased eddy current losses in the intermediate state, Maxwell and Strongin took the observation of a loss peak in χ_0'' and χ_H'' as evidence for filamentary superconductivity and developed an "effective conductivity " model to account for the behavior of χ' and χ''. The existence of two or more loss "peaks" was regarded as evidence of multiple superconducting phases being present in the alloy. This filamentary argument invoked considerable controversy, see Strongin et.al.[15].

When it was noted that that the changes in χ' and χ'' occur in the high field tail of the dc magnetization curve, their behavior was taken as reflecting magnetic hysteresis effects associated with the surface superconductivity of Saint-James and de Gennes[16]. Starting with the works by Paskin et al.[17], Fink and Barnes[18] and Fink[19], a myriad of papers appeared dealing with calculations of the magnitude of the peak in χ'' and its position with regard to the overall change in χ'. These calculations consider the behavior of minor hysteresis loops in terms of induced surface currents in the sheath. Subsequent refinements of this critical state model for the superconducting current sheath appeared and are discused in considerable detail by Rollins and Silcox[20]. Whatever model one cares to cite, it seems as though a peak in χ'' occurs whose magnitude is between $0.30(1/4\pi)$ and $0.43(1/4\pi)$ and that it occurs when $\chi' \cong 0.5(-1/4\pi)$. The reader is referred to the work of van der Klein et al.[21] for further experimental and theoretical details. Unless there have been developments of which this writer is unaware, the mechanism which gives rise to the "extra" loss in the superconducting sheet i.e. flux creep, flux flow etc. is still an open question[22].

THIN FILMS

Most of the above remarks are applicable to "bulk" samples i.e. where dimensions are large with regard to the penetration depth. Cody and Miller[23] used the ac self inductance technique to study magnetic transitions in "thin" films of Pb and Sn with thicknesses of 200 to 1200 nm. They worked with the dc magnetic field oriented parallel and perpendicular to the film surface and observed a loss peak in both field orientations. This is a comprehensive study of the ac response involving the effects of frequency, f, and amplitude of the ac measuring field. They found that the loss peak in their thicker films, while very pronounced at say f<100 Hz, decreased in magnitude with increasing frequency and was not observable above some "critical frequency". In general their data indicated that $\Delta R \approx \chi''_H(T)$ and $\Delta L \approx \chi'_H(T)$ were related in that $\Delta R(\max) = 0.32(2\pi f)\Delta L$ and that the peak occurred when $\Delta L = (0.5)\Delta L(NS)$. They felt that magnetic hysteresis models could not adequately account for all details of the magnetic response especially the fact that the peak in thicker films occurs in the reversible portion of the high field "tail" of the magnetization curve. They developed an "effective conductivity" model which could account for all observations, including the large harmonic content in the ac response which occurs in the vicinity of $\Delta R(\max)$. In this model, flux-flow resistivity plays a dominant role in determining the losses. They stress the usefulness of χ' and χ'' measurements in perpendicular and parallel dc fields on the same specimen as a means of determining microscopic parameters of the superconducting state.

Ishida and Mazaki[24] also use the ac mutual inductance technique to measure the zero field superconducting transitions in electrodeposited films of technetium, a 4d transition metal. The films with thicknesses in the 2 to 5 μm range had T_c values of about 7.5K. They observed multiple loss peaks in some of their films and followed the reasoning of Maxwell and Strongin in attributing this effect to sample inhomogeneities. They did not cite the work of Cody and Miller.

LOW-DIMENSIONAL AND PROXIMITY EFFECT SUPERCONDUCTORS

Ribault and coworkers[25] measured χ' and χ'' of crystals of the organic superconductor $(TMTSF)_2PF_6$ under a pressure of 12 kbars where $T_c \approx 1K$. They used an $|h^{ac}|$ of 0.03 Oe with f=68 Hz. and the ac field was perpendicular to the the high conductivity axis of the crystal. A single loss peak was observed in $\chi_O''(T)$ and $\chi_H''(T)$. These authors argue for "bulk" superconductivity and use the misnomer "ac Meissner Effect".

Oda et al.[26] measure χ' and χ'' of $(SN)_x$ crystals with $T_c \approx 250$ mK. No frequency effects were noted, therefore eddy current losses were minimal. Data were obtained for h^{ac} directed perpendicular and parallel to the b axis and a single loss peak was observed. A point to be made here is that the authors reported a "remarkable dependence of the susceptibility on $|h^{ac}|$". They concluded however that a complete Meissner effect was observed. Note well that they only observed complete ac shielding and not the Meissner effect. A subsequent paper[26] reports on a study of the loss peak in χ'' as a function of frequency, $|h^{ac}|$, its orientation with respect to the crystal axis and H^{dc}. The large sensitivity to $|h^{ac}|$ and relative insensitivity to H^{dc} led them to use a model in which the $(SN)_x$ sample is considered to consist of electrically isolated fibers weakly coupled via a network of Josephson junctions. Magnetic flux passing the junctions for each cycle of h^{ac} gives rise to an effective resistance, hence a peak in χ''. Based on this reasoning, they postulated an equivalent loop model in which the network of junctions is replaced by a single loop with a weak link and calculated how such a loop leads to changes in the measured mutual inductance, i.e changes in m' and m". In this way, they could account for the observed temperature dependences of χ' and χ''.

The concept of multiply-connected Josephson networks also appears in the work of Ishida and Mazaki[27] and Ishida et al.[28] on technitium impregnated in a porous alumina substrate to form a multiply-connected network of weak links. Direct observation of the wave form of the output of their Hartshorn mutual inductance bridge led them to propose a phenomenological "equivalent loop model". The magnetization loop of the model was analyzed by means of Fourier analysis. Measured behaviors for χ' and χ'' were consistent with those of Oda et al. From the shape of the χ'' peaks they conclude that the junctions involved were of the microbridge type as opposed to tunnel junctions. Their model produced harmonics consistent with those found by the theories discussed in the preceding sections.

Starting in about 1980, Oda and coworkers in Japan have used ac magnetic susceptibility measurements to study proximity-effect induced superconductivity in Cu[29]. Similar studies have been carried out for Cu and Ag[30,31] by A. Mota and coworkers in Switzerland.

These few selected categories should serve to illustrate the widespread use and utility of ac magnetic susceptibility measurements in studies of superconductors. This workshop will highlight areas in which ac magnetic susceptibility studies are currently playing an important role; namely, the area of granular superconductors in general, the high T_c oxides in particular, and magnetic spin systems.

REFERENCES

1. W. Meissner, and R. Ochsenfeld, Naturwissenschaften, **21**, 787 (1933).
2. N. Kurti, and F. E. Simon, Proc. Roy. Soc. (London), **A151**, 610 (1935).
3. D. Shoenberg, Proc. Camb. Phil. Soc. **33**, 559 (1937).
4. J. Daunt, Phil.Mag., **24**, 361 (1937).
5. D. Shoenberg, "Superconductivity", Cambridge University Press, Cambridge England, second edition 1952, p. 54.
6. R. P. Hudson, Proc. Phys. Soc. **A64**, 751 (1951).
7. J. G. Daunt, and C. V. Heer, Phys. Rev. **76**, 1324 (1949).
8. M. C. Steele and R. A. Hein, Phys. Rev., **92**, 243 (1953).
9. M. C. Steele, Phys. Rev., **87**, 1137 (1952).
10. R. A. Hein and R. A. Falge Jr., Phys. Rev., **123**, 407 (1961).
11. R. Serin, C. A. Reynolds and L. B. Nisbitt, Phys. Rev., **80**, 761 (1950).
12. T. F. Smith, R. V. Shelton, and J. E. Schriber, Phys. Rev., **B8**, 3479 (1973).

13. G. W. Hull, J. H. Wernick, T. H. Geballe, J. V. Waszczak, and J. E. Bernardini, Phys. Rev., **B24**, 6715 (1981).
14. E. Maxwell and M. Strongin, Phys. Rev. Letts., **10**, 212 (1963).
15. M. Strongin, E. Maxwell, and T. B. Reed, Rev. Mod. Phys., **36**, 165 (1964).
16. D. Saint-James and P. G. de Gennes, Phys. Letts., **7**, 306 (1963).
17. A. Paskin, M. Strongin, P. P. Craig, and D. G. Schweitzer, Phys. Rev. **137**, A1816 (1965).
18. H. J. Fink and L. J. Barnes, Phys. Rev. Letts., **15**, 792 (1965).
19. H. J. Fink, Phys. Rev. Letts., **16**, 447 (1969).
20. R. W. Rollins, and J. Silcox, Phys. Rev., **155**, 404 (1967).
21. C. A. M. van der Klein, J. D. Elen, R. Woolf, and D. de Klerk, Physica, **49**, 98 (1970).
22. J. P. Wagner and R. W. Rollins, J. Appl. Phys., **44**, 1778 (1974).
23. G. D. Cody and R. E. Miller, Phys. Rev., **173**, 481 (1968) also Phys. Rev. Letts., **16**, 697 (1966).
24. T. Ishida, and H. Mazaki, Phys. Rev., **B20**, 131 (1979).
25. M. Ribault, G. Benedek, D. Jerome, and K. Bechgaard, J. Physique, **41**, L-397 (1980).
26. Y. Oda, H. Takenaka, H. Nagano, and I. Nakada, Solid State Communications, **32**, 659 (1979), also Y. Oda, H. Takenaka, H. Nagano, and I. Nakada, Solid State Communications, **35**, 887 (1980).
27. T. Ishida, and H. Mazaki, Phys. Letts., **87A**, 373 (1982)
28. T. Ishida et al. Phys. Rev. **B29**, 1183 (1984).
29. Y. Oda, A. Sumiyama, and H. Nagasno, Jap.. J. of Appl. Phys., **22**, 464 (1983).
30. A. C. Mota, D. Marek, and J. C. Weber, Helv. Phys. Acta, **55**, 647 (1982).
31. P. Visani, A. C. Mota and A. Pollini, Phys. Rev. Letts., **65**, 1514 (1990).

MULTIPURPOSE CRYOSTAT FOR LOW TEMPERATURE MAGNETIC AND ELECTRIC MEASUREMENTS OF SOLIDS

C. Rillo, F. Lera, A. Badía, L.A. Angurel, J. Bartolomé, F. Palacio, R. Navarro
Instituto de Ciencia de Materiales de Aragón
CSIC-Universidad de Zaragoza, 50009 Zaragoza, Spain

A. J. van Duyneveldt
Kamerlingh Onnes Laboratorium, Rijks Universiteit te Leiden
Niewsteeg 18, 2311 SB Leiden, The Netherlands

ABSTRACT

Magnetization, ac susceptibility and electrical resistance measurements are performed in the temperature range between 1.2 and 300 K and fields from zero (earth field compensated) up to 5 T, by means of a multipurpose cryostat which has common cryogenics, temperature control system, data acquisition and controlling computer. The magnetic measurements are based on the determination of magnetic flux variation on a pick-up coil having two symmetric, oppositely wound sections, when the position of the sample is switched from the center of one subsection to the center of the other. For determining the magnetization the *emf* is integrated while for the ac susceptibility the mutual inductance between an exciting coil and the pick-up coil is detected. Voltage versus dc intensity as well as ac resistivity are determined by a four probe technique. Moreover, simultaneous magnetic and electric transport measurements may be performed, to our knowledge this being a novelty. Absolute values of the three magnitudes are measured with an accuracy of 1%. The sensitivity is 10^{-5} Oe cm^3 for the magnetization, 10^{-9} emu for the susceptibility and 10^{-8} Ω for the resistance. Illustrative measurements on high temperature superconductors, permanent magnets, as well as on other magnetic materials are included.

INTRODUCTION

Measurements of the differential (ac) susceptibility $\chi = \mathrm{d}M/\mathrm{d}H$, the magnetization, M, and the resistivity, ρ, are widely used for the study of magnetic and electric properties of materials. The presence of magnetic fields modify the transport properties of conductors and, conversely, electric currents influence the magnetic behaviour. The relationship between electric transport and magnetic quantities is specially important in superconductors and in permanent magnets, both from a fundamental point of view as well as for its applications. Usually these experiments are performed in different apparatus, making the correlation of data both difficult and time consuming. Consequently, the design of a multipurpose system for measuring χ, M and ρ sharing cryogenics, temperature control, data acquisition and controling computer is of great interest and furthermore the sinergy of the process may clarify some experimental uncertainties.

A great variety of mutual inductance methods are used in the study of magnetic systems[1-5]. However, all are based on the detection of the *emf* induced by a time varying magnetic flux in a pick-up coil, that contains the magnetic sample. Common ways to produce ***emf*** are by means of alternating fields (ac susceptometers), vibrating the sample/coil in an applied dc field (vibrating sample/coil magnetometer) or by extracting the sample from the coil (extraction magnetometer) [6].

Pick-up coils used in ac susceptometer bridges are often formed by two axially symmetric oppositely wound coils mounted in series. A primary coil concentric with the two secondaries carries a current that generates the alternating field. Ideally, in absence of a sample, the *emf* is zero and when a sample is introduced in one of the secondaries the imbalance is compensated with an inductance bridge or measured directly by synchronous detection.

Sensitivities of the order of 2×10^{-8} emu in the low frequency range are achieved [7]. One of the limiting factors to the sensitivity is the presence of a background signal that has to be subtracted. Different methods to balance the measurement system in the absence of a sample have been suggested. Between them are; ***a)*** The use of a matched, counterwound coil mounted in series[8], ***b)*** Recording the signal difference obtained when the sample is moved between the centers of the two sections of the pick-up coils [7] and ***c)*** Subtraction of an ac waveform equal to that of an empty pick-up coil, using a differential amplifier before the input to the lock-in amplifier [8].

From our experience of over two decades, we have found that sensitivities as high as 10^{-9} emu can be achieved with a simple combination of these three methods. The apparatus described in this paper has been optimized for ac susceptibility measurements at zero field in the temperature range from 1.5 to 350 K. Moreover, measurements in fields up to 5 T can be performed with the use of a superconducting magnet. The cryogenics, coil configuration, temperature and magnetic field control are described in the next two sections.

Moreover, the motion of the sample between two positions, that compensates the background signal in ac susceptibility, also enable the derivation of the dc magnetization by integration of the *emf* induced during the displacement. Furthermore, using the four point technique and the above detection electronics, ac or dc resistivity is determined, so magnetic and electric measurements are performed simultaneously. The multipurpose apparatus is described, whereas the final sections illustrate some experimental results obtained on high temperature superconductors (HTS), permanent magnets and other magnetic systems.

EXPERIMENTAL SET-UP

Cryogenic system

The cryogenic system, similar to that recently described by Deutz et al[7], is presented in Fig. 1. The 5 T superconducting magnet and the induction coils are immersed in a cryogenic liquid (He when operation of the magnet is needed) at atmospheric pressure. For experiments below 4.2 K, a vacuum rotatory pump and a manostat is used to stabilize the pressure and thereafter the He bath temperature. Two concentric borosilicate glass tubes provide a contact gas or vacuum space to allow cooling or heating of the sample. However, the sample space is filled with He gas at atmosferic pressure except for experiments at high temperatures in which a low pressure (≈10 mbar) is maintained to minimize liquid He boil off.

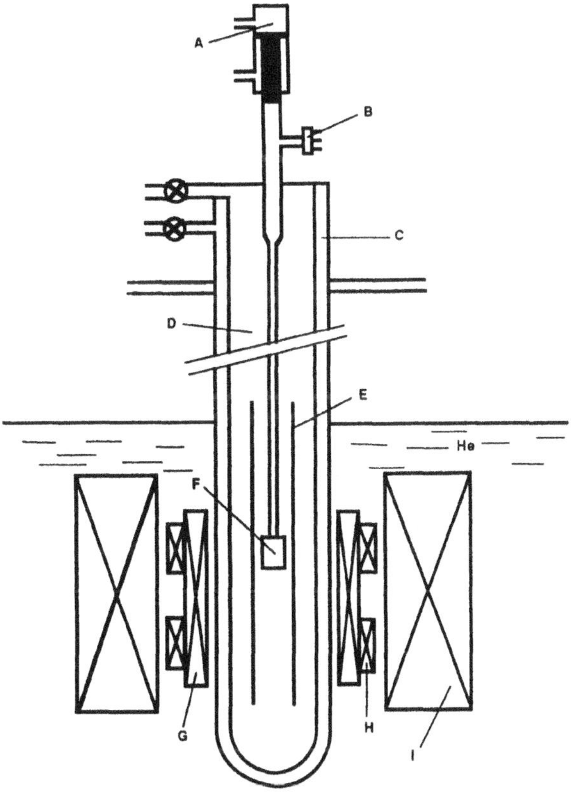

Fig 1. Schematic diagram of the cryostat. (A) Pneumatic cylinder for commuting the sample, (B) vacuum connector for transport measurements, (C) contact-gas space, (D) sample space, (E) coil-foil tube, (F) sample, (G) primary coil, (H) secondary coils, (I) superconducting magnet.

Table I

Characteristics of one primary and the two secondary coils used in the cryostat

	PRIMARY	SECONDARY	
		Section 1	Section 2
Copper wire diameter (mm)	0.15	0.15	0.15
Length (mm)	100	30	30
Number of turns	4 x 563	9 x 166	9 x 166+ΔN
Internal diameter (mm)	30.9		
Distance between centers (mm)		40	
Inductance (mH)	51	71	
Magnetic field in the center (Oe/mA)	0.55		
Resistance at 4.2 K (Ω)	1.9	2.3	
Resonance frequency (Hz)		330	

Coils

The mutual induction exciting and sensing set consist of one primary and two secondary coils wound in series opposition. Details are given in table I. The primary is wound on a teflon cylinder covered by capton foil; the secondary coils are wound on the primary. General Electric varnish and celotex cylindrical spacers are used to give dimensional stability to the assembly. A gold miniature four terminal connector is used for the electrical leads which are soldered with Cd.

One of the secondary sections is wound with additional turns, ΔN, to allow the compensation of the astatic pair. The mutual inductance set is constructed so that it can be separated from the supporting teflon cylinder by cooling to liquid nitrogen temperature. Afterwards, by small changes in the number of turns of one secondary, the background signal is minimized at room temperature and the coil set is attached to the external borosilicate glass tube. Two centering PVC pieces screwed on top and bottom of the superconducting magnet allow to fix the relative positions.

Sample holder

The sample holder is attached to a long rod made of a non-magnetic and low thermal conductivity alloy which is led through an O-ring to the attachment of a pneumatic piston, that can commute the sample's position. For transport measurements it has a four terminal vacuum connector in its upper extremity (Fig. 1).

For magnetic measurements, delrin holders are screwed to the lower part of the rod. A typical example is shown in Fig. 2 a. For electric current transport and simultaneous transport and magnetic measurements the lower part of the rod has a four pin terminal to

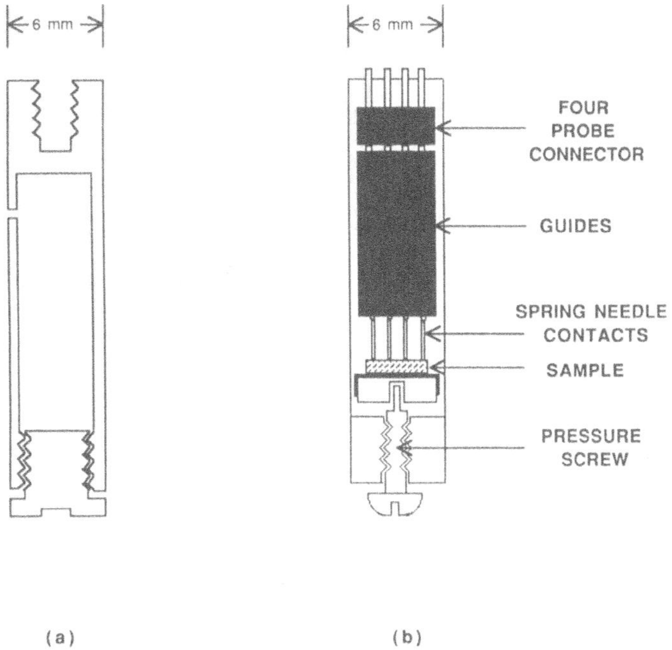

Fig 2. (a) Delrin sample holder for magnetic measurements.
(b) Sample support for resistivity measurements. Four equally spaced needles are used as constant pressure contacts with the sample. The current is fed through the external needles, and the voltage drop is measured across the inner ones.

which a typical sample holder, as shown in Fig. 2 b, is connected. The electric connections with the sample are made by means of mechanical pressure of gold spring needles or by silver paint contacts.

EXPERIMENTAL CONDITIONS

Magnetic field

A superconducting magnet immersed in the cryogenic bath enable to achieve dc magnetic fields, H, up to 5 T. The coil is energized with a *Hewlett Packard* power supply (model HP 6031A) implemented with a high stability option for inductive loads and a IEEE 488 interface. The source output is unipolar and for some applications (magnetic hysteresis loops) the magnetic field must be inverted. For that reason, a computer controlled switching (make before break) system is implemented [9].

A voltage source for persistent mode operation of the superconducting magnet is also included as well as a calibrated shunt resistance for measuring the current. The unit has also been provided with logical support for verifying safety conditions needed in the control of the superconducting magnet. When the power supply is operated as current source a minimum increment of 30 Oe can be generated. For low field measurements the power supply is used as voltage source and a resistance is placed in series with the superconducting magnet. In this way field increments as low as 1 mOe can be obtained.

Temperature measurement and control

The temperature control system consists of three main parts: a cryogenic bath to cool the sample, a vacuum system which enables to isolate the sample area from the bath for heating, and a microprocessor based temperature controller which reads and stabilizes the temperature through resistive thermometers and a heater (Fig. 3). The sample is in thermal contact with the cooling liquid (helium or nitrogen) by He exchange gas in the

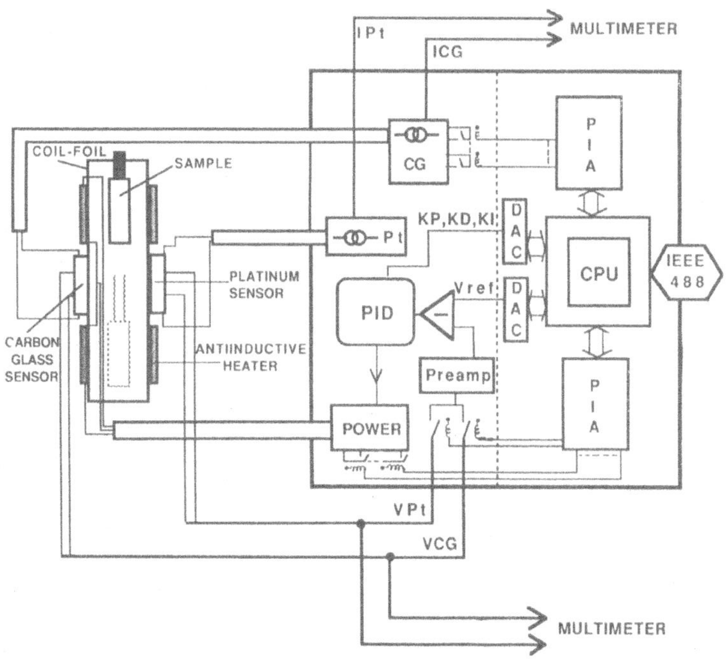

Fig 3. Schematic diagram of the temperature controlling system.The digital module consists of a CPU that sets PID parameters, two PIA integrated circuits that control relais for selecting the current in the carbon-glass thermometer, the controlling thermometer and the heating power range; and an IEEE 488 interface for comunication with the main computer. The analog module consits of two current sources, one for each thermometer, a PID module and a power heating unit. IPt and ICG are voltages proportional to the actual current in the thermometers. VPt and VCG are the voltages in the thermometers.

double wall container; regulation of the pressure with a vacuum system provides a control of heat leakage from the sample.

Temperatures between 1.2 and 4.2 K can be attained by controlling the pressure of the liquid helium bath with an Oxford Instruments manostat (model M26 vacuum regulator). In order to keep the sample space at constant temperature throughout the measurement, the thermometers and heaters are placed in a cylinder of vertical copper wires (coil-foil) with a length of 30 cm.

Two calibrated ohmic thermometers are used: carbon-glass and platinum. This allows optimum control throughout the whole range of temperature and field. For measurements at high magnetic fields we take advantage of the full calibration range of carbon-glass resistor (1.2 to 90 K) provided that its magnetoresistance is negligible. The Pt-thermometer is used from 90 to 350 K. At low fields the Pt-thermometer is used above 30 K provided it has higher sensitivity than the carbon-glass one.

Automated temperature controller

A PID analog temperature controller governed by microprocessor which is controlled by the main computer through IEEE 488 protocol (Fig. 3) has been designed and built [9]. The analog module includes two current sources (one for each thermometer), PID processing of the temperature error signal and a power unit which feeds the heater. The purpose of the digital module is to set PID parameters. The computer program selects the controlling and the reading sensor according to the experimental conditions. V_{ref} is the voltage corresponding to the temperature we want to achieve, for a known current through the thermometer. The current that feeds the Pt-thermometer is 1 mA. In the carbon-glass thermometer the current is varied from 50 nA to 300 μA to minimize self-heating. The PID constants are indicated by KD, KI and KP in Fig. 3. Finally there are four ranges of heating power, from 0 to 4 W. All the commands are computer controlled, allowing automatic operation over the whole range of temperature. From 1.2 to 30 K a temperature stabilization of 1 mK is maintained during the measurement. Above 30 K the stability becomes 10 mK and reaches 0.1 K above 300 K.

DESCRIPTION OF THE INSTRUMENT

A schematic block diagram of the measuring system is presented in Fig. 4. The sample is placed at the center of one of the secondary coils at given H and T values. Two voltage to current converters (VCC) that can generate alternating (i_{ac}), and dc (I_{dc}), currents are used for feeding current through the sample or the primary coil.

In the ac susceptibility measurements (χ_{ac}) a current i_{ac}, governed in frequency and amplitude by the lock-in amplifier oscillator voltage (the reference) is fed into the primary

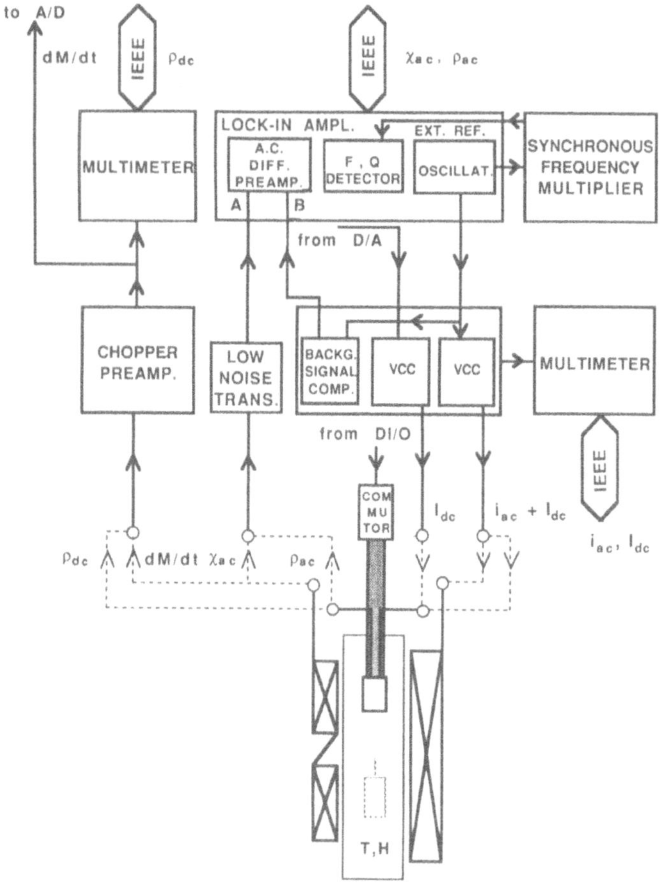

Fig 4. Schematic block diagram of the complete measuring system. For χ_{ac} measurements, $i_{ac}+I_{dc}$ is led to the primary coil. The emf induced on the secondary coils is filtered on the low noise transformer prior to enter the A input of the lock-in preamplifier. The background signal is substracted using the B input. Depending on the reference frequency selected with the synchronous multiplier, the nth harmonic is measured and transferred to the computer, for both positions of the sample. For resistivity measurements, the excitation current is driven trough the current leads into the sample. For ρ_{ac} measurements, the voltage drop is processed as in the ac susceptibility case. For ρ_{dc}, the amplified voltage is measured with a multimeter and transferred to the computer. In all cases, ac and dc currents are monitored with a multimeter. For M measurements, the emf induced by the sample displacement is amplified, digitized and transferred to the computer. The data are then numerically integrated to obtain the magnetization. The sample position is controlled with a pneumatically operated commutor. Lock-in amplifier: EG&G PAR 5301. Low noise transformer: EG&G PAR 1900. Multimeter: FLUKE 8842A. Chopper preamplifier: ANCOM 15C3A[9].

coil. In the absence of a sample, a background signal, due to non exact compensation of the two secondary sections, is obtained. The signal of the two secondaries in series is amplified by a low noise transformer and introduced in the A input of the lock-in differential preamplifier. A background compensation signal is generated by electronic amplitude and phase modulation of the reference and introduced in the B input. The preamplifier subtracts the A and B signals and, as a consequence, the sensitivity of the lock-in can be increased up to the noise level of the system (see below). When a sample is present in one of the secondary sections, a signal proportional to χ_{ac} is obtained at the output of the lock-in. Feeding a dc current through the primary, I_{dc}, we may compensate the Earth's magnetic field or study χ_{ac} in weak dc fields. A synchronous frequency multiplier [9] generates an external reference that is used for the harmonic analysis of the χ_{ac} signal.

The i_{ac} and I_{dc} values are read by a multimeter connected to the computer via an IEEE interface, while in-phase and out-of-phase readings of the lock-in are directly transferred to the computer via an IEEE interface.

For ac resistivity, ρ_{ac}, experiments the i_{ac} current is introduced in the current leads of the sample holder rod. The signal at the voltage leads is proportional to ρ_{ac} and detected by the lock-in. In this case the background signal compensation may be used as an offset when very small variations of ρ_{ac} have to be detected. For the determination of the dc resistivity, ρ_{dc}, the I_{dc} current is applied to the sample and the dc voltage is amplified by a low frequency chopper preamplifier and converted to a digital signal by a multimeter.

When the sample, having a magnetization M in the presence of a dc magnetic field H, is moved from the center of one secondary coil to the other, a low frequency signal proportional to dM/dt is induced. This signal is conducted to the chopper preamplifier and its output is fed into the A/D converter of an analog/digital board which transfers data to the computer in DMA mode. The magnetization is calculated by numerical integration.

Magnetic ac susceptibility measurements

The method is based on the lock-in detection of the mutual inductance changes produced in the secondary oppositely wound coils by the presence of a magnetic sample. Let us call M_1 and M_2 the mutual inductances between the primary and each of the secondary sections. When, in the absence of a sample, a current $i_{ac}(t) = i_0 \cos\omega t$ pass through the primary coil, $M_1 \neq M_2$ due to the non exact compensation of the coils and a background *emf*, $\varepsilon_B(t)$, is always present.

$$\varepsilon_B(t) = (M_1 - M_2)\, di_{ac}/dt \; = - (M_1 - M_2)\, i_0 \omega \sin\omega t$$

Furthermore, when a magnetic sample is introduced in one of the secondary coils the mutual inductance changes and a new term, $\varepsilon_s(t)$, appears. The total *emf* is given by:

$$\varepsilon_1(t) = \varepsilon_B(t) + \varepsilon_{s1}(t) \quad (1)$$

$$\varepsilon_{s1}(t) = - N_1 S_1 f \chi_{ac} \, dh_{ac}(t)/dt \quad (2)$$

where N_1 and S_1 are respectively the number of turns and the section of the first secondary; f the filling factor and $h_{ac}(t)$ the ac field generated by the exciting current $i_{ac}(t)$. χ_{ac} is the complex ac magnetic susceptibility, that in the linear aproximation is given by $\chi_{ac}=\chi'-i\chi''$, where χ' and χ'' are the in-phase and out-of-phase components, respectively. When the sample is placed in the other section of the secondary the induced *emf* will be

$$\varepsilon_2(t) = \varepsilon_B(t) - \varepsilon_{s2}(t)$$

where $\varepsilon_{s2}(t)$ follows expression (2) with subindex 1 replaced by 2. The lock-in amplifier measures the rms value of $\varepsilon_1(t)$ and $\varepsilon_2(t)$ and its difference is proportional to the ac susceptibility

$$\varepsilon_{1,rms} - \varepsilon_{2,rms} = \varepsilon_{s1,rms} + \varepsilon_{s2,rms} = C_\chi^{-1} h_0 \chi_{ac}$$

where $C_\chi = [(N_1S_1+N_2S_2)f\omega\,]^{-1}$ is a calibration factor, that represents the sensitivity of the system.

This method combines two compensation techniques: ***i)*** the use of a series matched counter-wound coil and ***ii)*** the position change of the sample from the center of one coil to the other. It gives reliable results when $\varepsilon_{s1}(t)$, $\varepsilon_{s2}(t) \geq \varepsilon_B(t)$. But in some cases, for small susceptibilities it occurs that $\varepsilon_{s1}(t)$, $\varepsilon_{s2}(t) << \varepsilon_B(t)$. In this case the sensitivity of the lock-in cannot be adapted to the low *emf* values of the sample and the signal to noise ratio becomes rather poor.

For calibration (determination of the constant C_χ) and phase setting (φ) of the lock-in amplifier, for proper determination of χ' and χ'', a paramagnetic compound ($Mn\ (NH_4)_2\ (SO_4)_2\ 6H_2O$) is used. In Fig. 5 a typical calibration run is presented; at low frequencies ($\omega/2\pi$ < 1MHz) φ is the phase which at zero dc field makes negligible χ'' of the paramagnetic salt and C_χ is obtained from $1/\chi'(T)$. In practice $\omega^{-1}C_\chi$ and φ are frequency dependent, due to stray capacitance effects on the secondary coils. These effects produce a resonance in the induced emf at frequency ω_r at which $\omega^{-1}C_\chi$ has a sharp peak and φ changes by 180º [10]. Therefore, both quantities C_χ and φ have to be determined for each measuring frequency, that should be different from ω_r. A calibrating point at 4.2 K is repeated each time that the whole apparatus is cooled from room temperature for a cycle of measurements. The calibration factor at a given frequency has a reproducibility and accuracy of 1% . Changes in φ of ≈ 0.1º are detected between different calibration processes, whereas variations in φ during a cycle of measurements are negligible, provided that the measuring frequency is sligtly different (10%) from ω_r.

The characteristics of the used mutual inductance coils were given in table I, being $\varepsilon_{B,rms} \approx 100\ \mu V$ for $i_0 = 2$ mA, $\nu = 120$ Hz, and $C_\chi = 3 \times 10^{-5}$ emu Oe/μV the calibration factor. Taking into account the output resolution of the lock-in amplifier (0.01%) the minimum detectable signal is 10 nV ($0.01\% \cdot \varepsilon_{B,rms}$), corresponding to 3×10^{-7} emu Oe. This is insufficient in some cases in which susceptibilities of the order of 10^{-6} emu or smaller have to be measured. So, before the phase sensitive detection, an additional background signal compensation is needed. A simple and inexpensive way of doing that is to utilize the differential capability of the lock-in preamplifier. A signal equal to $\varepsilon_B(t)$ times the low

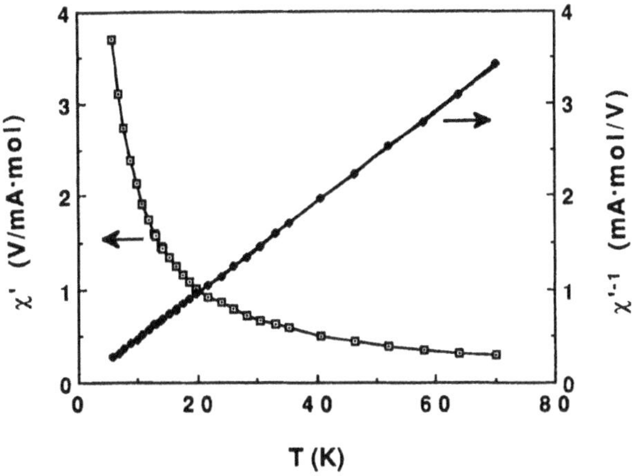

Fig 5. AC magnetic usceptibility and its inverse (ν = 120 Hz and h_0 = 1.1 Oe) of the paramagnetic Tutton salt $Mn(NH_4)_2(SO_4)_2.6H_2O$ used as calibrant compound. The phase is adjusted on the lock-in amplifier so that χ'' is zero. The calibration constant C_χ is obtained from the salt Curie constant , 4.375 emu/(K mol).

noise transformer ratio (≈100) (see Fig. 4), can be generated using the lock-in reference signal and a simple amplitude and phase modulation circuit. The block diagram of an electronic circuit to compensate the background is shown in Fig. 6 together with that of the VCC. The voltage from the lock-in oscillator controls the background signal compensation and the 20 mA VCC. The analog output of the computer controls the dc 2 A VCC. All the stages are standard operational amplifier circuits described elsewhere [10]. This method allows one to compensate the background signal at the input of the lock-in to a level which permits operating in the 1 μV full scale sensitivity (10 nV at the input of the low noise transformer). Consequently the resolution (0.01%) will be of the order of 1 pV, and, therefore, the minimum detectable signal will be determined by the noise.

To calculate the minimum equivalent voltage noise of the detection system (coils, low noise transformer and lock-in preamplifier) one has to consider the thermal voltage noise of the secondary coils $\varepsilon_{sn}(4.2\ K) \approx 20\ pV/\sqrt{Hz}$ and the equivalent voltage and current noise generators of the electronics. The dominant term is the noise voltage of the low noise transformer; i.e. $\varepsilon_{tn} \approx 0.4\ nV/\sqrt{Hz}$ at 120 Hz [10]. Then, the noise level and the resolution will be given by :

$$\chi_{\text{noise level}} = h_0^{-1} C_\chi\ \varepsilon_{tn} \approx h_0^{-1}(1 \times 10^{-8}\ \text{emu Oe}/\sqrt{\text{Hz}})\ .$$

Experimentally, the noise level attained using a high performance lock-in amplifier (EG&G model PAR5301) with 40 dB of dynamic reserve and a band-pass filter with Q =5 in the preamplifier stage, has been 10^{-9} emu with h_0= 10 Oe, in good agreement with the calculation. This ultimate high sensitivity is only achievable due to the three step of

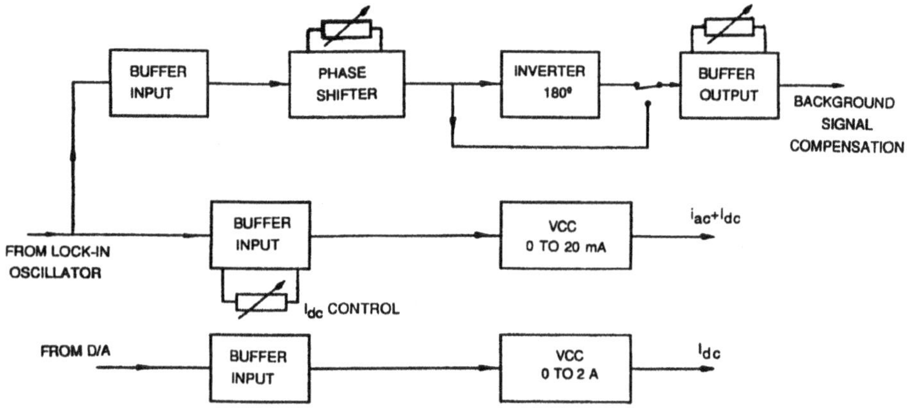

Fig 6. Block scheme diagram of the background signal compensation system and the VCC sources. The oscillator signal from the lock-in is phase and amplitude modulated.

background compensation: ***i)*** Use of a series matched counter-wound coil. ***ii)*** Recording of the signal difference between the centers of the two coil sections, ***iii)*** Electronic background compensation at intermediate signal level.

The system operates in the range from 10 Hz to 10 KHz, enabling frequency dependent studies of χ' and χ''. A phase locked-loop synchronous frequency multiplier can be connected to the lock-in oscillator output so that multiples of the fundamental frequency, ν, can be generated for harmonic analysis up to ninth order. This allows the study of non-linear effects that has become important in HTS ceramics. The method and calibration procedures for those measurements have been described elsewhere [11]. For a simultaneous recording of harmonics up to arbitrary higher order a *Hewlett Packard* dynamic analyzer (model HP 3562A) is used.

Magnetization measurements

For magnetization measurements it is common to place the sample in one of the secondary coils. As in the extraction method, the sample it is then pulled out of from one of the secondaries to the other, a process taking a certain time Δt. In the transient there is an induced *emf*, $\varepsilon(t)$, which is integrated between 0 and Δt. This measurement of the magnetic flux proportional to the magnetization of the sample is possible since the final position is out of the initial secondary coil, thus

$$\int_0^{\Delta t} \varepsilon(t)\, dt = - N S f M$$

where N is the number of windings. Outstanding to the sample simultaneous leaving the first secondary and entering the second (see Fig. 4), it is simple to see that the signal obtained after integration will be proportional to twice the magnetization .

Calibration is performed by measuring the spontaneous magnetization of a Ni standard 3 mm diameter sphere supplied by the Physikalisch-Technische Bundesanstalt. We obtain a value of 0.98×10^{-2} Oe·cm^3/μV·s. The magnetization curve $M(H)$ at 4.2 K of the calibrant (Mn $SO_4(NH_4)_2$ SO_4 $6H_2O$) for ac susceptibility is presented in Fig 7. From the low field slope we obtain a magnetic susceptibility that differs by 1 % from the tabulated value [12].

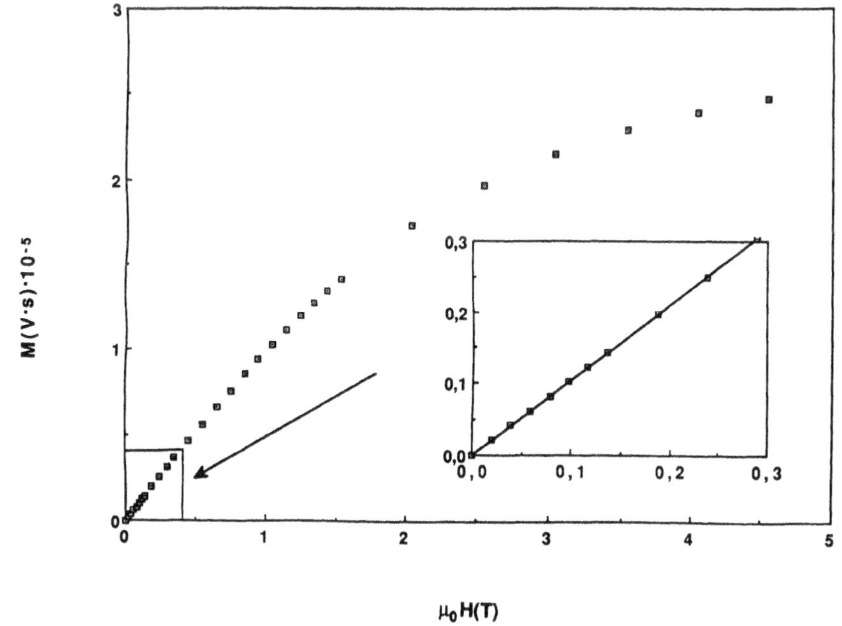

Fig 7. Magnetization curve M(H) at 4.2 K of the calibrant compound $Mn(NH_4)_2(SO_4)_2.6H_2O$. The calibration constant is obtained from the low field M(H) slope.

The equivalent input voltage noise of the chopper preamplifier in its equivalent noise bandwidth (≈10 Hz) is 1 nV, therefore, the calculated noise for an integration time of one second is of the order of 1×10^{-5} Oe cm^3, a value that coincides well with the measured noise.

Transport measurements

The four point method is employed for the measurements of the dc *V-I* characteristics and ac resistance. The contacts are made with gold plated steel pins provided with a spring so that the pressure at the contact point is constant. The sample has to be cut in the shape of a long bar (6 mm in length). The outside pin-points feed the current and the two intermediate ones allow the voltage drop measurement. (Fig. 2 b)

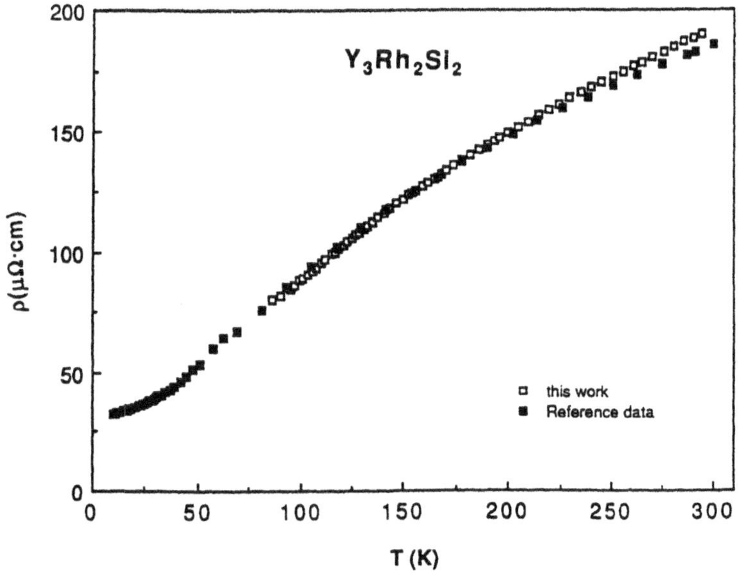

Fig 8. AC resistivity of a $Y_3Rh_2Si_2$ bar shaped sample. Our results (□) coincide well with those obtained in other laboratory (■).

The electrical measurement may be performed with dc or ac currents. For dc measurements a current of up to 2 A may be used. The voltage drop in the sample is amplified by the chopper preamplifier and digitized by a multimeter (see Fig. 4). For the ac voltage detection we use the same low noise lock-in amplifier system than in the susceptibility measurements. The in-phase and out-of-phase components of the signal are detected and analyzed. In Fig. 8 ac resistivity measurements on a $Y_3Rh_2Si_2$ bar shaped sample were compared with those obtained by other laboratory [13]. The results coincide within 1%. The sensitivity is limited by the noise voltage of the low noise transformer. For an applied current of 20 mA the calculated sensitivity is $1 \times 10^{-8}\ \Omega$ in a bandwith of 1 Hz, in agreement with the measured value.

MEASUREMENTS ON HIGH TEMPERATURE SUPERCONDUCTORS

Magnetic ac susceptibility, dc magnetization and resistivity provide valuable tools for the investigation of macroscopic properties of HTS. We have used our system to study these

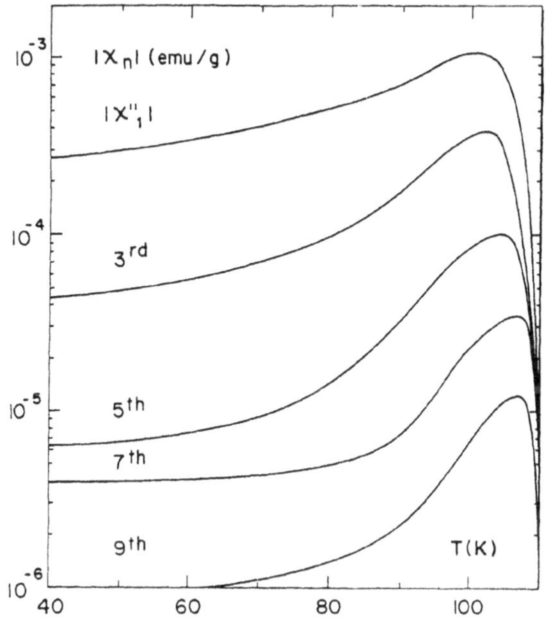

Fig 9. Modulus of the harmonic components $|\chi_3|$, $|\chi_5|$, $|\chi_7|$, $|\chi_9|$ as a function of the temperature for a high density sintered $Tl_2Ba_2Ca_2Cu_3O_X$ sample computed from the first nine harmonics ($\nu = 42$ Hz and $h_0 = 11$ Oe). For comparison $|\chi''_1|$ has been included.

materials, pioneering some techniques as the harmonic analysis of the ac susceptibility response, and below we present some examples.

Type II superconducting materials, under ac and dc magnetic fields higher than H_{c1} (lower critical field), show non linear behavior. For an exciting field $h(t)=H_{dc}+h_0\cos(\omega t)$ the resulting magnetization $m(t)$ is :

$$m(t)=M_0+h_0\sum_{n=1}^{\infty}\left[\chi'_n\cos(n\omega t)+\chi''_n\sin(n\omega t)\right] \qquad (3)$$

where χ'_n and χ''_n are the in-phase and out-phase components, respectively, of the complex ac susceptibility. For H_{dc}=0, there are only odd components, and for the case of HTS materials, the increase of $/H_{dc}/$ develops even ones [14].

When the sample is placed into an ideal sensing coil, the induced emf per unit volume is then

$$\varepsilon(t) = h_0 \sum_{n=1}^{\infty} C_{\chi_n}^{-1} n \left[\chi'_n sin(n\omega t) - \chi''_n cos(n\omega t)\right]$$

where C_{χ_n} are the callibration factors.

Following the procedures described elsewhere [11, 15], the χ'_n(T) and χ''_n(T) odd components of the harmonics up to ninth order of a high density sintered Tl-Ba-Cu-O sample, have been measured in thermal equilibrium between 40 and 115 K. True zero dc field (compensated) measurements for ν = 45 Hz and h_0 = 11 Oe were obtained, and the results for the modulus $\chi_n = (\chi'^2_n + \chi''^2_n)^{1/2}$ have been plotted in Fig. 9. For comparison, the out-of-phase component of the first harmonic $\chi_1''(T)$ has been included. Using equation 3 and the $\chi'_n(T, h_0)$ and $\chi''_n(T, h_0)$ $n \leq 9$ results we may represent $m(t,T)$ versus $h(t)$ deriving ac magnetization loops $m(h,T)$, which are depicted in Fig. 10. At low temperatures (40 K) and fields (11 Oe) the $m(h,T)$ cycles are Rayleigh shaped and there is good agreement with simple Bean critical state model predictions. Above 90 K more sofisticated critical state models are needed for the interpretation of the $\chi_n(T)$ and $m(h,T)$ results [16].

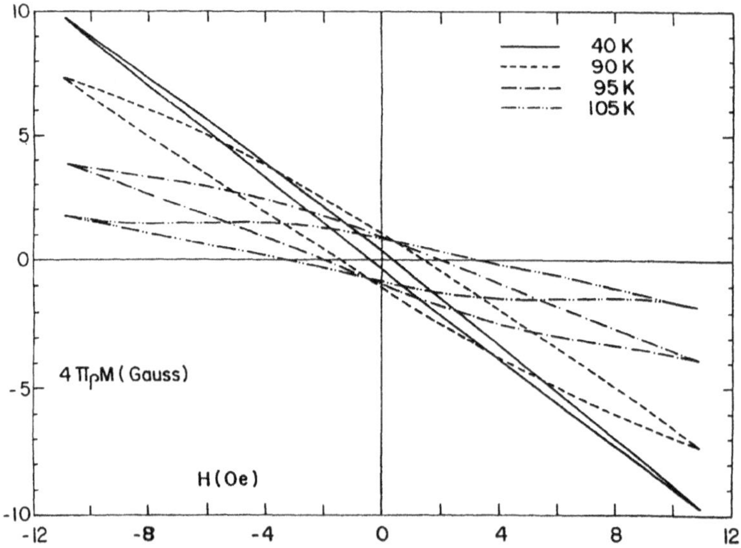

Fig 10. AC magnetization loops of a $Tl_2Ba_2Ca_2Cu_3O_X$ ceramic derived at different temperatures by addition of the first nine components of the harmonics, for h_0 = 11 Oe and ν = 42 Hz.

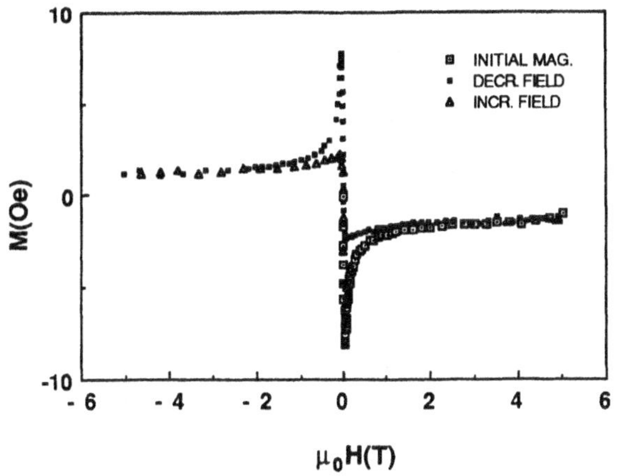

Fig 11. Hysteresis loop of an $YBa_2Cu_3O_{7-\delta}$ ceramic measured at 65 K.

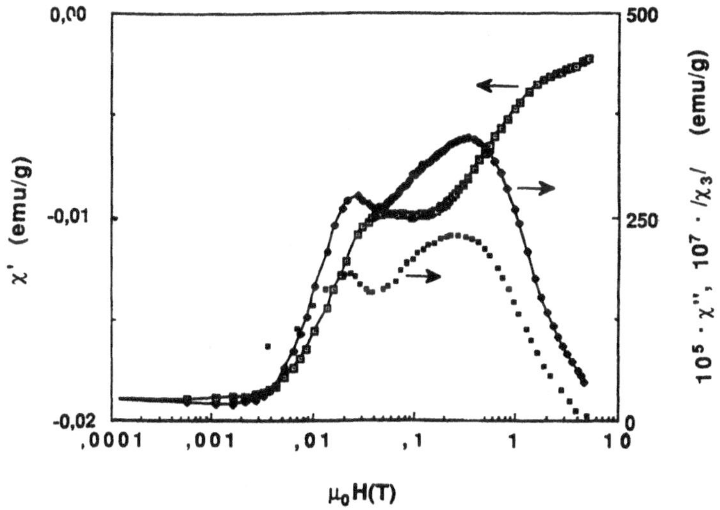

Fig 12. DC field dependence of (▫) χ_1' (emu/g), (♦) $10^5 \cdot \chi_1''$ (emu/g) and (▪) $10^7 \cdot |\chi_3|$ (emu/g) in an $YBa_2Cu_3O_{7-\delta}$ ceramic measured at 65 K (ν = 38 Hz and h_0 = 0.55 Oe).The magnetic field clearly separates the intra- and intergranular regimes. The sample was approximately a cylinder, diameter 2.5 mm, length 6 mm.

Harmonic analysis techniques have been of great importance for the understanding of the weak-link nature of oxide superconductors, being an appropriate test for theoretical models and a source of information to obtain microscopic parameters characterizing the samples [17]

The field dependence of the magnetization, ac linear susceptibility and its harmonics have also been studied in HTS samples. In Figs. 11 and 12 the measurements performed on a ceramic Y-Ba-Cu-O at 65 K are presented. From the magnetization hysteresis loop of Fig. 11 it is concluded that the irreversibility field is around 4 T, being difficult its exact determination from this type of experiment. The in-field ac susceptibility measurements clearly show a lower field regime, with full penetration at around 0.03 T, and a higher one indicating thet the grains are fully penetrated at around 0.32 T. Another interesting feature is the similarity between the out-of-phase component, χ_1'', and the amplitude of the third harmonic χ_3 (Fig. 12). It should be noted that χ_1'' and χ_3 are non-zero even at 5 T, indicating that the a.c. magnetization is irreversible, and therefore $H_{irr}(65\ K) > 5$ T. The onset of non linear effects in χ_{ac} (H,T) is an alternative experimental technique for the determination of $H_{irr}(T)$[18].

As an example of the sensitivity which may be achieved with the installation, measurements of $\chi_{ac}(T)$ in a *c*-axis oriented Y-Ba-Cu-O thin film are shown in Fig. 13. Two geometries have been recorded: with the exciting field parallel and perpendicular to the *c*-axis. In the last one, signals of the order of 10^{-7} emu have been obtained a situation in which the diamagnetic contribution of the sample holder signal has been measured and subtracted. As the penetration depth is of the order of the film dimensions the origin of these signals could be due to the misalignement of the film and the field.

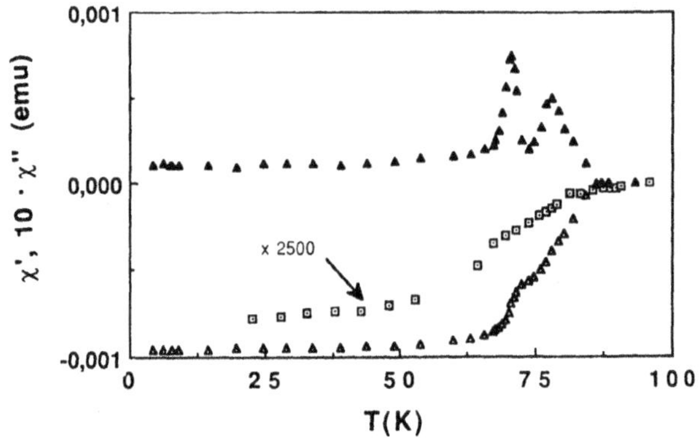

Fig 13. Temperature dependence of the ac susceptibility of a 300 nm c-axis oriented $YBa_2Cu_3O_{7-\delta}$ thin film. The c-axis is perpendicular to the face of the sample. Triangles (open for the in-phase and closed for ten times the out-of-phase component) show data measured with an ac field of 5.5 mOe applied parallel to the c-axis. Squares are used for the case in which an a.c. field of 11 Oe is applied perpendicular to the c-axis. The contribution of the sample holder has been subtracted.

In the course of our investigation on the HTS macroscopic properties we have inferred that studying the interplay of electrical and magnetic behavior of these materials should be of great interest. This arises from the controversial comparison of the results obtained for a macroscopic parameter which characterizes the samples; i.e. the critical current density J_c.

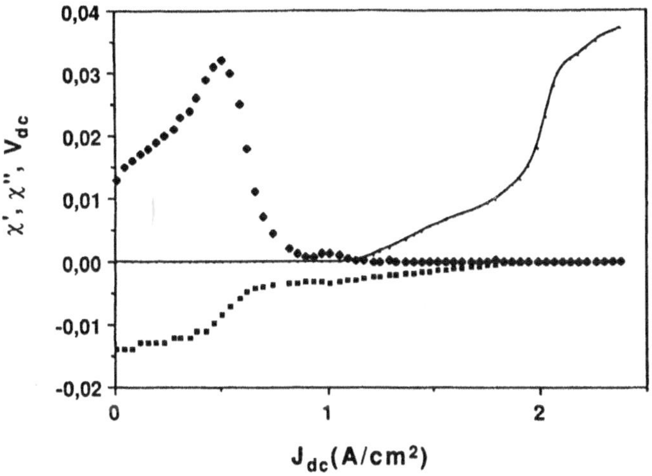

Fig 14. (▪) χ', (•) $10 \cdot \chi''$ in emu/g and $V_{dc}/50$ in mV simultaneous measurements in an $YBa_2Cu_3O_{7-\delta}$ ceramic. The temperature is kept constant at 84 K. An a.c. field of 0.55 mOe is applied parallel to the current. The voltage begins to increase almost in coincidence with the drop of χ'' to zero, but the material is superconductor until χ' becomes zero.

Electrical measurements directly determine the maximum transport current which the sample can bear in the non resistive state. However, the analysis of the magnetization or susceptibility data within critical state models also allows to derive values of J_c which may be different because it depend of other current trajectories. An experiment as we propose could unravel the mechanism of current distribution through a granular superconductor.

Simultaneously χ_{ac} components and V_{dc} have been recorded for a ceramic HTS sample through which an increasing dc current is led. In Fig. 14 it can be seen that the three quantities undergo significant changes for the same values of I_{dc}. Quantitative analysis of the data can be done in terms of critical state models [19] or of percolation models [20].

After the discovery of the high energy (*B-H*) product Nd compounds, the $RE_2Fe_{14}B$ series (RE= rare earth) have been the object of an extensive study. We have performed magnetic measurements on the pure and hydrogenated compounds, to study the Spin Reorientation Transitions (SRT) present in these compounds by competition of the anisotropies of the Fe and the RE sublattices.

A detailed study of the ac initial susceptibility of a $Ho_2Fe_{14}B$ single crystal was performed. It was measured along the [100], [110], and [001] directions (Fig. 15). The SRT present at T_{SRT}= 57.8 K give rise to an abrupt step-like anomalous increase in χ' for

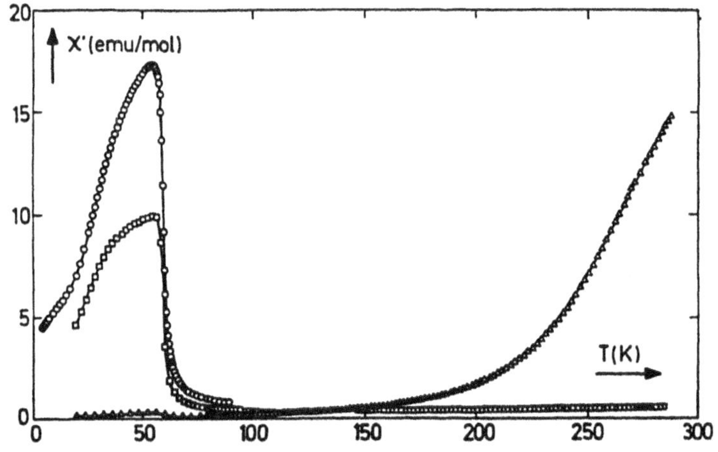

Fig 15. Magnetic susceptibility of $Ho_2Fe_{14}B$ single crystal with applied field along the (o) [110], ( ) [100] and (▽) [001] directions.

decreasing temperature. The results may be explained in terms of a reversible magnetic moment rotation produced by the ac magnetic field. Indeed, below T_{SRT} the transition towards the conical orientation sets in and the perpendicular component of *M* yields a significant contribution, giving rise to the strong increase observed both in the [110] and [100] directions. On the other hand, along the [001] direction no anomaly in $\chi'(T)$ is detected at T_{SRT}, but a strong continuous increase is manifest above 150 K. This increase is explained by the onset of narrow 180º domain wall motions induced by thermal excitations [21].

The resistance measurements as a function of temperature yield interesting information relative to the itinerant character of the 3d electrons of these compounds. Because of the low temperature domain of experimental accessibility, only the effects taking place in the Spin Reorientation Transition have been studied. Polycrystalline blocks of sintered $Nd_2Fe_{14}B$, as representative of a compound with second order SRT, and $Tm_2Fe_{14}B$, as the case with a first order one, were measured with the above described technique. No direct anomaly could be detected at first. However, after graphically evaluating $\partial R'/\partial T$, the anomaly was shown

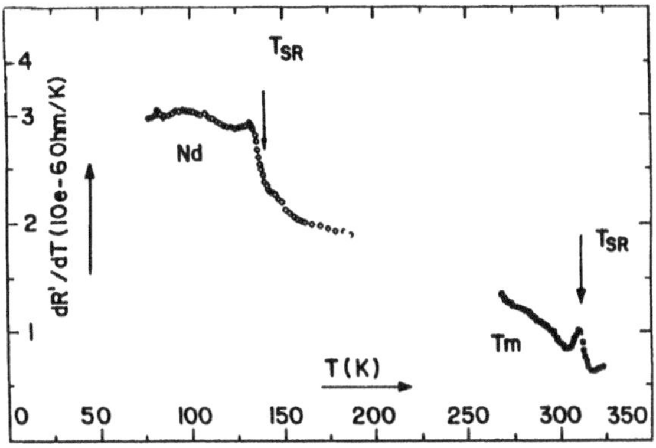

Fig 16. Temperature derivative of the resistance, $\partial R'/\partial T$.

to be present in each compound. In fact, for the Nd case, $\partial R'/\partial T$ presented a step-like anomaly, while in the Tm case, a small but definite peak was observed [22] (see Fig. 16). A possible explanation of this anomaly in $\partial R'/\partial T$ near T_{SRT} is the current dispersion caused by critical magnetic fluctuations.

MEASUREMENTS IN OTHER MATERIALS BELOW 4.2 K

As a representative example of the measuring capability of the instrument in the low temperature region we depict in Fig. 17 the zero-field (no screening of the earth's magnetic field) ac magnetic susceptibility of $CsMnF_4{\cdot}H_2O$ [23, 24]. The measurements show the rapid increase of the χ' values when the temperature decreases. The very sharp peak below 2 K is characteristic for the weak ferromagnetic nature of the magnetic alinement, which presents a maximum at 1.52 K with a half width of 0.3 K. The peak in χ' is accompanied by

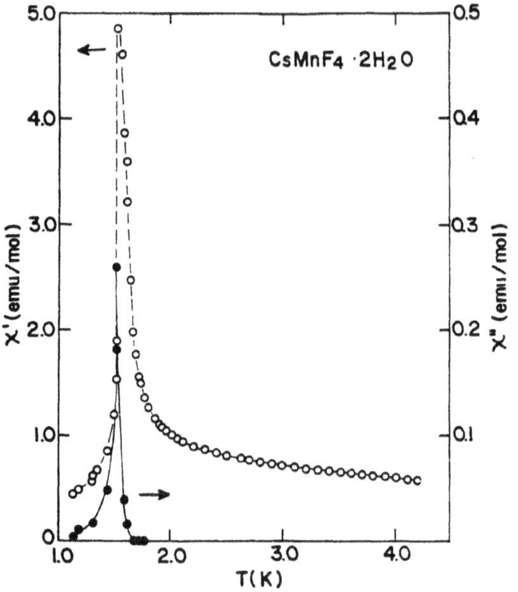

Fig 17. Temperature dependence of the a.c. susceptibility of $CsMnF_4$*H_2O

another one in the out-of-phase component, χ'', which presents a maximum at 1.49 K and a half width of 0.1 K. Such behavior of $\chi'(T)$ and $\chi''(T)$ is related to the absence of domain wall movements below T_c in weak ferromagnets.[25].

ACKNOWLEDGEMENTS

The authors would like to gratefully thank contributions from J.A. Rojo, J.A. Puértolas, F.J. Lázaro, P. Téllez, A. Camón, M. Andrés, C. Castro, L.M. García and M. Gabás to different projects in the developing process of the equipment described here. The research has been supported by several CAICYT and CICYT grants: MAT88-0174, MAT88-0268-CO2-02, MAT88-0152, MAT90-1174-CE, EC grants: SCI-0036-F and SCI-0389-C, and by the MIDAS Program (CICYT-REE-UNESA) no. 89/3797.

REFERENCES

1.- L.J.M. van der Klundert, C. de Rooji, M. Caspari, and L.C. van der Marel, Induction methods used in low temperature physics, Cryogenics 577:(1975).

2.- P.H. Muller, M. Schienle, and A. Kasten, Mutual inductance bridge for magnetic relaxation measurements at low temperatures, J. Mag. Mag. Mat. 28:341 (1982).

3.- M. Ocio and J. Hamman,Self-balanced bridge for automatic measurements of magnetic susceptibilities, Rev. Sci. Instrum. 56:1367 (1985).

4.- S. Ramakrishnan, S. Sudaram, R.S. Pandit, and Girish Chandra, An ac susceptometer from 1.5 to 300 K, J. Phys. E: Sci. Instrum. 28 650 (1985).

5.- R. A. Hein, ac magnetic susceptibility, Meissner effect, and bulk superconductivity, Phys. Rev. B 33:7539 (1986).

6.- S. Foner, Review of magnetometry, IEEE Trans. Magnetics 17:3358 (1981).

7.- A.F. Deutz, R. Hultsman, and F.J. Kranenburg, Automatic mutual inductance bridge for accurate ac susceptibility measurements from 1.2 to 300 K, Rev. Sci. Instr. 60:113 (1989)

8.- R.B Goldfarb and J.V Minervini, Calibration of ac susceptometer for cylindrical specimens, Rev. Sci. Instr. 55:761 (1984).

9.- Details about the electronic circuits can be obtained from the Servicio de Instrumentación Científica, Facultad de Ciencias, Universidad de Zaragoza, 50009 Zaragoza, Spain.

10.- C. Rillo, Ph. D. Thesis, I.S.B.N. 84-7733-083-9, University of Zaragoza (1986).

11.- F. Lera, Ph. D. Thesis, I.S.B.N. 84-7733-157-X, University of Zaragoza (1990).

12.- J.H. Landolt-Börnstein, Numerical data and functional relationships in science and technology, New series / editor in chief K.H. Hellwege, Berlin Springer-Verlag (1965).

13.- J.M. Rodríguez Fernández, R.J. López Sánchez, J.C. Gómez Sal, Dispositivo para medidas de resistividad eléctrica de alta precisión a bajas temperaturas en muestras metálicas cristalinas y amorfas, Anales de Física, serie B 82:203 (1986)

14.- K.H. Müller, J.C. Macfarlane, R. Driver, Nonlinear magnetic flux response in high temperature superconductors, Physica C 158:366 (1989).

15.- R. Navarro, F. Lera, C. Rillo, J. Bartolomé, $YBa_2Cu_3O_{7-\delta}$ low field diamagnetic properties: A multiharmonic analysis, Physica C 167:549 (1990).

16.- R. Navarro, F. Lera, A. Badía, C. Rillo, J. Bartolomé, W.L. Lechter, L.E. Toth, Critical current model analysis of inter- and intra- grain effects in a high density sinteres Tl-Ba-Cu-O ceramic, Physica C (Submitted).

17.- C. Rillo, L.A. Angurel, J. Bartolomé, J. Gonzalo, F. Lera, R. Navarro, A. Martínez, P. Téllez, On the sensitivity of high-T_c superconducting ceramics as magnetic field sensors, Sensors and Actuators A 25-27:775 (1991).

18.- L.A. Angurel, F. Lera, A. Badía, C. Rillo, R. Navarro, J. Bartolomé, J. Melero, J. Flokstra, R.P.J. IJsselsteijn, A.c. susceptibility harmonic analysis of the irreversibility lyne in an $YBa_2Cu_3O_{7-\delta}$ thin film, Proceedings of the E-MRS-ICAM'91 (in press).

19.- M.A.R. LeBlanc, Influence of transport critical current on the magnetization of a hard superconductor, Phys. Rev. Lett. 4:149 (1963).

20.- J. Aponte, H. Abache, and M. Octavio, Critical currents in bulk samples of $YBa_2Cu_3O_{7-\delta}$ and $DyBa_2Cu_3O_{7-\delta}$, Cryogenics 29:334 (1989).

21.- C. Rillo, J. Chaboy, R. Navarro, J. Bartolomé, D. Fruchart, B. Chenevier, A. Yaouanc, M. Sagawa, and S. Hirosawa, Dynamical susceptibility of $Ho_2Fe_{14}B$ single crystal: Spin rotation and domain wall motions, J. Appl. Phys. 64:5534 (1988).

22.- F. Lázaro, L.M. García, F. Luis, C. Rillo, J. Bartolomé, D. Fruchart, S. Miraglia, S. Obbade, and K. H. J. Buschow, Systematic ac susceptibility study of $(RE)_2Fe_{14}BH_x$ and $(RE)_2Fe_{14}CH_x$, EMMA , Dresden (1991), o J. Magn. Magn. Mat. in press

23.- F. Palacio and M. Andrés, Magnetic behavior of hydrated Tetrafluoromanganates (III) in "Organic and Inorganic Low Dimensional Crystalline Materials", P. Delhaes and M. Drillon ed.,Plenum Press, (1987).

24.- M. Andrés, Ph. D. Thesis, I.S.B.N. 84-7733-160-X, University of Zaragoza (1990).

25.- H.A Groenendijk, A.J. van Duyneveltdt, and R.D.Willet,Experimental study of the effect of domains on the ac susceptibility of the weak ferromagnet $(C_3H_7NH_3)_2MnCl_4$ Physica B+C101:320 (1980)

AC SUSCEPTIBILITY RESPONSES OF SUPERCONDUCTORS : CRYOGENIC ASPECTS, INVESTIGATION OF INHOMOGENEOUS SYSTEMS AND OF THE EQUILIBRIUM MIXED STATE

M. COUACH and A.F. KHODER

Centre d'Etudes Nucléaires de Grenoble
DRFMC/SPSMS/Laboratoire de Cryophysique
85 X - 38041 GRENOBLE Cedex - France

ABSTRACT

With the discovery of high T_c superconductors, ac magnetic susceptibility measurements have become widely used for sample characterization as well as for fundamental studies. The present article, intended to introduce the reader to these areas of research, emphasizes ; (a) the need for precise temperature determination, (b) geometrical concerns associated with quantitative measurements of χ' and χ'', the in-phase and out-of-phase components of the complex ac magnetic susceptibility $\chi^{ac} = \chi' + i\chi''$, (c) the use of an externally applied dc magnetic field to obtain additional information on inhomogeneous superconductors and the study of the mixed state. The magnetic responses of both single and multiphase systems of classical and of high T_c superconductors are treated in detail.

A. INTRODUCTION

The discovery of high T_c superconductors has renewed the interest and popularised the use of the ac susceptibility technique as a versatile method of characterization and investigation of the superconducting materials.

This technique has been widely used in the investigation of low T_c type I and type II superconductors [1,2,3,4,5] as well as other magnetic materials [6,7,8,9,10] where dynamic aspects (relaxation phenomena) could be of major importance. The usefulness of this technique and its relevance beyond its use as a characterization method can be severely limited by experimental difficulties such as : sensitivity and noise level, signal drift, temperature control and thermal equilibrium within the system of the sample, calibration and phase setting of the ac signal. All aspects which will be examined in some details. The paper is organized as follows. Section B is devoted to the description of the basic elements of the ac susceptibility coil assembly, bridge design and the set-up control of flux equilibrium. Cryogenic considerations in relation to the thermal equilibrium problem will also be emphasized.

Magnetic Susceptibility of Superconductors and Other Spin Systems
Edited by R.A. Hein *et al.*, Plenum Press, New York, 1991

Signal calibration which is a function of the measuring coils and sample geometry, is discussed and expressions for the filling factor α of some simple geometries are given.

In section C, χ_{ac} of selected examples of multiphase (inhomogeneous) superconductors are discussed. The importance of the application of a dc magnetic field and of a quantitative investigation of the differential susceptibility χ'_H, especially in the reversible mixed state, will be particularly emphasized in section D.

B. PRINCIPLES OF THE χ_{ac} MEASUREMENT

1. The mutual inductance technique

The ac susceptibility measurement is based on a mutual inductance technique where a primary coil and two secondary (oppositely wound) coils form the basic unit of the measuring circuitry. In the absence of a sample, which is usually centered in one of the secondary coils, the detection system should be ideally in equilibrium i.e. the net flux Φ_{net} accross the secondary coils is zero. In the presence of a sample the induced magnetization due to the ac primary field $h = h_0 e^{-i\omega t}$, will result in an off-balance signal of the secondary coils detection system given by :

$$|e| = \left| - \frac{d\Phi}{dt} \right| = \alpha \; n_s \; V \; \omega \; b_0 \; |\chi_m| \tag{1}$$

α : filling factor (see paragraph 5)
n_s : number of turn per unit length (meter) of a secondary coil
V : volume of the sample (m^3)
$b_0 = \mu_0 \; h_0$ magnetic induction (Tesla)
$\omega = 2 \pi f$, f frequency
e : induced voltage (volts)

with the above S.I. units, $\chi_m = \chi_{(measured)}$ is dimensionless.

In the case of demagnetizing effects

$$\chi_m = \frac{\chi}{1 + D\chi}$$

where χ is the effective susceptibility of the sample. For a superconductor $\chi = -1$ and $\chi_m = - \frac{1}{1 - D}$.

2. The coils assembly and the bridge design

The primary and secondary coils are wound on a cylindrical insulating holder or coil form. Secondary coils are first wound in opposition with an equal number of turns symmetrically vis-a-vis the center of the coil form. A long homogeneous test coil with known magnetic center is used to insure the symmetry of the secondary coils system and to adjust the number of turns to obtain perfect compensation of flux in the two coils $\left(e = - \frac{d\Phi}{dt} = 0\right)$. In the next step, the primary coil is wound over the secondaries. Such an assembly always displays an appreciable flux imbalance due to the inhomogeneity of the primary field over the secondary coils and to achieve perfect flux balance, an independent compensation coil is wound upon one of the secondary coils (the one which is not designed to contain the sample). The perturbation introduced by

Table I : Specifications of the coil assembly : the two secondary coils are separated by a distance of 10 mm.

	Primary coil	Secondary coil	Compensation coil
Diameter (mm)	21	16	21
Length (mm)	100	20	20
Number of turns	21800	3650	155
Inductance (H)	2,5	0,267	

the compensation coil on the excitation field is around 0.1 % and can be considered as negligeable for the detection coil containing the sample. Table I summarizes the parameters of a typical coil assembly (wire size 0.1 mm)..

When operating with an applied dc magnetic field superimposed on the ac measuring field, it is necessary to fix, very rigidly, the coil assembly to the support structure of the static field coil in order to avoid any relative displacement (mechanical vibrations).

Schematic representations of the bridge design and of the different apparatus used in connection with the primary, secondary and compensation coils is given in figure 1. A synthetizer HP 3326A with two synchronized outputs is used as a signal generator. One output is connected to a current amplifier and to an off balance circuitry delivering a controlled amplitude and phase current to the compensation coil. The second output of the synthetizer serves as a reference to the phase sensitive detector

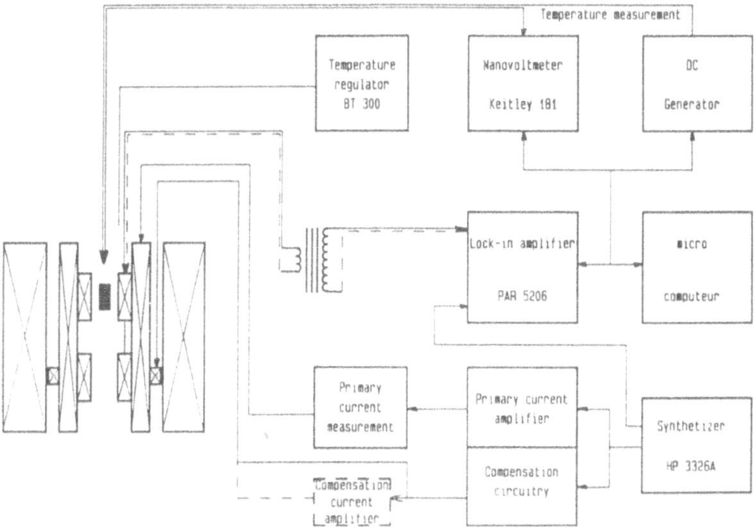

Figure 1 Block diagram of the ac susceptibility set-up

(PSD) PAR 5206 which is connected to the output of the secondary coils. Finally the PSD is connected to a micro computer. The temperature of the sample holder is controlled separately by a regulator acting on a resistance heater wound on a sapphire rod (Figures 2 and 3) which allows a continuous linear sweep in temperature with different rates ranging from 2.10^{-2} K/min to 2 K/min.

3. The set-up protocol and the adjustement of the flux equilibrium

Even without a sample being present, the coil system displays a flux imbalance which is compensated by the supplementary "compensation" coil. The current in the compensation coil is phase and amplitude controlled by means of an adequate compensation circuitry. The out-of equilibrium signal to be compensated $e = -\dfrac{d(\Phi_1 - \Phi_2)}{dt} \sim \iota\, \omega\, \mu_0\, h_p$ has an amplitude which increases with ω and h_p and a phase of 90° vis-a-vis the primary field h_p (no dissipation in the absence of a sample). This off-balance signal is thus used as a reference to adjust the rotation axis (phase setting) of the PSD. This allows one to avoid all the uncontrollable phase shifts which originate from the excitation circuitry (current amplifier) as well as from the detection circuitry (low noise transformer). After the axis rotation of the phase sensitive detector, the off-balance signal is compensated by the intermediary of the compensation coil.

Finally the validity of this adjustment may be verified, and eventually slightly corrected by the use of a superconducting test sample at 4.2 K for which the condition of null dissipation ($\chi'' = 0$) is always fulfilled in the superconducting state provided that the primary field $h_p < H_{c_1}$ (ex : Nb-Ti H_{c_1} (4.2 K) $\sim$ 400-500 Oe).

The application of a dc magnetic field is of great importance in the study of superconductors. In such cases, the introduction of a superconducting coil connected to a low impedance circuitry results, by mutual interaction with the primary coil circuit, in a reduction of the primary field amplitude and some phase shift. Flux considerations allows one to give an estimate of the back field h_s due to the shorted superconducting solenoid :

$$h_s \sim \alpha \frac{N_p}{N_s} \times \frac{S_p}{S_s} h_p$$

where

α : filling factor (see paragraph 5)
N_p : total number of turns of the primary coil
S_p : mean section of the primary coil
N_s : total number of turns of the superconducting coil
S_s : mean section of the superconducting coil

The effective primary field, as seen by the secondary coils and by the sample is $h_p - h_s$ and is not simply given by the primary driving current.

Preliminary experimental calibration is necessary to determine the actual field amplitude in the presence of the closed dc circuitry. This also can be done by the use of superconducting test sample. The phase adjustement is made by setting $\chi'' = 0$ in the superconducting state with closed dc circuitry but without the dc field. This adjustment should be revised if the frequency is changed.

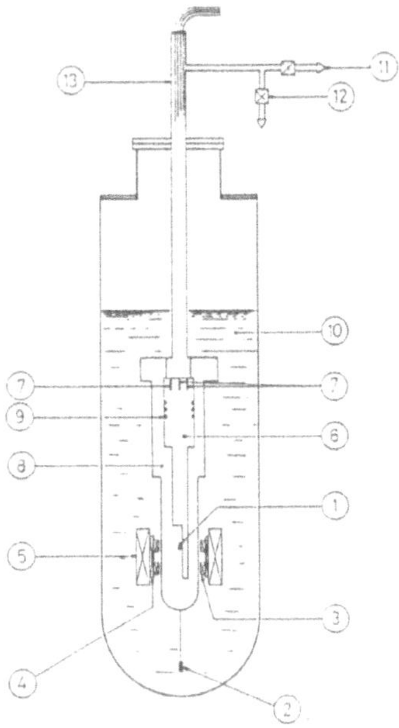

Figure 2 Schematic diagram of set-up A

(1) Sample
(2) Calibration sample
(3) Secondary coils
(4) Primary coil
(5) Superconducting magnet
(6) Sapphire
(7) Thermometers
(8) High vacuum
(9) Resistance heater
(10) Helium bath
(11,12) Pumping valves
(13) Moving tybe

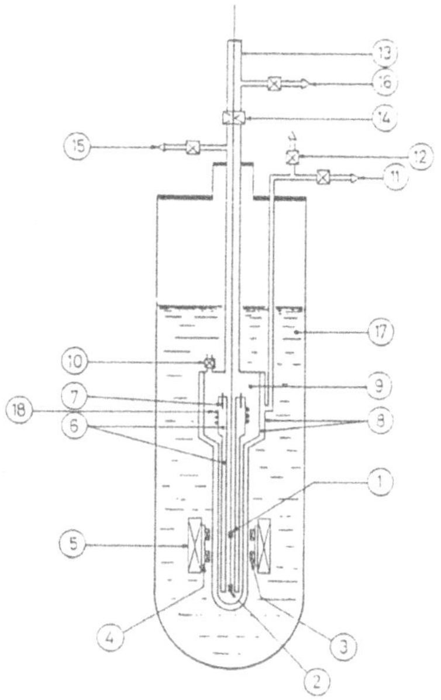

Figure 3 Schematic diagram of set-up B

(1) Sample
(2) Calibration sample
(3) Secondary coils
(4) Primary coil
(5) Superconductivity magnet
(6) Sapphire
(7) Thermometer
(8) Double walled glass tube
(9) Helium gas 0,1 mmHg
(10) Cold valve
(11,12,15,16) Pumping valves
(13) Assembly for introducing sample
(14) Isolation valve
(17) Helium bath

4. Coils and sample thermalization

To avoid a major contribution to any signal drift with temperature, the coil assembly is maintained at constant temperature in liquid helium. The temperature of the sample which is maintained at a fixed position in the center of one of the secondary coil can be controlled by two methods.

The first one (A) is to place the sample in direct physical contact with a sapphire holder [11] (figure 2) whereas in the second method (B) thermal contact with the sapphire is obtained by helium exchange gas (figure 3).

The principle of thermal coupling in A is to use the high level of thermal conductivity of pure sapphire [12]. Around 30 K, the thermal conductivity can reach value as high as 100 W $cm^{-1}K^{-1}$ and is no lower than 1 W $cm^{-1}K^{-1}$ in the temperature range 2-150 K. Moreover sapphire is an insulating material which can be obtained on a large scale with very low impurities level [13] and shows no appreciable magnetic contribution if choosen very pure. These two physical properties allow one to realize intimate thermal coupling of the sample inside the measuring coil while the heater and the thermometers are placed outside the coil system to minimize the effects of all extra temperature dependent susceptibilities. In order to compensate any residual magnetic contribution a reduced section of the sapphire enters the two secondary coils as shown in figure 2. The sample is coupled to the sapphire by silicon grease. The sapphire is contained in a small insulating cryostat in which high vacuum is maintained during the experiment. Thermal shielding is achieved by

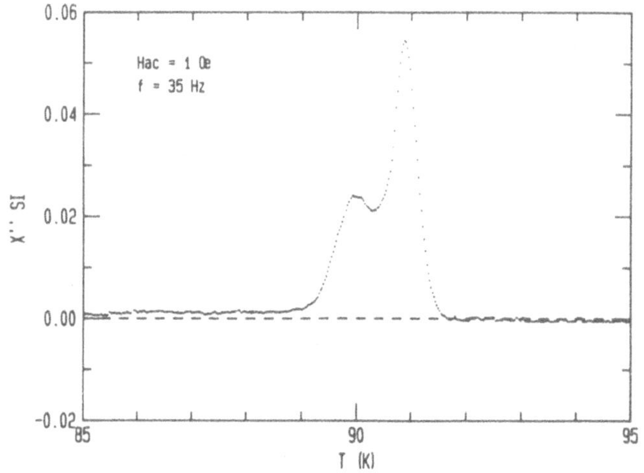

Figure 4 Transition of a single crystal $YBa_2Cu_3O_{7-\delta}$ (χ" component) with ac field parallel to the ab plane showing subtle details evidenced by a carefull choice of the rate change of temperature (0,5 K/min) and the acquisition time (2 sec).

introducing a glass tube between the sapphire and the insulating cryostat. Such an arrangement allows dynamical investigation in the temperature range 4-150 K with a maximum linear rate of 2 K/min to minimize the temperature gradient between the sample and the thermometer (carbon glass or Fe-Rh). This gradient is dependent on the temperature change rate. At 0.5 K/mn the uncertainty is no more than 0.05 K at 90 K. By this way, good thermalization of the sample is achieved which allows one to increase significantly the resolution in temperature by optimizing the acquisition time and the rate of the linear rise in temperature (figure 4,6). The main advantage of this type of coupling is to allow one to perform rapid experiments at 2 K/min with an uncertainty in temperature no more than 0.15 K and to study more precisely at a lower rate the domain of temperature where the interesting phenomena occur.

The main disadvantage of method (A) comes from the necessity to reheat the whole cryogenic system to room temperature when dismounting the sample. Such a system is suitable for long term studies but not for rapid characterization. For this last purpose set-up B was designed (figure 3) which allows one to directly introduce the sample inside the coil assembly. Helium exchange gas at a pressure of 0.1 mm Hg is used to keep the sample in thermal equilibrium with the machined sapphire tube which supports the heater and thermometers. The whole assembly is contained in a double walled glass tube. A typical precision in temperature obtained is 1 Kelvin at 1 K/min rate around 90 K.

Besides the possibility of directly introducing the sample from room temperature, method B is also capable of reaching temperatures as low as 1 K. This is accomplished by first permitting liquid helium to enter the inner double-wall cryostat by means of a cold valve, see fig. 3, and then reducing the vapor pressure over the liquid helium.

5. Calibration of the experiment

Calibration of the experiment is an important point to achieve quantitative measurements. The calibration of the response of the measuring coils i.e. determining the coefficient α of formula (1) can be done in two ways : first by using a good superconductor test sample of known dimensions with almost zero demagnetizing factor. α can be deduced from the condition $\chi' = -1$ in the superconducting state. Another method is to calculate α taking into account the geometrical dimensions of the secondary coils and of the sample [14]. In the case of very small samples with magnetization M, M can be approximated to that of a dipole and the flux of the dipole threading the measuring coil can be calculated

$$\Phi_m = \alpha \, \mu_0 \, M \, n$$

n number of turns by unit length

$$\alpha = \frac{L/R}{(1 + L^2/R^2)^{1/2}}$$

2L and 2R being respectively the length and the diameter of a coil.

The flux Φ_2 enclosed by the other secondary coil can be calculated in a similar way

$$\Phi_2 = \mu_0 \, M \cdot n \cdot \alpha_2$$

$$\alpha_2 = \frac{1}{2}\left\{\frac{\frac{L}{R}\left(3 + \frac{d}{L}\right)}{\left(1 + \frac{L^2}{R^2}\left(3 + \frac{d}{L}\right)^2\right)^{1/2}} - \frac{\frac{L}{R}\left(1 + \frac{d}{L}\right)}{\left(1 + \frac{L^2}{R^2}\left(1 + \frac{d}{L}\right)^2\right)^{1/2}}\right\}$$

d being the distance separating the two secondary coils.

As the two secondary coils are in opposition the net flux in the system is

$$\Phi_{net} = \mu_0 \, M \, n \, (\alpha - \alpha_2)$$

For a typical set of coil as detailed in table 1 one obtains

$$\begin{cases} \alpha = 0.78 \\ \alpha_2 = 0.05 \end{cases} \qquad \alpha_{net} = 0.73$$

In the case of a cylindrical sample of length 2a with radius $r \ll R$, the magnetization can be considered as a linear distribution of dipoles. One obtains a more complicated expression for $\alpha = \frac{R}{2a}(X - Y)$ with :

$$X = \left[\left(\frac{L + a}{R}\right)^2 + 1\right]^{1/2}$$

$$Y = \left[\left(\frac{L - a}{R}\right)^2 + 1\right]^{1/2}$$

In a similar manner a coefficient α_2 can be calculated to take into account the other secondary coil. For the last set of coils (table I) one obtains for a sample length $2a = 16$ mm $\alpha = 0.70$ and $\alpha_2 = 0.03$ $\alpha_{net} = \alpha - \alpha_2 = 0.67$. An excess error of 8 % is made by neglecting the correction due to the sample length. The calculation of α is in good agreement with the experimental determination.

6. Sensitivity problems

The principal problems encountered in χ_{ac} measurements are sensitivity, temperature dependent signal drift and noise.
The noise can be reduced by avoiding mechanical vibrations and by matching low impedance sources to high impedance measuring circuitry.

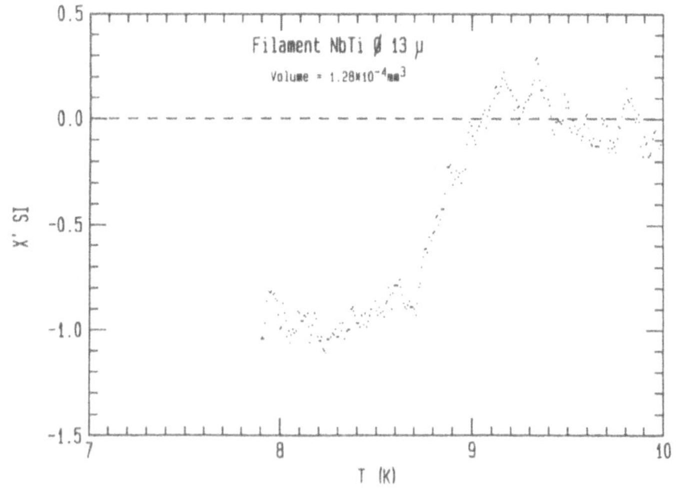

Figure 5 Superconducting transition of a Nb-Ti filament diameter 13 μm length 0.95 mm - ac field 30 Oe - 35 Hz

Although the background signal drift between 4 K and 150 K is more than 1.5 10^{-9} A.m^2 (for h_p = 30 Oe and f = 35 Hz), the sensitivity may reach the level of 1.5 10^{-10} A.m^2 in a reduced range of temperature such as that of a small ΔT around the critical temperature of a small superconducting particle ($\sim 10^{-4}$ mm^3) as shown in figure 5 where the noise level is about half the sensitivity ($V\chi \sim 2.4\ 10^{-12} m^3$, see formula (1) for the significance of the symbols).

C. MULTIPHASE SUPERCONDUCTING SAMPLES

Frequently, problem of homogeneity arises for many different reasons which results in a broadening of the superconducting phase transition and often in many well separated T_c's. In the search for new superconducting materials, identification of potential superconducting phases and a quantitative estimation of their ratio can be of great importance and correlations can be done with various structural characterizations (X rays, etc...).

On the other hand, different T_c's can originate for the same macroscopic phase due to either a lack of stoichiometry or the presence of structural defects. In this section we will give examples showing the efficiency of carefull ac susceptibility investigations in the clarification of such problems. Screening of minor superconducting phases by the major one should be, however, considered with much care. Practical solutions to avoid such situations will be discussed : powdering samples [5] is not the only way to detect the minor screened phases.

1. Case of A15 compound : V_3Si

Before the discovery of the new high T_c, the A15 superconductors have been intensively studied. They offer examples where superconductivity and structural instabilities were supposed to be in competition [15] or originating from the same electronic instability [16]. A good representative of this family is V_3Si which undergoes a structural phase transition from cubic to tetragonal symmetry just a few Kelvins above T_c ($T_c \sim 17$ K, $T_m \sim 22$ K). This crystallographic phase transition is known to be very sensitive to structural defects : point like or off stoichiometric defects. Only very good quality samples with resistivity ratio RRR > 25 are expected to be transforming.

This structural phase transition can be revealed by specific heat and static susceptibility anomalies [17,18], but in any cases the observation of such anomalies does not allow one to know the extent of the structural transformation and it was common to use terms as "non transforming" and "partially transforming" to characterize V_3Si samples.

In an earlier work [19] this problem was approached indirectly, by investigating the superconducting transition of V_3Si single crystals. Figure 6 displays the χ', χ'' components of χ_{ac} of such a single crystal. Well separated peaks in χ'' develop in a temperature range $\Delta T \sim 0.4$ K causing one to question the homogeneity, on large scale, of the single crystal. Additional insights on the large scale homogeneity problem have been obtained by γ rays diffraction experiments which revealed low angle (< 1°) mosaïc structure of the investigated single crystal which is actually formed from many grains with small relative disorientation (figure 7). A good correspondance between the γ rays diffraction spectrum (12 peaks) and the χ'' peaks (11 χ'' peaks and shoulders) results can be made ; suggesting that each χ'' peak is associated with superconductivity developing in one grain. Moreoever by investigating the evolution of the

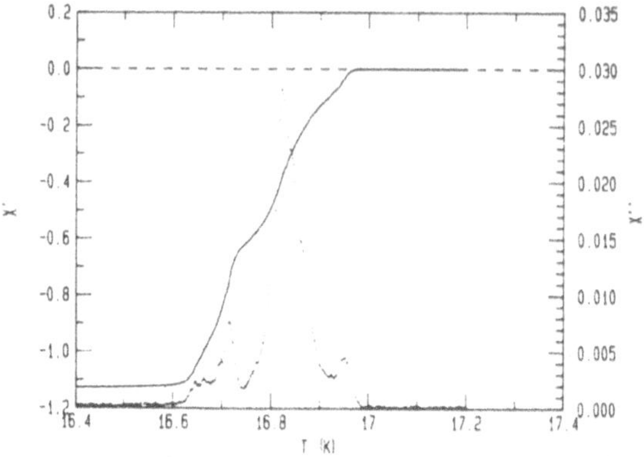

Fig. 6 χ' and χ'' components for a V_3Si single crystal with an ac field of 0.05 Oe at 35 Hz. Linear rate change of temperature 0.05 K/min, acquisition time 1.6 sec.

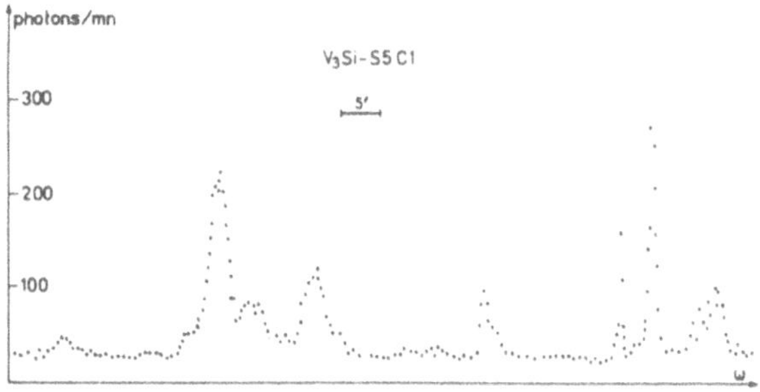

Fig. 7 γ-ray diffraction rocking curve revealing mosaic structure of the same single crystal of V_3Si.

χ'' peaks under static magnetic fields it was possible to separate the behavior of the transforming from non-transforming grains which have different "critical" field slopes $\frac{d\,H^*_{c_2}}{dT} \approx 1.85$ T/K for the transforming grains and $\frac{dH^*_{c_2}}{dT} \simeq 2.25$ T/K for the non transforming one.

More or less efficient shielding due to the relative orientation of the grains with the field may affect the magnitude of the transition. Moreoever the maximum value of the χ'' peak might appear very low in comparison to the χ' signal as theoretical consderations [20] predict the ratio $\chi''/\chi' \sim 0.2 - 0.4$. But in this complicated case each χ'' peak has to be compared with the correspondant χ' transition.

The major conclusion of such investigation was to show the efficiency of χ_{ac} data to separate superconducting phases with very close T_c's and to allow correlation in this case with structural data on the same sample. The use of dc fields, in this case, is of prime importance for obtaining additional insight in the identification of the superconducting phases.

2. Case of the high T_c superconductors

The same kind of problem arises in the case of the high T_c superconductors. It is now well known that oxygen off-stoichiometry plays a major role in determining T_c (i.e. $YBa_2Cu_3O_{7-\delta}$). On the other side, it is not always possible to grow a single phase sample because of the lack of sufficient knowledge of the phase diagram of these multicomponent compounds. Many T_c's are usually encountered in investigating these new superconductors. The oxygen inhomogeneous off-stoechiometry problem has been investigated in the case of early prepared $YBa_2Cu_3O_{7-\delta}$ single crystal grown by flux method and treated under 20 bars of oxygen. The mechanism of oxygen diffusion has been revealed to be very anisotropic [21] and a stair case structure has developed in this single crystal with oxygen step-like distribution along the C-axis [22].

The ac susceptibility investigation was performed along the two different crystallographic directions (⊥ and ∥ to c-axis). When the field is parallel to the basal plane, the imaginary susceptibility shows a distribution of peaks ranging between 64 ans 85 K (fig. 8) characterizing different superconducting phases. When the same field is applied along the c-axis we essentially observe a single χ'' peak at 80 K (fig. 9). The comparison of the different responses for H_{ac} parallel and perpendicular to the c-axis allows one to determine the geometrical arrangement of the different superconducting phases of the single crystal. This behaviour can be understood in detail if one considers a platelike structure

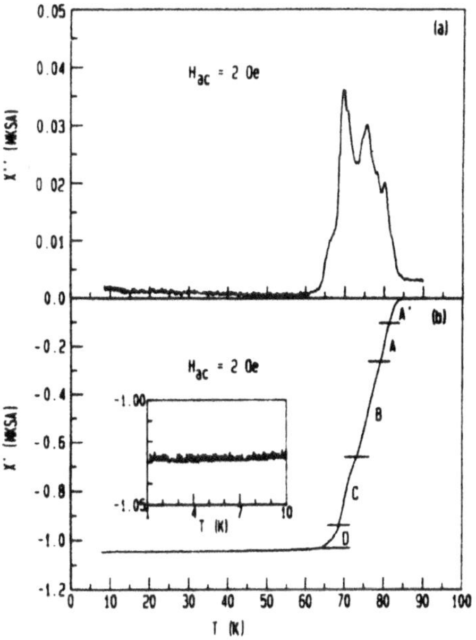

Fig. 8 $\chi''(T)$ and $\chi'(T)$ for H_{ac} (2 Oe) parallel to the ab-plane show different superconducting phases A', A, B, C, D.

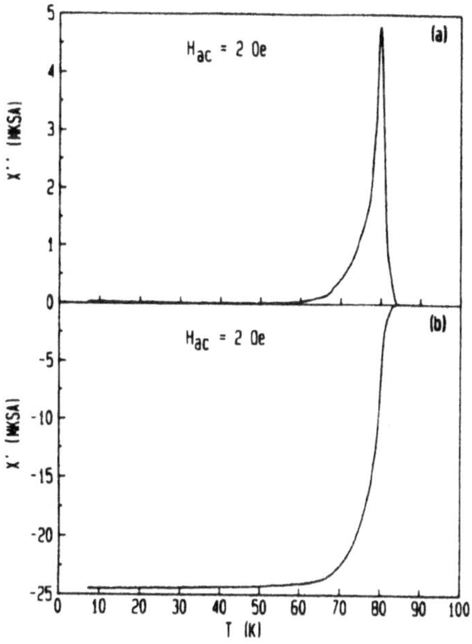

Fig. 9 $\chi''(T)$ and $\chi'(T)$ for H_{ac} (2 Oe) parallel to the c-axis show only the A superconducting phase.

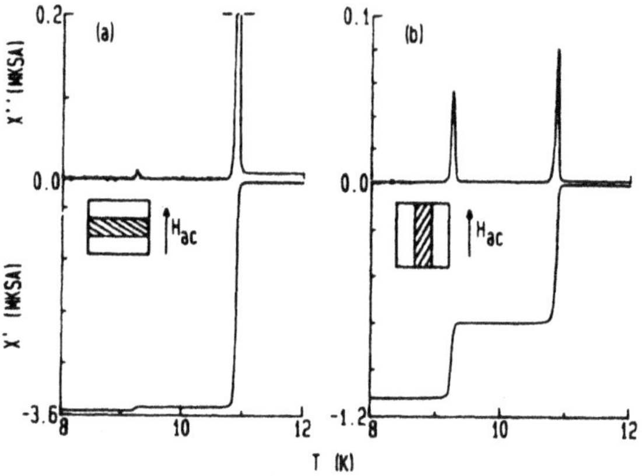

Fig. 10 Simulation of the screening effect with NbTi plate sandwiched between two NbZr plates (a) H_{ac} is perpendicular to the plates (b) H_{ac} is parallel to the plates.

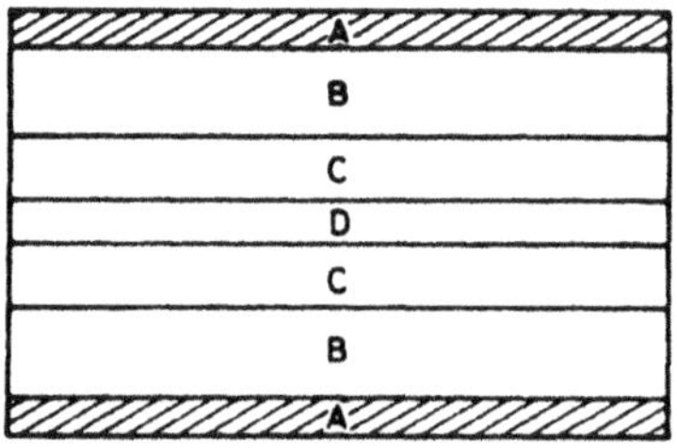

Fig. 11 ac susceptibility experiments performed with the ac field // and ⊥ c-axis allows one to give a schematic view of the platelet structure. The different superconducting layers A, B, C, D are perpendicular to the c-axis and correspond to decreasing T_c.

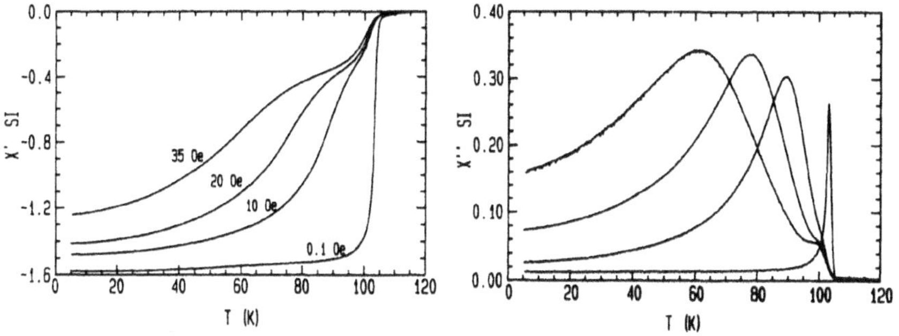

Fig. 12 ac suceptibility transition of a Thallium ceramic compound 1223 with increasing ac field amplitude allowing one to separate the granular and intergranular contribution.

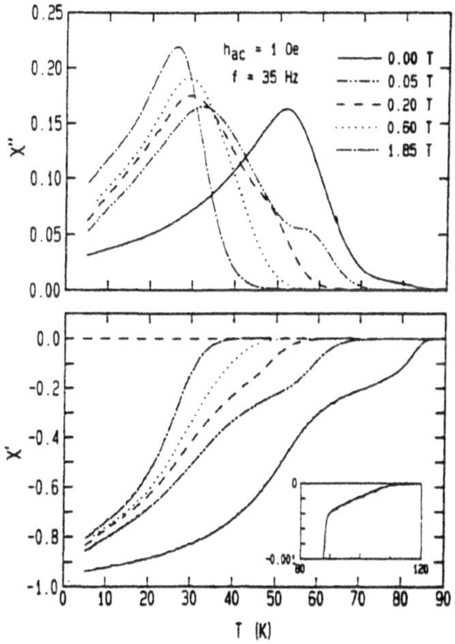

Fig. 13 ac susceptibility responses of a ceramic compound $Bi_2Sr_2CaCu_2O_{0.8}$ with an ac field of 1 Oe and static fields up to 1.85 T. Insert 2D fuctuations below 110 K.

(fig. 11) where the lowest T_c phases are sandwiched between two extreme layers of the $T_c \simeq 80$ K phase. Such a structure has been checked by an experimental simulation using a NbZr plate ($T_c = 10.9$ K) separated by a NbTi plate ($T_c = 9.3$ K) which gives the susceptibility response of fig. 10 and displays a striking similarity with results of figs. 8-9.

On the other hand, multiphase problem arises in the case of Bismuth and Thallium based cuprates. In the case of ceramic multiphase samples, the superconducting phase with lower T_c can be screened by the highest T_c phase due to intergranular shielding. In spite of powdering the samples to cut down the intergranular coupling, the presence of such coupling may still be detected by either applying a large dc field or/and working with large ac field amplitude.

Figure 12 shows in the case of a Thallium based compound a sequence of χ_{ac} behavior for different ac field amplitude ranging from 0.1 to 35 Oe. The low field behavior appears very sharp while it becomes more complex for larger amplitudes [31]. The granular response is hidden by the intergranular shielding at low amplitudes where the shielding currents are below the critical currents of the intergranular junctions at temperature very close to T_c of the grains [23].

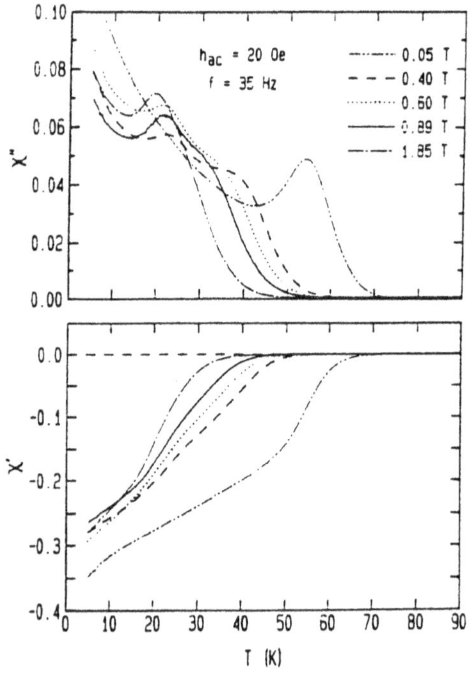

Fig. 14 The same Bi compound with an ac field of 20 Oe and static fields up to 1.85 T. The intergranular shielding is strongly reduced ($\chi' = -0,3$) for $h_{ac} = 20$ Oe.

Thus, increasing the amplitude of the alternating field h_{ac} results in a more rapid shift to lower temperatures of the intergranular response relatively to the granular one.

Figure 13 shows another sequence of χ_{ac} behavior at constant ac field amplitude of 1 Oe and for different dc fields up to 1.8 T in the case of multiphase Bi-based ceramic.

At low dc field (0.05 T) two χ'' peaks develop, one corresponding to the response of the grains per se and the second to the intergranular response. For highest field, the "grain" peak shifts rapidly to lower temperature due to the large reversibility of the mixed state for these Bi compounds (see further) and coalesce with the intergranular peak which appears to be less sensitive to the dc field effect.

Figure 14 shows the same sequence as figure 13 for a higher ac field amplitude (20 Oe). In this case the intergranular contribution is already shifted to very low temperature even at the lowest dc field and the χ'' peak for the intergranular shielding is no more observable in the temperature range investigated ($T \geqslant 4.2$ K). Meanwhile a third χ'' peak is revealed with the dc field increase. This new peak, absent in the case of low ac field amplitude, is obviously due to an another phase and has been revealed only when the intergranular shielding has been made ineffective by the conjunction of dc and ac fields. The effect of the ac field amplitude appears to be more critical for such an investigation.

Intergranular effect extensively studied with the appearance of high T_c superconductors are in fact not limited to this class of superconductors. Detailed studies performed on Chevrel phases compounds ($Pb\,Mo_6S_8$) have shown by ac susceptibility measurements the existence of intergranular effects clearly substantiated by comparison with specific heat data [24].

D. EQUILIBRIUM MIXED STATE INVESTIGATED BY ac SUCEPTIBILITY : THE DIFFERENTIAL PARAMAGNETIC EFFECT (DPE)

In the new high T_c superconductors all problems related to irreversibility and reversibility properties are of major importance for fundamental and practical purposes [25]. Each compound can be characterized in the (H, T) plane by an irreversibility (reversibility) line defining a domain where magnetic properties are reversible. Many methods [26] allow the investigation of the irreversibility line but among them ac susceptibility offers a reliable and versatile way to determine the position of such a line where the experimental time scale is well defined. Moreoever the extent in temperature of the reversible regime is easily measured by the temperature where shielding begin to occur [27].

In the reversible state the differential susceptibility $\chi'_H(T)$ is essentially governed by the equilibrium magnetization properties of the mixed state. Retardation (shielding) effects due to flux flow resitivity can be made negligeable by an adequate choice of the frequency i.e. by taking a sufficiently low value.

In this regime, $\chi'_H(T)$ can be developed as follows [4,32]

$$\chi'_H(T) = \frac{dM}{dH} - \left(\frac{1}{\mu_0}\frac{dB}{dH}\right)^3 \cdot \left(\frac{\rho_n}{\rho_f}\right)^2 \frac{a^4}{12\,\delta_n^4}$$

where a is the radius of the (cylindrical) sample, δ_n the normal state penetration depth, ρ_n and ρ_f are respectively the normal state and the flux flow resistivities. As $\frac{1}{\mu_0}\frac{dB}{dH} \sim 1$ and δ_n ($\delta_n(cm) \simeq 5000\ (\rho/f)^{1/2}$, ρ in Ω.cm) is of the order of 10 cm for $\rho_n = 10^{-4}$ Ω.cm and f = 10 Hz, the dynamical (negative) contribution to χ'_H becomes comparable to $\frac{dM}{dH}$ ($\sim 10^{-4}$) when $\frac{\rho_f}{\rho_n} \sim 10^{-3} - 10^{-2}$.

Thus the domain of reversibility, as it can be investigated by χ'_H, is extended nearly over the whole resistivity enlargement under magnetic field [28] or conversely it is possible to deduce the enlargement of the resistivity domain under the same field from the positive χ'_H.

As far as the dynamic contribution can be neglected, the χ'_H gives the equilibrium differential susceptibility

$$\chi'_H = \frac{dM}{dH}$$

According to the Abrikosov-Ginsburg-Landau theory, the equilibrium magnetization of the mixed state in the field H [29] can be written for the case $H_{c_1} \ll H \ll H_{c_2}$ (London regime)

$$M = \frac{B}{\mu_0} - H = - H_{c_1} \frac{\text{Ln}\left(H_{c_2}/H\right)}{\text{Ln}\ \kappa}$$

κ being the constant of Ginzburg-Landau. One deduces

$$\frac{dM}{dH} = \frac{H_{c_1}}{H\ \text{Ln}\kappa} \quad ; \quad H.\chi'_H = \frac{H_{c1}}{\text{Ln}\ \kappa} \tag{2}$$

For the case $H \sim H_{c_2}$ the theory provides

$$B = \mu_0\ (H + M) = \mu_0 \left(H - \frac{H_{c_2} - H}{(2\ \kappa^2 - 1)\ \beta_A}\right)$$

and gives

$$\frac{dM}{dH} = \frac{1}{(2\kappa^2 - 1)\ \beta_A} \tag{3}$$

At $T_{c_2}(H)$ one can expect from equation (3) with reasonable value of $\kappa \simeq 100$-300 and $\beta_A \simeq 1$ a step-like increase $\frac{dM}{dH} \sim 10^{-5}$. As it will be shown this step-like increase in the susceptibility is not observed although the sensitivity of the set-up is high enough. Many reasons can explain the absence of this step-like feature in $\chi'_H(T)$ (inhomogeneities, T_c distribution, κ temperature dependence,...).

On the other hand for the London regime $H_{c_1} \ll H \ll H_{c_2}$, one finds a positive $\chi'_H(T)$ as expected and which follows the 1/H functional dependence given by equation 2 [30].

The first experiments were performed on a bismuth based compound containing the major phase $Bi_2Sr_2CaCu_2O_8$ and already studied in section C.2. The order of magnitude of the χ' value expected in the positive region is aroud 10^{-4}-10^{-5} and experimental evidence of such an effect required high sensitivity experiments with an applied dc field ranging from 0 to 1.8 Tesla. The setting of the experiment was performed with great care according to the procedure described in section B. The $\chi'_H(T)$ results are displayed on figure 15 for fields up to 0.8 Tesla. Note that the 0.05 T curve displays a quite unexpected behaviour showing a small diamagnetic contribution between the positive $\chi'_H(H = 0.05\ T)$ region and the normal state. Diamagnetic fluctuations may be responsible for such a contribution otherwise it would imply the existence of a non reversible phase a quite unexpected result. For fields higher than 0.05 T the results are analyzed in the form $H.\chi'$ to test the validity of relation (2) (figure 15b). It is quite obvious from this figure that the functional dependence of $\chi'_H \sim \frac{1}{H}$ is followed in the reversible mixed state. Just below $T_{c_0} = 88$ K a deviation is observed and can be attributed to a vanishing contribution of diamagnetic fluctuations as mentioned above. The fact that the relation $H.\chi'$ is independent of H, allows one to show that the H_{c_1} temperature dependence is linear over a wide range of temperature and that a quantitative estimation of H_{c_1} can be achieved by taking a reasonable value of κ. Moreover for temperatures higher than that of the onset of shielding, χ'_H is frequency independent as expected (figure 16).

A second study concerns the investigation of the reversible mixed state properties of Thallium based compounds [31]. This study was performed on a ceramic compound containing the major phase 1223 ($T_{c_0} \simeq 106$ K) and a minor phase 2223 ($T_{c_0} \simeq 118$ K) already studied in Section C.2. A carefull analysis taking into account the reversible contribution of each phase allows one to show that the functional dependence $\chi' \sim \frac{1}{H}$ is well obeyed in a large range of temperature. The onset of shielding which stops the development of the reversible regime is thought to be that of the 2223 phase otherwise the 1223 phase would appear much less reversible than the 2223. Here too the fact that χ' obeys an 1/H dependence allows one to assert that the temperature dependence of H_{c_1} is linear over a significative range of temperature below T_c.

In the case of the $YBa_2Cu_3O_{7-\delta}$ phase, the reversible domain is much less extended for $\delta \sim 0$, than that of the Bi and Tl based compounds. As good quality single crystals are now available, in order to increase the domain of reversibility, oxygen deficiency single crystals were carefully prepared and well characterized by ac susceptibility [32]. The results for $\delta \simeq 0.48$ ($O_{6.52}$) obtained for magnetic fields parallel to the C-axis are displayed in figure 17. A large regime of reversibility is obtained showing a good verification of the 1/H functional dependence of χ'_H. As a consequence, very precise and direct determination of the H_{c_1} temperature dependence can be performed. Such data reveal an anomalous behaviour of H_{c_1} ($/\!/$ c) when the temperature is sufficiently lowered, which is attributed to proximity effect induced superconductivity in the normal layers [33]. For the case where the magnetic field is parallel to the ab plane, it is no longer possible to observe a DPE effect. So the reversible behaviour observed in the case of Bi and Tl compounds allows one to determine a mean H_{c_1} dependence resulting from very anisotropic contributions.

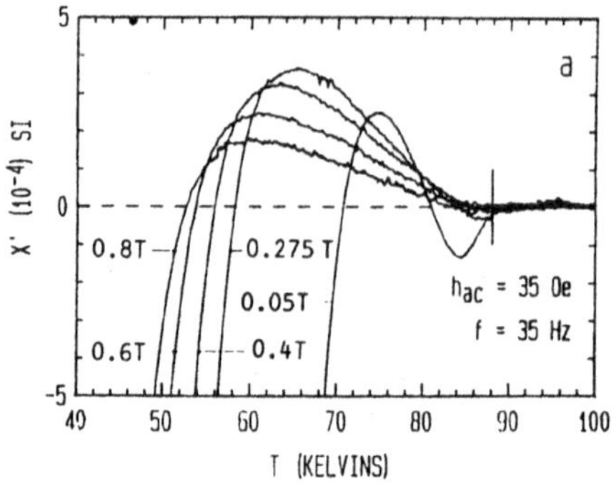

Fig. 15a ac susceptibility response of a ceramic compound $Bi_2Sr_2CaCu_2O_8$ showing a χ'_H DPE effect.

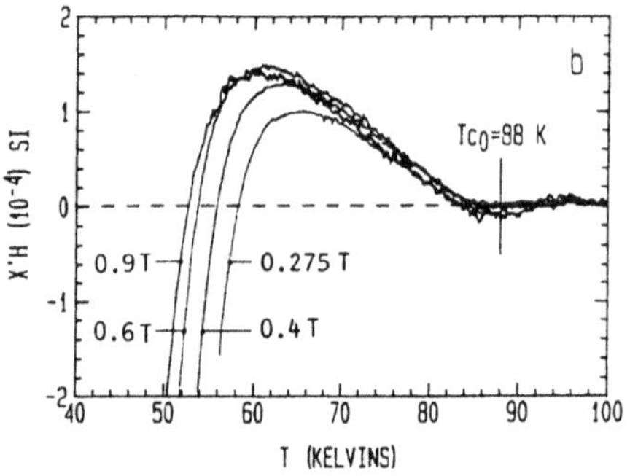

Fig. 15b The functional dependence 1/H of χ'_H for the same compound.

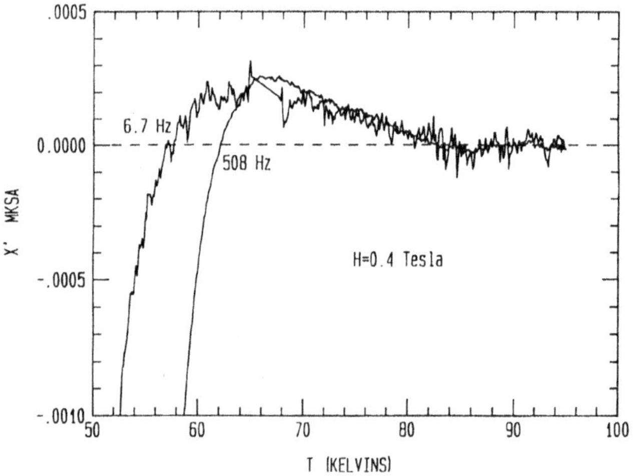

Fig. 16 Frequency effect on χ'_H for ac field of 2.5 Oe and static field of 0.4 Tesla. Onset temperature of shielding increases with frequency.

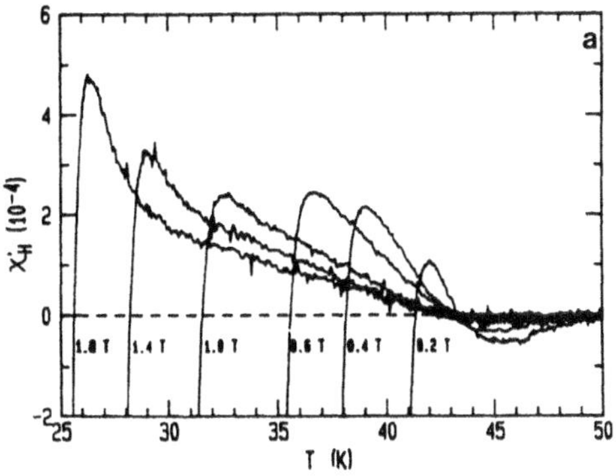

Fig. 17a ac susceptibility response of a single crystal (T_c onset ≃ 46 K) with ac field parallel to the c-axis. At low static field (0.2 - 0.4 T) diamagnetic fluctuations are still discernable.

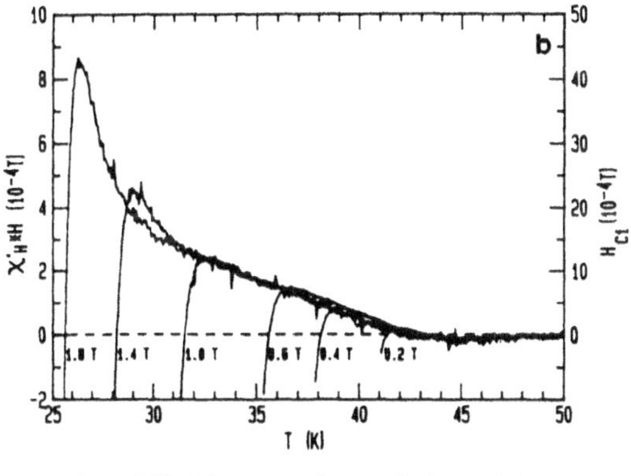

Fig. 17b The results of fig. 14a in the χ'_H representation showing the validity of 1/H dependance of χ'_H. The right hand scale gives $H_{c1}(T)$ for $\kappa = 150$. An anomaly of $H_{c1}(T)$ appears clearly at 31 K.

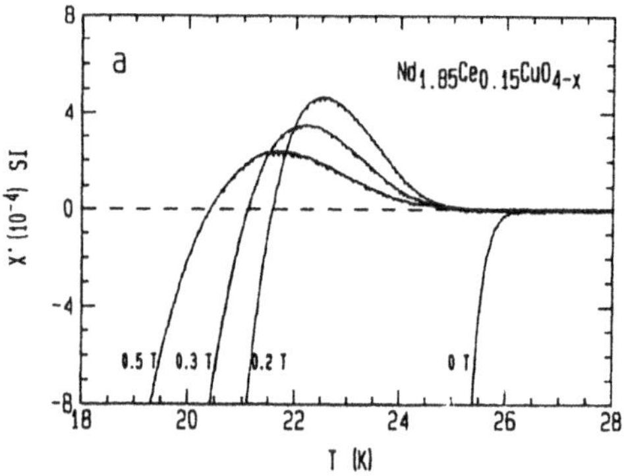

Fig. 18a The ceramic compound $Nd_{1.85}Ce_{0.15}CuO_{4-x}$ ($T_c \simeq 25$ K) exhibits positive χ'_H effect.

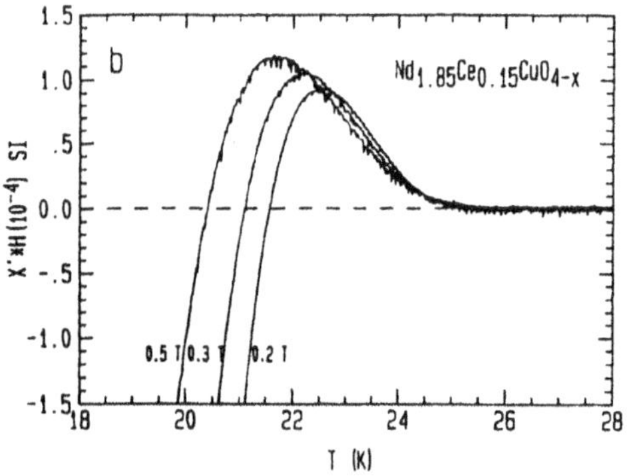

Fig. 18b The functional dependance $\chi'_H \sim 1/H$ is obeyed quite satisfactorily.

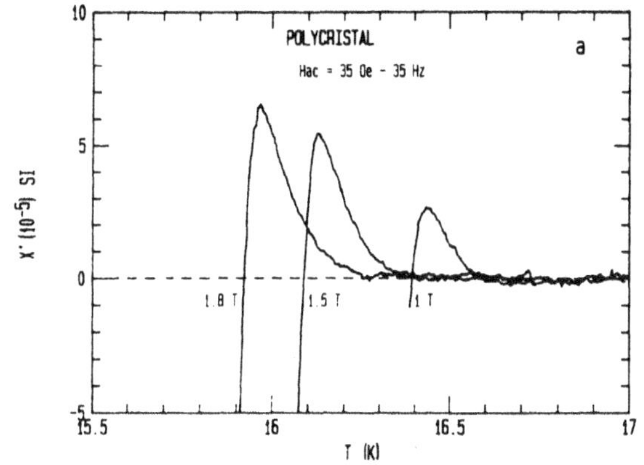

Fig. 19a The polycrystal V_3Si $T_c \simeq$ 17 K displays a positive χ'_H effect over a very reduced range of temperature.

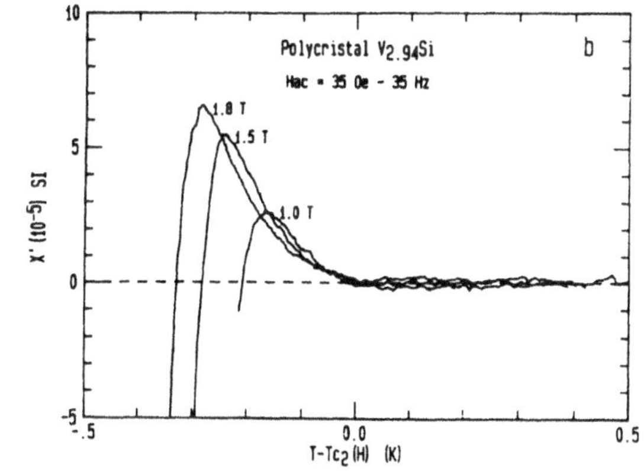

Fig. 19b The shape of the positive effect is field and temperature independent.

The ceramic compound $Nd_{1.85}Ce_{0.15}CuO_{4-x}$ although exhibiting a low $T_c \sim 25$ K, reveals also a large reversibility domain. This investigation is more difficult as important magnetic contributions exists in the normal state. Nevertheless by using a generalized Curie law to extrapolate the properties of the normal state, it was possible to prove the 1/H functional dependance of χ'_H as shown by figure 18. Whatever their critical temperatures, the new oxides families display extended domain of reversibility of the mixed state due to large anisotropy.

It is important to recall that investigation of DPE effect was firstly undertaken in detail by R.A. Hein [3] and J.R. Clem [4] to study the reversible mixed state of alloys. As shown by R.A. Hein classical type II superconductors may exhibit a more extended reversibility range when they are defect free. In the case of a transforming V_3Si compound with $T_c \sim 17$ K [34] the reversible range is limited to 1 K at 1.8 T (Fig. 19a). The functional dependence $\chi'_H \sim \frac{1}{H}$ is no longer valid and the χ'_H shape appears temperature and field independent suggesting an $1/(2\kappa^2 - 1)$ behaviour (Fig. 19b). Assuming $\kappa = 24$ [35], the theoretical step like increase is much larger ($\sim 10^{-3}$SI) than experimentally observed at 1.8 Tesla. So further investigations at higher fields and other frequencies are necessary to elucidate the true nature of this positive effect.

Nevertheless this effect raises a question with regard to determining the critical field slope $\frac{d H_{c_2}}{dT}$ by ac susceptibility measurements. It has been shown on Chevrel phase compound [36] that a determination by ac susceptibility yields a lower critical field slope when compared to specific heat determinations. In the present case of V_3Si, two methods can be used to achieve such a determination by taking either the onset of shielding (1.76 T/K) or the onset of the positive effect (2.65 T/K). As such a method can be tentatively tried and compared in detail with specific heat results, it seems to be no longer useful with high T_c superconductors.

Meanwhile the DPE effect appears in the case of the new high T_c superconductors to offer a powerful method of investigation of the reversible mixed state properties.

CONCLUSION

We have shown the usefulness of the ac susceptibility technique in the investigation of the superconducting materials. A large variety of problems, ranging from the detection of a minor superconducting phase and/or multiphase inhomogeneous samples to studies of the equilibrium, as well as nonequilibrium properties, of the mixed state, can be approached by this technique ; (semi) quantitative analysis can be done if parameters as frequency, ac field amplitude and dc field magnitude can be varied to control shielding and the χ'' peaks levels.

Temperature regulation and accuracy is crucial to resolve possible multitransitions in a narrow ΔT around an average T_c as it is often the case in single crystals : such subtle inhomogeneities can seldom be detected by other technique (like resistivity...).

The importance of varying the ac field amplitude has been underlined. The best resolution in the χ'' spectrum is obtained with low ac fields whereas higher ac fields (with or without application of a dc field) are often necessary with ceramics to lower the temperature where intergranular shielding is effective in order to reach an upper estimation of superconducting volume. Diamagnetic fluctuations may simulate a low fraction of high T_c phases. Their contribution to χ' can be distinguished by application of moderate static fields.

Finally ac susceptibility appears as a very efficient method to investigate the reversible mixed state whose extent in temperature can be increased by lowering the frequency, or increasing ac and dc field amplitude. A direct consequence of such investigations is to reach the H_{c1} temperature dependence and H_{c1} values if the Ginzburg-Landau constant κ is known. Such a determination is independent of weak links and demagnetizing effects.

ACKNOWLEDGMENTS

The authors would like to thank J.Y. Henry and F. Monnier from CENG-DRFMC for supplying Bi and Nd ceramics, YBCO single crystals and for extensive contribution to these experiments. The efficient collaboration on Thallium based compounds with Professor M.Th. Cohen-Adad, J.L. Jorda and R. Abraham from Laboratoire Physico Chimie Menérale II of University of Lyon I is gratefully acknowledged.

The authors are indebted to Professor R.A. Hein for critical and constructive reading of the manuscript.

REFERENCES

1. E. Maxwell and M. Strongin, Phys. Rev. Lett. 10, 212 (1963).
2. M. Strongin, E. Maxwell and T.B. Reed, Rev. Mod. Phys. 164 (1964).
3. R.A. Hein and R.L. Falge Jr., Phys. Rev. 10, 212 (1963).
4. J.R. Clem, H.R. Kerchner and S.T. Sekula, Phys. Rev. B, 14, 1893 (1976).
5. R.A. Hein, Phys. Rev. B, 33, 7539 (1986).
6. A.J. Van Duynevelt, J. Appl. Phys. 53, 11 (1982).
7. J.R. Owers-Bradley, Wen-Sheng Zhou and W.P. Halperin Rev. Sci. Instr. 52, 1106, (1981).
8. M. Ishizuka and Y. Tohi, Jap. J. Appl. Phys. 19, 639 (1980).
9. M. Ocio and J. Hamman, Rev. Sci. Instr. 56, 1367 (1985).
10. A.F. Deutz, R. Hulstman and F.J. Kranenburg, Rev. Sci. Instr. 60, 113 (1989).
11. M. Couach, A.F. Khoder and F. Monnier, Cryogenics, 25, 695 (1985).
12. A.M. de Goer, J. Phys. (France), 30, 389 (1969).
13. Sapphire supplied by RSA Le Rubis SA BP 16 - 38560 JARRIE FRANCE.
14. A.F. Khoder, M. Couach, Cryogenics, to appear (1991).
15. L.P. Gor'kov and O.N. Dorokhov, J. Low Temp. Phys. 22, 1 (1976).
16. J. Labbé, Phys. Rev. 158, 647 (1967).
17. J.P. Maïta, E. Bucher, Phys. Rev. Lett. 29, 931 (1972).
18. M. Pellizone, A. Treyvaud, Applied Physics, 24, 375 (1981).
19. A.F. Khoder, J. Labbé, M. Couach and J.P. Sénateur, J. Phys. (France) 47, 1223 (1986).
20. A.F. Khoder and M. Couach, In these proceedings;
21. S.H. Glarum, L.F. Schneemeyer and J.V. Waszczak, Phys. Rev. B 41, 1837 (1990).
22. M. Couach, A.F. Khoder, F. Monnier, J.Y. Henry and B. Barbara, Phys. Rev. B 38, 748 (1988).
23. K.H. Müller, Physica C, 159, 717 (1989).

24. D. Cattani, Thesis University of Geneva (1990).
25. J.G. Bednorz, M. Takashige and K.A. Müller, Europhys. Lett. 3, 379 (1987).
26. A.P. Malozemoff, Physical properties of High Temperature Superconductors I, Ed. by D.M. Ginsberg, World Scientific (1989).
27. J. Van der Berg, C.J. Van der Beek, P.H. Kes, J.A. Mydosh, M.J.V. Menken, A.P. Menovsky, Superc. Sc. Technol. 1, 249 (1989).
28. T.T.M. Palstra, B. Batlogg, R.B. Van Dover, L.F. Schneemeyer and J.V. Waszczak, Phys. Rev. B, 41, 6621 (1990).
29. M. Tinkham, Introduction to Superconductivity, McGraw-Hill (1975).
30. A.F. Khoder, M. Couach and J.L. Jorda, Phys. Rev. B 42, 8714, (1990).
31. M. Couach, A.F. Khoder, F. Monnier, J.L. Jorda, M.Th. Cohen-Adad and R. Abraham, High Temperature Superconductors in Materials Sciences Monographs, Vol. 70, p. 353-362, Elsevier Sci. Publ. Amsterdam 1991.
32. A.F. Khoder, M. Couach, F. Monnier and J.Y. Henry, Europhys. Lett. 15, 337, 1991.
33. T. Koyama, N. Takezawa and M. Tachiki, Physica C, 168, 69 (1990).
34. This sample was kindly supplied by J.P. Senateur, ENSIEG, BP 46, 38402 ST Martin d'Hères, France.
35. D. Eckert and K.H. Berthel, 479, Cryogenics, (1975).
36. D. Cattani, J. Cors, M. Decroux, B. Seeber and Φ Fisher, Physica C, 153-155, 461, (1988).

ALTERNATING–FIELD SUSCEPTOMETRY AND MAGNETIC SUSCEPTIBILITY OF SUPERCONDUCTORS

R. B. Goldfarb
National Institute of Standards and Technology
Boulder, Colorado 80303

M. Lelental
Eastman Kodak Company
Rochester, New York 14650

C. A. Thompson
National Institute of Standards and Technology
Boulder, Colorado 80303

ABSTRACT

This review critically analyzes current practice in the design, calibration, sensitivity determination, and operation of alternating-field susceptometers, and examines applications in magnetic susceptibility measurements of superconductors. Critical parameters of the intrinsic and coupling components of granular superconductors may be deduced from magnetic susceptibility measurements. The onset of intrinsic diamagnetism corresponds to the initial decrease in electrical resistivity upon cooling, but the onset of intergranular coupling coincides with the temperature for zero resistivity. The lower critical field may be determined by the field at which the imaginary part of susceptibility increases from zero. Unusual features in the susceptibility of superconductor films, such as a magnetic moment that is independent of film thickness and the variation of susceptibility with angle, are related to demagnetization. Demagnetizing factors of superconductor cylinders are significantly different from those commonly tabulated for materials with small susceptibilities. Rules for the susceptibility of mixtures with specific demagnetizing factors are used to estimate the volume fraction of superconducting grains in sintered materials. Common misunderstandings of the Meissner effect, magnetic units, and formula conversions are discussed. There is a comprehensive summary of critical-state formulas for slabs and cylinders, including new equations for complex susceptibility in large alternating fields. Limitations on the use of the critical-state model for deducing critical current density are listed and the meaning of the imaginary part of susceptibility is considered.

INTRODUCTION

The term "susceptibility" was originated by William Thomson (Lord Kelvin) in his annotated *Reprint of Papers on Electrostatics and Magnetism.*[1] He defined "the magnetic susceptibility of an isotropic substance [as] the intensity of magnetization acquired by an infinitely thin bar of it placed lengthwise in a uniform field of unit magnetic force." The specification of an infinitely thin bar eliminated the need to consider demagnetizing fields. The stipulation of a field of unit magnetic force defined susceptibility as the ratio of magnetization M (magnetic moment per unit volume) to magnetic field strength H. Thomson distinguished between susceptibility[2] and permeability, a term he devised to mean the ratio of magnetic induction B to H.

Magnetization and susceptibility measurements on superconductors detect signals, usually inductively, that have their origins in circulating persistent shielding currents, in addition to any magnetic properties of the material. We distinguish between eddy currents in normal metals, which decay with time, and shielding currents in superconductors, which do not. Susceptibility may be measured using direct or alternating magnetic fields, yielding the ac susceptibility or the dc susceptibility. For either, we define χ as the differential magnetic susceptibility $\mathrm{d}M/\mathrm{d}H$; we do not necessarily require that $\mathrm{d}H \rightarrow 0$. In dc susceptibility, the zero-field-cooled (ZFC) curve demonstrates flux shielding (flux exclusion) upon warming, and the field-cooled (FC) curve demonstrates the Meissner effect (flux expulsion) upon cooling. Whether measured upon warming or cooling, ac susceptibility (with no dc bias field) always measures shielding.

It is easier to define a superconductor as a material with zero electrical resistivity than it is to experimentally verify zero resistivity. A four-point measurement of resistivity involves the selection of current, voltage criterion, correction for thermoelectric voltages, contact geometry, and the effect of magnetic field, including the self-field of the current. The magnetic manifestation of zero resistivity is that a material is a superconductor if it exhibits perfect diamagnetic shielding; that is, its susceptibility χ is exactly -1 (in SI units, where numerical results must be corrected for any sample demagnetizing factor). Susceptibility is reminiscent of electrical conductivity σ, both functions of temperature T: $\chi(T)/\chi(0) \approx \sigma(T)/\sigma(0)$. In the normal state, both are small. In the superconducting state, both are large. Important variables are the magnitude of the measuring field and the definition of the critical temperature T_c in terms of the onset, midpoint, or end of the diamagnetic transition.

MEISSNER EFFECT

The Meissner-Ochsenfeld effect[3] is the expulsion of magnetic flux upon cooling a superconductor through T_c in a dc magnetic field or upon reducing the magnetic field through the upper critical field H_{c2} at constant temperature. Type-II superconductors will not exhibit a pronounced Meissner effect if they have good flux pinning (important for high critical current density J_c) in the mixed state. Thus, while a material that has a Meissner effect is a superconductor, the converse is not necessarily true. What is sometimes called the "Meissner state" in superconductors is better termed the shielding state, perfect diamagnetism.[4]

It has long been known that the Meissner effect is incomplete in many materials. Shoenberg stated that "for many of the element superconductors, it has not yet been possible to obtain a specimen which shows a complete Meissner effect...."[5] Tantalum, a type-I superconductor, is a case in point, with a Meissner effect of only 1%.[6] Shoenberg attributed the incomplete Meissner effect in tantalum to its "mechanical state rather than ... chemical impurities."[7] Years later, Alers *et al.* returned to this: "It is well known that for tantalum the Meissner effect is practically nonexistent because the metal freezes in all of the existing flux when it becomes superconducting.... Pure tantalum in bulk ... [is] made by sintering ... small flakes or grains.... Thus from a physical point of view, the metal is not homogeneous, and one can understand that the Meissner effect might not be realized by a metal of this physical make up."[8] The similarity between the morphologies of sintered tantalum and sintered Y–Ba–Cu–O (or any of its analogs) will not be lost on most readers. However, sintered materials are not unique in this respect. Type-II elements and alloys and other inhomogeneous superconductors,[5,6,9] including *melt-cast* tantalum,[10] similarly fail to show a significant Meissner effect. The Meissner effect is also incomplete in *single crystals* of $YBa_2Cu_3O_{7-\delta}$, which suggests intragrain pinning sites.[11]

It is experimentally found that, in weak fields, the Meissner effect approaches 100%,[11–15] a value defined by the ZFC susceptibility curve. This is not surprising; it is tautologous that FC *upon warming* is equivalent to ZFC when the measuring field is zero. If the FC curve were completely reversible for warming and cooling in the limit of zero field, the Meissner effect would of necessity approach 100%.

CRITICAL TEMPERATURE

In field–current-density–temperature (H–J–T) space, there is a critical surface, with axis intercepts H_c, J_c, and T_c, separating the superconducting and normal states. Its intersection with the $H-T$ plane may be regarded as H_c versus T or T_c versus H, and similarly for intersections with the $J-T$ and $J-H$ planes. For type-II superconductors the H-axis intercept is the upper critical field H_{c2}. The mixed state lies between H_{c2} and the lower critical field H_{c1}.

Measured as functions of temperature, transitions in resistivity ρ and susceptibility χ may be used to define T_c. Ideally, T_c should be determined at $J = 0$ and $H = 0$. However, measurements of ρ require some J and measurements of χ require some H. These are best kept small, unless the current and field dependences are specifically required. Electrical resistivity complements susceptibility. Resistivity is a one-dimensional measurement. A specimen will show zero resistivity if there is a single zero-resistance percolation path. A higher-T_c phase can mask the presence of a lower-T_c phase. Either of these cases could lead to erroneous conclusions regarding the microstructure of the specimen under study. Susceptibility is a two-dimensional measurement in the sense that a surface current sheath is required for full diamagnetism. The interior of the material could remain normal or have a lower T_c. Both measurements deceive because they do not probe the entire sample volume. "Onset" temperatures (defined as *onset upon cooling*, even if the measurement is made upon warming) occur with the first zero-resistance segment and the first zero-resistance current circuit, for ρ and χ respectively. In low-dimensional systems, fluctuation effects may obscure the onset of superconductivity.[16]

Granular and Multifilamentary Superconductors

Sintered high-T_c superconductors and composite low-T_c superconductors with closely spaced filaments exhibit two critical temperatures. One is intrinsic to the superconductor and the other is characteristic of the coupling between either grains[9,17–25] or filaments.[26] In such materials, the coupling component supports supercurrents and has its own effective T_c, J_c, H_{c1}, and H_{c2}. In multifilamentary niobium-titanium (Nb–Ti) and niobium-stannide (Nb_3Sn) superconductors, the coupling component is the normal-metal matrix and the coupling mechanism is the proximity effect.[26–29] The situation is less certain in sintered high-T_c compounds, but lack of stoichiometry at the grain boundaries could give rise to normal metal barriers[30–32] and proximity-effect coupling.[33–36] Another coupling mechanism in sintered materials is microbridges between grains.[37] Low-dimensional compounds, such as Nb_3Se_4, exhibit coupling attributed to superconductor-insulator-superconductor Josephson junctions.[38]

Because of the large change in shielded volume that occurs at T_c of the coupling component, there is a striking change in susceptibility. The change in resistivity, in comparison, is minor because the coupling component forms a small part of the conduction path. A crushed sintered sample yields isolated grains with only intrinsic characteristics.[18,24,25,35,39,40] Both intrinsic and coupling critical temperatures are

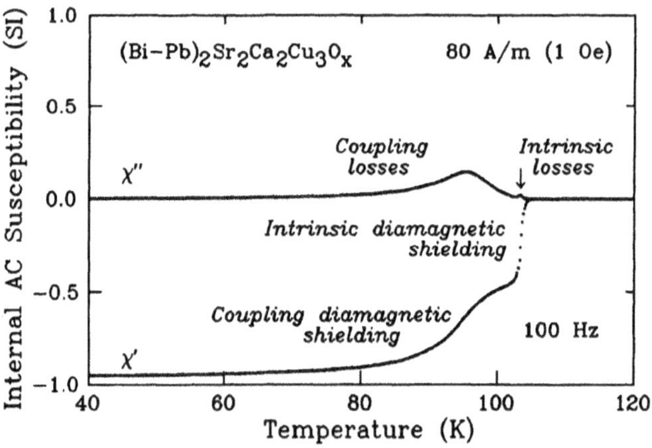

Fig. 1. Internal ac susceptibility for $(Bi{-}Pb)_2Sr_2Ca_2Cu_3O_x$. The measurement field, 80 $A \cdot m^{-1}$ rms at 100 Hz, is large enough to separate the intrinsic and coupling components. The real part does not extrapolate to −1 because the demagnetizing factor used to reduce the data was approximate.

field dependent, the latter more so.[17,19,23,40] This dependence may be examined with increasing ac measurement fields or dc bias fields.[41]

For high quality, strongly coupled, sintered superconductors, the two critical temperatures coincide for small measuring fields.[17] The coupling T_c is not depressed as much with increasing measurement field compared to a poor quality, weakly coupled sample. A typical measurement[42] for a sample of high quality, sintered, high-T_c $(Bi-Pb)_2Sr_2Ca_2Cu_3O_x$ is presented in Fig. 1. The real (χ') and imaginary (χ'') parts of internal susceptibility (corrected for demagnetizing factor) are shown as a function of increasing temperature, with the intrinsic and coupling segments identified. The measurement field of 80 $A \cdot m^{-1}$ rms is large enough to separate the two components. For a small measurement field of 0.8 $A \cdot m^{-1}$ rms, the critical temperatures overlap.

Figure 2 shows internal ac susceptibility curves for poor quality, weakly coupled, sintered $YBa_2Cu_3O_{7-\delta}$ measured in 0.8 $A \cdot m^{-1}$ and 80 $A \cdot m^{-1}$ rms. Even for the lower measurement field, the coupling T_c (90.3 K) is considerably below the intrinsic T_c (91.1 K). There is no intrinsic χ'' peak for this sample for the fields used.

Identification of Critical Temperature

In resistivity measurements, T_c is the temperature at which a percolation path is established. The corresponding temperature for magnetic susceptibility occurs when a bulk shielding path is established. This occurs at T_c of the coupling component, in particular at the onset of coupling. The distinction between the intrinsic onset and coupling onset is pertinent for samples with weak coupling and for measurements made in moderately large fields.

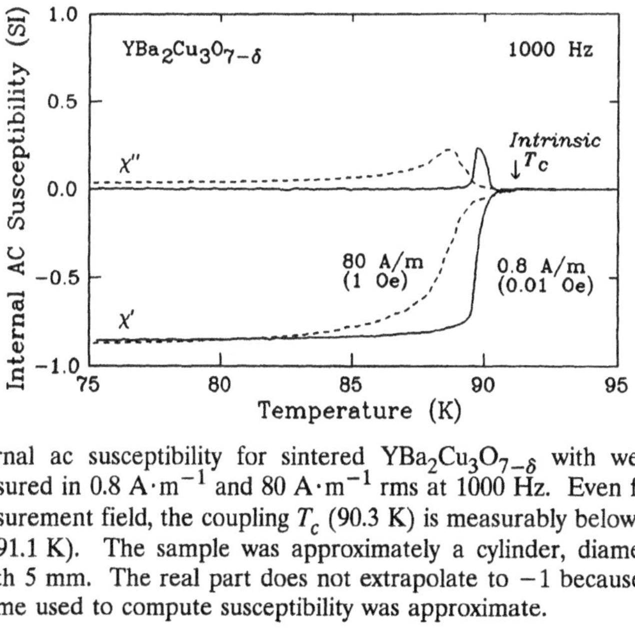

Fig. 2. Internal ac susceptibility for sintered $YBa_2Cu_3O_{7-\delta}$ with weak coupling measured in 0.8 $A \cdot m^{-1}$ and 80 $A \cdot m^{-1}$ rms at 1000 Hz. Even for the lower measurement field, the coupling T_c (90.3 K) is measurably below the intrinsic T_c (91.1 K). The sample was approximately a cylinder, diameter 0.9 mm, length 5 mm. The real part does not extrapolate to -1 because the sample volume used to compute susceptibility was approximate.

The critical temperature is sometimes taken as the midpoint of the diamagnetic transition and the width of the transition is quoted.[43] There are several problems with this. First, there are two transitions. Second, the widths of the transitions are field dependent. Third, a large part of the transition to full diamagnetism is due to coupling. Fourth, the complete intrinsic transition is often obscured by the coupling transition. Therefore, it is more useful to define the critical temperatures as the onset temperatures, although the precise onset temperatures are uncertain, particularly due to fluctuation effects.

Figure 3 shows ac susceptibility and ac resistance measured on a bar of $(Bi-Pb)_2Sr_2Ca_2Cu_3O_x$. To compare resistivity and susceptibility curves, we first

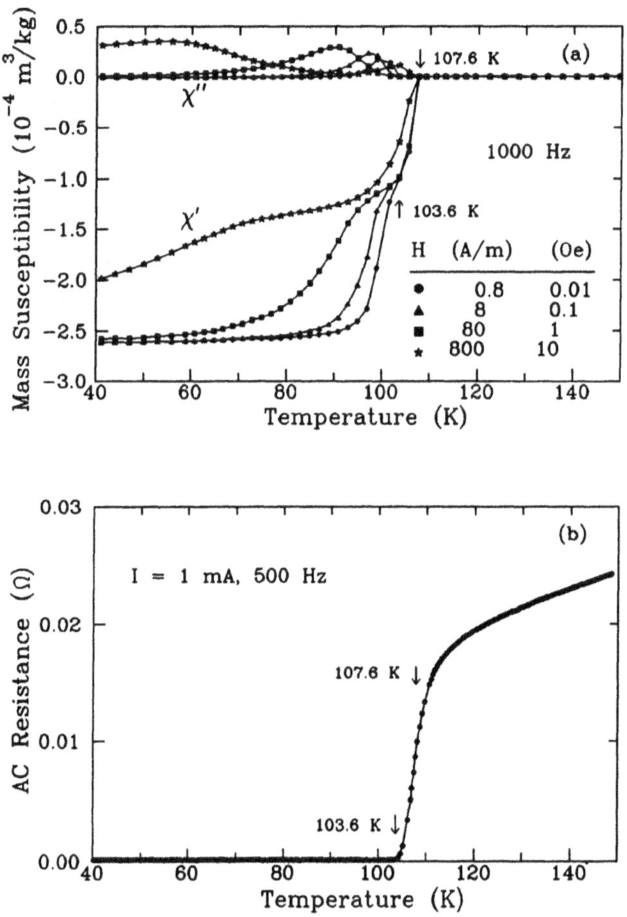

Fig. 3. Comparison of ac susceptibility and resistance as functions of temperature for $(Bi-Pb)_2Sr_2Ca_2Cu_3O_x$. (a) Mass susceptibility for four ac fields (rms values shown) at 1000 Hz. Intrinsic T_c is 107.6 K and, for the smallest measuring field, coupling T_c is 103.6 K. (b) Resistance for an ac measuring current of 1 mA rms at 500 Hz. The critical temperatures obtained from susceptibility are labeled.

calculate the self-field of the transport current used in the resistivity measurements. For a sample with circular cross section and uniform current density, the field inside the sample is $H(r) = rI/(2\pi a^2)$, from Ampère's circuital law, where r is the radial coordinate, I is the current, and a is the sample radius. The average field obtained by integration over the sample cross section (rather than over the radius) is $\langle H \rangle = I/(3\pi a)$. For our sample (actually of rectangular cross section with an effective radius $a \approx 0.7$ mm), the self-field of the transport current (1 mA) is negligible compared to any of the measuring fields, and the best comparison is with the lowest-field susceptibility curve. Intrinsic T_c is 107.6 K and coupling T_c for the lowest-field measurement is 103.6 K, as seen in Fig. 3(a). These temperatures are identified in the plot of resistance, Fig. 3(b).

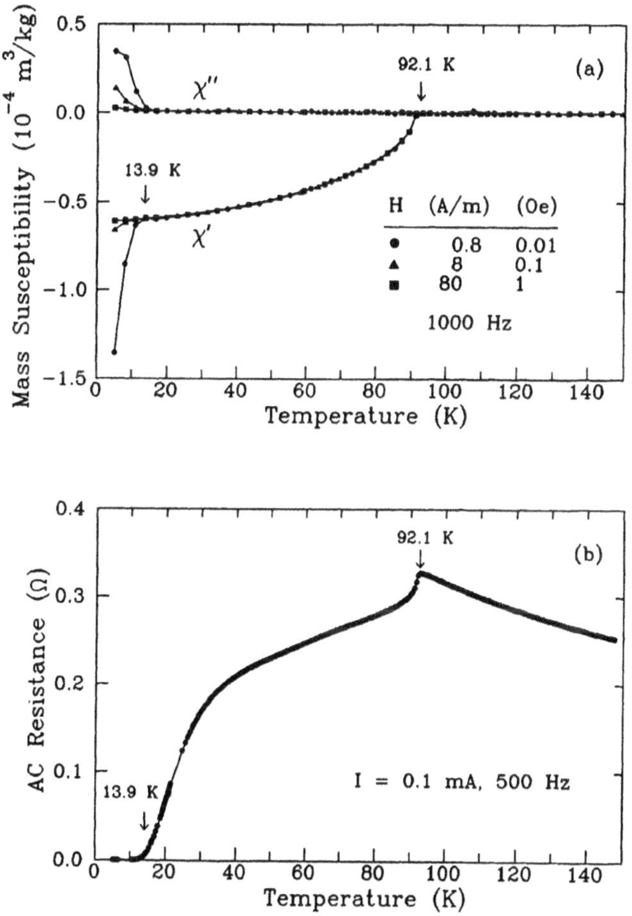

Fig. 4. Comparison of ac susceptibility and resistance as functions of temperature for $YBa_2Cu_3O_{7-\delta}$ with extremely weak intergranular coupling. (a) Mass susceptibility for three ac fields (rms values shown) at 1000 Hz. Intrinsic T_c is 92.1 K and, for the smallest measuring field, coupling T_c is 13.9 K. (b) Resistance for an ac measuring current of 0.1 mA rms at 500 Hz. The critical temperatures obtained from susceptibility are labeled. The intrinsic critical temperature is unambiguous for this sample, which is semiconducting in the normal state.

The positive slope of the resistance curve above T_c indicates a normal conductor and the intrinsic T_c does not correspond to a distinguishing feature. Coupling T_c coincides with the zero-resistance temperature.

Figure 4 shows ac susceptibility and ac resistance measured on a bar of sintered Y–Ba–Cu–O with extremely weak intergranular coupling. The comparison is informative. From Fig. 4(a), intrinsic T_c is 92.1 K but coupling T_c is only 13.9 K and very field dependent. In Fig. 4(b), there is a peak in the resistance curve at the intrinsic T_c where the material goes from the semiconducting state to the superconducting state. Resistance is zero at about the coupling T_c. In summary, the temperature for zero resistivity is related to the susceptibility coupling onset temperature, which is determined by the quality of intergrain coupling. The temperature for the initial drop in resistivity is related to the intrinsic onset temperature, which is determined by the quality of the grains. Other experimental studies are consistent with these conjectures.[44–49]

DEMAGNETIZING FACTORS

Demagnetizing factors are important for the understanding of the susceptibility of superconductors and especially films. In the equation for magnetic induction, $B = \mu_0(H + M)$, H is the internal field, equal to the external or applied field H_a corrected by the demagnetizing field H_d. The source of the demagnetizing field is taken to be magnetic poles on the surface of a magnetized specimen. In ellipsoids, the poles are distributed in such a way that all fields are uniform. These fields include H_a, H_d, and H, and the magnetization M. They are related vectorially by the equation $\boldsymbol{H} = \boldsymbol{H_a} + \boldsymbol{H_d} = \boldsymbol{H_a} - \mathsf{N}\boldsymbol{M}$, where N is the demagnetizing tensor. If $\boldsymbol{H_a}$ is along a principal axis of the ellipsoid, then $H = H_a + H_d = H_a - NM$, where N is the scalar demagnetizing factor. (If $\boldsymbol{H_a}$ is not along a principal axis of the ellipsoid and $\chi \neq 0$, $\boldsymbol{M}$ is uniform but not coaxial with $\boldsymbol{H_a}$, and the direction and magnitude of $\boldsymbol{M}$ depend on χ.) For ellipsoids of revolution (spheroids) N is a function of the aspect ratio γ of the ellipsoid (ratio of the polar axis to the equatorial axis) and is independent of susceptibility χ:[50,51]

$$N = (1 - \gamma^2)^{-1}[1 - \gamma(1 - \gamma^2)^{-1/2}\cos^{-1}\gamma] \qquad (\gamma < 1), \qquad (1a)$$
$$N = \tfrac{1}{3} \qquad (\gamma = 1), \qquad (1b)$$
$$N = (\gamma^2 - 1)^{-1}[\gamma(\gamma^2 - 1)^{-1/2}\cosh^{-1}\gamma - 1] \qquad (\gamma > 1). \qquad (1c)$$

Demagnetizing factors for cylinders have been examined in detail by Chen, Brug, and Goldfarb.[52] For cylinders, N is a function of γ (ratio of length to diameter) and also χ, which is assumed to be constant in the sample. With H_a along the cylinder axis, M and H_d are both nonuniform except in two cases. When $\chi = 0$, M is uniform. The approximation $\chi = 0$ is used for saturated ferromagnets, diamagnets, and paramagnets. When $\chi \to \infty$, H_d is uniform and equal to $-H_a$. The condition $\chi \to \infty$ applies to soft ferromagnetic materials. When $\chi = -1$, $(M + H_d)$ is uniform and equal to $-H_a$. [That is, $B = \mu_0(H_a + H_d + M) = 0$.] This applies to superconductors in the shielding state.

There are two types of demagnetizing factors for cylinders. The fluxmetric (or ballistic) demagnetizing factor N_f is the ratio $-\langle H_d\rangle_s/\langle M\rangle_s$, where $\langle\ \rangle_s$ indicates an

Table 1. Longitudinal and transverse magnetometric demagnetizing factors N_{mz} and N_{mx} for cylinders with $\chi = -1$, after Taylor.[53] γ is the ratio of length to diameter.

γ	N_{mz}	N_{mx}
0	1	0
0.25	0.6764	0.2136
0.5	0.5258	0.2928
1	0.3692	0.3669
2	0.2341	0.4237
4	0.1361	0.4596
∞	0	0.5

average over the center plane of the cylinder. The magnetometric demagnetizing factor N_m is the ratio $-\langle H_d\rangle_v/\langle M\rangle_v$, where $\langle\ \rangle_v$ indicates an average over the volume of the cylinder. Fluxmetric factors are used when magnetization is measured ballistically, with a short search coil closely wrapped around the center of a long sample. Magnetometric factors are used with magnetometers that sense the entire sample volume, such as vibrating-sample magnetometers, SQUID magnetometers, and ac susceptometers. Both N_f and N_m depend on γ and χ.

Experimental and theoretical work on demagnetizing factors for cylinders dates back to the 1880s. However, perhaps the only research ever published until recently for $\chi = -1$ was Taylor's paper on conducting cylinders.[53] He calculated polarizabilities, which we can convert to N_m, for several values of γ for both longitudinal and transverse fields. Our reduction of his results is given in Table 1.[52] Note that $N_{mx} + N_{my} + N_{mz} = 2N_{mx} + N_{mz} \geq 1$. The subscripts x, y, z indicate the orthogonal axes, with the applied field along z. (The sum of the three orthogonal magnetometric demagnetizing factors for cylinders equals 1 only when $\chi = 0$.) N_m and N_f for the complete range of γ and χ are given in Ref. 52. Values of N_m for $\chi = -1$ are different from those often tabulated for $\chi = 0$. For $\gamma = 1$, for example, $N_{mz} = 0.3116$ for $\chi = 0$. One caveat is that values of N_m for $\chi = -1$ are for superconductors in the shielding state. Superconductors in the mixed state do not have constant susceptibility, which is one of the basic assumptions in the derivation.

The measurement of susceptibility requires the application of H_a and the measurement of M. The susceptibility dM/dH_a is characteristic of the *sample* and may be termed the *external* susceptibility χ_{ext}. The *internal* susceptibility χ, characteristic of the *material*, is dM/dH. The two susceptibilities are related: $\chi = \chi_{ext}/(1 - N\chi_{ext})$ and $\chi_{ext} = \chi/(1 + N\chi)$. When ac susceptibility is measured, χ_{ext} is a complex quantity: $\chi_{ext} \equiv \chi'_{ext} + i\chi''_{ext}$. The internal susceptibility is also complex: $\chi \equiv \chi' + i\chi''$. When relating the two quantities χ and χ_{ext}, the real parts and the imaginary parts are separated, resulting in

$$\chi' = [\chi'_{ext} - N(\chi'^{2}_{ext} + \chi''^{2}_{ext})] \,/\, [N^2(\chi'^{2}_{ext} + \chi''^{2}_{ext}) - 2N\chi'_{ext} + 1], \tag{2a}$$

$$\chi'' = \chi''_{ext} \,/\, [N^2(\chi'^{2}_{ext} + \chi''^{2}_{ext}) - 2N\chi'_{ext} + 1]. \tag{2b}$$

This rule appears periodically in the literature.[54–56] (We should point out that an interesting artifact occurs in these equations for χ' and χ'' in terms of χ'_{ext}, χ''_{ext}, and N. When $N \approx 1$ and $\chi'_{ext} \approx 1$ and $\chi''_{ext} \approx 0$, as might occur for films in the normal state, χ' and χ'' diverge, causing severe scatter in χ' and χ''. Such values of χ'_{ext} and χ''_{ext} are not uncommon in actual measurements of thick films.)

Volume Fraction of Superconductor Grains

When the ideal ("X-ray") density of a superconductor is known, and when there are no nonsuperconducting phases present, the volume fraction of superconducting grains can be estimated from mass and volume measurements. Otherwise, susceptibility curves give some information on the volume fraction. When grains are fully coupled, the entire volume of a granular sample, including voids and nonsuperconducting phases, is shielded and $\chi = -1$. When grains are uncoupled, the inductive susceptibility signal represents a summation of shielding signals from many grains; voids and nonsuperconducting phases do not contribute.

Consider a sintered pellet of spherical superconductor grains, each with susceptibility $\chi_g = -1$, occupying a volume fraction ϕ in a medium of $\chi_v = 0$. Let the measuring magnetic field strength be large enough to decouple the grains. The demagnetizing factor of each sphere is $\frac{1}{3}$. Let the bulk pellet have a very different demagnetizing factor (0, for example). If the total internal susceptibility χ is based on the volume of the bulk pellet, including voids, can one deduce ϕ from the value of χ ($|\chi| \leq 1$)? When we first addressed this problem[17] we suggested that, for grains of unknown geometry, $\phi \approx |\chi|$.

The susceptibility of mixtures was discussed by Maxwell in his *Treatise*.[57] For perfectly conducting spheres in a nonmagnetic medium, the exact relationship is $\chi = -3\phi/(2 + \phi)$, or $\phi = -2\chi/(3 + \chi)$, where χ is the internal susceptibility of the mixture. One implication is that, for dense pellets ($\phi \rightarrow 1$), the effective demagnetizing factor for a susceptibility measurement is that of the *pellet*, not that of the grains, even when the grains are decoupled. (If demagnetizing fields are thought of as arising from surface magnetic poles, a dense mixture will have pole cancellation except at the surface of the pellet.) The effective susceptibility of granular superconductors, including the effect of magnetic penetration depth, has been examined recently.[58–60]

When the magnetic penetration depth λ is on the order of the grain size, a significant fraction of the grain volume does not contribute to the χ signal.[61] Typically, λ is on the order of 0.2 μm in high-T_c superconductors below $\frac{1}{2}T_c$.[62–68] For illustration,[69] a 100% dense sample composed of *uncoupled* plates of thickness 10 μm would have, for H in the plane of the plates, $\chi = -0.7$. The reduction in χ is especially severe near T_c, where λ becomes quite large: $\lambda(T)/\lambda(0) = [1-(T/T_c)^n]^{-1/2}$, where $n = 4$ in the two-fluid model,[70] but empirically $n \approx 1$ for Y–Ba–Cu–O.[62]

Therefore, to estimate the volume fraction of superconducting grains in a sintered sample containing voids and nonsuperconducting inclusions: (1) The grains should be decoupled by using measuring fields large enough to depress the coupling transition temperature. Thereby, voids and nonsuperconducting phases are not included in the shielded volume. (2) The grain dimensions should be significantly greater than λ so that a large fraction of the grain volume is shielded. Otherwise, λ should be included in the estimation. (3) The χ value used to estimate the volume fraction should be taken well below the intrinsic T_c of the grains. This avoids the increase in λ near T_c. Chen *et al.* have precisely modeled $\chi(T)$ and $\chi(H)$ curves and calculated the volume fraction of grains using λ and the critical-state model for both grains and matrix. They deduced the existence of grain *clusters* in some samples based on discrepancies between actual volume fractions and those computed from the susceptibility data.[71,72]

A destructive way to obtain the volume fraction of superconducting grains is to crush the sample pellet, collect all the powder, and use the original sample volume to compute χ. For loosely packed powder, the appropriate demagnetizing factor would be closer to that of a grain, typically approximated as a sphere. This is most effective when the crushing simply separates the grains from each other. If the grains were finely pulverized, their size may approach λ.

SUSCEPTIBILITY OF SUPERCONDUCTOR FILMS

To easily distinguish the superconducting and normal states, resistivity is best measured in specimens with at least one thin dimension. Susceptibility, in contrast, is best measured when there is a large sample volume. Susceptibility measurements on films thus present special problems and require some interpretation. In measuring the susceptibility of films there are considerations of adequate shielded volume, field orientation with respect to the film plane, demagnetizing-factor corrections, and film thickness compared to λ.

"The Absurdity of This Result ..."

The magnetic susceptibility of superconductor films measured in perpendicular fields presents a paradox arising from perfect diamagnetism and a demagnetizing factor N that approaches 1. For perfectly shielded superconductors, $\chi = -1$, so $\chi_{ext} = -(1 - N)^{-1}$. As superconductor films get thinner, $N \rightarrow 1$ and $\chi_{ext} \rightarrow -\infty$. In the early days of electromagnetism, Maxwell commented, "If the value of κ [susceptibility] could be negative and equal to $1/(4\pi)$ [in CGS units, 1 in SI units] we should have an infinite value of the magnetization in the case of a magnetizing force acting normally to a flat plate or disk. The absurdity of this result confirms what we said in Art. 428."[73]

Magnetization M is the measured magnetic moment m per sample volume V_s. As the superconductor gets thinner (with its cross sectional area constant), $\chi_{ext} \rightarrow -\infty$ because $V_s \rightarrow 0$, not because $m \rightarrow -\infty$.[74] Furthermore, as the film gets *very* thin and $N \approx 1$, flux immediately penetrates the film for any H_a and the superconductor is no longer in the shielding state.[75] But is not m linearly proportional to V_s or thickness t? If it were, M would be independent of t. We will show that, for a range of t, m remains constant for thick films of constant diameter d. Since a susceptometer pick-up coil voltage v is proportional m, this means that v is independent of t. The reason for this singular behavior is that, as $t \rightarrow 0$, $(1 - N)$ becomes proportional to t.

We model a circular superconductor film as an oblate ellipsoid, with major axes equal to the diameter d and minor axis equal to t. The ellipsoid volume $V_s = \frac{1}{6}\pi\gamma d^3$, where γ is the aspect ratio t/d. For small γ,[50,51] $N = 1 - \frac{1}{2}\pi\gamma + 2\gamma^2$. For $\gamma \leq 0.003$, the first two terms alone give N accurate to four significant figures. We use the linear approximation and $\chi_{ext} = m/(V_s H_a) = -(1 - N)^{-1}$ to get: $m = -V_s H_a/(1 - N) = -\frac{1}{3}d^3 H_a$ (*independent* of t)[76,77] and a gauge for superconductor films,

$$\chi_{ext} = -2/(\pi\gamma). \tag{3}$$

For computational convenience, we extrapolate these arguments to a flat cylinder: $V_s = \frac{1}{4}\pi\gamma d^3$ and $m = -\frac{1}{2}d^3 H_a$.

The magnetic moment of a superconductor arises from shielding currents which, for small applied fields, flow within a penetration depth λ of the surface. If we consider a cylinder of diameter d and height t, the magnetic moment of the current loop is $m = -\frac{1}{4}\pi d^2 I$, where I is the total shielding current. In terms of a current density J, $m = -\frac{1}{4}\pi d^2 J\lambda t$, which is equal to a constant from the previous discussion. (Strictly, the shielding current in a superconducting cylinder does not flow in a uniformly wide sheath on its circumference.) What happens when t gets too small? As t decreases J must increase until it equals the critical current density J_c. Any further decrease in t will result in flux penetration into the superconductor. From $m = -\frac{1}{2}d^3 H_a = -\frac{1}{4}\pi d^2 J\lambda t$, the critical thickness t_c is simply $2H_a d/(\pi\lambda J_c)$, which offers a way to determine J_c if λ is known. The analysis breaks down for thin films ($t \rightarrow \lambda$) for which there is an enhanced effective λ.[78]

We have experimentally verified some of these points with a series of seven superconducting Bi–Sr–Ca–Cu–O thick films. They were made by metallo-organic decomposition and had the usual granular characteristics. Their diameter was 3.22 mm and they ranged in thickness from 2.8 (± 0.2) μm to 0.37 (± 0.02) μm. Plots of χ'_{ext} as a function of temperature, measured in a field of 0.8 $A \cdot m^{-1}$, 1000 Hz, normal to their surface, were almost flat at low temperatures and indicated good diamagnetic shielding at 4.2 K for all except the 0.37-μm film. χ'_{ext} at 4.2 K ranged from −390 for the 2.8-μm film to −2080 for the 0.55-μm film. (χ'_{ext} was −2560 for the 0.37-μm film.) These values are about half those expected from $\chi_{ext} = -2/(\pi\gamma)$, which is not unreasonable considering the approximations involved. The pick-up coil voltages that gave rise to these χ'_{ext} ranged from 4.0 to 4.4 μV (almost constant). The 0.37-μm film voltage was smaller, 3.6 μV. The imperfect shielding for the 0.37-μm film suggests that $t_c \approx 0.4$ μm.

Dependence on Field Angle

When the field is perpendicular to the superconductor film plane there is more susceptometer signal than when the field is parallel. Aside from considerations regarding magnetic penetration depth, the reason is that, for the perpendicular orientation ($N \approx 1$), the applied field H_a is enhanced to give a large internal field, $H = H_a/(1 - N)$, and so are the magnetization, $M = -H_a/(1 - N)$, and the external susceptibility, $\chi_{ext} = -(1 - N)^{-1}$. (For magnetically soft, ferromagnetic films, in comparison, $\chi \rightarrow \infty$ and $\chi_{ext} \rightarrow N^{-1}$.)

Misalignment of a superconductor film in a magnetometer or susceptometer causes errors in the measurement of χ_{ext}. For example, Gyorgy found that χ_{ext} of a Nb film of dimensions 7.6 mm × 5 mm × 0.13 μm varied from −0.08, for H_a parallel to the film surface, to −0.17 for H_a at an angle of 0.5° from the film plane, to −18 for H_a at an angle of 8°.[79] Teshima *et al.* studied the angular dependence of the mixed-state magnetic hysteresis loop and concluded that the magnetization is perpendicular to the film plane for any direction of the applied field.[80]

We can examine the field-angle problem analytically. We start by modeling the film as an isotropic (constant χ), oblate ellipsoid. We use the relations

$$H_{a,\xi} = H_\xi + N_\xi M_\xi = H_\xi(1 + N_\xi \chi), \tag{4a}$$
$$M_\xi = \chi H_\xi = \chi H_{a,\xi} / (1 + N_\xi \chi) = H_{a,\xi} / (\chi^{-1} + N_\xi), \tag{4b}$$

where the subscript ξ denotes the three ellipsoid axes x, y, z. H_a is at an angle θ with respect to the normal to the film plane (z axis), so $H_{a,z} = H_a\cos\theta$ and $H_{a,x} = H_a\sin\theta$. For a superconductor in the shielding state, we take $\chi = -1$. Using $2N_x + N_z = 1$, we have

$$M_z = -H_a\cos\theta / (1 - N_z), \tag{5a}$$
$$M_x = -H_a\sin\theta / (1 - N_x) = -2H_a\sin\theta / (1 + N_z). \tag{5b}$$

But we measure M only in the axis of H_a, that is, $M_z\cos\theta + M_x\sin\theta$, so our measured susceptibility is

$$\chi_{ext} = -\cos^2\theta / (1 - N_z) - 2\sin^2\theta / (1 + N_z). \tag{6}$$

For superconductor films ($N_z \rightarrow 1$), a slightly out-of-plane H_a ($\theta < \frac{1}{2}\pi$) will cause M_z to dominate. The same analysis for magnetically soft, ferromagnetic ellipsoids ($\chi \rightarrow \infty$) gives

$$M_z = H_a\cos\theta / N_z, \tag{7a}$$
$$M_x = 2H_a\sin\theta / (1 - N_z), \tag{7b}$$
$$\chi_{ext} = \cos^2\theta / N_z + 2\sin^2\theta / (1 - N_z). \tag{8}$$

For ferromagnetic films, a slightly off-axis H_a ($\theta > 0$) will cause much of the magnetization to be in plane.

Multiple Phases

When susceptibility is measured as a function of temperature in perpendicular field, χ_{ext} curves appear strikingly different from the χ curves obtained after correcting for demagnetizing factor. Measurements made with the field perpendicular to the film plane give very large values of external susceptibility. Values of −1500 for χ'_{ext} are typical. However, when corrected for demagnetizing factor, any large negative value of χ'_{ext} will convert to $\chi' \approx -1$; there is negligible difference in χ' between $\chi'_{ext} = -1500$

and $\chi'_{ext} = -15$. Furthermore, huge errors in volume will barely affect the χ' result. The obvious way to avoid these difficulties is to measure χ with the field parallel to the plane of the film. However, this arrangement gives inadequate signal if the sample volume is insufficient or if the magnetic penetration depth is large relative to the film thickness. Furthermore, it is sometimes difficult to align films perfectly parallel to the applied field. If these problems arise, one has to resort to measurements in perpendicular field in which χ_{ext} is measured and χ is computed.

Superconducting Bi–Sr–Ca–Cu–O thick films, about 1 μm thick, were prepared by metallo-organic decomposition on single-crystal (100)-oriented MgO substrates.[81–83] By adjusting the heat treatment of the films, we varied the relative concentration of the two superconducting phases, $Bi_2SrCa_2Cu_2O_x$ ("2122," nominal $T_c = 85$ K) and $Bi_2Sr_2Ca_2Cu_3O_x$ ("2223," nominal $T_c = 110$ K). Each phase had both intrinsic and coupling components. The films were highly oriented, composed of platelets with c axes perpendicular to the film plane. The platelets were 10–30 μm wide and 0.2–0.3 μm thick. The films were characterized by X-ray diffraction, electrical resistivity, and ac susceptibility in parallel and perpendicular fields with the intent of ascertaining the relative concentration of the 2212 and 2223 phases.

The areas of the low-angle (002) X-ray diffraction peaks give a relative measure of the concentration of the two phases.[84] (Diffraction does not distinguish between intrinsic and coupling components.) For the (002) reflection, 50% of the signal comes from the top 0.16 μm of the film. This is important because phase segregation through the film thickness is likely. Resistivity provides evidence of two phases, but the lower-temperature 2212 phase will not be detected once the concentration of the higher-temperature 2223 phase exceeds the percolation threshold. Thus, the relative magnitudes of the resistivity transitions will give only a hint of the phase fractions.

For susceptibility, which is more useful, χ_{ext} or χ? As noted above, the demagnetization correction to obtain χ' from χ'_{ext} is nonlinear. Thus χ'_{ext} gives a better indication of the relative fractions of the different intrinsic phases. The coupling components should not be included in the determination; this might require measurements at several fields. (As expected, we found that the coupling component was tied to the "parent" phase. In single-phase 2212 samples, for example, there was no 2223 coupling component.) χ, highly nonlinear, is useful for highlighting the transition temperature of each phase, especially the higher-temperature 2223 phase, which is often obscured in χ_{ext}. One strategy for determining the phase fractions, not suitable for routine measurements, is to scrape the films into a powder and measure its susceptibility, thus eliminating the coupling components and reducing the demagnetizing factor problem. Perhaps the major utility of susceptibility measurements in granular superconductors, however, is for characterizing the quality of intergrain coupling by the field dependence of the coupling transition temperature.

In Table 2 we give values of the percentages of the 2212 and 2223 phases as inferred from resistivity, X-ray diffraction, and susceptibility, for one of the films while intact and after powdering. The presence of a small amount of $Bi_2Sr_2CuO_6$ was ignored except in the X-ray determinations, which do not sum to 100. As far as we

Table 2. Volume percentage of 2212 and 2223 phases in a Bi–Sr–Ca–Cu–O film as inferred from different measurement methods.

Morphology	Method	2212	2223
Film	ρ	5	95
Film	X-ray	12	82
Film	χ'_{ext}, $H\parallel$	75	25
Film	χ'_{ext}, $H\perp$	60	40
Powder	X-ray	32	47
Powder	χ'_{ext}	30	70
Powder	χ'	40	60

know, none of these determinations is "correct." Our intent is to illustrate the difficulty of the problem.

UNITS

Magnetic volume susceptibility χ is dM/dH in both SI and CGS. In terms of base units, χ is a dimensionless quantity. In CGS, however, χ is usually expressed as emu, $\mathrm{emu \cdot cm^{-3}}$, or $\mathrm{emu \cdot cm^{-3} \cdot Oe^{-1}}$. The designation "emu" is merely an indicator that electromagnetic units are in use; it is not a unit. The unusual χ units in CGS arise from ambiguity in the units for M.[85] In CGS, H is in Oe (dimensionally and numerically equivalent to G). Magnetization, when written as $4\pi M$, is in G. When magnetization is expressed simply as M (the magnetic moment m per unit volume) its units are $\mathrm{erg \cdot G^{-1} \cdot cm^{-3}}$ (conventionally expressed as $\mathrm{emu \cdot cm^{-3}}$), which are dimensionally but *not numerically* equivalent to G. Occasionally, CGS susceptibility is written as $4\pi\chi$ (dimensionless), which is equal to SI susceptibility (dimensionless).

Some of the symbols used in this paper and their associated SI units are M [$\mathrm{A \cdot m^{-1}}$], J [$\mathrm{A \cdot m^{-2}}$], a [m], H [$\mathrm{A \cdot m^{-1}}$], and W [$\mathrm{J \cdot m^{-3}}$]. To convert equations to CGS EMU, replace symbols for H by $(4\pi\mu_0)^{-1/2}H$, symbols for J by $(4\pi/\mu_0)^{1/2}J$, symbols for M by $(4\pi/\mu_0)^{1/2}M$, symbols for χ by $4\pi\chi$, and simplify as necessary. The symbols and their associated CGS EMU are M [$\mathrm{emu \cdot cm^{-3}}$], J [$\mathrm{abamp \cdot cm^{-2}}$], a [cm], H [Oe], and W [$\mathrm{erg \cdot cm^{-3}}$]. (Note that in CGS Gaussian units, J would be in $\mathrm{statamp \cdot cm^{-2}}$.) We avoid the use of "practical" or mixed units.

SUSCEPTIBILITY OF BULK SUPERCONDUCTORS, CRITICAL–STATE MODEL

In this section we review magnetic formulas for isotropic superconductors of several geometries in the Bean critical-state model.[86–88] Except for some of the high field susceptibility equations, most can be found in the literature in one form or another. Transport current, dc bias field, lower critical field, and surface barrier are all taken to be zero. H_a is the applied dc field. H_p is the full-penetration field, a function of the critical current density J_c (considered isotropic) and the cylinder radius (or slab half-thickness) a. M is the magnetization per unit volume of superconductor, which is equal to the half-width of the hysteresis loop ($\frac{1}{2}\Delta M$) for $H_a \geq H_p$. H_m is the maximum field for the hysteresis cycle, that is, the amplitude of the ac field. W is the hysteresis loss per unit volume per field cycle. χ' and χ'' are the real and imaginary parts of ac susceptibility. The equations for χ' and χ'' are derived from Fourier integrals of the magnetization[89] using the complete equations for magnetization as a function of field,[90] not the initial curve. By χ' and χ'' we mean the fundamental Fourier components χ_1' and χ_1''. As a check, we know from first principles[91,92] that $\chi'' = W/(\pi\mu_0 H_m^2)$ for any H_m. For $H_m \gg H_p$, the Bean hysteresis loop is almost rectangular, and $W \rightarrow 4\mu_0 H_m M$. A point worth emphasizing is that $\chi' \neq \bar{\chi}_{dc} \equiv M/H_a$. Only in the linear limit $H_m \ll H_p$ does $\chi' \rightarrow \bar{\chi}_{dc}$. Another point, applicable to the cylinder in transverse field, is that the susceptibilities are based on the applied field, and are not corrected for demagnetization.

Infinite Slab, Thickness 2a, Parallel H[93,94]

The equations below are for the initial magnetization curve ($0 \leq H_a \leq H_p$) and the *descending* portion of the hysteresis loop (which depends on whether $H_m \leq H_p$ or $H_m \geq H_p$). To get the *ascending* portion, replace M by $-M$ and H_a by $-H_a$ in the equations. For example, the ascending magnetization curve for $H_m \leq H_p$ is $M = -H_a - \frac{1}{4}(H_m^2 - 2H_a H_m - H_a^2)/H_p$, based on Eq. (11). To calculate χ' and χ'' we use $H_a = H_m \cos\theta$. For $H_m \geq H_p$, we integrate from $\theta = 0$ to $\cos^{-1}(1 - 2/x)$, and from $\theta = \cos^{-1}(1 - 2/x)$ to π, where $x \equiv H_m/H_p$. A useful trigonometric identity is $\sin[\cos^{-1}(1 - 2/x)] = (2/x)(x - 1)^{1/2}$.

$$H_p = J_c a \tag{9}$$

$$M = -H_a + \tfrac{1}{2}H_a^2/H_p \qquad (0 \leq H_a \leq H_p) \tag{10}$$

$H_m \leq H_p$:

$$M = -H_a + \tfrac{1}{4}(H_m^2 + 2H_a H_m - H_a^2)/H_p \qquad (-H_m \leq H_a \leq H_m) \tag{11}$$

$$W = \tfrac{2}{3}\mu_0 H_m^3/H_p \tag{12}$$

$$\chi' = -1 + \tfrac{1}{2}x \tag{13}$$

$$\chi'' = 2x/(3\pi) \tag{14}$$

$H_m \geq H_p$:

$$M = -\tfrac{1}{2}H_p + H_m - H_a - \tfrac{1}{4}(H_m - H_a)^2/H_p \qquad (H_m - 2H_p \leq H_a \leq H_m) \tag{15a}$$

$$M = \tfrac{1}{2}H_p \qquad (-H_m \leq H_a \leq H_m - 2H_p) \tag{15b}$$

$$W = 2\mu_0 H_m H_p - \tfrac{4}{3}\mu_0 H_p^2 \tag{16}$$

$$\chi' = \{(-1 + \tfrac{1}{2}x)\cos^{-1}(1 - 2/x) + [-1 + 4/(3x) - 4/(3x^2)]\,(x - 1)^{1/2}\} / \pi \tag{17}$$

$$\chi'' = (6/x - 4/x^2)/(3\pi) \tag{18}$$

Infinite Cylinder, Radius a, Axial H[93,95–98]

$$H_p = J_c a \tag{19}$$

$$M = -H_a + H_a^2/H_p - \tfrac{1}{3}H_a^3/H_p^2 \qquad (0 \leq H_a \leq H_p) \tag{20}$$

$$H_m \le H_p\colon\ M = -H_a + \tfrac{1}{2}(H_m^2 + 2H_aH_m - H_a^2)/H_p - \tfrac{1}{4}(H_m^3 + H_m^2H_a - H_mH_a^2 + \tfrac{1}{3}H_a^3)/H_p^2 \qquad (-H_m \le H_a \le H_m) \quad (21)$$

$$W = \tfrac{4}{3}\mu_0 H_m^3/H_p - \tfrac{2}{3}\mu_0 H_m^4/H_p^2 \quad (22)$$

$$\chi' = -1 + x - 5x^2/16 \quad (23)$$

$$\chi'' = (4x - 2x^2)/(3\pi) \quad (24)$$

$$H_m \ge H_p\colon\ M = -\tfrac{1}{3}H_p + H_m - H_a - \tfrac{1}{2}(H_m - H_a)^2/H_p + (H_m - H_a)^3/(12H_p^2) \qquad (H_m - 2H_p \le H_a \le H_m) \quad (25a)$$

$$M = \tfrac{1}{3}H_p \qquad (-H_m \le H_a \le H_m - 2H_p) \quad (25b)$$

$$W = \tfrac{4}{3}\mu_0 H_m H_p - \tfrac{2}{3}\mu_0 H_p^2 \quad (26)$$

$$\chi' = \{(-1 + x - 5x^2/16)\cos^{-1}(1 - 2/x) + [-19/12 + \tfrac{5}{8}x + 1/x - 2/(3x^2)]\,(x-1)^{1/2}\}\,/\,\pi \quad (27)$$

$$\chi'' = (4/x - 2/x^2)/(3\pi) \quad (28)$$

Infinite Cylinder, Radius a, Transverse H

This case has not been solved exactly, but a few approaches have been used successfully. The method of Carr *et al.* gives analytic equations in the limits of small and large fields[99–102] and is mostly numerical in between.[103] The hysteresis loss in SI units is $W = 256\mu_0 H_m^3/(9\pi^2 H_p)$ for $H_m \ll H_p$, and $W = \frac{8}{3}\mu_0 H_m H_p$ for $H_m \gg H_p$. The method of Zenkevitch *et al.*, presented below, gives reasonable equations for the full field range.[104–106] For M we preserved the form of the equations developed in Ref. *104*. In terms of H_p, the equations are simply twice those for the cylinder in axial H. Different functional forms are given by other authors.[107,108]

$$H_p = 2J_c a/\pi \quad (29)$$

$$M = -\tfrac{2}{3}H_p[1 + (H_a - H_p)^3/H_p^3] \qquad (0 \le H_a \le H_p) \quad (30)$$

$$H_m \le H_p\colon\ M = \tfrac{4}{3}H_p[1 + (H_m - H_a - 2H_p)^3/(2H_p)^3] - \tfrac{2}{3}H_p[1 + (H_m - H_p)^3/H_p^3] \qquad (-H_m \le H_a \le H_m) \quad (31)$$

$$W = \tfrac{8}{3}\mu_0 H_m^3/H_p - \tfrac{4}{3}\mu_0 H_m^4/H_p^2 \quad (32)$$

$$\chi'_{ext} = -2 + 2x - \tfrac{5}{8}x^2 \quad (33)$$

$$\chi''_{ext} = (8x - 4x^2)/(3\pi) \quad (34)$$

$$H_m \ge H_p\colon\ M = \tfrac{4}{3}H_p[1 + (H_m - H_a - 2H_p)^3/(2H_p)^3] - \tfrac{2}{3}H_p \qquad (H_m - 2H_p \le H_a \le H_m) \quad (35a)$$

$$M = \tfrac{2}{3}H_p \qquad (-H_m \le H_a \le H_m - 2H_p) \quad (35b)$$

$$W = \tfrac{8}{3}\mu_0 H_m H_p - \tfrac{4}{3}\mu_0 H_p^2 \quad (36)$$

$$\chi'_{ext} = \{(-2 + 2x - \tfrac{5}{8}x^2)\cos^{-1}(1 - 2/x) + [-19/6 + \tfrac{5}{4}x + 2/x - 4/(3x^2)]\,(x-1)^{1/2}\}\,/\,\pi \quad (37)$$

$$\chi''_{ext} = (8/x - 4/x^2)/(3\pi) \quad (38)$$

In Fig. 5 we plot χ' and χ'' as functions of H_p/H_m for the three cases. As we discuss in Ref. *89*, H_p is a good proxy for temperature, and the inverted abscissa in Fig. 5 is meant to indicate the temperature dependence of susceptibility. Note that, for the transverse-field case, $\chi'_{ext} \rightarrow -2$ as $x \rightarrow 0$. For this geometry the demagnetizing factor is $\frac{1}{2}$; after correcting for demagnetization, $\chi' \rightarrow -1$.

Critical Current Density

Transport J_c is measured directly using electrical techniques.[109] Alternatively, several magnetic methods may be used.[110] To estimate J_c from magnetic measurements, one could measure M (at $H_a \geq H_p$) and derive what has become known as "magnetization J_c." Alternatively, one could measure the reduction in field $\Delta H_a = 2H_p$ required to reverse the magnetization in the hysteresis loop.[111,112] Other magnetic methods are based on ac susceptibility,[55,113–116] usually with a dc bias field,[116–119] sometimes using harmonic analysis.[89,93,120–122] Curve fitting is often involved. For example, at the peak of χ'', $H_m = H_p$ for cylinders[113] and $H_m = \frac{4}{3}H_p$ for slabs[108] (Fig. 5). The dimensions of the sample are used in all these methods.

The equations that relate J_c to the magnetization are based on the assumption that J_c is a constant, *independent* of H_a. Apparently begging the question, the equations are often used to deduce J_c as a *function* of H_a. This is not necessarily a serious error provided that certain conditions are satisfied: (1) The sample is homogeneous and isotropic, although the critical-state model has been extended for anisotropic critical current densities.[123,124] (2) The sample has dimensions consistent with the model. However, in the fully penetrated state ($H \geq H_p$) in the Bean formalism, the magnetization is saturated even for finite dimensions. Thus, for example, the infinite cylinder equation for M can be used for a finite disk in perpendicular field.[80] (3) The field at which magnetization M is taken should be large enough ($H > H_p$) such that J_c is not a strong function of H_a.[90] Specifically, estimates of J_c from M at $H_a = 0$ are

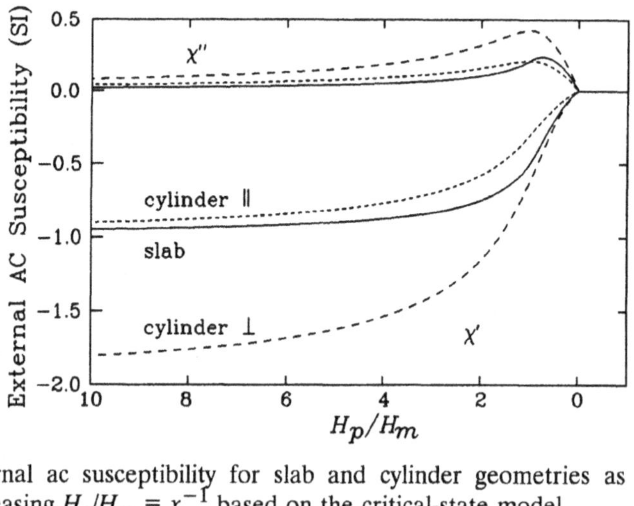

Fig. 5. External ac susceptibility for slab and cylinder geometries as functions of decreasing $H_p/H_m \equiv x^{-1}$ based on the critical-state model.

likely to be in error. (4) Flux vortices are well pinned; that is, there is no flux creep.[88] In practical type-II superconductors with high J_c, or in high-T_c superconductors at low temperature, this is generally true. (5) There is little contribution from surface barriers[125–127] and reversible magnetization.[127] This often applies at low temperatures, where the hysteresis loop symmetrically spans positive and negative M at high fields.

All these conditions for use of the critical-state model are seldom satisfied simultaneously. Furthermore, interpretation of the magnetization loop and application of the critical-state formulas are different for weakly linked and homogeneous samples. For weakly linked samples, such as sintered, high-T_c superconductors, the magnetization at high fields gives the intragrain critical current.[35] The average dimension of the grains should be used to deduce J_c. In low fields the grains are coupled. The correct dimension is that of the specimen and the deduced J_c is comparable to transport J_c. But as noted above, the critical-state model cannot be used accurately at low fields, and in any event, the coupled material is not homogeneous. For homogeneous samples, such as single crystals and samples with contiguous oriented grains, the basic critical-state model can be used if the field is applied such that shielding currents flow isotropically.

Lower Critical Field

In materials with intrinsic and coupling components, there will be two lower critical fields. Generally, H_{c1} for type-II superconductors corresponds to the field at which the initial magnetization-versus-field curve deviates from linearity. In practice, detection of this field value is difficult because the deviation may be subtle, especially if the critical current density is large. However, if the magnetic field is cycled, an area, indicative of hysteresis loss, will be traced out in the $M-H$ plane when $H > H_{c1}$.[128–130] This is effectively the field cycle used in ac susceptibility measurements, and hysteresis loss will appear as a positive imaginary part of susceptibility χ''.[98,131] Therefore, when the field amplitude used in a χ measurement exceeds H_{c1}, positive χ'' is expected and measurements of $\chi''(T)$ at constant H,[17,132] or $\chi''(H)$ at constant T,[133] may be used to deduce an upper limit for $H_{c1}(T)$. Equivalent to the $\chi''(T)$ method, the corresponding feature in $\chi'(T)$ or $M(T)$ is a departure from full diamagnetism.[129,134–136]

Susceptibility and its harmonics, measured as functions of H, reveal a distinct feature at H_{c1}.[122,128] In another method, H_{c1} is included in an expanded critical-state model and is a function of the remanent (trapped) magnetization.[137] In small specimens, such as thin films, one should be aware of the enhancement in H_{c1} that arises when the magnetic penetration depth is on the order of one of the sample dimensions.[27,61,69,138,139]

Interpretation of Peak in χ''

The critical-state model can account for many of the features in the temperature-dependent χ'' (and χ') without invoking any kind of loss mechanism or irreversibility other than magnetic hysteresis. The model may even be used to describe the behavior of the intergranular coupling component of sintered superconductors.[89,140] The following interpretation of the peaks in χ'' versus increasing T is based on the critical-state approach and can be applied to both the intrinsic and coupling components.[17,113,119] Susceptibility is measured in an ac field of constant magnitude. We divide the temperature scale into three ranges. (1) For $T \ll T_c$, the ac field causes shielding currents to flow on the surface of the sample and a line to be traced out in the $M-H$ plane. There is no hysteresis because $J < J_c$, $H < H_{c1}$, and $\chi'' = 0$. (2) For T somewhat below T_c, J_c and H_{c1} have decreased and shielding currents have to flow within the sample. The hysteresis loop in the $M-H$ plane has an area associated with it, and $\chi'' > 0$. The losses and χ'' attain their maximum values after supercurrents and penetrated flux reach the center of the specimen. (3) As T approaches T_c, J_c approaches 0, and the $M-H$ loop has collapsed. Even though H_{c1} also approaches 0, there is no area to the loop and no hysteresis loss; $\chi'' = 0$. This interpretation is in accordance with the expectations of the critical-state model, in which all energy losses are hysteretic and frequency independent.

Other loss mechanisms besides hysteresis may contribute to χ'' in superconductors. These losses may be classified as time-dependent or time-independent.[141] In particular, frequency effects may be explained by flux creep and flux flow.[142–144] Other frequency-dependent contributions to χ'' could be eddy currents of normal-state electrons in a two-fluid model[4,145–147] and vortex lattice viscosity and viscous damping.[4,148–150] One physical interpretation of the ac susceptibility of superconductors is in terms of BCS theory, the generation of supercurrents, and the establishment of the Meissner state.[151,152] Others have used a superconductor glass model and scaling with critical exponents.[153,154] The *time- and field-dependent* onset of irreversibility (an "irreversibility *plane*"), which may occur below the temperature of the χ'' onset, may be observed by the generation of odd harmonics of susceptibility.[89,122]

Occasionally the intrinsic χ'' peak is not apparent.[155] There are several cases where this is likely, with regard to granular high-T_c superconductors. (1) In well coupled materials, a small measuring field will cause the coupling peak to obscure the intrinsic peak. (2) There may be insufficient total grain volume. (3) Grain sizes may be on the order of λ.[113] (4) The grains may be superconducting only on their surface; the interior is normal, perhaps owing to deficient oxygenation, or superconducting only at a lower temperature. In this state, there would be insufficient superconducting volume and therefore a low level of losses. (5) In good quality grains,[145] H_{c1} may be large just below T_c. If H_{c1} only falls below H_a at T_c, no χ'' peak will be seen. That is, at the temperature that flux penetrates the grain, there is no longer any bulk pinning. A larger H_a will often elicit a measurable χ''.

Upper Critical Field

A plot of H_{c2} versus T is the same as a plot of T_c versus H. Following the distinction of intrinsic and coupling components, there are both intrinsic and coupling upper critical fields. Susceptibility can be used to deduce H_{c2} versus T (at high temperatures) in much the same way as it was used to determine H_{c1} versus T. At the onset temperature T_c there is an equivalence between the measuring field and H_{c2}.[25]

SUSCEPTOMETER DESIGN

Construction

Low-frequency ac susceptibility measurements are often made with a coaxial mutual-inductance coil system consisting of a primary excitation field coil, a secondary pick-up coil, and a secondary compensation coil (three-coil system). The two secondary coils, connected to a bridge circuit, have the same dimensions. The midpoint between them is concentric with the primary coil.[156–159] An alternative three-coil system, suitable for short or long samples, or when coil length, eddy currents, or temperature gradients are a problem, is with all coils concentric.[160] Here, the pick-up coil is close to the sample, but the compensation and field coils are not. To ensure balance between the pick-up and compensation coils, they are wound so that the mutual inductance of each with respect to the field coil is the same. Another three-coil system uses a large-bore field coil with side-by-side secondary coils.[161] If the compensation coil, or some other field compensation source, is omitted (two-coil system),[162] the measured quantity is ac permeability ($\mu = \mu' + i\mu''$), which is numerically related to ac susceptibility: $\mu = \mu_0(1 + \chi)$, $\mu' = \mu_0(1 + \chi')$, $\mu'' = \mu_0\chi''$, where μ_0 is the permeability of vacuum. If a single coil is used (one-coil system), one can relate the changes in inductance and resistance of the coil to ac χ. Calibration is readily achieved in any magnetometric system when the sample is contained within the pick-up coil.

At audio-frequencies, sensitivity is greatest for the susceptometer (three-coil) configuration. The ac susceptometer relies on inductive coupling between coils. A large number of turns on the pick-up coil increases the signal-to-noise ratio at low frequencies but causes capacitive coupling at high frequencies.[163] A typical instrument with 520 turns of 28 gauge wire (0.32 mm diameter) on the primary and 1340 turns of 38 gauge wire (0.10 mm diameter) on each secondary has capacitive coupling above about 5 kHz. The usable frequency range can be extended by reducing the number of turns.[164] Resonance methods, using a single coil[70,165–168] or a two-coil bridge,[169,170] may be used up to radio frequencies. A sample inserted in one of the coils causes a change in self inductance and phase that is related to the susceptibility. These methods are quite sensitive but have been often neglected. For metallic samples, eddy-current signals may present a problem at high frequencies.

Alternating-field susceptibility characterizes the shielding properties of superconductors, whether measured upon cooling or upon warming after zero-field cooling. Typically, measurements are made in zero dc field as a function of ac field or in small ac fields as a function of dc bias field.[89,117,118] The laboratory environment should not be ignored as a source of dc field, particularly in materials, such as sintered high-T_c superconductors, that are weakly coupled. For precise measurements it is desirable to surround the susceptometer with a high-permeability magnetic shield not too close to the coils.

The pick-up and compensation coils may be connected in series opposition or in parallel to the differential input of a lock-in amplifier. The lock-in amplifier may be used either as a null detector upon adjustment of standard inductors and resistors

(Hartshorn bridge),[171] or simply as an off-balance meter. An input band-pass filter should be used to attenuate harmonics with typical attenuation of more than 60 dB. In harmonic susceptibility measurements, filtering is used to isolate the harmonic of interest, as was done in Ref. *89*. A constant ac current source (transconductance amplifier) should be used to drive the alternating-field coil if the temperature of the coil changes during the measurement. If the pick-up coil and compensation coil are immersed in liquid helium, Johnson (thermal) noise will be reduced and the coil resistances will remain constant. A disadvantage of this arrangement is that the sample will be weakly coupled to the pick-up coil if a reentrant Dewar is used to control the temperature of the sample. We will describe a system in which the coil temperature changes with that of the sample. To maximize sample coupling, the pick-up coils are wound beneath the field coil. It is generally considered desirable to match the impedance of the pick-up coils to the input impedance of the lock-in amplifier. In practice, however, we have found that there is no advantage in terms of noise or sensitivity, and that the transformer may contribute phase shifts.

Inevitably, there will be mismatch between the pick-up and compensation coils. This can vary with temperature and over time as the coils contract and expand. Two-position susceptometers avoid this problem.[172–177] The sample is measured in each coil, with its position controlled by a sample rod and piston. The voltage signal attributable to the sample changes sign but the signal arising from coil imbalance does not. When the two measurements are subtracted and divided by two, the imbalance signal is removed. To minimize the imbalance and exploit the dynamic range of the lock-in amplifier, we use a trimming loop in series with the field coil. Its position is adjusted once to increase the coupling to either secondary coil. Its contribution to the measurement field is negligible. Sometimes a grounded "coil foil" (a sheet consisting of thin parallel strands of insulated copper wire) is used between the field coil and pick-up coils to reduce their capacitive coupling.[158] We have found that such a shield contributes little to the susceptometer performance. Another possible use for coil foil is for thermal stability.[178]

To help achieve an isothermal environment, we use a sapphire (Al_2O_3) tube as a coil form. Sapphire is a good thermal conductor and poor electrical conductor. Metal coil forms are not used because they can contribute eddy-current signals. Metal structural components and heaters are well separated from the coils for the same reason. The normal heat leak in the Dewar can be used to slowly warm the sample, or the temperature may be actively controlled. A resistance or semiconductor thermometer and its connecting wires are thermally anchored to the sapphire. (In systems designed for use with high dc bias fields, the thermometer should be relatively insensitive to field. Carbon glass is an example.) Our sample holder is designed in three identical sections so that, in both the upper and lower positions, each secondary coil detects the same holder material except for the sample itself. The holder is open at the sides for sample insertion and removal. The clearance between the sample holder and the sapphire tube is small. To reduce the possibility of damage to the coils from sample-rod motion if air becomes trapped and frozen, we use the release mechanism shown in Fig. 6. The O-ring releases at about 7 N force.

Phase Adjustment

The phase angle ψ of the lock-in amplifier used to detect the pick-up coil voltage must be adjusted to correctly separate the real and imaginary parts of susceptibility. The adjustment may be done before the measurement or by computation, after the measurement: $\chi'_{ext} = \chi'_{ext,0} \cos\psi + \chi''_{ext,0} \sin\psi$, $\chi''_{ext} = \chi''_{ext,0} \cos\psi - \chi'_{ext,0} \sin\psi$, where the 0 subscripts indicate the measured susceptibilities before adjustment of the phase angle.

Here is an opportunity for the experimentalist to use good judgement. The first guideline is that χ'' must never be negative. Phase adjustment is accomplished when samples are in low-loss states. Examples are superconductors in the shielding state (low temperature and low measuring field) and spin glasses[179] in the paramagnetic state. Precise phase adjustment is necessary to observe frequency shifts in the susceptibility curves.

The phase adjustment should be repeated for each sample and each measurement frequency but not for each measurement field. It is not practical to adjust phase for each measuring temperature, but, as temperature changes, the resistance of the susceptometer coils does change. This could cause some phase change in the mutual inductance bridge. To avoid this, the lock-in amplifier should be referenced to the voltage drop across a resistor in series with the primary coil.

On the secondary circuit, the input impedance of the lock-in amplifier is large enough to make any change in the coil resistance insignificant. If desired, however, wire with a residual resistivity ratio close to 1, such as brass,[180] could be used to wind the coils instead of copper magnet wire. (Wire with magnetic impurities should be avoided.) An added advantage would be the suppression of any eddy currents in the coils themselves. We have found copper wire to be satisfactory. Phase adjustment at different temperatures may be unavoidable if there are problems with eddy currents in structural elements.

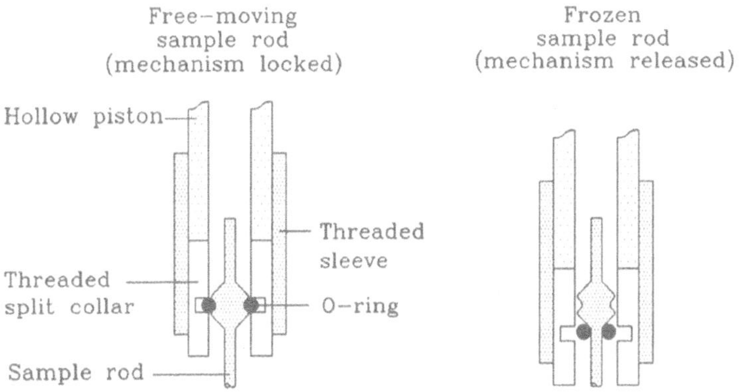

Fig. 6. Release mechanism used to decouple the top of the sample rod from the drive piston if there is excessive resistance to motion. The device is at room temperature, above the experiment Dewar.

With the use of computer controlled instruments, it is tempting to take as much data as possible in a single measurement session. Multiple measurement frequencies may be used at each temperature. It is usually not advisable to change magnetic field strength during the measurement of hysteretic materials because, at high fields, magnetic flux penetrates a superconductor and remains pinned. When measurements are continued at reduced fields, the pinned flux might wiggle around and contribute to the susceptometer signal. For a similar reason, ac measurements are usually best made upon warming, after cooling in zero field. Upon cooling through T_c, flux exclusion is often incomplete and pinned flux may remain in the sample. The effect is small and it would be of concern only in precise work.

SUSCEPTOMETER CALIBRATION

Analytical and Numerical Calibrations and Standards

A susceptometer pick-up coil can be calibrated analytically for spherical samples.[156] We use the dipole field of a uniformly magnetized sphere of magnetic moment $m = MV_s$, where M is the magnetization and V_s is the sample volume, and calculate the total flux Φ through a thin pick-up coil of radius a, length ℓ, and n turns.[181] We assume $M = \chi_{ext} H_a$ and $H_a = H_0 \sin 2\pi f t$, where χ_{ext} is the external susceptibility in SI units and H_0 and f are the amplitude and frequency of the applied field H_a. Finally, we use Faraday's law, $v = \mathrm{d}\Phi/\mathrm{d}t$, where v is the pick-up coil voltage, and get

$$\chi_{ext} = v_{rms}[(\tfrac{1}{2}\ell)^2 + a^2]^{1/2} \,/\, (nV_s \pi f \mu_0 H_{rms}), \tag{39}$$

where we now refer to the rms field and voltage and have ignored any sample voltage induced in the compensation coil. For a given pick-up coil, we assign a constant α to the quantity $[(\frac{1}{2}\ell)^2 + a^2]^{1/2} / (n\pi\mu_0)$. When measuring harmonic susceptibility, the harmonic frequency, not the field frequency, is used for f.[89]

It is also possible to calibrate susceptometers numerically for cylindrical samples with small susceptibilities ($\chi \approx 0$) or small demagnetizing factors ($N \approx 0$). Either will have almost uniform magnetization. The procedure models the sample as a solenoid and requires computation of the mutual inductance L^* between the model solenoid and the susceptometer pick-up coil.[182] Once L^* is known for the sample, we have $\chi_{ext} = v_{rms} / (L^* \ell 2\pi f H_{rms})$, where v_{rms} is the pick-up coil voltage, ℓ is the sample length, f is the frequency, and H_{rms} is the applied field. Usually L^* is calculated numerically, but if the pick-up coil is thin, L^* can be calculated analytically.[183]

Other calibration methods use standards. These include materials with known susceptibility, and magnetically soft ferromagnets ($\chi \rightarrow \infty$) and superconductors ($\chi = -1$) with known demagnetizing factors (such as spheres and cylinders). These are discussed in Ref. *182*. If cylinders are used, accurate values of N_m, corresponding to the standards' susceptibilities, are necessary.

A classical exercise is the calculation of the complex magnetic susceptibility of an isotropic, conducting sphere in a uniform ac magnetic field.[184,185] The apparent susceptibility is due to eddy currents, not to the magnetic properties of the material, much like the magnetic susceptibility of a superconductor arises from shielding currents. The real and imaginary parts of external susceptibility are calculated in terms of the sphere radius a and the skin depth δ,

$$\chi'_{ext} = \tfrac{9}{4}(\delta/a)[\sinh(2a/\delta)-\sin(2a/\delta)] \,/\, [\cosh(2a/\delta)-\cos(2a/\delta)] - \tfrac{3}{2}, \quad (40a)$$

$$\chi''_{ext} = \tfrac{9}{4}(\delta/a)[\sinh(2a/\delta)+\sin(2a/\delta)] \,/\, [\cosh(2a/\delta)-\cos(2a/\delta)] - \tfrac{9}{4}(\delta^2/a^2). \quad (40b)$$

In the limit of zero resistivity, $\delta \rightarrow 0$, $\chi'_{ext} \rightarrow -\tfrac{3}{2}$; using $N = \tfrac{1}{3}$ for a sphere, $\chi' \rightarrow -1$.

This result provides the basis for another calibration method that uses spheres of normal conductors such as copper. One requirement is knowledge of the temperature dependence of δ or, equivalently, resistivity ρ. Matthiessen's rule, expressed in terms of the residual resistivity ratio RRR $\equiv \rho(273\text{ K})/\rho(4\text{ K})$, is $\rho(T) = \rho_i(T) + \rho(273\text{ K})/\text{RRR}$, where ρ_i is the intrinsic resistivity. For copper, $\rho(273\text{ K}) = 1.543\times10^{-8}\ \Omega\cdot\text{m}$, and values of RRR range from 10 to 2000. Values of $\rho_i(T)$ are tabulated.[186] From $\rho(T)$ we calculate $\delta(T) = [\rho(T)/(\pi f\mu)]^{1/2}$, where f is the measurement frequency and $\mu = \mu_0$ for a nonmagnetic material. The point is that a copper sphere with known RRR will have predictable curves of χ'_{ext} and χ''_{ext} (or χ' and χ'') as functions of temperature and frequency. [If the ac susceptometer is already calibrated, this method can be used to measure $\rho(T)$.] In Fig. 7 we show χ'_{ext} and χ''_{ext} as functions of temperature for a copper sphere, $a = 3.088$ mm, at 10, 100, and 1000 Hz (points). The measurement field was 80 $\text{A}\cdot\text{m}^{-1}$ rms, although the susceptibilities are independent of field. The curves shown are the predicted χ'_{ext} and χ''_{ext} from the

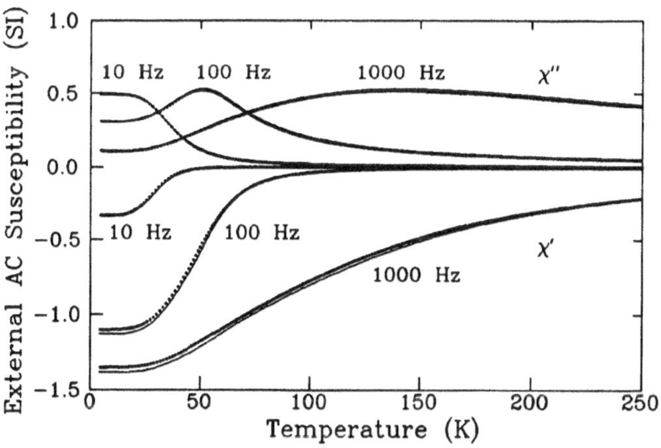

Fig. 7. Apparent susceptibility of a copper sphere as a function of temperature at 10, 100, and 1000 Hz based on eddy currents and skin depth. Points are measured susceptibilities; curves are calculated susceptibilities.

eddy-current equations. The curves overlap the data at 10 Hz. The adjustable parameters were the lock-in-amplifier phase angle ψ (for each frequency) and RRR (adjusted to 150). We have used the 1000-Hz data in lectures to show students an "onset of diamagnetism" near 300 K: a room-temperature superconductor!

Two related problems are for infinite cylinders in perpendicular and parallel fields.[146,164,184,187–191] The infinite-cylinder solutions, together with the demagnetizing factors in Table 1, suggest that calibrations can be done with finite-sized cylinders provided δ is small (that is, ρ is small and f is large).

SUSCEPTOMETER SENSITIVITY

The equation that describes the response of an ac susceptometer is $\chi_{ext} = \alpha v/(V_s f H_a)$, where χ_{ext} is the volume susceptibility in SI units, α is the calibration constant (a function of the pick-up coil geometry) $[\mathrm{A \cdot m^2 \cdot V^{-1} \cdot s^{-1}}]$, v is the pick-up coil voltage [V], V_s is the sample volume $[\mathrm{m^3}]$, f is the frequency $[\mathrm{s^{-1}}]$, and H_a is the magnetic field strength $[\mathrm{A \cdot m^{-1}}]$. The sensitivity of an ac susceptometer depends on α and on the precision in the output v_p of the ac voltmeter, typically a lock-in amplifier. Usually v is proportional to f, so $\alpha v_p / f$ is the magnetic moment precision $[\mathrm{A \cdot m^2}]$. (There is actually some degradation in voltage precision v_p at low frequencies such as 10 Hz due to $1/f$ noise.) For one of our susceptometers, $\alpha = 2.1546$ and $v_p/f = 2.5 \times 10^{-10}$, giving a moment precision of 5×10^{-10} $\mathrm{A \cdot m^2}$ (5×10^{-7} emu). For comparison, commercial vibrating-sample magnetometers are able to measure about 5×10^{-8} $\mathrm{A \cdot m^2}$ (5×10^{-5} emu) and commercial SQUID magnetometers can detect about 10^{-11} $\mathrm{A \cdot m^2}$ (10^{-8} emu). In principle, their sensitivity can be improved by increasing the pick-up coils' filling factor.[192,193] Alternating-gradient-force magnetometers can measure about 10^{-11} $\mathrm{A \cdot m^2}$ (10^{-8} emu).[194]

Our moment precision of 5×10^{-10} $\mathrm{A \cdot m^2}$ means the precision in χ_{ext} is $5 \times 10^{-10}/(V_s H)$. (That is, we can measure the susceptibility of a sample more precisely if we have a larger sample or use a larger measuring field.) For the favorable case of a sample sphere of diameter 5 mm measured in a field of 800 $\mathrm{A \cdot m^{-1}}$, the susceptibility precision would be 10^{-5} (SI). For a 3-mm-diameter sphere measured in 80 $\mathrm{A \cdot m^{-1}}$, the precision would be only 4×10^{-4} (SI). This assumes that both V_s and H are known exactly. Precision is not the same as accuracy, which depends on instrument calibration.

ACKNOWLEDGMENTS

We had valuable discussions with D.-X. Chen (Universitat Autònoma de Barcelona), I. B. Goldberg (Rockwell International Corporation), T. Ishida (Ibaraki University), and J. K. Krause (Lake Shore Cryotronics, Inc.). R. A. Hein (University of Washington) offered many thoughtful comments and suggestions. A sintered Y–Ba–Cu–O sample was provided by A. K. Mehrotra. D. L. Ried, R. L. Spomer, and R. J. Loughran assisted with data acquisition. T. W. Petersen, R. M. Gerrans, and R. W. Cross prepared the figures for publication. Work at the National Institute of Standards and Technology was supported by the Electronics and Electrical Engineering Laboratory.

REFERENCES

1. W. Thomson, "Reprint of Papers on Electrostatics and Magnetism," Macmillan and Co., London (1872), p. 472.
2. In his original paper, *Phil. Mag.*, ser. 4, **1**:177 (1851), Thomson used the term "inductive capacity" for susceptibility, which seems to associate susceptibility with the magnetic induction *B*. He abandoned this term by 1872, possibly on the suggestion of Maxwell. Maxwell used the term "magnetic inductive capacity" for *permeability* in "A Treatise on Electricity and Magnetism," first published in 1873.
3. W. Meissner and R. Ochsenfeld, *Naturwissenschaften* **21**:787 (1933).
4. R. A. Hein, *Phys. Rev. B* **33**:7539 (1986).
5. D. Shoenberg, "Superconductivity," Cambridge University Press, Cambridge, U.K. (1962), pp. 43-47.
6. F. G. A. Tarr and J. O. Wilhelm, *Can. J. Res.* **12**:265 (1935).
7. D. Shoenberg, *Proc. Cambridge Phil. Soc.* **33**:260 (1937).
8. P. B. Alers, J. W. McWhirter, and C. F. Squire, *in:* "Low-Temperature Physics," Nat. Bur. Stand. (U.S.), Circular 519 (1952), pp. 85-88.
9. M. B. Elzinga and C. Uher, *Phys. Rev. B* **32**:88 (1985).
10. R. A. Hein and R. L. Falge, Jr., *Phys. Rev.* **123**:407 (1961).
11. L. Krusin-Elbaum, A. P. Malozemoff, and Y. Yeshurun, *in:* "High-Temperature Superconductors," M. B. Brodsky, R. C. Dynes, K. Kitazawa, and H. L. Tuller, eds., Materials Research Society, Pittsburgh, *Symp. Proc.* **99**:221 (1988).
12. L. Krusin-Elbaum, A. P. Malozemoff, Y. Yeshurun, D. C. Cronemeyer, and F. Holtzberg, *Physica C* **153-155**:1469 (1988).
13. V. I. Aleksandrov, M. A. Borik, V. G. Veselago, V. V. Voronov, Yu. K. Voron'ko, P. A. Ivanov, G. V. Maksimova, V. E. Makhotkin, V. A. Myzina, V. V. Osiko, A. M. Prokhorov, V. M. Tatarintsev, V. T. Udovenchik, V. A. Fradkov, and M. A. Chernikov, *JETP Lett. (Suppl.)* **46**:S77 (1987) [*Pis'ma Zh. Eksp. Teor. Fiz. (Prilozh.)* **46**:90 (1987)].
14. V. V. Alexandrov, V. V. Borisovskii, T. A. Fedotova, L. M. Fisher, N. V. Il'in, O. K. Smirnova, I. F. Voloshin, M. A. Baranov, and V. S. Gorbachev, *Physica C* **173**:458 (1991).
15. S. Ruppel, G. Michels, H. Geus, J. Kalenborn, W. Schlabitz, B. Roden, and D. Wohlleben, *Physica C* **174**:233 (1991).
16. M. Tinkham, "Introduction to Superconductivity," Krieger Publishing, Malabar, Florida (1980), pp. 244-250.
17. R. B. Goldfarb, A. F. Clark, A. I. Braginski, and A. J. Panson, *Cryogenics* **27**:475 (1987).
18. D.-X. Chen, R. B. Goldfarb, J. Nogués, and K. V. Rao, *J. Appl. Phys.* **63**:980 (1988).
19. D.-X. Chen, J. Nogués, N. Karpe, and K. V. Rao, *Kexue Tongbao* (Beijing, English edition) **33**:560 (1988).
20. H. Mazaki, M. Takano, R. Kanno, and Y. Takeda, *Jpn. J. Appl. Phys.* **26**:L780 (1987).
21. T. Ishida and H. Mazaki, *Jpn. J. Appl. Phys.* **26**:L1296 (1987).
22. H. Mazaki, M. Takano, Y. Ikeda, Y. Bando, R. Kanno, Y. Takeda, and O. Yamamoto, *Jpn. J. Appl. Phys.* **26**:L1749 (1987).
23. J. García, C. Rillo, F. Lera, J. Bartolomé, R. Navarro, D. H. A. Blank, and J. Flokstra, *J. Magn. Magn. Mater.* **69**:L225 (1987).
24. R. Renker, I. Apfelstedt, H. Küpfer, C. Politis, H. Rietschel, W. Schauer, H. Wühl, U. Gottwick, H. Kneissel, U. Rauchschwalbe, H. Spille, and F. Steglich, *Z. Phys. B* **67**:1 (1987).

25. H. Küpfer, I. Apfelstedt, W. Schauer, R. Flükiger, R. Meier-Hirmer, and H. Wühl, *Z. Phys. B* **69**:159 (1987).
26. J. R. Cave, A. Février, Hoang Gia Ky, and Y. Laumond, *IEEE Trans. Magn.* **23**:1732 (1987).
27. C. M. Bastuscheck, R. A. Buhrman and J. C. Scott, *Phys. Rev. B* **24**:6707 (1981).
28. C. Ebner and D. Stroud, *Phys. Rev. B* **31**:165 (1985).
29. P. England, F. Goldie, and A. D. Caplin, *J. Phys. F: Met. Phys.* **17**:447 (1987).
30. Y. M. Chiang, J. A. S. Ikeda, and A. Roshko, *in:* "Ceramic Superconductors II," M. F. Yan, ed., American Ceramics Society, Westerville, Ohio (1988), p. 607.
31. S. E. Babcock, T. F. Kelly, P. J. Lee, J. M. Seuntjens, L. A. Lavanier, and D. C. Larbalestier, *Physica C* **152**:25 (1988).
32. P. Dubots and J. Cave, *Cryogenics* **28**:661 (1988).
33. D. K. Finnemore, R. N. Shelton, J. R. Clem, R. W. McCallum, H. C. Ku, R. E. McCarley, S. C. Chen, P. Klavins, and V. Kogan, *Phys. Rev. B* **35**:5319 (1987).
34. J. W. Ekin, A. I. Braginski, A. J. Panson, M. A. Janocko, D. W. Capone II, N. J. Zaluzec, B. Flandermeyer, O. F. de Lima, M. Hong, J. Kwo, and S. H. Liou, *J. Appl. Phys.* **62**:4821 (1987).
35. M. Suenaga, A. Ghosh, T. Asano, R. L. Sabatini, and A. R. Moodenbaugh, *in:* "High Temperature Superconductors," D. U. Gubser and M. Schluter, eds., Materials Research Society, Pittsburgh, **EA-11**:247 (1987).
36. D. C. Larbalestier, M. Daeumling, X. Cai, J. Seuntjens, J. McKinnell, D. Hampshire, P. Lee, C. Meingast, T. Willis, H. Muller, R. D. Ray, R. G. Dillenburg, E. E. Hellstrom, and R. Joynt, *J. Appl. Phys.* **62**, 3308 (1987).
37. T. Ishida and H. Mazaki, *J. Appl. Phys.* **52**:6798 (1981).
38. T. Ishida, K. Kanoda, H. Mazaki, and I. Nakada, *Phys. Rev. B* **29**:1183 (1984).
39. H. Mazaki and T. Ishida, *Jpn. J. Appl. Phys.* **26**:L1508 (1987).
40. Y. Oda, I. Nakada, T. Kohara, H. Fujita, T. Kaneko, H. Toyoda, E. Sakagami, and K. Asayama, *Jpn. J. Appl. Phys.* **26**:L481 (1987).
41. E. Babić, Ž. Marohnić, Đ. Drobac, M. Prester, and N. Brničević, *Physica C* **153-155**: 1511 (1988).
42. H. M. Ledbetter, S. A. Kim, R. B. Goldfarb, and K. Togano, *Phys. Rev. B* **39**:9689 (1989).
43. The material described in Fig. 2 was used in an interlaboratory comparison sponsored by the Defense Advanced Research Projects Agency (DARPA) in 1989. Participants were asked to determine T_c by susceptibility and report "the midpoint of the full inductive transition," presumably referring to the coupling transition. No guidance was given as to measuring field. The 16 participants reported T_c's ranging from 83 to 94 K. Most of the systematic differences in the measurements were likely due to different measuring fields used by the participants.
44. X. Obradors, C. Rillo, M. Vallet, A. Labarta, J. Fontcuberta, J. Gonzalez-Calbet, and F. Lera, *Physica C* **153-155**:389 (1988).
45. H. Morita, K. Watanabe, Y. Murakami, S. Kondo, Y. Obi, K. Noto, H. Fujimori, and Y. Muto, *Physica B* **148**:449 (1987).
46. H. Nobumasa, K. Shimizu, Y. Kitano, and T. Kawai, *Jpn. J. Appl. Phys.* **27**:L846 (1988).
47. C. E. Gough, *J. Physique Colloq.* **49**:C8-2075 (1988).
48. A. K. Sarkar, B. Kumar, I. Maartense, and T. L. Peterson, *J. Appl. Phys.* **65**:2392 (1989).
49. A. Mehdaoui, B. Loegel, and D. Bolmont, *J. Appl. Phys.* **66**:1497 (1989).
50. E. C. Stoner, *Phil. Mag.*, ser. 7, **36**:803 (1945).
51. J. A. Osborn, *Phys. Rev.* **67**:351 (1945).
52. D.-X. Chen, J. A. Brug, and R. B. Goldfarb, *IEEE Trans. Magn.* **27**:3601 (1991).

53. T. T. Taylor, *J. Res. Nat. Bur. Stand. (U.S.)* **64B**:199 (1960).
54. D.-X. Chen, "Physical Basis of Magnetic Measurements," China Mechanical Industry, Beijing (1985), pp. 139-140.
55. S. D. Murphy, K. Renouard, R. Crittenden, and S. M. Bhagat, *Solid State Commun.* **69**:367 (1989).
56. J. Ferreirinho, S. J. Lee, S. J. Campbell, and A. Calka, *J. Magn. Magn. Mater.* **88**:281 (1990).
57. J. Clerk Maxwell, "A Treatise on Electricity and Magnetism," 3rd Ed., Vol. 2, Clarendon Press, Oxford (1892), pp. 57-58 and pp. 476-477. Reprinted, Dover Publications, New York (1954).
58. A. M. Campbell, F. J. Blunt, J. D. Johnson, and P. A. Freeman, *Cryogenics* **31**:732 (1991).
59. R. Navarro and L. J. Campbell, unpublished, 1991.
60. T. C. Choy and A. M. Stoneham, *J. Phys.: Condens. Matter* **2**:939 (1990).
61. J. R. Clem and V. G. Kogan, *Jpn. J. Appl. Phys. Suppl.* **26-3**:1161 (1987).
62. T. Ishida and H. Mazaki, *Jpn. J. Appl. Phys.* **26**:L2003 (1987).
63. G. Aeppli, R. J. Cava, E. J. Ansaldo, J. H. Brewer, S. R. Kreitzman, G. M. Luke, D. R. Noakes, and R. F. Kiefl, *Phys. Rev. B* **35**:7129 (1987).
64. Y. J. Uemura, V. J. Emery, A. R. Moodenbaugh, M. Suenaga, D. C. Johnston, A. J. Jacobson, J. T. Lewandowski, J. H. Brewer, R. F. Kiefl, S. R. Kreitzman, G. M. Luke, T. Riseman, C. E. Stronach, W. J. Kossler, J. R. Kempton, X. H. Yu, D. Opie, and H. Schone, *Phys. Rev. B* **38**:909 (1988).
65. J. R. Cooper, C. T. Chu, L. W. Zhou, B. Dunn, and G. Grüner, *Phys. Rev. B* **37**:638 (1988).
66. R. J. Cava, B. Batlogg, R. B. van Dover, D. W. Murphy, S. Sunshine, T. Siegrist, J. P. Remeika, E. A. Rietman, S. Zahurak, and G. P. Espinosa, *Phys. Rev. Lett.* **58**:1698 (1987).
67. P. M. Grant, R. B. Beyers, E. M. Engler, G. Lim, S. S. P. Parkin, M. L. Ramirez, V. Y. Lee, A. Nazzal, J. E. Vasquez, and R. J. Savoy, *Phys. Rev. B* **35**:7242 (1987).
68. A. J. Panson, A. I. Braginski, J. R. Gavaler, J. K. Hulm, M. A. Janoko, H. C. Pohl, A. M. Stewart, J. Talvacchio, and G. R. Wagner, *Phys. Rev. B* **35**:8874 (1987).
69. A. C. Rose-Innes and E. H. Rhoderick, "Introduction to Superconductivity," 2nd Ed., Pergamon Press, Oxford, U.K. (1978), p. 93-97.
70. A. L. Schawlow and G. E. Devlin, *Phys. Rev.* **113**:120 (1959).
71. D.-X. Chen, Y. Mei, and H. L. Luo, *Physica C* **167**:317 (1990).
72. D.-X. Chen, A. Sánchez, T. Puig, L. M. Martínez, and J. S. Muñoz, *Physica C* **168**:652 (1990).
73. Ref. 57, p. 70. Maxwell probably meant to refer to Art. 429 in this quotation.
74. R. B. Goldfarb, *Cryogenics* **26**:621 (1986).
75. J. A. Cape and J. M. Zimmerman, *Phys. Rev.* **153**:416 (1967).
76. L. D. Landau, E. M. Lifshitz, and L. P. Pitaevskii, "Electrodynamics of Continuous Media," 2nd Ed., Pergamon Press, Oxford, U.K. (1984), p. 185.
77. V. F. Elesin, I. V. Zakharchenko, A. A. Ivanov, A. P. Menushenkov, A. A. Sinchenko, and S. V. Shavkin, *Supercond., Phys. Chem. Technol.* **3**:1376 (1990) [Sverkhprovodn., Fiz. Khim. Tekh. **3**: 1704 (1990)].
78. P. G. de Gennes, "Superconductivity of Metals and Alloys," Addison-Wesley, Redwood City, California (1989), pp. 60-63.
79. E. M. Gyorgy, *AT&T Bell Laboratories*, personal communication, 1990.
80. H. Teshima, A. Oishi, H. Izumi, K. Ohata, T. Morishita, and S. Tanaka, *Appl. Phys. Lett.* **58**:2833 (1991).
81. J. A. Agostinelli, G. R. Paz-Pujalt, and A. K. Mehrotra, *Physica C* **156**:1208 (1988).

82. M. Lelental, S. Chen, S.-Tong Lee, G. Braunstein, and T. Blanton, *Physica C* **167**:614 (1990).
83. D. Majumdar and M. Lelental, *Physica C* **161**:145 (1989).
84. T. Blanton, M. Lelental, C. L. Barnes, *Physica C* **173**:152 (1991).
85. R. B. Goldfarb, Magnetic units and material specification, *in:* "Concise Encyclopedia of Magnetic and Superconducting Materials," J. E. Evetts, ed., Pergamon Press, Oxford, U.K. (1992).
86. C. P. Bean, *Rev. Mod. Phys.* **36**:31 (1964).
87. H. London, *Phys. Lett.* **6**:162 (1963).
88. Y. B. Kim, C. F. Hempstead, and A. R. Strnad, *Phys. Rev.* **129**:528 (1963).
89. T. Ishida and R. B. Goldfarb, *Phys. Rev. B* **41**:8939 (1990).
90. D.-X. Chen and R. B. Goldfarb, *J. Appl. Phys.* **66**:2489 (1989).
91. Ref. *76*, pp. 204-205.
92. J. R. Clem, "AC Losses in Type-II Superconductors," Ames Lab. Tech. Rep. IS-M 280, Iowa State University, Ames, Iowa (1979), Eqs. (25). Reprinted with additions in: "Magnetic Susceptibility of Superconductors and Other Spin Systems," R. A. Hein, T. L. Francavilla, and D. H. Liebenberg, eds., Plenum Press, New York (1992).
93. C. P. Bean, *Rev. Mod. Phys.* **36**:31 (1964).
94. These slab equations for χ' and χ'' may be obtained by simplification of Eqs. (9) in: L. Ji, R. H. Sohn, G. C. Spalding, C. J. Lobb, and M. Tinkham, *Phys. Rev. B* **40**:10936 (1989).
95. C. P. Bean, *Phys. Rev. Lett.* **8**:250 (1962).
96. W. A. Fietz and W. W. Webb, *Phys. Rev.* **178**:657 (1969).
97. These axial cylinder equations for χ' and χ'' may be obtained by simplification of Eqs. (118)–(124) in Ref. *92*. Note that $2 \sin^{-1}(x^{-1/2}) = \cos^{-1}(1 - 2/x)$.
98. R. B. Goldfarb and A. F. Clark, *IEEE Trans. Magn.* **21**:332 (1985).
99. W. J. Carr, Jr., M. S. Walker, and J. H. Murphy, *J. Appl. Phys.* **46**:4048 (1975).
100. W. J. Carr, Jr., J. H. Murphy, and G. R. Wagner, *Adv. Cryo. Eng.* **24**:415 (1978).
101. W. J. Carr, Jr. and G. R. Wagner, *Adv. Cryo. Eng. (Materials)* **30**:923 (1984).
102. W. J. Carr, Jr., "AC Loss and Macroscopic Theory of Superconductors," Gordon and Breach, New York (1983), pp. 63-67.
103. M. Ashkin, *J. Appl. Phys.* **50**:7060 (1979).
104. V. B. Zenkevitch, A. S. Romanyuk, and V. V. Zheltov, *Cryogenics* **20**:703 (1980).
105. C. Y. Pang, P. G. McLaren, and A. M. Campbell, *Int. Cryo. Eng. Conf.* **8**:739 (1980).
106. J. V. Minervini, *Adv. Cryo. Eng. (Materials)* **28**:587 (1982).
107. M. N. Wilson, "Superconducting Magnets," Clarendon Press, Oxford, U.K. (1983), pp. 165-170.
108. K. V. Bhagwat and P. Chaddah, *Physica C* **166**:1 (1990).
109. J. W. Ekin, *Appl. Phys. Lett.* **55**:905 (1989).
110. D. M. Kroeger, C. C. Koch, and J. P. Charlesworth, *J. Low Temp. Phys.* **19**:493 (1975).
111. R. B. Goldfarb and A. F. Clark, *J. Appl. Phys.* **57**:3809 (1985).
112. J. R. Cave, P. R. Critchlow, P. Lambert, and B. Champagne, *IEEE Trans. Magn.* **27**:1379 (1991).
113. J. R. Clem, *Physica C* **153-155**:50 (1988).
114. J. Z. Sun, M. J. Scharen, L. C. Bourne, and J. R. Schrieffer, *Phys. Rev. B* **44**:5275 (1991).
115. F. Gömöry and P. Lobotka, *Solid State Commun.* **66**:645 (1988).
116. D.-X. Chen, J. Nogués, and K. V. Rao, *Cryogenics* **29**:800 (1989).
117. A. M. Campbell, *J. Phys. C: Solid State Phys.* **2**:1492 (1969).

118. R. W. Rollins, H. Küpfer, and W. Gey, *J. Appl. Phys.* **45**:5392 (1974).
119. H. Küpfer, I. Apfelstedt, R. Flükiger, C. Keller, R. Meier-Hirmer, B. Runtsch, A. Turowski, U. Wiech, and T. Wolf, *Cryogenics* **29**:268 (1989).
120. H. A. Ullmaier, *Phys. Stat. Sol.* **17**:631 (1966).
121. D. M. Kroeger, C. C. Koch, and W. A. Coghlan, *J. Appl. Phys.* **44**:2391 (1973).
122. A. Shaulov and D. Dorman, *Appl. Phys. Lett.* **53**:2680 (1988).
123. E. M. Gyorgy, R. B. van Dover, K. A. Jackson, L. F. Schneemeyer, and J. V. Waszczak, *Appl. Phys. Lett.* **55**:283 (1989).
124. R. L. Peterson, *J. Appl. Phys.* **67**:6930 (1990).
125. C. P. Bean and J. D. Livingston, *Phys. Rev. Lett.* **12**:14 (1964).
126. R. W. Rollins and J. Silcox, *Phys. Rev.* **155**:404 (1967).
127. D.-X. Chen, R. W. Cross, and A. Sánchez, unpublished, 1991.
128. S. A. Campbell, J. B. Ketterson, and G. W. Crabtree, *Rev. Sci. Instrum.* **54**:1191 (1983).
129. B. Loegel, A. Mehdaoui, and D. Bolmont, *Supercond. Sci. Technol.* **3**:504 (1990).
130. R. W. Cross and R. B. Goldfarb, *J. Appl. Phys.* **67**:5476 (1990).
131. J. R. Clem, *J. Appl. Phys.* **50**, 3518 (1979).
132. R. B. Goldfarb, A. F, Clark, A. J. Panson, and A. I. Braginski, *in:* "High Temperature Superconductors," D. U. Gubser and M. Schluter, eds., Materials Research Society, Pittsburgh, **EA-11**:261 (1987).
133. R. J. Loughran and R. B. Goldfarb, *Physica C* **181**:138 (1991).
134. E. Babić, Ž. Marohnić, Đ. Drobac, and M. Prester, *Int. J. Mod. Phys. B* **1**:973 (1987).
135. M. Avirovic, Ch. Neumann, P. Ziemann, J. Geerk, and H. C. Li, *Solid State Commun.* **67**:795 (1988).
136. L. Krusin-Elbaum, A. P. Malozemoff, Y. Yeshurun, D. C. Cronemeyer, and F. Holtzberg, *Phys. Rev. B* **39**:2936 (1989).
137. V. V. Moshchalkov, J. Y. Henry, C. Marin, J. Rossat-Mignod, and J. F. Jacquot, *Physica C* **175**:407 (1991).
138. E. W. Collings, A. J. Markworth, J. K. McCoy, K. R. Marken, Jr., M. D. Sumption, E. Gregory, and T. S. Kreilick, *Adv. Cryo. Eng. (Materials)* **36**:255 (1990).
139. S. Takács, *Czech. J. Phys. B* **33**:1248 (1983).
140. K.-H. Müller, *Physica C* **159**:717 (1989).
141. R. B. Goldfarb and R. L. Spomer, *Adv. Cryo. Eng. (Materials)* **36**:215 (1990).
142. M. Nikolo and R. B. Goldfarb, *Phys. Rev. B* **39**:6615 (1989).
143. T. T. M. Palstra, B. Batlogg, R. B. van Dover, L. F. Schneemeyer, and J. V. Waszczak, *Phys. Rev. B* **41**:6621 (1990).
144. J. H. P. M. Emmen, V. A. M. Brabers, and W. J. M. de Jonge, *Physica C* **176**:137 (1991).
145. E. Maxwell and M. Strongin, *Phys. Rev. Lett.* **10**:212 (1963).
146. G. D. Cody and R. E. Miller, *Phys. Rev.* **173**:481 (1968).
147. L. M. Fisher, N. V. Il'yn, and I. F. Voloshin, *Adv. Cryo. Eng. (Materials)* **36**:423 (1990).
148. F. Irie and K. Yamafuji, *J. Phys. Soc. Jpn.* **23**:255 (1967).
149. J. R. Clem, H. R. Kerchner, and S. T. Sekula, *Phys. Rev. B* **14**:1893 (1976).
150. S. Takács, F. Gömöry, and P. Lobotka, *IEEE Trans. Magn.* **27**:1057 (1991).
151. A. F. Khoder, *Phys. Lett.* **94A**:378 (1983).
152. A. F. Khoder, M. Couach, and B. Barbara, *Physica C* **153-155**:1477 (1988).
153. C. Lucchini, C. Giovannella, R. Messi, B. Lecuyer, L. Fruchter, and M. Iannuzzi *Phys. Stat. Sol. B* **157**: K123 (1990).
154. A. Gianelli and C. Giovannella, *Physica A* **168**:277 (1990).

155. R. A. Hein, H. Hojaji, A. Barkatt, H. Shafii, K. A. Michael, A. N. Thorpe, M. F. Ware, and S. Alterescu, *J. Supercond.* **2**:427 (1989).
156. W. R. Abel, A. C. Anderson, and J. C. Wheatley, *Rev. Sci. Instrum.* **43**:444 (1964).
157. W. L. Pillinger, P. S. Jastram, and J. G. Daunt, *Rev. Sci. Instrum.* **29**:159 (1958).
158. S. C. Whitmore, S. R. Ryan, and T. M. Sanders, *Rev. Sci. Instrum.* **49**:1579 (1978).
159. J. R. Owers-Bradley, Wen-Sheng Zhou, and W. P. Halperin, *Rev. Sci. Instrum.* **52**:1106 (1981).
160. D.-X. Chen, "Ballistic and Bridge Methods of Magnetic Measurements of Materials," China Metrology, Beijing (1990), pp. 526-572.
161. Ref. *107*, pp. 243-249.
162. D. Shoenberg, *Proc. Cambridge Phil. Soc.* **33**:559 (1937).
163. P. H. Müller, M. Schienle, and A. Kasten, *J. Magn. Magn. Mater.* **28**:341 (1982).
164. T. Ishida, K. Monden, and I. Nakada, *Rev. Sci. Instrum.* **57**:3081 (1986).
165. B. J. Dalrymple and D. E. Prober, *Rev. Sci. Instrum.* **55**:958 (1984).
166. D. G. Xenikos and T. R. Lemberger, *Rev. Sci. Instrum.* **60**:831 (1989).
167. J. N. Fox and J. U. Trefny, *Am. J. Phys.* **43**:622 (1975).
168. J. G. Elliott and W. Y. Liang, *Meas. Sci. Technol.* **1**:1351 (1990).
169. I. Maartense, *Rev. Sci. Instrum.* **41**:657 (1970).
170. I. Maartense, *J. Appl. Phys.* **53**:2466 (1982).
171. L. Hartshorn, *J. Sci. Instrum.* **2**:145 (1925).
172. A. J. de Vries and J. W. M. Livius, *Appl. Sci. Res.* **17**:31 (1967).
173. H. A. Groenendijk, A. J. van Duyneveldt, and R. D. Willett, *Physica B* **101**:320 (1980).
174. J. L. Tholence, F. Holtzberg, T. R. McGuire, S. von Molnar, and R. Tournier, *J. Appl. Phys.* **50**:7350 (1979).
175. A. J. van Duyneveldt, *J. Appl. Phys.* **53**:8006 (1982).
176. A. F. Deutz, R. Hulstman, and F. J. Kranenburg, *Rev. Sci. Instrum.* **60**:113 (1989).
177. A. K. Rastogi, *Jawaharlal Nehru University, New Delhi*, and J. L. Tholence, *Centre National de la Recherche Scientifique, Grenoble*, personal communication, 1985.
178. L. J. de Jongh, W. D. van Amstel, and A. R. Miedema, *Physica* **58**:277 (1972).
179. K. Baberschke, *Freie Universität Berlin*, personal communication, 1984.
180. F. R. Fickett, *in:* "Materials at Low Temperatures," R. P. Reed and A. F. Clark, eds., American Society for Metals, Metals Park, Ohio (1983), pp. 164-165.
181. The flux from an *ellipsoidal* sample through a *single-turn* pick-up coil is calculated in: M. Denhoff, S. Gygax, and J. R. Long, *Cryogenics* **21**:400 (1981).
182. R. B. Goldfarb and J. V. Minervini, *Rev. Sci. Instrum.* **55**:761 (1984). On page 763, second column, line 3 and Fig. 3, "L^*" should be "$2L^*$."
183. L. Cohen, *Bull. Bureau Standards* **3**:295 (1907).
184. Ref. *76*, pp. 205-207.
185. J. E. Zimmerman, *Rev. Sci. Instrum.* **32**:402 (1961).
186. R. A. Matula, *J. Phys. Chem. Ref. Data* **8**:1147 (1979).
187. R. M. Bozorth, "Ferromagnetism," Van Nostrand, Princeton, New Jersey (1951), pp. 775-776.
188. R. G. Chambers and J. G. Park, *Brit. J. Appl. Phys.* **12**:507 (1961).
189. M. D. Rosenthal and B. W. Maxfield, *Rev. Sci. Instrum.* **46**:398 (1975).
190. A. M. Ricca and S. Zannella, *IEEE Trans. Magn.* **23**:1422 (1987).
191. H. Zijlstra, "Experimental Methods in Magnetism," Vol. 2, North-Holland, Amsterdam (1967), pp. 72-79.
192. S. Foner, *Rev. Sci. Instrum.* **46**:1425 (1975).
193. J. S. Philo and W. M. Fairbank, *Rev. Sci. Instrum.* **48**:1529 (1977).
194. P. J. Flanders, *J. Appl. Phys.* **63**:3940 (1988).

AC INDUCTIVE MEASUREMENT: ITS APPLICATION TO THE STUDIES OF HIGH T_c SUPERCONDUCTIVITY

Q. Y. Chen

Honeywell Systems and Research Center
3660 Technology Drive
Minneapolis, MN 55418, U.S.A.

I. Introduction

Alternating current (ac) inductive method is a time-derivative or differential technique in which a modulated magnetic field is impressed on the sample through a primary coil and the response is then measured with a secondary (pickup) coil. If loss elements exist in the sample, the induced voltage (V_s) will involve a phase shift with respect to that (V_r) measured without the sample. This phase-shifted signal can be decomposed into two components: one in phase with V_r and the other in quadrature to it.[1,2] It has been widely used in superconductivity measurements[3–43] and, most controversial of all, in the determination of the so-called irreversibility lines for the high T_c superconductors[37–43] first observed using dc magnetization measurements.[44] Because of the extremely complex vortex dynamics and possible material imperfections in most high-temperature superconductors, the ac approach remains controversial in many aspects of data interpretation.

There are several perplexing questions regarding the magnetic measurement of superconductors: (1) Can ac magnetic susceptibility be used as an intrinsic physical property for type-II superconductors? If so, how can the frequency, field strength, sample size, and sample geometry dependencies of each measurement be properly normalized? Is there a scaling law? (2) Is dc susceptibility equivalent to the real part of ac susceptibility? (3) What is a mixed-state superconductor? A diamagnetic and near-perfect conductor? If so, can it be treated as an ultra-low-resistivity normal conductor using classical electrodynamics? (4) Can parallel and perpendicular fields be treated on a more or less equal footing so that theories available for parallel field, such as the critical state model, may be used for reasonable perpendicular field data interpretations?

Answers to these questions are of practical interest because magnetic measurements, especially the ac inductive methods, are noncontact, easy to implement, and cost-effective for basic superconductivity studies. Question 4 is particularly important for thin films and bulk samples of aspect ratio close to unity. It is a difficult subject that has been treated in several classic papers.[45-49] Useful discussions can also be found in the current proceedings.[50] Therefore I will restrict my remarks to the remaining three questions.

Following a review of the basic operational principles of the ac inductive method in order to clarify the underlying fundamental physics, the concepts of complex permeability and susceptibility will be developed using two typical cases: (1) the hysteretic regime where superconducting screening effects dominate and the critical state model is applicable, and (2) the vortex-motion regime where ac screening is effective and classical electrodynamics of magnetic diffusion is applicable. Experimental results are then reported. Finally, a scaling law will be discussed. The thermally activated flux creep and flux flow models will be used to evaluate the "ac irreversibility line" and its dependence on the field, frequency, and sample size in the flux-motion regime.

II. Principles of AC Inductive Measurements

The ac inductive method typically consists of a primary coil that generates an excitation signal and a secondary coil, wound around the periphery of the sample, that detects the rate of change of the linked magnetic flux density (B). B is the mean value of the flux density b(r) within the sample averaged over the entire sample volume. For a tubular sample with a secondary coil placed inside the hollow space, the signal detected would be the rate of change of flux density at the inner boundary (r = Ri) of the tube (i.e., $db/dt|_{r=Ri}$), since b(r<Ri) = b(r = Ri). Within the sample per se variations in b(r) arise due to the current density, J(r), induced by the applied magnetic field, H. Depending on the extent of penetration by the applied field, the current distribution J(r) and therefore the flux density b(r) could be nonuniform within the sample. $\underline{J}(\underline{r})$ and the penetrated field profile $\underline{b}(\underline{r})$ can be correlated as $\underline{\nabla} \times \underline{b}(\underline{r}) = \mu_o \underline{J}(\underline{r})$, where μ_o is the vacuum permeability and the underline denotes a vector. In the absence of transport current, $\underline{J}(\underline{r})$ is the magnetization current whereby a magnetic dipole moment $\underline{m}$ is defined. Written explicitly:

$$\underline{B} = V^{-1}\int_V \underline{b}(\underline{r})\, d^3r$$

One also defines the magnetization as:

$$\underline{M} = \underline{m}/V = (2V)^{-1}\int_V \underline{r} \times \underline{J}(\underline{r})\, d^3r \qquad (1)$$

Both B and M depend on the sample size and geometry for a nonuniform current distribution or flux density profile within the sample. The nonuniformity and size dependence may come from either the superconducting screening in the strong pinning (hysteresis) regime[5] or the ac screening in the flux-motion regime, or most probably a combination of both.

From electromagnetic theory we have $B = \mu_0(H + M)$ (in MKS units), where the magnetic susceptibility χ and permeability μ are defined such that $M = \chi H$ and $B = \mu_0\mu H$ ($\mu = 1 + \chi$). While the ac inductive technique is generally not restricted to specific sample geometries or pickup coil configurations, the conventional definitions of μ and χ are limited to the volume-averaged quantities. From Eq. (1), χ and μ are both sample size and shape dependent. In the next section, μ and χ are defined in order to illustrate the concepts of phase shift, complex permeability, complex susceptibility, and the related magnetic losses.

By differentiating $M = B/\mu_o - H$ on both sides with respect to time one obtains:

$$dM/dt = (\mu_o)^{-1} dB/dt - dH/dt \qquad (2)$$

The rate of change of magnetization for a sample placed within the secondary coil can be determined from the pickup voltages measured with sample (V_s) and without sample (V_r, r = reference), since from Faraday's emf law $V_r = -\kappa_r\mu_o dH/dt$ and $V_s = -\kappa_s dB/dt$, where κ_r and κ_s are the coil constants that depend on the turns and total area enclosed. In the case of a circular coil of N turns with radius a, we may write $\kappa_r = N\pi a^2$. κ_s also depends on the sample-to-coil volume ratio, which may be approximated by $\kappa_s = N\pi a^2$ if the sample is in close contact with the coil. Otherwise, κ_s includes a system coefficient associated with the sample–coil gap. Conventionally, an ideal superconductor of the same geometry is usually first measured to establish a calibration curve $V_{so}(T,H)$. The diamagnetic shift $V_s(T,H) - V_{so}(T,H)$ is often assumed to be proportional to the volume of the superconducting regions. In addition to practical difficulty, the use of this technique in determining bulk superconductivity has been addressed by Hein.[29] Calibrations using ferromagnetic materials of known susceptibility have also been employed by various researchers.[27,40] Putting aside questions regarding measurement standardizations, I will focus on discussing the size, field, and frequency dependences of the ac measurements and thus will assume proximity between sample and secondary coil; V_s and V_r will still be used to represent the sample and reference signals. Two cases will be discussed to illustrate the basic principles of the ac inductive measurements. First, the quasi-static approximation will be employed to treat the strong pinning (hard superconductor) case[5]; in this regime, the critical state model is applicable and the size dependence should not be neglected. Then the field, frequency, and sample size dependences of the measurements for the weakly pinned case applicable to the high T_c superconductors (diamagnetic conductors of ultra-low resistivity) will be treated.

III. Hysteresis Losses: Hard Superconductors

Hard superconductors represent an extreme case of strong flux pinning that typically exhibits hysteretic behaviors.[5] In response to an ac field $H(t) = H_{ac}\cos \omega t$, the B-field in the sample will show a phase lag ϕ behind the H-field. Characteristic of such hysteretic behavior is the distorted periodic waveform where neither B nor M can be expressed as a sinusoidal function of single frequency. However, as will become clear, the higher harmonics are not responsible for any power dissipations, and hence the fundamental term remains the key component for loss analyses. As such, for the sake of heuristics, the first harmonic analysis will be used to illustrate the concepts of phase shifts, complex permeabilities, and susceptibilities.

Induced voltages measured with and without a sample can be written as $V_S = -|V_S| \sin(\omega t - \phi)$ and $V_r = -|V_r| \sin \omega t$, respectively. While the origins of magnetic losses vary, the phase shift introduced here generally applies to a dissipative system.[1–2] Provided that the coil is kept away from its resonance frequency, the pickup voltage measured without sample is:

$$|V_r| = \mu_o \kappa_r \omega H_{ac} \tag{3}$$

Let the magnetization lag behind the applied field by an angle θ, then

$$M = \chi H = \chi H_{ac}\cos(\omega t - \theta) = H_{ac}(\chi \cos\theta \cos\omega t + \chi \sin\theta \sin\omega t) \tag{4}$$

Also, $$B = \mu_o \mu H = \mu_o \mu H_{ac}\cos(\omega t - \phi) = \mu_o H_{ac}(\mu \cos\phi \cos\omega t + \mu \sin\phi \sin\omega t) \tag{5}$$

and $$V_S = -|V_s| \sin(\omega t - \phi) = -(|V_s| \cos\phi \sin\omega t - |V_s| \sin\phi \cos\omega t) \tag{6}$$

Define $$\chi' = \chi \cos\theta \tag{7}$$

and $$\chi'' = \chi \sin\theta \tag{8}$$

Then from Eq. (4), $$M = H_{ac}(\chi' \cos\omega t + \chi'' \sin\omega t) \tag{9}$$

Obviously, $$\tan\theta = (\chi''/\chi') \tag{10}$$

and $$\chi = (\chi'^2 + \chi''^2)^{1/2} \tag{11}$$

By the same token, $$V_s' = |V_s| \cos\phi \tag{12}$$

and $$V_s'' = |V_s| \sin\phi \tag{13}$$

then $$\tan\phi = V_s'' / V_s' \tag{14}$$

and $$|V_s| = (V_s'^2 + V_s''^2)^{1/2} \tag{15}$$

By direct substitutions of $dB/dt = -V_s/\kappa_S$ and $dH/dt = -V_r/(\mu_o \kappa_r)$, as well as Eq. (3), (6), (9), (12) and (13) into Eq. (2),

$$\chi' = -1 + (\kappa_r/\kappa_s)V_S'/|V_r| \tag{16}$$

or $$1 + \chi' = (\kappa_r/\kappa_s)V_s'/|V_r| \tag{17}$$

and $$\chi'' = (\kappa_r/\kappa_s)V_s''/|V_r| \tag{18}$$

With Eq. (18)–(19), it then follows from Eq. (11) and (15) that

$$\tan\phi = V_s''/V_s' = \chi''/(1 + \chi') \tag{19}$$

Upon substitution of Eq. (11) into (20) we find:

$$\tan\theta/\tan\phi = 1 + 1/\chi' \tag{20}$$

Eq. (20) suggests that as χ' deviates greatly from its perfect extreme $\chi' = -1$ where $M = -H$ and $B = 0$, say $\chi' = -0.5$ in particular, one has $\phi = \theta \pm \pi$; that is, M and B or the effective current and the flux density are 180° out of phase. This is not surprising because in the quasi-static limit, just like in a static situation, when $\chi = -0.5$, $M = -0.5\,H$ and therefore $B = M + H = 0.5H = -M$. It is necessary to stress that, since magnetic quantities discussed here are volume averaged, the phase relationships should also be treated in the same contexts.

From Eq. (9), χ' is associated with the in-phase and χ'' with the out-of-phase signal with respect to the applied field. A complex susceptibility can be defined as

$$\tilde{\chi} = \chi' + i\,\chi'' = \chi \exp(i\theta) \tag{21}$$

such that
$$M = \mathrm{Re}\{\tilde{\chi} H_{ac} \exp(-i\omega t)\} = \chi H_{ac}\, \mathrm{Re}\,\{\exp{-i(\omega t - \theta)}\} \tag{22}$$

which is equivalent to Eq. (4). Using the similar procedure, one may also obtain the complex permeability $\tilde{\mu} = 1 + \tilde{\chi} = 1 + \chi' + i\,\chi''$.

V_s' and V_s'' can be measured with a phase-sensitive detector typically available on a lock-in amplifier. One could set the phase first and then measure V_s' and V_s'' directly (the X-Y mode), whereby the phase angle ϕ can be calculated using Eq. (14). This mode could sometimes result in extra errors if the sensitivity settings have to be changed in the process. Otherwise, one could also use the phase tracking (R-ϕ) mode in which the magnitude $|V_s|$ and the phase angle ϕ of the voltage are measured, whereby V_s' and V_s'' can be calculated using Eq. (12) and (13). R-ϕ mode usually requires more time for data acquisition than the X-Y mode. Depending on the instrument, the phase angle measured by a lock-in amplifier could range from –180° to 180° or from 0° to 360°. However, since only half of the full angular domain is needed for an inverse trigonometric function, the angular domain $\{\phi \mid 0 \le \phi \le \pi\}$ was chosen. As a result, the phase angle ϕ computed using the arctangent function, whose domain is $\{\phi \mid -\pi/2 \le \phi \le \pi/2\}$, must be adjusted by adding π when $\tan\phi$ is negative. This will shift the functional domain into the desired range. In comparing with the experimental data obtained from a ±180° lock-in amplifier, one has to bear in mind that ϕ, $\phi + \pi$, and $\phi - \pi$ all represent the same angle.

Based on Eq. (4) and (5), the energy dissipation, W, per unit volume per cycle can be calculated[67] as the area enclosed by the B-H (or M-H) loop for $H(t) = H_{ac}\cos\omega t$:

$$W = \oint B dH = \mu_o \oint M dH \tag{23}$$

$$= \pi\,\mu''\mu_o H_{ac}^2 = \pi\chi''\mu_o H_{ac}^2 \tag{24}$$

This is the area of an elliptic B-H (or M-H) hysteresis loop with semimajor axis $H_{ac}/\cos\beta$ and semiminor axis $\mu_o\mu'' H_{ac}\cos\beta$ (or $\mu_o\chi'' H_{ac}\cos\beta$) tilted at an angle $\beta = \tan^{-1}(\mu_o\mu')$ (or $\beta = \tan^{-1}(\mu_o\chi')$) versus the H-axis. The power dissipation per unit volume can also be calculated as $P = W/T = \omega W/2\pi = \mu''\omega\mu_o H_{ac}^2/2$. The hysteresis loop traced out using the ac technique should be the same as that done using the dc technique unless other ac loss mechanisms exist. By definition $\mu_{dc} = B/H$ ($\chi_{dc} = M/H$), while μ_{ac} (χ_{ac}) for H_{ac} superimposed on H_{dc} are associated with the above-mentioned semiaxes of a minor hysteresis loop. Generally speaking, the two techniques measure different quantities except for a few occasions in their magnetic histories; direct comparisons of the dc susceptibility with the real part of the ac susceptibility thus offer limited physical implications.

With the above analyses one can now write the full Fourier series of B(t) for an applied field of $H(t) = H_{dc} + H_{ac}\cos\omega t$ as:

$$B(t) = \mu_o\mu_{dc}H_{dc} + \mu_o H_{ac} \sum_n (\mu_n'\cos n\omega t + \mu_n''\sin n\omega t) \tag{25}$$

where $\chi_{dc}H_{dc}$ is the dc term included here to account for the major loop.[2] If the ac modulation is superimposed on a much stronger but slowly ramping dc field (i.e., $H = H_{dc} + H_{ac}\cos\omega t$), it would display a minor hysteresis loop of Rayleigh type[2] (owing to H_{ac}) superimposed on a major one (owing to H_{dc}) in a B-H or M-H curve. Earlier comments regarding the fundamental differences between ac and dc susceptibilities remain valid, although here the loops will no longer be simple ellipses. The higher order χ_n and μ_n terms indicate the extent of waveform distortion. The ac susceptibility (permeability) must be specified as a dynamic quantity at H_{dc}, where sample size, H_{ac}, and ω should also be noted. From Eq. (25):

$$\mu_n' = (\pi\mu_o H_{ac})^{-1} \int_0^{2\pi} B(t)\cos n\omega t \, d\,\omega t \tag{26}$$

and
$$\mu_n'' = (\pi\mu_o H_{ac})^{-1} \int_0^{2\pi} B(t)\sin n\omega t \, d\,\omega t \tag{27}$$

Similar analysis on $M[H(t)]$ at $H = H_{dc}$ gives

$$M(t) = \chi_{dc}H_{dc} + H_{ac} \sum_n (\chi_n' \cos n\omega t + \chi_n'' \sin n\omega t) \tag{28}$$

where

$$\chi_n' = (\pi H_{ac})^{-1} \int_0^{2\pi} M(t) \cos n\omega t \, d\omega t \tag{29}$$

and

$$\chi_n'' = (\pi H_{ac})^{-1} \int_0^{2\pi} \sin n\omega t \, d\omega t \tag{30}$$

It is worth noting that $\mu_1' = 1 + \chi_1'$ and $\mu_n' = \chi_n'$ for $n > 1$ while $\mu_n'' = \chi_n''$ for all n. In the meantime, it is also important to recognize from the orthogonality relations of the trigonometric functions that Eq. (24) remains unchanged in spite of the higher harmonic components. That is:

$$W = \pi\chi_1'' \mu_o H_{ac}^2 = \pi \mu_1'' \mu_o H_{ac}^2 \tag{31}$$

It suggests that although the waveform distortion results in a nonelliptical hysteresis loop, the enclosed area remains unchanged; therefore, only the fundamental signal carries the energy loss information. μ'' could be a function of applied field $(H_{dc} + H_{ac})$. The field dependence of hysteretic losses may vary from being proportional to H_{ac}^3 at low field (below full field penetration) to being proportional to H_{ac} at high field (beyond full field penetration) in accordance with the Bean critical state model.[5,51]

Knowing that $V_s = -\kappa_s dB/dt$, one may differentiate with respect to time on both sides of Eq. (25) and express $V_s(t)$ in Fourier terms:

$$V_s(t) = (\kappa_s/\kappa_r)|V_r| \sum_n n\,(\mu_n' \sin n\omega t - \mu_n'' \cos n\omega t) \tag{32}$$

$$= \sum_n (V_{s_n}' \sin n\omega t - V_{s_n}'' \cos n\omega t) \tag{33}$$

where we have used Eq. (3): $|V_r| = \kappa_r\mu_o\omega H_{ac}$. By direct comparisons of Eq. (32) and (33), one finds

$$\mu_n' = (1/n)\,(\kappa_r/\kappa_s)\, V_{s_n}' / |V_r| \tag{34}$$

and

$$\mu_n'' = (1/n)\,(\kappa_r/\kappa_s)\, V_{sn}'' / |V_r| \tag{35}$$

The fundamental components V_{s1}' and V_{s1}'' are the same as described in Eq. (17) and (18) where the subscripts were left out.

In principle, the higher harmonics can be measured in a similar fashion as the fundamental signal. Commercial lock-in amplifiers are typically equipped with a second harmonic mode that allows synchronous measurement of V_{s2}', V_{s2}'', and ϕ_{s2}. For the third and higher harmonics, phase-lock techniques usually have to be used to ensure the phase stability required for the synchronism between the input and output signals.[52]

Studies of higher harmonics have been widely used to explore the nonlinear magnetizations and associated hysteretic behaviors for high-temperature superconductors. Earlier works[5–23] based on the Bean critical state model failed to predict the even harmonic generations in the presence of a dc field. In this model, the critical current was assumed to be independent of magnetic field. Later efforts pioneered by Kim and coworkers,[53,54] which took the field dependence into consideration, were able to bridge the discrepancy, although others also successfully predicted similar results using magnetoresistive[55] and Josephson junction models.[56–59] While consensus seems to be lacking as to which model best describes the critical state, a field-dependent critical current of any kind should generate even harmonics in a dc bias field.[60–61]

IV. Flux Motions and Skin Effects: The Frequency and Size Dependencies

Eq. 24 shows that the energy loss per unit volume per period for a purely hysteretic system is independent of frequency. In reality, frequency-dependent susceptibility has been commonly observed for high-temperature superconductors. Below we will show that this can be understood in the general context of classical electrodynamics for a conductor of finite resistivity. The physics of the flux-motion-related ac screening effects for such conductors can be described using the magnetic diffusion equation[63–65]:

$$\partial \underline{b}/\partial t = (\rho_t/\mu_o)\nabla^2 \underline{b} = D_t \nabla^2 \underline{b}, \tag{36}$$

where ρ_t is the effective or total resistivity and $\rho_t/\mu_o = D_t$ is the magnetic diffusion coefficient. The basic physical meaning of Eq. (36) can be stated as the following. First, the varying applied magnetic field induces an electric field, which then results in an electric current, $\underline{J}$, inside a conductor of finite resistivity ρ. $\underline{J}$ and $\underline{b}$ are correlated through $\underline{\nabla} \times \underline{b}(\underline{r}) = \mu_o \underline{J}(\underline{r})$. The resistive dissipation per unit volume can be calculated as ρJ^2. The solution to Eq. (36) would thus carry a quadrature term in association with this loss. Cases for hollow cylinders,[63,66] solid cylinders,[2,63,64] spheres,[63,64] and planar sheets[2,63,64] have all been solved analytically for constant diffusion coefficients. Recently, approximations taking into account field-dependent D_t have also been reported.[67]

The diffusion coefficient may be assumed constant provided that the amplitude of the vortex movement is not too large, or the frequency is not too low, to allow one to treat the resistivity as an averaged value over one period of the oscillation. The resistivity introduced by the oscillating vortices then would simply depend on the root mean square (rms) value of the ac amplitude in addition to the dc field, if any, and is absorbed into the diffusion coefficient as a parameter. The resistivity ρ_t could originate from various sources; i.e., eddy current losses from electron scattering in the normal regions, field-driven flux flow losses, and thermally activated flux creep losses in those regions that contain vortices. Thus one may write $\rho_t(H,T) = \rho_e(T) + \rho_{ff}(H,T) + \rho_{tf}(H,T)$ where the subscripts denote the various contributions. In wiriting this equation, it is assumed that the normal regions are nonmagnetic so that the eddy current losses are independent of H. In principle, it is possible to use the applied field strength, frequency, and temperature as modulating parameters in order to discriminate various loss mechanisms; however, this will rely on the resistivity models used, which unfortunately still need to be developed and tested.

In studying magnetic diffusivity, it is useful to directly monitor the local flux density inside the material as a response to the changing external field. Hence it is interesting to use tubular samples. The ac field profiles in the superconductors and other metallic systems are known to be nonuniform and the phase relationships quite complex.[2] The use of tubular samples serves several purposes: (1) small changes in the pickup voltage can be better identified because of superconducting shielding; (2) flux dynamics and ac screening contributions can be observed directly; and (3) when the secondary coil is wound around the tube, this geometry mimics an inhomogeneous material composed of a dissipative cylindrical shell enclosing a virtually nonelectromagnetic core; for oxide superconductors requiring oxygenation through thermal annealing, this could represent a practical situation.

A calculation has been performed based on tubular geometries that essentially become a solid cylinder as the inner diameter of the tube is set to zero. The solution to Eq. (36) for a tube of infinite length, inner radius r_b, and outer radius r_a can be obtained by imposing the following two boundary conditions:[63]

$$\underline{\nabla} \times \underline{b}\,|_{r=r_b} = 0, \text{ or } \frac{db_z}{dr}\,|_{r=r_b} = 0 \tag{37}$$

and

$$b|_{r=r_a} = \mu_o H_{ac} = B_o \tag{38}$$

The first condition arises from the fact that there is no current inside the hollow space, while the second arises from the continuity of the tangential components of the two fields across the surface of the sample. Following the approach used by Smythe,[63] we obtained:

$$\frac{\tilde{b}(kr)}{B_o} = \frac{K_1(kr_b)\, I_o(kr) + I_1(kr_b)\; K_o(kr)}{K_1(kr_b)\, I_o(kr_a) + I_1(kr_b)\, K_o(kr_a)} \tag{39}$$

where $\tilde{b}(kr)$ is the complex local magnetic field at position r and $I_\nu(kr)$ and $K_\nu(kr)$ are, respectively, the modified Bessel functions of the first kind and the second kind of order ν. k is related to the skin depth $\delta = (2\rho_t/\mu_o\omega)^{1/2}$ by $k = (1 + i)/\delta$,[63–65] where $(i)^2 = -1$, and $k = (ip)^{1/2}$ in Smythe's notation,[63] which makes use of the identity $(i + 1)/\sqrt{2} = \sqrt{i}$. In what follows, the parameter $(\mu_o\omega/\rho_t)^{1/2}r = (p)^{1/2}r$ is denoted by x.

The sensed flux density with pickup coils placed in the bore will be $\tilde{b}(kr_b)$. If the coil is wrapped around the tube, the sensed flux density should be the volume average $\tilde{B}(k,r_a,r_b) = <\tilde{b}(kr)>_r$ given as:

$$\frac{\tilde{B}}{B_o} = \frac{1}{r_a^2}[r_b^2 \tilde{b}(kr_b) + 2\int_{r_b}^{r_a} \tilde{b}(kr)rdr]$$

$$= (\frac{r_b}{r_a})^2 \frac{\tilde{b}(kr_b)}{B_o} + \frac{2}{kr_a} \frac{K_1(kr_b)\, I_1(kr_a) - K_1(kr_a)\, I_1(kr_b)}{K_1(kr_b)\, I_o(kr_a) + I_1(kr_b)\, K_o(kr_a)} \tag{40}$$

In performing the integraton in Eq. (40), recurrence relations for I_o and K_o have been used.[63] Note that $I_1(r_b) \to 0$ as $r_b \to 0$. Eq. (39) and (40) then reduce to the forms for a solid cylinder.[24,63] We may write $\tilde{b} = b' + ib'' = b \exp(i\alpha(r))$ and $\tilde{B} = <\tilde{b}>_r = B' + iB'' = B \exp(i\phi)$ to distinguish the in-phase and the out-of-phase signals. The time harmonic fields b(t) and B(t) can be obtained, respectively, from

$$b(t) = \mathrm{Re}\{(\tilde{b} \exp(-i\omega t)\} \tag{41}$$

and

$$B(t) = \mathrm{Re}\{\tilde{B} \exp(-i\omega t)\} \tag{42}$$

Then

$$b(t) = b\, \mathrm{Re}\, \{\exp - i(\omega t - \alpha)\} \tag{43}$$

$$b' = b \cos \alpha \tag{44}$$

$$b'' = b \sin \alpha \tag{45}$$

$$b = (b'^2 + b''^2)^{1/2} \tag{46}$$

and

$$\tan \alpha = (b''/b') \tag{47}$$

Again, $\alpha(r)$ is defined in the interval $[0, \pi]$. B can also be expressed in similar form except for a different phase angle. Some numerical results using Eq. (36)–(47) are given below. The Bessel functions of complex arguments were computed using Kelvin's ber_v and bei_v functions.[68]

Fig. 1a and 1b show the calculated $b(kr_b)/B_o$ and B/B_o ($B_o = \mu_o H_{ac}$), respectively, as a function of $x_a = (\mu_o\omega/\rho)^{1/2} r_a = (1/2)^{1/2} r_a/\delta$. The abscissa can be treated as the varying outer radius r_a with $(\mu_o\omega/\rho)^{1/2}$ held constant, or as the varying $(\mu_o\omega/\rho)^{1/2}$ while holding r_a fixed. The scaled parameter x has special physical implications to be discussed later. First, assume fixed frequency and resistivity. Fig. 1a then depicts $b(r = 0)/B_o$ versus the sample radius r_a in unit of $\delta/(1/2)^{1/2}$ or $(\mu_o\omega/\rho)^{-1/2}$. The inset shows the phase angle between $b(r = 0)$ and B_o; we observe that the b-field phasor rotates with increasing radius over the entire angular domain while the in-phase and out-of-phase signals alter signs in an almost periodic fashion.

The volume-averaged flux density as a function of x_a, as shown in Fig. 1b, exhibits a trend similar to Fig. 1a except that the phasor rotation increases monotonically and eventually saturates at 45° at large x-values. This represents the case of a homogeneous cylindrical sample with a secondary coil wound on the periphery. Note that $|B|/B_o = \mu$, $B'/B_o = \mu'$, and $B''/B_o = \mu''$, as defined in the previous section. In comparing the data of Fig. 1b with that shown in Fig. 1a, we see that the drastic variation in local flux density has been smeared out as a result of the volume averaging. When the sample is small (i.e., for small x_a), much of the total induced screening current stays in phase with the applied field and the loss, represented by the magnitude of B″ (or b″) is thus low (Fig. 1b, $x_a \lesssim 2.5$). In this region, as the sample increases in size, the loss, represented by the magnitude of B″ or b″, increases accordingly. On the other hand, when the sample is large (Fig. 1b, $x_a \gtrsim 2.5$), the screening current is mostly confined within the skin depth and thus the effective losses would decrease as the sample size increases. The maximum $B'/B_o = \mu'' \approx 0.38$ occurs at $x_a \approx 2.5$ or $r_a \approx 1.77\delta$ when the ac screening current permeates an optimal fraction of the sample.

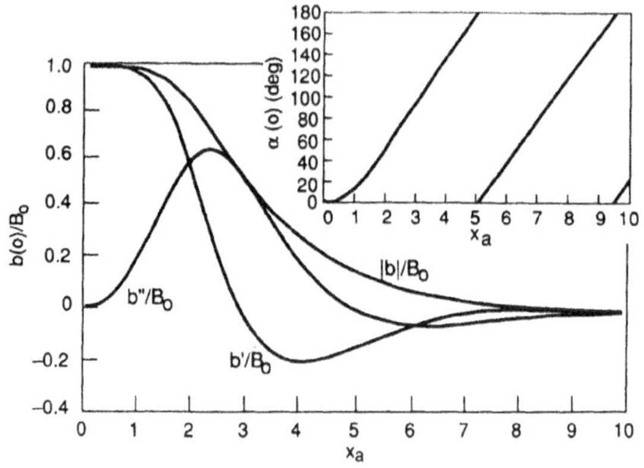

a. Calculated Local Field, $b(o)/B_o$ vs. x_a.
Inset is phase angle $\alpha(o)$ vs. x_a.

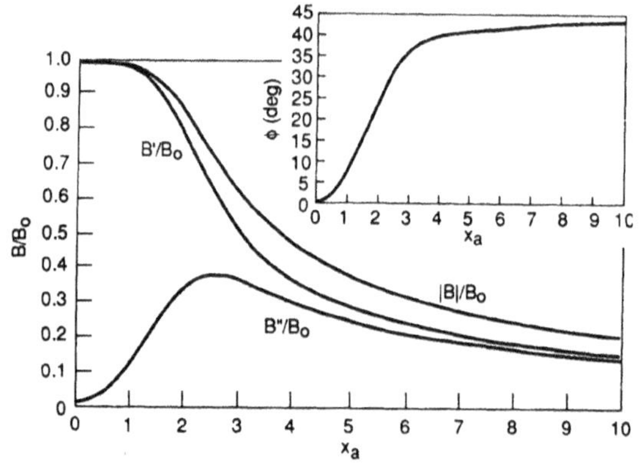

b. Calculated Permeability $\mu = (\mu'^2 + \mu''^2)^{1/2} = B/B_o$, $\mu' = B'/B_o$, and $\mu'' = B''/B_o$ vs. x_a.
Inset shows phase angle ϕ (deg).

Fig. 1. Solid Cylinder with $x_a = (\mu_o\omega/\rho)^{1/2} r_a$ (r_a = radius)

If we consider a specific sample size and let either frequency or resistivity vary, the B/B_o versus x plot then represents the ω or $\rho(B)$ dependent permeability where one may write $x_a = (\mu_o r_a^2/\rho)^{1/2}\omega^{1/2}$. It is interesting to note that $\tilde{\mu}(\omega) = B/B_o$ ($B_o = \mu_o H_{ac}$) given in Fig. 1b qualitatively obeys the Debye equation for a relaxation process:[37,69,70]

$$\frac{\tilde{\mu}(\omega)}{\mu(0)} \approx \frac{1}{1 + i\omega\tau} \tag{48}$$

where τ is the relaxation time for vortex motions. Because of the qualitative functional similarity, superconductors in an ac field can also be understood in the context of magnetic relaxation; the "forced and damped oscillators" tend to relax to their equilibrium state according to the relaxation equation $dB/dt = (\mu_o H\text{-}B)/\tau$, to which Eq. (48) is the solution. The peak μ'' occurs at $\omega\tau_m \approx 1$ as compared to that at $x_a^2 = x_m^2 = \omega(\mu_o r_a^2/\rho_m) \approx 6.25$ obtained from Eq. (40) (Fig. 1b). We may correlate $1/\tau_m \approx \omega_m$ where $\omega_m = \rho_m x_m^2/(\mu_o r^2)$.

Using a small ac signal superimposed on a dc field, the points (H_{dc},T) for peak μ'' have been used to determine the "irreversibility line" on the H–T phase diagram for high-temperature superconductors.[30] While this line should be a thermodynamic boundary between the vortex liquid and vortex solid (based on the notion of vortex melting), and thus should be an intrinsic property, its dependencies on sample size, ac amplitude, and frequency have been observed,[30,34,37–39,70–76] although opposite results also exist.[40] For the former, the line shifts toward higher temperature as the frequency increases and toward lower temperature as the sample size decreases or as the ac amplitude increases. These effects can be understood in the context of classical electrodynamics based on the flux-motion-induced resistive dissipations. As the frequency increases, the peak in μ'' will shift toward higher temperature where ρ is larger, because the μ'' maximum occurs at $(\mu_o\omega/\rho)^{1/2}r_a = 2.5$. By the same token, as the sample size decreases, the peak temperature will shift toward lower temperature, where ρ is smaller.[43]

The remaining issue is how to properly correlate the overall resistivity to the relevant physical parameters, such as temperature, applied ac field, critical current, pinning potential, and viscosity coefficient. Because of the multiple variable dependence, it is anticipated that more careful investigations on well-characterized materials will be needed before one can construct a universal model. The main point is that while the loss characteristics can be interpreted using various flux-motion resistivity models under different circumstances, they are governed by the same magnetic diffusion equation based on the scaling parameter $x = (\mu_o\omega/\rho)^{1/2}r$.

Fig. 2a–2c show the calculated flux density profile $b(kr)/B_o$ for three samples with x_a = 1, 2, and 5, respectively. First, we observe that the two boundary conditions are indeed satisfied: $b = B_o$ at $r = r_a$ (x = 1, 2, 5) and $db/dr = 0$ at $r = r_b = 0$ (x = 0). Second, we see that the quadrature signal develops at the expense of the in-phase signal as the radius x_a increases from 1 to 2. Meanwhile, the skin effect becomes increasingly significant and the quadrature begins to gain dominance at the inner part of the sample. Finally, as $x_a = 5$ the quadrature signal recedes while the in-phase signal reaches into the negative regime and the skin effect becomes prominent. The magnitude of the overall flux density on the axis has now been reduced to about 18% of the applied field.

Fig. 3a and 3b show the calculated field profiles $\{b(kr)/B_o \mid r_b \le r \le r_a\}$ for two hollow cylinders, both with the same outer radii $r_a = 20(\mu_o\omega/\rho)^{-1/2}$ but with different inner radii $r_b = 10(\mu_o\omega/\rho)^{-1/2}$ and $18(\mu_o\omega/\rho)^{-1/2}$. The radii were chosen so that the b(kr) values are comparable with those in Fig. 2. The abscissa here is $x = (\mu_o\omega/\rho)^{1/2}r$. The results (Fig. 2b vs. Fig. 3b and Fig. 2c vs. Fig 3a) suggest that as the sample is bored, $(\mu_o\omega/\rho)^{1/2}r_a$ must be increased such that a similar number of field lines can still be confined within the thickness. Experimentally, two situations would be of interest: (1) the flux density in the bore of a tube (i.e., $b(kr_b)$ based on Eq. (39)), and (2) the average flux density over the entire volume of the tube and the bore (i.e., B based on Eq. (40)). The latter could mimic an inhomogeneous sample with the region $r<r_b$ being electromagnetically insensitive, whereas the former provides a sensitive scheme for phase detection.

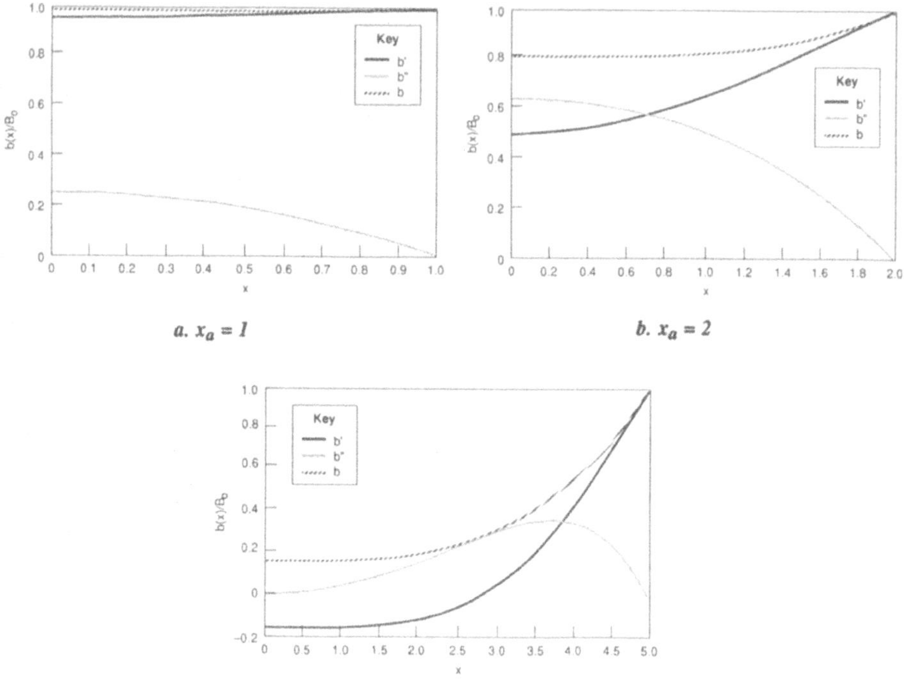

a. $x_a = 1$ b. $x_a = 2$ c. $x_a = 5$

Fig. 2. Calculated Local Field $b(x)/B_o$ vs. x for Solid Cylinders with Various x_a ($x_a = (\mu_o\omega/\rho)^{1/2}r_a$, $x = (\mu_o\omega/\rho)^{1/2}r$; r_a = radius, r = position)

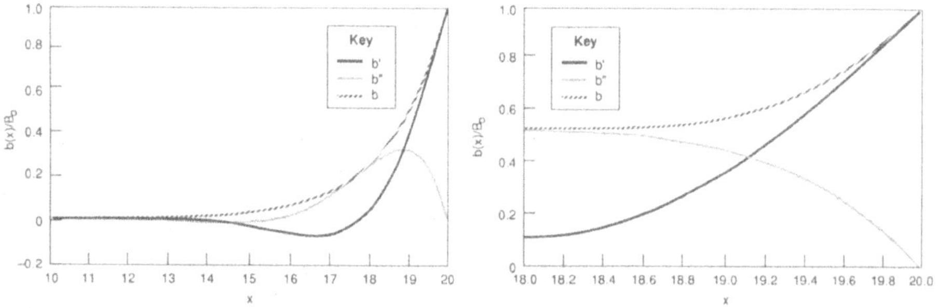

a. $x_b{:}x_a = 10{:}20$ b. $x_b{:}x_a = 18{:}20$

Fig. 3. Calculated Local Field $b(x)/B_o$ vs. x for Tubular Samples of Inner Radii r_b and Outer Radii r_a ($x_a = (\mu_o\omega/\rho)^{1/2}r_a$, $x_b = (\mu_o\omega/\rho)^{1/2}r_b$, $x = (\mu_o\omega/\rho)^{1/2}r$; $x_b < x < x_a$)

Fig. 4a depicts the calculated $b(kr_b)/B_o$ and the associated phase angle as a function of x_a in the range [0, 100] for a tube with $r_b/r_a = 2/3$. The phase angle varied from 0° to ±180°. Neither the in-phase nor the out-of-phase signal was restricted to a unipolar value. For the same sample, the calculated volume average flux density B/B_o and the phase angle are given in Fig. 4b. Note $B/B_o = \mu$, $B'/B_o = \mu'$ and $B''/B_o = \mu''$, and μ, μ' and μ'' are all >0. The results are similar to Fig. 1b except for some features. The shoulder exhibited here resembles a susceptibility versus temperature curve in the superconducting transition region. We found that as the inner radius approached zero, the shoulders were smeared out while the permeability curves converged to those of Fig. 1b for the solid cylinder. Here the phase angle could go up to 70°, well beyond the limiting value of 45° shown in Fig. 1b. The peak value for μ''(≈0.5) now occurs at $x = (\mu_o\omega/\rho)^{1/2}r_a \approx 5.5$ (or $r_a/\delta \approx 3.89$) as compared to 2.5 (or $r_a/\delta \approx 1.77$) for the solid cylinder. If we regard the tube as an inhomogeneous cylinder, this suggests that the peak for μ'' should occur at a lower temperature (where ρ is smaller) for such a sample as compared to a homogeneous sample of the same size measured at the same frequency and applied field. Furthermore, since ρ increases with temperature, a backward plot of Fig. 4b (with decreasing x_a) resembles the permeability vs. temperature curve; the shoulder in the transition region arises from the sample inhomogeneity discussed here and should not be mistaken for evidence of "phase transition" or "second phase impurity."

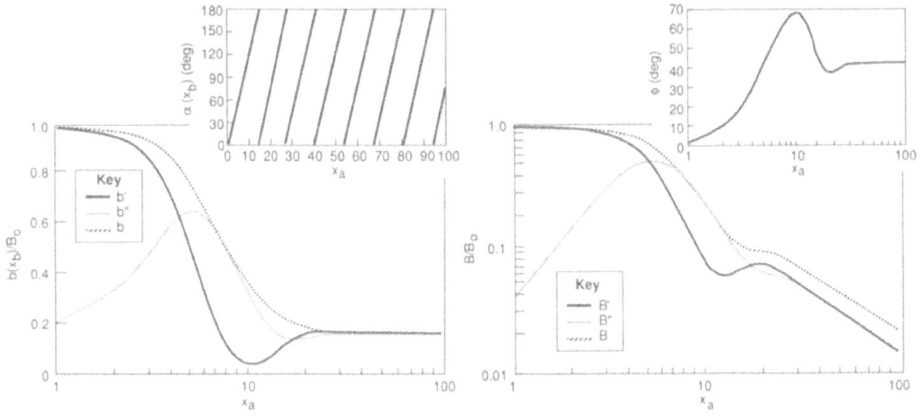

a. Calculated $b(x_b)/B_o$ vs. x_a. Inset is phase angle α (x_b) (deg).

b. Calculated Permeability $\mu = (\mu'^2 + \mu''^2)^{1/2} = B/B_o$, $\mu' = B'/B_o$, and $\mu'' = B''/B_o$ vs. x_a. Inset shows phase angle ϕ (deg).

***Fig. 4.** Tubular Sample (x_b:x_a = 67:100 (2:3))*

At the other extreme, for a very thin tube with 90.5% bore (or $r_b/r_a = 0.95$), Fig. 5a and 5b show that the volume-averaged signal is almost the same as that measured at the center. Although they could reach into the negative regime at higher x-values than presented here, in the present cases, the imaginary components are always positive up to x = 100 but the in-phase component $\mu' = B'/B_o$ has assumed some negative values. This amounts to a susceptibility $\chi' = -1 + \mu'$ less than –1. In taking account of the demagnetization effect, an uncertainty usually exists regarding the exact value of demagnetizaton factor. The effect discussed here further contributes to that uncertainty. Note that the peak for μ'' now occurs at $(\mu_o\omega/\rho)^{1/2}r_a \approx 30$ and the phase angle could rotate from 0° to ±180°. Under such circumstances the ac irreversibility line will be shifted toward lower temperature (to obtain smaller ρ) for a given sample size, ac amplitude, and frequency.

V. Experimental Setups and Sample Preparations

The ac measurements can be largely divided into the following functional categories: (1) shielding effectiveness measurements, (2) susceptibility measurements, (3) magnetization and critical current determinations using a small modulation field superimposed on a larger dc bias field, and (4) electrodynamic studies focusing on relative signal variations rather than elaborated calibrations. While the detailed experimental setup for each category varies, the general scheme is largely composed of a primary and a secondary coil, an ac current source, a phase-sensitive detector, a waveform analyzer (digital storage oscilloscope and spectrum analyzer), and a data acquisition system. The secondary coil can be either compensated (double coil) or uncompensated (single coil). The former uses two coils wound in opposite directions to null the signal in the absence of sample. As the sample is inserted, a "large" signal would be detected. This scheme has been commonly employed in ac magnetic susceptometry. The difficulty with this approach is the coil impedance drift over a wide range of measuring conditions. With the advance of computer technology, the

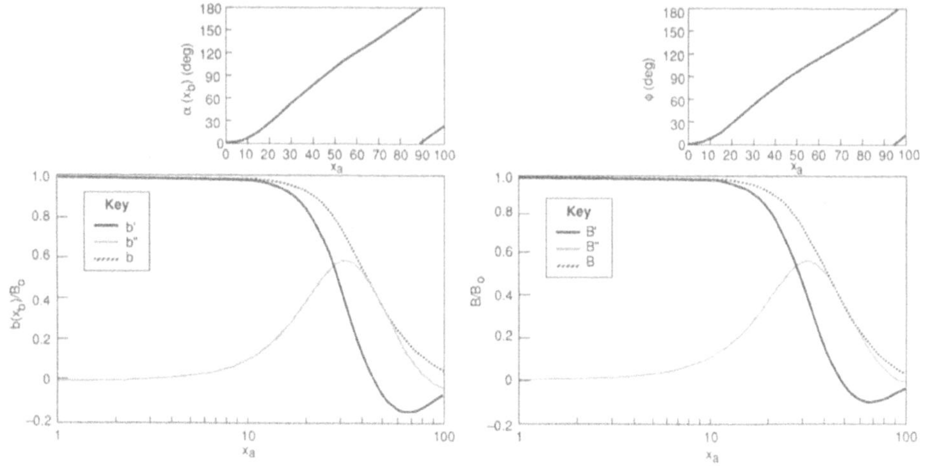

a. Calculated $b(x_b)/B_o$ vs. x_a. Inset is phase angle $\alpha\,(x_a)$ (deg).

b. Calculated Permeability $\mu = (\mu'^2 + \mu''^2)^{1/2} = B/B_o$, $\mu' = B'/B_o$, and $\mu'' = B''/B_o$ vs. x_a. Inset shows phase angle ϕ (deg).

Fig. 5. Tubular Sample (x_b:x_a = 95:100)

single-coil approach may become more favorable. Here calibration runs can be carried out over the range of interest from which polynomial data fitting can be established. The coil impedance variations over temperature can thus be better controlled. One frequently encounters data showing negative μ'' (χ'') below T_c or finite positive μ'' (χ'') above T_c. When this happens, the phase setting could be improper because the phase angle may drift as a result of coil impedance mismatch.

From a measurement sensitivity viewpoint, when the lock-in amplifier is set at large scale (low sensitivity) to detect the "large" signal from a compensated coil system, a small variation would be very difficult to recognize. On the other hand, measuring the shielding signal using a tubular sample allows for subtle flux-motion studies since the lock-in amplifier scale would be set at a much smaller scale (higher sensitivity). Fig. 6a shows the basic components of the ac measurements for an uncompensated coil system; here a tubular sample was used. For a compensated system, it was useful to introduce a phase shifter and a voltage divider between the two secondary coils, as shown in Fig. 6b, to fully null out the signal (i.e., by adjusting the phase as well as the amplitude). Flux density b(t) can be obtained by numerically integrating V(t) over the acquired digital waveform or via analog integration.

The frequency response of a measuring setup is important to the system performance. Fig. 7a shows a typical self-resonance at 250 kHz for a small coil roughly 2 mm in diameter wound with ≈500 turns of AWG40 magnet wire at 77 K. The setup performance becomes unreliable as the resonance frequency is approached. The resonance effect extends over a wide bandwidth into the lower frequency regime, which could limit the useful operational frequencies to the kHz range. Fig. 7b illustrates the shielding effectiveness for a sintered $YBa_2Cu_3O_{7-x}$ superconducting tube which is roughly 70 dB at 77 K and fairly constant up to 5 kHz for applied fields well below the penetration threshold.[5] Spurious pickups existed due to the geometric effect; the actual shielding effectiveness of this material can be as high as 160 dB or 10^8 at low frequencies (<1 kHz).[36] The decline in shielding effectiveness at higher frequencies was believed to be partly due to the coil resonance and partly due to the flux motion. To account for the frequency effects of flux motion, it is essential that a valid bandwidth be clearly defined, especially for low-level measurements.

Based on superior shielding effectiveness, we have focused mainly on the tubular sample in investigating the flux behaviors of the $YBa_2Cu_3O_{7-x}$ superconductors. The ceramic samples were sintered from powder at 950°C for 20 h, followed by annealing at

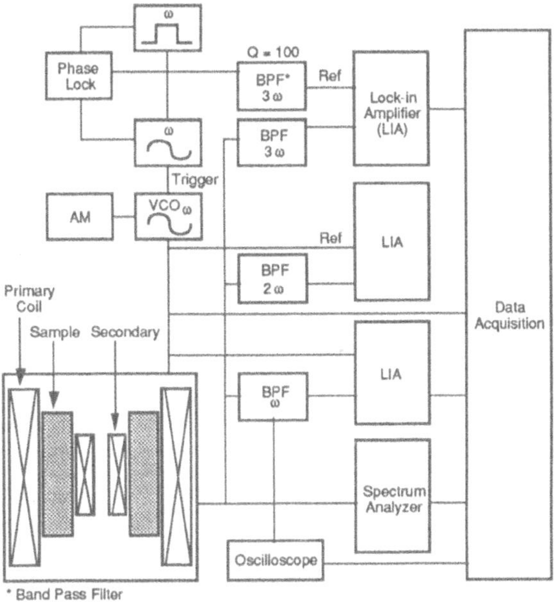

a. Basic Setup for ac Magnetic Measurements

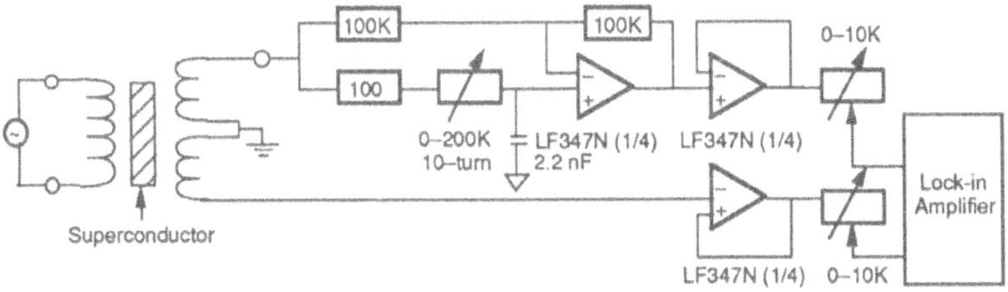

b. Signal-Nulling Electronics

Fig. 6. Uncompensated and Compensated Coil Systems

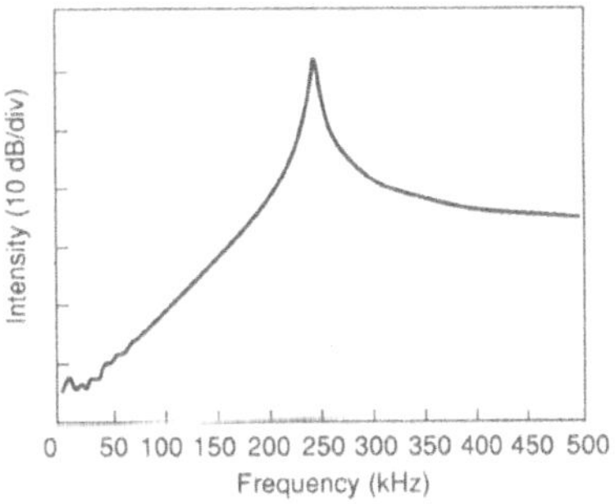

a. Typical Resonance Spectrum for Low Inductance Coils

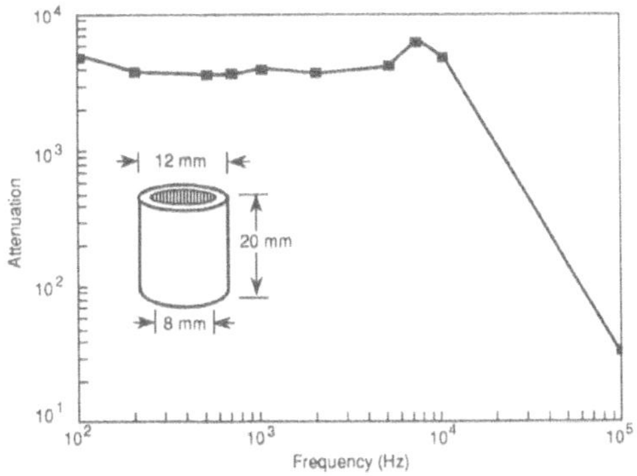

b. Typical ac Shielding Effectiveness for Tubular Samples ($YBa_2Cu_3O_{7-x}$ at 77 K for H_{ac} < Penetration (threshold) Field H_p)

Fig. 7. Coil Resonance Effects on ac Measurements

400–700°C for 4–20 h. The bulk densities of the samples averaged ≈89% of the theoretical density, or ≈5.6 g/cm^3. Typical onset T_c was 91 K with transition width 0.2–2 K according to the 10–90% magnetization criterion. Rod samples were roughly 1.5 mm in diameter and 10 mm long. Tubular samples, formed by isostatic pressing of calcined powder in a steel mold, were approximately 20 mm long, with a 12-mm outer diameter and an 8-mm inner diameter, or $r_b/r_a \approx 2/3$.

VI. Experimental Results

When the flux pinning is strong, the critical state model provides a simple picture for a field penetration and the associated critical current distribution in type II superconductors. In a tubular sample with a secondary coil placed inside the bore, we essentially measure the flux density at the inner surface since $b(r<r_b) = b(r = r_b)$. To the extent that the alternating field does not penetrate as deep as r_b, the coil will detect no change of flux. The response will therefore be flat. Fig. 8a shows the pickup voltage V(t) as well as the integrated waveform b(t) (both normalized to the peak value) due to a sinusoidal applied field at 200 Hz for a sample annealed at 550°C for 48 h (the position variables for V and b have been dropped for the sake of simplicity). The flat shoulder of V(t) (equivalent to the clipped portion of b(t)) represents the time period where the remanent field established by the previous half cycle of the ac field is still present and the opposite half cycle of flux has yet to enter. In light of the critical state model, this shoulder indicates the existence of flux pinning and hysteresis while its width reflects the strength of pinning force and the magnitude of the critical current. Certainly it also depends upon the thickness of the tube. In the present case, the flat shoulder represents approximately two-thirds of a period, a fraction equal to the ratio of the inner radius to the outer radius: r_b/r_a = 4 mm/6 mm. The flat shoulder observed here could become oblique or bear some features if rod samples are used instead; this is generally due to the uncompensated fundamental signal due to the empty space between the sample and coil. Typical of a hysteretic system, waveforms with the above-described shoulders contain higher harmonic terms, as discussed previously.

If there is no flux motion of any kind, the area enclosed by the hysteresis plot b(t) versus $H(t) = H_{dc} + H_{ac}\cos\omega t$ should represent the total hysteretic dissipation. Suppose that one ignores the phase lag of b(t) with respect to H(t), this area would become zero since no losses are incurred in the path. The phase shift is then added to account for the hysteretic behavior: the loop area would expand with the increasing phase angle, as predicted in Eq. (33) and (40) where phase information was contained in χ'' and μ''. If the b(t) – H(t) plot could not be adjusted in any way to zero out the loop area, other dissipative mechanisms must exist, as will be discussed next.

Fig. 8b–8d illustrate a series of normalized b–H hysteresis loops with increasing phase angle. The expanding loop reflects a larger phase shift and hence larger losses. Note that when the phase lag was set to zero, the plot in Fig. 8b still enclosed a finite, although small, area. This suggests the existence of certain flux motion losses (on the order of a few percent) in this hysteresis-dominated regime. The overall hysteresis loss should be measured with the volume-averaged field rather than the local field discussed here; the local field measurement is only employed here to take advantage of the phase sensitivity in recognizing distinctive loss mechanisms. The key point here is that for ac measurements where flux motions coexist with hysteresis, conclusions regarding loss mechanisms must be made cautiously.

Fig. 9a and 9b show the dc bias effects on the waveforms V(t) and b(t), respectively. H_{dc} of 10 G and 13.6 G were added to the 26.8 G (rms) ac field (i.e., $H_{ac} \approx 38$ G). In the positive half cycle where H_{dc} was added to H_{ac} to give an even higher field, the flat region became narrower. This narrowing could be due to the weakened flux pinning or to the deteriorated critical current as a result of increasing field. In the negative half cycle, the overall phase angle shifted consistently and the remanent field b(t) was essentially unaffected by H_{dc}, even though $H(t) \approx H_{dc} - H_{ac} \approx H_p$; the reason for this was that, according to the critical state model, H_{ac} rather than H_{dc} determines the current distribution in the sample for the strongly pinned case. Based on symmetry arguments, it is believed that the asymmetric narrowed flat region between the two half cycles arose from flux motions more than from

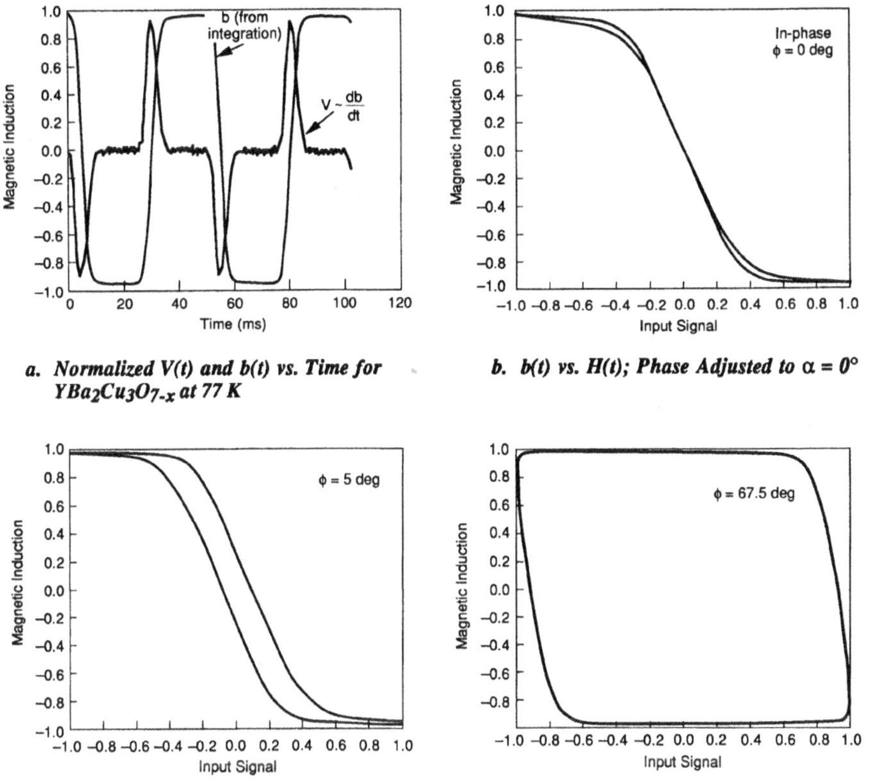

a. Normalized V(t) and b(t) vs. Time for $YBa_2Cu_3O_{7-x}$ at 77 K

b. b(t) vs. H(t); Phase Adjusted to α = 0°

c. b(t) vs. H(t); Phase Adjusted to α = 5°

d. b(t) vs. H(t); Phase Angle α = 67.5° (unadjusted)

Fig. 8. Illustrations of Remanent Field Using Tubular Samples (input waveform $H(t) = H_{ac} \cos \omega t$; $H_{ac} \approx 5$ G greater than the penetration field H_p; ω = 20 Hz)

the field-dependent critical current; otherwise the hysteresis plot in Fig. 8b would not have shown a linear region between saturations (flat regions). At higher temperature, this region was also found to be narrower for the same applied field and frequency. Using the phase-zeroing technique applied to Fig. 8b, larger flux motion losses were observed as H_{dc} increased.

Weakening of the pinning force was also observed as the frequency increased. Fig. 9c shows the narrower flat region for a 2000-Hz signal; it is understood that as the frequency increases the pinning effect becomes less effective and the flux-dynamics-related dissipative processes set in. Such frequency effect has also been observed in low-temperature superconductors, but at much higher frequency.[51]

In the flux-motion regime, the flux density in the bore of a tubular sample was calculated as a function of the scaling parameter $x_a = (\mu_o\omega/\rho)^{1/2} r_a$. These calculations showed that the phase angle could rotate over the entire angular domain of ±180°, in contrast to a solid cylinder where it was limited to 45° as far as the skin effect was concerned (Fig. 1b). In the purely hysteretic regime, it can be shown that such a phase shift for a solid cylinder[77] is bounded at 90° for $H_{ac} \gg H_p$. Here H_p is the applied ac field threshold at which the field penetration reaches the center of a cylinder (or the inner surface of a tube).

According to the Bean critical state model, $H_p = J_c r_a$ for the cylinder and $H_p = J_c(r_a - r_b)$ for the tube, where as usual r_a and r_b are the outer and inner radii, respectively. For a tubular sample, it will be shown that there is a transition from the hysteresis into the flux-motion regime. The oscillating local flux density in the tube in response to an alternating field

$H(t) = H_{ac} \cos \omega t$ can be divided into four regions for $H_{ac} > H_p$ ($b(r<r_b) = 0$ if $H < H_p$):

(1) $-H_{ac} < H < -H_{ac} + 2H_p$, where $b(t)/\mu_o = b(r<r_b)/\mu_o = -H_{ac} + H_p$ = constant,

(2) $-H_{ac} + 2H_p < H < H_{ac}$, where $b(t)/\mu_o = H(t) - H_p$,

(3) $H_{ac} - 2H_p < H < H_{ac}$, where $b(t)/\mu_o = H_{ac} - H_p$ = constant, and

(4) $-H_{ac} < H < H_{ac} - 2H_p$, where $b(t)/\mu_o = H(t) + H_p$.

These four segments enclose a hysteresis loop that is a limiting case (hysteretic losses only) to what is depicted in Fig. 8b–d. The corresponding induced waveforms in response to a sinusoidal excitation should consist of blips on flat shoulders. The in-phase and out-of-phase fundamental components of the local field ($b_{1\omega}$) in the tube can be found through direct Fourier expansion $b(t) = \Sigma_n[b'_{n\omega} \cos(n\omega t) + b''_{n\omega} \sin(n\omega t)]$. It follows that:

$$b'_{1\omega} = (\mu_o/\pi)\{-2(H_{ac} - H_p) \sin\gamma - H_{ac}(\gamma + \sin(2\gamma)/2\,) + 2H_p \sin\gamma\} \tag{49}$$

and

$$b''_{1\omega} = (\mu_o/\pi)\{2(H_{ac} - H_p)(1 + \cos\gamma) + (\cos\gamma-1)[H_{ac}(\cos\gamma + 1) - 2H_p]\} \tag{50}$$

where $\cos\gamma = -1 + 2H_p/H_{ac}$. The magnitude of the total field and its phase shift α with respect to the applied field are, as usual:

$$b_{1\omega} = [b'_{1\omega}{}^2 + b''_{1\omega}{}^2]^{1/2} \tag{51}$$

and

$$\tan \alpha_{1\omega} = b''_{1\omega}/b'_{1\omega} \tag{52}$$

Fig. 10 depicts the calculated α and the normalized local field b as a function of H_{ac}/H_p (>1) for a tubular sample based on Eq. (49)–(52). The local field increases linearly with the applied field for $H_{ac} \gg H_p$, whereas the phase shift reverses from approximately 80° at $H_{ac} \approx H_p$ to about 10° for $H_p/H_{ac} > 10$. This qualitatively agrees with the experimental results except for the exact magnitude of the phase angle, which is attributed to the emerging flux motions, as will be discussed below.

Fig. 11 illustrates the measured fundamental component $V_{1\omega}$ of the total pickup voltages and phase angles (ω = 200 Hz) under various H_{dc} for a tubular sample annealed for 16 h at 400°C. The slope of the curve below threshold arises from the leakage field (≈0.86μV/G) and may be subtracted in data processing. Above threshold the signal rose ≈70 dB (20 dB/decade) to saturation. The typical 70-dB shielding effectiveness has already been shown in Fig. 7b. The inset of Fig. 11 shows the phase angle shift of more than 90° as H_{ac} passes H_p. Eventually, the phase shift reversed as it increased further. Although the phase reversal was predicted in the critical state model, the shift of greater than 90° cannot be fully accounted for with hysteresis losses alone since this would have restricted the phase shift to less than 90° for $H_{ac} > H_p$. As H_{dc} increased, H_p shifted consistently toward lower H_{ac}. Sign alterations between the in-phase and the quadrature signal are implied in the large phase shift (i.e., $\alpha > 90°$). It was noted (see Fig. 4) that α as a function of $x_a = (\mu_o\omega/\rho)^{1/2}r_a$ varied periodically over the entire angular domain. Here one has the applied field as the varying parameter, which can be linked with the scaling parameter x_a through the field-dependent resistivity $\rho = \rho(b)$, as will be discussed next.

Fig. 12 illustrates $V_{1\omega}$ and α as a function of H_{ac} for a tubular sample annealed for 48 h at 700°C. As usual, the voltage went to saturation and the phase shift reversed as $H \gg H_p$. The interesting point here is that α completely reversed its sign with respect to the reference angle at $H<H_p$. We stress that the phase angle in the flux-motion regime may range from 0° to ±180° and the phase reversal simply reflects the local effect due to flux motion resistivity variations. Note that the initial phase lag before the turning point was only ≈30°, as compared with ≈95° for Fig. 11. Hence we have seen that as H_{ac} passes the threshold, flux motion resistivities, whether due to flux creep, flux flow, or eddy current, will become increasingly important.

In the scaling parameter $x = (\mu_o\omega/\rho)^{1/2}r$, frequency can also be varied. Fig. 13 shows the relative amplitude (dB) of the nth harmonics $V_{n\omega}$ (n = 2,3,...9), which has been normalized with respect to the fundamental signal to remove the linear frequency

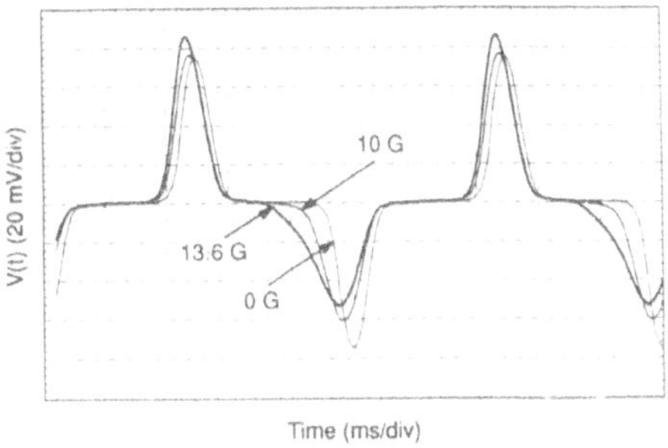

a. V(t) for $H_{dc} \approx$ 0, 10, and 13.6 G, ω = 200 Hz

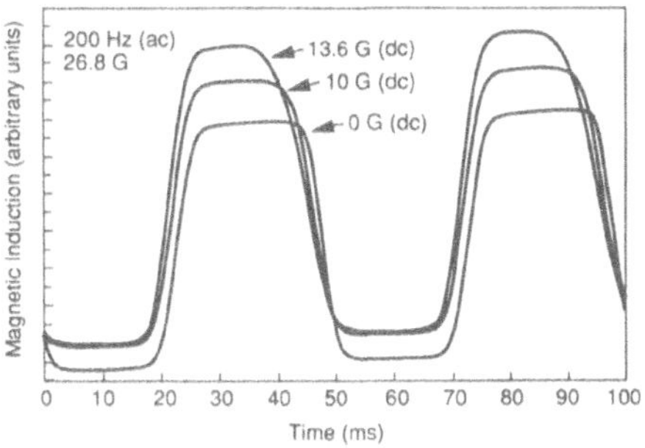

b. Corresponding b(t), ω = 200 Hz

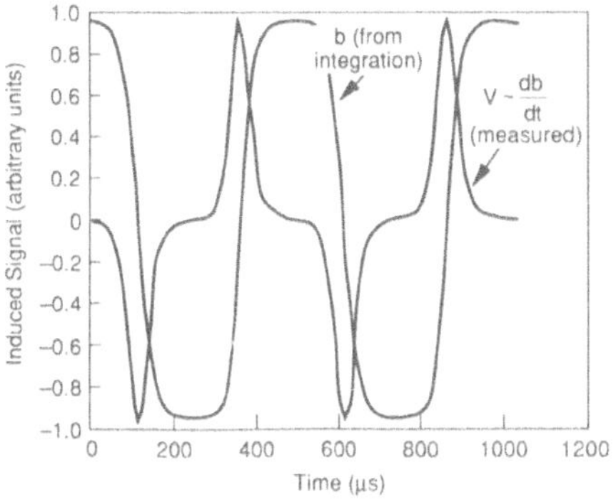

c. V(t) and b(t) with $H_{dc} \approx$ 0 for ω = 2000 Hz (evidence of weakening flux pinning as ω increases from 200 to 2000 Hz)

Fig. 9 The dc Bias Effects on the Remanent Fields ($H_{ac} \approx$ 5 G greater than the penetration field)

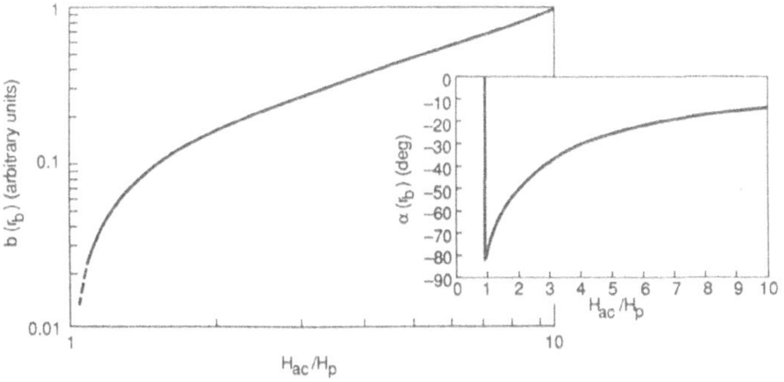

Fig. 10. Calculated Local Field b and Phase Angle α (r_b) (deg) vs. H_{ac} Based on Bean Model

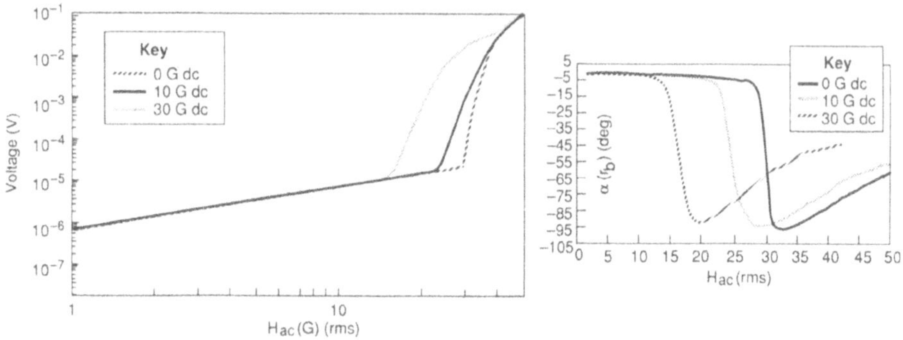

***Fig. 11.** Measured b and Corresponding Phase Angle α for $YBa_2Cu_3O_{7-x}$ at 77 K (three dc levels) and (r_b/r_a = 2/3)*

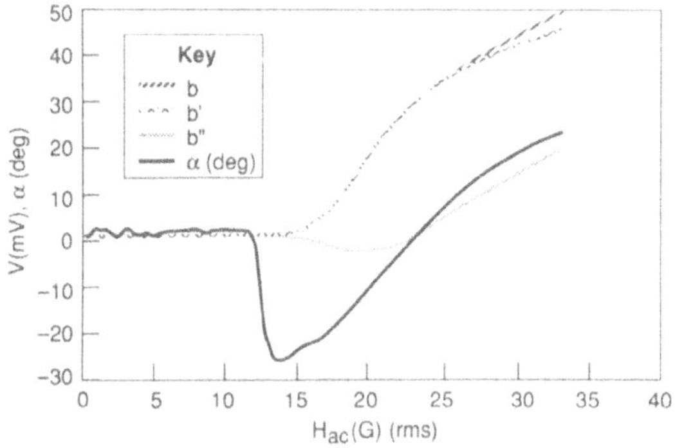

***Fig. 12.** Measured b, b′, b″ in Lock-in Voltages (mV) and the Phase Angle α (deg) for $YBa_2Cu_3O_{7-x}$ at 77 K. Sign alternation in b″ was observed (r_b/r_a = 2/3).*

dependence. These higher harmonics are the Fourier components contained in the distorted waveforms shown in Fig. 8 and 9 (with the same sample at roughly the same applied field). They were measured using an HP3582A low-frequency spectrum analyzer with long time averaging to improve the signal-to-noise ratio at low frequencies. The reason for the occurrence of even harmonics is unclear. Mathematically speaking, even harmonics should not occur for a periodic waveform $f(x + 2\pi) = f(x)$, i.e., to observe even harmonics one needs $f(x + \pi) \neq f(x)$. Only a trace amount of trapped field would be sufficient to generate a significant level of even harmonics; we were unable to eliminate the even harmonics through reverse biasing the sample, assuming that they did indeed arise from the spurious dc signal from the power supply or from the earth field. Zero-field cooling in a zero-gauss chamber to reduce the ambient field to around 10 mG did not seem to prevent it either.

The main point here is the remarkable evidence of skin effects for various odd harmonics, which agrees with the calculations in section IV. It is unfortunate that the associated phase data could not be obtained, but if one takes the 3-dB rolloff at 750 Hz as the peak position of the quadrature signal (note that the data represents the total magnitude of the local field), then for the sample with $r_b/r_a \approx 4$ mm/6 mm = 2/3, the peak position should occur at $x_a = (\mu_o\omega/\rho)^{1/2}r_a = 5.5$ (see Fig. 4). Thus it is estimated that $\rho \approx 0.7\mu\Omega$-cm at the applied ac field of approximately 25 G (rms). Since the normal resistivities of the $YB_2Cu_3O_{7-x}$ compounds were found to be on the order of mΩ-cm, this $\mu\Omega$-cm resistivity is believed to be caused mainly by the vortex motions. The corresponding magnetic diffusion coefficient is ≈ 224 cm^2/s. At 750 Hz the skin depth is around 1.54 mm as compared with the 2-mm sample wall thickness.

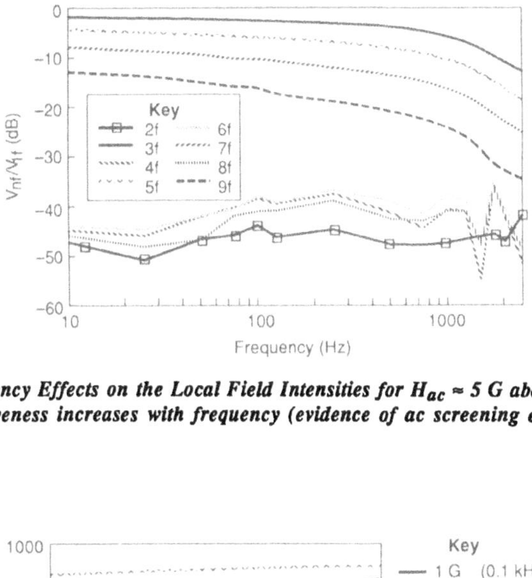

Fig. 13. Frequency Effects on the Local Field Intensities for $H_{ac} \approx 5$ G above H_p. Shielding effectiveness increases with frequency (evidence of ac screening effect).

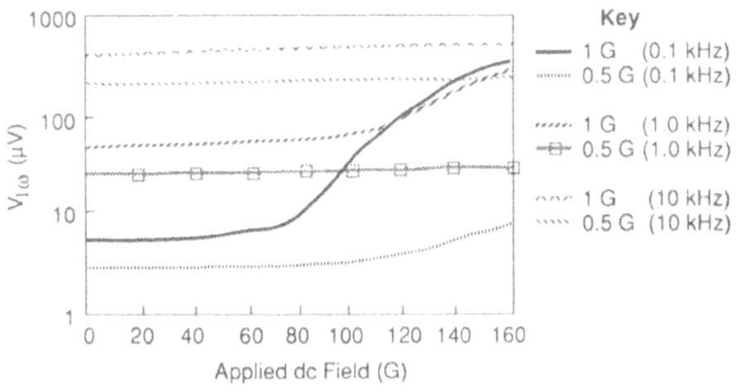

Fig. 14. Evidence of ac Screening Effects (shielding signal for various H_{ac} (rms) and H_{dc})

Fig. 14 shows the effects of H_{dc} and H_{ac} on the ac shielding effectiveness of a tvpical sintered $YB_2Cu_3O_{7-x}$ superconducting tube measured at 77 K (displayed as the pmkup voltage $V_{1\omega}$) as a function of H_{dc} (with H_{ac} and ω as the parameters). The data show that at low H_{dc}, the signals scale linearly with H_{ac} and frequency. In this regime the coil inside the tube is largely shielded from the applied field except for some leakage signal. The data can be grouped into two categories: (1) fixed ω, various H_{ac}, and (2) fixed H_{ac}, various ω. In the following, two extreme cases will be discussed.

In case 1, at 0.1 kHz, the shielding signal is a strong function of H_{ac} and H_{dc}. As H_{ac} increased from 0.5 to 1 G (rms), the shielding effectiveness of the sample was reduced to the extent that the alternating field was able to penetrate through the sample wall at a small H_{dc}. At this frequency, the skin effect was not effective and the applied field had enough time to sweep in and out of the sample volume.

In case 2, at 10 kHz, the overall shielding effectiveness was independent of H_{ac} up to the H_{dc} applied. Compared with the data at 0.1 kHz, in this case the increases in ρ due to the increasing field must have been overcome by the increase in ω, resulting in a reduced skin depth and therefore higher ac shielding capability. The 1-kHz signal showed a transition between these two extremes; thus this frequency dependence can be interpreted as due to the skin effect based on the classical electrodynamics of flux motion. In using the peak temperatures for the imaginary part of the ac susceptibilities to determine the ac irreversibility line,[30,34,37–39,70–76] the frequency effects suggest that such a line is in fact just a demarcation line (DL) separating different magnetic regimes in accordance with the dynamic responses of the flux vortices to external excitations. The true irreversibility line that defines the boundary between vortex solids and vortex liquids should be measured under isothermal and equilibrium conditions accessible only through true dc techniques (tentatively called the dc irreversibility line for distinction).

In association with these measurements, Fig. 15 shows the susceptibilities of a solid rod sample as a function of temperature measured with a conventional susceptometer (H_{ac} = 0.1 G, f = 1 kHz, and H_{dc} = 0–200 G). As the dc field was increased from 0 to 200 G, the peak of χ'' shifted toward lower temperature (89.8 K→84.8 K). The peak position (H,T) of χ'' is plotted in Fig. 16, which, as discussed above, is a demarcation line at 1 kHz. The screened lines mark the trend of the demarcation line as the frequency or the ac amplitude increases. The size effect can also be sketched in a similar fashion. As the size decreases, the

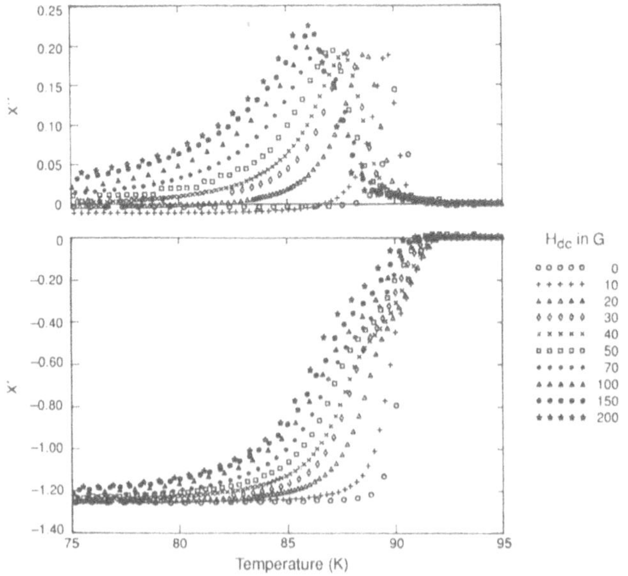

Fig. 15. Typical Field and Temperature Dependences of ac Susceptibilities for Rod Samples of $YBa_2Cu_3O_{7-x}$ for H_{ac} = 0.1 G, H_{dc} = 0 to 200 G, and ω = 1000 Hz

line shifts toward lower temperature. The ac amplitude has a larger effect on the position of the demarcation than does the dc field; a 30-G increase in H_{ac} could lead to a 10 K shift in peak temperature. The ac fields force the vortices to move periodically, and the energy losses are proportional to the amplitudes of the vortex displacements for such damped oscillations. The increases in energy losses due to the increases in the amplitude of vortex displacements are thus believed to be more significant than the decreases in thermal activation energy (or the increases in flux motion resistivity) caused by the dc field, based on the thermally activated flux creep model.

VII. Scaling Law for Flux Dynamics

Thus far, the discussion, which has emphasized the flux dynamics based on a scaling parameter $x = (\mu_o\omega/\rho)^{1/2}r$, leads one to conclude that the frequency and field dependences of magnetic measurements were accountable via this parameter. These arguments were particularly interesting for the x-value where the peak of the quadrature signal occurs and the energy losses were maximal (denoted as x_m; m stands for maximum μ''). For homogeneous samples, $x_m \approx 2.5$ as discussed in Fig. 1b. Once the sample radius is given, resistivity is the only unknown. Various models have been proposed to account for the vortex-motion-induced resistivity based on (1) the conventional viscous drag theories of flux flow, and (2) thermally activated flux creep. It will be illustrated that the scaling parameter could be used to substantiate a general theory in the flux-motion regime using specific $\rho(T,H)$ functionals.

The starting point for this discussion is the resistive transition regime based on the thermally activated process.[77] Basic to this process is the temperature- and field-dependent thermal activation energy $U_o(T,H)$. Following previous analyses conducted by various researchers using the Ginzburg-Landau theory, we take the activation energy functional as[30,79–80]:

$$U_o(T,B)/kT = A(1-t)^{3/2}/B \tag{53}$$

where A is a material constant to be determined experimentally (or to be fitted numerically, as will be discussed later), B is the flux density in the sample, and $t = T/T_c$ (T_c being the zero field transition temperature) is the reduced temperature. Eq. (53) will lead to an upward curvature for the H-T irreversibility line measured with a dc technique as well as the demarcation line measured with an ac technique. A simple power law is not applicable for a sintered polycrystalline $YBa_2Cu_3O_{7-x}$ superconductor measured with $H_{ac} = 0.1$ G (rms), as is the case shown in Fig. 16, but the general trend agrees semiquantitatively with experiments.

Based on the concept of phase slippage at the Josephson junctions, where vortices slip past one another over the activation energy barrier, the resistivity functional for $T_c/2<T<T_c$ (or $1/2 < t < 1$) was found[79] to be:

$$\rho_{tf}(T,B)/\rho_n = [I_o(U_o/2kT)]^{-2} \tag{54}$$

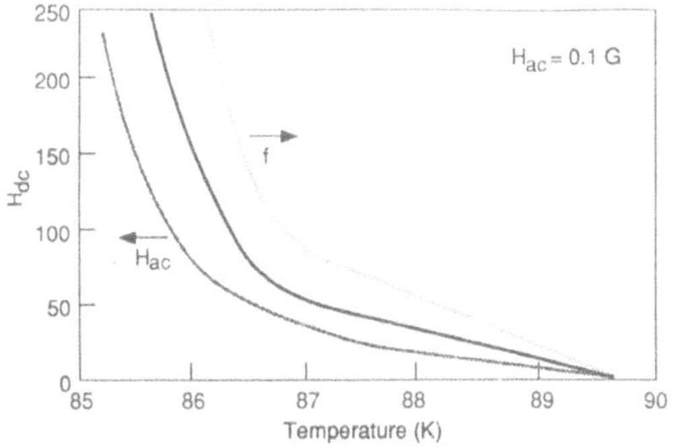

Fig. 16. The ac Irreversibility Line Determined from the χ" Peak Temperatures for $YBa_2Cu_3O_{7-x}$ rods. The screened lines sketch the general trend of line shifts as a function of H_{ac} and ω.

where ρ_n is the normal state resistivity and I_o is the modified Bessel function of order zero. Taking the peak temperature of the quadrature signal as the onset of irreversibility (i.e., $\rho_{tf}(T_m,H_m) = \rho_m$), and substituting Eq. (54) into the scaling parameter $x_m = (\mu_o\omega/\rho_m)^{1/2}r$, or $\rho_m = \omega/(x_m^2/\mu_o r^2) = \rho_{tf}(T_m,B_m)$, one finds:

$$\rho_{tf}(T_m,B_m)/\rho_n = \omega/\omega_o \tag{55}$$

where

$$\omega_o = \rho_n x_m^2/(\mu_o r^2) \tag{56}$$

ω_o is a characteristic frequency corresponding to the frequency of maximum dissipation when the sample is normal. For $r \approx 1$ mm, normal resistivity $\rho_n \approx 1 m\Omega$-cm, and $x_m \approx 2.5$, one obtains $f_o = \omega_o/2\pi \approx 10^7$ Hz. Using the experimental results from Palstra et al.,[73] where the single crystal was 1 mm × 0.2 mm × 0.01 mm, with the radius approximated as the geometric mean of the shorter dimensions ($r \approx 0.022$ mm) and $\rho_n \approx 0.1 m\Omega$-cm, one obtains $f_o \approx 10^9$ Hz. Fundamental to Eq. (55) is that as the frequency increases, the ac screening will become more effective and therefore the resistivity has to be increased to maintain the desired maximal dissipation. This is the theme of the scaling law being discussed here.

Combining Eq. (53)–(56) and assuming that $B_m \approx \mu_o H_m$,

$$\{I_o[A(1-t_m)^{3/2}/(2H_m)]\}^{-2} = \omega/\omega_o \tag{57}$$

where $t_m = T_m/T_c$. Eq. (57) can be solved through numerical iterations to obtain the ac irreversibility lines, with frequency and sample size as varying parameters. The ac amplitude effect is contained in H_m, where $H_m = H_{dc} + H_{ac}/(2)^{1/2}$. For superconducting $YB_2Cu_3O_{7-x}$ compounds (with $T \approx 90$ K), it was estimated[79] that $A \approx 3.5\ \beta J_{co}(0)$, where $J_{co}(0)$ is the zero-field critical current at 0 K on the order of 10^7 for single crystals (A in G, $J_{co}(0)$ in A/cm^2, and $\beta \approx 1$). B (in G) is the flux density in the sample. Two limiting cases will now be considered: $U_o \gg 2kT_m$ and $U_o \ll 2kT_m$.

For case 1 ($U_o \gg 2kT_m$), using $I_o(x) \approx \exp(x)/(2\pi x)^{1/2}$ for $x \gg 1$ and Eq. (57) one obtains:

$$(1-t_m)^{3/2}/[H_m/A] - \ln\{\pi(1-t_m)^{3/2}/[H_m/A]\} \approx \ln(\omega_o/\omega) \tag{58}$$

Note that the applied field has been scaled with respect to $A = 3.5\ \beta J_{co}(0) \approx 10^7$G for single-crystalline $YB_2Cu_3O_{7-x}$. At low temperature or high frequency, the logarithmic term on the left-hand side of Eq. (58) is much smaller than the first term and one can simply write:

$$(1-t_m)^{3/2}/[H_m/A] \approx \ln(\omega_o/\omega) \tag{59}$$

This relationship displays the logarithmic frequency dependence derived by previous researchers using transport current approaches.[30] Here we see that as frequency increases, $\ln(\omega_o/\omega)$ decreases and $(1-t_m)^{3/2}$ thus must also decrease (H_m kept constant). This means that t_m must increase and the line must shift toward higher temperatures in the H-T diagram. At constant frequency, the line obeys the $H \propto (1-t_m)^{3/2}$ law and concaves upward for $\omega \ll \omega_o$. Meanwhile, a reduction in radius of the sample will lead to a larger ω_o and consequently a line shifted toward lower temperatures, in agreement with the experimental results on thin-film samples.[43] When the measuring frequency $\omega \rightarrow \omega_o$, $t_m \rightarrow 1$, case 2 ($U_o \ll 2kT$) is in effect and Eq. (58) will no longer be applicable.

For case 2 ($U_o \ll 2kT_m$), using $I_o(x) \approx (1 + x^2/4)$ for $x \ll 1$, one obtains:

$$\rho_{tf}(T_m,H_m)/\rho_n \approx \{1 + [(1-t_m)^{3/2}/(4H_m/A)]^2\}^{-2} \tag{60}$$

Therefore

$$(1-t_m)^{3/2}/(4H_m/A) \approx [1 - (\omega/\omega_o)^{1/2}]^{1/2} \tag{61}$$

or

$$1-t_m \approx (4H_m/A)^{2/3}[1 - (\omega/\omega_o)^{1/2}]^{1/3} \tag{62}$$

Note that $t_m \rightarrow 1$ as $\omega \rightarrow \omega_o$; that is, as the frequency increases, the ac irreversibility line will shift toward the H_{c2} line and eventually merge with it. However, if $\omega > \omega_o$ the peak should fail to appear. Under such circumstances, an alternative value must be chosen for the scaling factor x_m (or characteristic frequency ω_o), but this will mean loss of the "peak" as a discriminator, which practically renders the technique unreliable.

Now consider the flux flow case. From conventional flux flow theory,[51] one has:

$$\rho_{ff}(T,H)/\rho_n \approx H/H_{c2}(T) \tag{63}$$

In the Ginzburg-Landau approximation, $H_{c2}(T) = H_{c2}(0)\,(1-t)$, where $t = T/T_c(H = 0)$; therefore:

$$\rho_{ff}(T,H)/\rho_n = H/[H_{c2}(0)(1-t)] \tag{64}$$

Since at $t = t_m$, $H = H_m$, one has, via Eq. 55, that:

$$H_m/[H_{c2}(0)(1-t_m)] \approx \omega/\omega_o$$

or

$$1\text{-}t_m \approx [H_m/[H_{c2}(0)](\omega_o/\omega) \tag{65}$$

Again, $H_m = H_{dc} + H_{ac}$. Increases in ac amplitude will thus shift the T_m toward lower temperatures. Decreasing the sample size will follow the same trend. Note that the frequency effect takes a different form than the thermally activated process, but in general, an increase in frequency will cause T_m to shift to higher temperature. Here the controversial power law for the ac irreversibility line $1-t \propto H^\alpha$ takes $\alpha = 1$.

VIII. Summary

The ac magnetic properties of high-temperature superconductors depend on the magnitude of the dc bias field as well as the amplitude and frequency of the alternating field. They also depend on the sample geometries and sizes. Metallurgical imperfections in the material could significantly affect the phase information and the peak position of quadrature signal on the H–T plane. In the regime where flux motion can be neglected, the critical state model was able to account for the basic characteristics of the hysteretic behaviors. In the weakly pinned regime, flux dynamics could be understood using classical electrodynamics for normal metals in conjunction with various flux motion resistivity models. Flux-motion-induced finite resistivity could lead to significant skin effect where the associated pickup voltage is in quadrature to the true magnetic contribution from magnetic hystereses, which could therefore result in misleading phase information. It is recommended that the ac susceptibility data be specified with sample geometry, sample size, measuring temperature, ac amplitude and frequency as well as dc field strength.

A scaling law of flux-motion resistivity based on the peak position of the quadrature component of the ac measurements has been formulated to account for the frequency, field, and sample size effects on the ac irreversibility lines. It was argued that the ac irreversibility line was a demarcation line separating different flux vortex regimes in accordance with the dynamic responses of vortices to external excitations. It is different from a dc irreversibility line that separates vortex liquid from vortex solid on an H–T phase diagram. Below the ac irreversibility line, pervasive vortex motions follow the ac field, sweeping in and out, whereas beyond it, only those within the skin depth are affected. The demarcation occurs at the (H,T) line, where the above two extremes are balanced and a maximal energy dissipation status is reached. Estimations of physical parameters are in semiquantitative agreement with the experimental data. Generally speaking, the ac technique is a dynamic measurement that is different from its dc counterpart. Comparisons between them can only be done at extremely low frequencies and low ac amplitudes in the strong pinning regime for the high T_c superconductors. Application of the thermally activated flux-motion and traditional flux flow models to the scaling law allows one to predict the general trends of ac amplitude, frequency, dc bias, and sample size effects on the ac measurements.

IX. Acknowledgments

Technical assistance and useful conversations with Drs. K.V. Rao (the Royal Institute of Technology,. Sweden), J.L. McArdle (3M Center), T.R. Werner, J.E. Lenz, and C.J. Torng (Honeywell) are appreciated.

This work is supported by the Defense Advanced Research Projects Agency through the Office of Naval Research under contract No. N00014-88-0753.

X. References

1. Very instructive treatments regarding eddy current losses and associated phase shifts can be found in K.L. Scott, *Proc. Inst. Radio Engineers* 18.10 (1930): 1750.
2. Rigorous classic treatments on eddy currents can be found in references quoted by R.M. Borzorth, *Ferromagnetism*. Princeton: D. Van Nostrand Co., Inc., 1951: 490, 769.
3. R.A. Hein and R.L. Falge, *Phys. Rev.* 123.2 (1961): 407.
4. E. Maxwell and Myron Strongin, *Phys. Rev. Lett.* 10.6 (1963): 212.
5. C.P. Bean, *Rev. Mod. Phys.* 36 (1964): 31.
6. J.E. Evetts, D. Dew-Hughes, and A.M. Campbell, Phys. Lett. 16.2 (1965): 113.
7. R.M. Easson and P.H. Halwiczka, *Phys. Lett.* 20.4 (1966): 333.
8. R.W. Rollins and J. Silcox, *Solid State Comm.* 4 (1966): 323.
9. B.W. Maxfield and E.F. Johnson, *Phys. Rev. Lett.* 16.15 (1966): 652.
10. J.I. Gittleman and B. Rosenblum, *Phys. Rev. Lett.* 16.17 (1966): 734.
11. P. Alais and Y. Simon, *Phys. Rev.* 158.2 (1967): 426.
12. R.W. Rollins and J. Silcox, *Phys. Rev.* 155.2 (1967): 155.
13. H.J. Fink, *Phys. Rev.* 161.2 (1967): 417.
14. R.M. Eason and P.H. Halwiczka, *Brit. J. Appl. Phys.* 18 (1967): 1237.
15. R.A. Hein and P.H.E. Meijer, *Phys. Rev.* 179.2 (1969): 497.
16. D.W. Deis, J.R. Gavler, C.K. Jones, and A. Patterson, *J. Appl. Phys.* 42.1 (1971): 21.
17. S.T. Sekula, *J. Appl. Phys.* 42.1 (1971): 16.
18. A.M. Campbell and J.E. Evetts, *Adv. Phys.* 21 (1972): 199.
19. J.F. Wagner and R.W. Rollins, *J. Appl. Phys.* 44.4 (1973): 1778.
20. F. Vidal, *Phys. Rev.* B8.5 (1973): 1982.
21. R.W. Rollin, H. Kupfer, and W. Gey, *J. Appl. Phys.* 45.12 (1974): 5392.
22. A. Feu and A. Trueba, *J. Appl. Phys.* 45.2 (1974): 2286.
23. D.J. Griffith, C.C. Koch, and J.P. Charlesworth, *Phil. Mag.* 33.3 (1976): 505.
24. J.C. Clem, *Phys. Rev.* B14.5 (1976): 1893.
25. T. Ishida and H. Mazaki, *Phys. Rev.* B20.1 (1979): 131.
26. B.J. Dalrymple and D.E. Prober, *Rev. Sci. Instrum.* 55.6 (1984): 958.
27. R.B. Goldfarb and J.V. Minervini, *Rev. Sci. Instrum.* 55.5 (1984): 761.
28. R.B. Goldfarb and A.F. Clark, *IEEE. Trans.* 21.2 (1985): 332.
29. R.A. Hein, *Phys. Rev.* B33.11 (1986): 7539.
30. A.P. Malozemoff, T.K. Worthington, Y. Yeshurun, and F. Holtzberg, *Phys. Rev.* B38.10 (1988): 7203.
31. D.G. Xenikos and T.R. Lemberger, *Rev. Sci. Instrum.* 60.5 (1989): 831.
32. D.X. Chen, J. Nogues, and K.V. Rao, *Cryogenics* 29.8 (1989): 800.
33. A.P. Malozeboff, in *Physical Properties of High Temperature Superconductors I*, D.M. Ginsberg (ed.). Singapore: World Scientific, 1989: 71.
34. K.-H. Muller, *Physica* C168 (1990): 585.
35. T. Ishida and R.B. Goldfarb, *Phys. Rev.* B41.13 (1990): 8937.
36. Q.Y. Chen, J.L. McArdle, T.R. Werner, and J.E. Lenz, *Appl. Phys. Lett.* 57.24 (1990): 2603.
37. L.T. Sagdahl, S. Gjolmesli, T. Laegreid, K. Fossheim, and W. Assmus, *Phys. Rev.* B42.10 (1990): 6797.
38. L. Civale, A.D. Marwick, M.W. McElfresh, T.K. Worthington, A.P. Malozemoff, and F.H. Holtzberg, *Phys. Rev. Lett.* 65.9 (1990): 1164.
39. D.-H. Wu and S. Sridhar, *Phys. Rev. Lett.* 65.16 (1990): 2074.
40. S.L. Shinde, J. Morrill, D. Goland, D.A. Chance, and T. McGuire, *Phys. Rev.* B41.13 (1990): 8838.
41. J.H. Claassen, M.E. Reeves, and R.J. Soulen, *Rev. Sci. Instrum.* 62.4 (1991): 996.
42. J. Emmen, V. Brabers, and W. de Jonge, *Physica* C176 (1991): 137.
43. L. Civale, T.K. Worthington, and A. Gupta, *Phys. Rev.* B43.7 (1991): 5425.
44. K.A. Muller, M. Takashige, and J.G. Bendorz, *Phys. Rev. Lett.* 58.11 (1987): 143.
45. M. Tinkham, *Phys. Rev.* 129.6 (1963): 2413.
46. A.L. Fetter, *Phys. Rev.* 159.2 (1967): 330.
47. F.E. Harper and M. Tinkham, *Phys. Rev.* 172.2 (1968): 441.
48. G.D. Cody and R.E. Miller, *Phys. Rev.* 173.2 (1968): 481–493 and 494–503.
49. A.L. Fetter, *Phys. Rev.* B22.3 (1980): 1200.
50. J. Claassen, this proceedings.
51. M. Tinkham, *Introduction to Superconductivity*. Malabar, FL: R.E. Krieger Pub. Co., 1975: 171.

52. P. Horowitz and W. Hill, *The Art of Electronics*. 2nd ed. Cambridge: Cambridge Univ. Press, 1989: 641.
53. Y.B. Kim, C.F. Hempstead, and A.R. Strnad, *Phys. Rev.* 129 (1963): 528.
54. P.W. Anderson and Y.B. Kim, *Rev. Mod. Phys.* 36 (1964): 39.
55. D.G. Xenikos and T.R. Lemberger, *Phys. Rev.* B41.1 (1990): 869.
56. T. Ishida and H. Mazaki, *J. Appl. Phys.* 52.11 (1981): 6798.
57. C.D. Jeffries, Q.H. Lam, Y. Kim, C.M. Kim, A. Zettle, and M.P. Klein, *Phys. Rev.* B39.16 (1989): 11526.
58. Q.H. Lam and C.D. Jeffries, *Phys. Rev.* B39.7 (1989): 4772.
59. C.D. Jeffries, Q.H. Lam, Y. Kim, C.M. Kim, L.C. Bourne, and A. Zettle, *Phys. Rev.* B37.16 (1988): 9840.
60. L. Ji, R. Sohn, G.C. Spalding, C.J. Lobb, and M. Tinkham, *Phys. Rev.* B40.16 (1989): 10936.
61. Q.H. Lam, Y. Kim, and C.D. Jeffries, *Phys. Rev.* B42.7 (1990): 4846.
62. T. Ishida and R.B. Goldfarb, *Phys. Rev.* B41.13 (1990): 8937.
63. W.R. Smythe, *Static and Dynamic Electricity*. 3rd ed. New York: McGraw-Hill, 1968: 198, 371.
64. L.D. Landau, E.M. Lifshitz, and LP. Pitaevskii, *Electrodynamics of Continuous Media*, 2nd ed. Oxford: Pergamon Press, 1984: 199–224.
65. J.D. Jackson, *Classical Electrodynamics*. 2nd ed. New York: John Wiley & Sons, 1976: 296, 472.
66. Y.B. Kim, *J. Appl. Phys.* 42.2 (1971): 550.
67. J.C. Clem, this proceedings.
68. M. Abramowitz and I. A. Stegun, *Handbook of Mathematical Functions with Formulas, Graphs, and Mathematical Tables*. New York: John Wiley & Sons, 1972: 348.
69. I. Bunget and M. Popescu, *Physics of Solid Dielectrics*. Amsterdam: Elsevier, 1984: 282.
70. V.B. Geshkenbein, V.M. Vinokur, and R. Fehrenbacher, *Phys. Rev.* B43.4 (1991): 3748.
71. R.B. Flippen and T.R. Askew, *J. Appl. Phys.* 67.9 (1990): 4515.
72. M. Nikolo and R.B. Goldfarb, *Phys. Rev.* B39.10 (1989): 6615.
73. T.T. Palstra, B. Batlogg, R.B. van Dover, L.F. Schneemeyer, and J.V. Waszczak, *Phys. Rev.* B41.10 (1990): 6621.
74. F. Gömöry and P. Lobotka, *Solid State Comm.* 66.6 (1988): 645.
75. S. Takacs and F. Gömöry, *Supercond. Sci. Technol.* 3 (1990): 94.
76. W.J. Gallagher, T.K. Worthington, T.R. Dinger, F. Holtzberg, D.L. Kaiser, and R.L. Sandstrom, *Physica* 148B (1987): 228.
77. M.R. Beasley, R. Labusch, and W.W. Webb, *Phys. Rev.* 181.2 (1969): 682.
78. P.H. Kes, J. Arts, J. van der Berg, C.J. van der Beek, and J.A. Mydoh, *Supercond. Sci. Technol.* 1 (1989): 242.
79. M. Tinkham, *Phys. Rev. Lett.* 61.14 (1988): 1685.
80. N.C. Yeh, *Phys. Rev.* B43.1 (1991): 523.

AC SUSCEPTIBILITY STUDIES OF SUPERCONDUCTING PROPERTIES UNDER HIGH HYDROSTATIC PRESSURE

J.S. Schilling[1], J. Diederichs[1], S. Klotz[1,2], and R. Sieburger[2]

[1]Department of Physics, Washington University, St. Louis, Missouri
[2]Sektion Physik, University of Munich, Munich, Germany

INTRODUCTION

Before revealing why the ac susceptibility technique is of particular importance in high pressure investigations, we would like first to briefly discuss the types of changes in the magnetic and superconducting properties which can occur when a solid is subjected to high pressures. The changes observed for the elemental solids across the periodic table are representative for all compounds and alloys. The only elemental solids which exhibit strong magnetic behavior, such as a Curie–law susceptibility or magnetic ordering, are solid oxygen, the 3d transition metals (Cr, Fe, Co, Ni), the 4f rare earth metals (Ce, Pr, Nd, Sm, Eu, Gd, Tb, Dy, Ho, Er, Tm, Yb), and the heavy 5f actinides (Cm, Bk, Cf, Es, Fm, Md, No) [1]. An atomic orbital will contribute to strong forms of magnetism only if the orbital is partially filled and does not overlap or hybridize too extensively with the orbitals of neighboring atoms. This explains why the above elements with their relatively well localized 3d–, 4f–, and 5f–orbitals are strongly magnetic. The remaining elements exhibit only weak forms of magnetism such as Pauli paramagnetism, Van Vleck paramagnetism, and Larmor diamagnetism.

Since subjecting a strongly magnetic solid to high pressures will bring its constituent atoms closer together, thus increasing nearest–neighbor orbital overlap, it is clear that pressure will normally weaken magnetism. It is thus no surprise that the Curie temperature of the weak itinerant ferromagnet $ZrZn_2$ falls rapidly to zero under pressure as the magnetism is destroyed [2]. In the rare earth elements, on the other hand, the 4f–orbitals are located so close to the nucleus that applying pressure at first leads to (multiple) valence changes, where an electron is squeezed from the 4f–orbital into a valence state, before the 4f–4f overlap becomes sufficient to destroy the magnetism. Such valence changes of themselves can cause a drastic variation in the magnetic properties, leading to such fascinating phenomena as valence fluctuations and heavy–fermion or Kondo–lattice behavior.

We now consider the superconducting state. What role do high pressure investigations play here? Firstly, the application of pressure has lead to the creation of superconductivity in a large number of elements and compounds, including the 16 elements Si, P, S, Ca, Sc, Ge, As, Se, Sr, Y, Sb, Te, Cs, Ba, Bi, and Lu, as well as the organic metals $(TMTSF)_2PF_6$ (the first organic superconductor, where $T_c \simeq 1.1$ K at 0.65 GPa) [3] and $(BEDT{-}TTF)_2Cu[N(CN)_2]Cl$ (the organic superconductor with the highest transition temperature $T_c \simeq 12.7$ K at 0.03 GPa) [4]. Secondly, a large pressure derivative of T_c, either positive or negative, is a signal that the system under study may well be capable of reaching higher values of T_c at ambient pressure, if only the system is suitably modified. The observation [5] of the very large pressure deriva-

Magnetic Susceptibility of Superconductors and Other Spin Systems
Edited by R.A. Hein *et al.*, Plenum Press, New York, 1991

tive $dT_c/dP \simeq +9$ K/GPa for $La_{2-x}Ba_xCuO_4$ with $T_c(0) = 32$ K led Wu et al.[6] to the substitution of the smaller Y for La and the discovery of the 90 K superconductor $YBa_2Cu_3O_{7-y}$. The third application of high pressure is as a continuous well–defined parameter to check theories of superconductivity. Comparing the pressure dependence of normal and superconducting properties can also give information on the mechanisms responsible for the superconductivity.

Many of the papers presented at this workshop consider the important information on the superconducting and magnetic state of solids which can be gained from a precise determination of the real and imaginary parts of the ac susceptibility as a function of temperature and applied magnetic field. In this paper we would like to emphasize that the ac susceptibility technique is of particular value in high pressure studies. There are two principle reasons for this which are evident to us. Firstly, many experimental techniques (e.g. electrical resistivity, thermopower, Hall effect) require high quality electrical contacts to the sample. Securing adequate contacts is particularly difficult under high pressure conditions because of the tiny size of many pressure cells, particularly those which are capable of reaching the highest pressures. The fact that the ac susceptibility technique does not require electrical connections directly to the sample is an advantage of inestimable value. Secondly, the volume of the sample is only a very small fraction of the volume of the pressure clamp. In the pressure range to 10 GPa (100 kbar or 100,000 atmospheres), which results in a change of sample volume by typically 5–10%, the ratio of these volumes is only approximately 1:100,000! Magnetic and superconducting properties are often measured in SQUID, vibrating–sample, and Faraday magnetometers. Using these techniques the properties of both the sample and the pressure cell are measured together, i.e. $M_{meas}(T,P,H) = M_{sample}(T,P,H) + M_{cell}(T,P,H)$. To determine M_{sample}, both M_{meas} and M_{cell} must be determined separately and subtracted from one another. Since M_{cell} is often much larger than M_{sample}, it is very difficult to determine the latter to high accuracy. The ac susceptibility technique, however, allows the pick–up coil to be wound directly around the sample itself in the high pressure environment, thus minimizing contributions from the pressure cell.

EXPERIMENTAL

There are two basic kinds of stress which can be applied to a solid: (1) hydrostatic stress, such as transmitted by a liquid or gas, which is everywhere normal to the surface of the sample and constant in magnitude, and (2) uniaxial stress which is applied to the sample in a particular direction. Whereas for a dense sample there is no limit to the magnitude of hydrostatic pressure it can withstand, only relatively small uniaxial stresses can be applied without permanently deforming the sample. The use of hydrostatic pressure carries with it the difficulty of totally containing the gaseous or liquid pressure medium, particularly when electrical leads are brought into the pressure chamber. A relatively simple pressure technique, termed "quasihydrostatic", utilizes a solid pressure medium such as steatite or NaCl. The nature of the applied "quasihydrostatic" pressure is neither purely hydrostatic nor purely uniaxial, but a combination of both. During the course of a quasihydrostatic pressure experiment the samples are exposed to shear stresses and cold worked. Any coil system embedded in the solid pressure medium will suffer a change in its geometric factors and a degradation in secondary–coil compensation. Quasihydrostatic pressure techniques are often used by groups striving to reach the highest pressures. Lotter and Wittig [7] have developed an interesting quasihydrostatic pressure cell to 10 GPa with vaseline as pressure medium where one secondary coil with 50 turns is located inside the pyrophyllite gasket with the second compensating secondary loop outside.

In a hydrostatic pressure experiment the pressure is changed at a temperature above the melting curve of the pressure medium. The pressure medium freezes upon cooling. However, the magnitude of the shear stresses on the sample generated by the freezing process are far smaller than those that occur in a quasihydrostatic experiment. To which pressures does a pressure medium remain fluid and capable of transmitting hydrostatic pressures? The answer is different for each substance and is strongly temperature dependent; however, the last fluid substance to become solid at room temperature (295 K) under pressure is liquid He at 11.8 GPa. Although He is solid above this pressure, it is still far softer than conventionally used pressure fluids such as 1:1 n–pentane iso–amyl or 4:1 methanol–ethanol.

In general, it is always preferable to use hydrostatic rather than quasihydrostatic high pressure techniques. Whether or not the two techniques lead to different experimental results depends on the phenomena and materials under investigation. In superconductors, either technique leads to the same pressure dependence of the superconducting transition temperature $T_c(P)$ in Pb [8], but not so in the A–15 compound V_3Si [9] or the thorium–phosphide–structure compound La_3S_4 [10] where T_c initially increases under hydrostatic pressure, but decreases under quasihydrostatic pressure!

The most commonly used hydrostatic pressure device is the so–called "piston—cylinder" technique [11] where two pistons in a bore compress a teflon bucket containing pressure fluid, coil system, sample, and manometer to pressures as high as 2 – 2.5 GPa. Since the teflon bucket has an inner diameter of typically 5–10 mm, the coil system for an ac susceptibility measurement can be easily placed inside. To avoid the sealing difficulties associated with bringing wires into the high pressure chamber, some groups move the coil system outside the high pressure region, thereby suffering a reduction in the signal/noise ratio and the ability to cleanly separate χ' and χ''. In a piston–cylinder device the pressure is normally changed by removing the pressure clamp from the cryostat, placing it under a hydraulic press, and changing the force applied to the pistons. The pressure can be estimated by dividing this force by the area of the piston ($P_{ex} = F/A$) or determined quantitatively in the low temperature region by using a superconducting manometer in the pressure cell to obtain P_{in}[12]. A plot of P_{ex} versus P_{in} for increasing and decreasing pressure shows a hysteresis which arises from internal friction in the pressure cell.

Helium–Gas Cell

A far more versatile technique for the pressure range 0–1.5 GPa is the He–gas pressure cell shown in Fig. 1 (Unipress, model GLC10) where the gas compressor and manometer (usually a manganin gage which can resolve a change in pressure of only 0.002 GPa) remain at room temperature and are connected to the Cu–Be pressure vessel in a cryostat at low temperature via a Cu–Be capillary tube [13]. The pressure can be changed at any temperature above the melting curve of He which is at 78 K for 1.5 GPa or 38 K for 0.5 GPa; below the melting curve the capillary freezes shut with solid He, thereby preventing any further flow of gas between the pressure vessel and the external gas compressor. The pressure vessel can remain in the cryostat except when changing samples or replacing a defective seal. Since up to 12 Cu wires can be brought into the pressure chamber, a wide variety of measurements can be simultaneously carried out on one or more samples, including ac susceptibility, electrical resistivity, Hall effect, and other transport properties.

The χ_{ac} coil system used in the present studies (see Fig. 1) consists of two compensated secondary coils (each 485 turns of 30 μm Cu wire) surrounded by the primary coil (740 turns of 60 μm Cu wire). The sample is placed in the center of one of the secondary coils. A Stanford Research model 530 dual phase lock–in amplifier is used to detect the signal from the compensated secondary coils utilizing the circuit diagram shown in Fig. 2. A highly stable constant–current oscillator (Optimation Inc., model RCD–9) sends an excitation current with several mA amplitude through R_1, L and the primary coil. The voltage drop across R_1 serves as a reference voltage for the lock–in and allows an accurate determination of the excitation current. An important part of the circuit is a variable impedance unit consisting of a variable inductance L coupled into the excitation current circuit, a variable resistor R_2, and a fixed capacitor C. In general, the signals from each of the two secondary coils will differ somewhat in magnitude and phase; a phase shift can occur if each secondary coil couples differently with metallic parts of the pressure cell lying nearby. The variable impedance unit allows one to generate a compensation signal with arbitrary magnitude (through variation of L) and phase (through variation of R_2) so as to exactly compensate the net signal from the secondary coils. A complete compensation allows the lock–in amplifier to be operated in an adequate sensitivity range. Since in the helium–gas cell the coupling between the secondary windings and the pressure vessel is minimal in the frequency range below 1500 Hz, and thus the individual signals from the two secondary coils are in phase, R_2 and C were removed from the circuit in Fig. 2. A small change in L represents a pure change in inductance which should only cause a signal change in the χ' channel. The internal phase–angle adjustment in the lock–in amplifier is utilized to meet this requirement. The temperature of the sample inside the pressure cell can be accurately determined by two pairs of calibrated Pt– and Ge–resistors located pairwise directly above and below the pressure cell.

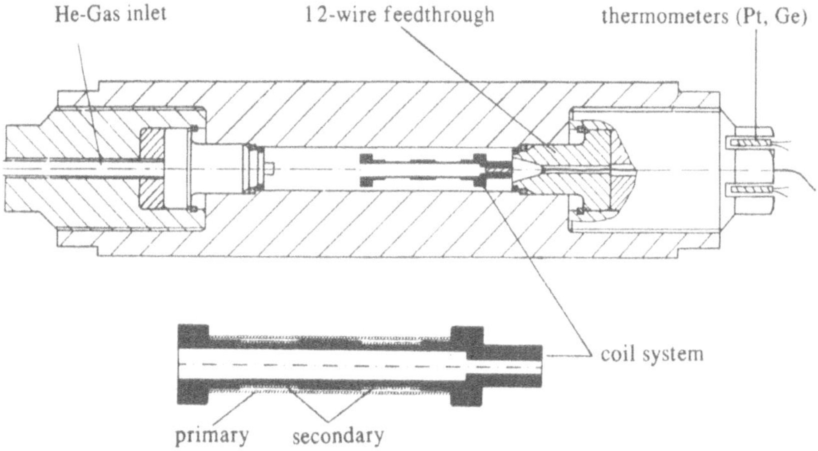

Fig. 1. He–gas pressure vessel to 1.5 GPa (7 mm I.D. and 28 mm O.D.) made from Cu–Be alloy by Unipress (see Ref. 13). An enlarged view of the 24 mm long coil system for 2 mm dia. samples is also shown.

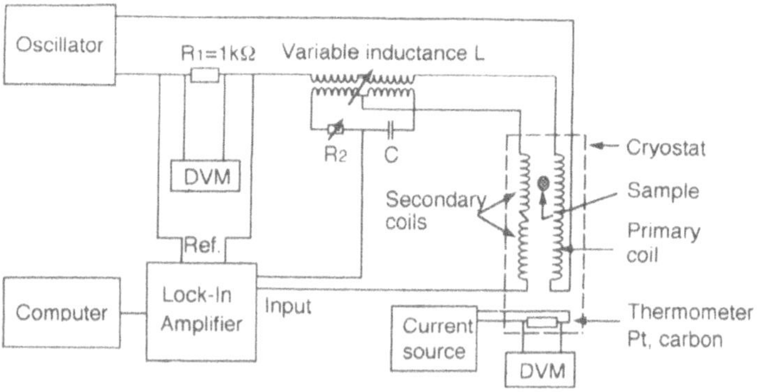

Fig. 2. Circuit diagram for ac susceptibility and temperature measurement systems. See text for details.

To test the coil system the temperature dependence of χ_{ac} at 0.5 Oe for a 1.85 mm dia. Pb–sphere was measured as a function of frequency, as shown in Fig. 3. At temperatures well below the superconducting transition temperature $T_c = 7.2$ K, a large negative inductive signal $\chi'(T)$ is observed which for low frequencies falls to nearly zero above T_c. At higher frequencies the sharp change in $\chi'(T)$ at T_c decreases in size because of the increasing importance of the shielding in the normal state. $\chi''(T)$ passes through a maximum near that temperature T_m where the applied field penetrates to the center of the sample. The fact that T_m shifts to higher temperatures with increasing frequency is also a result of the increased shielding at higher frequencies in the normal state. Both $\chi'(T)$ and $\chi''(T)$ are universal functions of the ratio of the sample dimension "a" to the screening length "δ", the maximum in $\chi''(T)$ occurring when $a/\delta = 2.41$ [14]. We have also calculated the absolute magnitude of $\chi'(T)$ and $\chi''(T)$ by taking into account the geometry of the coil system and the demagnetization factor (D=1/3) of a sphere [15]. Within the accuracy of this determination (~10%) the shielding effect is 100% so that we set $\chi'(5\text{ K}) = -1$ (SI units) in Fig. 3.

Azevedo et al.[16] have developed a useful technique in which the low–field ESR resonance of a single–crystal sample is measured by winding a radio frequency coil directly onto the sample mounted inside a He–gas pressure cell. The ESR signal and the measured absorption in the rf coil are strongly dependent upon the magnetic state of the sample. With this technique they were able to study the pressure dependence of both the superconducting and spin–density wave transitions in $(TMTSF)_2PF_6$.

Metal–Gasket Cell

One of the few drawbacks of the He–gas technique is the limitation of the pressure range to 1.5 GPa. To reach appreciably higher pressures it is necessary to press opposing anvils together. More than a decade ago Fasol and one of the present authors (J.S.S.) [17] modified the so–called "metal–gasket" technique, which was well known in conjunction with the diamond–anvil cell, to allow electrical leads to be brought through the gasket into the high pressure chamber, as shown in Fig. 4. The use of tungsten carbide anvils instead of diamonds allowed a scaling up of the cell by roughly an order magnitude. In this technique two WC–anvils press into a gasket made of a high tensile–strength alloy such as Cu–Be. Six Cu leads are brought into the 2mm dia. high pressure bore in the middle of the gasket. Although initially we used the cell to carry out measurements of the electrical resistivity, a miniature self–

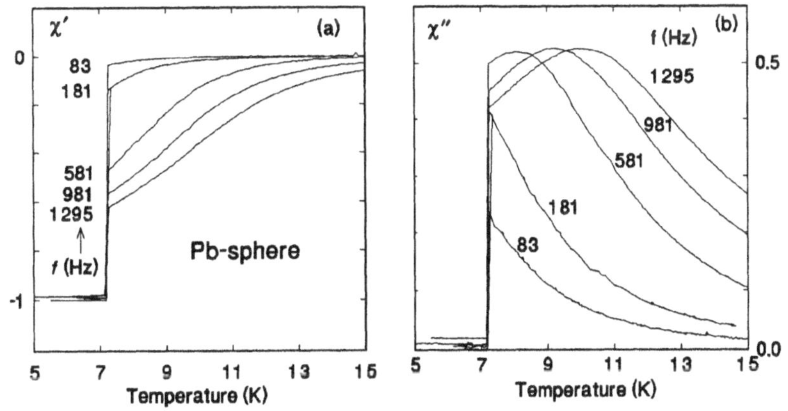

Fig. 3. Temperature dependence of the (a) real χ' and (b) imaginary χ'' parts of the ac susceptibility in SI units for a 1.85 mm dia. Pb—sphere in coil system from Fig. 1 as a function of frequency.

supporting coil system was soon developed which allowed ac susceptibility studies [18]. A 4:1 mixture of methanol–ethanol serves as pressure medium. As shown in Fig. 4, a small piece of Pb is placed next to the sample to serve as a superconducting manometer. The superconducting transition temperature of Pb decreases under pressure at the rate $dT_c/dP = -0.365$ K/GPa in the range 0 – 5 GPa [12]. The Cu wire in the primary coil winding can also be used as a resistive manometer [19] over a wide temperature range, as will be discussed in a future publication. The usable pressure range depends on a number of factors, including the gasket material and the thickness and dimensions of the WC–anvils. Pressures near 5 GPa have been reached over the entire temperature range below 300 K using a Cu–Be gasket. The use of a special steel alloy extended this range to 12 GPa at room temperature; however, failure occurred at low temperatures, presumably due to gasket embrittlement. Pressures over 13 GPa would start to indent the WC anvils and thus limit the pressure.

The self–supporting coil system used in this technique is also shown in Fig. 4. Since the coil system must be relatively flat to avoid getting squeezed by the anvils, the compensated secondary coils are wound on top of each other, instead of side–by–side as in Fig. 1, to save height. The coil system typically has the dimensions 0.8 mm O.D., 0.3 mm I.D. and 0.8 mm height with approximately 100 windings in the secondary coils and 30 windings in the primary coil; varnished Cu–wire with 30 μm dia. is

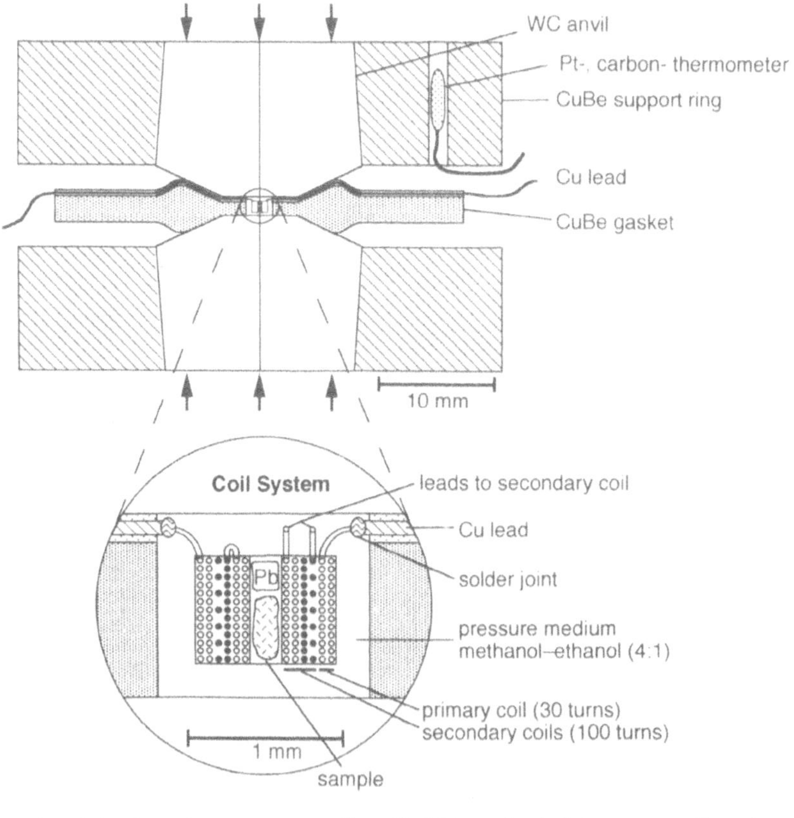

Fig. 4. Metal–gasket pressure cell to 10 GPa and miniature χ_{ac} coil system.

used for the windings. The lock–in amplifier used here is EG&G model 5210. For the frequency used (5.05 kHz) the coupling between the coil system and the metal gasket leads to a phase shift of appoximately 2°. This could be compensated for within the lock–in amplifier, so that a complete separation of $\chi'(T)$ and $\chi''(T)$ was possible.

Diamond–Anvil Cell

To appreciably extend the pressure range beyond 10 GPa, it is necessary to use anvil materials of superior hardness to tungsten carbide. Diamond is the hardest substance known, is transparent over a wide spectral range, possesses a very high thermal conductivity, and is electrically insulating which simplifies electrical insulation problems in the pressure cell. Unfortunately, the high cost of diamond dictates the miniaturization of the entire pressure cell. Although several groups have succeeded in bringing electrical leads into the pressure chamber [20], we are not aware that anyone has successfully mounted an χ_{ac} coil system inside the bore of a diamond–anvil gasket which typically has an initial diameter of 0.3 mm; under pressure the bore diameter decreases. Bringing electrical leads through the gasket region would considerably complicate, if not doom, any attempt to use liquid He as pressure medium. For these reasons we have chosen to place the entire χ_{ac} coil system outside the 3 mm dia. gasket, as shown in Fig. 5. The pressure technique itself is identical to that for the metal–gasket cell illustrated in Fig. 4. The diamond–anvil clamp in Fig. 5 was built nearly a decade ago by one of the authors (J.S.S.)[21]. The force pushing the diamond anvils together is provided by a double–diaphragm press [22] which can be pressurized to more than 200 atm. of He gas. The gasket (Cu–Be, Re, or a Ta–W alloy) is loaded at 2.0 K with superfluid helium before sealing off the pressure chamber by pressing the diamond anvils into the gasket. Pressure is only changed at room temperature or above to stay above the melting curve of He. The measurement of the R_1–fluorescence line of ruby chips located both inside and outside the pressure cell allows an accurate pressure determination at all temperatures to within 0.05 GPa [20]. In a test run where only a ruby chip, but no sample, was included in the He–loaded pressure cell, a pressure as high as 23 GPa was reached at room temperature before we lost courage. Much higher pressures should be possible.

As before, the compensated secondary coils are wound on top of one another. The extremely poor filling factor of a sample typically 0.1x0.1x0.02 mm^3 in a secondary coil with 3.5 mm dia. drastically complicates the compensation of the two secondary coils. Even though insulating epoxy is used as a coil form, the inductive coupling of the individual secondary coils with their metallic surroundings, including the metal gasket, is sufficiently different that the net signal exhibits a temperature dependence which is so large as to prevent the resolution of the superconducting transition of the sample. This temperature–dependent background can be significantly reduced by adjusting the resistor R_2 and the inductance L in Fig. 2. Due to the very strong signal from the metal gasket inside the coil system, it was not possible to phase separate χ_{ac} into its real and imaginary parts, but rather the magnitude of χ_{ac} is measured versus temperature. The superconducting transition is barely visible in the measured data, as seen in Fig. 6a. After subtracting off the temperature dependence of the background and expanding the vertical scale, the curve in Fig. 6b is obtained. The superconducting transition is clearly resolved. In fact, changes in T_c as small as 0.1 K at 90 K can be resolved.

RESULTS

Weak–Itinerant Ferromagnets

Before discussing the results of several recent high pressure experiments in the field of superconductivity, we would like to mention in passing one result from studies of magnetism which we feel exemplifies the ability of high pressure investigations to critically test theoretical predictions. As sketched in the Introduction, weak–itinerant ferromagnets such as $ZrZn_2$ or $TiBe_{2-x}Cu_x$ are situated very near the magnetic–nonmagnetic transition; for this reason the application of only very modest pressures is sufficient to destabilize the magnetic state and drive the Curie temperature to zero [2,23]. Such behavior was given a theoretical framework by the following approximate expression proposed by Wohlfarth: $d\ell nT_{curie}/dP \simeq -\alpha/T^2_{curie}$, where α is a slowly varying positive quantity [24]. According to this expression, the decline of T_{curie} under pressure should become more precipitous the smaller T_{curie} is, a trend which has received support from experiment.

Sc_3In is a well known weak–itinerant ferromagnet with $T_{curie} \simeq 5$ K. However, Gardner et al.[25] reported that T_{curie} increases under pressure for the single Sc_3In

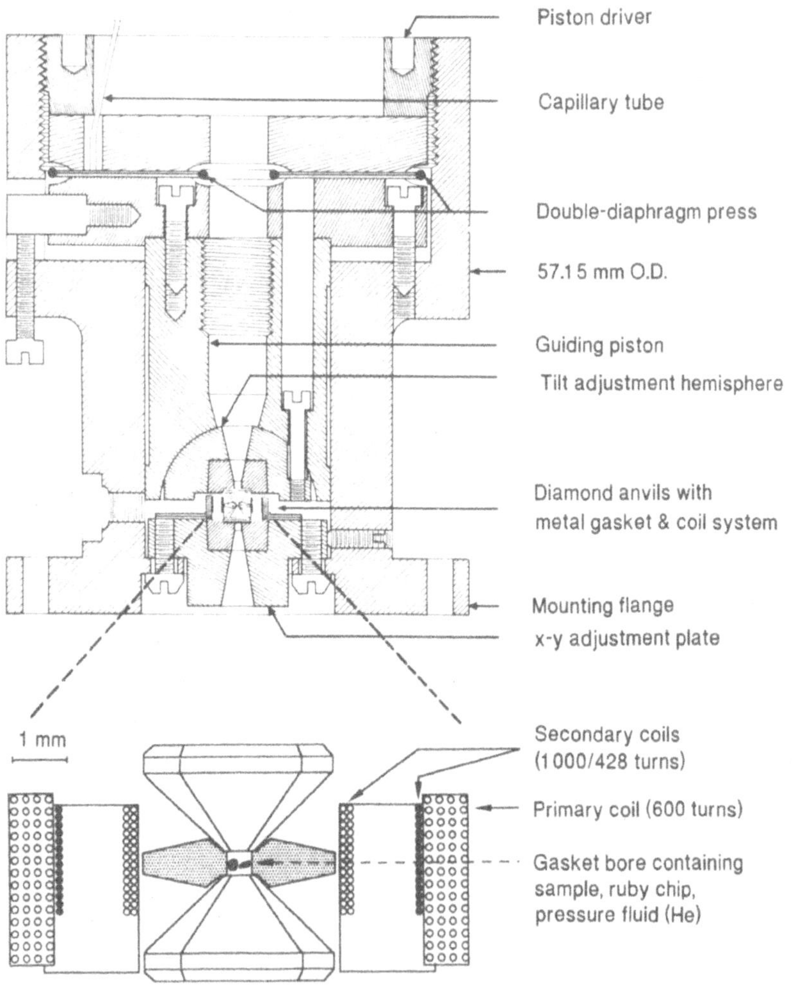

Fig. 5. Diamond–anvil pressure clamp for pressures above 10 GPa. Two 1/6 carat diamond anvils press into a 3 mm O.D. metal gasket with a 0.25 mm dia. bore containing sample, ruby manometer, and liquid He pressure fluid. An enlarged view of the primary(secondary) coil system using 30(16) μm dia. Cu–wire is also shown.

sample they investigated, which clashes with the above theoretical expectation. Utilizing the metal–gasket pressure cell shown in Fig. 4 for ac susceptibility and a Faraday balance magnetometer for magnetization studies, Grewe et al.[26] studied the pressure dependence of $\chi_{ac}(T)$ and M(T,H) on three well–characterized polycrystalline Sc_3In samples and found that both T_{curie} and the magnetization at a given temperature and field increase under pressure. Both results stand in direct contradiction to the predictions of Wohlfarth's theory and show that the validity of this theory is not as general as had been believed.

Conventional Superconductors

We now discuss several high pressure experiments in the field of superconductivity. In simple–metal superconductors like Pb, Hg, Sn, In, Al, Zn, and Ga, the superconducting transition temperature T_c decreases under pressure at the rate $dT_c/dP \simeq -0.2$ to -0.4 K/GPa [27]. In transition–metal superconductors, on the other hand, dT_c/dP can be either positive or negative and take on much larger magnitudes[28]. Why does T_c decrease under pressure for the simple metals? This can be most easily understood by considering a simplified form of the BCS formula: $T_c = \Theta \exp[-M\Theta^2/C]$, where Θ is the Debye temperature, M is the molecular weight, and C is a quantity which is only weakly pressure dependent for simple metals [27]. The Debye temperature increases under pressure as the lattice stiffens and the phonon spectrum is shifted to higher energies. Since the factor Θ^2 in the exponent of the above T_c formula overpowers the prefactor Θ, T_c should decrease as Θ increases. The rate of increase of the Debye temperature with pressure is given by $d\ell n\Theta/dP = \kappa[-d\ell n\Theta/d\ell nV] = \kappa\gamma$, where γ is the Gruneisen constant, κ is the compressibility, and V is the sample volume. For the above elements the Gruneisen constant takes on the values $\gamma \simeq +1.5 - 3$ [27] which means that the Debye temperature increases under pressure about twice as fast as the sample volume decreases. This all means that the superconducting transition temperature of simple metals decreases under pressure because of the increase of the Debye temperature, i.e. because the lattice stiffens under pressure. This is reasonable physically, since the positive ions in a stiffer lattice can

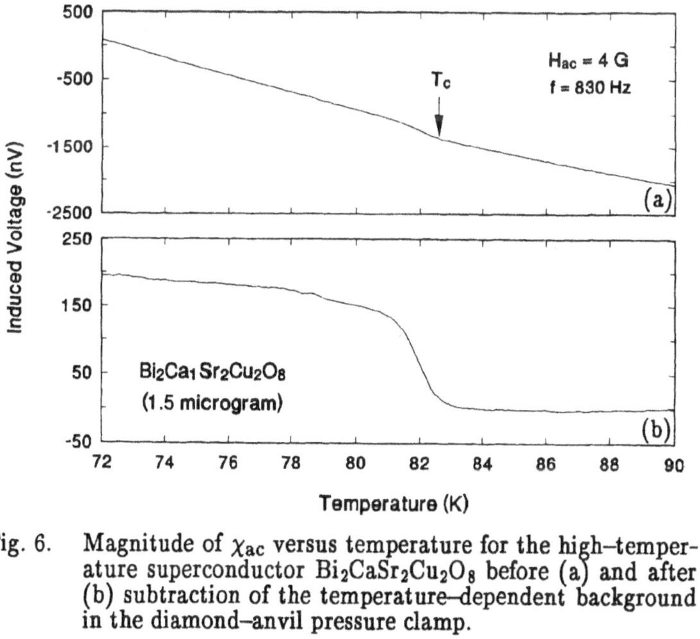

Fig. 6. Magnitude of χ_{ac} versus temperature for the high–temperature superconductor $Bi_2CaSr_2Cu_2O_8$ before (a) and after (b) subtraction of the temperature–dependent background in the diamond–anvil pressure clamp.

not approach each other as closely when an electron flies by as when the lattice is soft. Thus the behavior of T_c under pressure is further evidence that the coupling between electrons in a Cooper pair is electron–phonon in nature involving the crystal lattice.

Organic Superconductors

A decrease of T_c under pressure is also a hallmark of organic superconductors such as $(TMTSF)_2PF_6$ [3], $(TMTSF)_2ClO_4$ [29], and $(BEDT\text{–}TTF)_2I_3$ [30]. Is this an indication that here also the electron–phonon interaction leads to the superconductivity? To address this question we need to examine the magnitude of the pressure derivative of T_c. The results of the ac susceptibility measurements of Sieburger [31] obtained for $(BEDT\text{–}TTF)_2Cu(NCS)_2$ using the He–gas pressure system are shown in Fig. 7. The decrease of T_c with pressure is enormous, $dT_c/dP = -29$ K/GPa, in good agreement with the results of others [32]. Large pressure derivatives have been reported for many other organic superconductors [30]. Since the value of dT_c/dP for organic superconductors is approximately two orders of magnitude larger than for the simple metals, one might jump to the conclusion that a mechanism other than the electron—phonon interaction is responsible for the superconductivity in organic metals. In fact, the very large pressure derivative can be easily understood within an electron–phonon framework: (1) the compressibility $\kappa \simeq 0.1$ GPa^{-1} is approximately an order of magnitude larger than for simple metals, and (2) the Gruneisen parameter $\gamma \simeq 4-6$ is about twice as large as for simple metals. Both factors contribute to a larger pressure derivative of T_c for organic metals [33].

High Temperature Superconductors

Considering the relatively short five year time period since the discovery of high temperature superconductivity, there have been a large number of studies of the pressure dependence of T_c. Although it appears that the sign of dT_c/dP is normally positive for hole–doped and negative for electron–doped materials, and that the magnitude of dT_c/dP is inversely proportional to the value of T_c, there is a large scatter in the reported data [34]. This scatter may be due to differences in pressure technique, i.e. whether hydrostatic or quasihydrostatic, or to differences in the samples studied such as the concentration of oxygen or other components, defect density, whether polycrystalline or single–crystalline, etc. Indeed, for $YBa_2Cu_3O_7$ the value of dT_c/dP is known to vary markedly with the oxygen concentration [35]. The disparity in the values of the pressure derivative is well illustrated by an ac susceptibility study of

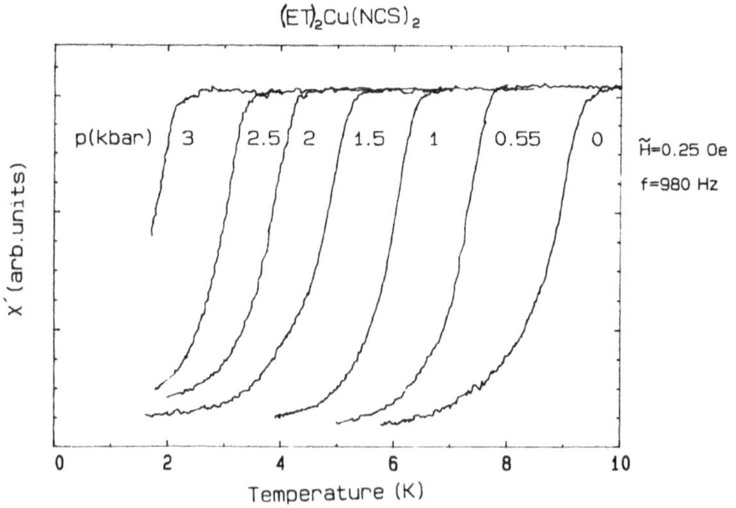

Fig. 7. Superconducting transition of $(BEDT\text{–}TTF)_2Cu(NCS)_2$ at different pressures in the He–gas system (1 GPa = 10 kbar).

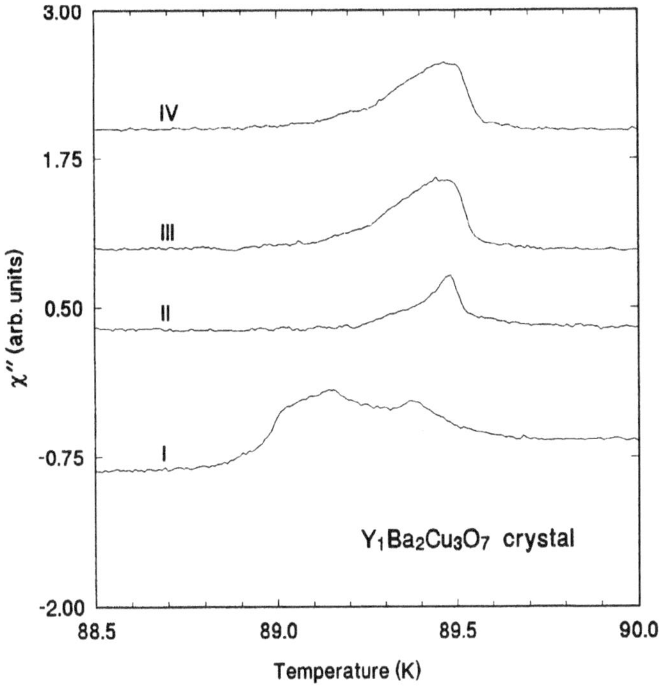

Fig. 8. Temperature dependence of the imaginary part χ'' of the ac susceptibility for a "single crystal" of $YBa_2Cu_3O_7$ before (I) and after (II, III, IV) it broke into three parts.

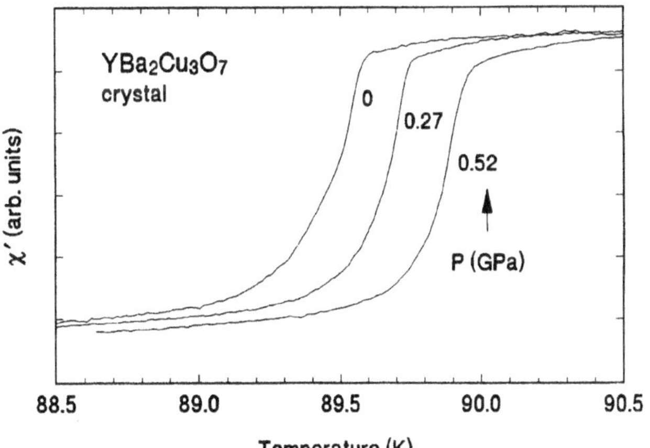

Fig. 9. Temperature dependence of χ' for a $YBa_2Cu_3O_7$ crystal at three different pressures in the He–gas system. See text.

Koch et al.[36] on several single crystals of $YBa_2Cu_3O_7$. For one crystal $T_c(P)$ increased initially with pressure, passing through a maximum near 5 GPa; for two others T_c decreased monotonically with pressure. Koch et al.[36] raised the possibility that the vaseline pressure medium could have exerted shear stresses on the sample under high pressure.

To try to "set things straight", one of the authors (R.S.) decided to use our best pressure technique and study $T_c(P)$ to 0.6 GPa on $YBa_2Cu_3O_7$ crystals using the He–gas system (see Fig. 1). First, a crystal was chosen and the ac susceptibility was measured at ambient pressure, yielding curve I in Fig. 8. A double–peak structure in $\chi''(T)$ is clearly seen. Under pressure the double–peak structure was observed to shift bodily to higher temperature at the rate +0.7 K/GPa. After this measurement the sample was removed from the pressure cell and found to have broken apart into three pieces. In Fig. 8 we also show the $\chi''(T)$–data on these three "subcrystals" as the curves II, III, and IV. It is seen that here the whole is not the sum of its parts! The lower–temperature peak in curve I is missing in the three other curves. We ascribe this peak to weak–link behavior between the three subcrystals before they separated. We next measured $T_c(P)$ on one of these subcrystals; the results are shown in Fig. 9. The unusually sharp superconducting transition is seen to shift to higher temperatures under pressure at the rate $dT_c/dP = +(0.7 \pm 0.2)$ K/GPa, exactly as found for the larger "mother" crystal above. Measurements at ten different values of the pressure taken with both increasing and decreasing pressure reveal $T_c(P)$ to vary in a completely reversible fashion. A further measurement on another crystal from a different batch gave an identical pressure dependence [37].

In Fig. 10 we display previously published data [38] on another $YBa_2Cu_3O_7$ crystal where we found T_c to remain constant under hydrostatic pressure, but $\chi''(T)$ to split reversibly into two peaks under pressure; the $\chi'(T)$–transition developed a shoulder under pressure. With increasing field amplitude the lower peak in $\chi''(T)$ shifts rapidly to lower temperatures and broadens [37]. While it is certainly possible that pressure induces a phase separation, perhaps due to oxygen reordering, on the

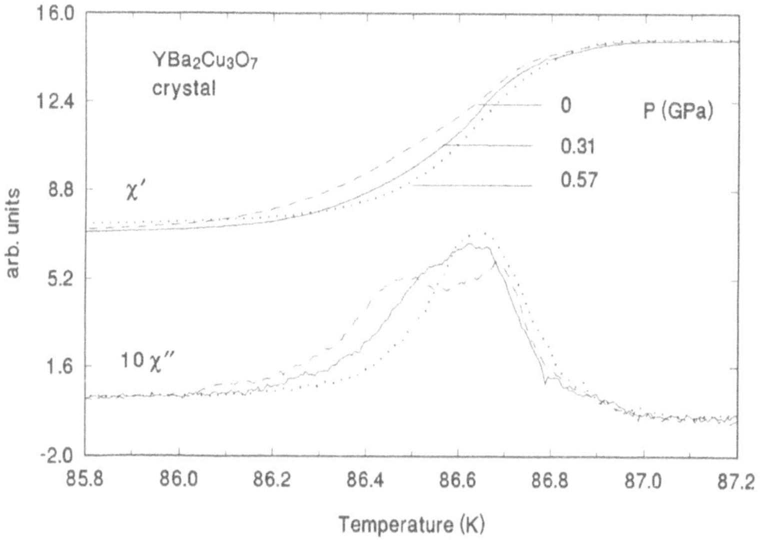

Fig. 10. Temperature dependences of the real and imaginary parts of the ac susceptibility of $YBa_2Cu_3O_7$ at three pressures in the He–gas system.

basis of the above studies it is tempting to conjecture that there are microcracks in the crystal which reversibly open and close as external pressure is applied. It is appropriate here to consider (Fig. 11) some beautiful experiments by Dian–lin et al.[39] on tiny (0.15x0.15x0.03 mm^3) $YBa_2Cu_3O_7$ crystals where a double–step structure in $\chi'(T)$ is seen to go away under pressure. The superconducting transition was measured by a modified ac bridge working at 5 MHz with an excitation field of less than 2 mG. To improve the sensitivity, both arms of the secondary coil system were placed in the pressure cell and tuned to resonance. The results in Fig. 11 are interpreted to mean that there is weak–link behavior even within a single crystal which apparently is suppressed under pressure, i.e. the contacts between the subcrystals in the sample improve under pressure. We will see below that similar behavior is also found for polycrystalline material where the grain boundaries form weak links.

To be able to identify systematics in $T_c(P)$ it would be of obvious benefit to expand the pressure region available. The diamond–anvil apparatus shown in Fig. 5 is capable of generating pressures one to two orders of magnitude higher than those accessible with the He–gas system. In Fig. 12 we show $\chi(T)$ for a $YBa_2Cu_3O_7$ single

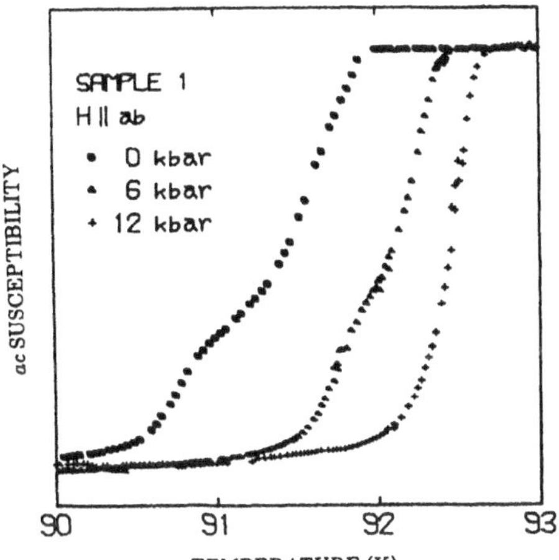

Fig. 11. Real part of the ac susceptibility versus temperature for a $YBa_2Cu_3O_7$ crystal at three pressures (1 GPa = 10 kbar) from Ref. 39. Under pressure the low–temperature step disappears.

crystal [40]. Under pressure T_c at first increases but then decreases. This nonmonotonic behavior is brought out more clearly in Fig. 13 where $T_c(P)$ is seen to pass through a maximum near 4 GPa [40]. This nonmonotonic pressure dependence offers a possible explanation for the widely differing initial pressure derivatives discussed above. Differences in sample composition may place the sample at different points on a generalized phase diagram. Indeed, T_c for $La_{2-x}Sr_xCuO_4$ is known to pass through a maximum as a function of the hole concentration [41]. Jorgensen et al.[42] have used structural data under pressure on $YBa_2Cu_3O_7$ to estimate that pressure leads to an increase in the hole concentration in the Cu–O planes. It would seem reasonable to predict that nonmonotonic behavior of $T_c(P)$ will be found for many, if not all, hole–doped and electron–doped oxide superconductors so that the sign of dT_c/dP is not a hallmark of either.

The following experiment was an unsettling experience for us, but the results are very relevant in the present context [43]. In Fig. 14 we show $T_c(P)$–data on

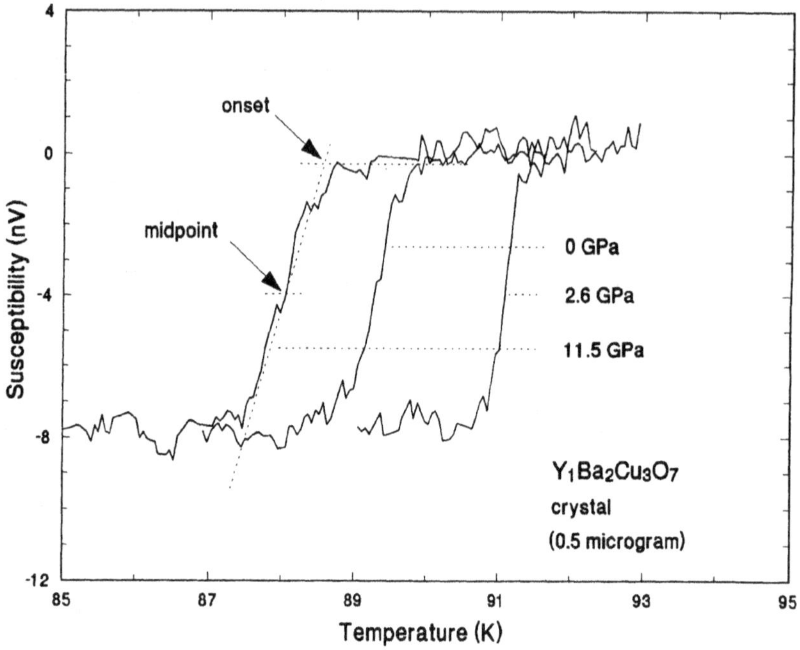

Fig. 12. Temperature dependence of ac susceptibility for a tiny crystal of $YBa_2Cu_3O_7$ at three pressures in the diamond–anvil clamp.

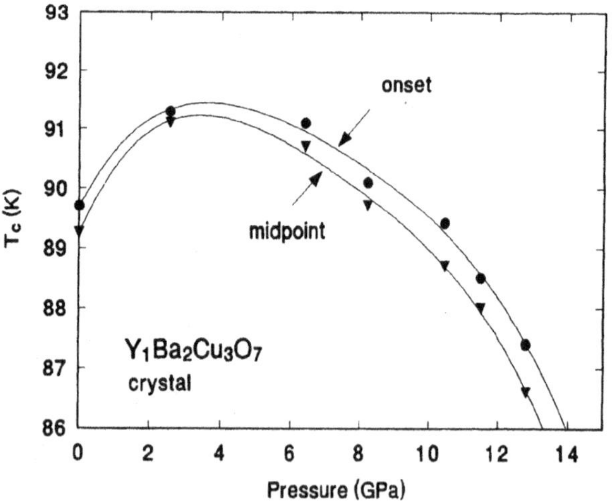

Fig. 13. Superconducting transition temperature of a $YBa_2Cu_3O_7$ crystal versus pressure in the diamond–anvil clamp.

$Tl_2Ba_2CuO_{6+y}$, a single–layer hole–doped oxide superconductor [44]. Not only does T_c decrease under pressure, it chooses to fall very rapidly! For the $T_c(P)$–data shown in Fig. 14 the pressure was always changed at room temperature. In the He–gas system used the pressure could have been changed at low temperatures, but we preferred not to do this since the cell is then more prone to leakage. Before a particular weekend, however, one of the authors (R.S.) was in a hurry and decided to save time by changing the pressure at a temperature only a few degrees above T_c (but always at a temperature well above the melting curve of He). To his astonishment, $T_c(P)$ remained constant whether the pressure was increased or decreased at low temperature. Warming back up to room temperature, however, caused T_c to jump back to a value along the $T_c(P)$–curve given in Fig. 14. Our first thought was that solid He was blocking the capillary, preventing the pressure in the cell from changing. Measurements using an internal superconducting manometer indicated that the pressure in the cell was changing as expected. But perhaps the most direct proof that the erratic behavior of T_c with pressure is actually an intrinsic property of the sample is given by the beautiful data in Fig. 15. Going from curve A to B the pressure has been increased at room temperature from 0.013 GPa to 0.13 GPa. T_c is seen to rapidly decrease, as expected from Fig. 14. The pressure was then increased to 0.34 GPa at low temperature (45 K), yielding the solid curve C. T_c did not change! Curves B and C lie almost on top of each other. The proof that the pressure on the sample in the pressure cell actually did increase between B and C is given by the appearance of a notch in curve C at 30.5 K. This notch marks exactly where the He pressure fluid froze, i.e. the melting curve of He for 0.34 GPa lies at 30.5 K. Upon releasing the pressure a bit to 0.32 GPa at room temperature, T_c decreases in a manner to agree with the $T_c(P)$–dependence in Fig. 14. Here again, the notch in $\chi'(T)$ near 30 K marks the decrease in the melting temperature of He at the slightly lower pressure.

It is thus apparent that for this Tl–compound the pressure dependence of T_c depends on the temperature where the pressure is changed. This is brought out clearly in Fig. 16 where the relative pressure derivative of T_c is plotted versus oxygen concentration [44]. We interpret these results in the following way. All compounds studied have excess oxygen atoms which are distributed on interstitial sites. As a function of pressure there may be a number of different ordered states into which the interstitial oxygen, or the other oxygen for that matter, would like to assume. If the pressure is

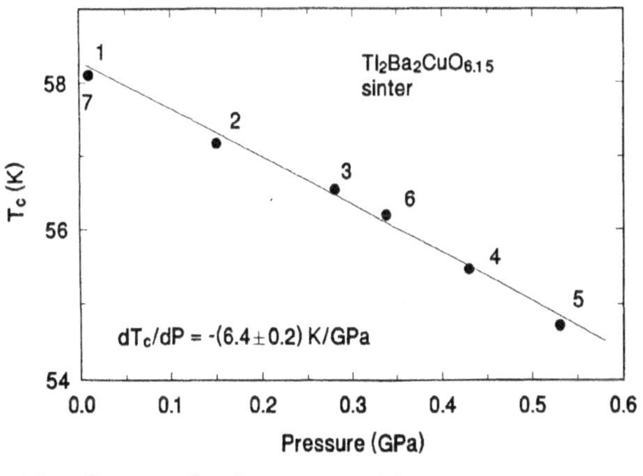

Fig. 14. Superconducting transition temperature for $Tl_2Ba_2CuO_{6.15}$ versus pressure in the He–gas system. Pressure was changed at 300 K. Numbers give order of measurement.

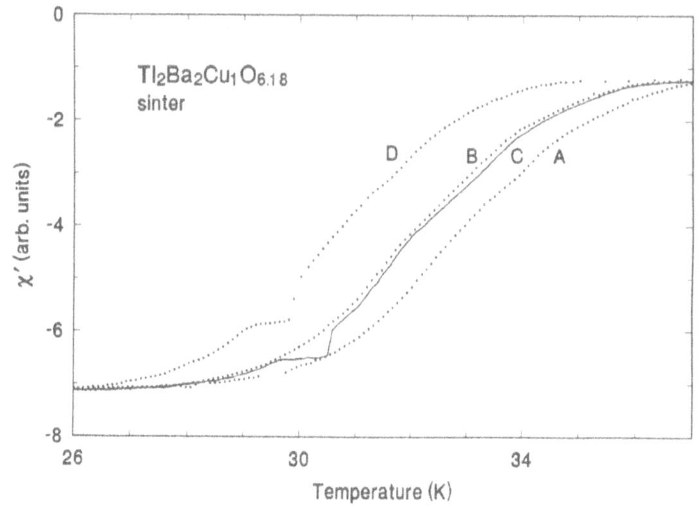

Fig. 15. Real part of the ac susceptibility for $Tl_2Ba_2CuO_{6.18}$ versus temperature at 0.013 GPa (A), 0.13 GPa (B), 0.34 GPa (C), and 0.32 GPa (D) in the He–gas system, in that order. See text for details.

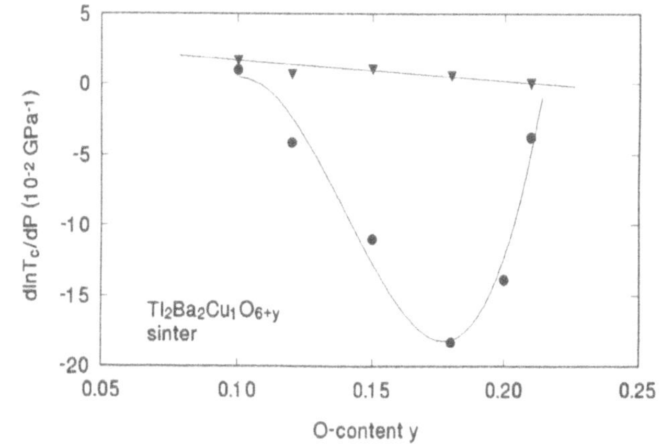

Fig. 16. Relative pressure derivative of T_c for $Tl_2Ba_2CuO_{6+y}$ versus oxygen content y. Pressure was changed either at 300 K (•) or at 45 K (▲).

changed at room temperature, there is enough thermal energy available that the oxygen sublattice can take on the ground state oxygen ordering arrangements. However, at low temperatures insufficient thermal energy is available to allow the oxygen atoms to reorder under pressure, so that T_c doesn't change significantly, but increases only slowly, which is the more normal behavior for hole–doped superconductors.

The next question to ask is the extent to which other systems might show similar effects. Single–crystalline $YBa_2Cu_3O_7$ [44] and polycrystalline $Bi_2Sr_2CaCu_2O_{8+y}$ [43] do <u>not</u> show any dependence of dT_c/dP on the temperature at which the pressure is changed, at least below 300 K. However, the results on $Tl_2Ba_2CuO_{6+y}$ are of obvious interest and further experiments should be carried out to determine the nature of the reordering which must be occurring.

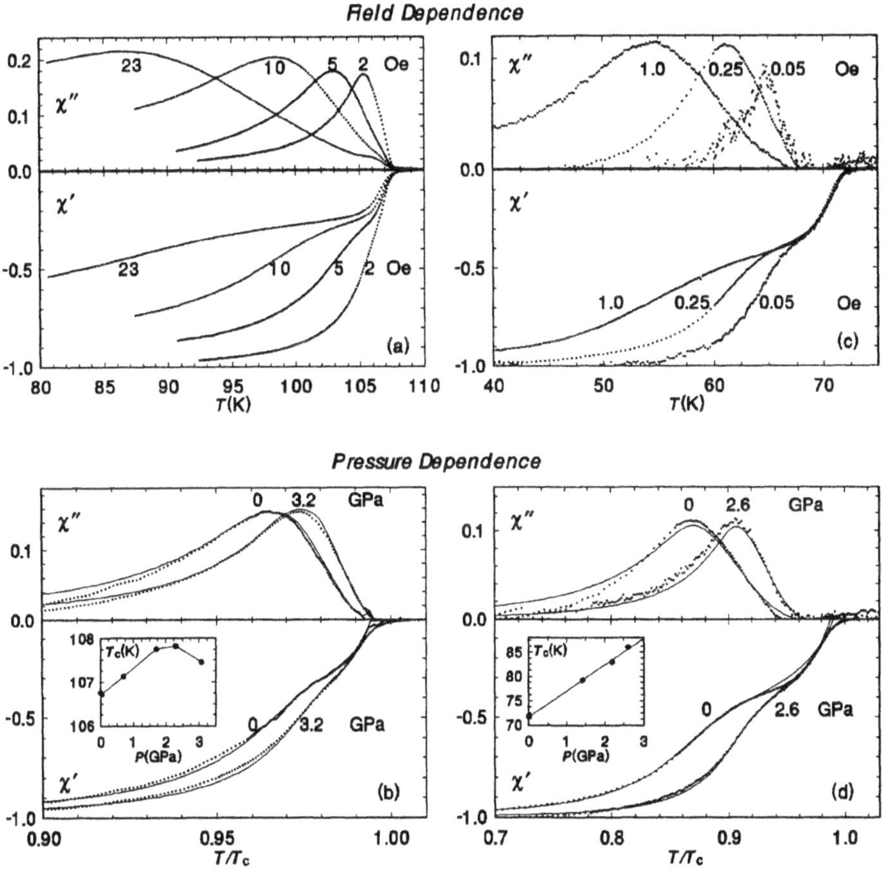

Fig. 17. Influence of the ac field amplitude H_{ac} (a,c) and pressure (b,d) on the temperature dependence of the real and imaginary parts of the ac susceptibility for $Tl_2Ba_2CaCu_2O_8$ (a,b) and $YBa_2Cu_4O_8$ (c,d). Numbers in (a,c) give values of H_{ac} in Oe. Data in (a) taken at 3 GPa and in (c) at ambient pressure. Data in (b,d) are plotted for two values of pressure in metal–gasket cell versus reduced temperature at H_{ac} = 2 Oe (b) and H_{ac} = 0.25 Oe (d). Dotted curves give data, solid lines are fits using Müller's model (Ref. 46). Insets show pressure dependence of the superconducting onset temperature T_c.

Figs. 8, 10, and 11 contain ac susceptibility data which give evidence that even in single crystals weak–link behavior can occur. The weak–link boundaries are presumably caused by planar defects, micro–cracks, or other structural imperfections. In polycrystalline materials it is widely accepted that the contacts between the individual grains, the grain boundaries, behave as weak–links so that a sintered material can be viewed as a dense array of Josephson–coupled strongly superconducting grains [45]. Because of the inability of a sufficient number of grain boundaries to carry a large density of supercurrent, the transport critical current density J_c for the material as a whole is very limited, becoming rapidly even weaker when a magnetic field is applied. A better understanding about why grain boundaries and other defects behave as weak–links could lead to processing changes to improve their current–carrying capacity. In addition, any technique, such as preferential grain orientation, which increases J_c would be of interest in its own right. Since the application of high pressure would be expected to improve the contact between the atoms on either side of a grain boundary, it might be anticipated that J_c would increase under pressure. We will see below that this, in fact, actually occurs.

The ac susceptibility technique is particularly well suited for studying J_c under high pressure conditions since it does not require electrical contacts to the sample under study. In addition, Müller [46] has developed a critical state model of superconducting sinters which gives in detail the temperature and magnetic field dependence of the real and imaginary parts of the ac susceptibility. The experimentalist need only fit theory to experiment to extract values of the intergranular critical current density J_c and the London penetration depth λ_g in a grain. The only input parameters are the mean grain size, the magnitude of the applied ac and dc magnetic fields, and the sample density.

In Figs. 17a and 17c we show the temperature dependence of $\chi'(T)$ and $\chi''(T)$ for sinters of $Tl_2Ba_2CaCu_2O_8$ and $YBa_2Cu_4O_8$ as a function of the applied ac field H_{ac} [47]. At intermediate values of the applied field the two–step $\chi'(T)$–behavior for both samples is clearly seen. As one cools through the superconducting transition, the grains are the first to be shielded from the applied flux, leading to the first step in $\chi'(T)$. At this point, however, flux still penetrates through the intergranular material, i.e. through grain boundaries, voids, planar defects, etc., since this material is not yet capable of carrying the supercurrent density J_c needed to shield the sample as a whole.

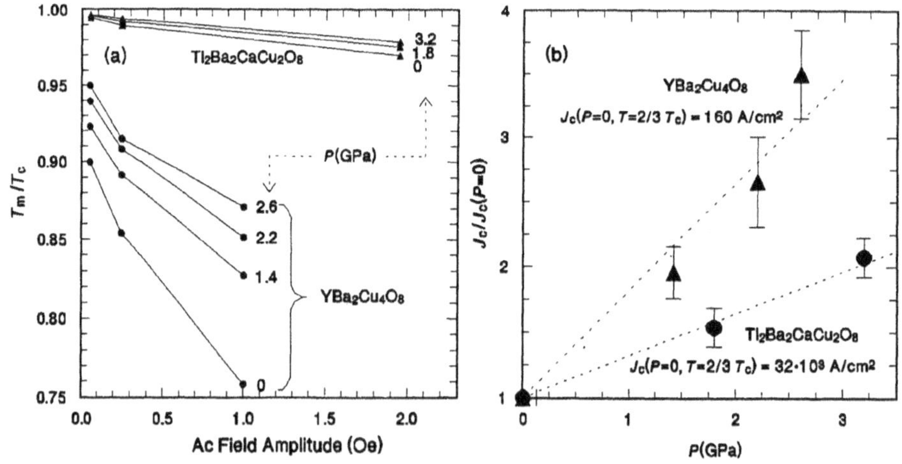

Fig. 18. (a) Dependence of the temperature T_m of the maximum in χ'' (see Fig. 17) divided by T_c versus ac field amplitude H_{ac} at different pressures in metal–gasket cell. Solid lines drawn through data at same pressure. (b) Pressure dependence of relative intergranular critical current density J_c.

This occurs at still lower temperatures where $J_c(T)$ has grown larger, leading to the second step in $\chi'(T)$. It is in this temperature region, where the intergranular supercurrents are being set up, that the largest losses in the sinter occur, leading to the maximum in $\chi''(T)$ at T_m seen in Fig. 17. Since $J_c(T)$ is weakened when a magnetic field is applied, the maximum in $\chi''(T)$ would be expected to shift to lower temperatures with increasing H_{ac}. This is clearly observed in Figs. 17a and 17c. The fact that the maximum in $\chi''(T)$ at T_m shifts much more rapidly to lower temperatures for $YBa_2Cu_4O_8$ than for $Tl_2Ba_2CaCu_2O_8$ implies that J_c must be much less for the former compound. The following analysis supports this conclusion.

The effect of pressure on the ac susceptibility is seen in Figs. 17b and 17d for the two sinters studied. Relative to the superconducting transition temperature T_c in the grains, the maximum in $\chi''(T)$ is seen to clearly shift to higher temperatures. From the above discussion it follows immediately that the application of pressure has sharply increased the intergranular critical current density J_c for both sinters. This conclusion from a cursory qualitative analysis of the raw data is supported by the detailed data fits to Müller's model [46] given by the solid lines in Figs. 17b and 17d. A simplified analysis of the pressure dependence of J_c is possible using Müller's expression $T_m/T_c = 1-C(J_c)^{-1/2}H_{ac}$, where J_c is the intergranular critical current density for zero applied field at a temperature 67% of T_c and C is a parameter given by Müller. This expression tells us that we can estimate J_c simply by determining the slope of a (hopefully) straight–line plot of T_m/T_c versus H_{ac}. The smaller the slope, the larger is J_c. The plots in Fig. 18a are not perfectly straight, but a good rough estimate of J_c can be made. As a function of applied pressure, the slope of the plots progressively decreases which implies that J_c increases under pressure. This is confirmed in Fig. 18b, where the relative pressure dependence $J_c/J_c(P=0)$ is plotted versus pressure. The values of J_c for zero applied field and pressure at a temperature 67% of T_c are also given in this figure. It is interesting to note that whereas J_c is more than two orders of magnitude greater for the Tl–compound, the increase of J_c with pressure is almost twice as large for the Y–compound. He who has little has more to gain! The increase of J_c with pressure is surprisingly large. Our hope is that understanding the reason(s) for this increase will lead to the synthesis of better sintered superconductors.

The estimated value of J_c for the Tl–sample at $67\%T_c$, or 72 K, is $3x10^4$ A/cm^2, a very respectable value for a sinter. An attempt to check this value by a direct measurement of the transport critical current density has not yet been successful due to the porosity of the sample. Müller et al.[48] report good agreement between the J_c–values from the ac susceptibility and transport measurements. A more detailed discussion of the above experimental results will be given elsewhere.

The $AuGa_2$ Dilemma Recalled

This review has primarily focussed on high temperature superconductivity since these materials are of intense current interest in the field. The highly anomalous behavior of T_c under pressure for $Tl_2Ba_2CuO_{6+y}$ was pointed out above. In our efforts to solve the puzzles of the present we should not forget that many mysteries from the past still remain unsolved. One of the most interesting high pressure transitions ever discovered was first reported by Schirber [49] on the intermetallic superconductor $AuGa_2$ where both a study of the Fermi surface and a sharp maximum in $T_c(P)$ near 0.6 GPa gave evidence for an isostructural Lifschitz or "electronic" transition. Even more remarkable is the report of Smith et al. [50] that for pressures above 1.5 GPa the superconducting state of $AuGa_2$ exhibits "a well resolved and reproducible supercooling of $\sim$ 5 mK for decreasing temperature and a marked differential paramagnetic effect (DPE) upon warming back through the transition." This anomalous behavior is evident in the measurements by Hein et al. [51], shown in Fig. 19, of the dependence of the ac susceptibility on applied dc magnetic field at a fixed temperature for two values of the pressure, one just below (0.57 GPa) and the other just above (0.60 GPa) the Lifschitz transition. It is not understood why a purely electronic transition should be able to set the scene for such unusual behavior as DPE and supercooling. To the knowledge of the authors, and at least one of the workshop coorganizers, this behavior under pressure remains unique in the field of superconductivity.

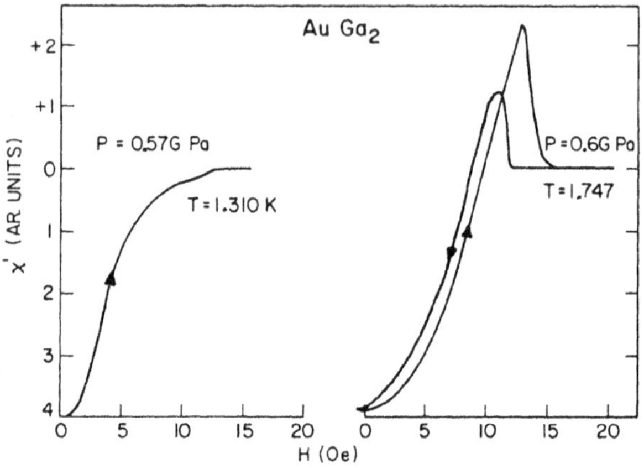

Fig. 19 In–phase component χ' of the ac magnetic susceptibility of $AuGa_2$ as a function of dc magnetic field showing the transition to superconductivity for two values of the applied pressure. This figure is adapted from Figs. 4 and 7 of Ref. 51. The sudden appearance of supercooling at the higher pressure is evident.

ACKNOWLEDGMENTS

The authors would like express gratitude to K. Andres, P. Müller, and other members of the Walther–Meissner–Institute for Low Temperature Research in Garching for fruitful discussions and for providing some of the excellent single–crystals used in the studies described here. Thanks are due W. Reith for preparing the thallium samples. Of particular importance for the analysis of our studies of the weak–link properties of ceramic superconductors was the Fortran program to fit our ac susceptibility data kindly provided to us by K.–H. Müller who is also a participant at this workshop. One of the authors (J.D.) acknowledges detailed discussion with J. Clem and V. Kogan.

REFERENCES

1. J.S. Schilling, in Physics of Solids under High Pressure, edited by J.S. Schilling and R.N. Shelton (North Holland, Amsterdam 1981)379; Physica 139&140B, 369(1986), and references therein.
2. R.C. Wayne and L.R. Edwards, Phys. Rev. 188, 1042(1969); T.F. Smith, J.A. Mydosh, and E.P. Wohlfarth, Phys. Rev. Lett. 27, 1732(1971).
3. D. Jerome, A. Mazaud, M. Ribault, and K. Bechgaard, J. Phys. Lett.(Paris) 41, L95(1980).
4. J.M. Williams, A.M. Kini, H.H. Wang, K.D. Carlson, U. Geiser, L.K. Montgomery, G.J. Pyrka, D.M. Watkins, J.M. Kommas, S.J. Boryschuk, A.V.S. Crouch, W.K. Kwok, J.E. Schirber, D.L. Overmyer, D. Jung, and M.–H. Whangbo, Inorgan. Chem. 29, 3272(1990).
5. C.W. Chu, P.H. Hor, R.L. Meng, L. Gao, Z.G. Huang, Y.Q. Wang, Phys. Rev. Lett. 58, 405(1987).
6. M.K. Wu, J.R. Ashburn, C.J. Torng, P.H. Hor, R.L. Meng, L. Gao, Z.J. Huang, Y.Q. Wang, and C.W. Chu, Phys. Rev. Lett. 58, 908(1987).
7. N. Lotter and J. Wittig (preprint); see also Ref. 36.

8. M.J. Clark, and T.F. Smith, J. Low Temp. Phys. 32, 495(1978); A. Eichler and J. Wittig, Z. Angew. Phys. 25, 319(1968).
9. T.F. Smith, J. Low Temp. Phys. 6, 171(1972).
10. R.N. Shelton, A.R. Moodenbaugh, P.D. Dernier, and B.T. Matthias, Mat. Res. Bull. 10, 1111(1975).
11. T.F. Smith, in High Pressure Science and Technology, edited by K.D. Timmerhaus and M.S. Barber (Plenum, N.Y., 1979)309.
12. A. Eiling and J.S. Schilling, J. Phys. F 11, 623(1981).
13. The pressure cell and electrical feed–through shown in Fig. 1 were designed and built by the company Unipress, ul. Sokolowska 29, 01–142 Warsaw, Poland.
14. L.D. Landau and E.M. Lifshitz, in Electrodynamics of Continuous Media (Addison–Wesley, Boston, 1960).
15. M. Couach, A.F. Khoder, and F. Monnier, Cryogenics 25, 695(1985).
16. L.J. Azevedo, J.E. Schirber, J.M. Williams, M.A. Beno, and D.R. Stephens, Phys. Rev. B 30, 1570(1984).
17. G. Fasol and J.S. Schilling 49, 1722(1978).
18. A. Eiling, J.S. Schilling, and H. Bach, in Physics of Solids under High Pressure, edited by J.S. Schilling and R.N. Shelton (North–Holland, Amsterdam, 1981)p. 385.
19. See Ref. 12 for a discussion of a Pb resistive manometer.
20. See, A. Jayaraman, Rev. Mod. Phys. 55, 65(1983) as well as, S.W. Tozer, PhD Thesis, The Johns Hopkins University (1986), and references therein.
21. J.S. Schilling, Mat. Res. Soc. Symp. Proc. 22, 79(1984).
22. W.B. Daniels and W. Ryschkewitsch, Rev. Sci. Instr. 54, 115(1983).
23. A.L. Giorgi, B.T. Matthias, G.R. Stewart, F. Acker, and J.L. Smith, Solid State Commun. 32, 455(1979).
24. E.P. Wohlfarth, J. Phys. C2, 68(1969); T.F. Smith, J.A. Mydosh, and E.P. Wohlfarth, Phys. Rev. Lett. 27, 1732(1971).
25. W.E. Gardner, T.F. Smith, B.W. Howlett, C.W. Chu, and A. Sweedler, Phys. Rev. 166, 577(1968).
26. J. Grewe, J.S. Schilling, K. Ikeda, and K.A. Gschneidner, Jr., Phys. Rev. B 40, 9017(1989).
27. N.H. March, in Advances in High Pressure Research, Vol. 3, edited by R.S. Bradley (1969).
28. T.F. Smith and R.N. Shelton, J. Phys. F 5, 911(1975); J.J. Hopfield, Physica 55, 41(1971).
29. K. Bechgaard, K. Carneiro, M. Olsen, and F.B. Rasmussen, Phys. Rev. Lett. 46, 852(1981).
30. J.E. Schirber, L.J. Azevedo, J.F. Kwak, E.L. Venturini, P.C.W. Leung, M.A. Beno, H.H. Wang, and J.M. Williams, Phys. Rev. B 33, 1987(1986).
31. R. Sieburger, H. Müller, C.–P. Heidmann, K. Andres, and J.S. Schilling (in preparation).
32. J. Schirber, E.L. Venturini, A.M. Kini, H.H. Wang, J.R. Whitworth, and J.M. Williams, Physica C 152, 157(1988).
33. See discussion and references in, B. Rothaemel, L. Forro, J.R. Cooper, J.S. Schilling, M. Weger, P. Bele, H. Brunner, D. Schweitzer, and H.J. Keller, Phys. Rev. B 34, 704(1986).
34. R. Griessen, Phys. Rev. B 38, 3690(1988); R.J. Wijngaarden and R. Griessen, in Studies of High Temperature Superconductors, ed. A.V. Narlikar (Nova Science Publisher, N.Y., 1989).
35. K. Murata, Y. Honda, H. Oyanagi, Y. Nishihara, H. Ihara, N. Terada, R. Sugise, M. Hirabayashi, M. Tokumoto, and Y. Kimura, MRS Int'l. Mtg. on Adv. Mats. 6, 682(1989).
36. U. Koch, N. Lotter, J. Wittig, W. Assmus, B. Gegenheimer, and K. Winzer, Solid State Commun. 67, 959(1988).
37. These measurements are part of the PhD thesis of R. Sieburger, University of Munich (1991).
38. R. Sieburger, J. Diederichs, H. Gossner, H. Neumaier, W. Reith, P. Müller, and J.S. Schilling, High Press. Res. 3, 117(1990).

39. Z. Dian–lin, L. Li, H. Jing, D. Hong–min, L. Shu–yuan, M. Bei–hai, C. Shao—chun, and B.J. Jin, Phys. Rev. B 41, 6692(1990).
40. S. Klotz, W. Reith, and J.S. Schilling, Physica C 172, 423(1991); this work is part of the PhD thesis work of S. Klotz, University of Munich.
41. J.B. Torrance, A. Bezinge, A.I. Nazzal, T.C. Huang, S.S.P. Parkin, D.T. Keane, S.J. LaPlaca, P.M. Horn, and G.A. Held, Phys. Rev. B 40, 8872(1989).
42. J.D. Jorgensen, B.W. Veal, A.P. Paulikas, L.J. Nowicki, G.W. Crabtree, H. Claus, and W.K. Kwok, Phys. Rev. B 41, 1863(1990).
43. This work is part of the PhD thesis of R. Sieburger, University of Munich (1991).
44. R. Sieburger and J.S. Schilling, Physica C 173, 403(1991).
45. J.R. Clem, Physica C 153, 50(1988).
46. K.–H. Müller, Physica C 159, 717(1989).
47. J. Diedrichs, W. Reith, B. Sundqvist, J. Niska, K.E. Easterling, and J.S. Schilling, Supercond. Sci. Technol. 4, S97(1991); this work is part of the Diplom thesis work of J. Diederichs, University of Munich (1990).
48. K.–H. Müller, B.W. Ricketts, J.C. Macfarlane, and R. Driver, Physica C 162&164, 1177(1989).
49. J.E. Schirber, Phys. Rev. B 6, 333(1972); ibid., Phys. Rev. Letters 28, 1127(1972).
50. T.F. Smith, R.N. Shelton, and J.E. Schirber, Phys. Rev. B 8, 3479(1973).
51. R.A. Hein, J.E. Cos, and R.W. McCallum, in "High–Pressure and Low–Temperature Physics," C.W. Chu and J.A. Woollam, editors (Plenum Press, N.Y., 1978)p. 419.

DC Magnetisation and Flux Profile Techniques

A.M.Campbell

IRC in Superconductivity, Madingley Road, Cambridge, U.K.

Abstract

Some general results on electromagnetism are used to define fields and magnetisation in superconductors. Experimental techniques for measuring ac and dc magnetisation are described. Treatment of the results to give critical currents is discussed and details of flux profile techniques given. It is shown that if the displacement of the flux lines is proportional to the force and a function of frequency the London equation is obtained with a complex penetration depth. This is derived and the inductive transition plotted through the reversibility line.

1. General Results on Magnetic Fields

The magnetisation we measure in a magnetometer is due to the field generated in a body by currents which do not cross its boundaries. The magnetic moment of a small current loop is $i\delta\mathbf{S}$ and the magnetic moment of a body is $\int i\delta\mathbf{S}$. This expression is most appropriate to magnetic materials in which the moments are local dipoles, but it can also be used when bulk currents are flowing, as in a superconductor. The most general expression for the magnetic moment is $\frac{1}{2}\int \mathbf{r}\times\mathbf{j}\,dv$ where $\mathbf{j}$ is the average local current density, including that from atomic dipoles (1).

Having defined the magnetic moment we then define a magnetisation **M** as the moment divided by the volume. At this point we must distinguish between two types of magnetisation. If the magnetisation is due to local dipoles the state of the material is determined by local fields, and we can relate **M** at any point to **B** at that point through material parameters which are independent of the size or shape of the sample. **B** is the average of the local magnetic field on a scale large enough to make the properties uniform and we can then

Magnetic Susceptibility of Superconductors and Other Spin Systems
Edited by R.A. Hein *et al.*, Plenum Press, New York, 1991

define a useful vector $\mathbf{H}=\mathbf{B}/\mu_o-\mathbf{M}$. However if the magnetisation is due to currents flowing on the scale of the sample size, as occurs in eddy current and superconducting situations, this cannot be done and it makes no sense to talk about a local value of M. We have currents in a non-magnetic material and $\mathbf{B}=\mu_o\mathbf{H}$. If these are not thermodynamic equilibrium currents they should be classified as transport currents.

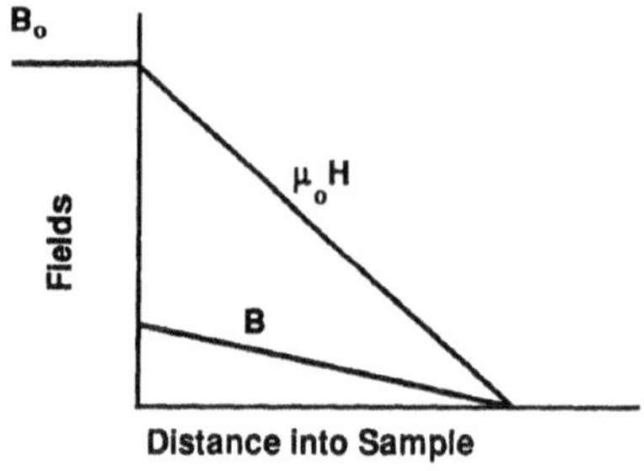

Fig. 1.
The gradients of B and H for constant J_c

Sometimes both types of current are present and this is of importance in granular materials. In these circumstances (which also include currents in ferromagnetic materials) we define H(B) as the external field in equilibrium with a flux density B in a long cylinder parallel to the field. This defines a B-H curve for the material. The transport current is defined as $\mathbf{J}=\mathrm{Curl}\mathbf{H}$ and it is this current which determines the driving force on the vortices, $\mathbf{J}\mathrm{x}\mathbf{B}$, not the gradient in B (2,3). Figure 1 illustrates the distinction. We take a material with a constant intergranular J_c in which about 70% of the volume is occupied by perfectly diamagnetic grains. The flux penetrates between the grains building up a current density J_c but the flux cannot penetrate into the grains. The flux density, B, is averaged over many grains. At the surface the equilibrium average flux density will be about 0.3 times the external flux density and it decreases linearly towards the centre. (The exact value will depend on the shape of the grains, it will be 0.3 for long grains parallel to the field.) The external field in equilibrium with this is about three times the value of B at any point and this gives the local value of H which decreases linearly from a surface value equal to the external field. We can now define an effective permeability $\mu=B/\mu_0 H$ which is related to the volume fraction of superconductor and the shape of the grains. This has been used in the calculations of Gömöry et. al. (4) and Müller (5), but it can be seen that if we treat the effective permeability as a measurable parameter, as is normally done, Maxwell's equations are exact without having to consider the details of the void network. i.e. $\mathbf{B}=\mu\mu_o\mathbf{H}$ and $\mathbf{J}=\mathrm{curl}\mathbf{H}$ where $\mathbf{J}$ is the intergranular transport current. It is in relating μ to the microstructure that uncertainties appear. A method of measuring μ is described in §4.1.

1.1. Measurement

Most measurements (apart from mechanical ones) involve measuring the voltage from a coil round the sample in an external field A voltage is generated by changing the magnetic moment of the sample, or by moving it. There are four geometries in which the voltage can be directly related to the magnetisation and these are shown in Figure 2.

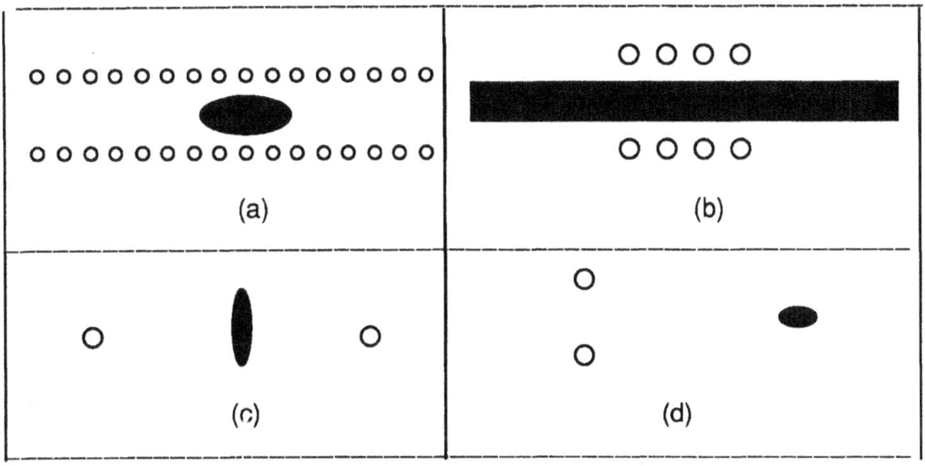

Fig. 2.
Four Different Coil Arrangements which Give Direct Magnetisation Measurements.

First we can have a small sample enclosed by a long search coil. Secondly we can have a coil wound round a cylinder well away from the ends. Thirdly we can have a small sample in a large current loop, and finally we can have a small sample a long way from a search coil. The common feature of all but the second is that if we pass a current into the search coil the the field across the sample is uniform. The second does not give the magnetic moment of the body, but the axial flux, and for a cylinder of arbitrary cross section this is directly related to the magnetisation per unit length well away from the ends.

We can relate the measured moment to the mutual inductance between the search coil and a coil of the same shape as the sample. Suppose a current i_C in the search coil produces a uniform field $\mathbf{B_s}$ at the sample. Then a current loop in the sample position of area $\boldsymbol{\delta}\mathbf{S}$ will have a flux through it of $\mathbf{B_s}.\boldsymbol{\delta}\mathbf{S}$. The mutual inductance M is therefore $\mathbf{B_s}.\boldsymbol{\delta}\mathbf{S}/i_C$. Suppose now we change the current in the loop in the sample i_S, and measure the voltage in the search coil keeping i_c constant. The voltage is $Mdi_s/dt=(di_s/dt)\boldsymbol{\delta}\mathbf{S}.\mathbf{B_s}/i_C.=kd\mathbf{m}/dt$ where k is a geometrical factor equal to the applied field produced at the sample per unit current in the search coil, Thus the voltage is directly proportional to the rate of change of total magnetic moment along the axis of the search coil.

In the case of the geometry of figure 2b we consider a long axial coil in the sample of area δS. (The cross section does not have to be circular). If it has n turns per unit length the mutual inductance is $\mu_o Nn\delta S$ where N is the number of turns on the search coil. If it carries a current i_s the induced voltage is $\mu_o Nn\delta S(di_s/dt)$. The magnetic moment per unit length of such a coil is $ni_s\delta S$. Hence the voltage from the sample is proportional to the magnetic moment per unit length. In this geometry (and only this one) the magnetic moment can also be determined from the difference between the external field and the mean B in the sample.

These are the geometries for accurate absolute values. They are also wasteful in that we can get a larger signal to noise ratio by making the search coil the same length as the specimen. For a perfectly diamagnetic sample the signal can be related to the magnetisation by using mutual inductance tables (6) for concentric coils. However as flux penetrates to the centre the coupling factor changes because the mutual inductance between the search coils and interior currents is not a simple function of the radius at which the currents flow, and even a calibration using a lead sample will no longer be strictly accurate. For practical purposes these corrections are not too important since most experiments compare similar samples in the same apparatus.

1.2. Losses

As well as an unambiguous measurement of magnetisation we can get an unambiguous measurement of the loss. For any change in the total magnetic moment $\boldsymbol{\delta m}$ we saw above that the voltage induced in a coil round the sample (in this case the drive coil, see Fig.7.) is $(\mu_o/i_d)\mathbf{H_o}.\mathbf{dm}/dt$ where $\mathbf{H_o}$ is the applied field at the sample due to the drive current i_d. Hence the power needed to keep $\mathbf{H_o}$ constant is $\mu_o\mathbf{H_o}.\mathbf{dm}/dt$ and the work done by the drive field on the sample is $\mu_o\mathbf{H_o}\boldsymbol{\delta m}$. Hence the work done is directly proportional to the change in total magnetic moment of the body in the direction of the applied field. In a dc measurement the loss during a complete cycle is the area of the hysteresis loop.

In an ac measurement the voltage due to the changing magnetisation can be expressed as a Fourier series If the applied field is $H_o \mathrm{Cos}\,\omega t$ and the voltage waveform from the search coil is turned into a magnetisation waveform by the appropriate calibration constant, we can write the magnetisation as

$$m=\Sigma\ (\ a_n \mathrm{Cos}\ n\omega t+b_n \mathrm{Sin}\ n\omega t)$$

The instantaneous power going into the sample is

$$\Sigma\ \mu_o H_o \mathrm{Cos}\ \omega t\ n\omega\ (\ -a_n\ \mathrm{Sin}\ n\omega t+b_n \mathrm{Cos}\ n\omega t)$$

Averaged over a cycle the loss is

$$1/2\ \mu_o\omega H_o b_1.$$

In other words it is proportional to the amplitude of the voltage at the fundamental frequency out of phase with the drive current. Hence to measure the loss we need to filter out higher harmonics.and pass only the fundamental frequency. The loss is then given by the out of phase signal in the search coil.

1.3. Voltage Level

It is important to have some idea of the voltage level involved. Suppose flux is flowing out of a cylinder of radius r at a rate $\dot{\phi}$(Figure 3.)

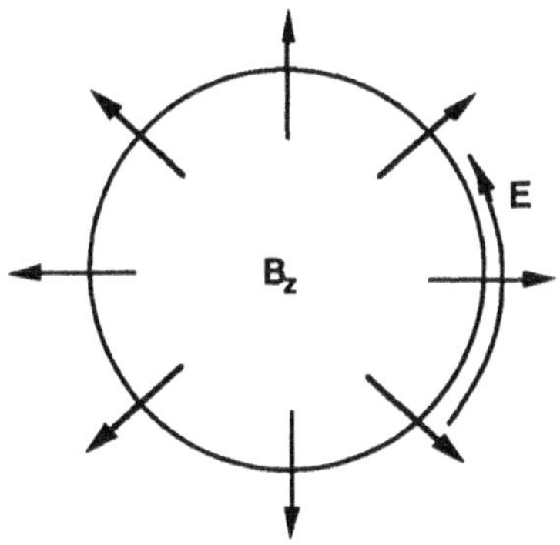

Fig. 3.
Flux perpendicular to the paper leaving a cylinder radially

From Faraday's law the surface field is given by $2\pi r\,E = \dot{\phi}$. The rate of change in magnetisation is $\mu_o\dot{M} = \dot{\phi}/\pi r^2$ so $E = \frac{1}{2} r\mu_o\dot{M}$. A similar electric field will be generated for samples of other shapes. For example if the magnetisation remains constant to 0.1mT over a minute, the electric field on the surface of a 10 micron particle must be less than 10^{-11} volts per meter In larger samples the minimum detectable voltage level will be larger in proportion to the size.

If the external field is changing we can find the electric field in a similar way from the Bean model. For example if we take a cylinder with full penetration of the field Faraday's law gives $2\pi r\,E = \pi r^2\dot{B}$. Hence $E = \frac{1}{2} r\dot{B}$. If flux has not penetrated to the centre of the sample it is the distane to the point where flux is stationary, the electric centre, which is used instead of the radius.

Whatever the model the electric field in a coil close to the sample has to be nearly equal to the electric field at the sample surface. The voltage detected is then $2\pi rEn$ where n is the number of turns and r the radius of the coil. This explains why inductive techniques explore lower voltage levels than resistance measurements. The use of a coil multiplies the EMF round the circumference by the number of turns giving several orders of magnitude amplification. For example if in a susceptibility measurement a lock-in amplifier measures 10 nanovolts from a 1000 turn coil with a 1mm. radius the average electric field at the surface is about 10^{-9} volts/meter. (If coils are balanced to give zero in the Meissner state are used the voltage gives the electric field at the sample surface instead of at the search coil).

1.4. Flux Displacement

Finally in this section on general electromagnetism we should put some figures to the amount of vortex movement that occurs. This is important to know because experiments can be divided into the low amplitude regime and the high amplitude regime. These are distinguished by the amplitude of the flux line movement in comparison with the distance required to build up a critical state, not the amplitude of the field change in Tesla. Although the distance the vortices move may appear to depend on the details of the vortex lattice, it is in fact independent of the size of the vortices and can be found from an essentially classical picture of lines of force. The quantised vortices behave and move in exactly the same way as Faraday imagined for his lines of force, or as would be drawn by a finite element magnetic field package on a computer. The only effect of the quantisation is to give a a fixed amount of flux to each line so that the spacing is determined uniquely by the magnetic field, rather than being merely proportional to it. We will therefore derive the result in two ways, firstly using a vortex lattice picture and secondly purely classical arguments.

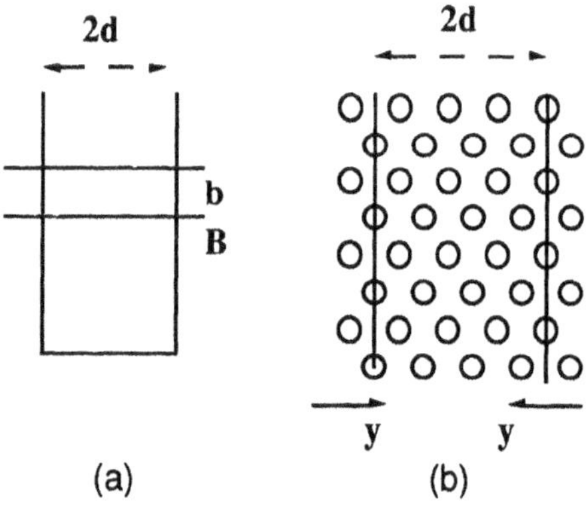

Fig. 4.
The flux displacement y for a rise in field b with no Pinning.

We take a slab of thickness 2d in a large external field B parallel to its faces. We assume no pinnning so that the internal field is equal to B. We then raise the external field by a small amount b. Flux enters the sample to raise the internal field by the same amount. The fields are shown in Fig.4a. If we look at the top surface of the slab we would see flux entering the slab by moving parallel to the x axis, as shown by the arrows in Fig 4b. The distance moved by the vortices at the surface is y, which is assumed much smaller than d. (A more rigorous treatment is found in §5.1. We find the distance moved, by equating the flux entering the sample in the two pictures. From Fig.4a the flux entering per unit length is 2bd. From Fig.4b all the flux within y of the surface enters, so the flux entering through the two surfaces is 2By. Since these must give the same flux it follows that 2bd=2By or y=bd/B. For a uniform increase in flux density the vortex displacement rises linearly from zero at the electric centre, and in general for slab geometry dy/dx=-b/B. A vector treatment is given in §5.1.

We can get the same result from the electric field. If B is increasing at a rate $\dot{B}$ the electric field at the surface is $\dot{B}d$. Since E=Bv, where v is the flux line velocity, v=$\dot{B}$d/B. or dy/dt=(db/dt)d/b. Hence for any small time interval y=bd/B This essentially classical argument gives an upper limit for the vortex displacement which is true for zero pinning. For a cylinder or sphere the displacement at the edge is half that of a slab since the electric field at the circumference is half that of a slab of thickness equal to the diameter. If pinning is significant the distance to the point where the flux stops moving should be used instead of the half width, as in calculating the electric field.

It is also possible to find the displacement directly from the search coil voltage independently of any model. We have an oscillating ripple of amplitude b, balance the search coils in the superconducting state, and calibrate the system by measuring the voltage in the normal state, V_o. Then in the superconducting state, when a voltage V is measured, V/V_o is the ratio of the flux entering to that which would enter in the normal state. If the circumference is c and the area A this is cBy/bA. Hence $y=(V/V_o)(b/B)(A/c)$. A/c is the half width for a slab and half the radius for spheres and cylinders. To take an example, if we can measure 10^{-3} of the calibration voltage in a sample of ten micron grains at an amplitude of 1 millitesla in a field of 1Tesla the vortex displacement we are detecting is 2.5×10^{-11} meters. It can be seen that susceptibility measurements can explore very small vortex displacements.

If the displacement is much less than the vortex spacing the movement is linear and reversible, and the critical state model (see §2) does not apply. The vortex displacement to build up a critical state is sometimes called the interaction distance and varies with field and sample. It can be measured experimentally from the force displacement curve described below (Fig.11) but a typical value is about a quarter the vortex spacing, in the region of 30 nanometers for typical fields. For 10 micron particles in a one Tesla field the upper limit to the displacement if the field is changed by 1mT is 10 nm. which is too low for the critical state model. If pinning is strong the relevant distance is the distance to the electric centre and in general since the search coil voltage is a direct measure of the flux entering the sample it also gives a measurement of the amplitude of vortex movement at the sample surface which is model independent. The only uncertainty is the need to know the effective sample size, which may be the grain size in high T_c materials rather than the sample size.

A final point of interest is that the vortex displacement gives a clearer meaning to the vector potential **A** than in classical materials. (See §5.1)

2. DC Magnetic Measurement Techniques

Let us look first at problems common to all types. These are mainly concerned with temperature and field control. The field distribution in the sample, and hence the magnetisation, is usually calculated from the Bean model. In terms of flux lines this arises from the fact that flux lines pushed into a sample experience a pinning force, building up a flux density gradient, and when the external field is reduced the pinning force acts in the opposite direction to prevent the exit of flux. This critical gradient in which the Lorentz force of the current is just balanced by the pinning forces is called the 'critical state'. Figure 5a shows the field distribution after a large increase in external field, and the effect of a small subsequent decrease. Since the amount of flux coming out is proportional to the square of the change in field the initial slope of the reverse magnetisation curve is perfectly diamagnetic, (apart from the effects of reversible movement, §4.1). It is therefore essential that there is no tendency to reverse the direction of the flux movement in the sample. This would bring the magnetisation down on the diamagnetic slope, which is very large, and small overshoots in temperature or field can bring the magnetisation to the opposite side of the hysteresis loop.

Figure 5b shows a typical magnetisation curve and the path taken when the field reverses. The field change required to cross to the other side of the hysteresis curve is approximately equal to the hysteresis and may be only a millitesla in a background field of several Tesla.

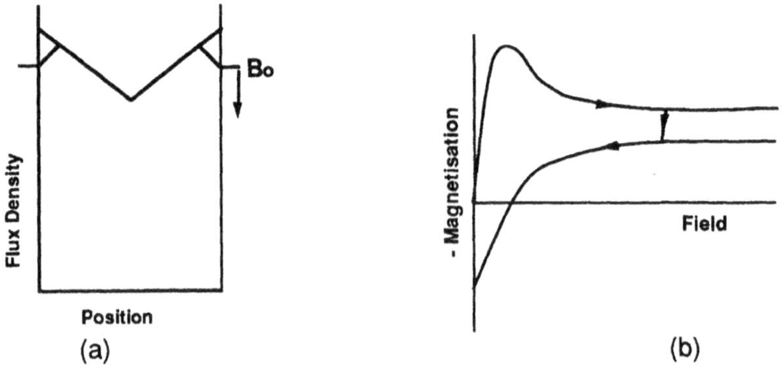

Fig. 5.
A very small change in external conditions makes a large change in magnetisation

Smaller overshoots will put a surface layer in the reverse critical state (Fig.5a) invalidating flux creep measurements until this layer has been wiped out. The size of overshoot which can be tolerated is that which keeps the flux movement at much less than a vortex spacing and this can be estimated from the results above. Although magnet controllers sometimes have a 'no overshoot' mode it is much rarer in temperature controllers so that to plot things as a function of temperature it is normally necessary to set a temperature well below the current value and let the machine drift down, or up, to the required value. This means that there is an inevitable temperature difference between indicated temperature and sample temperature which will depend on the nature of the sample. A technique has been developed to find the relevant time constants during a temperature sweep by imposing a ripple on the set temperature and measuring the amplitude and phase of the resultant ripple in the magnetisation.(7)

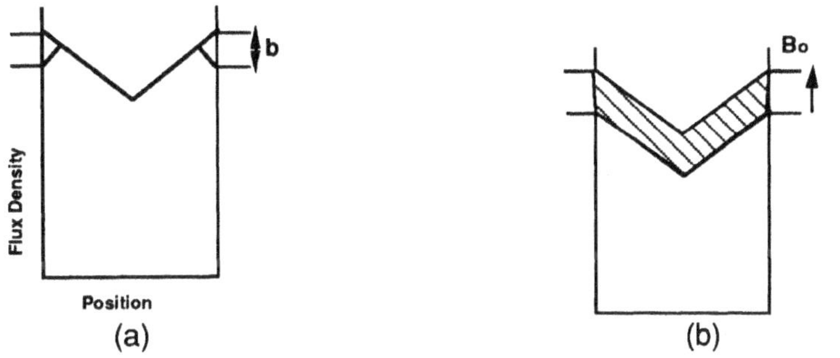

Fig. 6.
Showing how at the peak of a small amplitude ripple a swept field gives a large signal in the search coils

It is found that sweeping the field while measuring the susceptibility gives a very large signal which is difficult to interpret, but the effect may also occur to a smaller extent when sweeping the temperature, or when moving the sample through an inhomogeneous field. This is easily explained by the Bean model, and is illustrated in figure 6. In a steady external field the flux change due to the oscillating field is confined to the surface and the susceptibility is small (Fig.6a). In a swept field there are times near the peak of the cycle when flux moves right to the centre of the sample. (Fig.6b) During this section the hatched area of Fig.6b is filled with flux and a large spike appears on the waveform. If the field sweep is sufficiently fast that the external field never reverses direction the ac signal corresponds to complete flux penetration and the sample appears normal. (This only applies to the mixed state, in the Meissner state below H_{c1} the system is linear and the susceptibility is not changed by a swept field until vortices start to enter the sample).

2.1. *Moving Samples*

Measurements of dc magnetisation divide into two classes. Those in which the sample is moved and those in which it remains stationary. This is a crucial difference if the sample is hysteretic because all magnets are inhomogeneous and moving a sample can put it round a hysteresis loop, leaving it at a magnetisation which bears little relation to the magnetisation required. As shown in figure 5, if the width of the hysteresis loop is 1 mT then a field change of 1mT is sufficient to take the sample on to the return branch of the curve. Cooling in low fields can even lead to an apparent paramagnetic moment This is caused by the need to move a sample into a region of quite different field strength so that this flux density is trapped in the sample when it returns to its original position (8). However nearly all measurements which keep the sample stationary have great difficulty in measuring the magnetisation of type II superconductors. Over most of the field range this is likely to be less than 10mT in external fields of up 100T, so a resolution of 1 in 10^4 is needed. The high T_c materials pose particular problems in that an important part of the curve occurs between 0.1 and 10 mT and the rest of the magnetisation curve extends to over 100T. Few magnetometers can cope well with both régimes.

Perhaps the most popular magnetometer is a SQUID magnetometer. It is very sensitive and convenient to use, but we have found a number of problems when measuring hysteretic materials. This is because in the standard mode the sample is moved through 6 cms. to give a complete response function in the SQUID coils. There is always a very large drift in the SQUID signal so the software assumes a drift, and a standard shaped curve, and delivers a magnetic moment . This works well for the paramagnetic materials for which the system was designed. However the magnet inhomogeneity is such that measurements of hysteresis and the reversibility line require a much smaller scan length. This means that only a small proportion of the response curve is available, and since the magnetisation of the sample may not be uniform along its length, the curve shape may be anomalous and the software cannot cope. We have tried to get round this problem by putting the sample in a small coil which provides enough field to go round a small hysteresis loop, while the SQUID output is monitored directly. The large drift means that sophisticated signal analysis is needed to extract data and the results will be reported elsewhere in these proceedings (9).

The main alternative to a SQUID is a vibrating sample magnetometer which is slightly less sensitive, but has a much smaller movement and is therefore less afflicted by the problems of a SQUID in measuring hysteretic materials.

Although dc magnetometers are not ideal for hysteresis measurements, they are essential for measuring the reversible magnetisation and in particular for measuring B_{c2}.

2.2. *Stationary Samples*

Let us now look at techniques which do not involve sample movement. In the past physicists have used thermal demagnetisation and integration of the voltage round a coil on the sample as the sample demagnetises. Both require that the balance of the coils remain reasonably constant. Balancing search coils in a field is difficult because the effective size and field distribution of the magnet changes with field. The need with high T_c materials to make samples very hot to demagnetise them makes thermal demagnetisation unattractive. Both work well when the magnetisation is large compared with the external field, but in this régime other methods also work well. The challenge is to measure hysteresis, or reversible magnetisation, of less than a millitesla in an external field of over 10 Tesla and this will never be easy.

Force methods have not been used very much in the study of superconductors. This because to create a force requires a field gradient, and since the superconducting properties are very field sensitive, a field gradient will make interpretation difficult. In a Faraday balance the force is maximised by contouring the pole pieces to give a maximum of ∇B^2 which gives the maximum force in a linear material. If we are just looking for small amounts of magnetic material such a system can be very sensitive, since the balance can measure a microgram. The force is proportional to the field gradient multiplied by the magnetic moment, so that if the magnetisation decreases approximately linearly to zero at B_{c2} the maximum force is at half B_{c2}.

If we want to use the balance to measure a magnetisation curve we need to ensure the field variation across the sample is small compared with the applied field. This can be done by using a large superconducting solenoid for the main field, and generating a force with supplementary copper coils. Although we lose some sensitivity the system compares favourably with other types. The sensitivity can be increased by imposing a small oscillation on the coils and using a phase sensitive detector. A Faraday balance should be a very good way of measuring flux creep since it is not necessary to move the sample, and the force due to the dc magnetisation can be backed off backed off electronically.

Torque magnetometers have similar advantages. Here a constant field is appied to the sample and the torque required to turn it in the field is measured. There are three contributions to the torque in a linear material. First there is the shape effect proportional to χ^2 and the difference in demagnetising factors along the principal axes. For fully penetrated superconductors the effective value of χ^2 is very small, so this component is usually smallabove H_{c1}. The second component is the anisotropy torque which is proportional to to the difference in susceptibilities along the crystal axes, and is important in anisotropic materials such as layered superconductors. Finally there is the hysteresis torque due to the need to pull flux lines through the material. The relative size of these last contributions depends on the change in reversible magnetisation with angle, compared with the size of the hysteretic magnetisation. Torque magnetometry can not give the dc magnetisation curve.

Since the geometry of these experiments is usually a disc turning in a perpendicular field it is quite different from the conventional picture of a flux gradient in a long cylinder, as the currents induced are

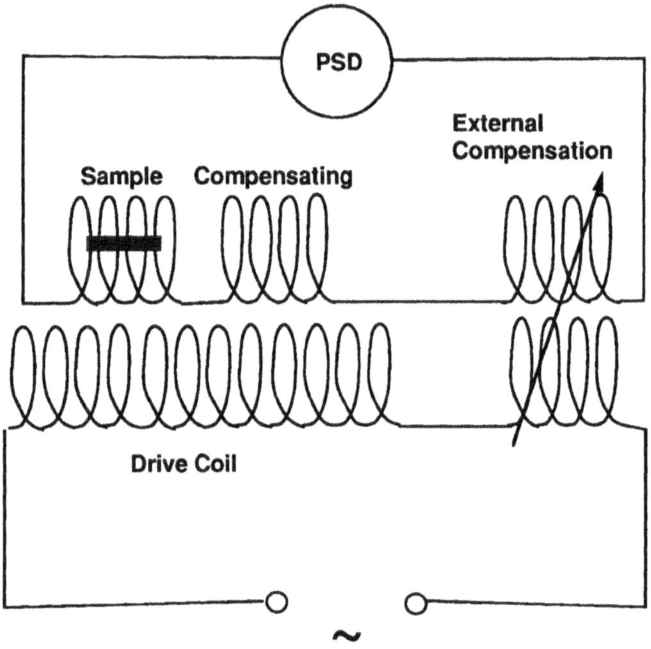

Fig. 7.
Typical coil systems for ac susceptibility measurements

perpendicular to those of the initial critical state. However it is likely that all geometries behave in a similar manner and torque magnetometers have been used with considerable success to measure flux creep. (10,11)

3.AC Magnetisation

Many of the remarks applied to dc magnetisation apply also to ac susceptibility measurements. The apparatus normally consists of a drive coil to provide an oscillating magnetic field and two search coils in opposition inside it. The drive coil should provide a uniform field at the sample position. Figure 7 is a schematic picture of a typical coil system.

The sample is inserted in one coil and the coils are balanced precisely with further compensation (active or passive) outside the cryostat. Normally a simple variable inductance can be used with one coil sliding in another. For sensitive measurements it will be necessary to compensate for the out of phase component as well as the main inductive signal. This can be done by one variable mutual inductance coupled to another fixed one, to give a 180 degree phase shift. A simple alternative is to slide a copper tube over the variable inductance coils. (It is not a good idea to derive this signal by tapping off a resisitive component of the drive circuit. This should be kept isolated from the low level signal circuit.)

For further sensitivity a transformer boosts the voltage, and to reduce mains pickup the compensating coils should divided in two and one turned round through 180 degrees.

For thin films a planar geometry is more convenient, but it leads to a variable field over the sample which makes quantitative conclusions more difficult to draw. Details will be found in reference (12).

The balance may done either with the sample normal, or with it in the Meissner state depending on the type of measurement. The coils are mounted on a long rod, or probe, which can be lowered into a variable temperature insert in a magnet. The limitations are random noise, changes in balance with field or temperature, non-linearity from the signal generator and any superconductors near the coils, and temperature control. The effect of nonlinearity depends on the nature of the compensating network. The search coil and compensating coil should be kept as identical as possible so that the balance is independent of frequency and harmonics are eliminated with the fundamental. Different parameters are limiting in different regimes and it is not possible to pick the best design for all purposes. We look at two extremes.

3.1. High Amplitude

Much of the work on Type II materials has been done on cast samples at helium temperatures. Samples are large so signal strength is not a problem. Large amplitudes are needed to drive the currents to the centre, so non linearity due to interaction with the magnet, or amplifier distortion, and heating of the drive coils are the main limitation. Use of a superconducting drive coil helps in keeping things cool, but adds to the nonlinearity. Relatively large probes are needed so there are large temperature lags and gradients if measurements are not made in a liquid cryogen. Also capacitative effects make high frequency operation difficult. Since balancing is usually done in the superconducting state the balance varies with sample so that there is little point in trying to achieve accurate compensation in the probe itself. (2,13,14)

3.2. *High Sensitivity*

The opposite approach is exemplified in a design by Cooper (15) for small single crystals at low amplitudes. In this design the search coils are outside the drive coils so that changes in their dimensions do not affect the balance. The drive coils are wound on quartz tubes to minimise thermal contraction and the pairs of coils are mounted side by side rather than axially since temperature gradients are less in this direction. Since the drive coils can be layer wound they can be made very accurately identical. This probe remains well balanced on cooling from room temperature to 4.2K and no outside compensation is needed.

3.3. *Moving Samples*

Commercial instruments frerquently use a system in which the sample probe contains neither thermometer nor search coils, but is a simple tube. The sample is then moved between the search coils. This has the disadvantage that the coils are very poorly coupled to the specimen has but several advantages. The search coils are kept in the constant temperature environment so that the balance does not change with sample temperature, and the problem of out of balance signals almost disappears.

A second major advantage is that the phase can be set accurately by balancing without the sample and then inserting the sample at zero field. Setting the phase correctly to measure losses in the presence of the inevitable background signal is difficult if the sample cannot be moved. The background signal may be of any phase and to cancel it out we must raise the sample above T_c. if the sample cannot be moved. However on cooling the phase of the background may not be the same as when it was balanced out. If there is a large diamagnetic signal a small inaccuracy in setting the phase will produce an apparent loss from the component of this which is sampled. It is a good idea to look for losses in two experiments. For large penetrations the coils should be balanced with no sample, while for small penetrations it should be balanced in the Meissner state. Oscillating the sample slowly up and down and using a second lock-in makes this type of susceptometer even more sensitive.

Under this heading we can put accoustic resonance techniques using a vibrating reed shaped sample, or a sample stuck to a vibrating spring (15,16). It should make no difference whether we move the sample or move the external field. Therefore these measurements measure the same thing as ac susceptometers although they tend to impose a fixed displacement rather than a fixed force on the vortex lattice. It is not yet clear whether they can cover a different regime or can be made more sensitive, but the geometry makes the results much more difficult to analyse.

3.4. *Calibration*

One of the problems in turning magnetisation into useful results is the calibration of the system. This is closely connected with the proportion of superconductor in the sample and is dealt with in some detail by Blunt et. al. (17). With conventional superconductors this was usually done by assuming the initial slope of the magnetisation curve to be unity. In ac measurements this is equivalent to taking the difference in signal between the normal and Meissner states to calibrate the system.

In high T_c materials this can cause problems. In a sintered sample the initial slope should correspond to the sample size. This requires that the amplitude should be such that currents do not exceed the intergrain critical current. When new materials appear they are of poor quality and the intergrain current may be very small. We have found an amplitude of less than 1μT is needed to get the full response from a cylinder of doped YBCO. (17)

In practice many magnetic measurements are carried out on powders or at fields which decouple the current between grains. We have made measurements on samples with diluted powder at increasing densities. It was found that for ground YBCO a demagnetising factor of 1.5, (i.e. a sphere), works well and up to 66% volume fraction (close random packing) the Lorenz correction gives a good fit to the calibration curve. The expression is $\chi=-3f/(2+f)$ where f is the volume fraction of superconductor. However decoupled grains in a sintered ceramic are denser than this and the magnetic response of such a system is difficult to model. It is usually bettter to use the weight of the sample and the density to get the volume of superconductor. For large penetrations the shape is not important.

3.5. The Superconducting Fraction

The response at low fields has often been used to assess the proportion of superconducting phases present. However the proportion of diamagnetism is very much influenced by two other factors. Firstly the decoupled grains look like spheres so that they appear to be 150% diamagnetic if large and reasonably well spaced. Counteracting this is the effect of a penetration depth comparable with the particle size which reduces the diamagnetism.

A low diamagnetic signal is usually interpreted as showing a small percentage of superconductor, but we believe this is more often caused by the combination of a large penetration depth and a small particle size. It is surprisingly difficult to find an experimental test which can distinguish between the two interpretations of imperfect ac diamagnetism. However it should be remembered that to show up as a diamagnetic phase mixed with a normal phase the scale of the microstructure must be larger than the penetration depth. Phase separation on this scale should be visible in a microscope and,if the diamagnetic signal is not too small, should be detectable by X rays.

It has sometimes been suggested that the field cooled magnetisation gives a measure of the proportion of superconductor. This is quite wrong. The magnetic moment on cooling in a field is a complex combination of the rate of change of J_c and H_{c1} with temperature. (18). When combined with magnet inhomogeneity it is even possible to get an apparent paramagnetic moment (8). This is caused by movement into a higher field region which traps flux on returning to the original position. It is difficult to see anything useful which has come out of field cooled measurements.

The calibration difficulty is part of a larger problem which is that if the material is subdivided it is very difficult to find out the size scale on which currents are flowing, and hence to get values for J_c. The ideal answer would be to break up the specimen into smaller pieces but small powders are difficult to handle and size, and the process can alter the material. The size scale is considered again below (see §4.2).

4. Treatment of Results

Whatever method is used the end result is a voltage proportional to the magnetisation of the sample. This can come from a minor hysteresis loop in a dc curve, but since the ac methods are more sensitive we concentrate on them. An interesting comparison of the techniques to be described is in reference (14).

There are three things we can do with this voltage. The first is to filter out harmonics and look at the real and imaginary parts. With a model assuming the parameters of the superconductor we can predict the results and adjust the parameters to fit the experiments. By filtering we improve signal to noise ratio and so long as the superconductor conforms to the model the parameters are useful measures(4,5). However the algebra is complex so if the superconductor does not behave as expected this may not be picked up in the analysis.

A second method was proposed by Bean (19) which is to measure the third harmonic of the signal. The big attraction of this is that almost nothing except a superconductor produces a third harmonic so that balancing coils are redundant, and even a search coil is not necessary since the third harmonic in the drive coil can be picked out from the much larger fundamental. The disadvantage is again that a simple model must be assumed to predict the third harmonic and it is difficult to cope with materials which do not obey the Bean model with constant J_c.

4.1. Total Flux Measurements

The other main technique is to measure the total flux put into the sample by the ac signal.(20,2) This can be done conveniently using a phase sensitive detectorwhich inverts the signal every half cycle and averages the resultant voltage. Figure 8 shows a typical wave form which would be generated by the Bean model, and the effect of inverting the signal by a phase sensitive detector set to pick up the in phase signal.

The output voltage is directly connected to the flux in the sample.

$$V=\int d\phi/dt = \phi(\pi)-\phi(0).$$

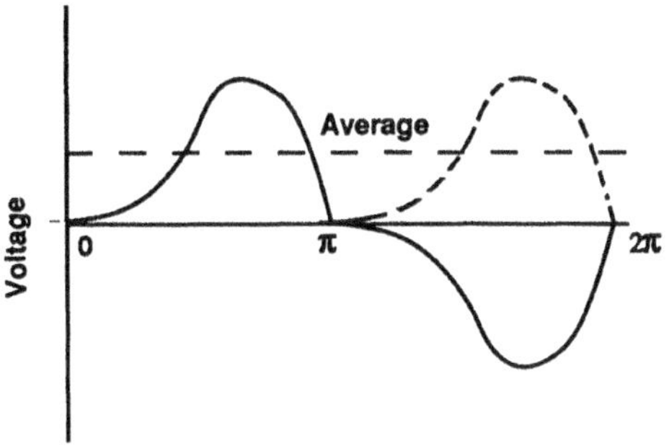

Fig. 8.
The voltage wave form with one half period inverted by the PSD

If the phase is set for maximum signal the lock-in signal is proportional to the difference in the flux in the sample at the maximum and minimum fields of the cycle, and at intermediate phases we get the flux at other points in the cycle. It is possible to trace out the complete hysteresis loop at constant amplitude by stepping the phase from 0 to 2π. (21).

An alternative technique is to record the complete wave form and integrate it numerically to get the flux at all points in the cycle 13). This generates a very large amount of information and it is frequently found that there are complexities which are difficult to explain. For example any model based on the critical state must give zero voltage at 0 and π, which are the extremes of the drive field cycle. However it is often found that the waveform is shifted in phase and that there is a finite voltage at these points which cannot be satisfactorily explained by flux creep.

The main advantage of measuring the flux is that is closely related to the physical model of the superconductor. For example if the material is inhomogeneous we can use it to measure the flux profile in a sample, that is to say the variation of B, and hence J_c, as function of depth from the surface

Figure 9a shows the field distribution if the surface layer has very high pinning. If we assume J_c is independent of B, which is accurate if we are imposing a small ripple on a large external field, then increasing the ampltiitude by δb increase the flux by $\delta\phi = x\delta b$ where x is the penetration of the signal into the sample. Hence by plotting the derivative of the signal with respect to amplitude, $d\phi/db$, against the

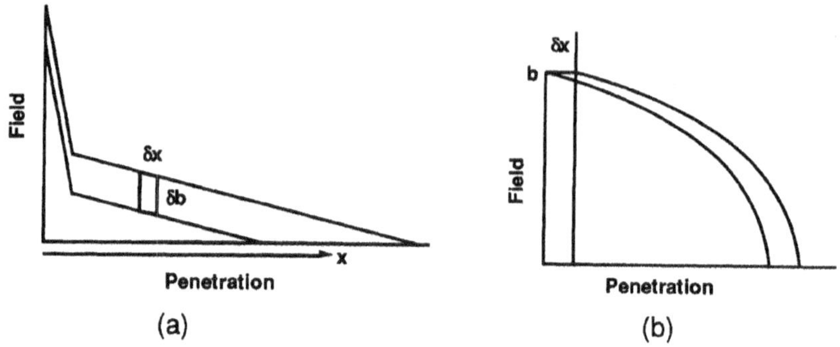

Fig. 9.
a) The change in flux for a sample in which J_c varies only with position
b)The change in flux for a sample in which J_c varies only with field.

amplitude we get a plot of the flux profile in the sample. (This expression can be easily modified for cylindrical geometry.)
The distance scale is calibrated by measuring the signal in the normal state, when the penetration is the sample halfwidth.

A different situation arises if the current density varies rapidly with field, but the sample is homogeneous (Fig.9b). This is likely to arise at very low or zero external field. Now there is a unique critical state profile given by the solution of $dB/dx=\mu\mu_o J_c(B)$. Suppose we have no external field and the amplitude of the ac field is b. If we raise the amplitude by δb the new profile is found by moving this universal curve into the sample until the value of b at the surface matches the external value. The extra flux is $\delta\phi=b\delta x=b\delta b(dx/db)=b\delta b/(\mu\mu_o J_c(b))$. Hence the derivative now gives us J_c as a function b, but again it also gives the penetration for a value of J_c averaged over the amplitude b.

In fact for almost any reasonable decay of the field with distance the value of $\delta\phi/db$ will give an estimate of the penetration of the field into the sample, so that it is always a useful quantity to plot. Figure 10 shows a number of possible results.

Figure 10a shows a typical plot in a conventional superconductor in a large applied field. At low amplitudes the response is linear, leading to an apparent penetration λ', independent of amplitude. (This is not the result of a finite London penetration depth, since the coils are balanced in the Meissner state so all penetrations start from λ inside the surface). This linear response is due to the reversible movement of flux lines in their potential wells and makes the superconductor appear like a London type superconductor with a large penetration depth which depends on the pinning. It is the cause of the broadening of the inductive transition in homogeneous materials at high fields. At larger amplitudes we get the linear gradient of the critical state, until the flux reaches the centre and the plot goes vertical since flux can penetrate no further. The signal here should be equal to that in the normal state.

Figure 10b shows a plot for a cast sample of A15 superconductor. Chevrell phases often give a similar plot. If not very carefully prepared these materials show many of the granular properties of high T_c materials. However the plot shows a marked contrast to 10c which is for a sintered high T_c material.

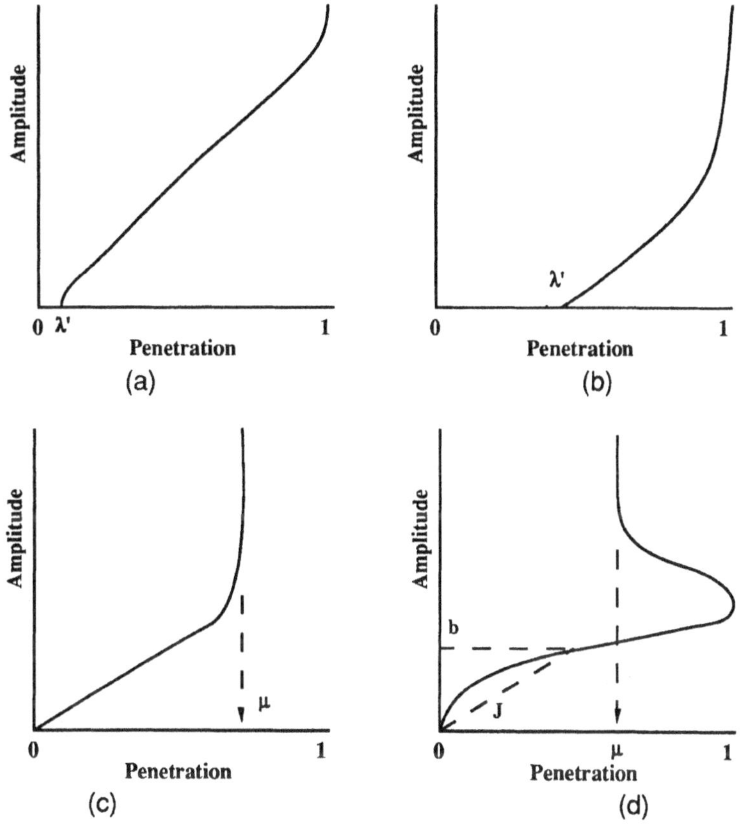

Fig. 10.
Some typical graphs of dϕ/db

In Fig.10b the large penetration in the limit of zero amplitude suggests that a significant volume fraction of the material is normal. There is some difficulty in understanding this type of curve but the continuous transition to full penetration could be produced by decoupled grains with the best material in the centre. This is likely to be caused by segregation during solidification leading to a smooth random variation of J_c through the material.

Figure 10c shows a typical plot for a high T_c material in zero field. The flux penetrates linearly but stops penetrating well before the centre. This is because we are measuring B, and although currents have reached the centre the diamagnetic grains exclude most of the flux. The effective permeability is given by the position of the vertical line, i.e if for example it is at 0.7 of the fully penetrated position in slab geometry the effective permeability defined in §1 is μ=0.7. By going to higher amplitudes, or higher external fields, we can plot the penetration into the grains as well. However there is always a very clear discontinuity, indicating a sharp distinction between grain boundary and bulk material. This is the clearest evidence of the sort of granularity that occurs in mixed oxides and is in marked contrast to the fuzzy transition to full penetration which is seen in poorly prepared metals and Chevrell phases.

Figure 10d shows the type of curve often obtained with sintered material in zero external field. The high gradient at low amplitudes might have been attributed to a surface barrier were it not for the 'nose' at full penetration. This 'nose' is the rapid increase in penetration as the steep flux fronts approaching from opposite sides of the sample meet in the middle meet, and it also occurs in susceptibility measurements (4).

Finally, in circumstances in which J_c is constant, we can get the full force displacement curve of the flux lines by plotting b db/ds against s, where b is the ripple amplitude and s the signal. The derivation will be found in reference (2). The shape is typical of many hysteretic systems such as domains in permanent magnets and plastic flow of dislocations. At low displacements the response is linear and reversible. The effect is like that of a large London penetration depth as discussed in above The depth depends on the pinning and is proportional to the initial slope of this graph. At larger displacements we start to unpin flux lines and we get hysteresis. Beyond a certain distance, sometimes called the interaction distance, the full critical state is built up and we get a frictional resistance to flux motion with a force BJ_c. It is this interaction distance, or maximum reversible displacement, which determines whether susceptibility measurements are in the high amplitude or low amplitude regime.

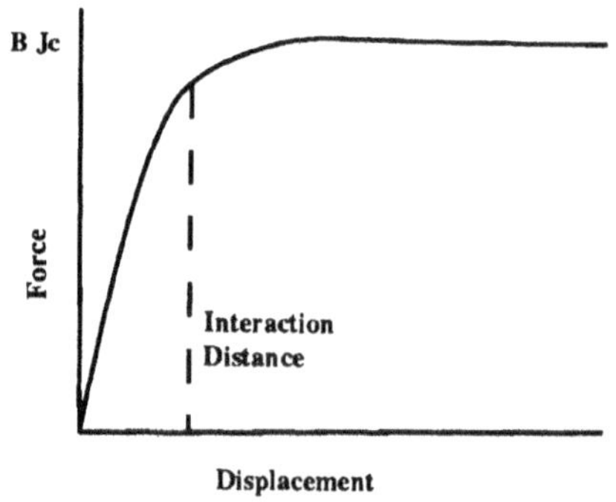

Fig. 11.

A Typical Force Displacement Curve. The linear slope gives λ' and the interaction distance is where irreversibility becomes important.

There are a number of problems in using these techniques in high T_c materials. Simple expressions are only obtained for large slabs and cylinders. If the particles are small compared with λ' it becomes necessary to fit parameters, although qualitative conclusions can still be drawn. The effect of having roughly spherical particles smooths out the curves and has not been considered in any detail. Finally there is an unknown length scale on which the currents flow which comes into all the formulae and which is normaly taken to be the grain size.

AC susceptibility measurements can be used to measure H_{c1} and are much more sensitive than looking for deviations from the straight line on a magnetisation curve. Loss and penetration increase rapidly when either the amplitude or the external field exceed H_{c1}. However the transition is never sharp since there is always some penetration into asperities and surface barriers can delay the entry. In powders small particle sizes increase the apparent H_{c1} so there are always problems in getting an accurate value.

4.2. Thin films

It is conventional to show the critical state as a flux gradient. However in materials with a demagnetising factor close to one the flux density is equal to the applied field and the driving force comes from a curvature of flux lines. Thermodynamics tells us that the driving force is still **BxJ** and the critical state equations still apply. There is no problem in finding the magnetisation for complete penetration, we just put a uniform current density J_c in a sheet and work out the magnetic moment. However the progress of the critical state from the edge to the centre is not straightforward and involves penetration from the flat faces rather than the edges. Numerical solutions have been published recently (22).

In thin films penetration from the flat faces has little meaning and on the scale of the vortices it is difficult to see how flux creep can occur if the vortex density remains constant. In spite of these conceptual difficulties, and the lack of experimental comparison between the critical state model in samples with large and small demagnetising effects, the fact that no major anomalies have been seen suggests that these worries are of mainly academic interest.

In a recent paper Angadi et. al. (23) have shown that in a sample of large demagnetising factor a unique length scale can be extracted. This technique could prove very useful in determining the granularity of materials. By measuring the slope of a minor hysteresis loop in a magnetisation experiment, or the susceptibility at low ampltudes, we can measure the demagnetising factor, and hence the ratio of diameter to thickness of the independent current carrying regions.

Since many measurements are made on thin films it is also important to consider the effect of putting the film at an angle to the field. In a superconductor **E=Bxv** so that the electric field is perpendicular to the magnetic field. This is not necessarily true in a normal conductor. For example a copper cylinder placed at an angle to an applied field will have currents generated with a component parallel to the field, so if we use a superconducting cylinder we must also get currents parallel to the field. The flux lines build up a force free configuration of helices and will try to cut each other. There is no satisfactory model for predicting longitudinal current flows in bulk samples and therefore we do not expect to explain the magnetisation of samples at an angle to the applied field until we can explain longitudinal currents.

However in thin films there is probably not enough room to generate any kind of force free configuration and we can model it with a flat mesh of thin filaments each with a critical current i_c. At full penetration all wires are carrying this current so the current distribution and magnetic moment are the same as if the field had been applied normal to the film. However the direction of the moment is perpendicular to the film. Now the magnetisation in a perfectly parallel field is J_c times the film thickness, while in a perpendicular field it is J_c times the film width which is hundreds of times larger. Thus measurements made in a field not exactly parallel to the film are essentially the perpendicular magnetisation multiplied by the sine of the misorientation angle. Only if the misalignment is less than the film thickness divided by the width will the parallel moment be greater than the component of the perpendicular moment. One way round this problem is to cut the film into narrow strips before making measurements but this technique does not seem to have been tried.

5. Shape Effects in the Linear Regime

5.1. The Generalised London Equation

The most physical picture of the behaviour of flux lines in superconductors is to express forces in terms of the displacement of the flux lines, **y**. Since a displacement along a flux line has no meaning, **y** is perpendicular to the local flux density **B**. The flux displacement can be directly related to the vector potential **A**. as follows (24).

Suppose the flux moves a small distance $\delta\mathbf{y}$ in a time dt.

Then:- $\mathrm{curl}\dot{\mathbf{A}} = \dot{\mathbf{B}} = -\mathrm{curl}\mathbf{E} = -\mathrm{curl}(\mathbf{B}\times\dot{\mathbf{y}})$.

Hence $\dot{\mathbf{A}} = -\mathbf{B}\times\dot{\mathbf{y}}$

and for small changes:-

$$\delta\mathbf{A} = -\mathbf{B}\times\delta\mathbf{y}$$

Thus the vector potential is directly related to the displacement of flux lines, which explains why it is also closely connected with the phase of the order parameter. We consider a material in a uniform applied field $\mathbf{B_o}$. We then impose a small change **b.** Displacements and the vector potential are measured as changes from this starting point. Since all displacements are assumed small on the scale of the sample:-

$$\mathbf{A} = -\mathbf{B_o}\times\mathbf{y}.$$

Two cases follow. We can assume the force is a general function of $\mathbf{y}$ and $\dot{\mathbf{y}}$. This leads to a modification of the critical state model.

However in this work we assume that the vortex displacement is much less than the vortex spacing so the problem becomes linear. We then assume a harmonic oscillation so that d/dt=-iω. The response can be written as a function of frequency, multiplied by the driving force.

$$\mathbf{y}=f(i\omega)\mathbf{B}\mathbf{x}\mathbf{J}.$$

Then $\mathbf{A}=-\mathbf{B_o}\mathbf{x}\mathbf{y}=-f(i\omega)\mathbf{B_o}\mathbf{x}(\mathbf{B_o}\mathbf{x}\mathbf{J})=-f(i\omega)(\mathbf{B_o}.\mathbf{B_o})\mathbf{J}+f(i\omega)(\mathbf{B_o}.\mathbf{J})\mathbf{B_o}$.

We consider only situations in which the current is perpendicular to **B**

Then $\mathbf{A}=-B_o^2 f(i\omega)\mathbf{J}=-B_o^2 f(i\omega)\mathrm{CurlCurl}(\mathbf{A})/\mu_o$.

That is $\mathbf{A}=\lambda_e^2$ Curl$\mathbf{A}$ where $\lambda_e^2=-B_o^2 f(i\omega)/\mu_o$. (1)

This equation has exactly the same form as the London equations except that the effective penetration depth λ_e is complex and a function of frequency. The effective penetration depth is B($_o^2$f(iω)/)$^{1/2}$. This equation also has the same form as the equation for penetration of microwaves into a superconductor and for the skin effect in a normal metal. We can therefore use solutions calculated for these phenomena to derive the magnetisation of a superconductor provided the force on the flux lines is a linear function of their displacement and speed.

5.2. *Susceptibility of Ellipsoids*

We can now write down the susceptibility of spheres and cylinders by using the the susceptibility worked out for a London superconductor. i.e. if the field in a thick slab decays from the surface as $\exp(-x/\lambda_e)$ we can immediately write down the moment per unit volume in a field H_0:-

For a Slab of half width a	$-H_0 \mathrm{Tanh}(a/\lambda_e)$
For a Longitudinal Cylinder radius a	$-H_0 I_1(a/\lambda_e)/I_0(a/\lambda_e)$
For a transverse cylinder	$-2H_0((2\lambda_e/a)I_1(a/\lambda_e)/I_0(a/\lambda_e)-1)$
For a Sphere of radius a	$-(3/2)H_0(1-(3\lambda_e/a)\ \mathrm{Coth}(a/\lambda_e)\ +3\lambda_e^2/a^2\ )$

Since the equation is the same as the eddy current equation we can make use of classical eddy current calculations for more complex shapes. The effective London penetration depth is iδ where δ is the classical skin depth.

We have established that for a linear response of the flux lines the London equations are obeyed so that we can deal with spheres and cylinders as easily as slabs. This is important as most high T_c .superconductors consist of grains for which the best approximation is a sphere.

6. The Linear Response Function and the Reversibility Line

6.1. The Effective Penetration Depth

We now consider the form of $f(i\omega)$. There are four linear forces on vortices. First there is the restoring force proportional to the displacement which depends on the potential well curvature. Secondly there is the viscous drag which the vortex experiences as it moves in the well. This depends on the flux flow resistivity. These have been combined by Coffey and Clem (25). Thirdly there is the Magnus force which gives rise to a small Hall effect and will be ignored. (2)

The fourth effect is the thermal activation of vortices which leads to the observed resistivity, as opposed to the flux flow resistivity calculated by Bardeen and Stephen.(26) This has been combined with the linear restoring force by Kes (27). A complete description of the reversibility line needs both resistivities, although it should be said that when the numbers are inserted most practical situations are dominated by one or other of the resistivities. Flux flow resistivity is dominant at low temperatures and high frequencies while linear thermal activation is dominant at high temperatures and low frequencies. Both regimes are different from that dealt with by Takacs and Gömöry (28) who were using amplitudes comparable to the applied fields so that vortex displacements were large, and a modification of the critical state model was needed.

We first consider the elastic displacement of the flux lines $\mathbf{y_e}$. The equation of motion is:-

$$\mathbf{B} \times \mathbf{J} = -\alpha \mathbf{y_e} - \eta_1 \dot{\mathbf{y}}_\mathbf{e}$$

where α is the force constant of the potential well and η_1 the viscous drag on the flux lines.

Hence $$\mathbf{y_e} = \mathbf{B} \times \mathbf{J}/(-\alpha + i\omega\eta_1).$$

Thermal activation of vortices out of wells leads to a plastic deformation, $\mathbf{y_p}$. The equation of motion is:-

$$\mathbf{B} \times \mathbf{J} = -\eta_2 \dot{\mathbf{y}}_\mathbf{p}$$

Hence $$\mathbf{y_p} = \mathbf{B} \times \mathbf{J}/i\omega\eta_2.$$

The total deformation is the sum of the plastic and elastic deformation. This was the assumption used by Kes and it is standard procedure in the theory of plastic flow in solids. It leads to a slight problem here in that the vortices cannot take part simultaneously in thermal hopping and an elastic oscillation. To take this into account requires a more detailed model so we assume that only a small proportion of the time is spent in the hopping process. This is true until the resistivity is close to that of the normal state.

With this assumption the total displacement is:-

$$\mathbf{y} = \mathbf{y_e} + \mathbf{y_p} = \mathbf{B} \times \mathbf{J}\{(1/(-\alpha + i\omega\eta_1) + 1/i\omega\eta_2)\}.$$

We see this fulfils the condition that the force on the flux lines is a function of the displacement and ω, so that the London equations are obeyed and from equation (1) the effective penetration depth is:-

$$\lambda^2 = -B^2/\mu_o\{(1/(-\alpha + i\omega\eta_1) + 1/i\omega\eta_2)\}.$$

For clarity we write this in terms of three lengths. $\lambda'=\sqrt{B^2/\mu_o\alpha}$, the pinning penetration depth, $\delta_1=B\sqrt{2/\mu_o\eta_1\omega}$, the skin depth based on the flux flow resistivity, and $\delta_2=B\sqrt{2/\mu_o\eta_2\omega}$,the skin depth based on the observed resistivity which includes flux creep.

The final result is:-

$$\lambda^2 = \frac{1}{\frac{1}{\lambda'^2} - \frac{2i}{\delta_1{}^2}} - \frac{\delta_2^2}{2i}$$

The best way to use this expression is to measure the susceptibility as a function of frequency and to adjust the parameters so that the expression gives the best fit to the experimental results.. However most experimental data involves measuring χ as a function of B or T. To fit this type of experiment needs a large number of assumptions on how the superconducting parameters vary with T/T_c and B. However by fitting typical variations of resistance, resistivity, and λ' in the mid temperature range we can reproduce the features of the reversibilty line.

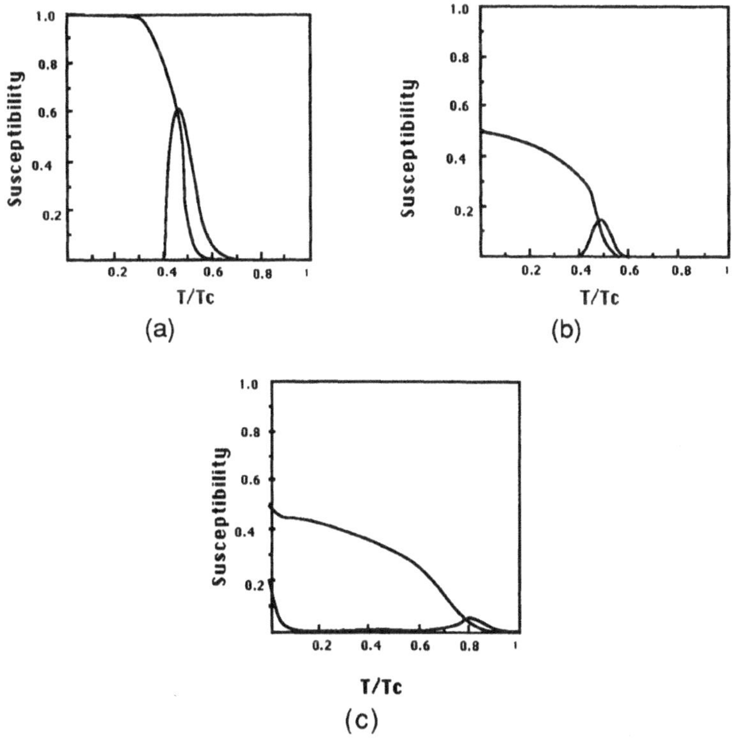

Fig. 12.

Three typical computed plots of the inductive transition

6.2. Plotting the Revesibility Line

Figure 12a is for a sample large compared with λ' at 400 Hz.so that at low temperatures it is perfectly diamagnetic and over most of the range the skin depth is larger than the sample size. However a low activation energy is assumed so that above about half Tc the material is completely reversible due to rapid flux creep. The inductive transition is narrow, it occurs at about half the critical temperature, which is where the skin depth is comparable to the sample size, and the loss peak is about half of the inductive change.
Figure 12b shows what happens if we reduce the pinning, or particle size, so that λ' is comparable to it. The transition occurs at the same temperature, but the inductive component is broadened and goes to a value less than perfect diamagnetism at low temperatures.
Figure 12c shows the effect of increasing the frequency to a megahertz. The low temperature diamagnetism is not changed except for a rise at very low temperatures. This is due to the assumption that the normal state resistivity continues to reduce linearly at low temperatures which cannot be true to absolute zero. The flux flow viscosity has little effect in the sample size assumed, 10 microns. However the transition is shifted to much higher temperatures because it occurs when then skin depth is comparable to the sample size. This means that both higher frequencies and larger grain sizes will move the peak of the loss curve to higher fields and temperatures. If the reversibility line is derived from dc magnetisation experiments it will be similarly shifted by an increase in grain size and this is an inevitable result of the critical state model. Since ac measurements of loss and hysteresis are much more sensitive than dc measurements the best definition of the reversibility line is the the peak of the loss curve as a function of frequency, extrapolated to zero frequency.

References

1) L.Landau and Lifschitz, The Electrodynamics of Continuous Media, Pergamon, Oxford, p123 (1960).
2) A.M.Campbell and J.E.Evetts, Adv. in Phys., **21**, 199., (1972)
3) R.P.Huebner, Magnetic Flux Structures in Superconductors, Springer Verlag, p122.
4) F.Gömöry, S.Takacs, P.Lobotka, K.Frölich, and V.Plechaacek, Physica C,**160**, 1, (1989)
5) K.H.Müller, Physica C , **159**, 717,(1989) and also these proceedings.
6) Grover, F.W., 'Inductance Calculations', Dover Publications Inc.,1973.
7) A.Perry and A.M. Campbell, these proceedings.
8) F.J. Blunt, A.R.Perry, A.M.Campbell, and R.S.Liu, Physica C,**175**,539, (1991).
9) A.Perry and A.M. Campbell, these proceedings.
10) C.W.Hagen and R.Griessen, Phys.Rev. Lett., **62**, 2857, (1989).
11) A. Hamzic, L.Fruchter, and I. Campbell, Nature, **345**, 515, (1990).

12) J.H.Claasen, IEE Trans. in Magnetics, **25**, 2233,(1989).
13) R.W.Rollins, H. Küpfer, and W. Gey, J. Appl. Phys., **45**, 5392, (1974).
14) D.M.Kroeger, C.C.Koch and J.P.Charlesworth, J.Low Temp. Phys., **19**, 493, (1975).
15) J.R.Cooper, L.Forro, G. Collin, and J.Y. Henry, Sol.State Comm.,**75**, 737, (1990)
16) E.H.Brandt, P.Esquinazi, and G.Weiss, Phys.Rev.Lett., **B62**, 751, (1991)
17) P.L.Gammel, L.F.Schneemeyer, J.V. Waszczak and D.J.Bishop, Phys.Rev. Lett., **61**, 1666, (1988).
18) A.M.Campbell, F.J.Blunt, J.D.Johnson, and P.A.Freeman, Cryogenics.,**31**, 732 (1991).
19) C.P. Bean, Rev. Mod. Phys.,**36,** 31, (1964).
20) A.M. Campbell, J. Phys. C., **2**, 1492, (1969)
21) S. Takacs, F.Gömöry, and P. Lobotka, IEEE Trans on magnetics, (Proc Appl. Superc. Conf Snowmass), **27**,1057. (1991).
22) R.Navarro and L.J. Campbell, to be published.
23) M.A.Angadi, A.D.Caplin, J.R.Lavery, and Z.X.Shen To be published in Physica C.
24) B.D.Josephson , Phys.Rev, A, **152**, 211.(1965)
25) M.W.Coffey and J.R.Clem, IEEE Trans on magnetics, (Proc Appl. Superc. Conf Snowmass), **27**,2136,. (1991).
26) J. Bardeen and M.J.Stephen, Phys. Rev. **140,** A1197, (1965).
27) P.Kes, Proc. Int. Conf on High T_c Superconductors, Garmisch Partenkirshen, (1990).
28) F.Gömöry, and S. Takacs, Proceedings of 6th. Int. Conference on Critical Currents, Cambridge, to be published in Supercon. Sci. and Technology,(1991).
29) A.M.Campbell and F.J.Blunt, Physica C, **172**, 253,1990.

STUDIES OF "NON-IDEAL" SUPERCONDUCTORS USING DC MAGNETIC METHODS

James R. Thompson

Solid State Division, Oak Ridge National Laboratory and Department of Physics, University of Tennessee, Knoxville.

D. K. Christen, H. R. Kerchner, L. A. Boatner, B. C. Sales, B. C. Chakoumakos, H. Hsu, J. Brynestad, D. M. Kroeger, R. K. Williams

Oak Ridge National Laboratory, P. O. Box 2008, Oak Ridge, TN 37831

Yang Ren Sun, Y. C. Kim, J. G. Ossandon

Dept of Physics, University of Tennessee, Knoxville TN 37996-1200

A. P. Malozemoff,* L. Civale, A. D. Marwick, T. K. Worthington, L. Krusin-Elbaum, and F. Holtzberg

IBM Watson Research Center, Yorktown Heights, NY 10598

INTRODUCTION

Among the most informative and facile methods for investigations of a superconductor are measurements of its static magnetization. The objective of this paper is to analyze some experimental features frequently observed in static (dc) magnetization studies of conventional and high-T_c superconductors. We shall discuss investigations employing measurement protocols in which the sample is cooled through the superconductive transition temperature in a finite magnetic field ("field-cooled") and compare this with zero-field-cooled studies. Also considered are reversible and irreversible materials; particle size effects; some effects of granular and multiply-connected materials; penetration depth studies of type II materials in the vortex state; and fine scale "multi-connected" materials formed by heavy ion irradiation that produces very significant enhancements of the critical current density J_c in the high temperature superconductor $Y_1Ba_2Cu_3O_7$. Many of these superconductors are "non-ideal" in that they may contain defects and inhomogeneities or may have small dimensions comparable with microscopic superconductive lengths. Practically speaking, however, such materials are often

Magnetic Susceptibility of Superconductors and Other Spin Systems
Edited by R.A. Hein *et al.*, Plenum Press, New York, 1991

encountered experimentally and can have very desirable physical properties, such as the high J_c values that developed in the ion-irradiated crystal. In the following, we assume that the reader is familiar with the fundamentals of superconductivity, which are thoroughly treated in many sources.[1-4]

While static magnetization measurements are quite useful, formidable problems of interpretation arise in studies of superconductors where the geometry is unknown and may be "irregular", e.g. very finely divided or multiply-connected. The difficulty is that the measured magnetic moment **m** of a superconductor arises from a distribution of volume and surface current densities **J**(**r**), with

$$\mathbf{m} = 1/(2c) \int dV\, [\mathbf{r} \times \mathbf{J}(\mathbf{r})]. \tag{1}$$

In the case of an inhomogeneous material with unknown geometry and volume, a sole measurement of its bulk moment provides information that is useful but also limited, since it is clearly not possible to reconstruct the distribution of currents **J**(**r**) from its integral.

I. EXPERIMENTAL METHODS FOR DC STUDIES

Most experimental methods used in static "magnetization" studies actually measure the magnetic moment "m" of a specimen via its associated field or the magnetic force on it. The most popular systems currently in use are the vibrating sample magnetometer (VSM) and the SQUID magnetometer, with some usage of force and torque magnetometers. Previously, induction methods with detection using a ballistic galvanometer, etc., were widely employed.

Sources of external magnetizing field H_{app} include magnets with continuously supplied currents flowing in nonsuperconducting coils (copper windings, Bitter solenoids, iron electromagnets, etc.) or in superconducting solenoids operated either with continuously supplied current or in a persistent mode. Implicit in the idea of static magnetization studies is that the sample experiences a magnetic field that is constant in time. However, most measurement methods entail some movement of the sample, in order to change its flux coupling to a measuring coil system and associated electronics. The motion is the order of 1 mm for a VSM and 10-50 mm for commercial SQUID-based instruments. Due to the spatial inhomogeneity inherent in all practical magnets, sample movement inevitably entails some temporal variation of H_{app}. The deleterious influence of this field excursion is most severe in two cases: (1) in high fields where the absolute variation of H_{app} is largest and (2) in fields that are nominally very low, so that residual field variations may constitute a large fraction of the total field. Each of these problems is exacerbated by sample geometries that tend to magnify changes in external field: thin film and "flat" geometries with large demagnetization factors are worst in this regard. Recent experience has shown that, without considerable care, widely used commercial instruments and others can generate field variations that are unacceptably large for study of irreversible superconductors.

Many external perturbations can produce large fractional changes in H_{app} when it is small. At levels near 1 Oe, these obviously include the earth's field. Others are magnetized construction materials such as structural and reinforcing steel in the building, welded "non-magnetic" stainless steel in cryogenic vessels; superconducting solders including at least one superconducting silver brazing alloy.[5] In the case of the increasingly popular SQUID-based magnetometers with superconducting magnets that remain in liquid helium for long periods (weeks to years), trapped flux in the magnet can generate fields that are inhomogeneous with non-axial components, according to indirect evidence from our laboratory. These small fields persist after deliberately quenching the superconducting magnet, using a scheme designed to eliminate most trapped flux. Other systems that use superconductive magnets are subject to these same problems, of course, but in more conventional usage, the magnets tend to be warmed above their respective T_c values fairly often, thus minimizing the problems of trapped fields. As mentioned, all of these effects are accentuated when the sample geometry is very oblate (e.g., thin films or plate-like crystals with H_{app} perpendicular to the surface).

Some of these considerations are obvious while others are less so. Considerable caution is appropriate when conducting and interpreting experiments that depend critically on a static field history to insure that the field on the sample actually remains stationary in time. Finally, it is encouraging that commercial equipment manufacturers are issuing technical application notes to users as various problems come to surface. All users of commercial instruments should review this information very carefully.

In general, electromagnetic units are used in this manuscript. Thus the magnetizing field H is measured in Oersted, which in free space is numerically equal to the flux density B that has units of Gauss, i.e., 1 Oe = 1 G in a vacuum. The unit of magnetic moment m is G-cm^3; we believe that it is most convenient and clear to identify the nebulous "emu" directly and solely with magnetic moment, so that 1 "emu" is 1 G-cm^3. Then the magnetization M = (sample moment)/(sample volume) has units of G. Finally, the volume susceptibility χ = (dM/dH) formally has units of G/Oe and is dimensionless. As is discussed below, one has in these units that $\chi = -1/(4\pi)$ for an ideal, completely diamagnetic body in which the flux density B = 0 everywhere. Consequently it is frequently convenient to reference the susceptibility of a superconductor to this ideal behavior and use the quantity $4\pi\chi$ which has a limiting value of -1. This we refer to as the "normalized susceptibility." It is <u>not</u> a ratio that is referenced to some limiting value, e.g., $\chi(T)/\chi(T{=}4.2K)$; in general, this latter practice is unfortunate, unless the reference value is stated explicitly.

Many methods and standards can be used to calibrate instrumental sensitivity. One convenient arrangement is the use of well-annealed, high-purity, relatively reversible spheres, one of Ni and one of Nb. In fields $H \geq 10$ kOe (≥ 1 Tesla), the ferromagnetic Ni sphere has a saturation magnetic moment that varies only weakly with field and temperature for $T \sim 295$ K. Under these conditions, the magnetic moment of a Ni sphere of 1 mm diameter is 0.256 G-cm^3. On the other hand, the low field magnetization of the superconducting Nb sphere provides a susceptibility standard that is temperature independent near 4.2 K. Such a sphere, 1 mm diameter, develops a diamagnetic moment of -0.625×10^{-3} G-cm^3 in an applied field of 10 Oe. Together, these two standards allow independent confirmation of the calibrations for field and for magnetic moment.

II. SUPERCONDUCTORS IN SMALL MAGNETIC FIELDS ($H_{eff} < H_{c1}$)

A. Meissner Effect

The Meissner Effect, discovered by Meissner and Ochsenfeld[6] in 1933, occurs when a superconductor spontaneously expels a static magnetic field from its interior, upon cooling through T_c. As emphasized by Hein,[7] the Meissner Effect is not amenable to measurement by ac methods, but rather by measurement of the static magnetic moment. To observe this phenomenon requires that magnetic flux move freely from the interior of a sample to its surface. Consequently, full flux expulsion is obtained only in materials with a low density of defects that inhibit motion of magnetic flux, a condition obtainable (with effort) in most elemental superconductors, but only with great difficulty in high-T_c cupric-oxide-based materials, for example.[8]

Shown in Fig. 1a are results of a low field, vibrating sample magnetometry study[9] on a high purity niobium sphere, sample Nb-A, plotted versus temperature for field-cooling (FC) in an applied field H_{app} = 11 Oe. Shown vertically is the normalized quantity $4\pi M/H_{eff}$ ($=4\pi\chi$, since we have $M \propto H$), where H_{eff}, defined below in Eq. 3, is the effective field, corrected for demagnetizing effects. Also shown is the response following zero-field-cooling (ZFC) to $T < T_c$, after which H_{app} = 11 Oe was applied. The two curves are identical (except for a small offset for clarity), demonstrating that this sample is highly reversible magnetically. Furthermore, the magnetization corresponds to having $B = H_{eff} + 4\pi M = 0$ in the interior, so that

$$M = (-1/4\pi)\, H_{eff}, \tag{2}$$

$$\text{where } H_{eff} = H_{app} - 4\pi DM \tag{3}$$

and D = 1/3 is the demagnetizing factor for a sphere. Magnetic reversibility in Nb-A sphere (4 mm diameter) was obtained after careful annealing and carbon removal by high temperature heat treatment in a controlled atmosphere, followed by a lower temperature surface oxidation[10] in air at 500 °C to eliminate pinning by surface defects.

These "textbook" results can be compared with an identical study in Fig. 1b on another Nb sample, sphere Nb-B. This second, high-purity sphere with similar diameter did not receive the surface oxidation treatment. As expected, its ZFC response was the same as that of Nb-A, since a magnetic moment was induced in each case by application of field while the sample was superconductive. In contrast, the FC responses were completely different, with no measurable expulsion of flux for sample Nb-B, as a consequence of vortices being pinned by defects. With weaker pinning and/or a lower density of defects, the fraction of flux expulsion can vary between zero and one (100%). These "well-known" results are presented as a stark counter-example to some reports in high-temperature superconductivity (HTSC), where on occasion the quantity ($-4\pi M/H_{eff}$) is equated to the volume fraction of superconductive material in a sample. The above results for the comparatively simple case of solid Nb demonstrate that considerable care is required when interpreting dc magnetization results. In the studies just cited, the FC measurements giving $-4\pi M/H_{eff} = 1$ in fact coincide with the true superconducting volume fraction (100%) of Nb-A, but this feature is hardly universal, as seen for sample Nb-B. Perhaps the strongest statement that can be made at this point is that a complete Meissner expulsion of flux is a sufficient but not necessary condition for the presence of bulk superconductivity.

B. Meissner Effect with Small Particles

Standard London theory for superconductivity[1,3] provides for a bulk Meissner Effect, via the differential equation

$$\nabla^2\mathbf{B} = (1/\lambda^2)\mathbf{B}. \tag{4}$$

Its solution in simple geometries gives $B = B_o \exp(-x/\lambda)$, where B_o is the flux density parallel to the surface of the superconductor and x is the depth within the superconductor. Here $\lambda(T)$ is the London magnetic penetration depth, given by $1/\lambda^2(T) = 4\pi e^2 n_s(T)/m^* c^2$ with n_s and m^* the density and effective mass of superconducting charge carriers,

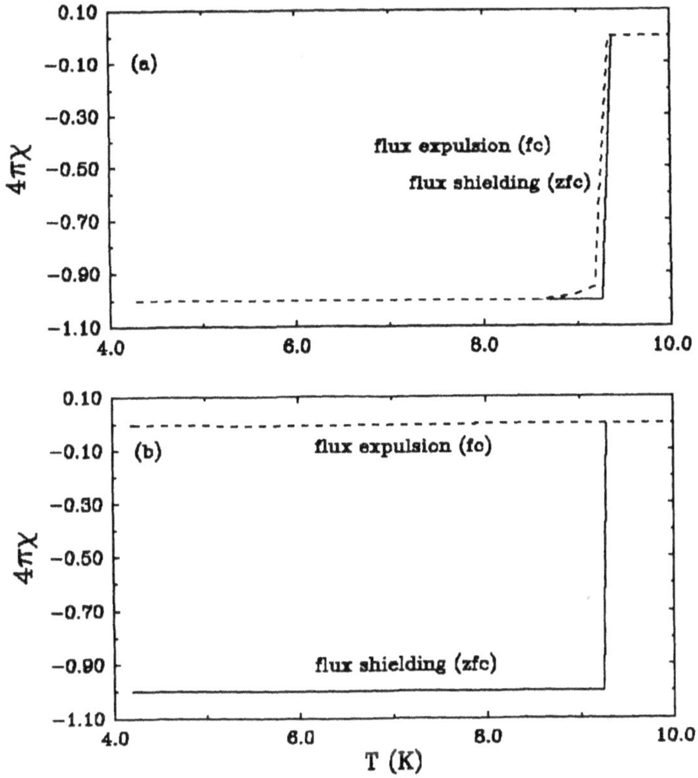

Fig. 1. The susceptibility $4\pi\chi = 4\pi M/H_{eff}$, normalized to equal -1 for an ideally diamagnetic body, versus temperature T for two high purity Nb spheres. The Meissner Effect was detected by field cooled (FC) measurements (dashed lines) in 11 Oe applied field. Zero field cooling (ZFC) measurements (solid lines) detected induced shielding currents. (a) The low pinning, "defect-free" sphere exhibited a complete Meissner expulsion of flux, while (b) the sphere with strong pinning expelled no flux. The ZFC shielding signals were identical, however.

respectively. Loosely speaking, the superconductor is penetrated by the field to a depth λ, which for $T << T_c$ is the order of 10^2 A for conventional materials and 10^3 - 10^4 A or greater for HTSC's.

Because of near-surface field penetration, the observed expulsion and shielding of magnetic field is less than perfect, even for samples that are uniformly superconductive. This point is illustrated in Fig. 2, a plot of normalized magnetization $4\pi M/(3H_{app}/2)$ versus reduced temperature $t = T/T_c$ for the material $Y_1Ba_2Cu_3O_7$. In this case, fine particles of the HTSC were produced by high temperature pyrolysis[11] of aerosols containing stoichiometric ratios of Y, Ba, and Cu cations in nitrate solutions. Pyrolysis near 950 °C in flowing O_2 carrier gas produced fine, mostly monocrystalline particles of $Y_1Ba_2Cu_3O_7$ that were superconductive as produced, with a magnetic onset T_c = 91 K. Scanning

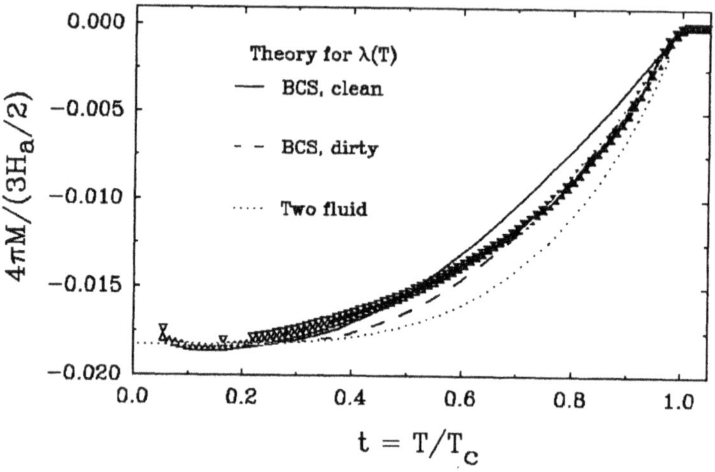

Fig. 2. Normalized magnetization vs temperature for assembly of small $Y_1Ba_2Cu_3O_7$ particles, with FC (▵) and ZFC (▿) data. Lines show modeling using London formulation for spheres, with three forms for the temperature dependence of the penetration depth λ(T); see text. The small response (~0.02) shows that the microscopic particles were nearly transparent to the 10 Oe field.

electron microscopy revealed approximately equiaxed particles with diameters of (0.1-1) μm. For the measurements presented, randomly oriented particles (mass = 0.354 g) were pressed with an applied pressure of 153 kg/mm^2 into a right circular cylinder (height = diameter = 4.76 mm), with no subsequent heat treatment. Consequently there was negligible connectivity between grains and the assembly approximated an array of independent spheres.

Figure 2 shows the normalized magnetization of the sample. Here $M = m/V_{sc}$ refers to the measured magnetic moment m divided by the volume V_{sc} of superconductor as calculated from the sample mass and material density. The FC and ZFC data are extremely similar, displaying a high degree of reversibility. Particularly noteworthy is the small magnitude of normalized magnetization, ~ 0.02, corresponding to a Meissner signal of about 2%. In other words, the particles were nearly transparent to the magnetic field as a consequence of their small size relative to λ.

London[1] has calculated the magnetic response for a sphere of radius R to account for the penetration of field. The results are that

$$-4\pi M = (3H_{app}/2)\,P(x), \qquad (5)$$

where $P(x) = 1 - (3/x)\mathit{coth}(x) + 3/x^2$ and $x = R/\lambda(T)$. (6)

Similar calculations[2] for some other regular geometries show that $-4\pi M$ is given by

(long cylinder, H ∥ axis) $-4\pi M = H_{app}[1 - (2/x)I_1(x)/I_o(x)]$ (7a) (long cylinder, H ⊥ axis) $-4\pi M = 2H_{app}[1 - (2/x)I_1(x)/I_o(x)]$ (7b)

(plate, H ∥ surface) $-4\pi M = H_{app}[1 - (1/x)tanh(x)]$, (7c)

where x is the ratio of cylinder radius or plate half-width to λ, and I_1 and I_o are modified Bessel functions of the first kind with order 1 and 0, respectively.

London expressions for $P(x(t)) = 4\pi M(t)/(3H_{app}/2)$ are plotted in Fig. 2 for comparison with experimental results. Three temperature dependencies have been used for $\lambda(t)$: weak coupling BCS calculations in the clean and dirty limits[12], and the empirical two fluid expression, $\lambda(0)/\lambda(t) = [1 - t^4]^{1/2}$. For each of these dependencies, the quantity $[\lambda(0)/\lambda(t)]^2$ is proportional to (1 - t) near T_c and is constant at low temperature; explicit examples of these and other theoretical temperature dependencies are shown in a later figure (Fig. 8). Now, the layered $Y_1Ba_2Cu_3O_7$ superconductor is anisotropic; for the two principal values[13-16] of the penetration depth tensor at t = 0, we use the values $\lambda_a(0) = 0.15\ \mu m$ and $\lambda_c(0) = 0.8\ \mu m$, corresponding to shielding by supercurrent flow in the Cu-oxide ab planes and along the c-axis, respectively. In Fig. 2, we model this anisotropy by taking $<P> = \{(1/3) P[R/\lambda_a] + (2/3)P[R/(\lambda_a\lambda_c)^{1/2}]\}$ and choosing $R = 0.12\ \mu m$, which is reasonably consistent with qualitative SEM observations. Deviations between the theory and experimental values are attributed to approximations in incorporating anisotropic effects; to the distribution of particle sizes; and to possible deviations from the BCS temperature dependence at lower temperatures.

From a perspective of characterizing superconducting materials, the main objective of this discussion is to demonstrate the strong influence that sample dimensions can have when they become comparable with the penetration depth λ. A recent example[17] of such influences, we suggest, is the observation of small diamagnetic ZFC and FC signals in potassium-doped C_{60} (Fullerene) having small, but unknown particle size and unknown homogeneity. In such cases, the normalized FC magnetization provides a lower bound on the fraction of material that is superconductive. More generally, this should be complemented by measurement of the ZFC magnetization. In studies of fine grained, sintered ceramic HTSC materials, we have frequently found that the two measures are quite similar in magnitude; the aerosol powders are a particularly pronounced example of this, as seen in Fig. 2. The relative reversibility arises, at least in part, from the fact that in low field measurements, the inter-vortex spacing can be substantially larger than the particle size. For example, with $H_{app} = 10$ Oe, the vortex-vortex separation is $\sim 1.4\ \mu m$. Qualitatively, the nucleation of vortices and their aggregate pinning is diminished considerably in low fields. In large applied fields, substantial magnetic irreversibility can develop, as was observed for the $Y_1Ba_2Cu_3O_7$ aerosol particles. However, a full discussion of these small particle effects in high fields lies beyond the intended scope of this report.

C. Meissner and Low Field Response: Multiply-Connected Geometry

To see the influence of multiply-connected geometry on the static magnetic response of a superconductor, let us consider a long, hollow, thick wall cylinder of inner radius r, outer radius R, and length L. The external field is applied parallel to the cylinder axis, as shown in Fig. 3. For a ZFC experiment, the results are relatively simple: application of a small field induces supercurrents that screen the entire geometrical volume of the cylinder, $V = \pi R^2 L$. Simply expressed, the low field ZFC magnetic moment is indistinguishable from that of a solid cylinder.

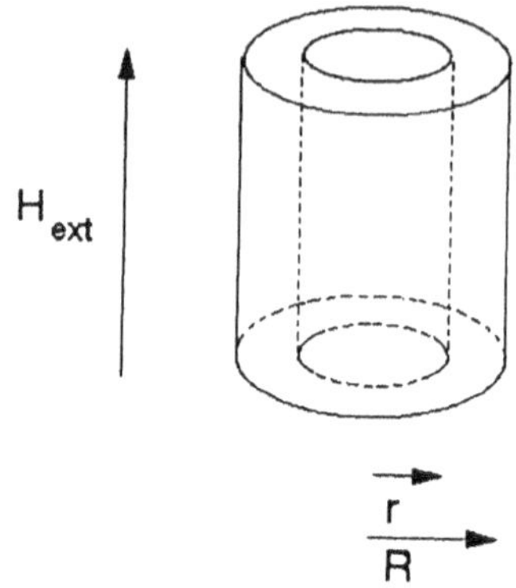

Fig. 3. This multiply connected geometry, a relatively simple case, consists of a thick wall cylinder of superconductor, with external field applied parallel to the cylinder axis.

The FC signal is smaller. Even if the superconducting material itself is defect-free and highly reversible, flux ϕ is trapped, with $\phi \geq H_{app}\ \pi r^2$. The maximum moment m that can be formed by flux expulsion lies between zero and $m = (-1/4\pi)\ H_{app}\pi(R^2-r^2)L$. If we have the simple case of a thin wall tube with $\lambda << (R-r) << R$, the overall behavior in low field will be indistinguishable from that of a solid superconductive body with strong pinning, with negligible flux expulsion in both cases.

Coffee et al.[18] investigated the field at the center of a hollow cylinder of the Type I material Pb, as a function of external field. The sample configuration was identical to that described above, except that Nb rings were fitted to each end of the tube to eliminate end effects. As one might expect, the external field was completely excluded ($B = 0$ inside) until it exceeded the critical field H_c for lead; for $H_{eff} > H_c$, the internal field tracked H_{app} linearly and reversibly; when it was reduced below H_c, that field was trapped within the cylinder. As noted by the authors, the overall response was roughly similar to that of a simply connected Type II cylinder with strong pinning. Of course, if the latter sample were a high-κ material, its behavior in the vortex state would differ considerably.

D. Granular Superconductors in Low Fields

Bulk ceramic samples of high temperature superconductors are generally considered to be granular superconductors.[19] They are polycrystalline, with grains that are weakly coupled together. This weak intergrain coupling manifests itself as an inability of the material to transport a significant supercurrent density from grain to grain, even though the current transport within individual grains can be quite good. This poor coupling between grains, which constitutes a major impediment to practical applications of ceramic superconductors, is generally attributed to a rapid attenuation of the superconducting order parameter in regions of imperfect material, such as grain boundaries. This attenuation occurs over a distance of the order of the coherence length ξ, which is very short in all high-T_c materials known to date, and consequently they are prone to have poor coupling across grain boundaries and other extended defects. Typically porous, bulk ceramic samples contain voids and perhaps other non-superconductive material that occupies from a few percent up to 50% of the apparent volume. Let us consider a ZFC measurement in which a cylindrical ceramic sample is cooled to low temperature and field H_{app} is applied. As before, this field induces surface currents, which are assumed for now to be smaller than

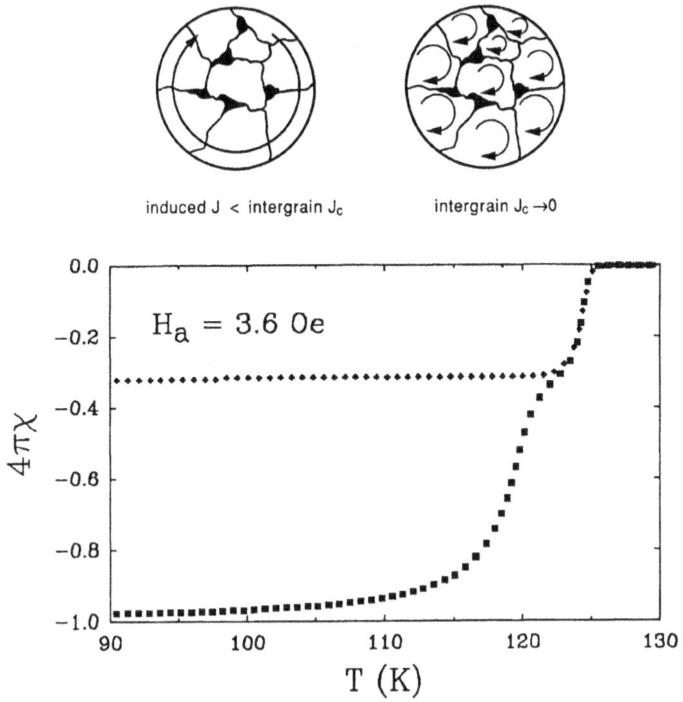

Fig. 4. The low field susceptibility $4\pi\chi$ (normalized to that of an ideally diamagnetic body) of polycrystalline, sintered $Tl_2Ca_2Ba_2Cu_3O_{10+\delta}$ vs T, in 3.6 Oe applied field. ZFC measurements (■) gave nearly complete shielding when supercurrents flowed between grains; see sketch at top left. At high temperature, grains became decoupled (see sketch at top right) and shielding decreased. Note that in the decoupled region, the ZFC and FC (◆) signals were nearly equal.

the intergrain critical current density $J_{c\text{-inter}}$ which is the Josephson critical current between grains or some analog thereof. In this case, the sample responds as a solid cylinder (or, for that matter, a hollow cylinder) of radius R and length L. We have in this low field case

$$(m/V_c) = M = (-1/4\pi)\, H_{eff}, \text{ where } H_{eff} = H_{app} - 4\pi DM, \tag{8}$$

$V_c = \pi R^2 L$ is the *geometrical* volume of the cylinder, and D is the effective, magnetometric demagnetization factor for a cylinder of radius R and length L. Zijlstra[20] has a convenient chart giving approximate values of D for cylinders with field parallel to the axis. (One should keep in mind, however, that the magnetizing field is *not* uniform within an arbitrarily shaped body and the value for D represents an average over the cylindrical volume.)

The normalized susceptibility $4\pi M/H_{eff} = 4\pi\chi$, as discussed in Section I, is shown in Fig. 4 as a function of temperature for a sintered, random polycrystalline cylinder of the high temperature superconductor $Tl_2Ca_2Ba_2Cu_3O_{10+\delta}$. A field of 3.6 Oe was applied parallel to the cylinder axis. For this sample, the geometrical density was 70 % of the theoretical density (7.04 g/cm^3), as calculated from the x-ray lattice parameters. At low

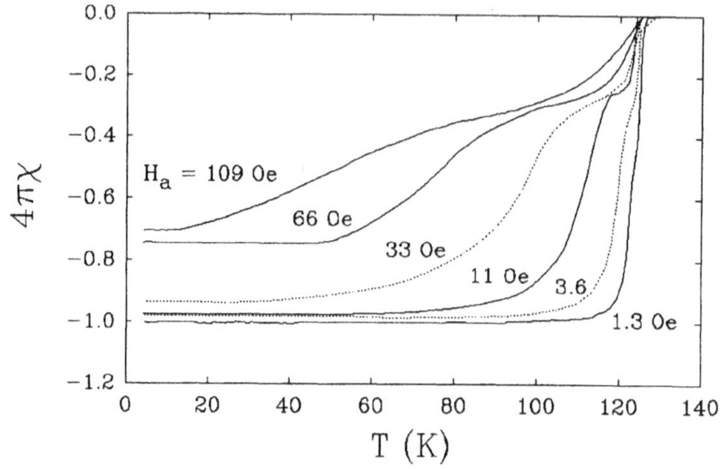

Fig. 5. Influence of applied field on grain decoupling. ZFC measurements of $4\pi M/H_{eff}$ for same sintered Tl-2223 sample as Fig. 3. Larger fields induced greater currents that caused grains to decouple at progressively lower temperatures. After decoupling, the grains produced the same shielding, ~30%, for all $H < H_{c1}$.

temperature, the ZFC signal (squares) gives $4\pi\chi = -0.98$, showing that virtually the entire volume was screened by induced surface currents. This situation is sketched in the accompanying drawing at the top of Fig. 4 (left side). With increasing temperature, however, this signal decreased substantially in magnitude near 120 K. This change arises from the reduction of $J_{c\text{-inter}}(T)$ with temperature, which eventually falls below the density of screening current, so that the grains become effectively isolated; see the drawing at the top of Fig. 4 (right side). The grains themselves are still superconducting, however, so that currents can flow on the "surface" of grains and shield their interior. For the case shown in Fig. 4, this corresponded to approximately 30 % of the geometrical volume of the sample. The penetrated volume of the cylinder has at least two components: (1) voids and (2) surface layers of grains that are penetrated to a depth of λ, as discussed above in part II B.

In the FC measurements shown as diamonds in Fig. 4, flux was expelled upon cooling and the magnetization reversibly tracked the ZFC signal near T_c. For temperatures below 121 K, the FC signal was nearly constant in magnitude, with flux being trapped between grains and penetrating their near surfaces.

For measurements in higher applied dc fields, the "knee" in the ZFC signal shown in Figs. 4 and 5 moved systematically and linearly[21] with H to lower temperature, with most other features unchanged. Again, the decoupling of grains is a consequence of the temperature dependence of weak intergrain currents. Larger applied fields induce larger screening currents, and those currents exceed the upper bound imposed by $J_{c\text{-inter}}(T)$ at progressively lower temperatures. This interpretation is consistent in that the signal at high temperature after decoupling, which corresponds to shielding by the grains only, was independent of field with $4\pi\chi \approx -0.3$, as expected so long as $H_{eff} < H_{c1}$ for the grains.

Figure 5 also shows that for sufficiently large applied field, e.g. 66 and 109 Oe, the intergrain current density (and perhaps H_{c1} for some crystallite orientations) was exceeded even at 4 K; consequently full screening of the geometrical volume was never achieved under these conditions.

An interesting and useful further analysis of the influence of granularity on the magnetic and transport properties of high-T_c superconductors has been given by Dersch and Blatter.[22] For a discussion of some other features found in dc magnetic studies of filamentary superconductive systems, the reader is referred to a recent paper by Collings et al.[23]

III. VORTEX STATE STUDIES of TYPE II MATERIALS: $H_{eff} > H_{c1}$

In magnetic fields that exceed the lower critical field H_{c1}, a type II superconductor is penetrated by vortices and enters the vortex or mixed state. As one increases the field applied to a material with numerous and/or strong pinning centers for vortices, their entry and motion into the sample is inhibited. Likewise, the movement of vortices out of the superconductor is inhibited upon decrease of the field. Consequently its magnetic response is hysteretic,[24] as illustrated in Fig 6 (top), a plot of magnetization M versus applied field H_{app} at 5 K for a $Bi_{1.8}Pb_{0.3}Sr_2Ca_2Cu_3O_{10+\delta}$ superconductor with $T_c = 106$ K. The measurements were made using a SQUID-based magnetometer from Quantum Design, Inc. This sample was a highly textured, pressed polycrystalline disk (mass = 0.169 g, diameter = 4.83 mm, thickness = 2.11 mm, effective demagnetizing factor D ≈ 0.54) with the c-axes nearly parallel to the pressing direction. The magnetizing field was applied parallel to the disk axis and hence perpendicular to the highly conductive Cu-O basal planes of the layered superconductor. The hysteresis seen in Fig. 6 is directly related to the critical current density J_c of the material, as will be discussed briefly in a later section. For now, let us note that the measured magnetization is a sum of at least two terms. The first is the equilibrium magnetization M_{eq} that is negative in sign and generated by microscopic supercurrents associated with individual vortices. This term is typically the order of 1-10 Gauss in magnitude and is discussed in the following section. The second, irreversible term M_{irr} is due to macroscopic, circulating supercurrents J_c, whose existence depends on pinning of vortices to prevent their motion. The magnitude of M_{irr} ranges from zero (for $J_c = 0$) to many tens of kiloGauss (for plate-like samples with high J_c's):

$$M = M_{eq} + M_{irr}.$$

In the opposite case where there is no effective flux pinning, vortices enter and leave the material easily and reversibly. This is illustrated in Fig. 6 (bottom), a plot M(H) for the same sample at 85 K. Here the effective pinning is very weak (the vortex lattice is "melted"), so that $J_c \rightarrow 0$ and $M_{irr} = 0$ also. Under these conditions, one can obtain the thermodynamic, equilibrium magnetization M_{eq}, which is related to the London penetration depth. The relationship between these two quantities is discussed in the next section.

A. Vortex State Penetration Depth Studies

Using the interaction energy between vortices and thermodynamic arguments, Kogan et al.[25] developed a theory in the London limit for the vortex state magnetization of a uniaxial, high-κ type II superconductor. In particular, the equilibrium M of a single crystal or its equivalent was related to applied field H_{app} as

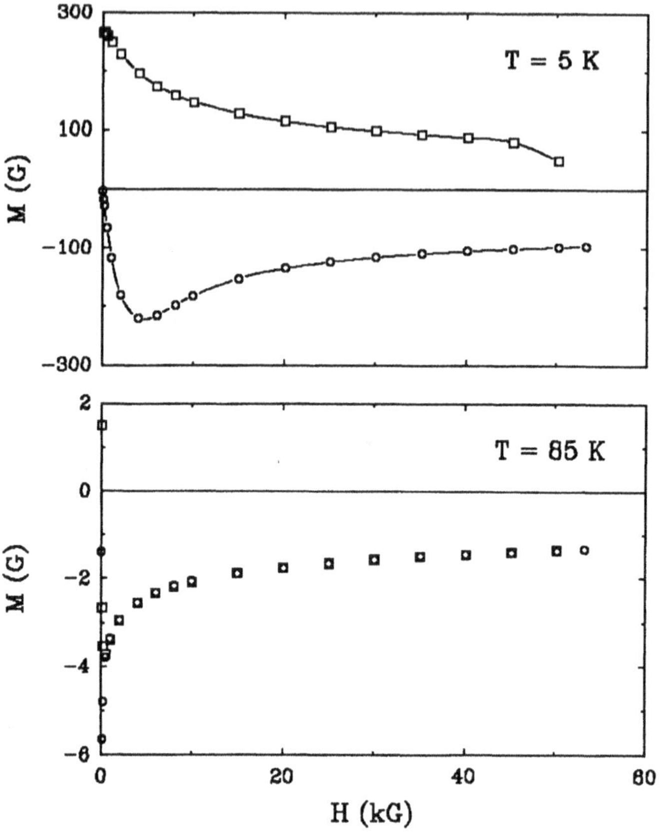

Fig. 6. Magnetization M vs applied field H at (a) 5 K and (b) at 85 K, for a highly textured polycrystalline disk of $Bi_{1.8}Pb_{0.3}Sr_2Ca_2Cu_3O_{10+\delta}$, with H ∥ c-axes for increasing (○) and decreasing (□) field history. Pinning caused substantial hysteresis at low temperatures, but the sample was highly reversible in a broad temperature range below T_c = 106 K.

$$M(H,T) = (\phi_o/32\pi^2\lambda_i\lambda_j)\ \ln[\beta H_{c2}(T)/H_{app}] \qquad (9)$$

where ϕ_o = 2.07 10^{-7} G-cm^2 is the flux quantum, β is a constant of order unity, and $H_{c2}(T)$ is the upper critical field corresponding to the applied field direction. Here λ_i is the magnetic penetration depth corresponding to field penetration that (in the Meissner state with B = 0 far from the surface) is screened by supercurrent flow in the i-th crystal direction. This relation is valid in the region $H_{c1}(T) << H_{app} << H_{c2}(T)$ where the magnetization is reversible and applies to uniaxial materials, which includes most high temperature superconductors containing copper oxide layers. In these cases, there are two independent eigenvalues in the penetration depth tensor, λ_a and λ_c. With the magnetizing field along the c-axis, one has in Eq. 1 that i=j=a such that M depends only on the shorter component λ_a; with the field in the basal a-b plane, one has i=a and j=c or visa versa, so that M depends on both λ_a and λ_c.

This theory predicts a logarithmic field dependence, with $M \propto ln(H)$. Figure 7, a semilog plot of M versus H, presents typical experimental data showing that this dependence is followed accurately. Shown is a set of isothermal magnetization data at various temperatures for a polycrystalline Bi-Sr-Ca-Cu-oxide sample containing two adjacent Cu-O layers with H ∥ c-axis. Since the individual crystallites were very flat platelets, pressing a thin sample with uniaxial pressure caused it to be highly textured. As thermally processed, this sample had T_c = 80.7 K (see Table I). The logarithmic slopes dM/d[*ln*(H)] from these and similar plots for other samples have been used to obtain values for $\lambda_a^2(T)$ (with H ∥ c) and, where possible, for the quantity $\lambda_{eff}^2(T) = \lambda_a(T)\lambda_c(T)$ (with H ∥ ab). Now, Ginzburg-Landau theory provides that $1/\lambda^2$ is proportional to (T_c-T) near T_c. Figure 8 illustrates this dependence, where $dM/d\,ln(H) \propto 1/\lambda^2 \propto n_s(T)$ is plotted versus T for a sample of magnetically aligned $Tl_2Ca_2Ba_2Cu_3O_{10+\delta}$ powder with H ∥ c-axes.[26] The composite, aligned[27] sample was produced by dispersing fine (~ 1 μm) monocrystalline powder in "45-minute" liquid epoxy and immersing it in a 50 kOe field at room temperature. Due to the anisotropy in the normal state magnetic susceptibility, the dispersed crystallites (6.5 % by volume) align with the c-axis parallel with the magnetizing field.

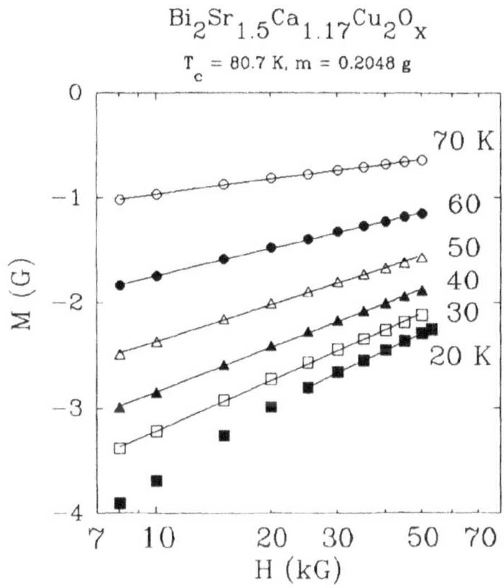

Fig. 7. The vortex state, equilibrium magnetization M vs field H on a logarithmic scale. The *ln*(H) dependence of Eq. 9 was followed accurately, except at the lowest temperature (20 K), where M was hysteretic at lower fields.

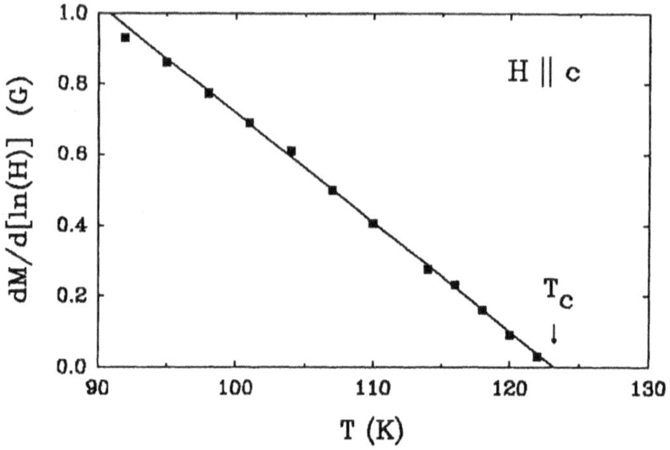

Fig. 8. A plot of logarithmic field derivative vs temperature T, for aligned $Tl_2Ca_2Ba_2Cu_3O_{10+\delta}$ powder with H ∥ c-axes. The quantity $dM/d\ln H \propto (1/\lambda^2) \propto n_s(T)$ varied linearly with T near T_c, as given by Ginzburg-Landau theory. The intercept where $(1/\lambda^2) = 0$ defines $T_{c,\lambda}$ appropriate to the penetration depth study.

Once cured, the epoxy matrix retained the c-axis alignment, as evidenced by X-ray diffraction studies showing only strong (00ℓ) reflections with rocking curve widths of $\leq 1°$. Figure 8 shows that the linear temperature dependence is quite well defined. We have observed[28] it in all cases investigated to date and used the linear variation to define an appropriate value of transition temperature $T_{c,\lambda}$ as the temperature at which the penetration depth diverged. In general, these values agreed relatively well with other measures of the transition temperature; some values are tabulated in Table I. For comparison with theoretical temperature dependencies for $\lambda(T)$, however, it is very important that the value of T_c be correct. Direct use of the magnetization data to determine $T_{c,\lambda}$ insures that a value appropriate to the penetration depth study is obtained.

Experimental results for the penetration depth $1/\lambda_a^2(T)$, which are denoted by symbols, are plotted versus reduced temperature $t = T/T_{c,\lambda}$ in Fig. 9 for two different materials. The data extend from t=1, near which the linear dependence on T is evident, down to $t \sim 0.3$, below which the magnetization became excessively hysteretic. Data are shown both for the textured, two-layer Bi-based cupric oxide material discussed above and for a small grained, random polycrystalline sample of $Y_2Ba_4Cu_8O_{16+\delta}$. In each case, the value of lambda at t=0, $\lambda_a(0)$, was used as the one free fitting parameter when comparing the data with various theoretical forms for $[\lambda(0)/\lambda(t)]^2$. These forms include weak coupling, BCS theory in the clean and dirty limits[12], the empirical, two-fluid model with $(1/\lambda^2) \propto (1-t^4)$, and strong coupling calculations by Rammer[29] in the clean and dirty limits, obtained by numerical solution of Eliashberg theory. The best fits to each of the theoretical forms are shown as lines in Fig. 9.

From Fig. 9, it is clear that weak coupling, clean limit theory provides an excellent description of the experimental results, which is visibly superior to that obtained for the other theoretical forms. Using the clean limit BCS temperature dependence, we obtain the value $\lambda_a(0) = 0.30\ \mu m$ for the two-layer Bi-Sr-Ca-Cu-Oxide sample. Results for this material and others, obtained by similar analyses, are tabulated in Table I. Where possible, studies with H ∥ ab have been used to estimate values for λ_c and these are included as well. Note, however, that measurements with H ⊥ c-axis are very sensitive[25] to any small

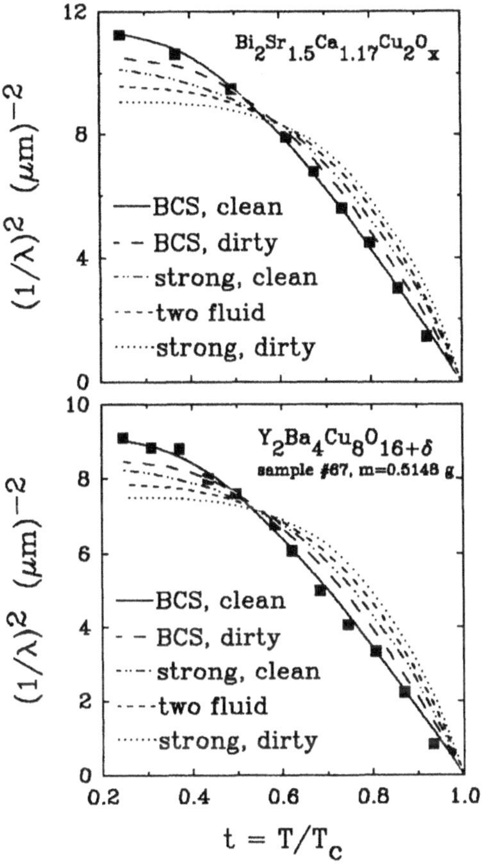

Fig. 9. The London magnetic penetration depth $1/\lambda^2$ vs reduced temperature $t = T/T_{c,\lambda}$, for (a) highly textured Bi-Sr-Ca-Cu-O with H $\parallel$ c-axes and for (b) fine grained, randomly oriented $Y_2Ba_4Cu_8O_{16+\delta}$. Lines are least squares fits to various theoretical forms for $1/\lambda^2$, showing that BCS theory in the clean limit represents the data best and rather accurately.

angular misalignment, unlike those with H $\parallel$ c. Consequently the values for λ_c should be regarded as lower bounds on this component of the penetration depth tensor.

For a random polycrystalline material such as the $Y_2Ba_4Cu_8O_{16+\delta}$ sample in Fig. 9, an angular average of Eq. 9 is required. The resulting expression is that[14,25]

$$dM/d\ln(H) = \phi_o/[32\pi^2\lambda_{avg}^2]\ [g(\gamma)/2] \text{ with} \tag{10a}$$

$$g(\gamma) = \gamma^{-1/3}\{\gamma + (\gamma^2-1)^{-1/2}\ln[(\gamma^2-1)^{1/2}+\gamma]\} \tag{10b}$$

where $\lambda_{avg} = (\lambda_a\lambda_a\lambda_c)^{1/3}$ and $g(\gamma)$ is a correction factor that depends on the superconductive anisotropy factor γ. The factor γ arises when anisotropy is incorporated into Ginzburg-Landau theory via a superconductive effective mass.[25] Then the components of the normalized mass tensor m_i are related by $\gamma = (m_c/m_a)^{1/2} = (\lambda_c/\lambda_a)$. Using the data for $Y_2Ba_4Cu_8O_{16}$ in Fig. 9, we can obtain values for $\lambda(0)$ if γ is known. Lacking an experimental determination of γ, we make the reasonable assumption that it has the same value[15] as in $Y_1Ba_2Cu_3O_7$, namely $\gamma = 5$. This gives $g(\gamma) = 3.2$ and implies that $\lambda_a(0) = 0.25\ \mu m$ and $\lambda_c(0) = 1.2\ \mu m$, as included in Table I.

One influence on these experiments may be "melting" of the flux lattice.[31] In one sense, this phenomenon facilitated many of the studies by widening the region of magnetic reversibility so that measurements could be extended to lower temperatures, thereby enabling us to make better comparisons with theories for $\lambda(T)$. According to Clem,[32] it is unlikely that a "melted" flux lattice has a significant effect on the determination of λ, since the vortex interaction energy in Kogan's formulation depends most strongly on the mean vortex spacing a ~ $(\phi_o/B)^{1/2}$. However, vortex core interactions, which are not included in the thermodynamic calculation, may lead to an overestimation[32] of ~15 % in the values of λ.

Table I. Experimental Values for the London Penetration Depths for Some High-T_c Superconductors.

COMPOUND	Preparation temperature, atmosphere	λ_a (μm) (except as noted)	$T_{c,\lambda}$ (K)	$T_{c,r}$ (K) resistive midpoint	$T_{c,m}$ (K) magnetic midpoint
$Bi_{1.7}Pb_{0.3}Sr_2Ca_2Cu_3O_{10}$	-----	0.22	105.9	108	105.5
$Bi_2Sr_{1.5}Ca_{1.17}Cu_2O_x$	840 C, 1% O_2	0.30	80.7	80.7	-----
Bi-2212 (Hoecht AG)	840 C, air	0.28	72.7	73.5	71.8
Bi-2212 (Hoecht AG)	840 C, 2% O_2	0.30	80.3	81.8	78.9
Bi-2212 single crystal	-----	λ_a=0.27 $\lambda_c \geq 3.7$	73.8	-----	-----
$Tl_2Ca_2Ba_2Cu_3O_{10+\delta}$ (aligned powder)	850 C, 1 Atm O_2	λ_a=0.17 $\lambda_c \geq 0.48$	122.8	-----	121.2
$Y_1Ba_2Cu_4O_8$ polyxtal. (assuming γ=5)	ref. 30.	λ_a=0.25 λ_c=1.2	80.3	-----	78.7

An overriding result of this work, however, is that weak coupling theory in the clean limit provides a decidedly superior description of the temperature dependence of $\lambda(T)$ in these measurements. Similar conclusions were reached in a study on $Y_1Ba_2Cu_3O_7$ and $Bi_2Sr_2CaCu_2O_8$ by Sreeparna Mitra et al.[14] These findings support both the earlier argument of Little[33] and the more recent theoretical work of Tsuei et al.[34] that the coupling in high temperature superconductors is weak. The latter authors use a model incorporating a van Hove singularity in the density of electronic states to account for both the high transition temperatures T_c and the compositional dependence of the isotope effect, all within a weak coupling framework.

An interesting aspect of these studies is that the vortex state magnetization is a bulk, volume-averaged quantity. In contrast, conventional Meissner state magnetic measurements of field penetration into a sample sense largely the near-surface properties of the superconductor, e.g., a single crystal of $Y_1Ba_2Cu_3O_7$. Reasonable consistency between these two types of measurements provides strong evidence that superconductivity in the compound is a *bulk* phenomenon and not a surface effect only. Indeed, such consistency is evident in penetration depth studies of YBCO using magnetization[14] and low field Meissner state[13] studies. These types of results complement other experimental studies that sense bulk superconductivity, such as a jump in specific heat at T_c and the observation of a flux line lattice by neutron scattering techniques.

B. Defect-induced Magnetic Hysteresis

As discussed at the beginning of section III, the presence of defects in a superconductor inhibits the motion of vortices and tends to "pin" them in energetically favorable locations. In addition to naturally occurring "pins," such regions can be created by many different methods. The technology of practical, high field superconductivity is devoted in considerable part to optimizing their size, distribution, and efficacy to obtain high, stable critical currents. The magnetic hysteresis, such as that shown in Fig. 6 (top), that is associated with pinning and the existence of critical currents provides a very useful and widely employed non-contact method to determine values for J_c. The critical state model was devised by Bean[35] and is described in the book of Tinkham.[3] In essence, the Bean model provides that $J_c(H,T)$ is proportional to $\Delta M = (M^- - M^+)$, where M^- (M^+) is the magnetization at temperature T measured in decreasing (increasing) field history, respectively. For the case of a cylinder of radius R (in cm) with field applied parallel to its axis, the J_c (in A/cm^2) is given by

$$J_c(H,T) = 15\ \Delta M(H,T)/R \tag{11}$$

with M in units of Gauss. Similar results apply to other geometries; e.g., for a sphere of radius R, the numerical factor[36] becomes 16.97. For a rectangular solid with field perpendicular to a face with sides $L_2 > L_1$, the "sandpile" model[24] provides that

$$J_c(H,T) = 20\ \Delta M/[L_1(1-L_1/3L_2)]. \tag{12}$$

For a sample of square section with $L_1 = L_2$, this expression reduces to Eq. 11 for a cylinder when one equates $R = L/2$.

Earlier in this paper, we discussed the trapping of flux in a multiconnected superconductor. This feature has been carried to an extreme, but extremely useful limit using ion irradiation methods. A single crystal of $Y_1Ba_2Cu_3O_7$ was irradiated[37] with 580 MeV Sn-ions, which created linear columns of damaged material nearly parallel to the c-

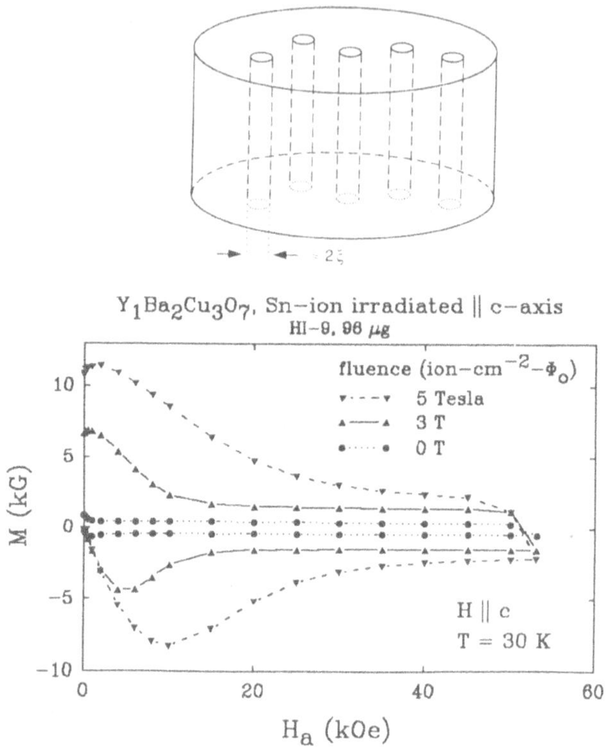

Fig. 10. Irreversible magnetization due to linear defects. Magnetization M(H) at 30 K for single crystal $Y_1Ba_2Cu_3O_7$, with H $\parallel$ c-axes. Ion irradiation with 580 MeV Sn-ions created linear damage columns $\parallel$ c-axes that pin vortices very effectively (see sketch). Hysteresis increased progressively with ion fluence.

axis. The diameter of these nonsuperconductive damage tracks is a few times ξ_a, the superconductive coherence length in the ab plane of Cu-O sheets. As such, a damage track is quite effective in pinning a vortex directed along the c-axis, in part because of the similarity of these dimensions. Most importantly, *both the damage columns and the vortices are linear structures,* which maximizes the pinning force when the two are collinear.

An example of increased pinning is given in Fig. 10, plots of M versus H $\parallel$ c at 30 K for a crystal at three levels of irradiation. The drawing at the top of the figure schematically shows columns of damaged material that extend from top to bottom of the plate-like crystal. In the plot itself, the circles denote data for the sample as prepared; here ΔM and J_c are relatively small and are produced by "accidental defects" incorporated during the growth process. The triangles show M following heavy ion irradiation, where the ion fluence, in units of (ions/cm^2), has been multiplied by the flux quantum $\phi_o = 2.07 \times 10^{-11}$ T-cm^2 to give it dimensions of flux density. For example, a fluence of 5 Tesla means that the area density of damage tracks is the same as the area density of vortices when B = 5 Tesla. It is evident that ion-induced defects make the crystal increasingly hysteretic, indicating substantial increases[37] in J_c. A Bean model calculation, using Eq. 12 with sample dimensions of L_1 = 0.80 mm and L_2 = 0.94 mm, shows that ΔM = 5 kG corresponds to $J_c = 1.8 \times 10^6$ A/cm^2. At this temperature, the improvement in J_c was significant. At still higher temperatures, e.g., 77 K, the effects were far more dramatic,[38] increasing the irreversibility field substantially.[37] In summary, this ion-irradiation study graphically demonstrates the central role of defects in controlling the magnetic hysteresis in a well characterized high temperature superconductor.

SUMMARY

Many features of static magnetization studies have been reviewed. While relatively straightforward in principle, considerable care is needed in practice when interpreting the results of these investigations. As demonstrated, the measurements are influenced by many variables, including geometry, sample size, granularity, and of course, defects that pin vortices. Studies of the dc magnetization can provide important insights into the properties of superconductors, especially when complemented by other investigations of structure, phase composition, and geometry.

A portion of the work of JRT and YRS was supported by the Science Alliance at the University of Tennessee, Knoxville. The research was sponsored by the Division of Materials Sciences, U.S. Department of Energy and technology development was funded by the Oak Ridge Superconducting Technology for Electric Energy Systems Program, Advanced Utility Concepts Division, Conservation and Renewable Energy Program, U. S. Department of Energy, both under contract DE-AC-05-84OR21400 with Martin Marietta Energy Systems, Inc.

REFERENCES

*APM is now at American Superconductor Corporation, 149 Grove Street, Watertown, MA 02172.

1. F.London, Superfluids, vol.1, (Dover, New York, 1961).
2. D. Schoenburg, Superconductivity (Cambridge Univ. Press, Cambridge, 1960).
3. M. Tinkham, Introduction to Superconductivity (McGraw-Hill, New York, 1975).
4. M. Tinkham and C. J. Lobb in Solid State Physics, vol. **42**, edited by H. Ehrenreich and D. Turnbull (Academic Press, New York, 1989), p. 91.
5. J. O. Thomson and J. R. Thompson, *Rev. Sci. Instrum.* ***49***, 1485 (1978).
6. W. Meissner and R. Ochsenfeld, *Naturwissenshaften* ***71***, 787 (1933).
7. R. A. Hein, *Phys. Rev. B.* ***33***, 7539 (1986).
8. L. Krusin-Elbaum, A. P. Malozemoff, Y. Yeshurun, D. C. Cronemeyer, and F. Holtzberg, *Physica C,* ***153-155***, 1469 (1988).
9. Y. C. Kim, "Anisotropies in the Magnetic Properties of Aligned High-T_c Superconducting $Y_1Ba_2(Cu_{1-x}Fe_x)_3O_z$ Powder Composites," thesis, The University of Tennessee, Knoxville, 1989 (unpublished).
10. H. R. Kerchner, D. K. Christen, and S. T. Sekula, *Phys. Rev. B.* ***21***, 86 (1980).
11. T. T. Kodas, E. M. Engler, V. Y. Lee, R. Jacowitz, T. H. Baum, K. Roche, S. S. P. Parkin, W. S. Young, S. Hughes, J. Kleder, and W. Auser, *Appl. Phys. Lett.* ***52***, 1622 (1988).
12. B. Muehlschlegel, *Z. Physik* ***155***, 313 (1959).
13. L. Krusin-Elbaum, R. L. Greene, F. Holtzberg, A. P. Malozemoff, and Y. Yeshurun, *Phys. Rev. Lett.* ***62***, 217 (1989).

14. Sreeparna Mitra, J. H. Cho, W. C. Lee, D. C. Johnston, and V. G. Kogan, *Phys. Rev. B* ***40***, 2674 (1989).
15. D. E. Farrell, B. S. Chandrasekhar, M. R. McGuire, M. M. Fang, V. G. Kogan, J. R. Clem, and D. K. Finnemore, *Phys. Rev. B* ***36***, 4025 (1987).
16. D. K. Christen, Y. C. Kim, S. T. Sekula, J. R. Thompson, J. Brynestad, B. C. Sales, L. A. Boatner, and J. D. Budai, *IEEE Trans. Mag.* ***25***, 2324-27 (1989).
17. A. F. Hebard, M. J. Rosseinsky, R. C. Haddon, D. W. Murphy, S. H. Glarum, T. T. M. Palstra, A. P. Ramirez, and A. R. Kortan, *Nature* ***350***, 600 (1991).
18. D. L. Coffey, W. F. Gauster, and H. E. Rorschach, Jr., *Appl. Phys. Lett.* ***3***, 75 (1963).
19. J. R. Clem, *Physica C* ***153-155***, 50 (1988).
20. H. Zijlstra, Experimental Methods in Magnetism, (North Holland, Amsterdam, 1967).
21. J. R. Thompson, J. Brynestad, D. M. Kroeger, Y. C. Kim, S. T. Sekula, D. K. Christen, and E. D. Specht, Phys. Rev. B **39**, 6652-59 (1989).
22. H. Dersch and G. Blatter, *Phys. Rev. B* ***38***, 11391 (1988).
23. E. W. Collings, A. J. Markworth, and K. R. Marken, Jr. in Cryogenic Materials '88: Superconductors, edited by R. P. Reed, Z. S. Xing, and E. W. Collings (International Cryogenic Materials Conference, Boulder, 1988), p. 61.
24. A. M. Campbell and J. E. Evetts, *Advances in Physics* ***21***, 199 (1972).
25. V. G. Kogan, M. M. Fang, and Sreeparna Mitra, *Phys. Rev. B* ***38***, 11958 (1988).
26. J. R. Thompson, D. K. Christen, H. A. Deeds, J. Brynestad, Y. C. Kim, S. T. Sekula, and J. Budai, *Phys. Rev. B* ***41***, 7293-6 (1990).
27. D. E. Farrell, B. S. Chandrasekhar, M. R. McGuire, M. M. Fang, V. G. Kogan, J. R. Clem, and D. K. Finnemore, *Phys. Rev. B* ***36***, 4025 (1987).
28. J. R. Thompson, D. K. Christen, Yang Ren Sun, B. C. Chakoumakos, B. C. Sales, H. R. Kerchner, J. G. Ossandon, and E. Sonder, *Physica B,* ***165-6***, 1453-4 (1990).
29. J. Rammer, *Europhysics Lett.* ***5***, 77 (1988).
30. R. K. Williams, J. Brynestad, T. J. Henson, D. M. Kroeger, G. C. Marsh, R. A. Padgett, and J. O. Scarborough, *J. Appl. Phys.* ***69***, 2426 (1991).
31. P. L. Gammel, L. F. Schneemeyer, J. V. Waszczak, and D. J. Bishop, *Phys. Rev. Lett.* ***61***, 1666 (1988).
32. J. R. Clem, private communication.
33. W. A. Little, *Science* ***242***, 1390 (1988).
34. C. C. Tsuei, D. M. Newns, C. C. Chi, and C. Pattnaik, *Phys. Rev. Lett.* ***65***, 2724 (1990).
35. C. P. Bean, *Phys. Rev. Lett.* ***8***, 250 (1962); W. A. Fietz and W. W. Webb, *Phys. Rev.* ***178***, 657 (1969).
36. J. R. Clem and V. G. Kogan, *Jpn. J. Appl. Phys.* ***26***, 1162 (1987).
37. L. Civale, A. D. Marwick, T. K. Worthington, M. A. Kirk, J. R. Thompson, L. Krusin-Elbaum, Y. R. Sun, J. R. Clem, and F. Holtzberg, *Phys. Rev. Lett.* ***67***, 648 (1991).
38. J. R. Thompson, Yang Ren Sun, D. K. Christen, H. R. Kerchner, A. P. Malozemoff, L. Civale, A. D. Marwick, L. Krusin-Elbaum, M. W. McElfresh, and F. Holtzberg, in *Physics and Materials Science of High Temperature Superconductors - II*, edited by R. Kossowsky, B. Raveau, and S. Patapis [NATO Advanced Studies Institute Series, Kluwer Academic Publishers, Dordrecht, The Netherlands, 1992), in press.

AC LOSSES IN TYPE-II SUPERCONDUCTORS

John R. Clem

Ames Laboratory–USDOE and
Department of Physics and Astronomy
Iowa State University, Ames, Iowa 50011

ABSTRACT

This paper discusses the physics of ac losses in type-II superconductors and in multifilamentary superconducting-normal composite conductors. As an introduction, the ac eddy-current losses in a homogeneous normal conducting cylindrical specimen subjected to an alternating applied magnetic field are briefly reviewed. The ac losses in a superconducting cylindrical specimen are discussed in detail next, with emphasis on the losses in a specimen subjected to an alternating applied longitudinal magnetic field. In general, dissipation occurs via both flux-flow losses, which are analogous to eddy-current losses in the normal state, and hysteretic losses, which arise from bulk and surface pinning as well as annihilation of vortices of the opposite sense. Tests are suggested by which the dominant loss mechanisms can be identified. Special considerations are noted for the case of the high-temperature superconductors. Finally, the losses in multifilamentary superconducting-normal composite conductors are discussed.

I. INTRODUCTION

The subject of ac losses in superconductors is of great interest to researchers involved in the field of applied superconductivity. Such losses are a key consideration in the design of composite conductors in a number of applications, including cables for long-distance power transmission and pulsed magnet coils for magnetic-confinement fusion (Tokamaks). The term ac losses has a somewhat different meaning to workers in different areas of applied superconductivity. To someone designing a power transmission line, ac losses may mean the losses generated by a 50 Hz or 60 Hz ripple field of small amplitude, say 0.1 T, whereas to someone designing a pulsed magnet coil the term may mean the losses in the coil as the field is ramped from zero to, say, 5 T in a fraction of a second. In this paper, I shall attempt to present a unified view of ac losses in superconductors. However, this is not a review paper, and I apologize for failing to cite many key references in this field.

First, let me make a few simple statements regarding the origin of ac losses. When a normal or superconducting metal is subjected to a time-varying applied magnetic induction $\vec{B}_a(t)$, which generates a net magnetic induction $\vec{B}$, an electric field $\vec{E}$ is generated according to Faraday's emf law. Accompanying $\vec{E}$ is a macroscopic current density $\vec{J}$, such that the local rate of energy dissipation per unit volume is $\vec{J}\cdot\vec{E}$.

Under alternating conditions with an angular frequency ω (frequency $\nu=\omega/2\pi$ and period $T=2\pi/\omega$), the time average of $\vec{J}\cdot\vec{E}$ integrated over the volume V of the specimen yields the time-averaged dissipated power P.

Poynting's theorem provides some useful relations. Recall that[1]

$$\frac{\partial u}{\partial t} + \vec{\nabla}\cdot\vec{S} = -\vec{J}\cdot\vec{E}, \tag{1}$$

where u is the energy density and $\vec{S}$ is Poynting's vector

$$\vec{S} = (c/4\pi)(\vec{E}\times\vec{H}) \quad \text{(Gaussian)} \tag{2a}$$

$$= (\vec{E}\times\vec{H}) \quad \text{(mks)}. \tag{2b}$$

Integrating Eq. (1) over the volume V of the specimen and making use of the divergence theorem to obtain an integral over the surface area A with inward normal $\hat{n}_{in}$, we obtain

$$\int_A d^2r\, \vec{S}\cdot\hat{n}_{in} = \int_V d^3r\, \vec{J}\cdot\vec{E} + \int_V d^3r\, \partial u/\partial t. \tag{3}$$

The left-hand side represents the net work per unit time done on the specimen by the fields, the first term on the right-hand side represents the rate of energy dissipation, and the second term on the right-hand side represents the rate of change of stored energy. Averaging Eq. (3) over the period T yields

$$P = T^{-1}\int_0^T dt \int d^3r\, \vec{J}\cdot\vec{E}$$

$$= T^{-1}\int_0^T dt \int_A d^2r\, \vec{S}\cdot\hat{n}_{in}. \tag{4}$$

When $\vec{J}\cdot\vec{E}$ is distributed rather uniformly over the volume, it is useful to introduce the time-averaged dissipated power per unit volume

$$P_V = P/V, \tag{5}$$

and, when $\vec{J}\cdot\vec{E}$ is localized at the surface, it is useful to make use of the time-averaged dissipated power per unit area

$$P_A = P/A. \tag{6}$$

It is also helpful to introduce the energy loss per cycle $W=PT$, the energy loss per unit volume per cycle $W_V=W/V=P_VT$, and the energy loss per unit surface area per cycle $W_A=W/A=P_AT$.

Consider a long, cylindrical specimen with a time-varying magnetic induction $B_a(t)$ applied parallel to its axis. Let $\bar{B}(t)$ denote the magnetic induction inside the specimen averaged over the volume V. As $B_a(t)$ sweeps through its cycle, $\bar{B}(t)$ lags behind, and the trajectory of $\bar{B}(t)$ versus $B_a(t)$ sweeps out a hysteresis loop of area A_H. Using Poynting's theorem to evaluate W_V, we obtain

$$W_V = A_H/4\pi \quad \text{(Gaussian)} \tag{7a}$$

$$= A_H/\mu_0 \quad \text{(mks)}. \tag{7b}$$

By consideration of the hysteresis loop, we find that the largest losses occur when the answer is "yes" to the following two questions:

1. Does a large amount of flux sweep in and out of the specimen each cycle?

2. Is $\bar{B}$ significantly out of equilibrium with B_a during the cycle?

Let us consider three examples: (a) In the case of *small screening currents*, although a large amount of flux sweeps in and out each cycle, $\bar{B}$ remains close to equilibrium with B_a during the cycle, the area A_H of the hysteresis loop is small, and the losses are small. (b) In the case of *modest screening currents*, a moderately large amount of flux sweeps in and out each cycle, $\bar{B}$ remains appreciably out of equilibrium with B_a during the cycle, the area A_H is relatively large, and the losses are large. (c) In the case of *large screening currents*, although $\bar{B}$ remains well away from equilibrium with B_a during the cycle, only a small amount of flux sweeps in and out each cycle, the area A_H is small, and the losses are small. Another helpful quantity to introduce, when considering a long cylindrical specimen of radius a in a time-varying longitudinal field, is the *complex permeability*[2] $\tilde{\mu}$. Suppose the applied magnetic induction has a dc component B_0 and a sinusodally varying component $b_0 \cos \omega t$:

$$B_a = B_0 + b_0 \cos \omega t\,. \tag{8}$$

The measured voltage V(t) in a single-turn coil wrapped tightly around the cylinder is

$$V = -c^{-1}\, d\Phi/dt \quad \text{(Gaussian)} \tag{9a}$$

$$= -d\Phi/dt \quad \text{(mks)}\,, \tag{9b}$$

where the magnetic flux through the coil is $\Phi = \pi a^2 \bar{B}$. [In Eqs. (8) and (9), as well as in many of the following equations, we suppress the time argument in quantities that depend upon the time t.] In the absence of the specimen, $\bar{B} = B_a$ and $V(t) = V_0 \sin\omega t$, where

$$V_0 = \pi a^2 b_0 \omega / c \quad \text{(Gaussian)} \tag{10a}$$

$$= \pi a^2 b_0 \omega \quad \text{(mks)}\,. \tag{10b}$$

In the presence of the specimen, we need convenient expressions for $\bar{B}(t)$ and V(t). This is accomplished by writing $\bar{B}$ as

$$\bar{B} = \langle \bar{B} \rangle_{time} + \bar{b}\,, \tag{11}$$

where $\bar{b}(t)$ is the deviation of $\bar{B}(t)$ from its time average, $\langle \bar{B} \rangle_{time}$. Because $\bar{b}(t)$ is periodic with period T, it can be expanded in a Fourier series as follows:

$$\bar{b} = \sum_{n=1}^{\infty} (\mu_n' \cos n\omega t + \mu_n'' \sin n\omega t) b_0 \tag{12}$$

$$\mu_n' = (\omega/\pi b_0) \int_0^T dt\, \bar{b} \cos n\omega t \tag{13a}$$

$$\mu_n'' = (\omega/\pi b_0) \int_0^T dt\, \bar{b} \sin n\omega t\,. \tag{13b}$$

The measured voltage V(t) is then

$$V = -(\pi a^2/c)\, d\bar{b}/dt \quad \text{(Gaussian)} \tag{14a}$$

$$= -\pi a^2\, d\bar{b}/dt\,. \quad \text{(mks)} \tag{14b}$$

Making use of Eq. (12), we obtain

$$V = V_0 \sum_{n=1}^{\infty} n(\mu_n' \sin n\omega t - \mu_n'' \cos n\omega t)\,, \tag{15}$$

where V_0 is the amplitude in the absence of the specimen, Eq. (10).

Frequently, lock-in techniques are used to measure only V_1, the component of V that varies with angular frequency ω:

$$V_1 = V_0(\mu_1' \sin \omega t - \mu_1'' \cos \omega t) \tag{16}$$

A *complex permeability* can be defined,

$$\tilde{\mu} = \mu' + i\mu'' = \mu_1' + i\mu_1''\,, \tag{17}$$

in terms of which the first term on the right-hand side of Eq. (16), the *out-of-phase* (with B_a) or *inductive* component, measures the real part (μ'), and the second term, the *in-phase, resistive*, or *lossy* component, measures the imaginary part (μ"). Note that

$$B_a = B_0 + \mathrm{Re}(b_0 e^{-i\omega t}) \tag{18}$$

$$V_1 = V_0 \mathrm{Re}(i\tilde{\mu} e^{-i\omega t})\,. \tag{19}$$

Here Re denotes the real part. In vacuum, μ'=1 and μ"=0, such that

$$V_1 = V_0 \sin n\omega t\,. \tag{20}$$

In a normal metal with static magnetic permeability μ, we have, at frequencies (typically less than 10^2 Hz) sufficiently low that the skin depth is much larger than the radius, μ'≈μ and μ"≈0, such that

$$V_1 \approx \mu V_0 \sin n\omega t\,. \tag{21}$$

In a superconductor in the Meissner state (where the internal magnetic induction is zero), μ'≈0 and μ"≈0, such that

$$V_1 \approx 0\,. \tag{22}$$

There is a close connection between μ" and the area A_H of the hysteresis loop, as can be seen by starting with the expression

$$A_H = \oint B_a d\bar{B} = \int_0^T (B_0 + b_0 \cos\omega t) \frac{d\bar{B}}{dt}\, dt\,. \tag{23}$$

Making use of Eqs. (11) and (12) and carrying out the time integration, we find that only the n=1, $\cos^2\omega t$ term survives, which yields

$$A_H = \pi b_0^2 \mu'' . \tag{24}$$

Thus, there is a simple geometric mnemonic for μ'': μ'' is simply the ratio of the area A_H of the hysteresis loop to the area of a circle of radius b_0. The loss per cycle per unit volume, expressed in terms of μ'' is

$$W_V = \pi b_0^2 \mu''/4\pi \quad \text{(Gaussian)} \tag{25a}$$

$$= \pi b_0^2 \mu''/\mu_0 . \quad \text{(mks)} \tag{25b}$$

II. NORMAL METALS

The basic equations needed for the calculation of ac losses in isotropic, magnetically reversible materials are

$$\vec{J} = \sigma_n \vec{E} = \rho_n^{-1} \vec{E} , \tag{26}$$

$$\nabla \bullet \vec{B} = 0 , \tag{27}$$

$$\nabla \bullet \vec{E} = 0 , \quad \text{(inside the metal)} \tag{28}$$

$$\vec{B} = \mu \vec{H} \quad \text{(Gaussian)} \tag{29a}$$

$$= \mu\mu_0 \vec{H} , \quad \text{(mks)} \tag{29b}$$

$$\nabla \times \vec{E} = -c^{-1} \partial\vec{B}/\partial t \quad \text{(Gaussian)} \tag{30a}$$

$$= -\partial\vec{B}/\partial t , \quad \text{(mks)} \tag{30b}$$

$$\nabla \times \vec{H} = (4\pi/c)\vec{J} \quad \text{(Gaussian)} \tag{31a}$$

$$= \vec{J} . \quad \text{(mks)} \tag{31b}$$

Note here that μ is dimensionless and that the displacement current can be neglected relative to the induced screening current at the frequencies typically used for ac loss measurements.

Combining the above equations yields the following diffusion equations for the magnetic flux density and the current density

$$\partial\vec{B}/\partial t = D_n \nabla^2 \vec{B} , \tag{32}$$

$$\partial\vec{J}/\partial t = D_n \nabla^2 \vec{J} , \tag{33}$$

where D_n is the *magnetic diffusivity in the normal state*,

$$D_n = \rho_n c^2/4\pi\mu \quad \text{(Gaussian)} \tag{34a}$$

$$= \rho_n/\mu\mu_0 . \quad \text{(mks)} \tag{34b}$$

For example, for high conductivity copper at low temperature, we have $\rho_n \approx 10^{-8}$ Ω-cm, $\mu \approx 1$, and $D_n \approx 1$ cm^2/sec.

In the presence of a sinusoidally varying applied magnetic flux density

$$\vec{B}_a = \mathrm{Re}(\vec{B}_{a\omega} e^{-i\omega t}) , \tag{35}$$

where $\vec{B}_{a\omega}$ is a constant, the induction inside the specimen can be expressed as

$$\vec{B}(\vec{r},t) = \mathrm{Re}[\vec{B}_\omega(\vec{r}) e^{-i\omega t}] , \tag{36}$$

where the diffusion equation [Eq. (32)] requires

$$-i\omega \vec{B}_\omega(\vec{r}) = D_n \nabla^2 \vec{B}_\omega(\vec{r}) . \tag{37}$$

In plane geometry, with the specimen occupying the space $x>0$, we obtain[3]

$$B_\omega(x) = B_\omega(0) e^{ikx} = B_\omega(0) e^{-x/\delta_n} e^{ix/\delta_n} \tag{38}$$

where

$$k = (1 + i)/\delta_n \tag{39}$$

and $\delta_n = (2D_n/\omega)^{1/2}$ is the *normal-state skin depth*

$$\delta_n = (\rho_n c^2/2\pi\mu\omega)^{1/2} \quad \text{(Gaussian)} \tag{40a}$$

$$= (2\rho_n/\mu\mu_0\omega)^{1/2} . \quad \text{(mks)} \tag{40b}$$

For example, for high conductivity copper at low temperature and $f=\omega/2\pi=60$ Hz, we have $\delta_n \approx 1$ mm.

The *normal-state magnetic flux diffusion time* τ_n is the characteristic time required for magnetic flux changes to diffuse in to the middle of a normal-metal specimen. If $2L$ is a characteristic specimen dimension, we have, simply by dimensional arguments, $D_n \sim L^2/\tau_n$ or $\tau_n \sim L^2/D_n = 2(L/\delta_n)^2/\omega$. Note that when $\omega\tau_n \ll 1$, $\delta_n \gg L$. Similarly, when $\omega\tau_n \sim 1$, $\delta_n \sim L$, and when $\omega\tau_n \gg 1$, $\delta_n \ll L$.

For a cylinder of radius a, it is convenient to assign the constants of proportionality such that $\omega\tau_n=1$ corresponds to $\delta_n=a$. We then have the following relations

$$\delta_n = (2D_n/\omega)^{1/2} , \tag{41}$$

$$a = (2D_n\tau_n)^{1/2} , \tag{42}$$

$$\tau_n = a^2/2D_n , \tag{43}$$

$$\omega\tau_n = a^2/\delta_n^2 , \tag{44}$$

as well as

$$\tau_n = 2\pi\mu a^2/\rho_n c^2 \quad \text{(Gaussian)} \tag{45a}$$

$$= \mu\mu_0 a^2/2\rho_n . \quad \text{(mks)} \tag{45b}$$

For example, for high conductivity copper at low temperature with a radius a~1 mm, we have $\tau_n \sim 10^{-2}$ sec.

Consider now a normal cylinder of radius a in a parallel magnetic field. Let

$$B_a = B_0 + \text{Re}(b_0 e^{-i\omega t}) . \tag{46}$$

Inside the cylinder, we have at radial coordinate ρ

$$B(\rho,t) = \mu B_0 + b_1(\rho,t) \tag{47}$$

$$b_1(\rho,t) = \text{Re}[b_{1\omega}(\rho)e^{-i\omega t}] , \tag{48}$$

where $b_{1\omega}$ obeys Eq. (37). The solution of this equation is[4,5]

$$b_{1\omega}(\rho) = \mu b_0 J_0(k\rho)/J_0(ka) \tag{49}$$

where J_0 is the Bessel function of order zero and

$$k = (1 + i)/\delta_n . \tag{50}$$

Averaging Eq. (47) over the cylinder cross section yields

$$\bar{B}(t) = \mu B_0 + \bar{b}_1(t) \tag{51}$$

$$\bar{b}_1(t) = \text{Re}(\bar{b}_{1\omega} e^{-i\omega t}) \tag{52}$$

$$\bar{b}_1 = \tilde{\mu} b_0 , \tag{53}$$

where the *complex permeability* $\tilde{\mu}$ is

$$\tilde{\mu} = \mu' + i\mu'' = \mu 2J_1(ka)/(ka)J_0(ka) . \tag{54}$$

The quantities μ' and μ" also can be written as

$$\mu'/\mu = f_1(x) = f_0(x)\cos\beta(x) \tag{55}$$

$$\mu''/\mu = f_2(x) = f_0(x)\sin\beta(x) , \tag{56}$$

$$x = a\sqrt{2}/\delta_n = (2\omega\tau_n)^{1/2} , \tag{57}$$

where, in terms of the modulus M_v and phase Θ_v of the Kelvin functions,[6]

$$f_0(x) = 2M_1(x)/xM_0(x) , \tag{58}$$

$$\beta(x) = \Theta_0(x) - \Theta_1(x) + 3\pi/4. \tag{59}$$

Plots of f_1, f_2, f_0, and tanβ versus x are shown in Fig. 1.

The hysteresis loop is easily analyzed as follows. The ac component of the applied magnetic flux density is $b_a = b_0 \cos \omega t$, and the induced ac component of the average magnetic flux density in the specimen is

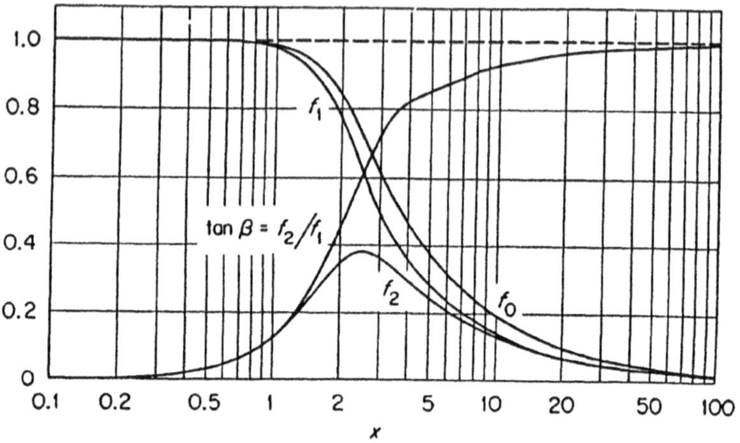

Fig. 1. Plots of the auxiliary functions used in the analysis of the complex permeability $\tilde{\mu}=\mu'+i\mu''$ of a normal cylinder, $f_1=\mu'/\mu$, $f_2=\mu''/\mu$, $f_0=|\tilde{\mu}|/\mu_0$, and $\tan\beta=f_2/f_1=\mu''/\mu'$, defined in Eqs. (55)–(58), versus $x=a\sqrt{2}/\delta_n=(2\omega\tau_n)^{1/2}$.

$$\bar{b}_1(t) = \mu b_0 f_0 \cos(\omega t-\beta)\ . \tag{60}$$

The hysteresis loop is thus an ellipse with area $A_H=\pi b_0^2\mu f_0\sin\beta$. Referring to Fig. 2, we see that

$$\mu' = \mu f_1 = \mu f_0\cos\beta \tag{61}$$

is the slope of the line MN and that

$$\mu'' = A_H/\pi b_0^2 = \mu f_2 = \mu f_0 \sin\beta\ . \tag{62}$$

At low frequencies, such that $\omega\tau_n<<1$, $a<<\delta_n$ and $x<<1$, we have $f_1\approx f_0\approx 1$ and $f_2\approx\beta\approx x^2/8=(a/2\delta_n)^2=\omega\tau_n/4<<1$. (See Fig. 1.) This is the case of *small screening currents.* A large amount of flux sweeps in and out each cycle but $\bar{B}$ remains close to equilibrium during the cycle, and the losses are small. (See Fig. 2.)

At intermediate frequencies, such that $\omega\tau_n\sim 1$, $a\sim\delta_n$, and $x\sim 1$, $f_1\sim f_2\sim f_0$. This is the case of *modest screening currents.* A moderately large amount of flux sweeps in and out each cycle, and $\bar{B}$ remains appreciably out of equilibrium with B_a during the cycle. The losses are maximized when $\omega\tau_n=3.2$, $a=1.8\ \delta_n$, and $x=2.5$, where $\beta=0.55$, $f_1=0.62$, $f_2=0.38$, and $f_0=0.72$. At high frequencies, such that $\omega\tau_n>>1$, $a>>\delta_n$, and $x>>1$, we have $f_1\approx f_2\approx f_0/\sqrt{2}\approx\sqrt{2}/x=\delta_n/a=(\omega\tau_n)^{-1/2}<<1$ and $\beta\approx\pi/4=0.79$. This is the case of *large screening currents.* Although $\bar{B}$ remains far from equilibrium with B_a during the cycle, only a small amount of flux sweeps in and out during the cycle, and the losses are small.

Shown in Fig. 3 is the local time-averaged rate of energy dissipation per unit volume at a distance ρ from the axis of the cylinder,

$$d(\rho) = T^{-1}\int_0^T \vec{J}\cdot\vec{E}\,dt. \tag{63}$$

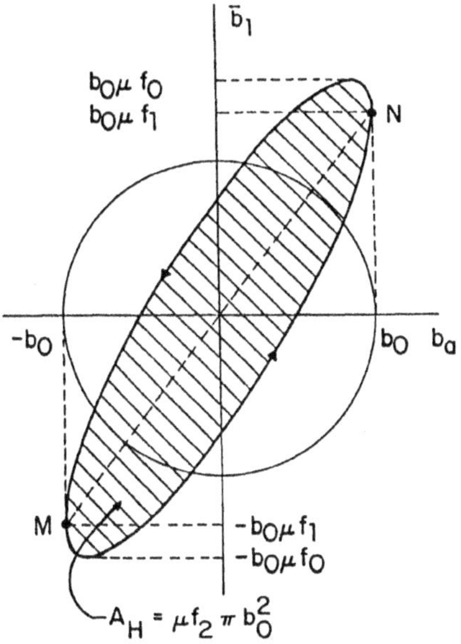

Fig. 2. Hysteresis loop for a normal cylinder, described by Eqs. (52)–(62). Note that $\mu''=A_H/\pi b_0^2$, where A_H is the area of the hysteresis loop (crosshatched) and πb_0^2 is the area of the circle of radius b_0. (For this figure, the values $\mu=2$, $x=2.4$, $\beta=0.52$, $f_0=0.75$, $f_1=f_0\cos\beta=0.65$, and $f_2=f_0\sin\beta=0.38$ were used, such that $\mu''=\mu f_2=0.75$.)

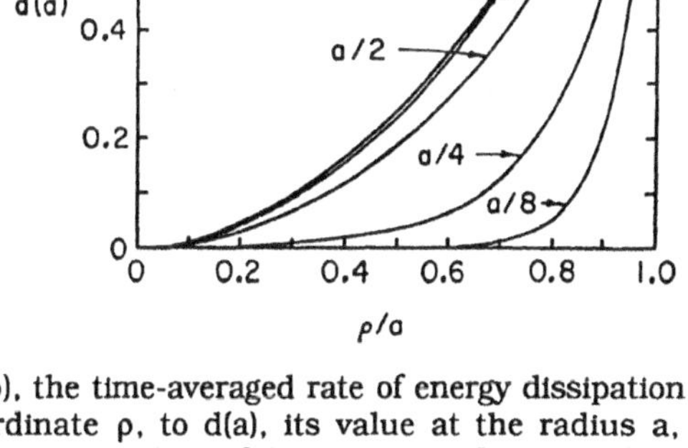

Fig. 3. Ratio of $d(\rho)$, the time-averaged rate of energy dissipation per unit volume at radial coordinate ρ, to $d(a)$, its value at the radius a, in a normal-metal cylinder for various values of the skin depth δ_n [Eq. (65)].

In Eq. (63), $\vec{J}$ and $\vec{E}$ have only azimuthal components, J_ϕ and E_ϕ. One obtains from Eqs. (26), (31), and (49)

$$d(\rho) = \mu(b_0^2/8\pi)\omega\,|J_1(k\rho)/J_0(ka)|^2 \quad \text{(Gaussian)} \tag{64a}$$

$$= \mu(b_0^2/2\mu_0)\omega\,|J_1(k\rho)/J_0(ka)|^2 . \quad \text{(mks)} \tag{64b}$$

Figure 3 is a plot of

$$d(\rho)/d(a) = |J_1(k\rho)/J_1(ka)|^2 \tag{65}$$

versus ρ/a. For $a \ll \delta_n$ or $\omega\tau_n=(a/\delta_n)^2 \ll 1$, both E_ϕ and J_ϕ are proportional to ρ, and $d(\rho)$ is proportional to ρ^2:

$$d(\rho) \approx \mu(b_0^2/8\pi)\omega(\rho^2/2\delta_n^2) \quad \text{(Gaussian)} \tag{66a}$$

$$\approx \mu(b_0^2/2\mu_0)\omega(\rho^2/2\delta_n^2) . \quad \text{(mks)} \tag{66b}$$

In the opposite limit, for $a \gg \delta_n$ or $(\omega\tau_n)=(a/\delta_n)^2 \gg 1$, E_ϕ and J_ϕ are large only within the normal-state skin depth δ_n of the surface.

III. EDDY-CURRENT LOSSES IN TYPE-II SUPERCONDUCTORS

We next consider the electrodynamic behavior of ideal, type-II superconductors and ignore, for the present, the possibility of flux pinning. Here, the ac losses can be called *eddy-current losses, flux-flow losses*, or *viscous losses*. In many respects, the behavior of an ideal, type-II superconductor is similar to that of a normal metal except that the linear relations between $\vec{E}$ and $\vec{J}$ [Eq. (26)] and between $\vec{B}$ and $\vec{H}$ [Eqs. (29a) and (29b)] do not hold. Instead, it is found that, when $\vec{E}$ and $\vec{J}$ are parallel,

$$\vec{E} = \rho_f(B)\vec{J} , \tag{67}$$

where ρ_f is the flux-flow resistivity,[7] and the thermodynamic magnetic field $\vec{H}$ and magnetic flux density $\vec{B}$ in thermodynamic equilibrium are[8]

$$\vec{H} = H_{eq}(B)\hat{B} \tag{68}$$

or

$$\vec{B} = B_{eq}(H)\hat{H} , \tag{69}$$

where $\hat{B}$ and $\hat{H}$ are unit vectors. The dependences of ρ_f and H_{eq} upon the flux density B are sketched in Fig. 4. The symbols H_{c2} and H_{c1} denote the upper and lower critical fields, respectively, and B_{c2} denotes the upper critical flux density. Note that the equilibrium flux density is $B=0$ for $-H_{c1} \leq H \leq H_{c1}$. The nonanalytic B dependence of ρ_f and H_{eq} makes the electrodynamic equations nonlinear, such that analytic solutions no longer can be obtained for arbitrarily large amplitudes of the applied magnetic field.

For small amplitudes b_0 of the applied magnetic induction, however, a linearization procedure can be used to solve analytically for the electrodynamic response of the superconductor.[5] Let the magnetic field and induction applied parallel to a cylindrical type-II superconductor of radius a be

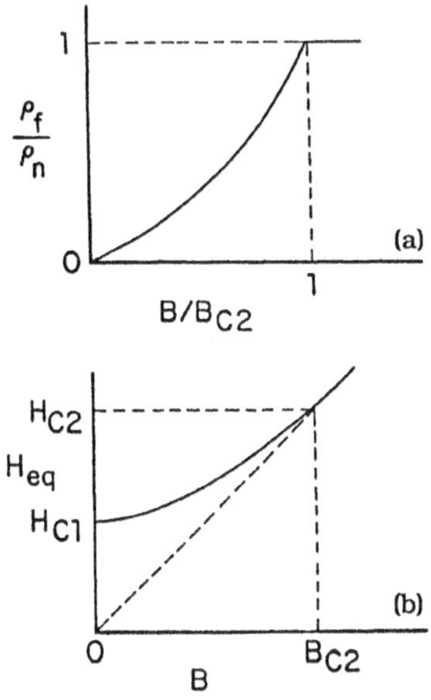

Fig. 4. Sketches of flux-density-dependent functions for type-II superconductors: (a) flux-flow resistivity ρ_f/ρ_n versus B/B_{c2} and (b) H_{eq} versus B.

$$H_a = H_0 + h_0 \cos\omega t \tag{70}$$

$$B_a = B_0 + b_0 \cos\omega t \,, \tag{71}$$

where $h_0 \ll H_0$ and $b_0 \ll B_0$. Inside the superconductor the resulting macroscopic flux density is

$$B(\rho,t) = B_{eq}(H_0) + b_1(\rho,t) \,, \tag{72}$$

where ρ is the radial coordinate; the corresponding spatial average over the cross section is

$$\bar{B}(t) = B_{eq}(H_0) + \bar{b}_1(t) \,, \tag{73}$$

where $|b_1(\rho,t)| \ll B_{eq}(H_0)$ and $|\bar{b}_1(t)| \ll B_{eq}(H_0)$. Then $b_1(\rho,t)$ obeys the same equations as in the normal state, <u>except</u> for the replacement of the normal-state resistivity ρ_n by the flux-flow resistivity ρ_{f0} and the replacement of the static normal-state permeability μ by the differential permeability of the mixed state μ_0', which is defined below:

$$\rho_n \to \rho_{f0} \tag{74}$$

$$\mu \to \mu_0' \,. \tag{75}$$

The flux-flow resistivity ρ_{f0} is defined by

$$\rho_{f0} = [\rho_f(B)]_{B=B_{eq}(H_0)} \tag{76}$$

and the dimensionless differential permeability μ_0' by

$$\mu_0' = [dB_{eq}(H)/dH]_{H=H_0} \quad \text{(Gaussian)} \tag{77a}$$

$$= \mu_0^{-1}[dB_{eq}(H)/dH]_{H=H_0} \quad \text{(mks)} . \tag{77b}$$

Here the prime denotes the derivative with respect to H, and the subscript 0 denotes evaluation at $H=H_0$. The dependences of ρ_{f0}/ρ_n and μ_0' upon H_0 are sketched in Fig. 5.

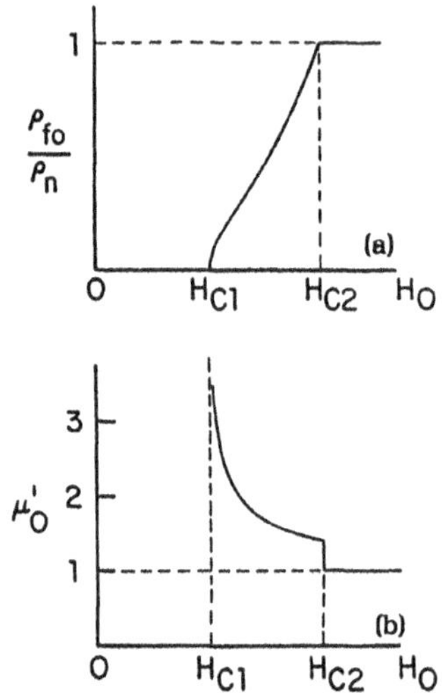

Fig. 5. Sketches of (a) ρ_{f0}/ρ_n [Eq. (76)] and (b) μ_0' [Eqs. (77a) and (77b)] versus H_0 for a type-II superconductor.

In Sec. II we discussed the normal-state magnetic diffusivity D_n, the normal-state skin depth δ_n, and the normal-state flux diffusion time τ_n. In the mixed state of the type-II superconductor, the corresponding quantities are the *flux-flow magnetic diffusivity*,

$$D_{f0} = \rho_{f0}c^2/4\pi\mu_0' \quad \text{(Gaussian)} \tag{78a}$$

$$= \rho_{f0}/\mu_0'\mu_0 \text{ , (mks)} \tag{78b}$$

the *flux-flow skin depth*,

$$\delta_{f0} = (2D_{f0}/\omega)^{1/2} , \tag{79}$$

$$\delta_{f0} = (\rho_{f0}c^2/2\pi\mu_0'\omega)^{1/2} \quad \text{(Gaussian)} \tag{79a}$$

$$= (2\rho_{f0}/\mu_0'\mu_0\omega)^{1/2} , \quad \text{(mks)} \tag{79b}$$

and the *flux-flow flux diffusion time*,

$$\tau_{f0} = a^2/2D_{f0} . \tag{80}$$

Shown in Fig. 6 are sketches of D_{f0}, δ_{f0}, and τ_{f0}, normalized to their values in the

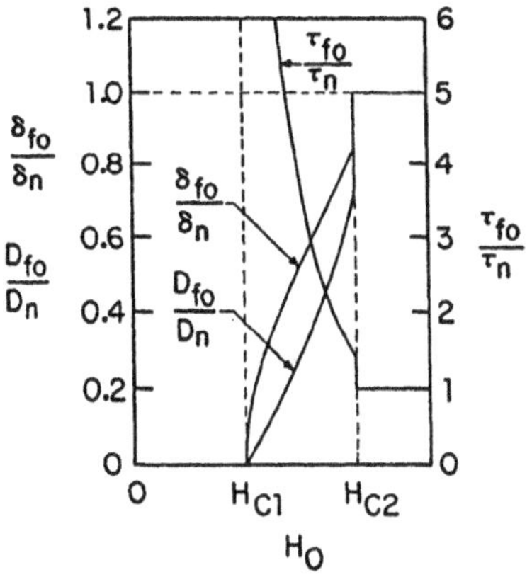

Fig. 6. Sketches of D_{f0}/D_n and $\delta_{f0}/\delta_n=(D_{f0}/D_n)^{1/2}$ (left scale) and $\tau_{f0}/\tau_n=D_n/D_{f0}$ (right scale) versus H_0 for a type-II superconductor. See Eqs. (78)–(80).

normal state just above H_{c2}, versus H_0. Note that D_{f0} and δ_{f0} vanish and τ_{f0} becomes infinite at $H_0=H_{c1}$. Moreover, at H_{c2} all three quantities exhibit discontinuities, which arise from the discontinuity of μ_0'.

In terms of these flux-flow quantities, Eq. (73) can be expressed as[5]

$$\bar{B}(t) = B_{eq}(H_0) + \mathrm{Re}(\tilde{\mu} b_0 e^{-i\omega t}) \,, \tag{81}$$

where $\tilde{\mu}=\mu'+i\mu''$ and

$$\mu'/\mu_0' = f_1(x) = f_0(x)\cos\beta(x) \tag{82}$$

$$\mu''/\mu_0' = f_2(x) = f_0(x)\sin\beta(x) \,, \tag{83}$$

where f_0, f_1, f_2, and β are the same functions as in Eqs. (55)–(58), except that here

$$x = a\sqrt{2}/\delta_{f0} = (2\omega\tau_{f0})^{1/2} \,. \tag{84}$$

Note that x now depends strongly not only upon frequency but also upon the applied dc field H_0 via the dependence of ρ_{f0} upon H_0.

Just as the loss per cycle per unit volume of a normal cylinder can be calculated from Eqs. (25) and (56), the loss per cycle per unit volume of an ideal, type-II superconductor can be calculated from Eqs. (25) and (83):

$$W_V = (\pi b_0^2/4\pi)\mu_0' f_2(a\sqrt{2}/\delta_{f0}) \text{ (Gaussian)} \tag{85a}$$

$$= (\pi b_0^2/\mu_0)\mu_0' f_2(a\sqrt{2}/\delta_{f0}) \,. \text{ (mks)} \tag{85b}$$

For fixed b_0 and ω, but varying H_0, we see that W_V is maximized when $\mu''=\mu_0' f_2$ is maximized, where both μ_0' and δ_{f0} depend upon H_0.

Illustrating the expected qualitative dependence of μ' and μ'' upon H_0 at different

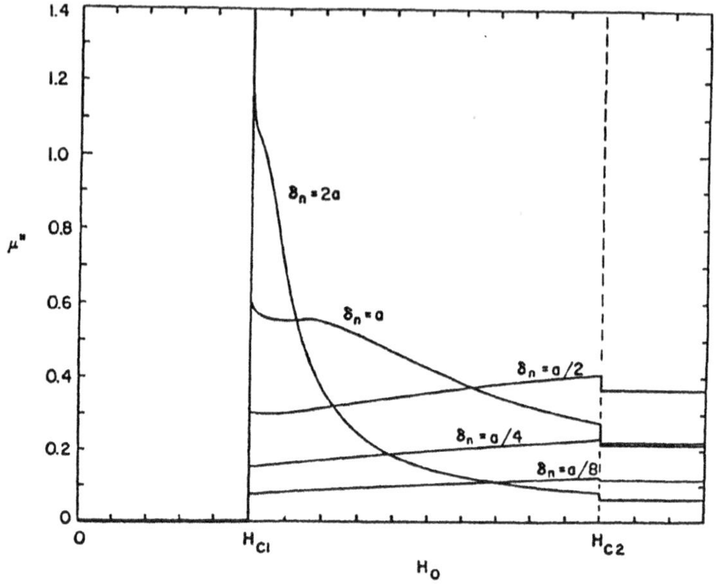

Fig. 7. Theoretically calculated [Eqs. (82)–(89)] curves of μ' versus H_0 at various frequencies and corresponding normal-state skin depths δ_n for eddy-current losses.

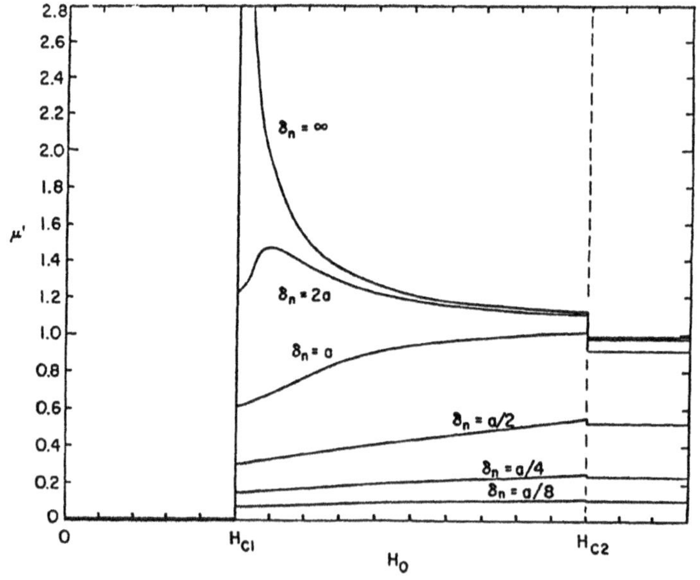

Fig. 8. Theoretically calculated [Eqs. (82)–(89)] curves of μ" versus H_0 at various frequencies and corresponding normal-state skin depths δ_n for eddy-current losses.

values of the frequency, Figs. 7 and 8 exhibit curves of μ' and μ" computed from Eqs. (82) and (83) with the crude model approximations of Eqs. (86)-(89).

$$\rho_{f0} = [B_{eq}(H_0)/B_{c2}]\rho_n \,, \tag{86}$$

$$B_{eq}(H_0) = B_{c2}[(H_0^2 - H_{c1}^2)/(H_{c2}^2 - H_{c1}^2)]^{1/2} \,, \tag{87}$$

$$dB_{eq}(H_0)/dH_0 = B_{c2}H_0/[(H_0^2 - H_{c1}^2)(H_{c2}^2 - H_{c1}^2)]^{1/2} \,, \tag{88}$$

$$\mu_0'(H_0) = H_{c2}H_0/[(H_0^2 - H_{c1}^2)(H_{c2}^2 - H_{c1}^2)]^{1/2} \,. \tag{89}$$

According to Eq. (89),

$$\mu_0'(H_{c2}) = [1-(H_{c1}/H_{c2})^2]^{-1} > 1 \,, \tag{90}$$

and the magnitude of the discontinuity of μ_0' at H_{c2} is $[(H_{c2}/H_{c1})^2-1]^{-1}$.

Measurements of μ' and μ" in nearly ideal, highly reversible samples of NbTa alloys were reported in Ref. 5. The measured dependences of μ' and μ" upon H_0 at different frequencies for two samples with different Ta concentrations are shown in Figs. 9 and 10. In these samples, the dissipation is dominated by flux flow over a wide range of fields. Flux-pinning effects, however, play an important role near H_{c1}. As expected when viscous losses are dominant, both μ' and μ" exhibit strong frequency dependence. The expected discontinuity in μ' and μ" at H_{c2} is smeared out over $2h_0$, the field resolution width of the experiment. The experiments of Ref. 5 confirm that in nearly ideal type-II superconductors, the complex permeability components μ' and μ" can be calculated using the theoretical approach described here.

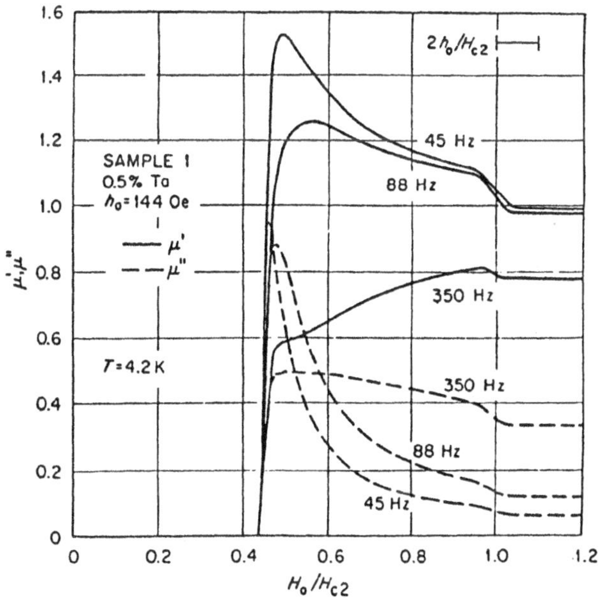

Fig. 9. Measurements [Ref. 5] of the real and imaginary parts of the ac permeability as a function of the dc applied magnetic field H_0 for a NbTa alloy containing 0.5 at. % Ta. H_{c2}(4.2K) = 2.87 kOe.

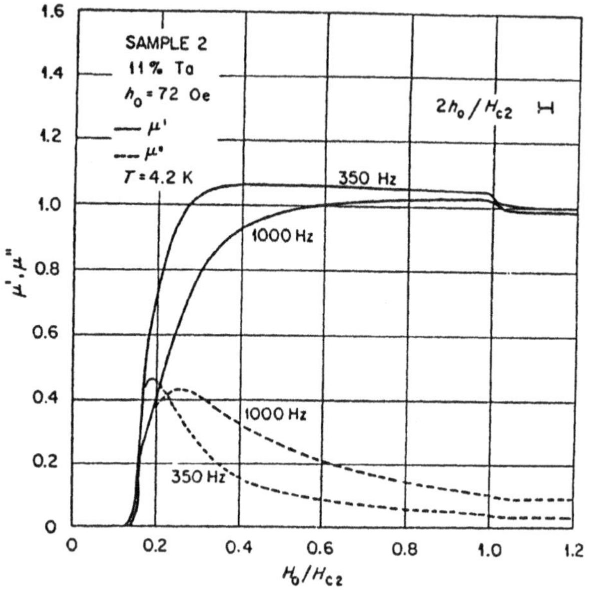

Fig. 10. Measurements [Ref. 5] of the real and imaginary parts of the ac permeability as a function of the dc applied magnetic field H_0 for a NbTa alloy containing 11 at. % Ta. $H_{c2}(4.2K)$ = 4.42 kOe.

In the conventional superconductors, the linear relation [Eq. (67)] between $\vec{E}$ and $\vec{J}$ holds at low temperatures only for samples containing very few pinning centers or for average driving current densities well above the critical current density J_c. On the other hand, in the high-temperature superconductors there often is a wide range of temperatures T near the critical temperature T_c where thermally assisted flux flow[9,10] occurs, resulting in a linear relation $\vec{E} = \rho_{th}\vec{J}$, but where the thermally activated resistivity ρ_{th} is generally a more complicated function of B and T than assumed, for example, in Eq. (86). The physics of the time-dependent behavior of the magnetic flux, however, remains exactly as discussed above, except for the simple replacement of ρ_f by ρ_{th}.

In hysteretic type-II superconductors, in which flux pinning plays an important role, another theoretical approach to the losses must be used, which is discussed in the following section.

IV. BULK-PINNING HYSTERETIC LOSSES IN TYPE-II SUPERCONDUCTORS

As is seen from Eq. (85), the loss per cycle per unit volume W_V in an ideal, pinning-free, type-II superconductor depends strongly upon the frequency. In the presence of pinning, magnetic hysteresis plays a significant role, and if hysteretic losses dominate, the loss per cycle W becomes independent of frequency. It can be argued as follows that hysteretic losses dominate when the electric field obeys $E \ll \rho_f J_c$, where ρ_f is the flux-flow resistivity and J_c is the critical current density. Assuming that, for $J > J_c$, $E = \rho_f(J - J_c)$, as discussed by Kim and Stephen,[7] the rate of energy dissipation per unit volume can be written as

$$JE = E^2/\rho_f + J_c E . \tag{91}$$

The first term on the right-hand side of Eq. (91) represents the eddy-current, flux-flow, or viscous losses discussed in Sec. III. The second term, which describes the hysteretic losses, gives the rate of heat generation per unit volume near the pinning centers that impede vortex motion. By comparing these two terms, we see that the hysteretic losses dominate when $E \ll \rho_f J_c$, that is, when the electric field generated by vortex motion is at a relatively low level, close to the foot of the curve of $E=\rho_f(J-J_c)$ versus J. This condition also requires that J not greatly exceed the critical depinning current density J_c.

It is perhaps surprising that a complete theory does not yet exist for the calculation of both hysteretic and eddy-current losses when $\vec{J}$ has a component locally parallel to $\vec{B}$. In this case, it is possible that flux-line cutting may occur[11-13] or that the vortex array may split up into domains.[14] The theory for hysteretic losses accounting for flux-line cutting has been discussed in Refs. 15 and 16, but in this paper I discuss only the situation where the theory is best established, the case for which the current density $\vec{J}$ is always locally perpendicular to $\vec{B}$.

The most important quantity determining the hysteretic losses is the critical current density J_c, which at constant temperature is a function only of the magnetic flux density B. Usually, $J_c(B)$ is a monotonically decreasing function of B, decreasing from its maximum value $J_c(0)$ at B=0 to $J_c=0$ at $B=B_{c2}$. The origin and properties of $J_c(B)$ are discussed in much greater detail in Ref. 10.

The basic equations needed for the computation of hysteretic but otherwise isotropic superconductors, for which $\vec{J}$ is perpendicular to $\vec{B}$, are

$$\vec{E} = \rho_f(\vec{J} - \vec{J}_c)\,, \tag{92}$$

$$\vec{J} = J_c\hat{E}\,, \tag{93}$$

where $J_c>0$ and $\hat{E}=\vec{E}/E$,

$$\vec{E} = \vec{B} \times \vec{v}/c \quad \text{(Gaussian)} \tag{94a}$$

$$= \vec{B} \times \vec{v}\,, \quad \text{(mks)} \tag{94b}$$

where $\vec{v}$ is the local vortex velocity,

$$\nabla \cdot \vec{B} = 0\,, \tag{95}$$

$$\nabla \cdot \vec{E} = 0\,, \ \text{(inside)}\,, \tag{96}$$

$$\vec{H} = H_{eq}(B)\hat{H}\,, \tag{97}$$

or

$$\vec{B} = B_{eq}(H)\hat{B}\,, \tag{98}$$

where $\hat{B}=\vec{B}/B$ and $\hat{H}=\vec{H}/H$,

$$\nabla \times \vec{E} = -c^{-1}\partial\vec{B}/\partial t \quad \text{(Gaussian)} \tag{99a}$$

$$= -\partial\vec{B}/\partial t\,, \quad \text{(mks)} \tag{99b}$$

$$\nabla \times \vec{H} = (4\pi/c)\vec{J} \quad \text{(Gaussian)} \tag{100a}$$

$$\nabla \times \vec{H} = \vec{J}\,. \quad \text{(mks)} \tag{100b}$$

At low frequencies when $E \ll \rho_f J_c$ and $J \approx J_c$, the *critical state equation* holds:

$$|\nabla \times \vec{H}| = (4\pi/c) J_c \quad \text{(Gaussian)} \tag{101a}$$

$$= J_c\,. \quad \text{(mks)} \tag{101b}$$

The equations (92)–(101) need to be solved simultaneously with the corresponding equations for the fields outside the specimen, subject to the boundary conditions at the surface, one of which is that the normal component of $\vec{B}$ be continuous. If there is no surface barrier (no barrier to vortex entry and exit, no surface pinning), then there is an additional boundary condition that the tangential component of $\vec{H}$ be continuous. For example, if the specimen is a long cylinder subjected to an externally applied parallel magnetic field $\vec{H}_a$, the boundary condition at the surface is $H_a=H_{eq}(B)$. On the other hand, if there is an appreciable surface barrier [barrier to vortex entry and exit, surface pinning not accounted for in $J_c(B)$], then the boundary condition on the tangential component of $\vec{H}$ is more complicated. For example, if the specimen is a long cylinder subjected to an externally applied parallel magnetic field $\vec{H}_a$, the boundary condition at the surface is $H_a=H_{en}(B)>H_{eq}(B)$ if the surface is unstable to the *entry* of vortices or $H_a=H_{ex}(B)<H_{eq}(B)$ if the surface is unstable to the *exit* of vortices. There exists a range of applied fields $H_{ex}(B)<H_a<H_{en}(B)$ for which the specimen remains in a metastable state, and neither entry nor exit of vortices occurs. The surface boundary conditions and the expected B dependence of $H_{en}(B)$ and $H_{ex}(B)$ are discussed at greater length in Refs. 17 and 18.

The nonlinearities introduced into Eqs. (92)–(101) and the boundary conditions via the nonlinear B dependence of J_c, H_{eq}, H_{en}, and H_{ex} make it impossible to obtain analytic solutions for the flux-density profiles and ac losses for arbitrarily large ac field amplitudes. In this case numerical solutions are required. A description of how to obtain such solutions in slab geometry is given in Ref. 18. Many useful results can be obtained, however, by linearizing the equations when the ac field amplitude is quite small.

Consider the hysteretic losses in a type-II superconducting cylinder of radius a in a parallel applied field,

$$H_a = H_0 + h_0 \cos\omega t \tag{102}$$

$$B_a = B_0 + b_0 \cos\omega t\,, \tag{103}$$

where $h_0 \ll H_0$ and $b_0 \ll B_0$. Assume no barrier to vortex entry and exit. The response of the superconductor at radial coordinate ρ is

$$H(\rho,t) = H_0 + h_1(\rho,t) \tag{104}$$

$$B(\rho,t) = B_{eq}(H_0) + b_1(\rho,t)\,, \tag{105}$$

where $|h_1| \ll H_0$ and $|b_1| \ll B_{eq}(H_0)$. Denote quantities that are generated by h_0 and b_0 by subscripts 1. Then

$$\vec{E}_1 = E_1 \hat{\phi}\,, \tag{106}$$

$$\vec{J}_1 = J_1 \hat{\phi}\,, \tag{107}$$

where $\hat{\phi}$ is the unit vector in the azimuthal direction ($\hat{\phi}=\hat{z}\times\hat{\rho}$). Starting from Eqs. (92)–(101), the resulting linearized equations inside the superconductor are, suppressing the space and time arguments ρ and t in E_1, J_1, h_1, and b_1,

$$E_1 = \rho_{f0}(J_1 - J_{c0}), \qquad J_1 > J_{c0} \tag{108a}$$

$$= 0, \qquad -J_{c0} < J_1 < J_{c0} \tag{108b}$$

$$= \rho_{f0}(J_1 + J_{c0}), \qquad J_1 < -J_{c0} \tag{108c}$$

where ρ_{f0} is defined in Eq. (76), and

$$J_{c0} = [J_c(B)]_{B=B_{eq}(H_0)}, \tag{109}$$

$$\rho^{-1}\partial(\rho E_1)/\partial\rho = -c^{-1}\partial b_1/\partial t \quad \text{(Gaussian)} \tag{110a}$$

$$= -\partial b_1/\partial t, \quad \text{(mks)} \tag{110b}$$

$$-\partial h_1/\partial\rho = (4\pi/c)J_1 \quad \text{(Gaussian)} \tag{111a}$$

$$= J_1, \quad \text{(mks)} \tag{111b}$$

$$b_1 = h_1[dB_{eq}(H)/dH]_{H=H_0}, \tag{112}$$

$$b_1 = \mu_0' h_1 \quad \text{(Gaussian)} \tag{113a}$$

$$= \mu_0' \mu_0 h_1, \quad \text{(mks)} \tag{113b}$$

where the dimensionless differential permeability μ_0' is defined in Eq. (77).

When $|E_1| << \rho_{f0}J_{c0}$, the *critical state equation*,

$$\partial h_1/\partial\rho = \pm(4\pi/c)J_{c0} \quad \text{(Gaussian)} \tag{114a}$$

$$= \pm J_{c0}, \quad \text{(mks)} \tag{114b}$$

determines the critical-state profiles of H_1 and B_1 versus ρ. Let us introduce H_b^* as the field step at the surface needed to push the critical state profile to the center of the specimen. Then, according to Eq. (114),

$$H_b^* = (4\pi/c)J_{c0}a \quad \text{(Gaussian)} \tag{115a}$$

$$= J_{c0}a. \quad \text{(mks)} \tag{115b}$$

For example, if J_{c0}~10^5A/cm^2 and a~0.1 mm, $\mu_0 H_b^*$ ~ 0.1 T = 1 kG. Let us also introduce L_p as the depth of penetration of a critical-state profile with an applied field step h_0. Then,

$$L_p = (c/4\pi)h_0/J_{c0} \quad \text{(Gaussian)} \tag{116a}$$

$$= h_0/J_{c0}. \quad \text{(mks)} \tag{116b}$$

For example, if $\mu_0 h_0$~0.1 T = 1 kG and J_{c0}~10^5A/cm^2, L_p~0.1 mm. The critical-state profiles of $h_1(\rho,t)$ versus ρ have a slightly different appearance depending upon whether $h_0<H_b^*$ ($L_p<a$) or $h_0>H_b^*$ ($L_p>a$). These are shown in Fig. 11a and b, respectively, as the

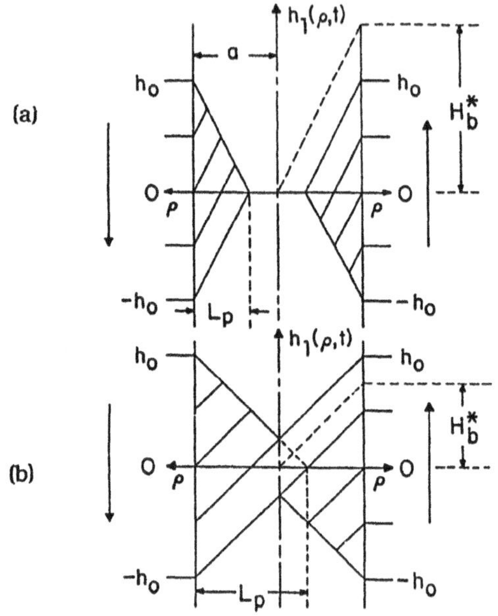

Fig. 11. Magnetic field profiles for bulk pinning, $h_1(\rho,t)$ versus ρ, for (a) $h_0<H_b^*$ ($L_p<a$) and (b) $h_0>H_b^*$ ($L_p>a$). The left side shows the cases for H_a decreasing with time and the right side shows the cases for H_a increasing.

straight-line segments obeying the boundary condition $h_1(\rho,t) = h_0\cos\omega t$. With the help of Eq. (113), we easily obtain $b_1(\rho,t)$ and the volume average

$$\bar{b}_1(t) = (2/a^2) \int_0^a d\rho\rho\, b_1(\rho,t) \,. \tag{117}$$

Substitution of $\bar{b}_1$ into Eqs. (13a) and (13b) with n=1 yields, after a tedious calculation of the Fourier integrals, the complex permeability $\tilde{\mu}=\mu'+i\mu''$, where

$$\mu'/\mu_0' = g_1(x) \tag{118}$$

$$\mu''/\mu_0' = g_2(x) \,. \tag{119}$$

In contrast to Secs. II and III, here we have

$$x = h_0/H_b^* = L_p/a \,, \tag{120}$$

and

$$g_1(x) = x(1 - \frac{5}{16}x) \,, \qquad 0 \le x \le 1 \tag{121a}$$

$$= 1 + \frac{2}{\pi}[(-\frac{1}{2} + \frac{x}{2} - \frac{5x^2}{32})\,\theta + (-\frac{2}{3x} + 1 - \frac{7x}{8} + \frac{13x^2}{48})\sin\theta$$

$$+ (-\frac{1}{4} + \frac{x}{4} - \frac{x^2}{12})\sin 2\theta + (-\frac{x}{24} + \frac{x^2}{48})\sin 3\theta$$

$$+ (-\frac{x^2}{384})\sin 4\theta] \,, \qquad x \ge 1 \tag{121b}$$

where

$$\theta(x) = 2\sin^{-1}(x^{-1/2}) , \tag{122a}$$

$$\theta(1) = \pi , \tag{122b}$$

$$\theta(\infty) = 0 , \tag{122c}$$

$$g_2(x) = \frac{4}{3\pi}\, x(1 - \frac{x}{2}) , \qquad 0 \le x \le 1 \tag{123a}$$

$$= \frac{4}{3\pi}\,\frac{1}{x}\,(1 - \frac{1}{2x}) . \qquad x \ge 1 \tag{123b}$$

The functions g_1 and g_2 are plotted as functions of x in Fig. 12. Note that g_2 has its maximum,

$$g_2^{max} = 2/3\pi = 0.21 , \tag{124}$$

at $x=h_0/H_b^*=L_p/a=1$. Note also that, in the limit of very weak pinning, $H_b^* \to 0$, $L_p \to \infty$, $x \to \infty$, $g_2 \to 0$, $g_1 \to 1$, such that $\tilde{\mu} \to \mu_0'$.

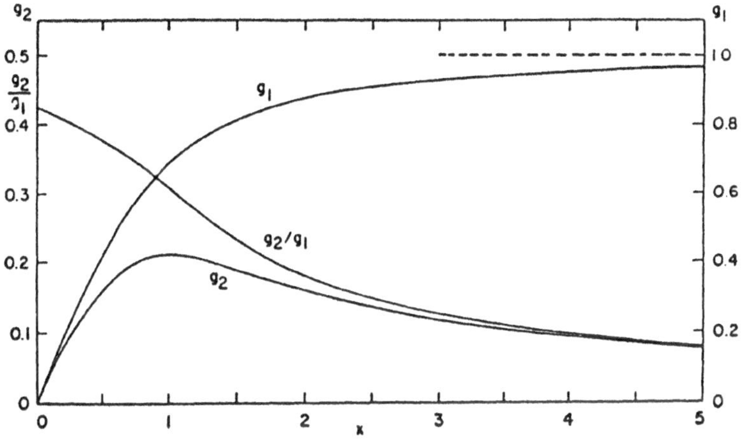

Fig. 12. Plots of the functions $g_1=\mu'/\mu_0'$ (right scale) and $g_2=\mu''/\mu_0'$ (left scale) [see Eqs. (118)–(124)] versus $x=h_0/H_b^*=L_p/a$, as well as their ratio $g_2/g_1=\mu''/\mu'$ (left scale), for bulk-pinning hysteretic losses.

Both μ' and μ'' [Eqs. (118) and (110)] depend strongly upon H_0 via their dependence upon both μ_0' and $x=h_0/H_b^*=L_p/a$, which, via Eq. (115) or (116), involves the dependence of J_c upon H_0. To illustrate the expected behavior of μ' and μ'' upon H_0, Figs. 13 and 14 exhibit theoretically calculated values of μ' and μ'', computed using the models of Eqs. (86)–(89) and the following model for J_c,

$$J_c(B) = J_c(0)(1 - B/B_{c2})/(1 + 3B/2B_{c2}) . \tag{125}$$

Thus, the H_0 dependence of x is determined by

$$x = x_0[J_c(0)/J_c(B)] , \tag{126}$$

where

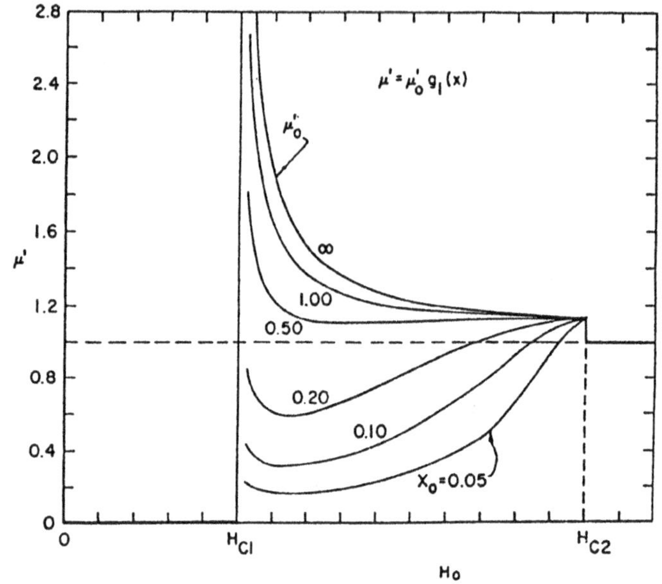

Fig. 13. Theoretically calculated [Eqs. (118)–(127)] curves of μ' versus H_0 at various values of x_0 [Eq. (127)] for bulk-pinning hysteretic losses.

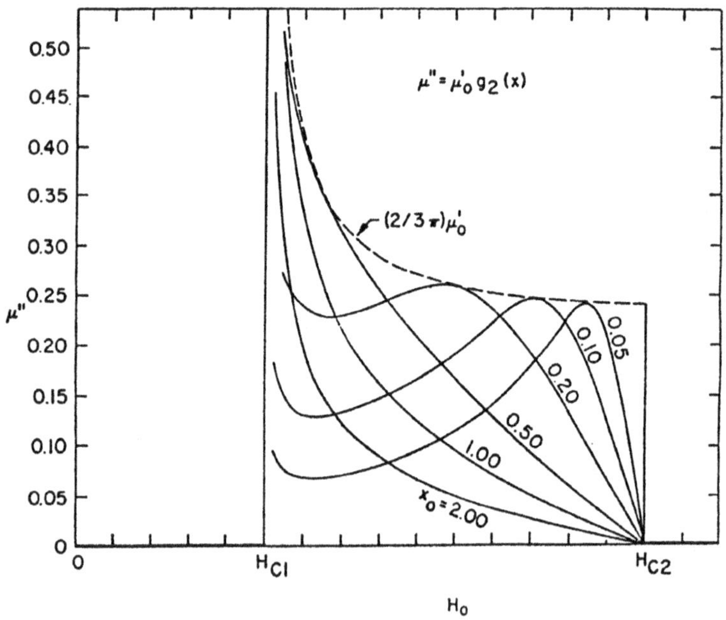

Fig. 14. Theoretically calculated [Eqs. (118)–(127)] curves of μ'' versus H_0 at various values of x_0 [Eq. (127)] for bulk-pinning hysteretic losses.

$$x_0 = ch_0/4\pi J_c(0)a \quad \text{(Gaussian)} \tag{127a}$$

$$= h_0/J_c(0)a\,. \quad \text{(mks)} \tag{127b}$$

$J_c(B)$ is given by Eq. (125) and $B=B_{eq}(H_0)$ by Eq. (87).

Experimental measurements of μ' and μ'' in plastically deformed niobium were reported in Refs. 19 and 20. That the complex permeability was dominated by hysteretic effects, rather than eddy-current effects, was shown by the lack of any frequency dependence of μ' and μ''. The experiments also revealed dependences of μ' and μ'' upon H_0 and h_0 very similar to those predicted in Figs. 13 and 14, with structure at $H_0=H_{c1}$ and H_{c2}.

According to Eqs. (25), (119), and (120), the energy loss per cycle per unit volume is

$$W_V = (\pi b_0^2/4\pi)\mu_0' g_2(x) \quad \text{(Gaussian)} \tag{128a}$$

$$= (\pi b_0^2/\mu_0)\mu_0' g_2(x)\,, \quad \text{(mks)} \tag{128b}$$

where $x=h_0/H_b^*=L_p/a$. For fixed b_0 and ω, W_V is maximized when $\mu''=\mu_0' g_2$ is maximized. Both μ_0' and x depend upon H_0, but, because the H_0 dependence of μ_0' is slow except near $H_0=H_{c1}$, the maximum loss occurs when $x\approx 1$, $h_0\approx H_b^*$, or $L_p\approx a$.

When $x=h_0/H_b^*=L_p/a<<1$, it is appropriate to use the small-argument approximation,

$$g_2(x) \approx (4/3\pi)x, \qquad x << 1\,. \tag{129}$$

Then $W_A=W_V(a/2)$ is proportional to h_0^3:

$$W_A = ch_0^3\mu_0'/24\pi^2 J_{c0} \quad \text{(Gaussian)} \tag{130a}$$

$$= 2h_0^3\mu_0\mu_0'/3J_{c0}\,. \quad \text{(mks)} \tag{130b}$$

Note that in this case the losses are confined to a narrow region of thickness $L_p<<a$ close to the surface.

For the case of bulk ceramic high-temperature superconductors, there is an important complication arising from the granularity of these materials. As stressed in Ref. 21, rather than there being only one parameter, J_c, describing hysteretic losses, there are two. These are J_{cg}, the intragranular critical current density (g = grain), and J_{cJ}, the intergranular critical current density (J = Josephson). Since J_{cJ} often is orders of magnitude smaller than J_{cg}, this leads to the well-known weak-link problem, which severely limits the transport critical current densities of bulk high-temperature superconductors. This problem can be avoided only by dealing with single crystals or with materials in which the microstructure has been strongly textured.

An elementary analysis of the hysteretic loss per cycle has been carried out in Ref. 22, which treats cylindrical samples of radius R containing grains modeled as cylinders of radius R_g. Analytic results are worked out under the assumption that the critical current densities J_{cg} and J_{cJ} can be taken to be slowly varying functions of B over the range of fields seen by the specimen during each cycle. Nevertheless, the expressions are very complicated, primarily because there are now two dimensionless parameters, $x_g = h_0/H_{bg}^* = L_{pg}/R_g$ for the intragranular behavior (characterized by J_{cg})

and $x_J = h_0/H^*_{bJ} = L_{pJ}/R$ for the intergranular behavior (characterized by J_{cJ}). As the temperature is varied and the dc bias field H_0 is held fixed, granular materials sometimes exhibit two peaks[21,23] in μ", one near T_c given by the condition that $x_g = 1$, and one at lower temperatures given by the condition that $x_J = 1$, as can be understood from Fig. 12.

A complete analysis of the ac susceptibility accounting for hysteretic effects has been developed by K.-H. Müller and coworkers (Refs. 24-26). A description of much of this work is given by Müller elsewhere in this volume.

V. SURFACE LOSSES IN TYPE-II SUPERCONDUCTORS

In Secs. III and IV we considered situations in which the losses in a type-II superconductor were dominated by either eddy-current losses or bulk-pinning hysteretic losses. Also introduced in Sec. IV were the quantitites $H_{en}(B)$ and $H_{ex}(B)$, the values of a parallel field at which, in the presence of a surface barrier, for a given value of B just inside the surface, the surface is unstable to entry and exit of vortices, respectively. For most of the results of Sec. IV, however, it was assumed that there was no barrier to vortex entry and exit and that $H_{en}=H_{ex}=H_{eq}$. In the present section, we consider the situation for which the measuring frequency is so low that eddy-current losses are negligible ($\delta_{f0}>>a$) and the bulk critical current density is so small that bulk-pinning hysteretic losses are negligible ($L_p>>a$), the only appreciable losses being due to energy dissipation at the surface as vortices cross the surface barrier.[18]

Consider a type-II superconducting cylinder of radius a in a parallel applied field,

$$H_a(t) = H_0 + h_0\cos\omega t \tag{131}$$

$$B_a(t) = B_0 + b_0\cos\omega t\,. \tag{132}$$

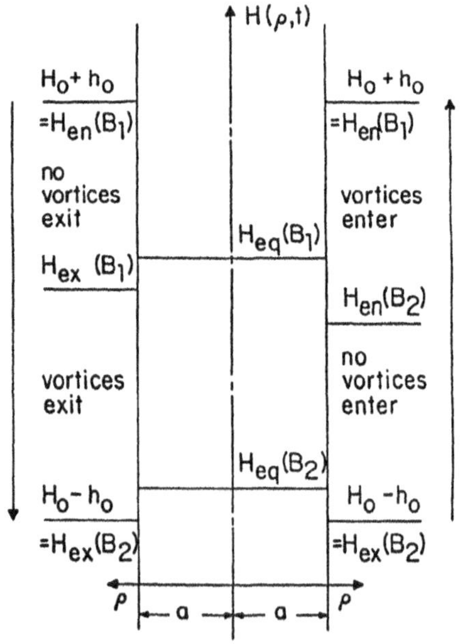

Fig. 15. Magnetic field profiles with a surface barrier but no bulk pinning, H(ρ,t) versus ρ, for the applied field $H_a(t)$ [Eq. (131)] decreasing (left side) and increasing (right side).

If $H_{ex}(B) \le H_a \le H_{en}(B)$ or, equivalently, $2h_0 \le H_{en}(B) - H_{ex}(B)$ for a given flux density B inside the cylinder, no vortices pass through the surface and no losses occur. For excursions of H_a not satisfying these conditions, vortices pass in and out through the surface during the cycle, and the magnetic field profiles are as sketched in Fig. 15. Suppose the applied field has just reached its maximum value H_0+h_0, which has produced a flux density B_1 and a magnetic field $H_{eq}(B_1)$ inside the cylinder. Because vortices have been entering the specimen, the applied field must obey $H_0+h_0=H_{en}(B_1)$. As the applied field decreases, no vortices come out of the specimen and B remains equal to B_1 so long as H_a remains in the metastable range $H_{ex}(B_1) \le H_a \le H_{en}(B_1)$. As H_a decreases below the value $H_{ex}(B_1)$, vortices exit the specimen, maintaining an internal flux density B obeying $H_{ex}(B)=H_a$. When H_a reaches its minimum value H_0-h_0, (assume for simplicity, that $h_0<H_0$), B reaches its minimum value B_2, determined by $H_{ex}(B_2)=H_0-h_0$. The corresponding magnetic field is $H_{eq}(B_2)$. As the applied field increases, no vortices enter the specimen and B remains equal to B_2 so long as H_a remains in the metastable range, $H_{ex}(B_2) \le H_a \le H_{en}(B_2)$. As H_a increases above the value $H_{en}(B_2)$, vortices enter the specimen, maintaining an internal flux density B obeying $H_{en}(B)=H_a$. When H_a reaches its maximum value H_0+h_0, B reaches its maximum value, determined by $H_{en}(B_1)=H_0+h_0$. The corresponding magnetic field is $H_{eq}(B_1)$.

Let us consider B>0 and define

$$\Delta H_{en}(B) = H_{en}(B) - H_{eq}(B) \tag{133}$$

$$\Delta H_{ex}(B) = H_{eq}(B) - H_{ex}(B) \tag{134}$$

$$H_s^*(B) = \frac{1}{2}\,(H_{en}-H_{ex}) = \frac{1}{2}\,(\Delta H_{en}+\Delta H_{ex})\,, \tag{135}$$

which set the scale for magnetic fields when a surface barrier is present. The calculation of the complex permeability taking into account the B dependence of $H_{en}(B)$ and $H_{ex}(B)$ is quite complicated and requires numerical solutions.[18] To obtain the main qualitative features, however, it is sufficient to linearize.

Assume in Eqs. (131) and (132) that $h_0 \ll H_0$ and $b_0 \ll B_0$, such that the internal response of the superconductor can be written as

$$H = H_0 + h_1 \tag{136}$$

$$B = B_{eq}(H_0) + b_1\,, \tag{137}$$

where $|h_1| \ll H_0$ and $|b_1| \ll B_{eq}(H_0)$. We also assume, for simplicity, that to good approximation

$$dH_{en}/dB = dH_{ex}/dB = dH_{eq}/dB \tag{138}$$

for B near $B_{eq}(H_0)$. The characteristic field scale is set by

$$H_{s0}^* = [H_{s0}^*(B)]_{B=B_{eq}(H_0)}\,. \tag{139}$$

The resulting hysteresis loop of $\bar{b}_1=b_1$ versus $b_a=b_0\cos\omega t$ is shown in Fig. 16. For $h_0 \le H_{s0}^*$, the specimen remains in the metastable state and no flux enters or leaves the specimen.

Expressed as a function of time during the period $0 \le \omega t \le 2\pi$,

$$b_1 = \mu_0' h_1 \quad \text{(Gaussian)} \tag{140a}$$

$$\quad = \mu_0' \mu_0 h_1 \quad \text{(mks)} \tag{140b}$$

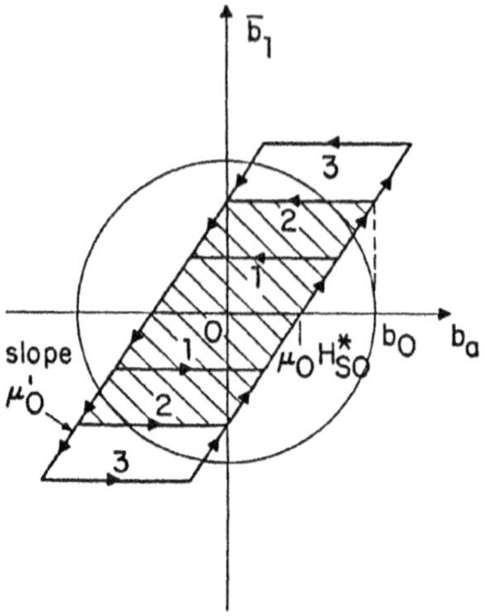

Fig. 16. Hysteresis loop expected with linearization for a surface barrier but no bulk pinning: (1) $h_0=(3/2)H_{s0}^*$, (2) $h_0 = 2H_{s0}^*$, and (3) $h_0=(5/2)H_{s0}^*$. Maximum μ'' occurs when $h_0=2H_{s0}^*$; μ'' is then the ratio of the crosshatched area to the area of the circle of radius $b_0=\mu_0 h_0$.

is obtained from

$$h_1 = h_0 - H_{s0}^* , \qquad 0 \le \omega t \le \theta , \tag{141a}$$

$$= h_a + H_{s0}^* , \qquad \theta \le \omega t \le \pi , \tag{141b}$$

$$h_1 = -h_0 + H_{s0}^* , \qquad \pi \le \omega t \le \pi + \theta , \tag{141c}$$

$$= h_a - H_{s0}^* , \qquad \pi + \theta \le \omega t \le 2\pi , \tag{141d}$$

where

$$h_a = h_0 \cos\omega t. \tag{142}$$

To assist in the calculation of the Fourier integrals, we have introduced an auxiliary phase angle θ, not to be confused with that used in Eq. (122a), via

$$\cos\theta = 1 - 2u \tag{143}$$

$$u = H_{s0}^*/h_0 \le 1 . \tag{144}$$

Substitution into Eqs. (13a) and (13b) yields

$$\mu'/\mu_0' = F_1(u) = 1 - (\theta - \sin\theta\cos\theta)/\pi , \tag{145}$$

$$\mu''/\mu_0' = F_2(u) = (\sin\theta)^2/\pi = (4u/\pi)(1-u) , \tag{146}$$

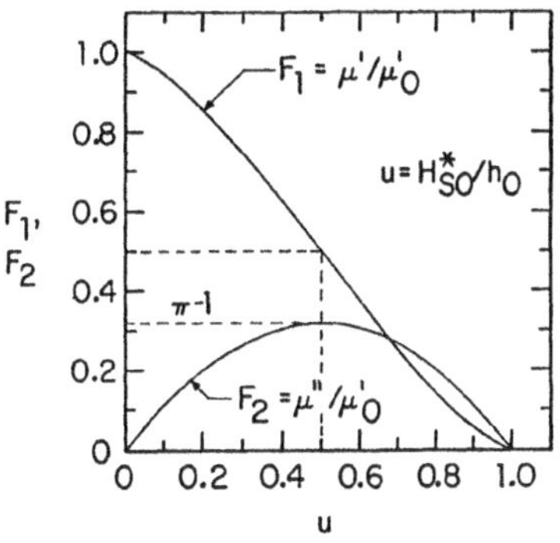

Fig. 17. Plots of the functions $F_1=\mu'/\mu'_0$ and $F_2=\mu''/\mu'_0$ [See Eqs. (143)–(146)] versus $u=H^*_{s0}/h_0$ for surface-barrier hysteretic losses.

which are plotted in Fig. 17. Note that F_2 is maximized when $\theta=\pi/2$, $u=1/2$, and $h_0=2H^*_{s0}$. The maximum value of F_2 is

$$F_2^{max} = F_2(0.5) = 1/\pi = 0.32\,, \tag{147}$$

and the corresponding value of F_1 is $F_1(0.5)=0.50$. In the limit $H^*_{s0}\to 0$, $u\to 0$, $F_2\to 0$, and $F_1\to 1$.

From Fig. 16 it is easily seen that the area of the hysteresis loop is

$$A_H = 4H^*_{s0}(h_0-H^*_{s0})\mu'_0 \quad \text{(Gaussian)} \tag{148a}$$

$$= 4H^*_{s0}(h_0-H^*_{s0})\mu'_0\mu_0^2\,, \quad \text{(mks)} \tag{148b}$$

such that

$$\mu'' = A_H/\pi b_0^2 = (4/\pi)(H^*_{s0}/h_0)(1-H^*_{s0}/h_0)\mu'_0\,, \tag{149}$$

as found in Eq. (146). The energy loss per cycle per unit volume W_V can be obtained from Eqs. (25) and (146) or (149).

VI. INTERPLAY OF THE VARIOUS LOSS MECHANISMS IN TYPE-II SUPERCONDUCTORS

In Secs. III-V we considered idealized situations in which the low-frequency losses were dominated by just one of the three possible loss mechanisms (eddy-current, bulk-pinning, or surface-pinning losses). Summarized in Table I are the key results of the three analyses.

From a theoretical point of view, it is unfortunate that the three loss contributions are *not simply additive*. There is a strong interaction among the three mechanisms which makes it extremely difficult to calculate the total losses in the not uncommon case when all three mechanisms are of comparable importance. As an example of how the mechanisms interact with each other, note that screening of the interior of the superconductor by eddy currents can reduce the amount of magnetic flux moving in and out of the specimen each cycle and hence reduce the energy dissipated via bulk pinning. Similarly, surface screening currents accompanying a surface barrier can produce an analogous effect.

TABLE I. Summary of the key results for the three loss mechanism in type-II superconductors. The symbols are defined in Secs. III-V.

Loss mechanism	Eddy currents (viscous drag, flux flow)	Bulk pinning	Surface pinning (surface barrier)
Characteristic measure	δ_f or τ_f	H_b^* or L_p	H_s^*
Maximum in μ''/μ_0' versus H_0 seen when	$\delta_f = 0.56a$ $\omega\tau_f = 3.2$	$h_0 = H_b^*$ $L_p = a$	$h_0 = 2H^*$
Are μ' and μ'' dependent upon			
H_0?	Yes	Yes	Yes
ω?	Yes	No	No
h_0?	No	Yes	Yes

At sufficiently low frequencies that eddy-current effects are negligible and only bulk-pinning and surface-pinning hysteretic losses remain, the interplay of the latter two mechanisms can be treated on a common footing. As has been shown in Ref. 18, it is possible to account completely for all the effects of each mechanism upon the other and to identify how much of the total hysteretic losses is dissipated at interior bulk-pinning sites and how much at the surface. On the other hand, it is a much more difficult problem to theoretically describe the interaction of eddy-current losses and hysteretic losses, and a complete theory of how to separate these two contributions remains to be developed.

VII. NORMAL-SUPERCONDUCTING COMPOSITES

For reasons of thermal, electrical, and mechanical stability, the high-current wires or cables used in large-scale applications of superconductivity generally involve normal-superconducting composites. A wire typically contains thousands of long, superconducting filaments, with diameters of the order of 1 μm, imbedded in a matrix of normal metal. When the wire is subjected to time-varying currents and magnetic fields, a large fraction of the ac losses is due to the Joule losses from currents induced in the normal matrix. Accordingly, much attention has been given to the development of composite wire designs that minimize these matrix losses. Twisting the wire, for example, is very effective in reducing the matrix losses of a wire in a time-varying transverse magnetic field. An extensive literature exists regarding the ac matrix losses in normal-superconducting composites, Refs. 27-49 representing only a fraction of the papers published on this topic. In this paper, I do not attempt to critically review the field but merely present a few of the principal findings, reexpressed in terms of the theoretical framework established in Sec. II.

A. Composite in a Time-Varying Parallel Applied Magnetic Field

Consider first a long, normal-superconducting composite wire of radius a placed in a sinusoidally time-varying parallel magnetic field. The matrix loss mechanism of interest is the Joule loss from induced currents flowing through the normal matrix, perpendicular to the superconducting filaments. Carr[27,28] has shown that it is convenient to analyze these losses by performing local averages of the fields and currents over linear distances characterized by the spacing between filaments. This results in the equation,

$$\vec{J} = \sigma_\perp \vec{B} = \rho_\perp^{-1} \vec{E}, \tag{150}$$

for the corresponding induced macroscopic current density. Here, $\rho_\perp$ is the *effective continuum transverse resistivity*,[27,28] which is expressible in terms of the normal matrix resistivity ρ_m and superconducting volume fraction α as follows. In the case of *no contact resistance* between the normal and superconducting components, the currents flow as much as possible through the superconductor and

$$\rho_\perp = \rho_m \frac{(1-\alpha)}{(1+\alpha)} < \rho_m. \tag{151}$$

On the other hand, in the case of *high contact resistance* between the components, the currents flow in such a way as to avoid entering the superconductor and

$$\rho_\perp = \rho_m \frac{(1+\alpha)}{(1-\alpha)} > \rho_m. \tag{152}$$

Carr also derived an equation similar to Eq. (150) relating the macroscopic magnetic induction $\vec{B}$ and magnetic field $\vec{H}$, which are parallel to the filaments in this example,

$$\vec{B} = \mu_\parallel \vec{H} \quad \text{(Gaussian)} \tag{153a}$$

$$= \mu_\parallel \mu_0 \vec{H}. \quad \text{(mks)} \tag{153b}$$

Here, $\mu_\parallel$ is the effective permeability, where ($\mu_m \approx 1$ in the normal matrix) $\mu_\parallel \approx 1$, if the filaments are fully penetrated by magnetic flux, or $\mu_\parallel \approx 1-\alpha$, if the filaments screen out the applied field.

The basic equations needed for the calculation of the ac matrix losses are then essentially Eqs. (26)–(31) given in Sec. II for a normal metal, except that Eq. (26) is replaced by Eq. (150) and Eq. (29) by Eq. (153). Thus, the results of Sec. II can be applied here but with the replacements

$$\rho_n \to \rho_\perp \tag{154}$$

$$\mu \to \mu_\parallel \tag{155}$$

$$D_n \to D_\parallel \tag{156}$$

$$\delta_n \to \delta_\parallel \tag{157}$$

$$\tau_n \to \tau_\parallel \tag{158}$$

where

$$D_\parallel = \rho_\parallel c^2/4\pi\mu_\parallel \quad \text{(Gaussian)} \tag{159a}$$

$$= \rho_\parallel/\mu_\parallel\mu_0, \quad \text{(mks)} \tag{159b}$$

$\delta_\parallel = (2D_\parallel/\omega)^{1/2}$, and $\tau_\parallel = a^2/2D_\parallel$. The resulting ac loss per cycle per unit volume is then [Eqs. (25) and (65)]

$$W_V = (\pi b_0^2/4\pi)\mu_{\parallel} f_2(x) \quad \text{(Gaussian)} \tag{160a}$$

$$= (\pi b_0^2/\mu_0)\mu_{\parallel} f_2(x)\ , \quad \text{(mks)} \tag{160b}$$

where f_2 is defined in Eq. (59) and plotted in Fig. 1, and

$$x = a\sqrt{2}/\delta_{\parallel} = (2\omega\tau_{\parallel})^{1/2}\ . \tag{161}$$

Note that, for fixed b_0, W_V has its maximum when x=2.5.

B. Composite in a Time-Varying Perpendicular Applied Magnetic Field

Consider next a normal-superconducting composite wire of radius a and length 2L, placed in a sinusoidally time-varying perpendicular magnetic field, as sketched in Fig. 18a. Assuming that the filaments remain fully superconducting, we see that the electric field component parallel to the filaments must vanish and that the electric field vector must always be perpendicular to the filaments. Accordingly, currents in the normal matrix are perpendicular to the filaments [see Fig. 18b]. Currents through the superconductor flow along the filaments as supercurrents (without generating a parallel electric field) provided the current does not exceed the critical depinning current for transverse vortex motion across the filament. Magnetic flux penetrates to the middle of the specimen only by diffusing in from the ends of the specimen through the normal matrix [Fig. 18b]. According to the anisotropic continuum model of Carr,[27,28] the macroscopic current density is

$$\vec{J} = \vec{J}_{\perp} + \vec{J}_{\parallel}\ . \tag{162}$$

Here, $\vec{J}_{\parallel}$ is the average current density, flowing as supercurrents ($\vec{E}_{\parallel}$=0) provided $J_{\parallel}<\alpha J_c$, where J_c is the critical current density for the superconducting filaments and α is the superconducting volume fraction; $\vec{J}_{\perp}$ is the transverse current density obeying

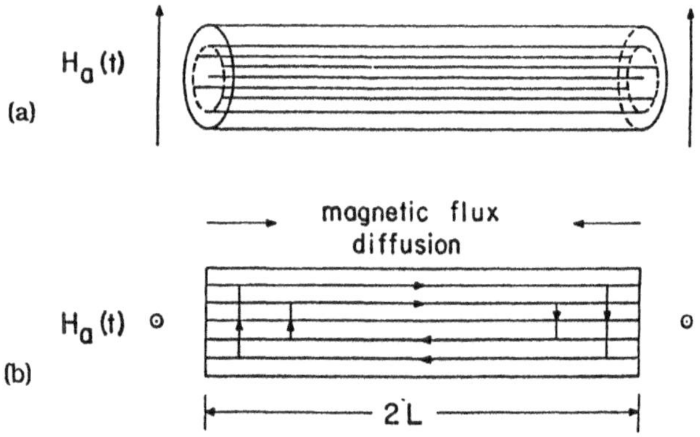

Fig. 18. Untwisted normal-superconducting composite wire of length 2L in a time-varying perpendicular applied magnetic field, increasing in the vertical direction (arrows): (a) Side view, showing internal superconducting filaments. (b) Top view, with arrows inside the wire showing directions of supercurrents flowing along the filaments and normal (coupling) currents flowing perpendicular to the filaments through the normal matrix. Arrows outside the wire show the direction of magnetic flux diffusion through the normal matrix from the ends toward the middle of the wire.

$$\vec{J}_\perp = \sigma_\perp \vec{E}_\perp = \rho_\perp^{-1} \vec{E}_\perp , \tag{163}$$

where $\rho_\perp$ is defined in Eqs. (151) and (152).

According to Carr,[27,28] the relation between $\vec{B}$ and $\vec{H}$ in this geometry is

$$\vec{B} = \mu_\perp \vec{H} \quad \text{(Gaussian)} \tag{164a}$$

$$= \mu_\perp \mu_0 \vec{H} , \quad \text{(mks)} \tag{164b}$$

where $\mu_\perp \approx 1$, if the filaments are fully penetrated by magnetic flux, or

$$\mu_\perp \approx (1-\alpha)/(1+\alpha) < 1, \tag{165}$$

if the filaments screen out the applied field.

The basic equations needed for the calculation of the ac matrix losses are then essentially Eqs. (26)–(31) given in Sec. II for a normal metal, except that Eq. (26) is replaced by Eq. (163) and Eq. (29) by Eq. (164). Thus, some of the results of Sec. II can be applied here, but with the replacements

$$\rho_n \to \rho_\perp \tag{166}$$

$$\mu \to \mu_\perp \tag{167}$$

$$D_n \to D_\perp \tag{168}$$

$$\delta_n \to \delta_\perp , \tag{169}$$

where

$$D_\perp = \rho_\perp c^2/4\pi\mu_\perp \quad \text{(Gaussian)} \tag{170a}$$

$$= \rho_\perp/\mu_\perp\mu_0 \quad \text{(mks)} \tag{170b}$$

and $\delta_\perp = (2D_\perp/\omega)^{1/2}$. Because of the difference in geometry, however, the characteristic magnetic flux diffusion time is

$$\tau_\perp = L^2/2D , \tag{171}$$

which involves the length 2L, rather than the diameter 2a.

On the basis of our previous examination of eddy-current losses, we expect the ac loss per cycle per unit volume to be given by an expression of the form

$$W_V = (\pi b_0^2/4\pi)\mu_\perp f_{2\perp}(x) \quad \text{(Gaussian)} \tag{172a}$$

$$= (\pi b_0^2/\mu_0)\mu_\perp f_{2\perp}(x) , \quad \text{(mks)} \tag{172b}$$

where

$$x = L\sqrt{2}/\delta_\perp = (2\omega\tau_\perp)^{1/2} . \tag{173}$$

and $f_{2\perp}$ is a dimensionless function similar to f_2 [Eq. (59) and Fig. 1]. That is, we expect W_V to reach its maximum at a frequency such that $x \sim 1$. Such a behavior indeed has been observed experimentally by Kwasnitza.[50]

C. Twisted Composite in a Time-Varying Perpendicular Applied Magnetic Field

Consider finally a long normal-superconducting composite wire of radius a and twist pitch l_t, placed in a sinusoidally time-varying perpendicular magnetic field, as sketched in Fig. 19. A complex pattern of currents and fields is generated in the twisted geometry under time-varying conditions.[48] Because the helical superconducting filaments are then electrodynamically coupled and induced currents flow from one filament to another through the normal matrix, the corresponding losses are usually referred to as *coupling losses*, rather than eddy-current losses. The most important aspects of the behavior of a twisted composite are revealed by consideration of just a single pair of helical filaments, shown in Fig. 19. Comparison of Fig. 19 with Fig. 18 reveals that the electrodynamic behavior of a twisted segment of length $l_t/2$ is very similar to that of an untwisted composite of length $2L=l_t/2$. Indeed, according to Carr,[27,28] the basic equations are then essentially Eqs. (26)–(31) except with the replacements of Eqs. (166)–(170). The appropriate characteristic diffusion time, however, is the time

$$\tau_t = (l_t/4)^2/2D_\perp \, , \tag{174}$$

which is simply Eq. (171) with 2L replaced by $l_t/2$.

Thus, on the basis of our previous examination of eddy-current losses, we expect the ac loss per cycle per unit volume to be given by an expression of the form

$$W_V = (\pi b_0^2/4\pi)\mu_\perp f_{2t}(x) \quad \text{(Gaussian)} \tag{175a}$$

$$= (\pi b_0^2/\mu_0)\mu_\perp f_{2t}(x) \, , \quad \text{(mks)} \tag{175b}$$

where

$$x = (l_t/4)\sqrt{2}/\delta_\perp = (2\omega\tau_t)^{1/2} \tag{176}$$

and $f_{2t}(x)$ is a dimensionless function similar to f_2 [Eq. (59) and Fig. 1]. That is, we expect W_V to reach its maximum at a frequency such that $x \sim 1$. Such a behavior has been found experimentally by Kwasnitza,[37] who observed exactly this scaling law behavior with a maximum W_V occurring when (assuming $\mu_\perp=1$) $x=1.6$.

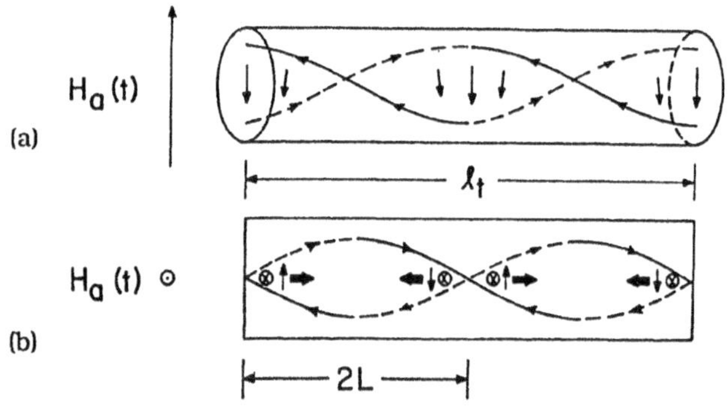

Fig. 19. Twisted normal-superconducting composite wire of twist pitch l_t in a time-varying perpendicular applied magnetic field, increasing in the vertical direction (arrows). Only a segment of length l_t and only two of the many helical superconducting filaments are shown. Light arrows inside the wire show the directions of induced supercurrents flowing along the filaments and normal (coupling) currents flowing perpendicular to the filaments through the normal matrix. (a) Side view. (b) Top view. Heavy arrows show the direction of magnetic flux diffusion through the normal matrix from the crossover regions, where flux is generated, toward the middle of a loop of length $l_t/2$.

VIII. SUMMARY

In this paper, I have discussed the physics of low-frequency ac losses in type-II superconductors and in multifilamentary superconducting-normal composite conductors. In type-II superconductors, there are two basic loss mechanisms: *Eddy-current (viscous, flux-flow) losses* (Sec. III), for which the loss per unit volume per cycle W_V is frequency dependent, are important in ideal type-II superconductors and in nonideal type-II superconductors at high frequencies; *hysteretic losses* (Secs. IV and V), for which W_V is independent of frequency, are important in nonideal type-II superconductors at low frequency. The hysteretic losses may be further subdivided into *bulk-pinning* (Sec. IV) and *surface-pinning losses* (Sec. V), the bulk-pinning losses including bulk annihilation losses[18] from the annihilation of vortices of the opposite sense and the surface-pinning losses including the effects of a surface barrier.[17,18] A complete theory for the interaction of eddy-current and hysteretic losses (Sec. VI) remains to be developed. An additional complication occurs in bulk, ceramic high-temperature superconductors. Because they are granular, they have at least two distinct current densities:[21] an intragranular J_{cg} and an intergranular J_{cJ}. In normal-superconducting multifilamentary composites, another loss mechanism becomes important: *coupling losses* (Sec. VII), which can be thought of as being very similar to normal eddy-current losses (Sec. II) but in a highly anisotropic medium.

A few simple concepts help to unify the broad field of ac losses in superconductors (Sec. I): The *loss per cycle per unit volume* W_V is proportional [Eq. (7)] to the *area of the hysteresis loop* A_H [Eq. (23)]. For a given amplitude of the ac magnetic field, both W_V and A_H are proportional [Eq. (25)] to μ'', the imaginary part of the dimensionless complex permeability $\tilde{\mu}=\mu'+i\mu''$ [Eq. (17)]. In all the cases considered where a single loss mechanism is dominant, μ'' can be expressed as

$$\mu'' = \mu_{eff} f_2(x) \, , \tag{177}$$

where μ_{eff} is an appropriate dimensionless effective permeability, $f_2(x)$ is a dimensionless function of x which has its maximum when $x \sim 1$, and x is an appropriate dimensionless ratio. Table II gives a summary of the appropriate dimensionless ratios x, aside from numerical constants of order unity, for the situations considered in this paper.

TABLE II. Summary of the appropriate dimensionless ratios x, aside from numerical constants of order unity, for the situations considered in this paper.

Situation	x	Eq.
Eddy currents in normal metal	a/δ_n	(59)
Eddy currents (flux flow, viscous drag) in type-II superconductors	a/δ_f	(84)
Bulk pinning	$h_0/H_b^*=L_p/a$	(120)
Surface pinning	H_{s0}^*/h_0	(144)
Composite in parallel field	$a/\delta_{\parallel}$	(161)
Composite in perpendicular field	$L/\delta_{\perp}$	(173)
Twisted composite in perpendicular field	$l_t/\delta_{\perp}$	(176)

ACKNOWLEDGMENT

Ames Laboratory is operated for the U.S. Department of Energy by Iowa State University under Contract No. W-7405-Eng-82. This work was supported by the Director for Energy Research, Office of Basic Energy Sciences.

REFERENCES

1. J. D. Jackson, "Classical Electrodynamics," Wiley, New York, (1962), p. 197.
2. R. W. Rollins and J. Silcox, Phys. Rev. 155:404 (1967).
3. J. D. Jackson, "Classical Electrodynamics," Wiley, New York (1962), p. 238.
4. R. G. Chambers and J. G. Park, Brit. J. Appl. Phys. 12:507 (1961).
5. J. R. Clem, H. R. Kerchner, and S. T. Sekula, Phys. Rev. B 14:1893 (1976).
6. Handbook of Mathematical Functions, Ed. by M. Abramowitz and I. A. Stegun, U.S. Natl. Bur. Stand. Appl. Math. Series No. 55 (U.S. GPO, Washington, D.C., (1967), p. 382.
7. See, for example, Y. B. Kim and M. J. Stephen, in Superconductivity, Ed. by R. D. Parks, Marcel Dekker, New York (1969), Vol. 2, Chap. 19, pp. 1107-1165.
8. See, for example, A. L. Fetter and P. C. Hohenberg, in Superconductivity, Ed. by R. D. Parks, Marcel Dekker, New York (1969), Vol. 2, Chap. 14, pp. 817-923.
9. D. Dew-Hughes, Cryogenics 28:674 (1988).
10. P. H. Kes, J. Aarts, J. van den Berg, C. J. van der Beek, and J. A. Mydosh, Supercond. Sci. Technol. 1:242 (1989).
11. D. G. Walmsley, J. Phys. F 2:510 (1972).
12. A. M. Campbell and J. E. Evetts, Adv. Phys. 21:199 (1972).
13. J. R. Clem, J. Low Temp. Phys. 38:353 (1980).
14. T. Ezaki, K. Yamafuji, and F. Irie, J. Phys. Soc. Japan 40:1271 (1976).
15. J. R. Clem, Phys. Rev. B 26:2463 (1982).
16. A. Perez-Gonzalez and J. R. Clem, Phys. Rev. B 32:2909 (1985).
17. J. F. Clem, in Low Temperature Physics-LT13, Ed. by K. D. Timmerhaus, W. J. O'Sullivan, and E. F. Hammel, Plenum, New York (1974), Vol. 3, p. 102.
18. J. R. Clem, J. Appl. Phys. 50:3518 (1979).
19. C. A. M. van der Klein, J. D. Elen, R. Wolf, and D. de Klerk, Physica 49:98 (1970).
20. C. A. M. van der Klein, Ph.D. Thesis, Leiden, 1974.
21. J. R. Clem, Physica C 153-155:50 (1988).
22. J. R. Clem, "AC Losses in the New High-Temperature Superconductors," Report No. EPRI EL-6277, March 1989, Electric Power Research Institute, 3412 Hillview Avenue, Palo Alto, CA 94304.
23. B. Renker, I. Apfelstedt, H. Küpfer, C. Politis, H. Rietschel, W. Schauer, H. Wühl, U. Gottwick, H. Kneissel, U. Rauchschwalbe, H. Spille, and F. Steglich, Z. Phys. B 67:1 (1987).
24. K.-H. Müller, B. W. Ricketts, J. C. Macfarlane, and R. Driver, Physica C 162:1177 (1989).
25. K.-H. Müller, Physica C 168:585 (1989).
26. K.-H. Müller, IEEE Trans. Magn. 27:2174 (1991).
27. W. J. Carr, Jr., J. Appl. Phys. 45:929 (1974).
28. W. J. Carr, Jr., J. Appl. Phys. 45:935 (1974).
29. W. J. Carr, Jr., J. Appl. Phys. 46:4043 (1975).
30. W. J. Carr, Jr., in Proc. Sixth Symposium on Engineering Problems and Fusion Research, (1975), p. 152.
31. W. J. Carr, Jr., Phys. Rev. B 11:1547 (1975).
32. W. J. Carr, Jr., IEEE Transactions on Magnetics MAG-13, 192 (1977).
33. W. J. Carr, Jr., M. S. Walker, and J. H. Murphy, J. Appl. Phys. 46:4048 (1975).
34. R. Hancox, Proc. IEE 113:1221 (1966).
35. K. P. Jüngst and G. Ries, IEEE Transactions on Magnetics MAG-13, 527 (1977).
36. K. Kwasnitza, Cryogenics 17:613 (1977).
37. K. Kwasnitza, Cryogenics 17:616 (1977).
38. K. Kwasnitza and I. Horvath, Cryogenics 14:71 (1974).
39. J. R. Miller and S. S. Shen, IEEE Transactions on Magnetics MAG-13, 534 (1977).
40. G. H. Morgan, J. Appl. Phys. 41:3673 (1970).

41. J. H. Murphy, IEEE Transactions on Magnetics MAG-13, 525 (1977).
42. G. Ries, IEEE Transactions on Magnetics MAG-13, 524 (1977).
43. J. P. Soubeyrand and B. Turck, IEEE Transactions on Magnetics MAG-15, 248 (1979).
44. B. Turck, Rev. Phys. Appl. 11:369 (1976).
45. B. Turck, IEEE Transactions on Magnetics MAG-13, 548 (1977).
46. M. S. Walker, J. H. Murphy, and W. J. Carr, Jr., IEEE Transactions on Magnetics MAG-11, 309 (1975).
47. M. N. Wilson, in 1972 Applied Superconductivity Conference (IEEE, New York, 1972), p. 385.
48. M. N. Wilson, C. R. Walters, J. D. Lewin, and P. F. Smith, J. Phys. D 3:1517 (1970).
49. J. J. Wollan, IEEE Transactions on Magnetics MAG-13, 544 (1977).
50. K. Kwasnitza, private communication.

EARLY THEORIES OF χ' AND χ'' OF SUPERCONDUCTORS : THE CONTROVERSIAL ASPECTS

A.F. KHODER and M. COUACH

Centre d'Etudes Nucléaires de Grenoble
DRFMC - SPSMS - Laboratoire de Cryophysique
85 X - 38041 GRENOBLE CEDEX FRANCE

ABSTRACT

The transition from the normal to the superconducting state is characterized by the disappearance of the electrical resistance of a superconductor below its transition temperature, $T_c(H)$, which, as indicated, is a function of magnetic field and the appearance of the Meissner effect i.e. the expulsion of any magnetic field (below some critical value) out of the bulk of the superconductor. This Meissner effect is considered as the hallmark of bulk superconductivity while a drop of the resistance can, at best, serve only as an indication of a superconducting phase transition.

The shielding of ac magnetic field is frequently misinterpreted as a Meissner effect which results in an overestimation of the superconducting volume fraction in investigated samples. On the other hand, questions are raised about the physical interpretation of the complex ac susceptibility ($\chi_{ac} = \chi' + i\chi''$) behavior around T_c, mainly the significance of the χ'' peak(s) ; is this the expected behavior of a normal conductor with sudden increase of the conductivity or it is more tightly linked with the normal to superconducting phase transition.

In this paper, we will review the different interpretations and discuss the controversial aspects of the χ_{ac} behavior around T_c.

A. INTRODUCTION

The ac susceptibility technique is now becoming one of the most common "tools" used in the search for and the study of high T_c superconductors. It offers a rapid identification of (possible) T_c's in multiphase material and allows the investigation of other macroscopic properties of superconductors such as :

H(T) irreversibility lines [1], critical current densities [2, 3] intergranular as well intragranular contributions [4, 5, 6], etc...

Magnetic Susceptibility of Superconductors and Other Spin Systems
Edited by R.A. Hein *et al.*, Plenum Press, New York, 1991

The aim of this paper is to review the controversial aspects related to the zero field $\chi_{ac}(T)$ interpretations and the physical conclusions which can be drawn from e.g. the peaks in $\chi''(T)$ vs T behavioir at $T \leqslant T_c$ [7]. From here on the fact that χ is a function of T will not be explicitly shown.
It is largerly accepted that χ'_{ac} is equivalent to the zero field cooling (ZFC) dc magnetization measurement of superconductors giving the shielding (exclusion) aspect of the problem in constrast to the expulsion, or Meissner effect, obtained by field cooling the samples (FC).

The shielding may originate from the (perfect) conductivity of the sample as well from the Meissner effect and can not be used, as such, as an indication of bulk superconductivity nor as a quantitative measurement of the volume fraction of this (bulk) superconductivity.

Complete shielding may also originate from a thin superconducting sheath on the sample surface and/or from the development of superconducting multifilamentary structure [8, 9] above T_c of the bulk which results in an effective increase of the conductivity $\langle\sigma\rangle$ of the medium leading to a dissipation peak χ''_{peak} and a high shielding level. Wall and domain boundaries can also be suspected of being particularly effective regions for superconductivity nucleation [10] resulting in the enhancement of $\langle\sigma\rangle$ and the appareance of shielding. Thus, the use of χ_{ac} to detect and characterize bulk superconductivity is matter of passionate debate and the conflicting interpretations appear to suffer the same prejudice [7].

Before going further, we should separate the actual problem into two main aspects :

1. Whether or not χ_{ac} reflects bulk superconductivity ?
   The length scale is fixed by λ, the magnetic field penetration depth.

2. If a positive answer is given to (1), the next step is whether or not χ_{ac} can be used to make a quantitative estimate of the bulk superconducting volume fraction ?

The paper is organised as follows : we first recall the main experimental facts related to χ_{ac} behavior, the models are treated in a next section and expressions of χ_{ac} are given for the different simple sample geometries. The shielding by a (very) thin superconducting layer ($D \ll \lambda$) is shown to be impossible unless unrealistic conditions are fulfilled.

B. THE EXPERIMENTAL FACTS

At low frequency low amplitude ac magnetic field $H_{ac} = H_0\, e^{-i\omega t}$, we can summarize the experimental behavior of the real (χ') and imaginary (χ'') components of χ_{ac} upon decreasing the temperature through T_c as follows :

1. χ' drops to a negative value in a monotonic way which is given by $\chi' = -1/(1-D)$ where D is the demagnetizing factor (complete shielding).

2. Simultaneously with the χ' decrease, χ'' generally shows a bell shape dependence on the temperature. The occurence of the χ'' peak is missed in some cases [7, 8, 9]. This fact is, among others,

responsible of the controversy about the physical meaning of the χ'' peak (bulk or non-bulk superconductivity).

3. The temperature of the χ'' peak, T_p, is frequency and field amplitude dependent : T_p increases with frequency while it decreases with an increase in the amplitude of the ac measuring field, $|H_{ac}|$.

4. The onset temperature T_{on} of the χ' decrease (or χ'' increase) is much less sensitive to frequency and field amplitude than is T_p and apart from experimental errors, T_{on} can be considered as f and $|H_{ac}|$ independent.

5. The χ'' peak level does not seem to scale with the χ'_s level in the superconducting state (χ'' peak/$|\chi'_s| \sim \alpha$) although $\alpha \sim 0.2$ - 0.4 in most cases. Moreover, in many cases, the χ'' peak level shows a dependence on $|H_{ac}|$.

6. Harmonic generation at the transition region.

These experimental facts are well established in the case of homogeneous samples but the stituation becomes more complicated when multiphase superconducting samples are investigated and/or different large scale coherent phase transitions are present i.e. the strong (intragranular) and weak (intergranular) superconductivity of the high T_c ceramics [4, 5, 6]. In such cases, the behavior of $\chi'(T)$ is still monotonic with inflections and "plateau", while the $\chi''(T)$ behavior becomes very complex ; many peaks can be resolved (or missed) and in such situations great care should be paid to shape and geometrical aspects [11] as well as to H_{ac} amplitude effects.
The χ'' peak level deserves close attention. Although this level is a function of the sample's shape, we will subsequently show that it can be used to help separate screening effects from bulk superconducting properties.

C. THE MODELS

The zero field ac susceptibility has been interpreted by several models [8, 12, 13, 14, 15]. We will start with models based on the normal metal behavior where the decreasing of the resistivity ρ to a null value over a temperature range ΔT around T_c results in the (un)usual behavior of χ' and χ''.

1. The Maxwell-Strongin filamentary model [8, 9]

Maxwell and Strongin argued that the χ'' peak is due to the development of superconductivity as a filamentary network (Mendelssohn sponge !) or thin flakes so as to increase the average conductivity $\langle\sigma\rangle$ while at the same time permitting the magnetic field to penetrate. A homogeneous growing of the superconducting phase will not allow such penetration of the magnetic field and thus the dissipation peak can not develop. The χ'' peak is, thus, the hallmark of filamentary superconductivity and it develops, as for normal metals, because of the increasing of the electrical conductivity with decreasing temperatures ; a more detailed discussion of this behavior will be presented in section C.3. In a footnote Maxwell and Strongin claim [8], that the London two fluid model also results in a χ'' peak at T_c when λ is still comparable to the sample size but this too has been shown to be incorrect [15] and will be discussed later.

2. The Eddy current model [12]

Another interpretation of the χ'-χ'' behavior was given by Cody and Miller [12] ; on the basis of the eddy current model. This model is also based on the theory of normal metal shielding of quasi-static fields [16].

As this theory is a linear response one, it is obvious that the $T_p(H_{ac})$ dependence and harmonic generation can not be accounted for within this approach. To explain the harmonic generation, Cody and Miller [12] invoke a supplementary hypothesis : "the resistivity that occurs in the Eddy current equations is clearly nonlinear". This suggestion is very important as it includes an element which is related to the superconducting phase.

3. Back to the χ_{ac} behavior of normal metal : few simple cases

Before going further, it will be helpful to reproduce the results of the electrodynamic theory of normal metal in terms of their polarisability or ac susceptibility [16].

The theory is based on the following equations

$$\text{rot } H = J \tag{1}$$

$$\text{rot } E = - \partial B/\partial t \tag{2}$$

$$\text{and div } B = 0 \tag{3}$$

The equation giving the average local magnetic field H within the conductor can be obtained if the appropriate equations governing the magnetic ($B = \mu_0 H$) and transport ($J = \sigma E$) properties are introduced :

$$\Delta H = \mu_0 \sigma \frac{\partial H}{\partial t} \tag{4}$$

which is a diffusion like equation for H with the diffusion constant being

$$D = (\mu_0 \sigma)^{-1} = \frac{\rho}{\mu_0} \tag{5}$$

where ρ is the resistivity.

If we take for H the time dependance $H(r,t) = H(r)\, e^{-i\omega t}$ we obtain :

$$\Delta H(r) + K^2 H(r) = 0 \tag{6}$$

where $K^2 = \dfrac{i \omega \mu_0}{\rho} = \dfrac{2i}{\delta^2}$ with $\delta = \sqrt{2\rho / \mu_0 \omega}$ being the normal skin depth (in pratical units where ρ is given in Ω.cm, $\delta \simeq 5000 \sqrt{\rho/f}$).

Depending on the geometrical aspects of the problem, which determine the boundary conditions for H, many cases can be considered and we will only give the final (text book) [16] expressions of the χ_{ac} susceptibility for the simplest geometrical forms.

a. Case of spherical sample (radius a)

$$\chi_{ac} = -\frac{3}{2}\left[1 - \frac{3}{K^2 a^2} + \frac{3}{K a} \operatorname{ctg} K a\right] = \chi' + i\chi'' \qquad (7)$$

This expression can be separated into the real and imaginary parts :

$$\chi' = -\frac{3}{2}\left[1 - \frac{3}{2}\frac{\delta}{a}\frac{\sinh(2a/\delta) - \sin(2a/\delta)}{\cosh(2a/\delta) - \cos(2a/\delta)}\right] \qquad (8)$$

and

$$\chi'' = \frac{9\,\delta^2}{4\,a^2}\left[-1 + \frac{a}{\delta}\frac{\sinh(2a/\delta) + \sin(2a/\delta)}{\cosh(2a/\delta) - \cos(2a/\delta)}\right] \qquad (9)$$

for $\delta/a \gg 1$

$$\chi' \sim -\frac{4}{105}\left(\frac{a}{\delta}\right)^4 \quad \text{and} \quad \chi'' \sim \frac{1}{5}\left(\frac{a}{\delta}\right)^2 \qquad (10)$$

for $\delta/a \ll 1$

$$\chi' \sim -\frac{3}{2}\left[1 - \frac{3}{2}\frac{\delta}{a}\right] \quad \text{and} \quad \chi'' \sim \frac{9}{4}\frac{\delta}{a} \qquad (11)$$

The χ'' peak occurs for $a/\delta \approx 2.4$ and its height is ≈ 0.54.

Figure 1 shows the χ' and χ'' behavior as function of a/δ. One should remark that for low shielding ($a/\delta < 1$) the χ_{ac} modulus is represented by the dissipation component χ''.

b. Case of cylindrical sample (radius = a) in longitudinal configuration (H_{ac} // to the cylinder axis) :

$$\chi_{ac} = -1 + \frac{2}{Ka}\frac{J_1(Ka)}{J_0(Ka)} \qquad (12)$$

where $J_n(Z)$ are Bessel functions.

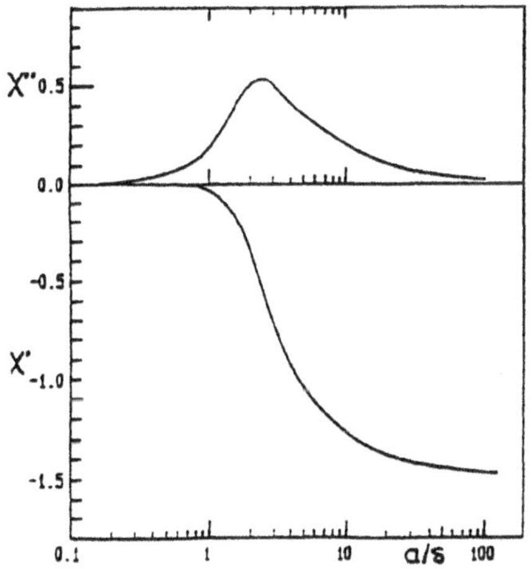

Fig. 1 : χ' and χ'' as function of a/δ for a spherical sample

In this case, analytical separation of the real and imaginary parts of χ_{ac} is not possible. Fortunately the function $F(Z) = 2 J_1(Z)/Z J_0(Z)$ can be written as continued fractions of the form [17] :

$$F(Z) = \frac{1}{1 -} \frac{1/4 \; Z^2/1.2}{1 -} \frac{1/4 \; Z^2/2.3}{1 -} \tag{13}$$

which is reducible to :

$$F(Z) = \frac{1}{Q_1} \quad \text{with} \quad Q_n = 1 - \frac{1/4 \; Z^2/n(n+1)}{Q_{n+1}} \tag{14}$$

Numerical solution converges rapidly for each Z^2 value by an adequate choice of $n \gg 1$ with the trial function $Q_{n+1} = 1$.

For $\delta/a \gg 1$ we get

$$\chi' \approx - \frac{1}{12} (a/\delta)^4 \quad \text{and} \quad \chi'' \approx \frac{1}{4} (a/\delta)^2 \tag{15}$$

and

for $\delta/a \ll 1$ we get

$$\chi' \approx - \left(1 - \frac{\delta}{a}\right) \quad \text{and} \quad \chi'' \approx \delta/a \tag{16}$$

The χ'' peak occurs for $a/\delta \approx 1.8$ and its height is 0.38.

c. Case of cylindrical sample (radius = a) in transverse configuration (H_{ac} ⊥ to the cylinder axis)

In this case, the ac susceptibility is enhanced by a factor of 2 relatively to the above case $\chi_{ac\perp} = 2 \chi_{ac/\!/}$.

d. Case of slab geometry (H_{ac} // to the slab surface) of thickness = 2a

$$\chi_{ac} = -1 + \frac{1}{Ka} \operatorname{tg} Ka \tag{17}$$

Giving

$$\chi' = -1 + \frac{\delta}{2a} \frac{\sinh(2a/\delta) + \sin(2a/\delta)}{\cosh(2a/\delta) + \cos(2a/\delta)} \tag{18}$$

and

$$\chi'' = \frac{\delta}{2a} \frac{\sinh(2a/\delta) - \sin(2a/\delta)}{\cosh(2a/\delta) + \cos(2a/\delta)} \tag{19}$$

For $\delta/a \gg 1$

$$\chi' \approx - \frac{8}{15} a^4/\delta^4 \quad \text{and} \quad \chi'' \approx \frac{2}{3} a^2/\delta^2 \tag{20}$$

and

for $\delta/a \ll 1$

$$\chi' \approx -1 + \frac{\delta}{2a} \quad \text{and} \quad \chi'' \approx \delta/2a \tag{21}$$

The χ'' peak occurs for $a/\delta \approx 1.12$ and its height is 0.42.

In all the cases, the χ'' peak develops as a compromise between the increasing of the local absorbed power density ($\sim 1/2\ \sigma E^2 \sim 1/2\ \rho\ J^2$) and the decreasing of the field penetration depth into the material. In other words, the available effective volume of the material where dissipation is physically possible decreases (the skin depth limits the field extension into the material). Although this statement is a simplification of the actual behavior of χ'', the notion of the available effective volume should be kept in mind when analysing the χ'' behavior, specially the peak level. By way of constrast, it is the apparent geometrical volume which is more important for the χ' level reached in full shielding cases. It is remarkable that in all these cases at low shielding ($a/\delta \ll 1$), while the χ' level can be hardly observable, the χ'' level ($\sim \sqrt{|\chi'|}$) can be used as a direct measurement of ρ

$$\chi'' \sim f/\rho$$

At strong shielding χ' and χ'' fulfill the following relationship $\chi' - \chi'' = \chi_0'$ where χ_0' is the level of χ' for perfect shielding ($\delta = 0$).

These remarkable behaviors of χ' and χ'' for low and high shielding levels should help in the verification of the experimental techniques and procedures.

As an illustration, figure 2 shows the normal state ($T > T_c$) $\chi''(T)$ and its frequency dependence around the martensitic transition temperature, T_m, of a V_3Si crystal. The slope discontinuity at T_m

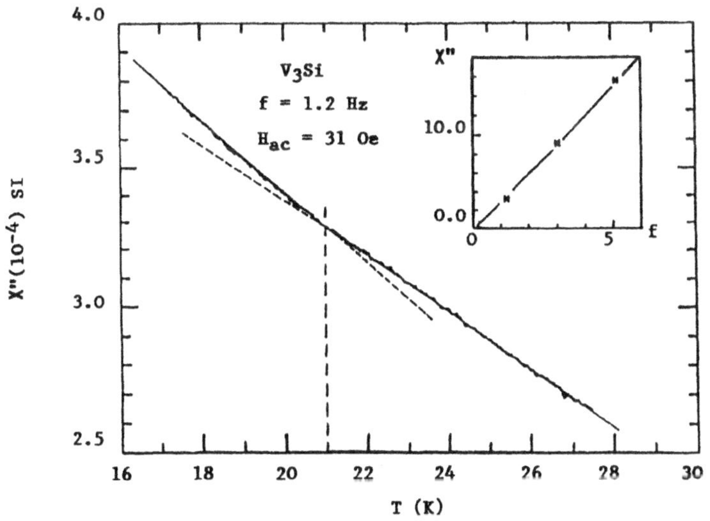

Fig. 2 The χ'' anomaly at T_m, the martensitic transition temperature for V_3Si crystal which correlates with resistivity, anomaly [18,19]. The inset shows the χ''-f linear relationship.

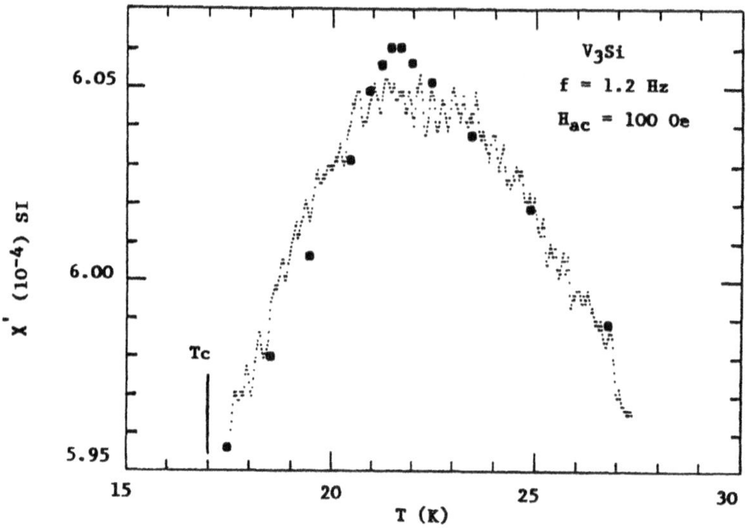

Fig. 3 The residual χ' scales with the dc susceptibility as measured by SQUID susceptometer [20] in the region of the martensitic phase transition temperature of a V_3Si crystal.

reflects similar behavior of the resistivity [18, 19] and one should also note the linear dependence of the χ'' level on frequency [inset].
The χ' level which should be associated with the χ'' level is negligible and the residual frequency independant signal is due to the paramagnetic susceptibility. A comparison with χ_{dc} measured by SQUID susceptometer [20] is given in figure 3.

The a/δ values at which χ'' peaks occur as well as the levels of χ'' peak, the χ'_0 values for complete shielding and the χ'' peak/χ'_0 ratios for the examples discussed above are summarized in table 1. No standard solutions exist for other geometries and the demagnetizing effects can not be limited to the enhancement of the χ'_0 level.

Table 1

	Sphere radius = a	cylinder radius = a H ∥ axis	Cylinder radius = a H ⊥ axis	Slab ∥ H thickness D = 2 a
(a/δ) peak	2.4	1.8	1.8	1.12
χ'' peak	0.54	0.38	0.76	0.42
χ'_0 $(\delta = 0)$	− 1.5	− 1	− 2	− 1
χ'' peak / $\lvert\chi'_0\rvert$	0.36	0.38	0.38	0.42

4. B.C.S. based intrepretation [15]

The physical meaning of the χ'' peaks has been raised again by one of the authors [15] on the basis of the B.C.S. theory. In this theory coherence effects play a major role in the determination of the electrical conductivity components ($\sigma = \sigma_1 + i\sigma_2$) both of which should be considered in the solution of the field distribution into a superconductor. In the quasi-static limit, the imaginary part σ_2 is simply related to the penetration length λ by the relationship [18] $\mu_0\ \sigma_2\ \omega = 1/\lambda^2$ and if we neglect the real component, the equation giving H(r) within the superconductor reduces to a well known one :

$$\Delta H - \frac{1}{\lambda^2} H = 0 \tag{22}$$

No dissipation can result form such a situation and $\chi'' \equiv 0$. As we have discussed, Maxwell and Strongin [8] suggested that a two fluid expression of σ would result in a very narrow, experimentally unobservable, χ'' peak. This suggestion, as was shown in an earlier work [15], is incorrect. In fact, in a two fluid model σ can be expressed by

$$\sigma = \sigma_{n1} + i\sigma_2 \tag{23}$$

where

$$\sigma_{n1} = (1 - X)\ \sigma_n < \sigma_n$$

X being the relative number of condensed electrons ($X = |\Psi_0^2|$) and σ_2 is related to λ as given before.

In the field equation :

$$\Delta H + K^2\ H = 0 \tag{24}$$

K^2 is now given by :

$$K^2 = i\ \omega\ \mu_0\ \sigma_{n1} - \frac{1}{\lambda^2} \tag{a}$$

or

$$K^2 = \frac{2\ i}{\delta_{n1}^2} - \frac{1}{\lambda^2} \tag{b}$$

with

$$\delta_{n1}^2 = \frac{2}{\mu_0\ \omega\ \sigma_{n1}} > \delta_n^2 \tag{c}$$

Analytical expressions for the χ_{ac} components are the same as in the case of a normal conductor with the new expression for K^2. In this case, it is easy to show that χ'', when λ becomes finite, decreases from its normal state value and there is no possibility at $T < T_c$ of having χ'' develop a peak. This expected and obvious result can be checked by developing any of the χ'' expression for $|Ka| \ll 1$ e.g. for cylindrical sample this gives [15]

$$\chi' = -\ a^2/8\lambda^2 + \frac{a^4}{48}\left(1/\lambda^4 - 4/\delta_{n1}^4\right) \tag{25}$$

$$\chi'' = a^2/4\delta_{n1}^2\ (1 - a^2/3\lambda^2) < a^2/4\delta_{n1}^2 < \chi''_n \tag{26}$$

The χ'' value is decrased by the effects of λ and $\delta_{n1}(> \delta_n)$.

On the other hand, an increase of χ'' will result if we consider σ_1 given by B.C.S. where, due to coherence effects, σ_1 increases over its normal state level below T_c :

$$\sigma_1/\sigma_n = 2 \int_{\Delta}^{\infty} \frac{E^2 + \Delta^2}{E^2 - \Delta^2} \left(\frac{-\partial f}{\partial E}\right) dE \qquad (27)$$

The divergence of the B.C.S. density of state at $E = \Delta$ results in a divergence of σ_1, but many reasons can be invoked to prevent this divergence.

Tinkham [21] has also shown that at low frequencies ($\hbar\omega \ll \Delta$) the behavior of σ_1 and σ_2 are related. Based on the validity of the Kramers-Kronig relationships and taking into account the behavior of these conductivities at high frequencies ($\hbar\omega \gg \Delta$), he found that if $\sigma_1 = A\delta(\omega)$, then $\sigma_2 = -2\ A/\pi\omega$. This results in $1/\lambda^2 = 2\ \mu_0\ A/\pi$. The behavior of σ_1 and σ_2 is reflected by λ and again λ emerges as the relevant length scale for the behavior of χ'' which will go through a maximum close to T_c as $\lambda \to \lambda(0) \ll a$. This maximum occurs because χ'' is given by the same analytical expressions as for the normal conductor but with a different expression of the "wave vector" K. It is obvious that no χ'' peak should be expected if the experimental conditions are such that this peak has already occured in the normal state. This B.C.S. based approach can not explain the dependence of the χ'' peak temperature, T_p, on $|H_{ac}|$ since the BCS theory is a linear response one. But again, as was suggested by Cody and Miller [12], there is the aspect of the critical current density (depairing current) which should be taken into account. When the χ'' peak occurs, the current density $J \sim H_{ac}/a$ is typically of the order of $10\ A/cm^2$ for $H_{ac} \sim 1$ Gauss and $a \sim 1$ mm this density could be much higher than the intrinsic critical current density close to T_c. This also explain non-linearity which results in harmonic generation.

To conclude, the B.C.S. based approach to χ', χ'' behavior needs the incorporation of another property of the superconducting state (the existence of critical or depairing current density) to explain the non linear aspects of the χ_{ac} behavior. Formally, the difference between thge BCS approach and the eddy-current one lies in the expression of the wave vector K. In the normal state :

$$K^2 = \frac{2i}{\delta_n^2} \qquad (28)$$

and in the superconducting state

$$K^2 = \frac{2i}{\delta_1^2} - \frac{1}{\lambda^2} \qquad (29)$$

The residual skin depth δ_1 plays the role of δ_n while the presence of λ serves to enhance the decrease in the available effective volume ; a quantity which otherwise is determined by δ_n (or δ_1). A result of the B.C.S. based model is that χ'' values are always less than those expected for a normal metal (where $\delta_1 = \delta_n$). In other words, the χ'' peak level observed at the superconducting phase transition should be smaller than that which would develop due to simply increasing σ_n of the normal metal. This has been proved in the case of low screening and it is easy to show its validity in the high screening situation. The χ_{ac} limit in the case of cylindrical sample is given by :

$$\chi_{ac} = -1 + \frac{2i}{Ka} \tag{30}$$

and

$$K^2 = \frac{2i}{\delta_1^2} - \frac{1}{\lambda^2} \tag{31}$$

can be developed to give

$$K = \alpha + i\beta$$

with $\beta = n/\delta_1$ and $\alpha = \dfrac{1}{n\,\delta_1}$ where $n = n\,(\lambda_1\,,\ \delta_1) > 1$.

The imaginary part χ'' is given by :

$$\chi'' = \frac{2/n}{\dfrac{1}{n^2} + n^2} \times \frac{\delta_1}{a} \tag{32}$$

since

$$\frac{2/n}{\dfrac{1}{n^2} + n^2} < 1 \tag{33}$$

This gives

$$\chi'' < \delta_1/a \tag{34}$$

δ_1/a is the expected level of χ'' in the normal state when $\delta_n = \delta_1$ $(\lambda \to \infty)$.

So, the χ'' peak level, for a given geometry, is reduced from its value expected on the basis of the eddy current model due to λ. The exact value reached by χ'' peak will depend on the $(\sigma_1\,,\ \sigma_2)$ or $(\delta_1\,,\ \lambda)$ relationship which can vary from one superconducting compound to another. We will discuss further the χ'' peak level dependence on the H_{ac} amplitude and on the frequency which is observed in many cases.

5. Other models [13, 14]

Other phenomenological models have been proposed to interpret the χ', χ'' behavior at the superconducting phase transition of quasi-one dimensional superconductors such as polysulfur Nitride $(SN)_x$ [13], transition metal chalcogenides Nb_3X_4 [22], and an artificially prepared multiconnected Josephson network [14].

In the weakly coupled filamentary superconductors model of Oda and al [13], the shielding current flows accross Josephson junctions to form a current loop which expands as the temperature decreases. Expressing the mutual inductance of the coil system as $M = M' - iM''$, their model yields :

$$M' = - AL\ (\omega L)^2 \ /\ [R^2 + (\omega L)^2] \tag{35}$$

$$M'' = + ALR\ (\omega L)\ /\ [R^2 + (\omega L)^2] \tag{36}$$

where L and R are respectively the self-inductance and resistance of an equivalent loop.

M'' has a maximum (= AL/2) at $R = \omega L$ and $M' = M_0'$ for the complete shielding (R = 0) has the value - AL, thus M'' peak/M_0' = 0.5 and M'' is symmetric in the M'' vs M' representation.

Another model, the multiconnected Josephson network model proposed by Ishida and Mazaki [14], explains the χ' and χ'' behavior as well as harmonic generation by performing a Fourier analysis of the induced magnetization m(t) which can be deduced from the experimentally observed e.m.f. signal e(t). The "expected" m(t) wave form is jusitifed on the basis of an equivalent superconducting loop which involves weak links as in the Oda and al model [13].

The two models, althought the second one [14] is actually a presentation of "modelized" experimental data, provide a good basis for the understanding of χ' vs χ'' behavior due to the (weak) intergranular coupling in the high T_c superconductors.

D. CAN A THIN SUPERCONDUCTING SURFACE LAYER (THICKNESS ≪ λ) SIMULATE (IN PART) THE DIAMAGNETIC SUSCEPTIBILITY OF A BULK SUPERCONDUCTOR ?

This question has been given a positive answer by Kittel at al. [23] in a recent paper thus supporting the commonly held reservations about the utility of χ_{ac} data.

The example of an insulating cylindrical sample of radius R covered with a thin layer of thickness D (≪ R) of superconducting material (D ≪ λ) considered by Kittel et al [23] has also been treated by De Gennes [24] and the solution which correspond to thermodynamic equilibrium is such that $h_i \sim h_0$ (h_i the internal field and h_0 the applied field). In this case the number of fluxoids are allowed to adjust to reach this equilibrium state. In the case, treated by Kittel et al. [23], the field h_i is given by [24] :

$$h_i = \frac{2\ h_0\ \lambda}{R\ \sinh D/\lambda} + \frac{\Phi_f}{\pi\ R^2} \tag{37}$$

where Φ_f is the contribution due to the fluxoids whose number is maintened constant.

Neglecting $\Phi_f/\pi R^2$ and considering $D/\lambda \ll 1$ we get ;

$$h_i = h_0 \frac{2\ \lambda^2}{RD} \tag{38}$$

which is the result of Kittel et al [23] and $h_i \ll h_0$ (substantial screening) if

$$D \gg \frac{2\,\lambda^2}{R} = \lambda \cdot \frac{2\,\lambda}{R} \tag{39}$$

Obviously there is the possibility of satisfying this last inequality and having, simultaneously, $D \ll \lambda$ as was argued by Kittel et al. If now we return to the screening current density in this situation we have

$$J = \frac{h_0 - h_i}{D} \approx \frac{h_0}{D} \approx \frac{h_0}{N\,\lambda\,(2\lambda/R)} \tag{40}$$

where $N \gg 1$.

The intrinsic limitation of J is given by ;:

$$J_c \sim \frac{H_{c1}}{\lambda} \quad \text{(type II superconductor)} \tag{41}$$

Thus

$$\frac{h_0}{N\lambda\,(2\lambda/R)} < \frac{H_{c1}}{\lambda} \quad \text{or} \quad h_0 < N \times \frac{2\lambda}{R} H_{c1} \tag{42}$$

if we take $N \approx 10 - 100$ and $\lambda/R \sim 10^{-5}$ this gives a limitation for the applied field amplitude :

$$h_0 < (10^{-3} - 10^{-4})\,H_{c1}$$

Although, Kittel el al [23] suggest that this problem of the current densitycan be avoided by working with low enough field amplitude $|h_0|$, the limitation placed on $|h_0|$ is rather stringent. Since H_{c1} is temperature dependent, variations in $|h_0|$ will yield apparent transition temperatures which are very sensitive to $|h_0|$. For example, if the superconducting layer has a $H_{c1}(T=0) \cong 10^3$ G then a measuring field of $|h_0| = 1$ G could depress the apparent transition temperture to temperatures as low as 0 K. Clearly, even for very high T_c, the $|\chi'|$ will be a function of T and $|h_0|$.

In the other case, where the system is allowed to reach thermodynamic equilibrium (adjustable phase) the internal field $h_i = h_0$ and the cylindrical superconducting thin layer will behave electrodynamically as a thin slab with very low, if any, screening, when $D \ll \lambda$. $e'_L/e'_C \sim \left(\frac{D}{\lambda}\right)^4 \cdot \left(\frac{D}{R}\right)$ where e'_L and e'_C are the in-phase signals for the layer respectively in the slab and the cylindrical geometries.

E. THE χ'' PEAK LEVEL, ITS FIELD AMPLITUDE DEPENDENCE AND THE SHIELDING EFFECTS

We have seen that, within an eddy current interpretation, the $\chi''\,(\text{peak})/|\chi'_0|$ ratio is a function of the sample shape. However, this ratio is decreased by the effect of λ which is an important factor in reducing the available effective volume for dissipation to occur. Other factors can limit this physical volume and thus reduce further the χ'' peak/$|\chi'_0|$ ratio such as voids and normal inclusions, superconducting inclusions in a matrix of highest T_c, intergranular shielding, etc...

In high T_c ceramics a second dissipative peak in $\chi''(T)$, just below T_{co}, is frequently observed to evolve with increasing $|H_{ac}|$. This peak is referred to as the intragranular loss peak [25,26] and shifts to lower temperatures accompanied by an increase in magnitude, i.e. χ''_{intra}(peak) increases with increasing values of $|H_{ac}|$. A natural explanation of this behavior is taht with increasing values of $|H_{ac}|$ the intergranular shielding breaks down when the induced currents required to shield out H_{ac} exceed the critical currents of the intergranular weak links ; consequently, the field can now penetrate into the bulk of the sample and currents induced in the grains per se contribute to the dissipation. In such cases, the saturated χ'' peak level becomes meaningful and the problem is not to prove the bulk nature of the superconductivity (the existence of intergranular contribution is an a posteriori proof) but to make quantitative estimation of the superconducting volume percentage. In the present example, increasing the field amplitude is similar to grinding the sample without introducing uncontrolable mechanical modifications. So, an anomalously low χ'' peak level along with a high shielding, i.e. χ' level, should be considered as the first indication of a reduced effective superconducting volume (we presume that signals are normalised to the apparent geometrical volume). Depending on the physical situation, the actual superconducting volume can be studied by eliminating the suspected elements responsible for the shielding.

F. EXPECTED BEHAVIOR OF χ_{ac} IN THE CASE OF TYPE-I SUPERCONDUCTOR

Type-I superconductors are characterized by a Ginzburg-Landau parameter $\kappa = \lambda/\xi$ lower than unity ($< 1/\sqrt{2}$) i.e. in the region of the magnetic field penetration, the order parameter amplitude is strongly reduced. The role of λ in limiting the available effective volume is much more important than the residual depth δ_1 and in spite of the strong shielding a χ'' peak can hardly be expected at the limit of a very low field amplitude. However, in reality, samples posses a non sero demagnetization factor which gives rise in type I superconductors to an intermediate state ; a state which exists over a small temperature range ΔT below T_{co} which is a function of $|H_{ac}|$. The intermediate state consists of a mixture of normal and superconducting regions of macroscopic size known as "domains". The dynamics of such domains can result in a χ'' dissipation peak as in the case of type-II superconductors in the mixed state [27, 28]. The χ'' peak temperature (T_p) should be, in this case, strongly dependent upon the frequency and magnitude of the ac measuring field and the χ'' peak level should approach the expected values from the eddy-current model (see table 1).

This explanation of the χ'' peak origin in type-I superconductors as being due to intermediate state dynamics, induced when, for a fraction of the ac cycle, $|H_{ac}|$ exceed the thermodynamical critical magnetic field, $H_c(T)$ can also be advanced in the case of type II superconductors in the mixed state. In this case, one is concerned with $|H_{ac}|$ exceeding $H_{c1}(T)$ rather than $H_c(T)$. This is certainly the case in most situations where $(\mu_0 H_{ac})S \gg \Phi_0$ where S is the sample or grain cross section and/or in a region near T_c where $H_{c1}(T) < |H_{ac}|$.

In such cases, vortices nucleate and move within the sample in a reversible or irreversible way. This scenario of χ', χ'' behavior based on the possibility of drawing the material to the intermediate state (type-I) or to the mixed state (type-II) assumes naturally the bulk nature of the superconducting phase transition. It should be complemented by further investigation of χ'_H and χ''_H under dc magnetic field, but this is another important domain.

CONCLUSION

We have discussed the χ' and χ'' behavior of superconducting materials at $T < T_c$ and have shown that the normal metal electrodynamic approach, although being a fundamental basic approach, should be generalised to incorporate the specific aspects of the superconducting state. This introduces a new expression for the wave vector of the ac field within the material.

The occurence of a peak in χ'' is associated with the notion of the available effective volume where dissipation can physically occur in constrast to the shielding component χ' for which only the apparent geometrical volume is relevant.

The magnitude of the peak in χ'' and its dependence on the ac field amplitude are important for estimating the volume percentage of superconductivity and, thus, should be investigated carefully.

REFERENCES

1. A.P. Malozemoff, in Physical Properties of High Temperature Superconductors I, ed. by D.M. Ginsberg, World Scientific (1989), and references therein.
2. A.M. Campbell, J. Phys. C2, 1492 (1969).
3. R.W. Rollins, H. Küpfer and W. Gey, J. Appl. Phys., 45, 5392 (1974).
4. D.X. Chen, J. Nogues and K.V. Rao, Cryogenics, 29, 800 (1989).
5. J.R. Clem, Physica C, 153-155, 50 (1988).
6. K.H. Müller, J.C. McFarlane and R. Driver, Physica C, 158, 366 (1989) and K.H. Müller, Physica C, 159, 717 (1989).
7. R.A. Hein, Phys. Rev. B, 33 7539 (1986).
8. E. Maxwell and M. Strongin, Phys. Rev. Lett., 10, 212 (1963).
9. M. Strongin and E. Maxwell, Rev. of Modern Phys., P164, Janv. 1964
10. J.L. Genicon, J.P. Modondanon, R. Tournier, E.D. Dahlberg and D.G. Hinks, J. Magn. & Magn. Mater., Vol. 54-57, p. 1545 (1986).
11. M. Couach, A.F. Khoder, F. Monnier, B. Barbara and J.Y. Henry, Phys. Rev. B, 37, 748 (1988).
12. G.D. Cody and R.E. Miller, Phys. Rev. 173, 481 (1968).
13. Y. Oda, H. Takenaka, H. Nagano and I. Nakada, Sol. St. Com. 35, 887 (1980).
14. T. Ishida and H. Mazaki, J. Appl. Phys. 52, 6798 (1981).
15. A.F. Khoder, Doctorat d'Etat Thesis-Grenoble (1983) and Phys. Lett. A 94, 378 (1983).
16. L. Landau et E. Lifshitz, Electrodynamique des milieux continus, Ed. Mir (Moscou 1969), p. 245, English Version Ed. by Pergamon Press.
17. Handbook of Mathematical Functions, Ed by M. Abramowitz and I.A. Stegun, National Bureau of Standard (Washington 1972), p. 363.
18. G. Fasol, J. Schilling and B. Seeber in Proceedings of LT15, J. Phys. (France), C6, 410 (1978).
19. K.H. Berthel and B. Pietrass, J. Phys. (France), C6, 383 (1978).
20. M. Pellizone and A. Treynaud, Appl. Phys. 24, 375 (1981) and M. Pellizone Thèse Univ. de Genève, (1982).

21. M. Tinkham, Introduction to superconductivity, Mc Graw-Hill, Kogakusha Ltd (1975), p. 59
22. T. Ishida, K. Kanoda, H. Mazaki and I. Nakada, Phys. Rev. B 29 1183 (1984).
23. C. Kittel, S. Fahy and S.G. Louie, Phys. Rev. B 37, 642 (1988).
24. P.G. de Gennes, Superconductivity in Metals and Alloys, W.A. Benjamin (New-York 1966), p. 194.
25. R.B. Goldfarb, A.F. Clark, A.I. Braginski and A.J. Panson, Cryogenics 27, 475 (1987).
26. T. Ishida and H. Mazaki, Jap. J. of Appl. Phys. 26, L1296 (1987).
27. J.R. Clem, H.R. Kerchner and S.T. Sekula, Phys. Rev. B 14, 1893 (1976).
28. C.P. Bean, Rev. Mod. Phys. 36, 31 (1964).

DETAILED THEORY OF THE MAGNETIC RESPONSE OF HIGH-TEMPERATURE CERAMIC SUPERCONDUCTORS

Karl-Heinz Müller

CSIRO Division of Applied Physics

Sydney 2070, Australia.

INTRODUCTION

High-temperature superconductors produced by sintering consist of anisotropic grains. The grains are separated by grain boundaries which act as Josephson weak-links. The response of such a granular superconductor to a weak magnetic field is determined by the diamagnetic response of the superconducting grains and the pinning of Josephson vortices which form along grain boundaries, producing an intergranular field. For strong magnetic fields Abrikosov-like vortices form inside grains and generate an intragranular field.

In this paper I apply the concept of the critical state to both the intergranular and intragranular magnetic fields. The dependence of the intergranular pinning force density on the local field is estimated and a critical state model formulated. The resulting flux dynamics is compared with experimental data. It is shown that measurements of the time-derivative of the total flux which threads a sample when an ac field is applied agree well with the model prediction. Intergranular and intragranular ac losses and ac susceptibilities are calculated. The effect of intergranular flux creep is incorporated into the critical state model and explains the observed frequency dependence of the ac susceptibility. Finally the ac susceptibility of a two-phase granular superconductor is modelled.

JOSEPHSON NETWORK

High-temperature ceramic superconductors produced by sintering consist of superconducting grains with a diameter of typically 10μm. Because of the small coherence length of these materials, $\xi(T{=}0) \simeq 1$ nm, the grain boundaries act as Josephson junctions which are interconnected and form a complicated Josephson (or weak-link) network[1]. For an applied magnetic field less than the Josephson lower critical field, H_{c1J}, the current induced along the outer surface region of the network is strong enough to restrict field penetration to a depth λ_J, the Josephson penetration depth, and the bulk of the granular superconductor is in the Meissner-state, i.e. the magnetic induction inside the superconductor is zero if the sample size is much greater than λ_J. For a regular array of Josephson junctions with cubic grains of side length $2R_g$ and $\lambda_J >> R_g$ one obtains[2]

$$\lambda_J(T) = \left(\frac{\Phi_o}{4\pi R_g J_o(T)\mu_o\mu_{eff}}\right)^{1/2} \quad \text{and} \quad H_{c1J} = \frac{\Phi_o}{4\pi\mu_o\lambda_J^2\mu_{eff}}\ln(2\lambda_J/R_g) \tag{1}$$

Here $\Phi_o = 2.07 \times 10^{-15}$ Vs is the flux quantum, $\mu_o = 1.256\times 10^{-6}$ VsA^{-1}m^{-1} is the magnetic

Magnetic Susceptibility of Superconductors and Other Spin Systems
Edited by R.A. Hein *et al.*, Plenum Press, New York, 1991

constant, J_o the junction critical current density and μ_{eff} the effective permeability[2] resulting from the diamagnetic superconducting grains. For example, if $J_o = 10$ Acm^{-2}, $R_g = 2$ μm and $\mu_{eff} = 0.5$ one obtains $\lambda_J = 37$ μm and $H_{c1J} = 10^{-2}$ Oe. A simple expression for μ_{eff} was derived by Clem[2] where in order to avoid the complication of demagnetizing factors the granular structure of the superconductor was approximated by cylindrical grains of radius R_g aligned along the z-direction parallel to the applied field (see fig. 1). In this case one obtains

$$\mu_{eff} = f_n + f_s F(R_g, \lambda_g) \quad , \tag{2}$$

where f_n is the volume fraction of non-superconducting material (grain boundaries, non-superconducting phases and voids), f_s is the volume fraction of superconducting material ($f_s = 1 - f_n$) and F is the fractional deviation of a grain from the complete Meissner-state due to London-type flux penetration where λ_g is the London penetration depth of grains (neglecting anisotropy effects). The fractional deviation, $F(R_g, \lambda_g)$, of a cylindrical grain from the complete Meissner-state is[2]

$$F(R_g, \lambda_g) = \frac{2\overline{I}_1(R_g/\lambda_g)}{(R_g/\lambda_g)\overline{I}_o(R_g/\lambda_g)} \quad , \tag{3}$$

where $\overline{I}_o$ and $\overline{I}_1$ are modified Bessel functions of the first kind.

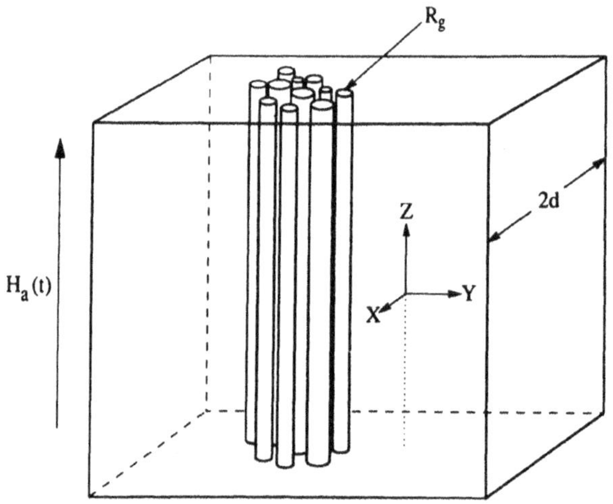

Fig. 1. Piece of slab of thickness $2d$. The superconducting grains are approximated by cylinders of radius R_g, aligned along the z-direction parallel to the applied field, H_a.

INTERGRANULAR CRITICAL STATE

When the applied magnetic field, H_a, exceeds H_{c1J} it becomes energetically favourable for the Josephson currents to form vortices along grain boundaries. The penetration depth of the magnetic field into the Josephson network depends on how efficiently the Josephson vortices are pinned. The pinning sites are grain boundary intersection points or voids and precipitates between grains. The microscopic magnetic field which results from these vortices can be used to define a macroscopic intergranular magnetic field, $\vec{H}_J(\vec{x})$, by averaging spatially over a sufficiently large number of vortices at position $\vec{x}$. Similar to Abrikosov vortices in conventional superconductors, the magnetic field profile $\vec{H}_J(\vec{x})$ can be determined by a critical state equation[3,4,5]. The critical state model assumes that locally everywhere inside the superconductor the Lorentz force density, $\vec{F}_{LJ}$, is equal to the maximum pinning force density, $\vec{F}_{PJ}$. The Lorentz force density is given by

$$\vec{F}_{LJ}(\vec{x}) = \vec{J}_{cJ}(\vec{x}) \times \vec{B}_J(\vec{x}) \quad , \tag{4}$$

where $\vec{J}_{cJ}(\vec{x})$ is the macroscopic intergranular critical current density and $\vec{B}_J(\vec{x})$ the macroscopic intergranular magnetic induction. Using Maxwell's equation

$$\vec{J}_{cJ} = curl\ \vec{H}_J \quad , \tag{5}$$

and setting $F_{LJ} = F_{PJ}$ one obtains the intergranular critical state equation

$$\mid (curl\vec{H}_J) \times \vec{B}_J \mid = F_{PJ}(B_J) \quad . \tag{6}$$

Equation (5) takes a simple form if one considers a superconducting sample in the shape of a large slab or a long cylinder and applies the external magnetic field parallel to the slab or the cylinder axis (z-direction). One obtains from eq.(6)

$$\mid \frac{dH_J(x)}{dx} \mid = \frac{F_{PJ}}{\mid B_J(x) \mid} \quad , \tag{7}$$

where x is the coordinate perpendicular to the slab or the radial coordinate of a cylinder. H_J

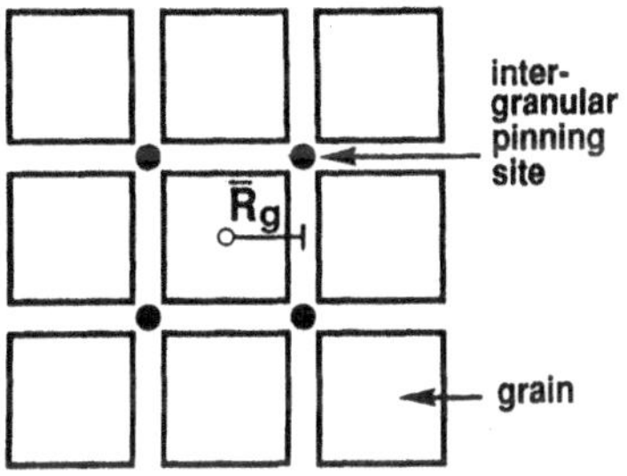

Fig. 2. Intergranular flux line pinning sites in an idealized regular array of Josephson junctions between cubic grains, representing a granular superconductor.

and B_J are the magnetic field and magnetic induction in z-direction. For $\mid H_J \mid >> H_{c1J}$ the magnetic induction B_J can be approximated by

$$B_J = \mu_{eff}\mu_o H_J \quad . \tag{8}$$

INTERGRANULAR FLUX PINNING

Lobb et al.[6] have shown that the depth of the pinning potential, U, of an isolated Josephson vortex in a regular two-dimensional array of square-like grains is proportional to the coupling strength, $E_J = \Phi_o I_J / 2\pi$, where I_J is the Josephson critical current across a grain boundary. The pinning sites are positioned at grain boundary intersection points as indicated in fig. 2 and the range of the potential is R_g. One can speculate that in the case of magnetically overlapping Josephson vortices, the Josephson current becomes suppressed by the average intergranular magnetic field, $H_J(x)$, produced by all the other vortices and that I_J varies with H_J in the form of a Fraunhofer diffraction pattern as for a single, isolated junction.

Because the grains in a real material vary strongly in size, the averaged Josephson current no longer possesses pronounced dips like the Fraunhofer pattern but has the form of an envelope function

$$I_J(H_J) = I_{0J}\frac{H_o}{\mid H_J \mid + H_o} \quad . \tag{9}$$

Here, I_{0J} is the maximum Josephson current across a junction and $\mu_o H_o = \Phi_o/A_J$, where A_J is the field penetrated junction area, $A_J = 2\lambda_g 2R_g$. The average intergranular pinning force density, F_{PJ}, is given by

$$F_{PJ} = \frac{U(T, H_J)}{R_g V_b} \quad , \tag{10}$$

where V_b is the volume which contains one pinning site with an attached flux line. Assuming the intergranular flux line correlation length is R_g, one finds for $B_J < \Phi_o/4R_g^2$ that $V_b = 2R_g\Phi_o/\mid B_J \mid$. Expressing the depth of the intergranular pinning potential, U, in terms of the Josephson junction current, i.e. $U = \beta E_J = \Phi_o I_J/2\pi$, and using eq.(9), one obtains

$$F_{PJ} = \frac{\beta I_{oJ} \mid B_J \mid}{4\pi R_g^2}\frac{H_o}{\mid H_J \mid + H_o} \quad . \tag{11}$$

The temperature dependence of F_{PJ} is determined by the dependence of the maximum Josephson current I_{OJ} and H_o on temperature. The factor β ($\beta < 1$) in eq.(11) takes into account effects of inhomogeneities of junction coupling strenghts and irregularities in a realistic Josephson network[7]. Substituting $\mu_{eff}\mu_o H_J$ for B_J in eq.(11) results in a pinning force density which is independent of the local magnetic field if $H_J >> H_o$. Using eq.(11) the critical state equation, i.e. eq.(7), becomes

$$\left| \frac{dH_J}{dx} \right| = J_o\frac{H_o}{\mid H_J \mid + H_o} \quad , \tag{12}$$

where $J_o = \beta I_{0J}/4\pi R_g^2$. A critical state equation as in eq.(12) was used by Kim et al.[9] in a phenomenological approach to describe the magnetic properties of certain conventional type II superconductors. Equation (12) does not account for the mutual vortex repulsion at higher fields which makes it difficult for vortices to use optimally the existing pinning sites. This will cause a further depression of F_{PJ} with H_J which has been neglected in this investigation.

INTERGRANULAR FLUX DYNAMICS

Solutions of eq.(12) for an applied ac magnetic field, $H_a(t) = H_{ac} \cos \omega t$, parallel to a cylinder (or slab) are given in the appendix A1. The plotted solutions in fig. 3 show the intergranular field profiles during the first half of the ac cycle where Josephson vortices move out of the material. Here, R is the radius of the cylinder. The figure illustrates that the most essential feature of the critical state is irreversibility, where the magnetic field inside the sample lags behind the applied field in a complicated manner.

The time-derivative $d\Phi/dt$ of the total flux Φ which threads a sample can be determined by measuring the induced voltage , V_{ind}, from a pick-up coil wound tightly around a cylindrical sample. One obtains

$$V_{ind} = -N\frac{d\Phi}{dt} \quad , \tag{13}$$

where N is the number of turns of the pick-up coil and

$$\Phi = 2\pi \int_0^R B_J(x)\, x\, dx \tag{14}$$

or, using eq.(8)

$$\Phi = \mu_{eff}\mu_o 2\pi \int_0^R H_J(x)\, x\, dx \quad . \tag{15}$$

Contributions to $d\Phi/dt$ from regions where $\mid H_J \mid < H_{c1J}$ and for which the critical state is not defined can be neglected as long as $H_{ac} >> H_{c1J}$. Analytical expressions for V_{ind} using

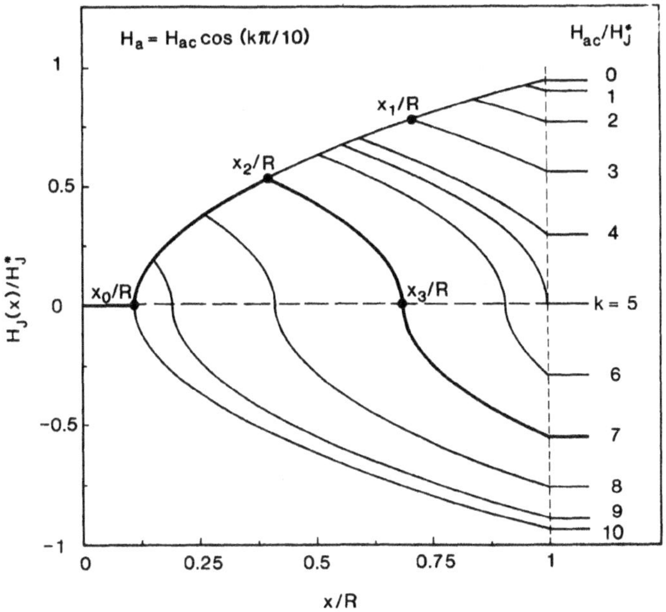

Fig. 3. Typical intergranular magnetic field profile $H_J(x)/H_J^*$ inside a cylinder during the first half of an ac cycle (k=0,...10). The penetrating field, H_J^*, is defined by Eq.(24).

the magnetic field profiles of appendix A1 are given in appendix A2. Similar results can be obtained for a large slab of thickness $2d$. In this case one has

$$\Phi \sim \mu_{eff}\mu_o \int_{-d}^{d} H_J(x)\, dx \ . \tag{16}$$

Figures 4 and 5 show the experimental and calculated results[10] at 77 K for a 50 mm long Y-Ba-Cu-O ceramic cylinder of radius $R = 1.65$ mm. The pick-up coil is wound directly on the sample and consists of 50 turns of 25 μm diameter wire. A 130 Hz ac magnetic field with variable amplitude generated by a solenoid is applied parallel to the axis of the sample. Good agreement between experiment and theory is found as long as the applied field is less than about 60 Oe. The data of fig. 5b cannot be fitted using the above model because for stronger applied fields flux penetrates into the grains where Abrikosov-like vortices are formed. The movement of these vortices significantly modifies $d\Phi/dt$. Equation (12) can also be solved when an additional dc field, H_{dc}, is applied

$$H_a(t) = H_{ac} \cos \omega t + H_{dc} \ . \tag{17}$$

Solutions for $H_J(x)$ are given in appendix A3. For the calculation of $d\Phi/dt$ the expressions for $H_J(x)$ of appendix A3 are used. The analytical expressions for $d\Phi/dt$ are similar to the ones given in appendix A2. Figure 6 shows the calculated results[11] which agree well with experimental data (not shown). As can be seen, an additional dc field breaks the $V_{ind}(t) = -V_{ind}(t + \pi/\omega)$ symmetry because for $H_{dc} \neq 0$ one has $H_J(t) \neq -H_J(t + \pi/\omega)$. To determine the spectral components, $C(n) = 20\, log(V_n/V_1)$, the Fourier components of V_{ind} are calculated numerically by evaluating integrals of the form

$$V_n = ([\int_0^{2\pi/\omega} V_{ind}(t)\, cos(n\omega t)\, dt]^2 + [\int_0^{2\pi/\omega} V_{ind}(t)\, sin(n\omega t)\, dt]^2)^{1/2} \ . \tag{18}$$

A comparison can be made with the experimental result obtained from a spectrum analyser. Figure 7 shows an example of measured spectral components up to n=50 and fig. 8 displays the corresponding computation. Good agreement is found up to the highest harmonics shown.

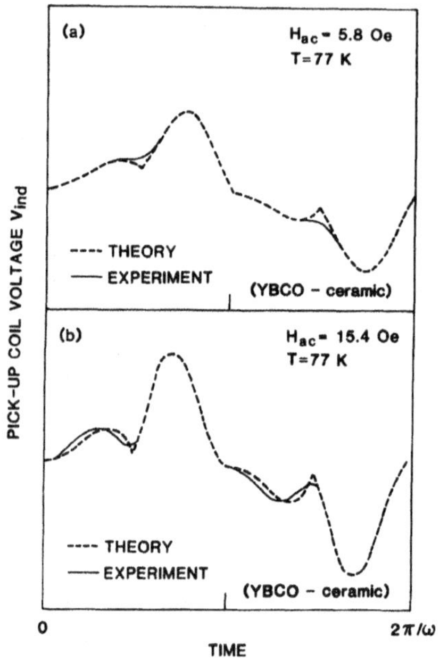

Fig. 4. Experimental and calculated pick-up coil voltage, V_{ind}, during a full ac cycle at 77 K. Here, $J_o\mu_o H_o = 116$ TAm^{-2} and $\mu_o H_o = 10^{-5}$ T.

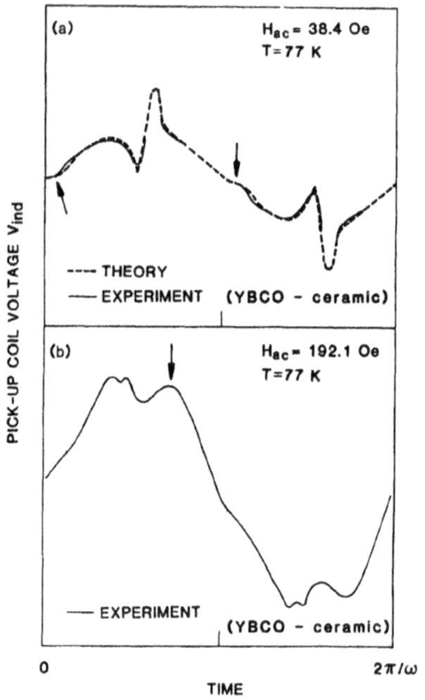

Fig. 5. As in fig. 4, but for different values of H_{ac}.

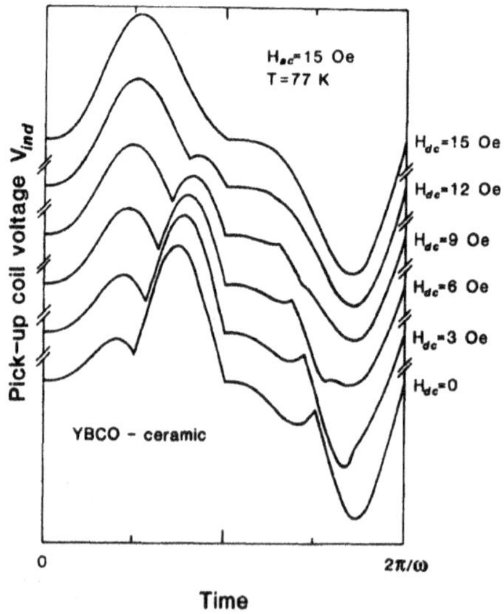

Fig. 6. Calculated pick-up coil voltage, V_{ind}, during an ac cycle for different additional dc fields, H_{dc}.

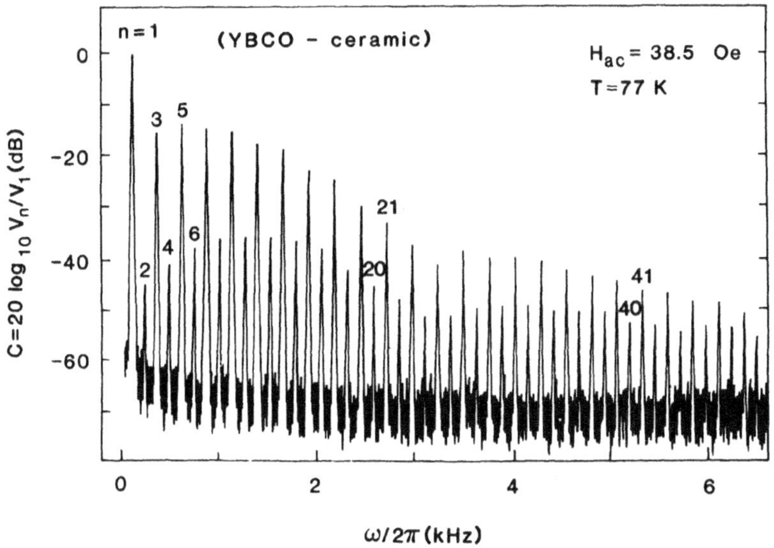

Fig. 7. Spectral components, $C(n)$, up to $n = 50$, for a Y-Ba-Cu-O ceramic long cylinder at 77 K driven by an ac magnetic field of amplitude 38.5 Oe at 130 Hz in the presence of a small dc field of 0.4 Oe ($\mu_o H_{dc} = 4 \times 10^{-5}$ T).

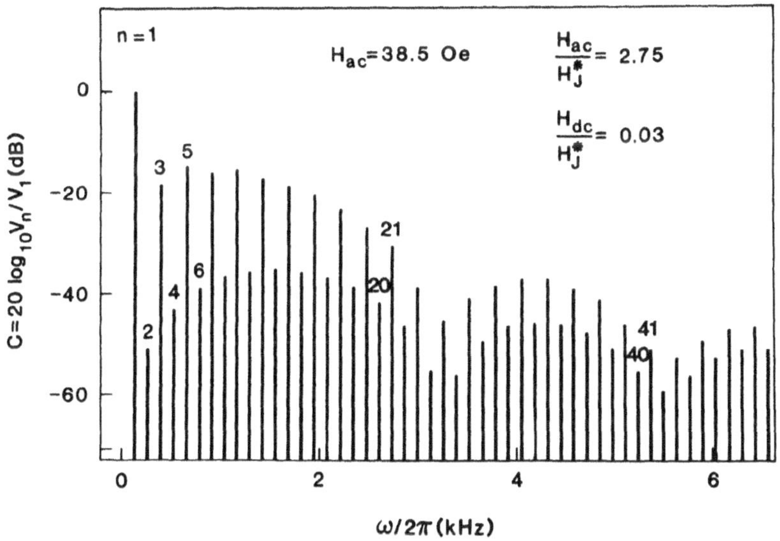

Fig. 8. Calculated spectral components, reproducing the experimental data of fig. 7. Here, $J_o \mu_o H_o = 116 \text{TAm}^{-2}$ and $\mu_o H_o = 10^{-5}$ T.

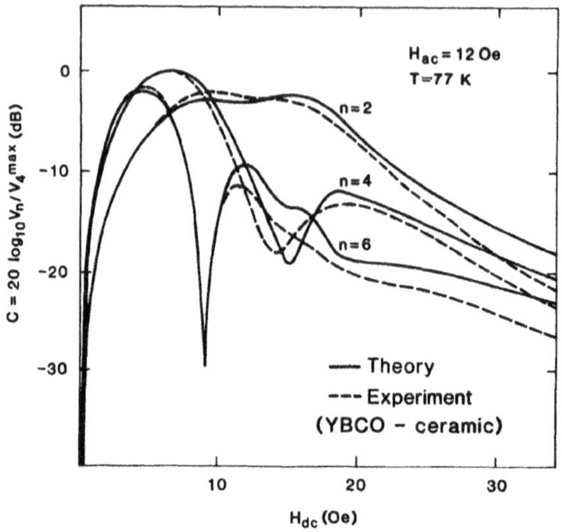

Fig. 9. Measured and calculated second, fourth and sixth harmonics as a function of the applied dc magnetic field, H_{dc}.

Several harmonics were analysed as a function of the applied dc field, H_{dc}. As shown in fig. 9 the experimental data and the calculated result agree well[12].

INTRAGRANULAR CRITICAL STATE

For $H_J(x) > H_{c1g}$, where H_{c1g} is the lower critical field of grains (neglecting anisotropy) Abrikosov-like vortices enter the grains. Because of an intrinsic pinning mechanism and pinning at twin boundaries and other defects, a critical state model can also be used to describe the magnetic field distribution inside grains. The intragranular magnetic field distribution $H_g(r,x)$ inside a cylindrical grain with radial coordinate r located at a position x inside the sample is described by the critical state equation

$$\left| \frac{dH_g(r,x)}{dr} \right| = \frac{F_{Pg}(B_g)}{|B_g(r,x)|} \tag{19}$$

with boundary condition

$$H_g(r = R_g, x, t) = H_J(x,t) \quad . \tag{20}$$

AC LOSS

The ac loss, W, is the energy dissipated in the superconductor per unit volume per cycle and is given by (see Clem, J. this volume)

$$W = -\oint \overline{B}(H_a)dH_a \quad , \tag{21}$$

where "$\oint$"indicates integration over H_a for a full ac cycle and $\overline{B}$ is the total magnetic induction (intergranular and intragranular contributions) spatially averaged over the cross-sectional area of the sample. The integral in eq.(21) represents the area of the $\overline{B}(H_a)$ hysteresis loop. For a slab one obtains

$$\overline{B} = \frac{\mu_{eff}\mu_o}{2d}\int_{-d}^{d} H_J(x)\,dx + \frac{(1-\mu_{eff})\mu_o}{R_g^2 d}\int_{-d}^{d}\int_0^{R_g} H_g(r,x)\,r\,dr\,dx \quad , \tag{22}$$

where the first term is the intergranular contribution and the second the intragranular one. In the limit of an applied field, H_a, close to H_{c2g} (upper critical field of grains) eq.(22) must result in $\overline{B} = \mu_o H_a$ which substantiates the factor $(1 - \mu_{eff})$ in front of the second term of eq.(22). Much experimental data is reasonably well described by assuming for the intragranular pinning force density

$$F_{Pg} = J_{cg} B_g \tag{23}$$

with J_{cg} independent of B_g (Bean model[3]).

Figure 10 compares experimental and calculated intergranular and intragranular ac losses as a function of the ac amplitude, $\mu_o H_{ac}$, ranging from 10^{-4} Tesla to 10 Tesla[13]. The intragranular critical state equation was solved by using the approximation $B_g = \mu_o H_g$, which makes the result for the intragranular part reliable only for $H_{ac} >> H_{c1g}$. The penetrating

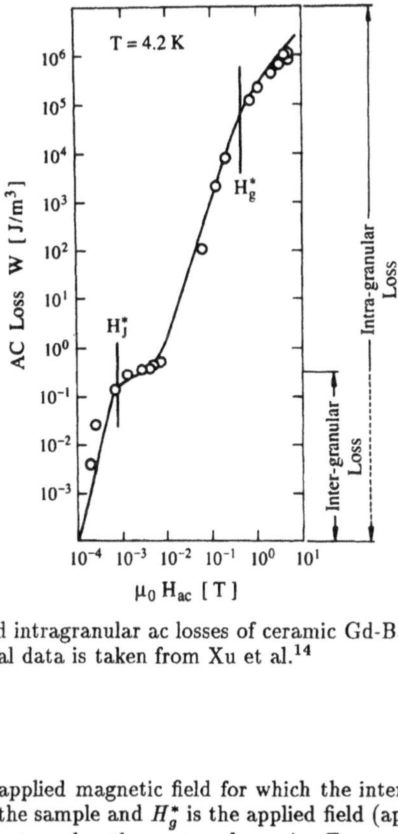

Fig. 10. Intergranular and intragranular ac losses of ceramic Gd-Ba-Cu-O as a fuction of H_{ac} at 4.2 K. The experimental data is taken from Xu et al.[14]

field H_J^* in fig. 10 is the applied magnetic field for which the intergranular field distribution just reaches the centre of the sample and H_g^* is the applied field (applied to a grain) for which the intragranular profile just reaches the centre of a grain. From eq.(12) one finds

$$H_J^* = -H_o + (H_o^2 + 2RJ_oH_o)^{1/2} \tag{24}$$

and from Eqs. (19) and (23)

$$H_g^* = R_g J_{cg} \quad . \tag{25}$$

The intergranular ac loss strongly depends on an additional dc field. A dc magnetic field causes the hysteresis loop to become asymmetric and the loop area to decrease. This is because the gradient $\mid dH_J/dx \mid$ decreases when H_J increases, diminishing the irreversibility of $\overline{B}(H_a)$. Figure 11 shows the experimental ac loss data as a function of the additional dc field, H_{dc}, for different magnetic field amplitudes, H_{ac}. Figure 12 shows the calculated result which agrees well with the experimental data[15].

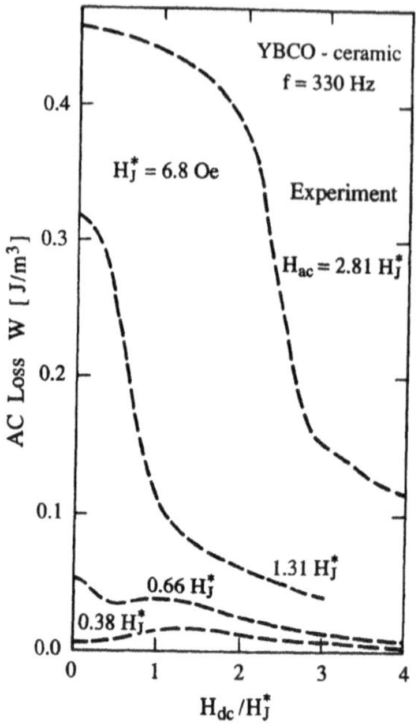

Fig. 11. Measured intergranular ac loss versus the applied dc field, H_{dc}/H_J^*, for different ac amplitudes, H_{ac}/H_J^*.

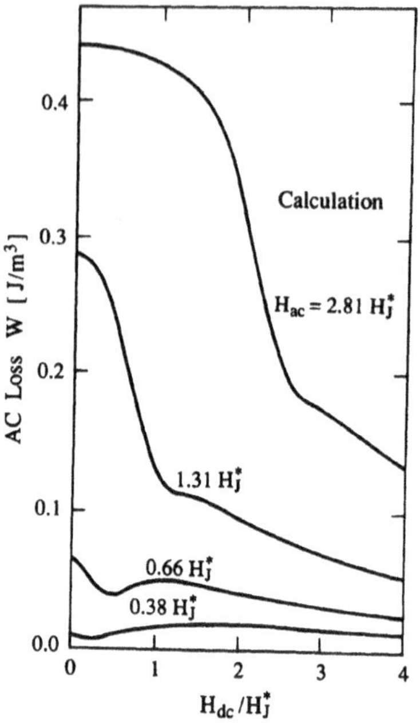

Fig. 12. Calculated intergranular ac loss versus the applied dc field, H_{dc}/H_J^*, for different ac amplitudes, H_{ac}/H_J^*.

AC SUSCEPTIBILITY

The ac loss, W, is related to the imaginary part of the ac susceptibility, χ_1'', by the simple expression

$$\chi_1'' = \frac{W}{\pi \mu_o H_{ac}^2} \quad . \tag{26}$$

The subscript 1 refers to the fundamental of the ac susceptibility. In general, χ_n' and χ_n'' with n=1,2,3... are defined by decomposing $\overline{B}$ into its Fourier components

$$\overline{B} = \mu_o H_{ac} \sum_{n=1}^{\infty} [(\chi_n' + \delta_{1n})\, cos(n\omega t) + \chi_n''\, sin(n\omega t)\,] \tag{27}$$

Here, $\delta_{1n} = 1$ if $n = 1$ and zero otherwise. Equation (26) can be verified by inserting eq.(27) into eq.(21) and using $dH_a = -\omega H_{ac} sin(\omega t)dt$. From eq.(27) one obtains for the susceptibilities χ' and χ'' (neglecting the subscript 1)

$$\chi' = -1 \ + \ \frac{\omega}{\pi \mu_o H_{ac}} \int_o^{2\pi/\omega} \overline{B}(t)\, cos(\omega t)\, dt \tag{28}$$

and

$$\chi'' = \frac{\omega}{\pi \mu_o H_{ac}} \int_o^{2\pi/\omega} \overline{B}(t)\, sin(\omega t)\, dt \ . \tag{29}$$

$\overline{B}(t)$ depends on H_{ac}, H_{dc}, f_n, R, R_g, $\lambda_g(T)$, $J_o(T)$ and $J_{cg}(T)$. The expressions used are $\lambda_g(T) = \lambda_g(0)(1-(T/T_c)^4)^{-1/2}$, $J_o(T) = J_o(0)(1-T/T_c)^2$, corresponding to SNS type of junctions and $J_{cg}(T) = J_{cg}(0)(1-T/T_c)$. $\overline{B}$ consists of an intergranular and an intragranular contribution, $\overline{B} = \overline{B_J} + \overline{B}_g$, (see eq.(22)) and therefore the susceptibility has an intergranular and an intragranular part.

$$\chi' = \chi'_J + \chi'_g \tag{30}$$

and

$$\chi'' = \chi''_J + \chi''_g \quad . \tag{31}$$

Figure 13 shows the experimental ac susceptibility of a typical ceramic Y-Ba-Cu-O sample as a function of temperature for different H_{ac} at zero dc magnetic field. Figure 14 displays the calculated results which are obtained by solving the eqs.(12) and (19) with the appropriate boundary conditions evaluating $\overline{B}$ from eq.(22) and finally determining χ' and χ'' numerically from eqs.(28) and (29). The model parameters used are discussed in ref. 17. The calculated positions of the intergranular and intragranular peaks, at temperatures T_m^J and T_m^g, depend on $J_o(T)$ and $J_{cg}(T)$ as illustrated in fig. 15.

The χ''_J-peak occurs at a temperature T_m^J at which $H_{ac} \simeq H_J^*(T_m^J)$. The electrically measured critical current density is equal to the spatially averaged critical current density $< J_{cJ} >$. For a slab one finds

$$< J_{cJ}(T) >= \frac{H_J^*(T)}{d} \quad . \tag{32}$$

It is important to notice that this equation was derived by using $curl\ \vec{H}_J = J_{cJ}$ and no explicit form of any critical state equation was employed. This means that eq.(32) holds for the Bean, Kim or any other critical state model. For a long cylindrical sample one finds[11,18]

$$< J_{cJ}(T) >= \frac{2}{\sqrt{3}} \frac{H_J^*(T)}{R} \quad . \tag{33}$$

Because of Eqs.(32) and (33), fig. 15 can be viewed as a plot of $< J_{cJ}(T) >$ and $< J_{cg}(T) >$.

Effect of Intergranular Flux Creep

The position of the χ''_J-peak at T_m^J shifts slightly to lower temperatures with increasing frequency of the applied ac magnetic field. In the following it is shown that incorporating the effects of flux creep into the critical state model can explain the observed frequency dependence of the ac susceptibility[19,20]. Using the Anderson flux creep model[4] and equating the Lorentz force density, F_{LJ}, to the maximum pinning force density, F_{PJ}, one obtains

$$F_{PJ} = \frac{kT}{V_b r_p} arcsinh[\frac{\nu_h}{2\nu_o}\ exp\ \frac{U(T,H_J)}{kT}] \quad , \tag{34}$$

where $U(T, H_J)$ is the average pinning potential, r_p the potential range which is equal to R_g, ν_o the attempt frequency for hopping and ν_h is the "minimal-observable" flux-line hopping rate of Josephson vortices. In the case that $U(T, H_J) >> kT$ only the forward hopping term of the flux creep model is essential and one finds

$$F_{PJ} = \frac{1}{V_b r_p}[U(T,H_J) + kT\ ln(\nu_h/\nu_o)] \quad . \tag{35}$$

Because $\nu_h << \nu_o$ the second term in eq.(35) is negative and F_{PJ} is negative between T^* and T_c. T^* is defined by $U(T^*, H_J) = -kT\ ln(\nu_h/\nu_o)$. Equation (35) is invalid for $T \geq T^*$ as well as for temperatures slightly below T^*. In this temperature range the less transparent eq.(34) has to be used. From eqs.(10) and (11) one obtains for the intergranular pinning potential

$$U(T, H_J) = U_o(T)\frac{H_o}{|H_J| + H_o} \tag{36}$$

where $U_o(T)$ has the form

$$U_o(T) = U_o(0)(1 - T/T_c)^\alpha \tag{37}$$

with $U_o = \beta E_J(T = 0)$ where E_J is the Josephson coupling strength defined above. For SIS grain boundary junctions $\alpha = 1$ and for SNS-type junctions $\alpha = 2$.

For a pinning force density, F_{PJ}, in the form of eq.(34), the critical state equation is extremely difficult to solve. If for $T > T^*$ the quantity H_o is greater than H_{ac} (H_{ac} is usually quite small for T_J^m close to T_c) one can substitute kT by $kT(H_{ac}/\sqrt{2} + H_o)/(|H_J| + H_o)$ and one obtains an expression for J_o which can be used in eq.(12)

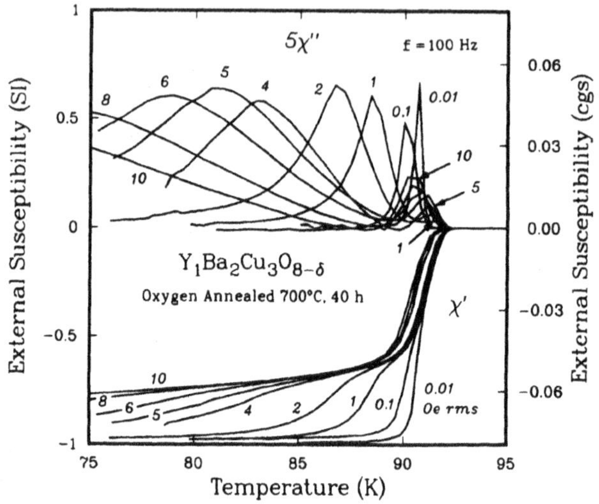

Fig. 13. Experimental data χ' and $5\chi''$ from ref. 16. The numbers labelling the curves are the r.m.s. values of the measuring ac field in Oe.

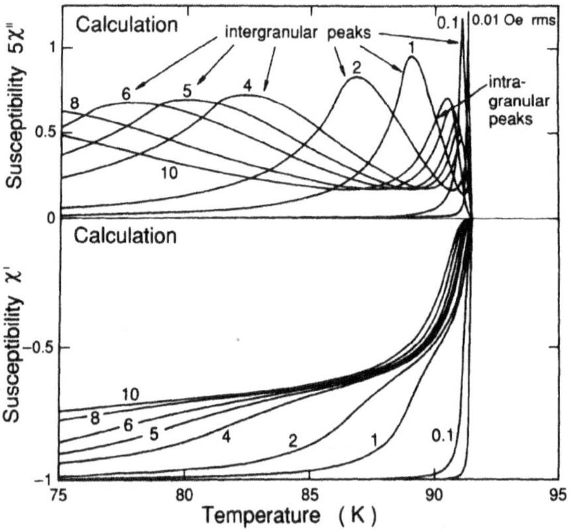

Fig. 14. Calculated susceptibility χ' and $5\chi''$ as a function of temperature. The χ'' curves show a broad intergranular peak at lower temperature and a narrower intragranular peak near T_c.

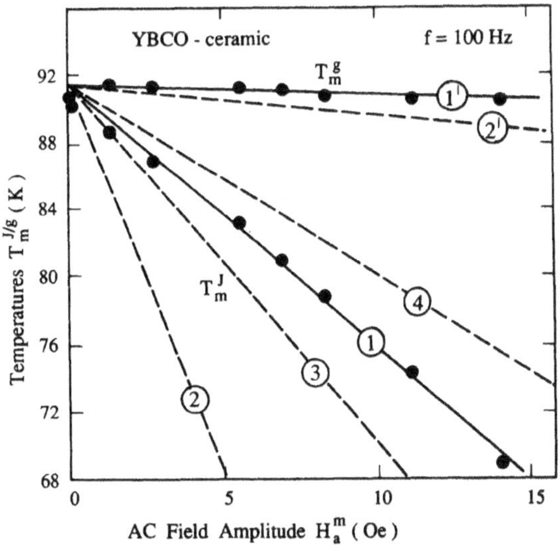

Fig. 15. Intergranular and intragranular χ''-peak temperatures T_m^J and T_m^g as a function of H_{ac} for different values of $J_o(0)\mu_o H_o(0)$ where $H_o(0) = 10^{-5}T$. (1) $J_o(0)\mu_o H_o(0) = 6.05 \times 10^3$ TAm^{-2}, (2) 7.24×10^2 TAm^{-2}, (3) 3.5×10^3 TAm^{-2}, (4) 1.21×10^4 TAm^{-2}. Parameters for (1') and (2') are discussed in ref. 17. The solid points are experimental data from ref. 16.

$$J_o = \frac{H_{ac}/\sqrt{2} + H_o}{H_o} \frac{kT}{2\Phi_o R_g^2} arcsinh[\frac{\nu_h}{2\nu_o} exp \frac{U(T, H_{ac}/\sqrt{2})}{kT}] \quad . \tag{38}$$

It is important to notice that the above J_o is independent of the local magnetic field H_J and the critical state profiles are those given in the appendix A1 and A3. An reasonable estimate for ν_o is 10^9 s^{-1} (see ref. 20). The "minimal-observable" hopping rate, ν_h, is the inverse of the measuring time if the applied magnetic field is a dc field. When an ac field is applied to a cylinder of radius R one obtains[20]

$$\nu_h \simeq \frac{8\Phi_o H_{ac} R_g^2}{\pi RkT}\omega \quad . \tag{39}$$

Figure 16 shows the experimental susceptibility as a function of temperature for a Bi-(Pb)-Sr-Ca-Cu-O ceramic sample at different amplitudes and frequencies. Figure 17 shows the calculated result. The parameters are discussed in ref. 22. It is important to notice that χ_J'' is almost zero between T^* and T_c where flux creep causes reversible behaviour. Because the pinning potential decreases with increasing field, the temperature T^* shifts to lower temperature with increasing amplitude. No reversible regime is seen in fig. 14 where the calculation does not consider the effect of flux creep. Figure 18 shows the effective pinning force density, $\alpha_J = J_o\mu_o H_o$ used in the calculation of fig. 17. The effective pinning force density becomes negligibly small above T^* which causes the reversible magnetic flux behaviour.

AC Susceptibility in Two-Phase Samples

The ac susceptibility has been modelled for a two-phase Bi-(Pb)-Sr-Ca-Cu-O sample where the low-T_c phase has a transition temperature T_{c1} and the high-T_c phase a transition temperature T_{c2}. A two-phase material consists of two intertwined Josephson networks where one network is composed of Josephson junctions between grains of the low-T_c phase and the other network of Josephson junctions of the high-T_c phase. If the two networks are only coarsly intertwined i.e. large regions exist which contain only one grain type, one can neglect

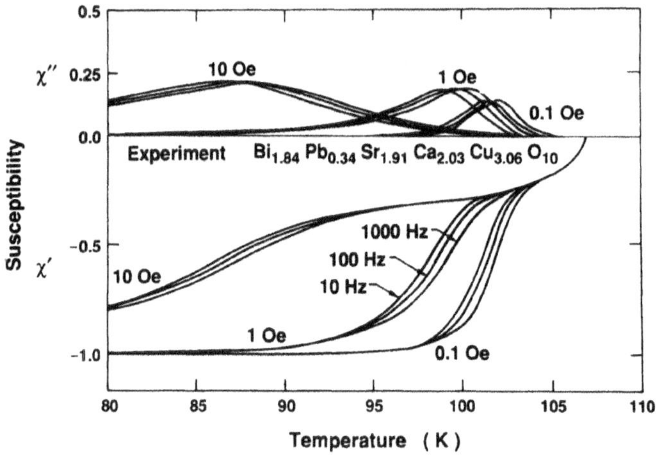

Fig. 16. Experimental data of χ' and χ'' versus temperature for a ceramic Bi-(Pb)-Sr-Ca-Cu-O sample from ref. 21.

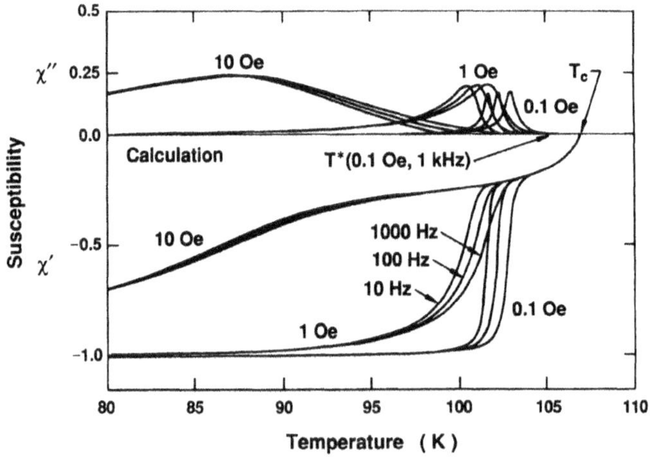

Fig. 17. Calculated susceptibility χ' and χ'' versus temperature for ceramic Bi(Pb)SrCaCuO. The parameters used are given in ref. 22.

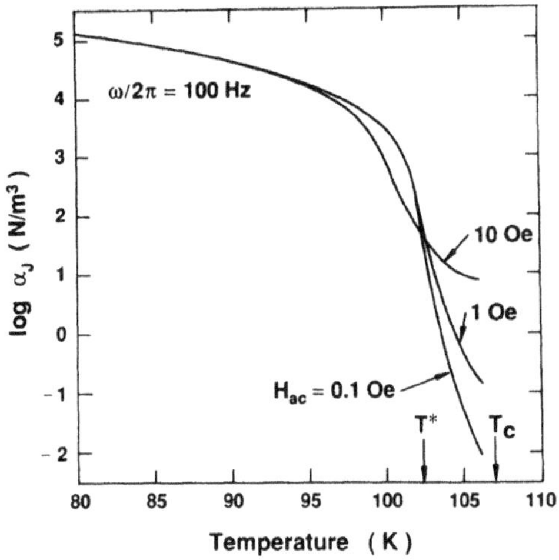

Fig. 18. Intergranular effective pinning force density, $\alpha_J = J_o \mu_o H_o$, used to calculate the susceptibility in fig. 17, as a function of temperature.

the coupling between the two networks. In this case the average intergranular pinning potential for $T_{c1} \leq T \leq T_{c2}$ is

$$U(T, H_J) = f_{s2}\, U_2(T, H_J) \tag{40}$$

and for $0 \leq T \leq T_{c1}$

$$U(T, H_J) = f_{s1}\, U_1(T, H_J) + f_{s2}\, U_2(T, H_J) \quad , \tag{41}$$

where f_{s1} and f_{s2} are the volume fractions of the low-T_c and high-T_c phases, respectively, and U_1 and U_2 are the corresponding pinning potentials (see eq.(36))

$$U_i(T, H_J) = U_{oi}(T) \frac{H_{oi}(T)}{\mid H_J \mid + H_{oi}(T)} \quad ; \; i = 1, 2 \quad . \tag{42}$$

Here,

$$H_{oi} = \frac{\Phi_o}{4\mu_o \lambda_i(T) R_g (1 - f_v^{1/2})} \quad ; \; i = 1, 2 \quad , \tag{43}$$

where f_v is the volume fraction of voids and non-superconducting material and λ_i the London penetration depths of the two different types of grains

$$\lambda_i(T) = \lambda_i(0)(1 - (T/T_{ci})^4)^{1/2} \quad ; \; i = 1, 2 \quad . \tag{44}$$

Assuming SNS-type junctions one has

$$U_{oi}(T) = U_{oi}(0)(1 - T/T_{ci})^2 \quad ; \; i = 1, 2 \quad . \tag{45}$$

For the effective permeability one obtains for $T_{c1} \leq T \leq T_{c2}$

$$\mu_{eff} = f_v + f_{s1} + f_{s2} \frac{2\overline{I}_1(R_g/\lambda_2)}{(R_g/\lambda_2)\overline{I}_o(R_g/\lambda_2)} \tag{46}$$

and for $0 \leq T \leq T_{c1}$

$$\mu_{eff} = f_v + f_{s1} \frac{2\overline{I}_1(R_g/\lambda_1)}{(R_g/\lambda_1)\overline{I}_o(R_g/\lambda_1)} + f_{s2} \frac{2\overline{I}_1(R_g/\lambda_2)}{(R_g/\lambda_2)\overline{I}_0(R_g/\lambda_2)} \quad . \tag{47}$$

Figure 19 shows the experimental data of $\chi'(T)$ and $\chi''(T)$ for a Bi-(Pb)-Sr-Ca-Cu-O sample[23]. The transition temperatures are $T_{c1} \simeq 65$ K and $T_{c2} = 107$ K. Figure 20 shows the calculated

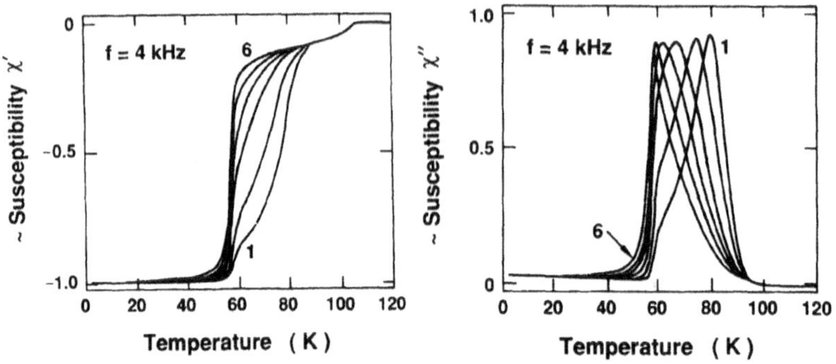

Fig. 19. Experimental susceptibility χ' and χ'' versus temperature for a two-phase Bi-(Pb)-Sr-Ca-Cu-O ceramic sample from ref. 23. The ac field amplitudes, H_{ac}, are 1: 0.16 Oe, 2: 0.4 Oe, 3: 1.0 Oe, 4: 1.6 Oe, 5: 2.4 Oe, 6: 3.6 Oe.

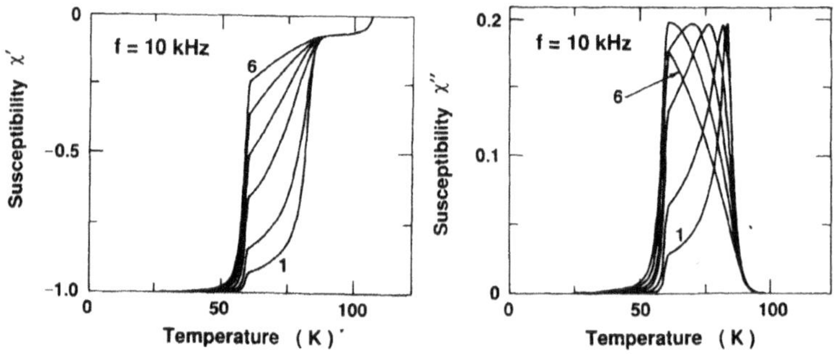

Fig. 20. Calculated susceptibility versus temperature for a two-phase sample. The parameters are $f_v = 0.6$, $f_{s1} = 0.3$, $f_{s2} = 0.1$, $R_g = 4\ \mu$m, $d = 1$ mm, $\omega/2\pi = 10$ kHz, $U_{01}(0) = 200$ eV and $U_{02}(0) = 0.35$ eV.

susceptibility which agrees well with the experimental data of fig. 19. A reversible region below T_{c2} exists which is about 15 K wide.

CONCLUSIONS

The concept of an intergranular and intragranular critical state is well suited to describe in detail the observed magnetic response of polycrystalline high-temperature superconductors. The intergranular flux line pinning mechanism suggests a Kim-like critical state model. The magnetic behaviour of grains is reasonably well described by a simple Bean critical state model. Because the intergranular shielding current decreases strongly with increasing field, even harmonics appear in the ac magnetic response when a dc magnetic field is present. Susceptibility measurements as a function of temperature and ac field amplitude allow the determination of parameters which characterise a granular superconductor. These parameters are the volume fraction of the non-superconducting material, the pinning force density and the intergranular critical current density. The weak freqnency dependence of the magnetic response in a weak applied field can be attributed to intergranular flux creep. Flux creep causes the appearance of a temperature regime just below T_c, where the magnetic flux behaviour is reversible. The temperature below which irreversibility occurs does not coincide with the temperature at which χ'' peaks. The critical state model concept can be extended to describe multiphase samples.

APPENDIX A1

The critical state equation (i.e. eq.(12)) for the boundary condition $H_J(R) = H_a = H_{ac}\, cos(\omega t)$ is solved by variable separation and subsequent intergration

$$\int_{H_J(x)}^{H_a} (| H_J | + H_o) dH_J = \pm J_o H_o \int_x^R dx' \quad . \tag{A1.1}$$

In the case of a slab, R has to be replaced by d.
One obtains for $H_J > 0$

$$H_J(x) = -H_o + [H_o^2 + H_a^2 + 2H_oH_a \pm 2H_oJ_o(R-x)]^{1/2} \quad , \tag{A1.2}$$

where the $-$ sign corresponds to vortices moving into the sample and the $+$ sign to vortices moving out of the sample.
One obtains for $H_J \leq 0$

$$H_J(x) = H_o - [H_o^2 + H_a^2 - 2H_oH_a \pm 2H_oJ_o(R-x)]^{1/2} \quad , \tag{A1.3}$$

where the $+$ sign is for antivortices (negative H_J) moving out of the sample and the $-$ sign for antivortices moving into the sample.
With these four different solutions (eqs.(A1.2) and (A1.3)) the critical state profile inside a long cylinder (or slab) can be constructed.

When $H_{ac} \leq H_J^*$, where H_J^* is the penetrating field one obtains for the first quarter of the ac cycle ($0 \leq \omega t \leq \pi/2$) (corresponding to k = 0 to 5 in fig. 3) :

$$H_J(x) = -H_o + [H_o^2 + H_a^2 + 2H_oH_a + 2H_oJ_o(R-x)]^{1/2} \quad \text{for} \quad x_1 \leq x \leq R \quad , \tag{A1.4}$$

$$H_J(x) = -H_o + [H_o^2 + H_{ac}^2 + 2H_oH_{ac} - 2H_oJ_o(R-x)]^{1/2} \quad \text{for} \quad x_o \leq x \leq x_1 \tag{A1.5}$$

and

$$H_J(x) = 0 \quad \text{for} \quad 0 \leq x \leq x_o \tag{A1.6}$$

where

$$x_o = R - (H_{ac}^2 + 2H_oH_{ac})/2H_oJ_o \tag{A1.7}$$

and

$$x_1 = R - (H_{ac}^2 - H_a^2 + 2H_o(H_{ac} - H_a))/4H_oJ_o \quad . \tag{A1.8}$$

(See fig. 3 for definitions of x_o and x_1.)
For the second quarter of the ac cycle one finds :

$$H_J(x) = H_o - [H_o^2 + H_a^2 - 2H_oH_a - 2H_oJ_o(R-x)]^{1/2} \quad \text{for} \quad x_3 \leq x \leq R \quad , \tag{A1.9}$$

$$H_J(x) = -H_o + [H_o^2 - H_a^2 + 2H_oH_a + 2H_oJ_o(2R-x)]^{1/2} \quad \text{for} \quad x_2 \leq x \leq x_3 \quad , \tag{A1.10}$$

$$H_J(x) = -H_o + [H_o^2 + H_{ac}^2 + 2H_oH_{ac} - 2H_oJ_o(R-x)]^{1/2} \quad \text{for} \quad x_o \leq x \leq x_2 \quad , \tag{A1.11}$$

and

$$H_J(x) = 0 \quad \text{for} \quad 0 \leq x \leq x_o \quad , \tag{A1.12}$$

where

$$x_2 = \frac{3}{2}R - \frac{1}{4H_oJ_o}(H_{ac}^2 + H_a^2 + 2H_o(H_{ac} - H_a)) \tag{A1.13}$$

and

$$x_3 = R - \frac{H_a^2 - 2H_oH_a}{2H_oJ_o} \quad . \tag{A1.14}$$

(See fig. 3 for definitions of x_2 and x_3.) The second half of the ac cycle ($\pi/2 \leq \omega t \leq \pi$) is obtained from the symmetry relation $H_J(x,t) = -H_J(x, t - \pi/\omega)$.

For $H_{ac} > H_J^*$ the value of x_o is always negative whereas x_1, x_2 and x_3 can be negative. The above equations for $H_J(x)$ are still valid in this case for positive x.

APPENDIX A2

Analytical expressions for the change of the total flux with time ($d\Phi/dt = \dot{\Phi}$) for a long cylinder.

As an example $\dot{\Phi}$ is calculated for $H_{ac} \simeq 2.5\ H_J^*$ (see eq.(24)) in which case x_o and x_2 are negative.

(i) For $0 \leq \omega t \leq \omega t_1$, where t_1 is determined by

$$x_1 = 0 = R - [H_{ac}^2(1 - cos^2\omega t_1) + 2H_oH_{ac}(1 - cos\ \omega t_1)]/4H_oJ_o \quad (A2.1)$$

one obtains via eq.(15)

$$\dot{\Phi} = \mu_{eff}\mu_o 2\pi \frac{d}{dt}\int_{x_1}^{R}(-H_o + [H_o^2 + H_a^2 + 2H_oH_a + 2H_oJ_o(R - x)]^{1/2})\ x\ dx \quad (A2.2)$$

or

$$\dot{\Phi} = \mu_{eff}\mu_o 2\pi[H_o x_1 \dot{x}_1 - \quad (A2.3)$$

$$\frac{1}{H_oJ_o}\{\dot{H}_a(H_a + H_o)^2 - ((H_o + H_a)^2 + 2H_oJ_o(R - x_1))^{1/2}((H_a + H_o)\dot{H}_a - H_oJ_o\dot{x}_1)\}]$$

(ii) For $\omega t_1 \leq \omega t \leq \pi/2$ one obtains

$$\dot{\Phi} = \mu_{eff}\mu_o 2\pi \frac{d}{dt}\int_{o}^{R}(-H_o + [H_o^2 + H_a^2 + 2H_oH_a + 2H_oJ_o(R - x)]^{1/2})\ x\ dx \quad (A2.4)$$

or

$$\dot{\Phi} = \mu_{eff}\mu_o \frac{2\pi}{H_oJ_o}[-\dot{H}_a(H_o + H_a)^2 + ((H_o + H_a)^2 + 2H_oJ_oR)^{1/2}(H_a + H_o)\dot{H}_a] \quad (A2.5)$$

(iii) For $\pi/2 \leq \omega t \leq \omega t_2$, where t_2 is defined by

$$x_3 = 0 = R - [H_{ac}^2 cos^2\omega t_2 - 2H_oH_{ac}cos\ \omega t_2]/2H_oJ_o \quad (A2.6)$$

one obtains

$$\dot{\Phi} = \mu_{eff}\mu_o 2\pi \frac{d}{dt}\{\int_{o}^{x_3}(-H_o + [H_o^2 - H_a^2 + 2H_oH_a + 2H_oJ_o(2R - x)]^{1/2})\ x\ dx \quad (A2.7)$$

$$+ \int_{x_3}^{R}(H_o - [H_o^2 + H_a^2 - 2H_oH_a - 2H_oJ_o(R - x)]^{1/2})\ x\ dx\}$$

or

$$\dot{\Phi} = \mu_{eff}\mu_o \frac{2\pi}{H_oJ_o}[-2H_o^2J_o x_3\dot{x}_3 - \quad (A2.8)$$

$$(H_o^2 - H_a^2 + 2H_oH_a + 2H_oJ_o(2R - x_3))^{1/2}(\dot{H}_a(H_o - H_a) - H_oJ_o\dot{x}_3) +$$

$$(H_o - H_a)\dot{H}_a(H_o^2 - H_a^2 + 2H_oH_a + 4H_oJ_oR)^{1/2} +$$

$$(H_o - H_a)^2\dot{H}_a + ((H_o - H_a)^2 - 2H_oJ_o(R - x_3))^{1/2}((H_a - H_o)\dot{H}_a + H_oJ_o\dot{x}_3)]$$

(iv) For $\omega t_2 \leq \omega t \leq \pi$ one obtains that $\dot{\Phi}$ is given by eq.(A2.8) with $x_3 = \dot{x}_3 = 0$

(v) For $\pi \leq \omega t \leq 2\pi$ one finds $\dot{\Phi}(t) = -\dot{\Phi}(t - \pi/\omega)$.

APPENDIX A3

Solutions of the critical state equation for the boundary condition $H_J(R) = H_a = H_{ac}cos\ \omega t + H_{dc}$, where H_{dc} is an additional dc field.

I. For $H_{dc} \leq H_{ac}$:

When $0 \leq \omega t \leq \omega t_1$ (time interval (i) in fig. 21 a) where t_1 is defined by

$$0 = -H_o + [(H_o + H_a(t_1))^2 + 2H_oJ_oR]^{1/2} \tag{A3.1}$$

one obtains :

$$H_J(x) = -H_o + [(H_o + H_a)^2 + 2H_oJ_o(R-x)]^{1/2} \quad \text{for} \quad x_1 \leq x \leq R \tag{A3.2}$$

$$H_J(x) = -H_o + [(H_o + H_{ac} + H_{dc})^2 - 2H_oJ_o(R-x)]^{1/2} \quad \text{for} \quad x_o \leq x \leq x_1 \tag{A3.3}$$

and

$$H_J(x) = 0 \quad \text{for} \quad 0 \leq x \leq x_o \tag{A3.4}$$

where

$$x_o = R - [(H_{ac} + H_{dc})^2 + 2H_o(H_{ac} + H_{dc})]/2H_oJ_o \tag{A3.5}$$

and

$$x_1 = R - [(H_{ac} + H_{dc})^2 - H_a^2 + 2H_o(H_{ac} + H_{dc} - H_a)]/4H_oJ_o \tag{A3.6}$$

When $\omega t_1 \leq \omega t \leq \pi$ (time interval (ii) in fig. 21 a) one obtains :

$$H_J(x) = H_o - [(H_o - H_a)^2 - 2H_oJ_o(R-x)]^{1/2} \quad \text{for} \quad x_3 \leq x \leq R \tag{A3.7}$$

$$H_J(x) = -H_o + [H_o^2 - H_a^2 + 2H_oH_a + 2H_oJ_o(2R-x)]^{1/2} \quad \text{for} \quad x_2 \leq x \leq x_3 \tag{A3.8}$$

$$H_J(x) = -H_o + [(H_o + H_{ac} + H_{dc})^2 - 2H_oJ_o(R-x)]^{1/2} \quad \text{for} \quad x_o \leq x \leq x_2 \tag{A3.9}$$

and

$$H_J(x) = 0 \quad \text{for} \quad 0 \leq x \leq x_o \tag{A3.10}$$

where

$$x_2 = 3R/2 - [(H_{ac} + H_{dc})^2 + H_a^2 + 2H_o(H_{ac} + H_{dc} - H_a)]/4H_oJ_o \tag{A3.11}$$

and

$$x_3 = R - H_a(H_a - 2H_o)/2H_oJ_o \tag{A3.12}$$

When $\pi \leq \omega t \leq \omega t_2$ (time interval (iii) in fig. 21 b) where t_2 is defined by

$$0 = H_o - [(H_o - H_a(t_2))^2 + 2H_oJ_oR]^{1/2} \tag{A3.13}$$

one obtains :

$$H_J(x) = H_o - [(H_a - H_o)^2 + 2H_oJ_o(R-x)]^{1/2} \quad \text{for} \quad x_4 \leq x \leq R \tag{A3.14}$$

$$H_J(x) = H_o - [(H_{dc} - H_{ac} - H_o)^2 - 2H_oJ_o(R-x)]^{1/2} \quad \text{for} \quad \overline{x}_3 \leq x \leq \overline{x}_4 \tag{A3.15}$$

$$H_J(x) = -H_o + [H_o^2 - (H_{dc} - H_{ac})^2 + 2H_o(H_{dc} - H_{ac}) + 2H_oJ_o(2R-x)]^{1/2} \tag{A3.16}$$

$$\text{for} \quad \overline{x}_2 \leq x \leq \overline{x}_3$$

$$H_J(x) = -H_o + [(H_o + H_{ac} + H_{dc})^2 - 2H_oJ_o(R-x)]^{1/2} \quad \text{for} \quad x_o \leq x \leq \overline{x}_2 \tag{A3.17}$$

and

$$H_J(x) = 0 \quad \text{for} \quad 0 \leq x \leq x_o \tag{A3.18}$$

where

$$x_4 = R - [(H_{dc} - H_{ac})^2 - 2H_o(H_{dc} - H_{ac}) - H_a^2 + 2H_oH_a]/2H_oJ_o \tag{A3.19}$$

$$\overline{x}_4 = R - (H_{dc} - H_{ac})(H_{dc} - H_{ac} - 2H_o)/2H_oJ_o \tag{A3.20}$$

$$\overline{x}_3 = R - [(H_{dc} - H_{ac})^2 - 2H_o(H_{dc} - H_{ac})]/2H_oJ_o \tag{A3.21}$$

and

$$\overline{x}_2 = 3R/2 - [H_{ac}^2 + H_{dc}^2 + 2H_oH_{ac}]/2H_oJ_o \tag{A3.22}$$

When $\omega t_2 \leq \omega t \leq \omega t_3$ (time interval (iv) in fig. 21 b), where t_3 is defined by

$$0 = -H_o + [H_o^2 + H_a^2(t_3) + 2H_oH_a(t_3) - 2H_oJ_o(R - \overline{x}_3)]^{1/2} \tag{A3.23}$$

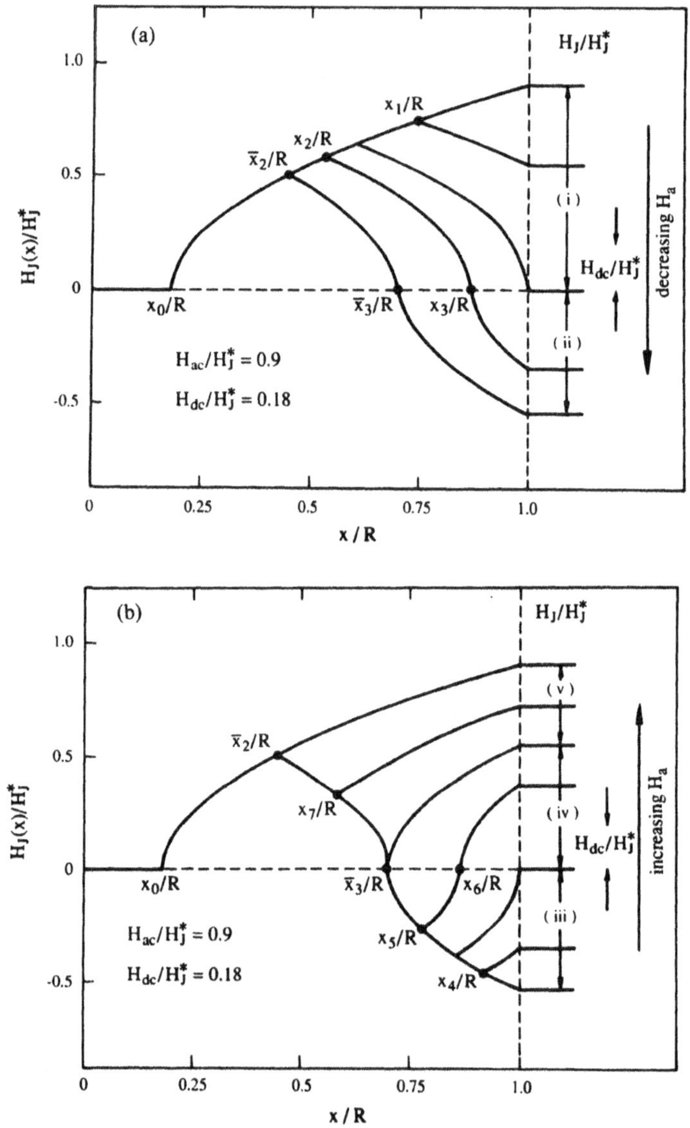

Fig. 21. (a) Critical-state field profiles inside a long cylinder of radius R during the first half of an ac cycle in the presence of an additional dc field. (b) Magnetic field profiles during the second half of an ac cycle in the presence of an additional dc field.

one obtains :

$$H_J(x) = -H_o + [(H_o + H_a)^2 - 2H_oJ_o(R-x)]^{1/2} \quad \text{for} \quad x_6 \le x \le R \tag{A3.24}$$

$$H_J(x) = H_o - [(H_o + H_a)^2 + 2H_oJ_o(2R-x)]^{1/2} \quad \text{for} \quad x_5 \le x \le x_6 \tag{A3.25}$$

$$H_J(x) = H_o - [(H_{dc} - H_{ac} - H_o)^2 - 2H_oJ_o(R-x)]^{1/2} \quad \text{for} \quad \overline{x}_3 \le x \le x_5 \tag{A3.26}$$

$$H_J(x) = -H_o + [H_o^2 - (H_{dc} - H_{ac})^2 + 2H_o(H_{dc} - H_{ac}) + 2H_oJ_o(2R-x)]^{1/2} \tag{A3.27}$$

$$\text{for} \quad \overline{x}_2 \le x \le \overline{x}_3$$

$$H_J(x) = -H_o + [(H_o + H_{ac} + H_{dc})^2 - 2H_oJ_o(R-x)]^{1/2} \quad \text{for} \quad x_o \le x \le \overline{x}_2 \tag{A3.28}$$

and

$$H_J(x) = 0 \quad \text{for} \quad 0 \le x \le x_o \tag{A3.29}$$

where

$$x_6 = R - (H_a^2 + 2H_oH_a)/2H_oJ_o \tag{A3.30}$$

and

$$x_5 = 3R/2 - [(H_{dc} - H_{ac})^2 - 2H_o(H_{dc} - H_{ac}) - H_a^2 - 2H_oH_a]/4H_oJ_o \tag{A3.31}$$

When $\omega t_3 \le \omega t \le 2\pi$ (time interval (v) in fig. 21 b) one obtains :

$$H_J(x) = -H_o + [(H_o + H_a)^2 - 2H_oJ_o(R-x)]^{1/2} \quad \text{for} \quad x_7 \le x \le R \tag{A3.32}$$

$$H_J(x) = -H_o + [H_o^2 - (H_{dc} - H_{ac})^2 + 2H_o(H_{dc} - H_{ac}) + 2H_oJ_o(2R-x)]^{1/2} \tag{A3.33}$$

$$\text{for} \quad \overline{x}_2 \le x \le x_7$$

$$H_J(x) = -H_o + [(H_o + H_{ac} + H_{dc})^2 - 2H_oJ_o(R-x)]^{1/2} \quad \text{for} \quad x_o \le x \le \overline{x}_2 \tag{A3.34}$$

and

$$H_J(x) = 0 \quad \text{for} \quad 0 \le x \le x_o \tag{A3.35}$$

where

$$x_7 = 3R/2 - [H_a^2 + 2H_oH_a + (H_{dc} - H_{ac})^2 - 2H_o(H_{dc} - H_{ac})]/4H_oJ_o \tag{A3.36}$$

II. For $H_{dc} > H_{ac}$:

When $0 \le \omega t \le \pi$ one obtains :

$$H_J(x) = -H_o + [(H_o + H_a)^2 + 2H_oJ_o(R-x)]^{1/2} \quad \text{for} \quad x_1 \le x \le R \tag{A3.37}$$

$$H_J(x) = -H_o + [(H_o + H_{ac} + H_{dc})^2 - 2H_oJ_o(R-x)]^{1/2} \quad \text{for} \quad x_o \le x \le x_1 \tag{A3.38}$$

and

$$H_J(x) = 0 \quad \text{for} \quad 0 \le x \le x_o \quad . \tag{A3.39}$$

When $\pi \le \omega t \le 2\pi$ one obtains :

$$H_J(x) = -H_o + [(H_o + H_a)^2 - 2H_oJ_o(R-x)]^{1/2} \quad \text{for} \quad x_8 \le x \le R \tag{A3.40}$$

$$H_J(x) = -H_o + [(H_{ac} - H_{dc} - H_o)^2 + 2H_oJ_o(R-x)]^{1/2} \quad \text{for} \quad \overline{x}_1 \le x \le x_8 \tag{A3.41}$$

$$H_J(x) = -H_o + [(H_o + H_{ac} + H_{dc})^2 - 2H_oJ_o(R-x)]^{1/2} \quad \text{for} \quad x_o \le x \le \overline{x}_1 \tag{A3.42}$$

and

$$H_J(x) = 0 \quad \text{for} \quad 0 \le x \le x_o \tag{A3.43}$$

where

$$x_8 = R - [H_a^2 + 2H_oH_a - (H_{ac} - H_{dc})^2 - 2H_o(H_{dc} - H_{ac})]/4H_oJ_o \tag{A3.44}$$

and

$$\overline{x}_1 = R - H_{ac}[H_{dc} + H_o]/H_oJ_o \quad . \tag{A3.45}$$

REFERENCES

1. V. V. Bryksin, A. V. Goltsev and S. N. Dorogovtsev, Theory of interacting Josephson junctions (Josephson lattices), J. Phys.: Condens. Matter 2:6789 (1990).
2. J. R. Clem, Granular and superconducting-glass properties of the high-temperature superconductors, Physica C 153–155:50 (1988).
3. C. P. Bean, Magnetization of hard superconductors, Phys. Rev. Lett. 8:250 (1962).
4. P. W. Anderson, Theory of flux creep in hard superconductors, Phys. Rev. Lett. 9:309 (1962).
5. F. Parodi and R. Vaccarone, The critical state and flux dynamics in squid arrays, Physica C 173:56 (1991).
6. C. P. Lobb, D. W. Abraham and M. Tinkham, Theoretical investigation of resistive transition data from arrays of superconducting weak links, Phys. Rev. B 27:150 (1983).
7. M. Tinkham and C. J. Lobb, Physical properties of the new superconductors, Solid State Phys. 42:91 (1989).
8. P. G. De Gennes, "Superconductivity of Metals and Alloys", W. A. Benjamin, Inc., New York (1966).
9. Y. B. Kim, C. F. Hempstead, A. R. Strnad, Critical persistent current in hard superconductors, Phys. Rev. Lett 9:306 (1962).
10. K.-H. Müller, Josephson vortices and flux penetration in high-temperature superconductors, Physica C 158:69 (1989).
11. K.-H. Müller, J. C. Macfarlane, B. W. Ricketts and R. Driver, Role of flux pinning in high-temperature superconductors, Aust. J. Phys. 42:413 (1989).
12. K.-H. Müller, J. C. Macfarlane and R. Driver, Nonlinear magnetic flux response in high-temperature superconductors, Physica C 158:366 (1989).
13. K.-H. Müller, AC losses in high-temperature superconductors, IEEE Trans. on Magnetics 27:2174 (1991).
14. Y. Xu, W. Guan and Z. Zeibig, AC losses in $RE\,Ba_2Cu_3O_{7-y}$ superconductors, Appl. Phys. Lett. 54:1699 (1989).
15. K.-H. Müller and A. J. Pauza, Intergranular ac loss in high-temperature superconductors, Physica C 161:319 (1989).
16. R. B. Goldfarb, A. F. Clark, A. I. Braginski and A. J. Panson, Evidence for two superconducting components in oxygen-annealed single phase Y-Ba-Cu-O, Cryogenics 27:475 (1987).
17. K.-H. Müller, AC susceptibility of high-temperature superconductors in a critical state model, Physica C 159:717 (1989).
18. K.-H. Müller, B. W. Ricketts, J. C. Macfarlane and R. Driver, Intergranular flux pinning in high-temperature superconductors, Physica C 162–164:1177 (1989).
19. K.-H. Müller, Frequency dependence of ac susceptibility in high-temperature superconductors: Flux creep and critical state at grain boundaries, Physica C 168:585 (1990).
20. K.-H. Müller, M. Nikolo and R. Driver, Flux pinning at grain boundaries in Bi-(Pb)-Sr-Ca-Cu-O ceramic superconductors, Phys. Rev. B 43:7976 (1991).
21. M. Nikolo, unpublished data.
22. K. H. Müller, S. J. Collocott, R. Driver and N. Savvides, AC susceptibility of granular superconductors, Supercond. Sci. Technol. 4:S325 (1991).
23. A. K. Sarkar, I. Maartense, B. Kumar and L. T. Peterson, Effect of sintering temperature on the superconducting properties of Bi-(Pb)-Sr-Ca-Cu-O ceramics, Supercond. Sci. Technol. 3:199 (1990).

EXPONENTIAL CRITICAL-STATE MODEL FIT FOR INTERGRANULAR AC SUSCEPTIBILITY OF SINTERED HIGH-T_c SUPERCONDUCTORS

Alvaro Sanchez and Du-Xing Chen
Physics Department, Universitat Autònoma Barcelona
08193 Bellaterra, Barcelona, Spain

Introduction

Ac susceptibility techniques have been widely used in the research of sintered high-T_c superconductors. In general, these are low-field techniques since the ac field produced by a small coil cannot be large. However, for many sintered high-T_c superconductors, even in such low fields the field-dependent intergranular magnetic properties can already be studied quite sufficiently. The study of these properties has been recognized as being important both practically and theoretically; they are directly related to the values and mechanisms of a very important quantity for applications, the transport critical-current density J_c.

To deduce the relationship between low-field ac susceptibility and the critical-current density, the contribution of intergranular supercurrents has to be separated from the total measured susceptibility. This is possible since the supercurrents start to penetrate into the grains at a rather high field so that their contribution to the total susceptibility is a constant at low fields. Although different terminologies have been applied in the literature, (for example, in Refs. 1-3) for the purpose of separation, this constant can be simply written as $-f_g$, where f_g is an effective grain volume fraction. Most authors considered this f_g to be less than the real volume fraction of the grains owing to the London penetration.[1,2,4,5] However, careful analysis of the experimental data led to further conclusions that this effective f_g is often larger, and beside the London penetration the grain-cluster structure plays an important role on it.[6-8] We will further describe the concept of this separation in the present work.

If we measure the complex susceptibility $\chi = \chi' - j\chi''$ as a function of ac field amplitude H_m, the result will give two χ'' peaks. The first low-field peak is due to the intergranular supercurrents, and the second high-field one is from the penetrated intragranular supercurrents. If these two peaks are well separated with a minimum very close to zero in between, then f_g can be easily determined accurately.[8] However, in many cases such a low minimum does not exist (the intergranular J_c is not very low and the sample is not thin enough) or is not available (limited by the maximum field of the susceptometer, or when only temperature dependence at some given H_m is measured). In these cases, we have to apply the critical-state model to the intergranular supercurrents with a correct $J_c(H_i)$ function. By means of this model, the low-field maximum χ'' point is fitted and f_g is determined simultaneously.[3] We will introduce a simple and rather accurate graphic technique which allows one to do this.

To obtain a correct $J_c(H_i)$ function is not easy, and this is in fact one of the main purposes of our low-field susceptibility research. The actual $J_c(H_i)$ must be a rather complicated function, different from sample to sample, but for materials research it will be very useful to have a simple common one which can be a reasonably good approximation for most samples.

Magnetic Susceptibility of Superconductors and Other Spin Systems
Edited by R.A. Hein *et al.*, Plenum Press, New York, 1991

This function was thought to be expressed by Kim's model.[3,4] Our recent study showed that the exponential model is a better one.[9,10] We now recommend this model. Some examples will be given to show the accuracy of using this model to determine f_g and J_c.

Low-field susceptibility separation

Sintered high-T_c superconducting samples consist of type-II superconducting grains which are superconductively coupled by Josephson junctions (for simplicity, we will not talk about the grain-clusters). At low fields, supercurrents only flow through the grain surface layers and the junctions among the grains. Since the surface-circulating supercurrents of the grains completely shield the grain cores the contribution of the grains to the magnetic response of the sample can be described by an effective grain volume fraction f_g; the partial susceptibility χ_g within this fraction is -1 and contributes $-f_g$ to the total susceptibility. The sample-circulating supercurrents flow through a network made of the grain surface layers and the junctions connecting them. This network is named the matrix. At very low fields, its susceptibility is in general not -1 (for example, due to damage of the sample surface during cutting). However, if it is stable and equal to $-1 + f_n$, where f_n is a positive number much smaller than 1, then we can introduce such an effective volume fraction $f_m = 1 - f_g - f_n$ for the matrix that its partial susceptibility χ_m equals -1 at very low fields. The sample surface layer will have a volume fraction f_n with a zero susceptibility. Based on this analysis, the total complex susceptibility χ of the sample can be written as

$$\chi = f_m \chi_m - f_g, \tag{1}$$

where χ_m is the partial complex susceptibility for the matrix, and the partial susceptibility for the grains and the sample surface layer have been taken to be -1 and 0, respectively, owing to the above definitions of three effective volume fractions.

Using Eq. (1) we can convert the measured χ to χ_m by

$$\chi'_m = (\chi' + f_g)/f_m, \tag{2}$$

$$\chi''_m = \chi''/f_m. \tag{3}$$

As discussed above, Eqs. (1-3) are valid for samples which have a stable low-field χ' close to -1 with $\chi'' = 0$. In our experience, the χ_m in this case can be rather well described in terms of the critical-state model. To use Eqs. (2) and (3), very accurate measurements and demagnetizing corrections are required. This is not easy. However, for many practical cases we can normalize the very-low-field χ to be -1 so that $f_n = 0$ and $f_m = 1 - f_g$. Thus, Eqs. (2) and (3) are simplified to

$$\chi'_m = (\chi' + f_g)/(1 - f_g), \tag{4}$$

$$\chi''_m = \chi''/(1 - f_g). \tag{5}$$

We will introduce our fitting technique based on the critical-state model and Eqs. (4) and (5).

It should be emphasized that although "partial" susceptibilities can be termed for the grains and the matrix when we separate χ into two contributions, the meaning of the "partial" is not the same for the two cases. f_g is the volume fraction of the grain cores defined above in a conventional meaning, that is, if there were no contribution from the matrix (all the junctions were broken) χ would be f_g times the partial susceptibility χ_g. However, if we could turn off all the grain-circulating supercurrents so removing the contribution of the grains, χ would not be $(1 - f_g)$ times the partial susceptibility χ_m but equal to χ_m itself. This is a consequence of the superconducting shielding effect. The important concept here is that although we call χ_m the partial susceptibility of the matrix, it is not for the volume where intergranular supercurrents flow, nor the volume remaining after removing the grain cores, but for the whole volume of the sample, including the grain cores and voids. Also, the local and averaged intergranular J_c values deduced from χ_m are the current densities with respect to a continuous cross section of the macroscopic volume supercurrent path.

Exponential critical-state model

The critical-state model assumes that in hard superconductors the penetrated volume supercurrents flow with a density equal to the critical-current density $J_c(H_i)$, where H_i is the local internal field. For the intergranular supercurrents discussed here, J_c decreases rather quickly with H_i. Considering the junctions to be narrow with random size and orientation distributions this kind of field dependence can be understood. For such a matrix, $J_c(H_i)$ has a finite value at zero field and often shows a power-law decay at high fields.[9,11] Its accurate expression will be complicated. However, for practical purposes such as J_c determinations, simple two-parameter models can give good approximations. We have found that the exponential model is the best one in many cases.

The advantages of the exponential model are: (1) it covers a wide field dependence ranging from J_c equal to a constant to a δ-function type dependence so that the most important characteristic quantity, the maximum χ'' due to intergranular supercurrents, can be fitted for all the samples as long as the critical-state model is applicable; (2) since its $J_c(H_i)$ is close to the actual one, it can often fit the field-dependent χ' and χ'' curves better than other two-parameter models in a wide field range; (3) analytic solutions for the critical-state loops have been derived and the susceptibility computations for different J_c parameters and sample dimensions can be made quickly.

The exponential model is written as

$$J_c(H_i) = J_0 \exp(-|H_i|/H_0), \tag{6}$$

where J_0 is the $J_c(H_i)$ value when $H_i = 0$, and H_0 is a positive field constant. The critical-state $M(H)$ curves have been calculated from this model for an infinite column of rectangular cross section $2a \times 2b (a \leq b)$. The results for $a/b = 1$ can also be used for cylinders.[12]

On the initial curve, a full-penetration field H_p is obtained to be

$$H_p = H_0 \ln(1 + p), \tag{7}$$

which is defined as the applied field at which the supercurrent penetrates to the sample center. The parameter p appearing in Eq. (7) is defined as

$$p = J_0 a / H_0. \tag{8}$$

An ideal example of partial intergranular dc $M_m(H)$ loops for a sintered 123-YBCO high-T_c superconductor, well fitted by the exponential model, is given in Fig. 1.[10]

After Fourier analysis of the hysteresis loops, χ' and χ'' as functions of a/b, p, and H_m/H_p, where H_m is the maximum field for the loop, are obtained. A characteristic point is the maximum in χ''. Figure 2 gives the calculated results of χ''_{max} and the corresponding χ' and H_m/H_p (symboled by *) for $a/b = 0$ and 1 as functions of p.

Exponential model fit for ac susceptibilities

For any model with two adjustable parameters, the model fit for the χ' and χ'' data at a given temperature can be done for one characteristic point only, which should be chosen to be

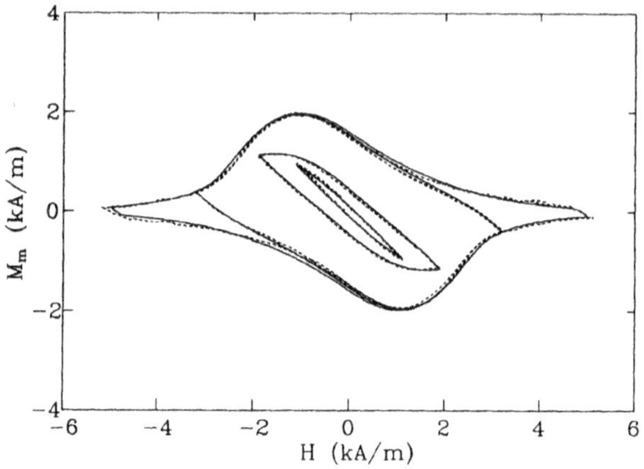

Figure 1: Partial intergranular dc $M_m(H)$ loops measured at 76 K for a sintered 123-YBCO sample. The curves give the exponential model fit.

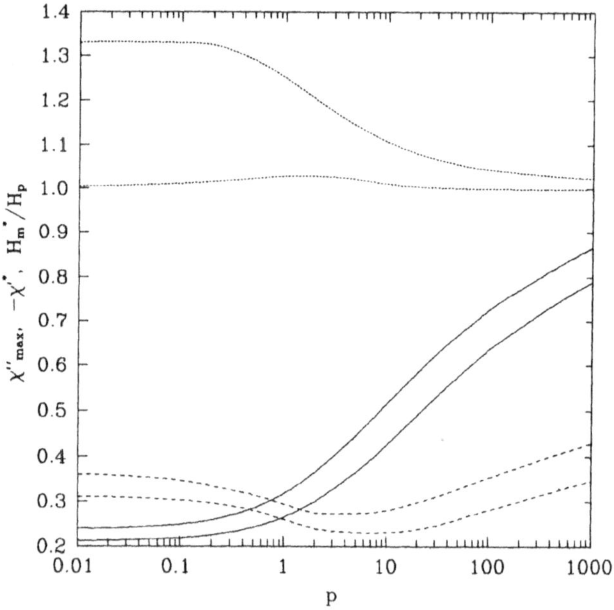

Figure 2: χ''_{max} and the corresponding χ^* and H_m^*/H_p as functions of p. Solid, dashed, and dotted lines are for χ''_{max}, χ'^*, and H_m^*/H_p, respectively. In each pair of curves, the upper one is for $a/b = 0$, and the lower one for $a/b = 1$.

the point where χ'' takes its maximum. Two model parameters are obtained by this one-point fit. With the parameters determined, the χ' and χ'' vs. H_m curves can be calculated using the same model. If the calculated curves are in agreement with the measured data then we confirm the model to be good, giving a correct $J_c(H_i)$ within the experimental error. If there is some disparity between them, then the fit is erroneous, although the point is well fitted. Therefore, it is very important to choose a proper model to make this fit. As explained above, the exponential model can be a good choice.

The exponential model fit can be done graphically just by using Fig. 2 and performing a simple iterative process. We give two examples below to introduce the fitting procedures. The samples are 123-YBCO superconductors studied in Ref. 8. The sample dimensions and the experimentally determined quantities are given in Table 1.

Understanding the quantities in Fig. 2 as for the matrix and denoting the ith iteration result by i in parentheses, the fitting procedures are written as follows:

(1) Starting with $\chi_m'^{*(0)} = -0.25$ (the starting value is arbitrary; this value is optimum for quick iteration for most samples), substituting it and the experimental χ'^{*} into Eq. (4), $f_g^{(1)}$ is obtained. (2) Substituting $f_g^{(1)}$ and the experimental χ''_{max} into Eq. (5), $\chi''^{(1)}_{m,max}$ is obtained. (3) With $\chi''^{(1)}_{m,max}$ and a/b, making a linear interpolation between the curves for $a/b = 0$ and 1, $p^{(1)}$ and $\chi_m'^{*(1)}$ are obtained using Fig. 2 again by linear interpolation on a/b. (4) Repeating procedure (1) but using $\chi_m'^{*(1)}$, $f_g^{(2)}$ is obtained. (5) Repeating procedure (2) but using $f_g^{(2)}$, $\chi''^{(2)}_{m,max}$ is obtained. (6) Repeating procedure (3) but using $\chi''^{(2)}_{m,max}$, $p^{(2)}$ and $\chi_m'^{*(2)}$ is obtained. (7) $H_m^*/H_p^{(2)}$ is obtained from $p^{(2)}$ using Fig. 2 by linear interpolation on a/b. For most samples such two-step iteration can already give a final result.

Table 2 lists the iteration parameters for the two samples. The final results are compared in Table 3 with the further refined results by direct exponential model calculations. Figs. 3a and 3b give the experimental χ'' and χ' vs. H_m data and the exponential model fit based on the f_g given in Table 1.

From Table 3, comparing the graphic technique and the direct computation, we see that using the former we can get f_g very close to the direct calculating results with a difference less than 0.3%. Another parameter that can be determined rather accurately (with error less than 1.2%) by this technique is H_p. In contrast, p, thus H_0 and J_0, are less accurate with a maximum error about 10%. However, all the differences between both results are only from the reading error and the linear interpolation error in the graphic measurements. It is easy to reduce both errors by making a larger figure and putting more curves between $a/b = 0$ and 1.

We can also see in Table 3 that compared with the accurate results of f_g, the direct computation using the exponential model for the one point gives a value 0.009 smaller for sample 1 but 0.003 greater for sample 2. When the one-point fitting approach used in this work gives f_g very close to the accurate one, all the other fitting parameters obtained by different techniques are also very close (sample 2).

Kim's model does not work well for these two samples, as described in Ref. 8. The same graphic approach using Kim's model gives f_g always less than the accurate value with an error up to -0.02. When p is large and $a/b \approx 1$, the H_p determined by Kim's model has an error up to $+7\%$.

J_c calculation from fitting parameters

One of the main purposes of the critical-state model fit for ac susceptibility data is

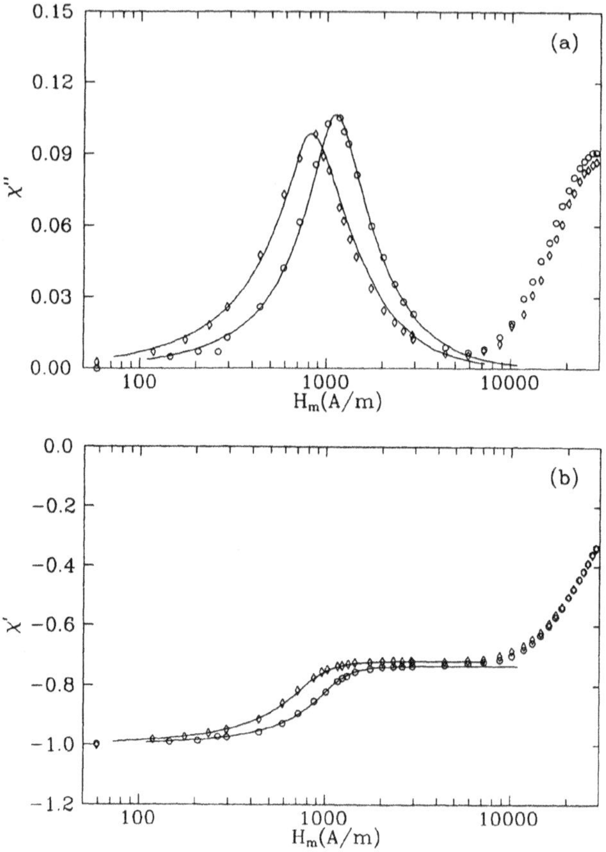

Figure 3: Exponential model fit of χ'' (a) and χ' (b) vs. H_m curves for samples 1 (◇) and 2 (○).

Table 1: Sample dimensions, the measured χ''_{max}-point parameters, and the accurately determined f_g.[8]

Sample	$2a$(mm)	$2b$(mm)	χ''_{max}	χ'^{*}	H^*_m(A/m)	f_g
1	1.29	2.60	0.0983	−0.782	830	0.718
2	2.50	2.60	0.105	−0.797	1100	0.733

Table 2: Iteration results.

Sample	$f_g^{(1)}$	$\chi''^{(1)}_{m,max}$	$\chi'^{*(1)}_m$	$f_g^{(2)}$	$\chi''^{(2)}_{m,max}$	$\chi'^{*(2)}_m$	$p^{(2)}$	$H^*_m/H_p^{(2)}$
1	0.709	0.338	−0.255	0.707	0.335	−0.255	2.1	1.12
2	0.729	0.387	−0.23	0.736	0.398	−0.23	7	1.03

Table 3: Comparison among the graphic results (first), the computer refined results (second), and the computed results using accurate f_g (third).

Sample	f_g	p	H_p(A/m)	H_0(A/m)	J_0(MA/m^2)
1	0.707/0.709/0.718	2.1/2.3/2.7	741/732/725	655/613/554	2.13/2.19/2.24
2	0.736/0.736/0.733	7/7.2/7.3	1,068/1,078/1,090	514/512/515	2.88/2.95/3.01

to obtain the intergranular J_c. Since the $J_c(H_i)$ function type has been fixed to be the exponential one in this work, the final $J_c(H_i)$ is readily obtained after the one-point fit by substituting the fitting values of J_0 and H_0 in Table 3 into Eq. (6). If the agreement between the calculated curves and the χ'' and χ' vs. H_m data is very good, as are the results for sample 2 given in Figs. 3a and 3b, then the $J_c(H_i)$ thus obtained is the actual one with some error owing to the limited resolution and accuracy of the measurements. If the agreement is not very good, then other types of $J_c(H_i)$ functions have to be tested. For an arbitrary $J_c(H_i)$, numeric critical-state model calculations have to be done, which will be introduced elsewhere.

In contrast with the local $J_c(H_i)$, a volume averaged J_c can be obtained from H_p rather accurately. This average J_c is calculated by

$$\langle J_c \rangle = H_p/a. \tag{9}$$

If $a/b << 1$, this $\langle J_c \rangle$ should be the same as the transport $\langle J_c \rangle$ measured at zero applied field; larger a/b should give a $\langle J_c \rangle$ which is smaller than the transport $\langle J_c \rangle$.[3] For the samples shown above we have $a/b > 0.49$, so that the $\langle J_c \rangle = 1130$ and 860 kA/m^2 calculated using Eq. (9) should be smaller than the transport $\langle J_c \rangle$.

The error in the $\langle J_c \rangle$ calculated from Eq. (9) is due only to a and H_p. H_p is determined by the experimental $H_m(\chi''_{max})$ and the H_m/H_p at the χ''_{max} point calculated from the critical-state model. This H_m/H_p is not equal to 1 in general. Without the critical-state model calculation for orthorhombic sample geometry, a ratio of 1 is often assumed in the literature; this will introduce an error up to about 30% in the $\langle J_c \rangle$ determination. Even with orthorhombic-geometry calculations and using the technique described in this work, some error can still exist owing to the erroneous $J_c(H_i)$ function. Using Kim's model, for example, this error can be up to about +10% if the actual J_c decreases with H_i more quickly than H_i^{-1}. However, such error is reduced to within 3% if the exponential model is used. In most cases, using the exponential model gives an error less than 2%, as can be seen in this work. All the conclusions made here are based on our latest work on the theoretical critical-state susceptibility spectra calculation.[13]

When different models yield quite different values of f_g and H_m^*/H_p, one must be concerned about the model choice for accurate $\langle J_c \rangle$ determination by ac susceptibility measurements. A judgement for the validity of the model can be made by comparing the f_g determined by the one-point model fit, on one hand, and by the most elaborate technique, described in Ref. 8, on the other hand; if one is close to the other then the model is valid. When the exponential model is used, successful results are often found for thick samples of sintered high-T_c superconductors. Consequently, the $\langle J_c \rangle$ determined by using this model is also accurate. For thin samples, a good agreement between both f_g's is not met very often. To estimate the error in $\langle J_c \rangle$ in this case, more caution has to be taken.

Conclusion

To fit the low-field intergranular ac susceptibility data of sintered high-T_c superconductors by the critical-state model, a proper choice of the $J_c(H_i)$ function is important. For most practical cases, the exponential model is satisfactory. The model fit to the experimental data should be done for one point where the low-field χ'' takes maximum. A simple graphic fitting technique has been introduced for this purpose. The effective grain volume fraction determined by this technique is usually accurate within 0.01 if the exponential model is used. Consequently, the $\langle J_c \rangle$ determination by this model is often more accurate than other models. In most cases, an error of 2% in $\langle J_c \rangle$ can be expected.

Acknowledgments

We thank R. W. Cross for his help with the hysteresis loop measurements, and T. L. Francavilla for comments. One of us (DXC) thanks the Ministerio Español de Educación y Ciencia for supporting his visit.

References

[1] J. R. Clem, Physica C **153-155**, 50 (1988).
[2] H. Dersch and G. Blatter, Phys. Rev. B **38**, 11391 (1988).
[3] D.-X. Chen, J. Nogues, and K. V. Rao, Cryogenics **29**, 800 (1989).
[4] K.-H. Muller, Physica C **159**, 717 (1989).
[5] D.-X. Chen, A. Sanchez, J. Nogues, and J. S. Muñoz, Phys. Rev. B **41**, 9510 (1990).
[6] D.-X. Chen, Yu Mei, and H. L. Luo, Physica C **167**, 317 (1990).
[7] D.-X. Chen, J. S. Muñoz, T. Puig, A. Sanchez, M. T. Aurell, L. M. Martinez, F. Lopez-Aguilar, and J. Costa-Quintana, in *Transport properties of superconductors*, R. Nicolsky ed. (World Scientific, Singapore, 1990), p. 198.
[8] D.-X. Chen, A. Sanchez, T. Puig, L. M. Martinez, and J. S. Muñoz, Physica C **168**, 652 (1990).
[9] A. Sanchez, D.-X. Chen, J. S. Muñoz, and Y.-Z. Li, Physica C **175**, 33 (1991).
[10] D.-X. Chen, R. W. Cross, and A. Sanchez, Phys. Rev. B, in press.
[11] R. J. Peterson and J. W. Ekin, Physica C **157**, 325 (1989).
[12] D.-X. Chen, A. Sanchez, and J. S. Muñoz, J. Appl. Phys. **67**, 3430 (1990).
[13] D.-X. Chen and A. Sanchez, J. Appl. Phys., in press.

PHENOMENOLOGICAL MODEL FIT TO INTRAGRANULAR AC SUSCEPTIBILITY OF SINTERED HIGH-T_c SUPERCONDUCTORS

Du-Xing Chen and Alvaro Sanchez
Physics Department, Universitat Autònoma Barcelona
08193 Bellaterra, Barcelona, Spain

Introduction

For sintered high-T_c superconductors the field dependence of the complex ac susceptibility $\chi(=\chi'-j\chi'')$ shows two χ'' peaks. The first low-field peak with maximum χ''_{max1} is due to sample-circulating intergranular supercurrents, while the second high-field peak with χ''_{max2} is due to intragranular supercurrents. For the intergranular susceptibility, the critical-state model works quite well. We have described how to fit the low-field χ by using an exponential critical-state model.[1] To analyse the high-field χ other factors, in addition to the critical currents, have to be taken into consideration. Recently, a simple model for type-II superconductors involving the volume supercurrents, the equilibrium magnetization, and the surface barrier for flux entry and exit has been developed,[2] providing a tool to analyse the high-field hysteresis loops of sintered high-T_c superconductors. In the present paper we introduce a technique to analyze the intragranular ac susceptibility of these materials based on a similar model.

Model

In Fig. 1, we show some dc hysteresis loops of a sintered 123-YBCO sample measured at 76 K with a vibrating sample magnetometer. The four cycles were recorded continuously in order of decreasing H_m, the maximum field. Since it was not easy to precisely control the applied field produced by the electromagnet, the given loops are not very symmetric. However, several features can be clearly observed from these data: (1) The loop with the largest H_m has a tilted Z shape with a narrow waist. (2) Both sides of the middle part of the loops are almost parallel, as long as the H_m is not very small. (3) Most sections of the loops with larger H_m collapse except for the regions where the reverse curves start, which are roughly parallel to the middle part with some rounding. (4) There is a little expansion in the central region. (5) The slope of the loop for the smallest H_m is less than the slope of the middle part of the loops with larger H_m.

Feature (4) is due to the intergranular supercurrents; their contribution to the susceptibility at higher H_m can be neglected. In feature (5), the slope of the loop at small H_m should be close to $-f_g$, the effective volume fraction of the grains[1] determined by low-field χ measurements; and the higher slope in the middle part of the high-H_m loops suggests that there is another grain volume fraction g that is smaller than f_g. f_g may be attributed to grain clusters and g may be ascribed to the individual grains.[2] From features (1-3) we can simplify the high-H_m loop to be composed of a number of linear sections, as shown in Fig. 2. The middle part and the initial reverse sections of the loop have a slope $-\tan\alpha = -g$; the top

Magnetic Susceptibility of Superconductors and Other Spin Systems
Edited by R.A. Hein *et al.*, Plenum Press, New York, 1991

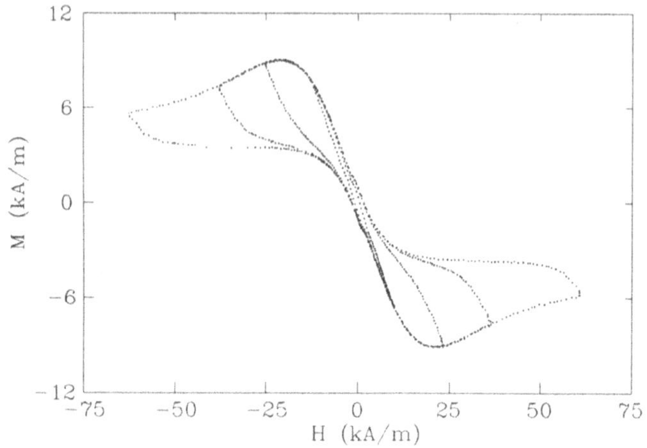

Figure 1: Experimental dc hysteresis loops for a 123-YBCO high-T_c superconductor.

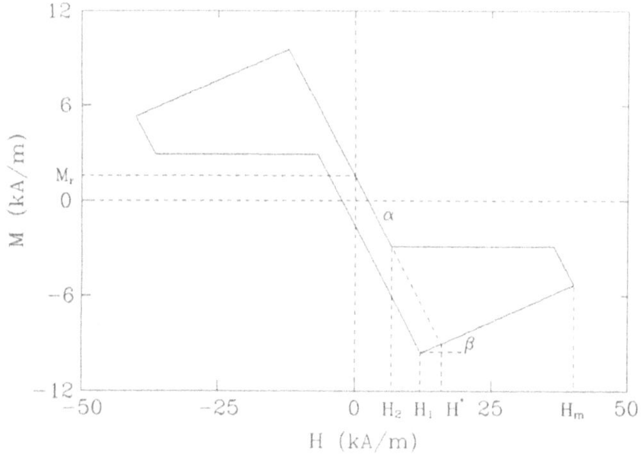

Figure 2: Modeling loop. The loop parameters are those in Table 2.

and the bottom sections have a slope $\tan\beta = f$. Three characteristic fields H_1, H_2, and H^* and the remanence M_r are also defined in the figure. According to Ref. 2, H_1 and H_2 can be defined in terms of physical parameters

$$H_{c1g} = (H_1 + H_2)/2, \tag{1}$$

$$\Delta H_0 = (H_1 - H_2)/2, \tag{2}$$

where H_{c1g} is an effective lower critical field for the grains, and ΔH_0 is the extra field due to the surface barrier, assumed to be the same for flux entry and exit, when the surface flux density $B \approx 0$.

This modeling is assumed to be valid for $H_m > H^*$. For a sinusoidal ac field $H(t) = H_m \cos \omega t$, χ'' is defined by

$$\chi'' = \oint H \mathrm{d}M/\pi H_m^2, \tag{3}$$

where the integration is equal to the area of the loop. Using this definition and the loop given in Fig. 2, we obtain

$$\chi''(H_m) = AH_m^{-2} + BH_m^{-1} + C \tag{4}$$

The coefficients A, B, and C are functions of the parameters defined in Fig. 2:

$$A = \{(f+g)(H^{*2} - H_1^2) + (1 + f/g)fH^{*2} - 2(1 + f/g)[(f+g)H_1 - gH_2 + 2M_r]H^*\}/\pi, \tag{5}$$

$$B = 2(1 + f/g)[(f+g)H_1 - gH_2 + 2M_r]/\pi, \tag{6}$$

$$C = -(1 + f/g)f/\pi, \tag{7}$$

where

$$H^* = H_1 + 2M_r/(f+g). \tag{8}$$

We will show below that Eq. (4) is appropriate for fitting some data on the field dependence of the ac susceptibility of sintered high-T_c superconductors for H_m around and higher than $H_m(\chi''_{max2})$. For lower fields, the loops are more complicated. We have noticed that $g < f_g$, which suggests that flux starts to penetrate into the grains (or grain clusters) at a field lower than H_1. In other words, the second χ'' rise starts from 0 at a field $H_1' < H_1$. The details of how the loop changes are not easily modeled, but we can make a reasonable approximation for χ''. Similar to the Rayleigh law for the low-field susceptibility of ferromagnets or hard superconductors,[3,4] we assume, for H_m larger than and close to H_1', the area of the loop to be

$$\oint H \mathrm{d}M = D\pi(H_m - H_1')^2 H_m \tag{9}$$

where D is a constant. Substituting Eq. (9) into Eq. (3) we obtain

$$\chi''(H_m) = D(H_m - H_1')^2 H_m^{-1} = D(H_m - 2H_1' + H_1'^2 H_m^{-1}) \tag{10}$$

This equation is used to fit the low-field side of the second χ'' peak.

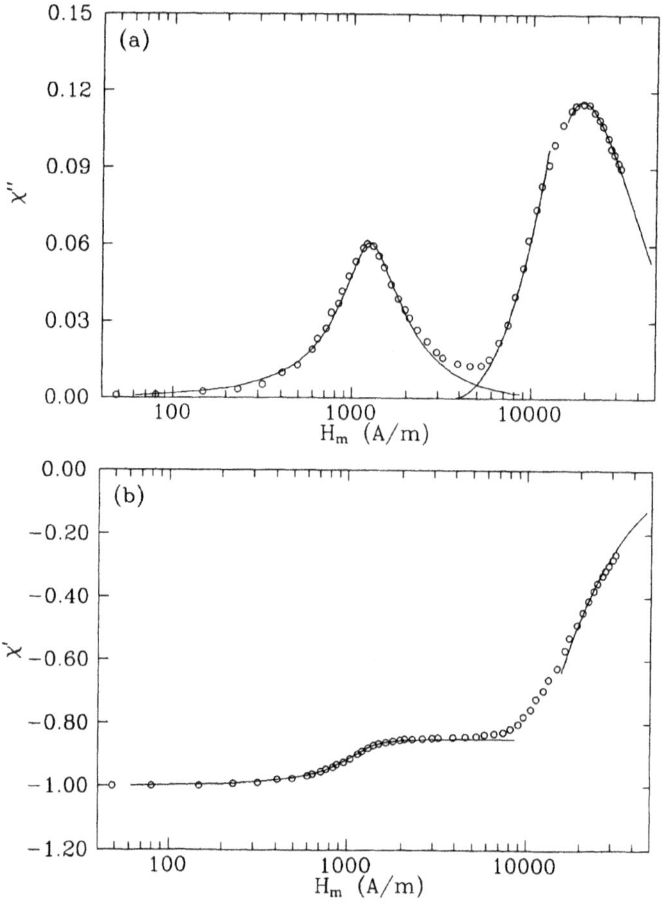

Figure 3: Fitting result.

Table 1: Experimental data and fitting parameters for the second χ'' peak.

χ''_{max2}	χ'	H_m(A/m)	A(MA^2/m^2)	B(A/m)	C	D(μm/A)	H'_1(A/m)
0.115	−0.487	19,000	−63.52	6,686	0.06	21	3,700

Table 2: The loop parameters obtained from high-field susceptibility fit.

H_1(A/m)	H_2(A/m)	M_r(A/m)	g	f
12137	6780	1566	0.66	0.153

Procedures for fitting susceptibility

For testing the power of our model, we theoretically fit some data given in Ref. 5. We converted the original permeability data to χ' and χ'' which are normalized so that $\chi' = -1$ at low fields. Figs. 3a and 3b show the experimental χ'' and χ', respectively, as functions of H_m for a sintered 123-YBCO sample studied in Ref. 5. The sample was cylindrical with radius $a = 1.5$ mm, containing grains with an estimated average radius of $a_g = 10\mu$m.

The χ'' vs. H_m curve for $H_m \geq H_m(\chi''_{max2})$ is fitted using Eq. (4). The fitting parameters A, B, and C are given in Table 1. The rising part of the second χ'' peak is fitted by Eq. (10), resulting in the D and H'_1, listed in Table 1. The fitting parameters A, B, and C together with the value of χ' corresponding to χ''_{max2} are further used to be fitted by the loop parameters using Eqs. (5), (6), and (7). The results have been found to be unique; they are listed in Table 2. Two sectional fitting curves for the second χ'' peak are drawn in Fig. 3a. A χ' vs. H_m curve in the high-H_m region obtained by Fourier analysis of the fitting loops is given in Fig. 3b.

The low-field χ''_{max1} and its corresponding χ' and H_m are fitted using the exponential critical-state model (Ref. 1). The fitting parameters are listed in Table 3. The susceptibility curves calculated from these parameters are given in Figs. 3a and 3b.

We see in Fig. 3a that all three theoretical sections of the χ'' curves are rather well in coincidence with the data points. Adding the inter- and intragranular χ'' together, our final result will reproduce almost perfectly the experimental data. The low field χ' curve also agrees well with the experimental data. For the high-field fit, however, a crossover can be seen. This is obviously due to our over-simplified modeling loop in Fig. 2. If the rounding sections of the real loops shown in Fig. 1 are taken into consideration, this crossover should disappear.

Thus, to fit the field dependence of the ac susceptibility of a sintered high-T_c superconductor over a wide field range, three regions have to be considered separately. In the low-field range for the intergranular susceptibility superimposed on a constant intragranular contribution, the exponential critical-state model is used (see Ref. 1). In the high-field range for the intragranular susceptibility around and above the χ'' maximum, Eq. (4) is used. For field values in the middle, i.e., for the rising part of the second χ'' peak, Eq. (10) is used.

Table 3: Exponential model fitting parameters for the low-field susceptibility data.

χ''_{max1}	χ'	H_m(A/m)	f_g	p	H_p	J_0(MA/m^2)	H_0(A/m)
0.0604	−0.890	1,212	0.85	8	1,230	2.99	560

Deduction of superconducting properties

Using the data given in Table 2 and Eqs. (1) and (2), the calculated values of H_{c1g} and ΔH_0 are 9,460 A/m and 2,680 A/m, respectively. All the published data on H_{c1} of single crystals of high-T_c superconductors are actually defined somewhat as H_1. However, they are much smaller than the value given here. The reason may be that the superconducting quality of the grains is much better than that of large single crystals.[2]

To determine the minimum H_{c1} of the grains from ac susceptibility data, the conventional way is to determine an H_m where the second χ'' starts. This field is actually H_1'. Our model gives a unambigous method to determine this point.

Many authors have used $H_m(\chi''_{max2})$ to stand for the full penetration field H_{pg} of the grains and estimate the average J_c of the grains using Bean's model:[4,6]

$$\langle J_{cg}\rangle = H_m(\chi''_{max2})/a_g \tag{11}$$

Using the above data we obtain $\langle J_{cg}\rangle = 1{,}900\ \mathrm{MA/m^2}$. However, this value is overestimated since the equilibrium magnetization and the surface barrier have not been considered. We have tested that if letting $M_r = 0$ (this quantity is due to J_c) and keeping the other loop parameters listed in Table 2 the same, the $H_m(\chi''_{max2})$ is 20,000 A/m, 5% *higher* than the experimental result. Actually, the higher the intragranular J_c the lower the $H_m(\chi''_{max2})$, provided this field to be on the same order of H_{c1g} and ΔH_0.

Therefore, with our results, a more correct way to calculate J_c is[6]

$$\langle J_{cg}\rangle = 3M_r/a_g g \tag{12}$$

This gives $\langle J_{cg}\rangle = 710\ \mathrm{MA/m^2}$. In Eq. (12), dividing by g changes the loop into the partial loop for the grains.[7]

The parameter f is also important. It characterizes the field dependence of J_{cg} and ΔH; they are constant if $f = 0$, and their field dependence becomes stronger when f is increasing. These two quantities are not separated in our model. However, the field dependence of J_c is not expected to be very strong in the grains, so that f will mainly indicate the field dependence of ΔH.

We notice that g is smaller than f_g in this sample. This is typical for many sintered high-T_c superconductors.[2] From the comparison between g and f_g we can get some information on the grain performance, grain orientations, and superconducting coupling among the grains.

Conclusion

The high-field hysteresis loop of sintered high-T_c superconductors at not too low temperatures are modeled to have a tilted Z shape with linear sections. From this loop the function $\chi''(H_m)$ has a simple form with an H_m^{-2}-, an H_m^{-1}-, and a constant term. This function is used to fit χ'' vs. H_m data in a field region around and above the second χ'' maximum. For the rising part of the second χ'' peak, a modified Rayleigh-law formula with an H_m-, an H_m^{-1}-, and a constant term is used to fit data. Since the equilibrium magnetization and surface barrier effects have been considered, the superconducting properties deduced from such a model fit for high-field ac susceptibility data are quite different from the commonly used approaches. An important new finding is that the H_m corresponding to the high-field χ'' maximum, which has been commonly related to the intragranular J_c and used for J_c determination in the literature, has little to do with this J_c.

Acknowledgments

We thank R. W. Cross for his help with the hysteresis loop measurements, and R. B. Goldfarb and T. L. Francavilla for comments. One of us (DXC) thanks the Ministerio Español de Educación y Ciencia for supporting his visit to the Universitat Autònoma de Barcelona.

References

1 A. Sanchez and D.-X. Chen, in this volume.
2 D.-X. Chen, R. W. Cross, and A. Sanchez, Phys. Rev. B (1991), in press.
3 S. Chikazumi, *Physics of Magnetism* (John Wiley & Sons, Inc., 1964) p. 296.
4 C. P. Bean, Rev. Mod. Phys. **36**, 31 (1964).
5 Q. H. Lam, Y. Kim, and C. D. Jeffries, Phys. Rev. B **42**, 4846 (1990).
6 D.-X. Chen and R. B. Goldfarb, J. Appl. Phys. **66**, 2510 (1989).
7 D.-X. Chen, A. Sanchez, T. Puig, L. M. Martinez, and J. S. Muñoz, Physica C **168**, 652 (1990).

CRITICAL CURRENT DENSITIES FROM ac SUSCEPTIBILITY DATA

Željko Marohnić and Emil Babić

Institute of Physics and Department of Physics
The University of Zagreb, POBox 304
41000 Zagreb, Croatia, Yugoslavia

ABSTRACT

A method for the accurate determination of bulk critical currents (J_s) of high temperature superconductors (HTS) via high resolution ac susceptibility measurements is described. It is based on the critical-state model, but its results, unlike those obtained from other methods (e.g. magnetization hysteresis or voltage versus transport current, i.e. V-I curves), are free from the effects of trapped field or self-field, thus it yields the true $J_s(H)$ and $J_s(T)$ variations. This method is employed for the determination of the J_s vs H variation at 78K in a $YBa_2Cu_3O_{7-x}$ ceramic sample. Furthermore, the applicability of the critical-state model to the investigation of J_s in sintered HTS samples has been verified over a wide range of applied magnetic fields (H<4,400A/m). Interesting effects occuring as a consequence of the partial cancellation of dc and ac magnetic fields acting on a sample have been observed. The importance of this method for the investigation of the strength and nature of the intergrain coupling in ceramic HTS samples is briefly disscused.

1. INTRODUCTION

It is well known that the transport critical current density (J_s) of sintered $YBa_2Cu_3O_{7-x}$ (thereafter YBaCuO) samples is some orders of magnitude lower than that of the corresponding monocrystals or epitaxial thin films[1]. Such poor performance of ceramic high temperature superconductors (HTS) is a result of weak intergrain coupling throughout the material. The

Magnetic Susceptibility of Superconductors and Other Spin Systems
Edited by R.A. Hein *et al*, Plenum Press, New York, 1992

quality of intergrain coupling depends on numerous factors associated with sample preparation (such as the purity of the components, the details of synthesis and sintering, the metal and oxygen stoichiometry, the grain size, size distribution and eventual texturing, etc). In spite of a large effort, the origin of a weak intergrain coupling is not well understood. Therefore the accurate determination of the macroscopic (bulk) critical currents in polycrystalline HTS samples is of considerable scientific (the understanding of the coupling mechanism) and technological (the achivement of higher J_s) interest.

In practice either transport measurements (V-I curve[2-4]) or magnetic (magnetization and ac susceptibility[5-7]) measurements are employed for the determination of bulk critical currents in sintered HTS samples. The direct determination of J_s (i.e. V-I curve) is somewhat uncertain due to the problems related to measurements of low voltages on short samples, the self-heating effects at higher currents, etc. In addition, because of the inhomogeneity inherent to ceramic materials, J_s obtained from V-I curves reflects only the onset of dissipation (probably at the weakest part of the sample) and thus cannot be correlated with the bulk properties (for instance, resistivity[6-8]) of the sample. Because of this J_s is often determined from the width of the magnetization hysteresis curve (irreversible magnetization) by the use of the Bean critical-state model[9]. This model is also used for the determination of J_s from ac susceptibility measurements. Whereas for YBaCuO thin films[10] and monocrystals (or grains) the validity of the Bean model for the determination of J_s has been established, to our knowledge no such proof for the intergranular J_s exists.

However, even assuming that all the technical and conceptual problems mentioned above are solved, the V-I curves and magnetization hysteresis loops cannot yield actual $J_s(H=0,T)$ value because both measurements are performed in a finite magnetic field (self-field and trapped field in the case of the transport and magnetization measurement respectively). Because of the sensitivity of J_s on the magnetic field in ceramic HTS samples, reduced J_s values[11] and as importantly the incorrect $J_s(H)$ and $J_s(T)$ dependences[8] have been reported.

In this paper we describe a method based on high resolution ac susceptibility measurements capable of giving correct $J_s(H=0,T)$ values for ceramic samples and thus suitable for the determination of true (intrinsic) $J_s(H)$ and $J_s(T)$ dependences. In addition, we present evidence that, over a wide range of applied fields, the critical-state model but with a field dependent J_s is applicable to ceramic HTS samples.

2. EXPERIMENTAL

The YBaCuO pellet was prepared in the usual way[12]. The average grain size of the ceramic was about 10μm. A plate-like sample with the dimensions $12x2.6x0.8mm^3$ was cut from the central part of the pellet. The surfaces of the sample were slightly polished. Transport measurements (performed immediately before ac susceptibility investigation) revealed T_{co}=90K, resistivity at 95K = 1mΩcm, the resistance ratio R_{295}/R_{95}=2.2 and transport critical current density (78K, H=0) of about $45A/cm^2$.

In order to achieve the highest possible resolution and sensitivity in our measurements a standard induction method for the measurements of ac susceptibility[13] was somewhat modified. A schematic block diagram of the full arrangements is shown in Fig.1. The secondary pick-up coils (connected in series opposition) were very carefully balanced so that the voltage signal for empty coils was much lower than that of the sample, and hence can be accurately substracted from the measured signal (V). In order to achieve the highest possible sensitivity in the measurements at low fields (≤60A/m) an additional pick-up coil was used for the compensation of the real part of the signal (V^{Re}). A two-phase lock-in amplifier (SR-510), connected to a personal computer, was employed for the detection, storage and elaboration of the in-phase (V^{Re}) and out-of-phase (V^{Im}) components of the signal. In the measurements of V^{Re} a precision ratio transformer (with the resolution 1 part in 10^5) was employed in order to "zero" the measured signal at each applied ac magnetic field. For the lowest fields the averaging over up to 200 data points (performed by computer) was used in order to eliminate the influence of the background noise. The effects of ambient laboratory fields (both static and ac) were reduced several orders of magnitude by the use of a magnetic shield, made out from Metglas[R] 2714 tape. The above modifications and procedures permitted measurements of the voltage signal with a resolution of 1 part in 10^4. Such resolution enables the accurate determination of the variation of V^{Re} with the amplitude of the ac field at very low field values (≤60A/m). Such resolution is vital for the determination of critical current density, J_s, with the method described later on in this paper.

All measurements were performed at a frequency of 28.5Hz with ac fields up to 4,000A/m. At low fields the signal from the pick-up coils was first amplified with a low-noise transformer (1:100); whereas at higher fields it was fed directly into a lock-in amplifier. (This change did not affect the dependence of signal (V) on the field amplitude.) The superposed dc magnetic field (0-4,400A/m) was controled by means of the precision (Zenner) current source (0-100mA) connected in parallel with the ac current amplifier. At our, relatively low operating frequency neither amplitude nor phase of the ac field was affected when the dc current source was employed. The ac current through the primary coil (thus magnetic field) was measured by means of the potential difference

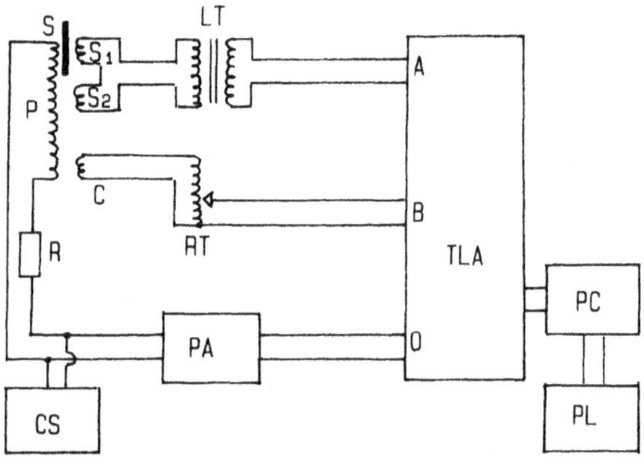

Fig.1. Block scheme of the arrangement: S-sample, P-primary, S_1, S_2- secondaries, C-compensation coil, R-standard resistor, CS-dc current source, PA-power amplifier, RT-ratio transformer, LT-low noise transformer, TLA-two phase lock-in amplifier, A,B-inputs, O-reference output, PC-computer and PL-plotter.

developed across a 1Ω standard resistor. The sample was mounted on a saphire sample holder which ensured both, a good thermal contact of the sample and no contribution of the sample holder to the measured signal. The set-up was calibrated by the use of a pure Nb sample as reported earlier[14]. Measurements reported in this paper were performed at 78K (liquid nitrogen). This temperature can be kept constant for prolonged periods of time and at the same time ensured bulk superconductivity of the sample.

3. DESCRIPTION OF METHOD AND RESULTS

The temperature dependence of the critical current density, ($J_s(T)$), in polycrystalline HTS samples is often determined via the temperature dependence of the intergrain maximum in the imaginary part, $\chi''(T)$, of the ac susceptibility[5,7]. This method is based on the Bean model[9] which predicts that the maximum in $\chi''(T)$ occurs when the magnetic field penetrates to the center of the sample. In particular, for a long cylinder (radius R) when the applied magnetic field (H_A) reaches the value of full penetration (H_P) the critical current density is determined by $H_P=RJ_s$.

Although simple, for practical realization, this method does not give reliable results for $J_s(T)$ of sintered HTS samples. First, at lower temperatures the intergrain peak in χ'' is broadened and reduced so that the actual position of its maximum is more difficult to determine. Secondly at lower temperatures a larger amplitude of the magnetic field is required in order to reach a maximum in χ'', hence J_s is measured in progressively larger magnetic fields on lowering the temperature. Since the bulk critical current of ceramic HTS samples depends rather strongly on the magnetic field[11,8] this method yields lower value of $J_s(T)$ than that which would be obtained in true zero-field. Furthermore, measured J_s depends on the cross section of the sample. As mentioned in the Introduction, a similar problem also appears in the determinatiom of $J_s(T)$ from the magnetization hysteresis loop or from V-I curves. The consequence of these facts is that in all above methods the measured (effective) critical current density depends on the geometry of the sample, in particular on thickness and radius for the plates and cylinders respectively[11].

By using the results of ac susceptibility measurements in very low ac fields superposed on dc fields, one can avoid the above problems and obtain the correct $J_s(H)$ dependence. For a thin plate (with thickness a) and low magnetic fields ($H_A<H_P$) the Bean model predicts (see Appendix) the following behaviours for the magnitudes of the real (χ') and imaginary (χ'') parts of the ac susceptibility:

$$\chi' = -1 + (H_A/2H_P)$$
$$\chi'' = (4/3\pi)(H_A/2H_P) \qquad (1)$$

Here $H_P=aJ_s(0)/2$ and the thin plate geometry is selected because of the simpler behaviour of the magnetization compared to that of a cylinder.

Accordingly all our measurements were performed on a sample in a shape approximating a thin plate, for which the above expressions are applicable.

When using the expressions (1) one should take into account the following limitations:

i) If J_s is independent of the magnetic field, equations (1) hold for all amplitudes of magnetic field up to $H_A=H_P$. However for ceramic HTS one should use the lowest possible amplitudes of the magnetic field (those at which data of sufficient accuracy can still be obtained) in order to ensure that for the selected H_A the dependence of J_s on H ($H<H_A$) can be neglected.

ii) The Bean model is developed for homogenous superconductor, hence in sintered samples one has to take into account their granularity. In particular, when the transport or macroscopic shielding currents are investigated the applied magnetic field should be small enough so as not to significantly affect the magnetization of the grains (i.e. $H_A<H_{c1g}$ where H_{c1g} is the lower critical field of the grains[5] in that particular sample).

For the ceramic (granular) material magnetic susceptibility can be roughly subdivided into two parts:

$$\chi' = (1 - \alpha)\chi'_g + \alpha\chi'_s \tag{2}$$

where α is the volume fraction of the intergranular material and χ'_g and χ'_s refer to grains and intergranular material respectively. We note that the properties of the material between the grains (weak links) determine $J_s(H,T)$.

If the condition in ii) is fulfilled one may approximate the state of the grains with perfect diamagnetism ($\chi'_g=-1$). Then the expression (2) becomes

$$\chi' = -(1 - \alpha) + \alpha\chi'_s \tag{3}$$

The dependence of χ' on the amplitude of the applied ac magnetic field for the sample at 78K is shown in Fig.2. After a strong initial variation, χ' tends to saturate at intermediate magnetic fields and then again decreases at larger magnetic fields ($H_A\geq H_{c1g}$). The above diagram (Fig.2) defines the range of low fields ($H_A\leq 1{,}000$A/m) for the given sample. Accepting as reasonable assumptions that $\chi'(H_A\to 0)=-1$ (i.e. complete diamagnetic shielding as verified by the comparison with a pure

Nb sample at 4.2K[14]) and that in sufficiently large fields $\chi_s=0$, one can determine α. From the data in Fig.2 we have found $\alpha=0.23$ for that particular sample. This value was obtained by the extrapolation of the solid line in Fig. 2 to $H_A=0$ i.e. we took into account that similar to the initial penetration of the magnetic field into the intergranular space, small magnetic fields penetrate into the grains. However the magnetic field penetration depth into the grains for small fields at 78K is, in addition to the difference in geometrical factor, smaller than that into

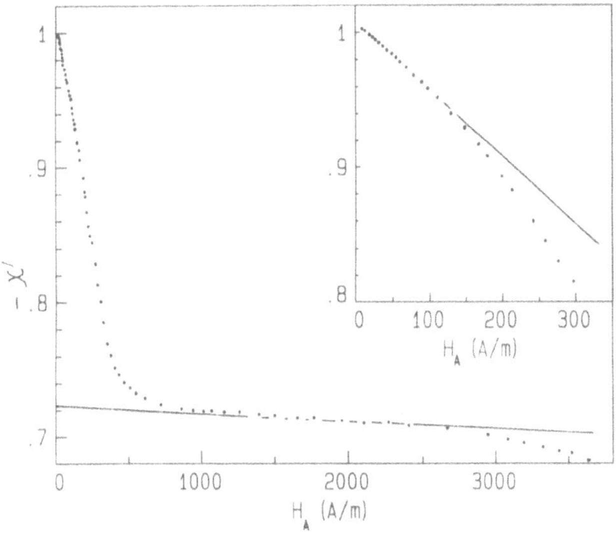

Fig.2. Variation of a real part of ac susceptibility (χ') with the amplitude of ac field for YBaCuO sample at 78K. The inset: the same but for low fields only.

the intergranular space by a factor of the order of ratio of the inter- and intragrain critical currents (Fig.2). Because of this, for the sample shown in Fig.2 at 78K, one can safely assume that the background contribution from the grains does not change with field. Accordingly, we have not taken into account the slope in the region of saturation (Fig.2) which would, in any event, have no significant effect on the agreement between the experimental results and calculated ones. Similarly, a small error appears in our determination of α, because we neglected completely χ'_s for fields of about 1,000A/m (i.e. we assumed that the magnetic fields of this magnitude destroy completely the intergrain couplings). This is apparently not exact because the transport measurements indicate some residual J_s even at much larger fields[2]. Although at such large fields the residual J_s is some orders of magnitude lower than that in zero-applied field, its effect on α can be investigated in future experiments.

We note that the value of α is central for all quantitative analyses performed later on in this paper and hence the approximations involved in its estimation warrant the discussion given above. In particular, α is used for the appropriate normalizations of the measured real and imaginary signals (V^{Re} and V^{Im} respectively) which are required in order to apply correctly the Bean model to a given case. Hence all results in the following figures are normalized in this manner. This procedure although of utmost importance both for the analysis of ac susceptibility and magnetization[15] is usually ignored. We also note that all factors which affect the intergrain couplings (such as the purity of components, the details of synthesis and sintering, the grain sizes and distributions, etc.) also affect α. Hence α has to be determined in the way depicted in Fig.2 for every sample and at each temperature for a given sample[15].

Assuming the validity of the Bean model in a given case one can apply the expressions (1) to the normalized ac susceptibility data. The dependences of χ' and χ'' on H_{ac} for the lowest magnetic fields are shown in Fig.3. It can be seen that the experimental results obey Eq. (1). The slopes of the straight lines in Fig.3 yield (according to the Eqs. (1) and (3)) $aJ_s(0)\cong 672$A/m and 675A/m from the data for χ' and χ'', respectively. The deviations of the data from the straight lines occuring already for $H_A \geq 60$A/m indicate a very strong dependence of J_s on H. However, practically identical values of $aJ_s(0)$ obtained from χ' and χ'' allow the use of only one of these quantities for the future determination of the variations of J_s with H. The experimental convenience makes χ' an obvious choice (the signal is considerably larger and much less sensitive to the eventual small drift of the phase during the measurements).

The determination of $J_s(H)$ is performed in the same way as of $J_s(0)$ (Fig.3) but with χ' data obtained with a dc magnetic field of preselected magnitude (H_D) applied to the sample. The results of these measurements are shown in Fig.4. $aJ_s(H_D)$ is determined from the slope of a linear part of $\chi'(H_A,H_D)$ vs. H_D variation. The variation of J_s with H_D (obtained by inserting the thickness of the sample a = 0.8mm into $aJ_s(H_D)$ is shown in Fig.5. Rather low $J_s(0)\cong 85$A/cm^2 at 78K as well as a very strong dependence of J_s on H_D clearly show that the sample used for testing this method for the determination of actual $J_s(H)$ and $J_s(T)$ dependences had rather weak intergrain coupling.

The above results for $J_s(H_D)$ are obtained from the analysis of low

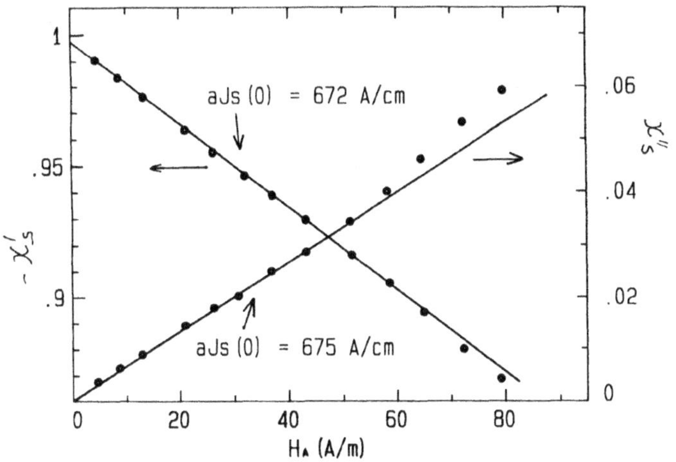

Fig.3. Variations of χ' and χ'' with the amplitude of the ac field for zero dc field. Straight lines are fit through low field data.

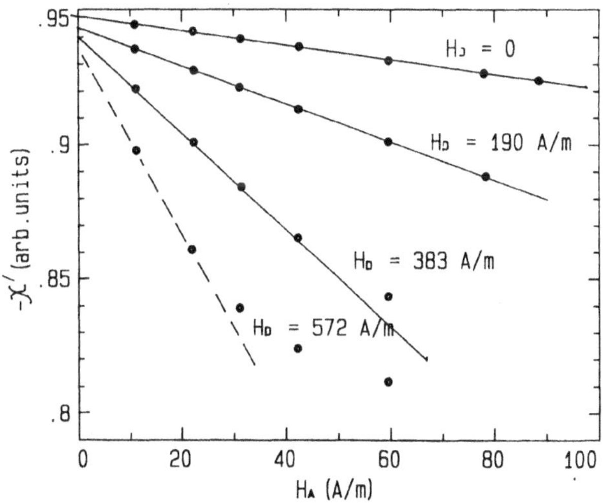

Fig.4. Variation of χ' with the amplitude of the ac field, with dc field as a parameter.

field ($H_A \rightarrow 0$) χ' and χ'' data under the assumption that the critical-state model is applicable to sintered YBaCuO samples. Although the low field data support this hypothesis (validity of Eq.(1)), a more detailed verification of the applicability of the critical-state model to sintered HTS can only be made by extending the analysis to higher fields. Over a wider range of the magnetic field, the variations of V^{Re} and V^{Im} with H_A are not linear (as illustrated in Fig.2 for H_D=0) and non-monotonic

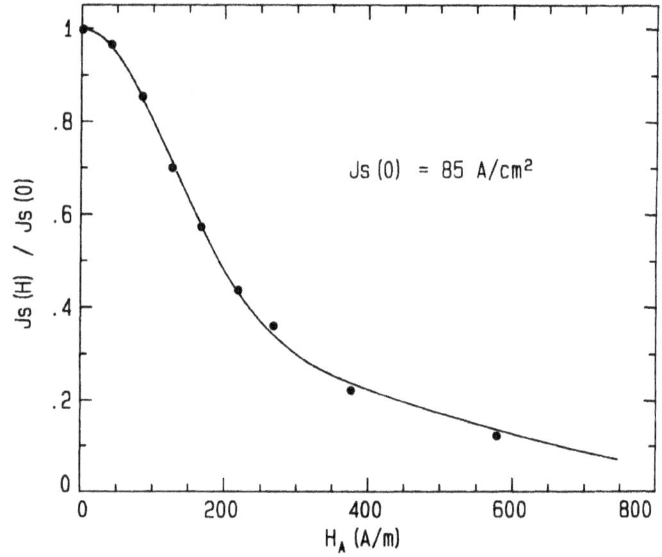

Fig.5. Variation of critical current density (J_s) with dc field: ( ) experimental data, (-) fitting curve.

functions of H_D. Because of this we have compared the results of our measurements with those calculated from the critical-state model by using the measured J_s(H) variation (Fig.5) which was obtained from the low H_A-data only. Since in these calculations no adjustable parameters were used, rather good agreement between the measured and calculated values of χ' (Fig.6) and χ'' (Fig.7) seems to confirm the applicability of the critical-state model to the analysis of the magnetic properties of these materials.

The differences between the results of calculations and the experimental data (more pronounced in Fig.6) stem mainly from the difference between the actual shape of the sample and that assumed in our calculations (a parallelepiped instead of an infinite plate, and the neglect of the demagnetizing factor of the actual sample). These effects can in principle be taken into account, but the calculations in that case would be considerably more involved. We deemed such a refinement unneccessary at this stage. We note that the calculations shown by the full line in Figs. 6 and 7 were performed numerically by means of a personal computer. A more detailed account of these calculations is given in the Appendix.

Rather complex behaviour of $\chi''(H_A,H_D)$ shown in Fig.7 is very interesting. In particular at larger values of H_D two maxima of $\chi''(H_A)$ appear, whereas a single maximum occurs for $H_D=0$. For J_s independent of magnetic field, the Bean model predicts a single maximum in $\chi''(H_A)$ (i.e. the largest hysteresis loss) for $H_A=(4/3)H_P$ ($H_A=H_P$ for a cylinder) and χ'' independent on H_D. The theoretical dependence of $\chi''(H_A)$ for $J_s=85A/cm^2$ (assumed to be independent of the applied field) is shown by full line in Fig.8. For this calculation we used the expression (1) for $H_A<H_P$ and

$$\chi''(H_A) = (2/\pi)\ (1 - (2H_P/H_A) + (1/3)(2H_p/H_A)^2) \qquad (4)$$

for $H_A>H_P$. For the sake of comparison, experimental results for $H_D=0$ (dots) are also shown in Fig.8. We note a considerable deviation of this calculation from the experimental results in the region of larger fields which is a consequence of the decrease of J_s with the increasing amplitude of magnetic field. When a dc field is present the first maximum (Fig. 7) is shifted to lower ac fields (because H_P decreases with decreasing J_s). The appearence of the second maximum at even higher fields seems unusual since it reflects an increase in J_s. We note however that this maximum occurs around $H_A \cong H_D$ and in this case the total external field (H_o) on the sample for $\omega t \cong \pi$ is zero, which causes an increase of J_s (hence loss) during a short time in which $H_o=H_D-H_A \cong 0$. A similar behaviour is also observed for $\chi'(H_A)$ at higher fields (Fig.6). In this case weak maxima occur when $H_o \cong 0$. We note that such behaviours would not be expected within the framework of the Bean model if the dependence of J_s on H was weak for the range of magnetic field used in this study. Therefore, strong dependence of J_s on H associated with the interplay of H_A and H_D is responsible for the unusual behaviours of the ac susceptibility in ceramic HTS samples.

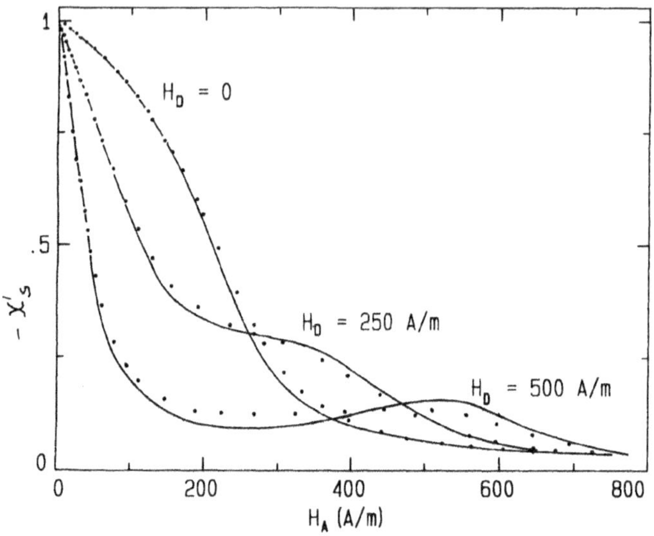

Fig.6. Variation of χ' in large ac and dc magnetic fields:( ) measured, (-) calculated (see Appendix).

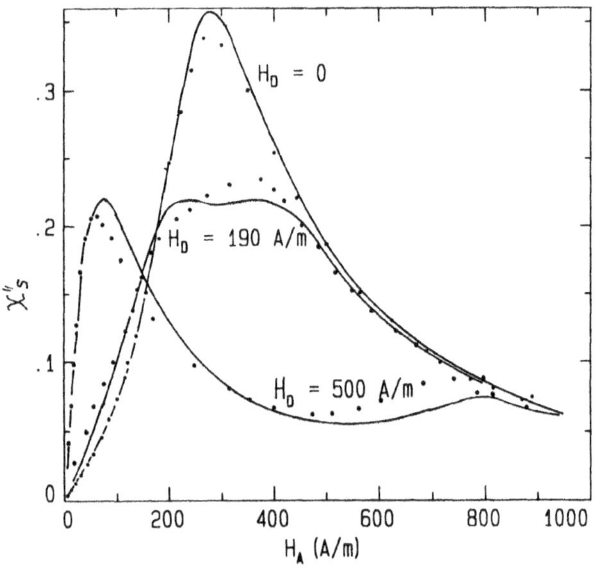

Fig.7. Variation of χ'' in large ac and dc fields: ( ) measured, (-) calculated (see Appendix).

Apparently the method described above is also applicable to single crystals and thin films but care has to be taken with regard to anisotropy and the demagnetizing factor. (Of course in that case the calculations would be different from those given in the Appendix.) For single crystals the possibility of using larger fields (J_s is much less dependent on $H^{15,16}$) may partially cancel the difficulty associated with the small mass of the sample. Our method has the advantage with respect to transport property measurements (V-I curve) in that heating problems as well as those associated with any changes in the structure of the sample during

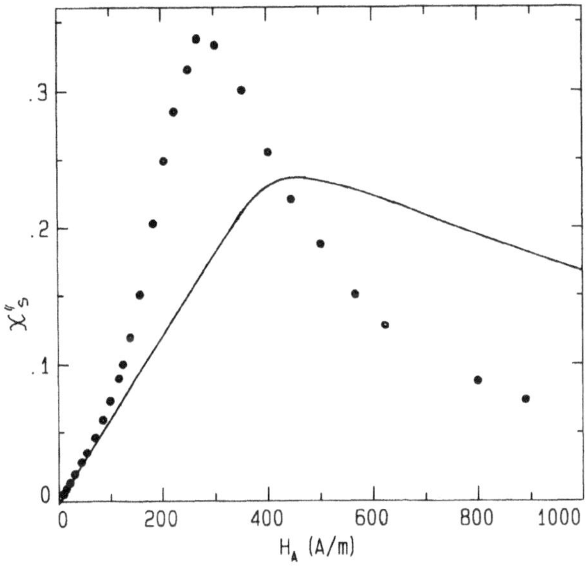

Fig.8. Variation of χ'' with the amplitude of the ac field in zero dc field ( ). Full line: calculation for $J_s(H)=J_s(0)$.

the contact preparation are avoided. Furthermore this method yields true $J_s(0)$ which does not depend on criteria employed in the determination of J_s as is the case in transport measurements[17]. Finally we note that the effectiveness of the method for polycrystalline material relies on the macroscopic homogeneity of the sample. This means in particular, that the average grain size of ceramic sample should be much smaller than its thickness. The homogeneity of the sample can of course be verified by reducing its dimensions.

4. CONCLUSION

A method for accurate determination of bulk critical currents (J_s) of sintered HTS samples from the result of high-resolution ac susceptibility is developed. The application of this method to one sintered YBaCuO sample (in a form of plate) yielded the J_s vs H variation at 78K which is free from spurious magnetic field effects (the trapped field and self-field in the case of magnetization and transport measurements respectively). Since the method is based on the critical state model, the applicability of that model to the studied samples (hence ceramic HTS samples in general) has been verified over a wide range of applied magnetic fields. Taking care of the anisotropy and the demagnetizing factor the method (but not the present calculations) may also be applied to single crystal and thin film HTS samples.

The availability of a method which yields correct $J_s(H)$ and $J_s(T)$ variations is of considerable interest for the understanding of the nature of intergrain coupling in sintered HTS samples. Because of an extreme sensitivity of J_s on H, effects of the self-field (V-I curves) and trapped fields (magnetization hysteresis) drastically modify[8] the magnitude of $J_s(0)$, as well as $J_s(H)$ and $J_s(T)$ variations deduced from transport and magnetization measurements[2-5]. Hence any intercomparison of the values of $J_s(0)$ obtained by these techniques is ambigous. Moreover the values of $J_s(0)$ obtained on the same sample and with the same technique change on changing its crossection[11]. As a consequence the comparison of magnitudes and variations of J_s with H and T obtained by the above techniques with the specific model predictions[18,19,20] can hardly help the understanding of the nature of intergrain coupling (thus J_s limitations) in ceramic HTS samples. In contrast to that, our method is free of these difficulties provided the sample is macroscopically homogeneous and that its shape remains self-similar upon changes in cross section. Therefore this method may have considerable scientific, and probably tehnological relevance. Indeed recent applications of this technique to the investigation of the nature of intergrain junctions in low-J_s ceramic YBaCuO sample[21] as well as the study of the sample size and grain size dependence of J_s in specially prepared ceramic samples of the same compound[22] seem to support that view. Finally we note that knowledge of the true $J_s(H,T)$ combined with the validity of the critical-state model enables the prediction of J_s obtained from transport and magnetization measurements on the same sample[23]. In particular it enables one to predict whether magnetization or transport measurements will give higher J_s values at a given H and what fraction of $J_s(0,T)$ it would be.

APPENDIX

According to the critical-state model, a change in the external magnetic field induces shielding currents of density equal to J_s which decreases the magnetic field inside the sample. The relation between the critical current density, $J_s(H)$, and the internal magnetic field is:

$$\vec{\nabla}\times\vec{H} = \vec{J}_s(H) \qquad (1)$$

In the simplest case of thin plate this relation becomes:

$$dH/dx = \pm J_s(H) \qquad (2)$$

where H is in z-direction and x is along the thickness of the plate. The magnetization of a thin plate with thickness a in an external magnetic field H_0 is:

$$M(H_0) = (2/a)\int_0^{a/2} H(x)dx - H_0 \qquad (3)$$

If $J_s(H)$ is known (for instance Fig.5), one can use the numerical integration of Eq.(2) in order to obtain the field profile within the sample, H(x), at a given external field amplitude H_0.

In the ac techniques of measurements (see Experimental) the quantity detected is the derivative of magnetization with time dM/dt (not the magnetization itself). A long thin sample with the crossection area S, situated in the compensated secondary coil with N turns, will cause the induced voltage:

$$V(t) = -\mu_0 NS(dM/dt) \qquad (4)$$

If the magnetic field applied to the sample has a general form $H_0=H_D+H_A\cos\,\omega t$, Eq.(4) can be written as:

$$V(t) = H_A\omega\mu_0 NS(dM/dH_0)\sin\,\omega t \qquad (5)$$

where we have set $dM/dt = (dM/dH_0)(dH_0/dt) = -\omega H_A \sin\,\omega t\,(dM/dH_0)$.
From Eq.(5) it is clear that any analysis of such measurements requires

the knowledge of time dependence of dM/dH_0. Because of this, let us consider the magnetic field profile within a superconducting sample in a form of long thin plate with thickness a. In the upper parts of Figs. 9a and 9b two characteristic field profiles for an external field of the form $H_0=H_D+H_A\cos\phi$ and ϕ in the range $0\leq\phi=\omega t\leq\pi$ are shown. (There, as in the case of Eq.(2), H_0 is along the sample and x is across the thickness.). In Figs. 9a and 9b, because of symmetry, only the right half of the sample is shown. It can be seen that, in general, the field profile consists of two

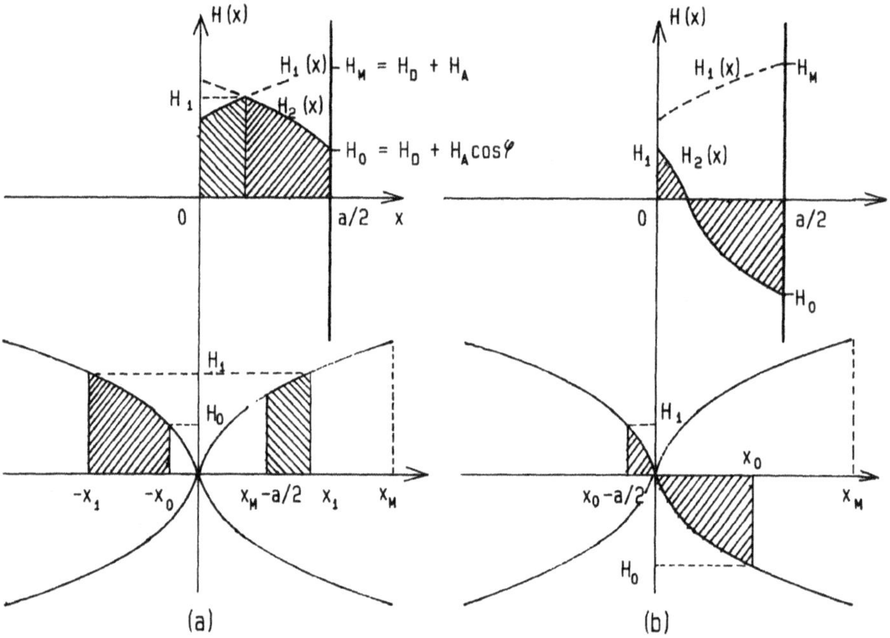

Fig.9. Examples of field profiles in thin superconducting plate. For details see Appendix.

curves ($H_1(x)$ and $H_2(x)$) which are, according to the critical-state model, parts of the universal function H(x) (shown in the lower parts of Figs. 9a and 9b) obtained by integration of Eq.(2) using the experimentally deduced $J_s(H)$ shown in Fig. 5. The shadowed areas, both in the upper and lower parts of Fig.9, are equal to the value of integral in Eq.(3). The integration limits are different in the two cases, but the value of integral is the same. The change of integration limits is associated with a translation of curves $H_1(x)$ and $H_2(x)$ (see Fig.9) performed in order to facilitate future calculations of dM/dH_0. Accordingly, the integral in Eq.(3) can be written as:

$$-(2/a)\int_{-x_1}^{-x_0} H(x)dx + (2/a)\int_{x_M-a/2}^{x_1} H(x)dx \qquad \text{(from Fig.9a)}$$

$$-(2/a)\int_{-x_0-a/2}^{-x_0} H(x)dx \qquad \text{(from Fig.9b)}$$

The integration limits are functions of characteristic fields (Fig.9): $x_0=x(H_0)$, $x_M=x(H_D+H_A)$ and $x_1=x(H_1)$, where H_1 is defined by $H_1(x)=H_2(x)$. Taking into account that $H(0)=0$ and $H(-x)=-H(x)$, it is seen from Fig.9a that $x_1=(x_M+x_0)/2$. By using the identity $dM/dH_0=(dM/dx_0)/J_s(H_0)$ it is simple to calculate the derivative of magnetization with external field H_0. The expression which includes both cases illustrated in Figs. 9a and 9b, is given by

$$dM/dH_0 = \frac{2}{aJ_s(H_0)}(H(x_1)-H(x_0)) - 1 \qquad (6)$$

Here x_1 is defined in the following way:

$$x_1 = (x_M+x_0)/2, \qquad \text{for } x_0 \geq x_M - a$$
$$(7)$$
$$x_1 = x_0 + a/2, \qquad \text{for } x_0 < x_M - a$$

In essentially the same way one obtains the expression for dM/dH_0 in the case of $\pi \leq \phi \leq 2\pi$:

$$dM/dH_0 = -\frac{2}{aJ_s(H_0)}(H(x_2)-H(x_0)) - 1 \qquad (8)$$

where x_2 is defined as

$$x_2 = (x_m+x_0)/2, \qquad \text{for } x_0 \leq x_m + a$$
$$(9)$$
$$x_2 = x_0 - a/2, \qquad \text{for } x_0 > x_m + a$$

and $x_m=x(H_D-H_A)$.

Having found suitable expressions for dM/dH_0 (Eqs. (6) and (8)), which, according to Eq.(5), determine the induced voltage V(t) at the secondary coils, we have to examine how the particular detecting instrument processes this voltage. As mentioned in the experimental section we have been using a two-phase lock-in amplifier SR510. This instrument, for a periodic signal V(t) with frequency ω at the input, gives at the output the RMS value of the first term (proportional to $\sin\omega t$ and $\cos\omega t$) in the Fourier series:

$$V' = \frac{1}{\sqrt{2}\,\pi} \int_0^{2\pi} V(t)\sin\phi\, d\phi$$

$$V'' = \frac{1}{\sqrt{2}\,\pi} \int_0^{2\pi} V(t)\cos\phi\, d\phi \qquad (10)$$

where $\phi=\omega t$.

By inserting the results from Eqs.(6)-(9) into Eq.(5) and transforming the integrals in Eq.(10) to ones with integration limits of 0 and π, one can describe the output readings from the lock-in amplifier in terms of the results of our calculations:

$$\frac{V'}{C} = -1 + \frac{2}{\pi a}\int_0^{\pi} \frac{H(x_1)-H(x_2)}{J_s(H_0)} \sin^2\phi\, d\phi \qquad (11)$$

$$\frac{V''}{C} = \frac{2}{\pi a}\int_0^{\pi} \frac{2H(x_0)-H(x_1)-H(x_2)}{J_s(H_0)} \sin\phi\cos\phi\, d\phi \qquad (12)$$

where $C=(1/\sqrt{2})NSH_A\omega$. The quantities V'/C and V''/C are usually called the real and imaginary part of susceptibility (χ' and χ'', respectively).

As will be shown below, the integrals in Eqs.(11) and (12) can be easily analytically solved if J_s is independent of the magnetic field. However, in the more general case of a field dependent J_s, numerical integration will, in general, be required.

For the results of our calculations presented in Figs. 6 and 7 we used the pre-calculated and computer stored functions H(x) and the inverse one x(H). These functions have been obtained by the use of an analytical expression for $J_s(H)$ which in turn was deduced by fitting the experimental $J_s(H_D)$ data obtained from the measurements in low ac field (H_A) to a suitable function (Fig.5). In this way, by using the simple formulae for numerical integration we have been able, with little expense of computer time, to calculate the predicted values of the measured quantities (V', V'') for arbitrary combinations of H_A and H_D.

In the case of a low ac field ($H_A \ll H_p$) but for arbitrarily large dc field (for ceramic samples $H_D < H_{c1g}$) one can safely assume $J_s(H) \cong J_s(H_D)$.

Because of this, $H(x)$ is a linear function of x ($H(x) \cong J_s(H_D)x$) and the following relations hold:

$$H(x_1) = H_D + H_A(\cos\phi+1)/2 \qquad (13)$$
$$H(x_2) = H_D + H_A(\cos\phi-1)/2 \qquad (14)$$

Accordingly the integrals in Eqs.(11) and (12) take on the simple forms:

$$\chi' = -1 + \frac{H_A}{\pi H_p}\int_0^{\pi} \sin^2\phi \, d\phi = -1 + \frac{H_A}{2H_p} \qquad (15)$$

$$\chi'' = \frac{H_A}{\pi H_p}\int_0^{\pi} \cos^2\phi \sin\phi \, d\phi = \frac{4}{3\pi}\frac{H_A}{2H_p} \qquad (16)$$

where $H_p = aJ_s(H_D)/2$. Eqs.(15) and (16) provide a proof of Eq.(1) presented in the main body of the text.

Assuming that $J_s(H)$ remains constant even at somewhat larger ac fields ($H_A \cong H_p$) one can predict the magnitude of H_A at which the maximum of χ'' occurs. In the above case, Eq.(16) remains valid up to $H_A \cong H_p$. For still larger fields, $H_A > H_p$, according to Eqs.(7) and (9), χ'' remains to be determined by the integral in Eq.(16) but with the upper limit of integration being determined by $\cos\phi_1 = 1 - 2H_p/H_A$ which is equivalent to $x_o = x_M - a$ or $x_o = x_m + a$.

In this way one obtains:

$$\chi'' = \frac{2}{\pi}\left(1 - 2H_p/H_A + (1/3)(2H_p/H_A)^2\right) \qquad (17)$$

By using the criterion for a maximum of the function χ'' (Eq.(17)) one finds:

$$H_A = (4/3)H_p \qquad (18)$$

As specified above this result can be used only in the special cases where $J_s(H)$ can be regarded as independent on the magnetic field. In particular this means that Eq.(18) is acceptable only if $H_D \gg H_A$ i.e. when $J_s(H_D \pm H_A) \cong J_s(H_D)$. However Eq.(18) can be applied even for $H_D=0$, but only for very thin ceramic HTS samples. In that case H_p is a very low field so that one can assume $J_s(H_p) \cong J_s(0)$ and therefore use the maximum of χ'' for determination of $J_s(0)$ as outlined in the text.

ACKNOWLEDGEMENT

We thank to Drs.Jovica Ivkov, Josip Horvat and Đuro Drobac for useful discussions.

REFERENCES

1. P.Chaudhary, R.H.Koch, R.B.Laibowitz, R.R.Mcguire and R.J.Gambino, Phys.Rev.Lett. 58:2684 (1987)
2. J.W.Ekin, A.I.Braginski, A.J.Panson, M.A.Janocko, D.W.Capone, M.J. Zaluzec, B.Flandermeyer, O.F.de Lima, M.Hong, J.Kwo and M.S.Liou, J.Appl.Phys. 62:4821 (1988)
3. J.E.Ewetts and B.A.Glowacki, Cryogenics 28:641 (1988)
4. R.Meisels, S.Bungre and A.D.Caplin, J.Less Comm.Met. 151:83 (1989)
5. E.Babić, Đ.Drobac, J.Horvat, Ž. Marohnić and M.Prester, J.Less Comm. Met. 151:89 (1989)
6. F.Küpfer, I.Apfelstedt, W.Shauer, T.Wolf and H.Wühl, Physica C 153-155:367 (1988)
7. F.Gomory and P.Lobotka, Sol.State Commun. 66:645 (1988)
8. E.Babić, M.Prester, Ž.Marohnić, T.Car, N.Biškup and S.A.Saddiqui, Sol. State Commun. 72:753 (1989)
9. C.P.Bean, Rev.Mod.Phys. 36:31 (1964)
10. A.Kapitulnik B.Oh, M.Naito, K.Char, A.D.Kent, N.Missert, E.Hellman, S.Arnason, J.W.P.Hsu, M.R.Hahn, P.Rosenthal, R.Barton, M.R.Beasley, T.H.Geballe and R.H.Hammond, Int.J.Mod.Phys. B 1:779 (1987)
11. R.B.Stephens, Cryogenics 29:399 (1989)
12. E.Babić, Ž.Marohnić, M.Prester and N.Brničević, Phil.Mag.Lett. 56:91 (1987)
13. Đ.Drobac and Ž.Marohnić, in: "Rapidly quenched metals", S.Steeb and H.Warlimont, eds., Elsevier Science Publ.B.V., Amsterdam, (1985) 1133.
14. E.Babić, Ž.Marohnić, Đ.Drobac, M.Prester and M.Brničević, Int.J. Mod. Phys. B1 :973 (1987).
15. V.Calzona, M.R.Cimberle, C.Ferdeghini, M.Putti and A.S.Sirri, Physica C, 157:425 (1989).
16. E.Babić, Ž.Marohnić, Đ.Drobac, M.Prester and N.Brničević, Physica C 153-155:1511 (1988).
17. J.W.Ekin, Appl.Phys.Lett. 55:905 (1989).
18. J.R.Clem, Physica C 153-155:50 (1988) and Refs. therein.
19. R.L.Peterson and J.W.Ekin, Phys.Rev.B 37:9848 (1988).

20. T.Y.Hsiang and D.K.Finnemore, Phys.Rev.B 22:154 (1980)and Refs.therein
21. E.Babić, M.Prester, Đ.Drobac, Ž.Marohnić and N.Biškup, Phys.Rev.B 43:1162 (1991).
22. E.Babić, M.Prester, Đ.Drobac, Ž.Marohnić, P.Nozar, P.Stastny, F.C.Matacotta and S.Bernik, submitted to Phys.Rev.B (1991)
23. Ž.Marohnić, unpublished (1991).

RESPONSES OF HIGH T_c SUPERCONDUCTORS TO VARIOUS COMBINATIONS OF AC AND DC MAGNETIC FIELDS

Fedor Gömöry

Inst. of Electrical Engineering, Slovak Academy of Sciences

842 39 Bratislava, Czech and Slovak Federal Republic

1. Introduction

In the study of classical as well as high T_c superconductors, magnetic measurements are often used.[1-4] In these experiments, the response of the sample to an applied magnetic field with the magnitude constant or changing in time is observed. In the former case one speaks about a DC measurement while in the later about an AC measurement, respectively. Sometimes the combination of AC and DC fields can be very helpful, too. Because it is common practice to use different terminology in reporting results of such experiments, we will define the basic terms used in what follows. First of all, to avoid problems caused only by the irregular shape of the sample, we consider a slim cylinder with a height much greater than its radius R, placed in a homogeneous magnetic field B_{ext} parallel to the axis. In this case the field acting on the sample is equal to B_{ext}. However, one should not neglect the fact that the magnetization inside the superconducting sample is in general not homogeneous. Then the quantity, essential for all experiments where the magnetic flux produced by supercurrents flowing in the sample is measured, is the mean value of flux density inside the sample

$$\bar{B} = A^{-1}.\int_A B(A)\ dA = \Phi/A \tag{1.1}$$

Here, A is the cross-section of the sample and Φ denotes the magnetic flux present in the sample. Due to the analogy with bodies of homogeneous magnetization, the superconducting samples are often characterized by the

Magnetic Susceptibility of Superconductors and Other Spin Systems
Edited by R.A. Hein *et al.*, Plenum Press, New York, 1991

magnetization M, meaning simply the difference between the value of the external field B_{ext} and the mean value of flux density over the cross-section of the sample

$$M = \bar{B} - B_{ext} = \Phi/A - B_{ext} \tag{1.2}$$

According to this relation, a measurement of Φ at known A and B_{ext} allows us to determine the magnetization of the sample.

Another thing we must clearly delineate is which measurements are to be considered DC and which measurements are to be considered AC. As mentioned above, for DC measurements the external magnetic field should be constant during the whole experiment. Then the only "pure DC" experiment is the measurement on the sample cooled in a DC field switched on at the temperature exceeding T_c - the so called field-cooled (FC) experiment. Exclusively in this way the presence of the Meissner effect in the sample can be tested to confirm that the sample exhibits superconductivity. In all the other types of measurements shielding currents are induced in the sample by the nonzero time derivative of B_{ext}, which allows one to detect the zero resistance state. Consequently, the non-DC class includes also the exepriments that are usually reported as DC ones: the zero-field-cooled (ZFC) and the switch -off or remanence (REM) measurements, respectively. Indeed, in these experiments the field remains constant during the registration of the magnetic moment of the sample, but this moment was first induced by a sudden change of the external magnetic field. Thus this moment is produced by shielding rather than Meissner currents. The comparative FC versus ZFC magnetization measurement is a very nice experiment. Particularly because it allows one to compare shielding with the Meissner effect on the same sample at nearly identical conditions.

Shielding currents induced by nonzero dB_{ext}/dt decrease with typical time constants, exceeding days in the case of low temperature superconductors at liquid helium temperature, and on the order of minutes in the case of high T_c superconductors at 77 K. Thus we call a measurement an AC measurement if the driving field changes periodically in time with a period T_0 substantially lower than these time constants. The magnetization process can be well described by the magnetization loop, i.e. the $M(B_{ext})$ dependence at constant temperature T (Fig.1). Sometimes a DC field is superimposed upon the AC one to study its influence on the properties of the sample. The combination of AC and DC fields in an AC measurement is problematical in that the DC field can not be measured directly, but the consequences of its presence are recognizable. This is a case of great interest and we will discuss some effects observed on the high T_c superconductors under the action of such fields in section 4. In the next

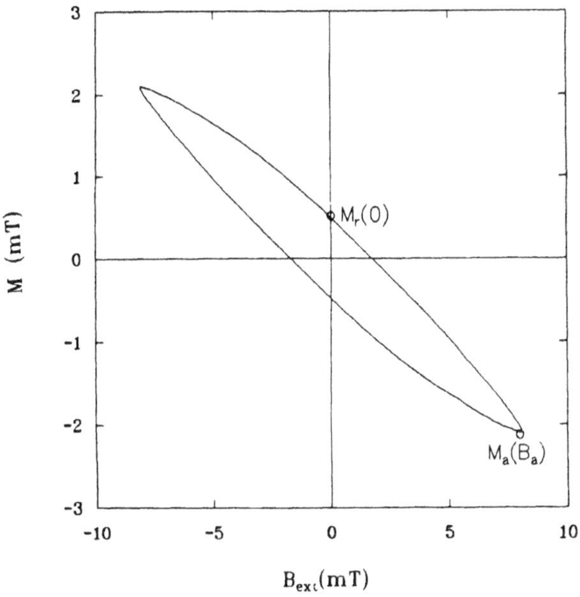

Fig. 1: Magnetization loop of a high T_c superconductor.

section we discuss briefly the physical quantities commonly used in discussions of an AC magnetization measurement.

2. AC susceptibility measurements

To process the information contained in a magnetization loop, several simplifications are often made. It is possible to study the whole loop with all its peculiarities,[5,6] but in most cases the loop may be characterized by a set of parameters, related to its significant features. We shall now describe a widely used approach of characterizing the loop by complex AC susceptibility.[7-13]

Let us suppose the harmonic driving field

$$B_{ext} = B_{ac} = B_a \cos(\omega t) \tag{2.1}$$

with the frequency $f = \omega/2\pi = 1/T_0$. One can define two dimensionless quantities

$$\chi' = 2\,(T_0 B_a)^{-1} \int_0^{T_o} \bar{B}(t)\cos(\omega t)\,dt - 1 \qquad (2.2a)$$

$$\chi'' = 2\,(T_0 B_a)^{-1} \int_0^{T_o} \bar{B}(t)\sin(\omega t)\,dt \qquad (2.2b)$$

called the real and the imaginary part of the complex AC susceptibility. What is the physical meaning of these two parameters?

Remember that harmonic functions are orthogonal within the time interval identical with the period T_0, i.e. the mean value of the product $\sin(p\omega t)\sin(q\omega t)$, (where p and q are integers), is 1/2 if p=q while zero in all other cases. The same result is valid for the product of two cosine functions, the product of sine and cosine functions exhibits zero mean value in every case. If we express the time dependence of $\bar{B}$ in the form of the Fourier series

$$\bar{B}(t) = \sum_n B_n \cos(n\omega t + \varphi_n) \qquad (2.3)$$

only the first harmonic - with the amplitude B_1 and the phase φ_1 - can contribute to the AC susceptibility components χ', χ''. Inserting (2.3) into (2.2) yields:

$$\chi' = 2\,(T_0 B_a)^{-1} \int_0^{T_o} B_1 \cos(\omega t + \varphi_1)\cos(\omega t)\,dt - 1 = ((B_1 \cos\varphi_1)/B_a) - 1 \qquad (2.4a)$$

$$\chi'' = 2\,(T_0 B_a)^{-1} \int_0^{T_o} B_1 \cos(\omega t + \varphi_1)\sin(\omega t)\,dt = -(B_1 \sin\varphi_1)/B_a \qquad (2.4b)$$

The last two expressions show why the phase sensitive detector (PSD) is an ideal device for the complex AC susceptibility measurement. The physical meaning of χ' and χ'' can be found if one remembers the basic energy relations for a magnetized body. The energy which is converted into heat during one cycle of magnetization is

$$Q = \mu_0^{-1} \oint B_{ext}\, d\bar{B} = \mu_0^{-1} \int_0^{T_o} B_{ext}(t)\, d\bar{B}(t)/dt\, dt$$

The $d\bar{B}(t)/dt$ term yields a nonzero contribution to the integral only for the first harmonic

$$d\bar{B}(t)/dt\big|_{1st\ harmonic} = -B_1\omega\,\sin(\omega t+\varphi_1)$$

and the loss per cycle can be calculated as

$$Q = -\mu_0^{-1}B_1B_a\omega \int_0^{T_o} \cos^2(\omega t)\ \sin\varphi_1\ dt = -\pi\ \mu_0^{-1}B_aB_1\sin\varphi_1 \tag{2.5}$$

Comparing this result with (2.4b) one easily obtains the relation

$$Q = -2\pi\ \chi''\ (B_a^2/2\mu_0) \tag{2.6}$$

giving the physical sense of the imaginary part of AC susceptibility. The heat has in our notation a negative sign which means that χ'' is a positive number.

Similar procedures must be carried out to reveal the meaning of χ'. Here we start from the expression for the magnetic field energy inside the sample, averaged over one cycle of magnetization

$$\bar{W}_m = (\mu_0T_0)^{-1} \int_0^{T_o} B_{ext}(t)\ \bar{B}(t)\ dt = (\mu_0T_0)^{-1}B_aB_1 \int_0^{T_o} \cos(\omega t)\ \cos(\omega t+\varphi_1)\ dt$$

Similarly, as in the previous case, the orthogonality of harmonic functions was taken into account. After evaluation of the last integral we obtain

$$\bar{W}_m = B_aB_1\cos\varphi_1/(2\mu_0) \tag{2.7}$$

In the high T_c superconductors the normal state resistivity is large enough to allow one to neglect the eddy current effects, and the magnetic susceptibility can be taken as zero. Then, at $T>T_c$ it is $B_1 = B_a$ and $\cos\varphi_1 = 1$, and the averaged value of magnetic energy is $\bar{W}_{nor} = B_a^2/(2\mu_0)$. When the sample becomes superconducting the energy values will differ by

$$\Delta W = \bar{W}_m - \bar{W}_{nor} = B_a^2/(2\mu_0)[((B_1\cos\varphi_1)/B_a)-1] = \chi'\ B_a^2/(2\mu_0) \tag{2.8}$$

Because the magnetic field is expelled from the superconductor, ΔW is negative resulting in negative value of χ'.

As we have seen, both parts of the complex AC susceptibility characterize the energy exchange between the sample and the source of the AC field. The real part reflects the screening properties expressed as a difference between the energy (of magnetic field in the sample) in the nonsuperconducting and the superconducting states, respectively. The imaginary part corresponds to the amount of the AC field energy converted to heat. Both these quantities depend on the whole course of the magnetization process rather then on some significant points, and they can be used in thermodynamic considerations. However, I recommend the use of another set of parameters characterizing the magnetization loop (see section 3). These behave similarly as χ' and χ'', but instead of energies they are related to magnetizations.

3. Measurements of the amplitude and remanent AC susceptibility

Let us consider a phase sensitive detector (PSD) having the following features:[14-16] (i) the whole frequency spectrum of the input signal can be processed without filtering, (ii) the mixer signal is a square wave and (iii)there are two output channels of the PSD, differing in the shift of the mixer square wave. Then, the input voltage u(t) is multiplied by the functions

$$r_s(t) = \begin{cases} +1 & \text{if} \quad 0 \le t < T_0/2 \\ -1 & \text{if} \quad T_0/2 \le t < T_0 \end{cases}$$

$$r_c(t) = \begin{cases} +1 & \text{if} \quad 0 \le t < T_0/4 \quad \text{or} \quad 3T_0/4 \le t < T \\ -1 & \text{if} \quad T_0/4 \le t < 3T_0/4 \end{cases}$$

and the output channels of the PSD give the voltages

$$S' = (T_0)^{-1} \int_0^{T_0} u(t)\, r_s(t)\, dt \quad ; \quad S'' = (T_0)^{-1} \int_0^{T_0} u(t)\, r_c(t)\, dt$$

Let the voltage u_i induced in one turn of the pick-up coil, wrapped tightly around the sample, be the PSD input signal. Because this is simply the time derivative of the flux Φ inside the sample, the relation

$$\int u_i(t)\, r(t)\, dt = \int -(d\Phi/dt)\, r(t)\, dt = -\int r(t)\, d\Phi$$

holds for the both r_s and r_c, replaced here symbolically by r. The result of the integration is the sum of Φ values at the moments when the mixer functions change the sign:

$$S' = -[\Phi(T_0/2) - \Phi(0) - \Phi(T_0) + \Phi(T_0/2)]/T_0 =$$

$$= -2\,[\Phi(T_0/2) - \Phi(0)]/T_0 \qquad (3.1a)$$

$$S'' = -[\Phi(T_0/4) - \Phi(0) + \Phi(T_0) - \Phi(3T_0/4) - \Phi(3T_0/4) + \Phi(T_0/4)]/T_0 =$$

$$= -2\,[\Phi(T_0/4) - \Phi(3T_0/4)]/T_0 \qquad (3.1b)$$

The last expressions can be further simplified if we consider a symmetrical hysteresis loop like that in the Fig.1. In such a loop, according to Eq.(1.2) and Eq.(2.1), the following relations are valid:

at $t=0$: $\quad B_{ext} = B_a \quad ; \quad M = M_a = \Phi(0)/A - B_a$

at $t=T_0/4$: $\quad B_{ext} = 0 \quad ; \quad M = M_r = \Phi(T_0/4)/A$

at $t=T_0/2$: $\quad B_{ext} = -B_a \quad ; \quad M = -M_a = \Phi(T_0/2)/A + B_a$

at $t=3T_0/4$: $\quad B_{ext} = 0 \quad ; \quad M = -M_r = \Phi(3T_0/4)/A$

Then

$$S' = -(2A/T_0)\,[-M_a - B_a - M_a - B_a] = (4A/T_0)\,[M_a + B_a]$$

$$S'' = -(2A/T_0)\,[\,M_r + M_r] = -(4A/T_0)\,M_r \qquad (3.2)$$

As we can see from these relations, the present mode of using the PSD allows us to obtain information about the magnetization of the sample at two significant moments: at $t=0$ or $T_0/2$, when the field is maximal, and at $t=T_0/4$ or $3T_0/4$, when it is zero. To facilitate comparison with the complex AC susceptibility, we normalize the output signals of the PSD by the S' value measured with the sample in the normal state ($M_a = 0$); $S'_{nor} = (4A/T_0)B_a$. The normalized output signals are

$$S'/S'_{nor} = 1 + M_a/B_a \quad ; \quad S''/S'_{nor} = - M_r/B_a$$

These expressions, although derived for a one turn pick-up coil, are valid for pick-up coils with an arbitrary number of turns, i.e. the above normalization cancels out all instrumental constants. Thus, we can characterize the magnetization loop by two dimensionless quantities

$$\chi_a = M_a/B_a = S'/S'_{nor} - 1 \quad ; \quad \chi_r = M_r/B_a = - S''/S'_{nor} \qquad (3.3)$$

that could be well called amplitude susceptibility (for χ_a) and remanent

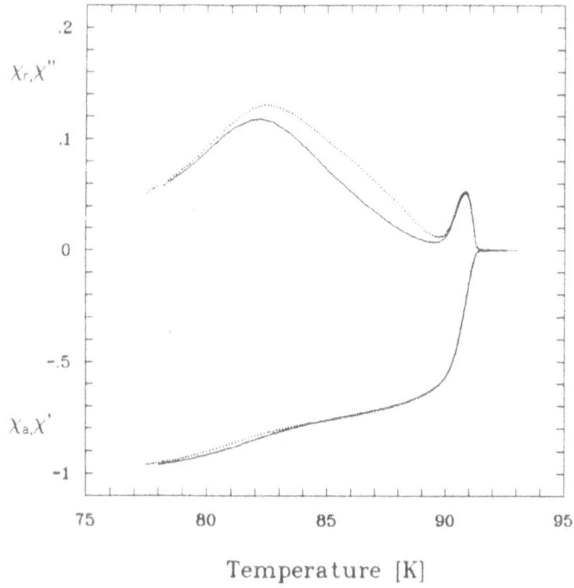

Fig.2: Comparison of the true AC susceptibility components χ', χ'' (solid lines) with the "wide band AC susceptibilities" χ_a, χ_r (dotted lines). Ref.17, sample C, $B_a = 0.5$ mT, f= 333 Hz.

susceptibility (for χ_r). What are the main features of these "wide band" susceptibilities? They are similar to their "narrow band" counterparts χ', χ'' in the sense that at temperatures $T>T_c$ $\chi'=\chi''=\chi_a=\chi_r=0$; while for $T \ll T_c$ $\chi'=\chi_a= -1$ and $\chi''=\chi_r= 0$. In the intermediate temperature region $\chi' \simeq \chi_a > 0$ and $\chi'' \simeq \chi_r > 0$. The temperature dependence of χ_a is similar to that of χ' and χ_r resembles χ''. This is illustrated in Fig.2, where χ', χ'', χ_a and χ_r are plotted as measured on the same polycrystalline sample of $YBa_2Cu_3O_x$ [Ref.17, sample C]. Notice the coincidence of the onset temperatures as well as slight differences in the temperatures T_m at which the χ'' and χ_r curves reach their maxima, and in the magnitude of the maxima. These differences are partially caused by the nonlinearity of the magnetic field profiles inside the sample, resulting in a difference between the penetration field and the field at which χ'' or χ_r reach the maxima. Theoretical calculations for a cylindrical sample using the simplest Bean model[18] with j_c= const. give $\max(\chi'') = 2/(3\pi) = 0.212$, while $\max(\chi_r) = 1/4 = 0.25$. Corresponding values for both remaining quantities are $\chi'(T_m) =$

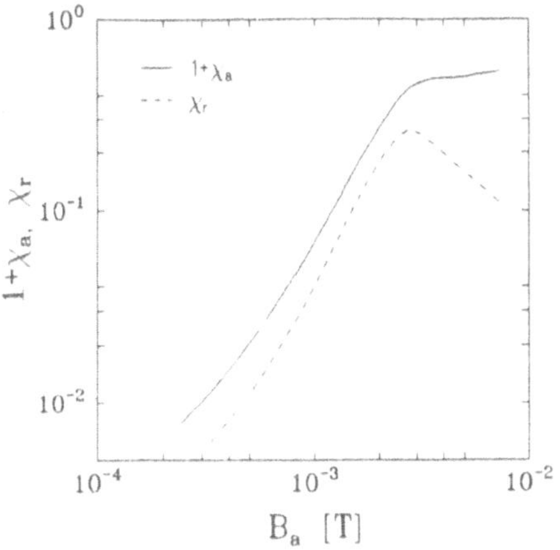

Fig.3: Typical dependence of the "wide band AC susceptibilities" χ_a, χ_r on the AC field amplitude B_a. Ref.17, sample B, T= 77 K, f= 123 Hz. In order to use a logarithmic scale, $1+\chi_a$ is plotted instead of $\chi_a<0$.

$-1/4 = -0.25$ and $\chi_a(T_m) = -1/3 = -.333$, respectively. The maxima in Fig.2 do not reach these values due to the presence of two types of screening currents in the sample.[19]

Using χ_a, χ_r instead of χ', χ'' has another advantage: i.e. a comparison with Campbell's method[14] is possible thanks to the fact that in Campbell's method the measured quantity is the in-phase signal of the PSD working in the wide band mode (in our notation S'). From this value measured at different B_a one can construct the magnetic field profile inside the sample at given temperature T. Then it is easy to check some points $\chi_a(T,B_a)$ obtained by both methods (i.e. variable T at constant B_a, variable B_a at constant T) to see if they coincide.

The out-of-phase component χ_r can be utilized as well:[16] it is quite useful to plot - at given temperature - both χ_a and χ_r as functions of B_a. To allow the use of logarithmic coordinates, $1+\chi_a$ is plotted instead of χ_a which is negative. In Fig.3 such a plot is given for another polycrystalline $YBa_2Cu_3O_x$ sample [Ref.17, sample B]. In the range of amplitudes used, the intergrain currents prevail in the shielding of the sample, therefore only one peak is seen in the $\chi_r(B_a)$ dependence. In this way, using the "wide band AC susceptibility parts" χ_a, χ_r instead of the true AC susceptibilities χ', χ'' allows us to perform with the same arrangement diverse experiments: (i) $\chi_a(T)$, $\chi_r(T)$ measurement with B_a as a parameter (ii) $\chi_a(B_a)$, $\chi_r(B_a)$ measurement with T as a parameter, $\chi_a(B_a)$ obtained in this experiment can be used directly for constructing the magnetic field profile according to Campbell's method.

There is another reason why such comparable measurements are rather unusual : Campbell's method was developed to study superconductors in a small AC magnetic field superimposed upon much higher DC background fields, while in more common AC susceptibility experiment only ac AC field is used.[20] The situation in the intermediate region, i.e. when AC and DC fields are comparable, will be discussed in the following section.

4. AC measurements at AC and DC fields of comparable amplitude

In the simple geometry chosen here to avoid the complications not related to the magnetization process itself (i.e a slim cylinder in a homogeneous external field parallel to its axis), the 1st Maxwell equation has the form

$$\partial B(r,t)/\partial r = \mu_0 j(r,t,B(r,t)) \qquad (4.1)$$

where r is the radial coordinate. This differential equation has to be solved with the boundary condition $B(R) = B_{ext}$, R being the radius of the cylinder. Having solved Eq.(4.1) one can insert B(r) into (1.1) to calculate $\bar{B}(t)$. In the case of a symmetrical driving field $B_{ext}(t+T_0/2) = B_{ext}(t)$, only two values - B(0) and $B(T_0/4)$ - are necessary to calculate χ_a and χ_r. Otherwise we need two more values, $B(T_0/2)$ and $B(3T_0/4)$. In this way, calculations of the "wide band AC susceptibilities" χ_a and χ_r are completely finished. However, if one is interested in calculating the AC susceptibility parts χ' and χ'', the complete time dependence of j(r,t) is essential in determining $\bar{B}(t)$. As a next step, the integrations (2.2) must be done to obtain χ' and χ''. Comparing the procedure having as a result χ_a and χ_r with that yielding χ' and χ'' one can clearly see that the comparison between the theory and the experiment is much simpler if we use the "wide band AC susceptibilities" χ_a and χ_r instead of the true ones i.e. χ', χ''.

The key point in solving Eq.(4.1) is the appropriate choice of j(r,t). Suppose that our sample is spatially homogeneous in the sense that it has the same properties on the surface as in the center, and let us ignore the rate dependent effects like flux creep and flux flow. Then the macroscopic picture (with an exception of an outer shell with the thickness given by the London penetration depth, λ_L, containing the Meissner currents) is given by the macroscopic critical current density $j_c(B(r))$ whose magnitude depends on the local value of magnetic field B(r) and its orientation ($\mp$) is determined by the most recent local nonzeroelectrical field. The quantity characterizing this picture is the penetration field

$$B_p = \mu_0 \int_0^R j_c(r)\, dr$$

One of the approaches widely used is to superimpose on the AC field a DC field B_d exceeding many times B_p. Then the local value of B(r) differs only slightly from B_d and $|j_c(r)|$ can be taken constant. On the other hand, in typical AC susceptibility experiments no DC field is used. In this case Eq.(4.1) must be solved in a self-consistent manner and solving it starts with some tentative $j_c(B)$ dependence. Different solutions, well suited for high T_c superconductors, have been found recently.[21-24]

Only few data are reported from experiments on high T_c superconductors in the intermediate region $B_d \approx B_a$. Therefore, we have tested in this domain the properties of various polycrystalline $YBa_2Cu_3O_x$ samples differing in their densities as well as grain sizes.[17] The samples were cylindrical, with 1 mm radius and 12 mm height. The quantities measured in our study were the amplitude AC susceptibility χ_a and the remanent AC susceptibility χ_r as function of temperature (at constant B_a and B_d) or as function of the AC field amplitude (at constant T and B_d), respectively. Along with these two dependences, field profiles have been constructed using the same data. There have been observed several unusual effects, present in some extent in the experiment on every sample and, therefore, of general validity. To illustrate these effects some appropriate curves measured on different samples have been chosen. Effects worth mentioning are the following: (i) In the $\chi_a(T)$ dependence, measured at certain combinations of AC and DC fields, a temperature interval exists, where $\chi_a(T,B_d)<\chi_a(T,0)$ - see Fig.4. This means that the shielding of the sample under these circumstances is better when the DC field is applied than if $B_d=0$. The remanent flux is, as expected, smaller. (ii) On the curves linking B_a with the normalized penetration depth δ/R obtained

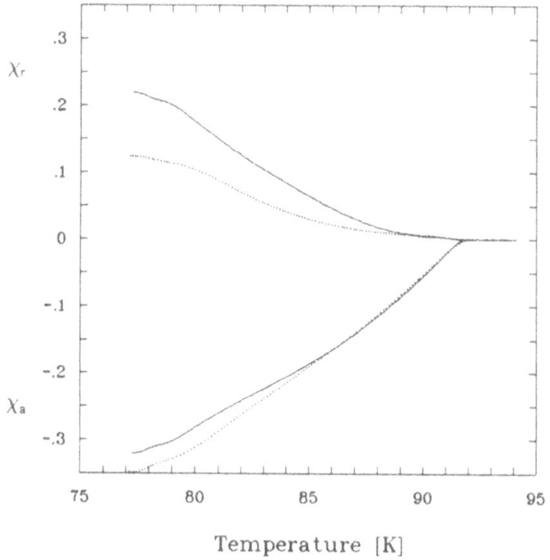

Fig.4: Temperature dependence of the "wide band AC susceptibilities" χ_a, χ_r. The curves differ in DC field superimposed on the AC field: solid lines - B_d= 0; dotted lines - B_d= 2 mT. Ref.17, sample B, f= 123 Hz, B_a= 2 mT.

following Campbell's method, some portions with negative slope $d(\delta/R)/dB_a$ appear - see Fig.5, solid line. Curves in Fig.5 were obtained by processing raw experimental data without any smoothing or correcting procedures in order to avoid possible introduction of artefacts. It is without doubt, that in a certain interval of AC field amplitudes the shielding is better at higher B_a. Data of Fig.5 show that for an applied DC field of 10 mT the curve exhibits the form typical for the high T_c superconductor.[25-28] (iii) The dependence of the remanent AC susceptibility χ_r on B_a exhibits at some combination of AC and DC fields complicated behavior with two maxima, while at other combinations only one

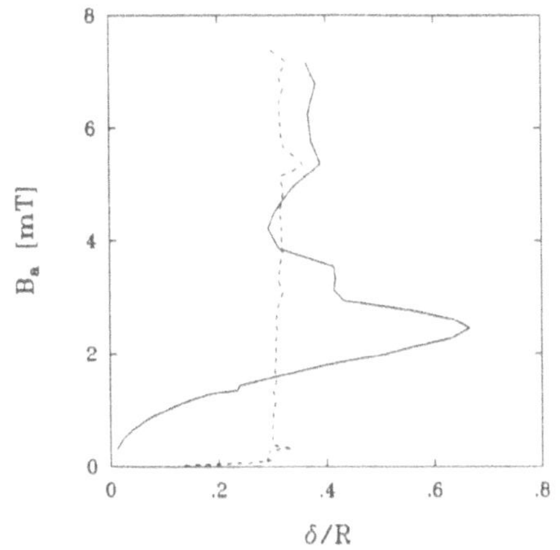

Fig.5: Relation between the normalized penetration depth δ/R and the AC field amplitude B_a at the superimposed DC field of 10 mT (dotted line) and without it (solid line). Ref.17, sample B, 77 K, 123 Hz.

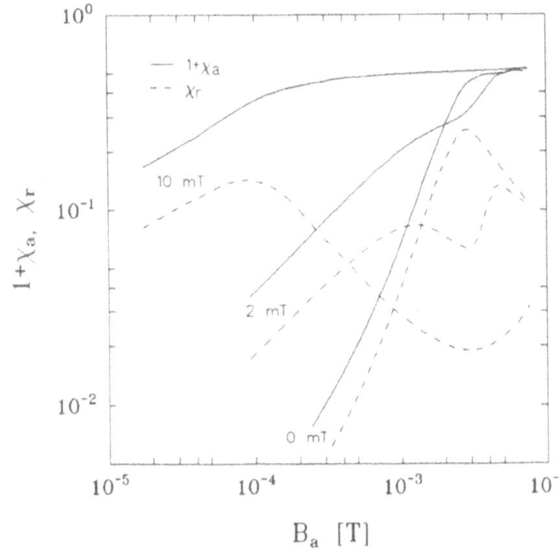

Fig.6: Amplitude dependence of the "wide band AC susceptibilities" χ_a, χ_r at 77 K, 123 Hz and superimposed DC fields indicated in the graph. Ref.17, sample B.

maximum is present. This is illustrated in Fig.6, by the dependence measured at B_d= 2 mT. The curves measured on this sample at B_d=0 as well as B_d= 10 mT exhibited only one intergrain maximum. According to the two-component model of high T_c superconductor[21] one expects one maximum due to intergrain currents and another maximum due to intragrain ones. However, at 77 K the AC field in the given interval of interest is too weak to penetrate the grains, and all the maxima can be ascribed to the intergrain currents only.

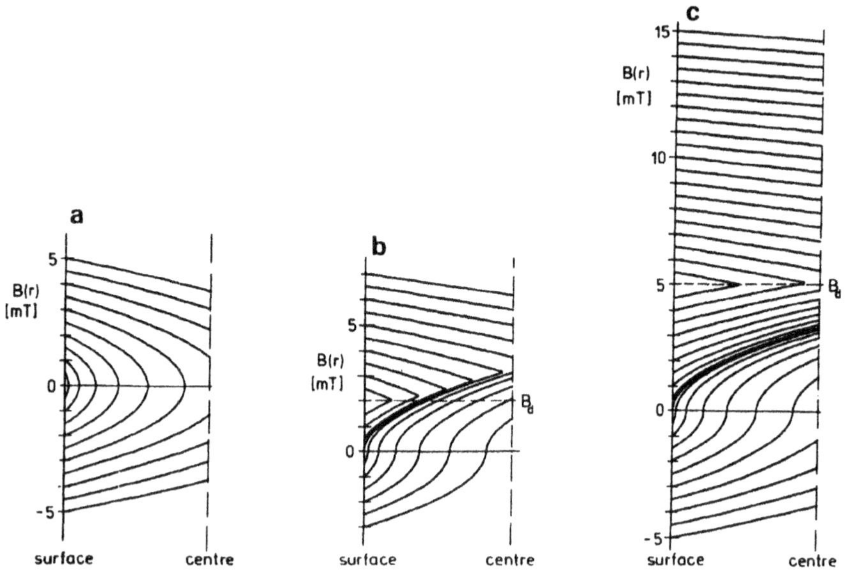

Fig.7: Calculated profiles of the magnetic field at ωt=0 (upper parts) and at $\omega t=\pi$ (lower parts) inside the cylindrical sample with R= 1 mm at different values of the AC field amplitude and various superimposed DC fields: a/ B_d= 0, b/ B_d= 2 mT, c/ B_d= 5 mT. Kim's model of $j_c(B)$ dependence with B_0= 5×10^{-5} T, α= 4×10^3 N/m^3.

All these strange effects have been explained taking into account the strong dependence of the intergrain critical current density on the local magnetic field B(r), a feature confirmed by other experiments on high T_c samples as well.[29-37] The principal understanding can be gained following the three cases of AC field penetration illustrated in Fig.7. Here, the magnetic field profiles have been calculated by numerically solving solving Eq.(4.1), omitting the surface currents and using Kim's model[38] to describe $j_c(B(r))$ dependence:

$$j_c(r) = \alpha/(B_0 + B(r))$$

where α and B_0 are constants and $j_c(r)$, $B(r)$ are local values of the critical current density and the magnetic field, respectively. To express this dependence other models can be used as well,[32,33] but the main feature remains: At B_d comparable with B_a a region inside the sample exists where the local $B(r)$ is nearly zero. In this case the value of the critical current density averaged over the sample's cross-section increases dramatically, causing the appearance of the effects (i) and (ii) mentioned above. The effect (iii) can be explained when we encounter the existence of two characteristic penetration fields at these circumstances: the first penetration field is reached when the profile first reaches the center of the sample (nearly 1.8 mT in Fig.7c), the second one appears when the lower part of the profile, corresponding to $\omega t = \pi$, crosses notably the $B(r) = 0$ line. Increasing the DC field to values exceeding the AC field amplitude, the $j_c(r)$ dependence for a homogeneous sample approaches a constant value and the second penetration field can no more be reached. Then the results obtained by Bean can be qualitatively applied to check the experimentally observed dependence. Analytical solutions for the profiles obtained using a simplified Kim's model with $B_0 = 0$ can be found in Ref.39.

There is another feature seen in our data, typical for the high T_c samples: The $\chi_{a,r}(B_a)|_{T,B_d}$ and $\chi_{a,r}(T)|_{B_a,B_d}$ dependences differ with respect to the sequence of cooling the sample and DC field application. In the $\chi_{a,r}(B_a)$ experiments the DC field was applied for $T>T_c$ and the sample cooled to $T<T_c$. In this case a higher B_a was needed to reach the center of the sample than in the case of cooling the sample in zero field and applying B_d with the sample in the superconducting state. Similar observations have been done in the $\chi_{a,r}(T)$ experiments: the intergrain maximum was reached at a higher temperature when the sample was cooled in DC field than in the case when it was first cooled and the DC field applied at $T \ll T_c$. This effect is illustrated in Figs. 8 and 9, respectively. In Fig.9 one can see that the intragrain maximum (close to T_c) is not affected by the manner in which the DC field is applied. The difference in sample response with regard to the principal (intergrain) maximum is more noticeable in porous samples with large grains than in the fine-crystalline BiSrCaCuO[40], and it vanishes in single crystals.[4] Thus it

should be ascribed to the influence of intragrain currents on intergrain ones. A possible explanation is as follows: During the cooling in DC field the grains expel the magnetic flux producing a self field directed oppositely to B_d. Then the actual field inside the sample is lower than B_d due to intragrain currents. This does not succeed when one cools the sample in zero DC field, and switches the field on at low temperature, inducing mostly the intergrain currents shielding the grains from the DC

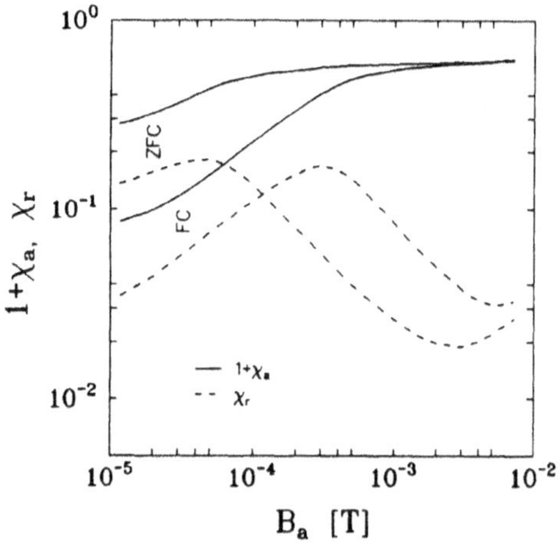

Fig.8: Influence of the succession of cooling and DC field switching on the "wide band AC susceptibilities" χ_a, χ_r. B_d= 5 mT was applied before (FC) and after (ZFC) cooling the sample from 95 K to 77 K (the temperature of measurement). Ref.17, sample B, 123 Hz.

field. In this case the grains do not contribute notably to the self field and the AC field penetrates more easily into the sample. The self field produced by the grains is weaker also in samples with small grain size due to so called "magnetic invisibility effect",[41] i.e. penetration of the magnetic field in the outer shell of the grain with a thickness given by the London penetration depth λ_L, resulting in decrease of the magnetic flux expelled from the sample. In this way, the small grains became "invisible" in magnetic measurements.

Up to now only AC measurements have been discussed, with DC fields applied in various ways. An interesting question is, if it is possible to do some comparison between the results of such experiments and those obtained by DC magnetization measurements.

5. Comparison of AC and DC experiments

To facilitate comparisons, we express the results of DC experiments in dimensionless quantities

$$\chi_{ZFC} = M_{ZFC}/B_d \qquad ; \qquad \chi_{FC} = M_{FC}/B_d$$

that can be reasonably called ZFC and FC susceptibilities, respectively. The ZFC susceptibility is a special case of the amplitude AC

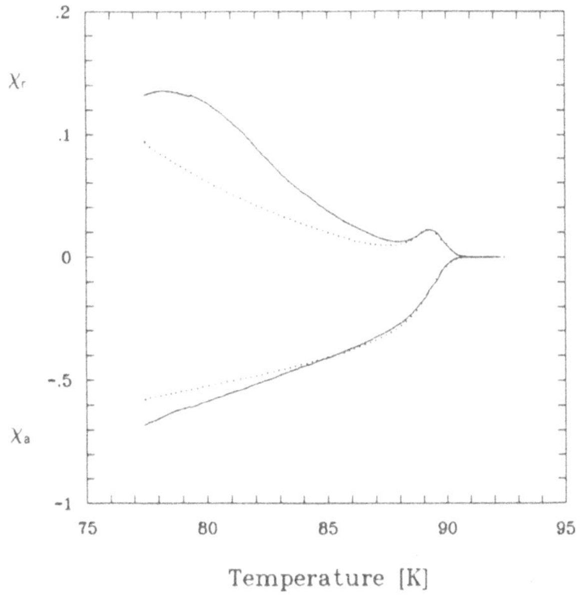

Fig. 9: Influence of the succession of cooling and field switching on the "wide band AC susceptibilities" χ_a, χ_r. Full lines - the curves measured at warming up the sample cooled to 77 K in DC field of 5 mT; dotted curves - the same but cooled in zero DC field with successive switching on the DC field at 77 K. Ref. 17, sample B, 123 Hz.

susceptibility χ_a, because it corresponds to the point (B_a, M_a) on the magnetization loop (see Fig. 1). The FC susceptibility has no AC analogue, because it reflects the Meissner effect rather than the shielding effect observed in the experiments with nonzero dB_{ext}/dt. Among the comparisons of ZFC and FC susceptibilities, the determination of the irreversibility line, i.e. the line dividing the (B_d, T) plane into two regions is of great interest.[42] In the reversible region $\chi_{ZFC} = \chi_{FC}$, while in the irreversible region $\chi_{ZFC} < \chi_{FC}$. It is difficult to determine the irreversibility line directly from the DC magnetization measurements, therefore AC measurements are very often used to localize it.[43]

In order to check the validity of this approach, a comparison of the results obtained on the same sample in DC measurements using a vibrating sample magnetometer with those given by the wide band AC susceptibility measurement as described in section 3 was done. Samples of BiSrCaCuO

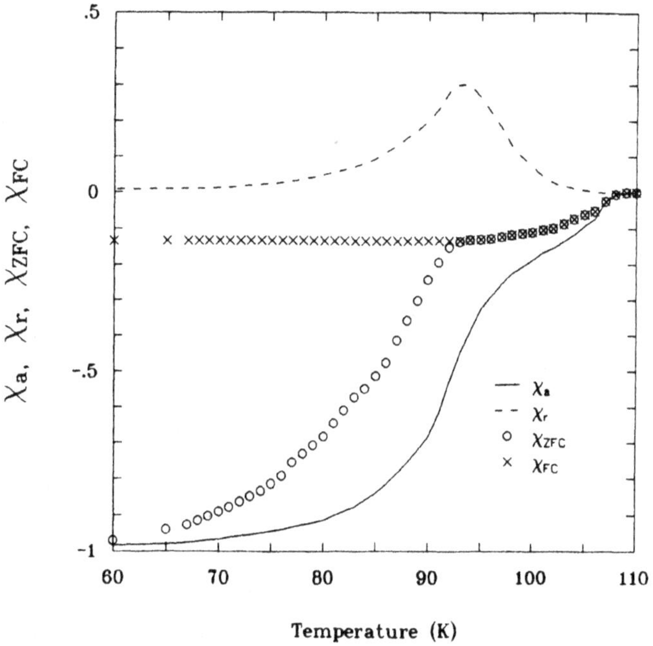

Fig. 10: Comparison of the "wide band AC susceptibilities" χ_a, χ_r with the DC susceptibilities χ_{ZFC}, χ_{FC}. BiSrCaCuO polycrystalline sample with ~75% of the 110 K phase and ~25% of the 80 K phase.

chosen for this study were in the form of a parallelepiped with dimensions $1x3x6\ mm^3$ cut from the polycrystalline pellets prepared by the solid state reaction from fine powders.[40] In sample A, the content of 110 K phase is nearly 100%, while the sample B contains 25% of the 80 K phase. In Fig. 10 the comparison of $\chi_a, \chi_r, \chi_{ZFC}$ and χ_{FC} measured on sample B at $B_a = B_d = 1$ mT is given. The temperature T_m, at which χ_r reaches its maximum coincides well with the temperature T_{irr}, corresponding to the irreversibility line. However, one should not forget that this sample is multiphased i.e. two superconducting components and therefore must ascertain if this coincidence is true also in the case of aone-component sample, i.e. sample A. Results of experiments with this nearly single phase sample reveal that the rule $T_m = T_{irr}$ is not valid for this sample. In Fig. 10, where the curves obtained at $B_a = B_d = 1$ mT are given, a two-step transition from normal to superconducting state is observable on the χ_a, χ_{ZFC} and χ_{FC} curves. The first step near T_c can be ascribed to the transition of the

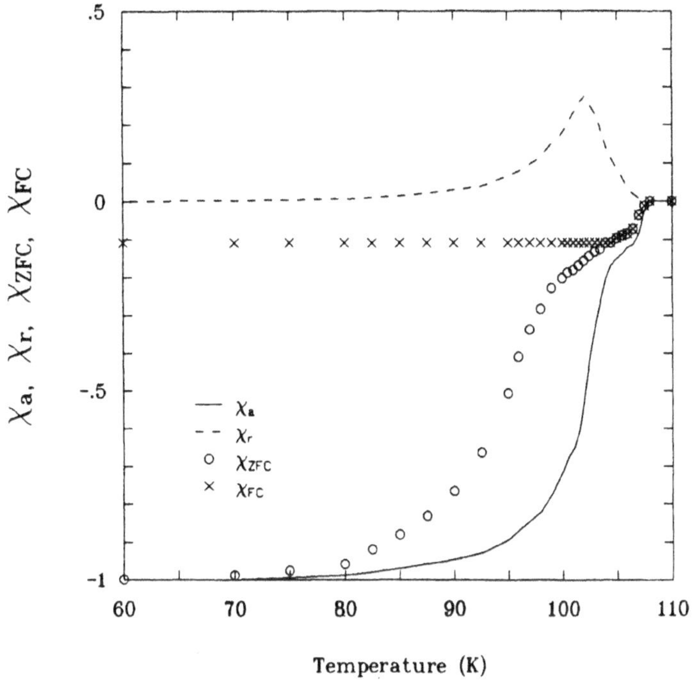

Fig. 11: Comparison of the "wide band AC susceptibilities" χ_a, χ_r with the DC susceptibilities χ_{ZFC}, χ_{FC}. BiSrCaCuO polycrystalline sample with nearly 100% content of the 110 K phase. $B_a = B_d = 1$ mT.

grains, connected with the expulsion of nearly 11% of the magnetic flux from the sample's volume. Remanent susceptibility χ_r, representing the magnetic flux trapped in the sample at the time when the amplitude of the AC field is zero, starts to increase with some delay with respect to the other susceptibilities. It becomes noticeable, when the temperature falls below T_{irr} (the temperature of splitting between χ_{ZFC} and χ_{FC}), but the maximum of χ_r is reached at the temperature T_m somewhat below T_{irr}. A probable explanation is that in this sample the irreversibility starts when the intergrain weak links become superconducting, and the maximum of remanent flux is reached when the whole sample volume is filled by the intergrain critical current density. This idea can be further supported by data shown in Fig. 12, where the susceptibility data measured on the same homogeneous sample A at a higher magnetic field $B_a=B_d=$ 8.5 mT are presented. The division of the transition into two steps is not so clearly visible here, but the completion of the first step (that is supposed to correspond to the transition of the grains) roughly corresponds to the

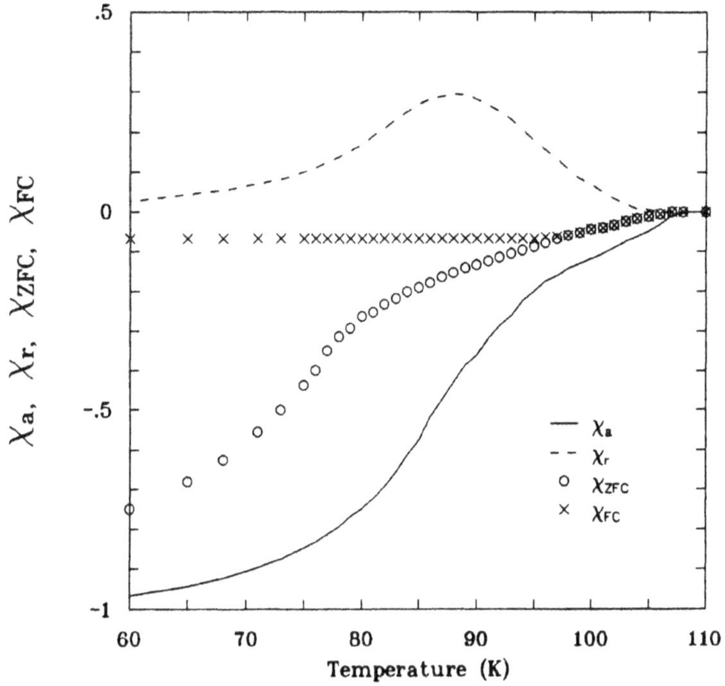

Fig. 12: Comparison of the "wide band AC susceptibilities" χ_a, χ_r with the DC susceptibilities χ_{ZFC}, χ_{FC}. BiSrCaCuO polycrystalline sample with nearly 100% content of the 110 K phase. B_a = B_d = 8.5 mT.

irreversibility temperature T_{irr}. The difference between the temperature T_m(at which the remanent susceptibility χ_r reaches its maximum), and T_{irr} is nearly 10 K in this experiment. Thus a reconsideration of the dependences observed on the inhomogeneous sample B, presented in Fig. 10, should be done. In this sample, according to the X-ray analysis, the 80 K phase consists of grains with size much larger than that of the 110 K phase. At temperatures over 80 K, they behave as non-superconducting voids, screened by the intergrain supercurrents connecting the grains of 110 K phase. According to the classical flux creep model,[44] it is possible to expect shorter relaxation times for these currents due to weaker pinning by the intergrain loops with a smaller current carrying capability or larger characteristic space scale. Thus at $T>T_{irr}$, where only the reversible part of screening currents is discernable in the DC measurement, the irreversible screening currents with short time constant are observable only under AC conditions. The existence of such currents at $T>T_{irr}$ follows from the nonzero remanent AC susceptibility χ_r in this temperature interval. The completely reversible shielding for both DC and AC regimes is observed - similarly as in the case of the homogeneous sample A - only in a narrow interval near T_c, corresponding to the shielding of the 110 K phase grains itself. Other explanations are possible as well and the results of these measurements indicate that the comparison of DC and AC measurements on high T_c samples[45,46] can stimulate further improvements of the models describing the behavior of these materials. Of course, similar experiments should be done on better characterized, single crystal samples.

6. Conclusions

In the paper an approach is presented that allows one to perform direct comparison of experimental results obtained by measuring the response of high T_c superconductors on various combinations of AC and DC fields. It consists in representing results of the AC measurements in terms of "wide band" AC susceptibilities, that express the magnetization of the sample at the maximum and at the zero value of AC field, i.e. χ_a and χ_r, respectively. Interesting results obtained by various AC methods can then be directly compared. Theoretical treatment of the results are simpler and can be more easily understood. Another attractive feature of this method is that it allows one to compare results of AC measurements with those obtained by DC measurement.

The utility of the described approach was demonstrated by the analysis of experimental data obtained on various polycrystalline samples. Characteristic features such as the existence of two types of supercurrents - the intergrain and the intragrain ones, the strong dependence of the intergrain critical current density on the magnetic field, and the presence of flux creep with the time constants much shorter than those typical for the low temperature superconductors has been illustrated.

Acknowledgement

Preparation of this paper was stimulated and its presentation sponsored in part by the organizers of S^4-ONR-Workshop'91. The author would like to express his appreciation for their help and in particular give a special thanks to R.A.Hein for his efforts which enabled the author to participate in the workshop.

The paper summarizes many experimental results obtained during the course of a long term research program that was consistently supported by many of the authors colleagues from the Institute of Electrical Engineering of the Slovak Academy of Sciences. Very important contributions were made by P.Lobotka who constructed the apparatus and carried out many of the experiments, and by L.Cesnak and S.Takács who suggested several helpful interpretations of the data.

Most of the samples have been kindly provided by V.Plecháček from the State Research Institute for Materials, Prague. The VSM measurements presented in section 5 were carried out with the help of G.K.Nikolaides at the Royal Institute of Technology in Stockholm, during the authors stay at the Department of Solid State Physics sponsored by the Royal Swedish Academy of Technical Sciences.

References

1. H. Ullmeier, Irreversible Properties of Type II Superconductors, Springer-Verlag, Berlin, 1975
2. M.N. Wilson, Superconducting Magnets, Claredon Press, Oxford, 1983
3. A.P. Malozemoff, in Physical Properties of High Temperature Superconductors, ed.D.M. Ginsberg, World Scientific Publishing Co., Singapore, 1989
4. F. Gömöry, Thermochimica Acta **174**, 299, (1991)
5. S. Senoussi, M. Oussena and S. Hadjoudj, J.Appl.Phys. **63**, 447, (1988)
6. A.K. Grover, C. Radhakrishnamurty, P. Chaddah, G. Ravi Kumar and G.V. Subba Rao, Pramana **30**, 569, (1988)
7. E. Maxwell and M. Strongin, Phys.Rev.Lett. **10**, 212, (1963)
8. M. Couach, A.F. Khoder and F. Monnier, Cryogenics **25**, 695, (1985)
9. R.A. Hein, Phys.Rev.B **33**, 7539, (1986)
10. T. Ishida and H. Mazaki, Jpn.J.Appl.Phys **26**, L1296,(1987)
11. R.B. Goldfarb, A.F. Clark, A.I. Braginski and A.J. Panson, Crygenics **27**, 475, (1987)
12. J. Garcia, C. Rillo, F. Lera, J. Bartolome, R. Navarro, D.H.A. Blank and J. Flokstra, J.Magn.Magn.Mater. **69**, L225, (1987)
13. F. Gömöry and P. Lobotka, Solid State Comm. **66**, 645, (1988)
14. A.M. Campbell, J.Phys. C **2**, 1492, (1969)
15. P. Dubots and J. Cave, Cryogenics **28**, 661, (1988)
16. F. Gömöry, Solid State Comm. **70**, 879, (1989)
17. F. Gömöry, S. Takács, P. Lobotka, K. Fröhlich and V. Plecháček, Physica C **160**, 1, (1989)
18. C.P. Bean, Rev.Mod.Phys **36**, 31, (1963)
19. J.R. Clem, Physica C **159**, 717, (1989)
20. B. Loegel, A. Mehdaoui and D. Bolmont, Supercond. Sci. Tech. **3**, 504, (1990)

21. K.-H. Müller, Physica C **168**, 585, (1990)
22. D.-X. Chen and R. Goldfarb, J.Appl.Phys. **66**, 2489, (1989)
23. P. Berg and G. Kozlowski, Mod.Phys.Lett B **3**, 1163, (1989)
24. D.-X. Chen, A. Sanchez, T. Puig, L.M. Martinez and J.S. Muñoz, Physica C **168**, 652, (1990)
25. H. Küpfer, I. Apfelstedt, R. Flükiger, C. Keller, R. Meier-Hirmer, B. Runtsch, A. Turowski, U. Wiech and T. Wolf, Cryogenics **28**, 650, (1988)
26. H. Küpfer, I. Apfelstedt, R. Flükiger, C. Keller, R. Meier-Hirmer, B. Runtsch, A. Turowski, U. Wiech and T. Wolf, Cryogenics **29**, 268, (1989)
27. P.A. Godelaine and M. Ausloos, Solid State Comm. **73**, 759, (1990)
28. B. Ni, T. Munakata, T. Matsushita, M. Ivakuma, F. Funaki, M. Takeo and K. Yamafuji, Jap.J.Appl.Phys **27**, 1658, (1988)
29. D.-X. Chen, J. Nogues and K.V. Rao, Cryogenics **29**, 800, (1989)
30. K.-H. Müller, J.C. MacFarlane and R. Driver, Physica C **158**, 69, (1989)
31. A.M. Campbell and F.J. Blunt, Supercond.Sci.Technol. **3**, 450, (1990)
32. S. Senoussi, C. Aguillon and P. Manuel, Physica C **175**, 202, (1991)
33. G. Ravi Kumar and P. Chaddah, Phys.Rev.B **39**, 4704, (1989)
34. S.F. Wahid and N.K. Jaggi, Physica C **170**, 395, (1990)
35. A.M. Campbell and F.J. Blunt, Physica C **172**, 253, (1990)
36. T. Ishida and R.B. Goldfarb, Phys.Rev.B **41**, 8937, (1990)
37. M. Forsthuber, G. Hilscher and F. Ludwig, Topics in Solid State Science **99**, 69, (1990)
38. Y.B. Kim, C.F. Hempstead and A.R. Strnad, Phys.Rev.A **139**, 1163, (1965)
39. F. Gömöry, S. Takács and P. Lobotka, in Studies of High Temperature Superconductors, Vol. 6, ed. A.Narlikar, Nova Science Publishers, New York, 1990
40. V. Plecháček and F. Gömöry, Solid State Comm. **73**, 349, (1990)
41. J.R. Clem and V.R. Kogan, Jap.J.Appl.Phys. **26**, Suppl. 26-3, 1161, (1987)
42. A.P. Malozemoff, L. Krusin-Elbaum, D.C. Cronemeyer, Y. Yeshurun and F. Holtzberg, Phys.Rev.B **38**, 6490, (1988)
43. L. Civale, A.D. Marwick, M.W. McElfresh, T.K. Wothington, A.P. Malozemoff, F.H. Holtzberg, J.R. Thompson and M.A. Kirk, Phys.Rev.Letters **65**, 1164, (1990)
44. M.R. Beasley, R. Labusch and W.W. Webb, Phys.Rev. **181**, 682, (1969)
45. G. Shaw, S.D. Murphy, Z-Y. Li, A.M. Stewart and S.M. Bhagat, IEEE Trans.Magn. **25**, 3512, (1989)
46. S.L. Shindé, J. Morril, D. Goland, D.A. Chance and T.M. Guire, Phys.Rev.B **41**, 8838, (1990)

NONLINEAR A.C. SUSCEPTIBILITY RESPONSE

NEAR THE IRREVERSIBILITY LINE

L. Civale, T.K. Worthington, L. Krusin-Elbaum and F. Holtzberg

IBM, Thomas J. Watson Research Center
Yorktown Heights, NY 10598-0213

INTRODUCTION

The analysis of the response to a.c. magnetic fields has been one of the most popular techniques to study the magnetic properties of conventional type II superconductors. The essence of the method is to immerse the sample in an a.c. magnetic field generated by a coil, and to monitor its response either by the change in the impedance of the same coil, or by the signal induced in a second coil magnetically coupled to the sample. The a.c. field is characterized by its frequency and amplitude. The response of the superconductor can be studied as a function of temperature and the value of a d.c. magnetic field superimposed on the a.c. excitation. Very often, a harmonic a.c. field is used, and both the out of phase and in phase components of the response signal are detected. These components are related to the real and imaginary parts of the samples ac magnetic susceptibility, χ' and χ'' respectively. (Note: the phase is with respect to the phase of the a.c. field, χ' measures the amount of shielding and χ'' the amount of energy loss due to the induced current.) If the sample is a type II superconductor, then depending upon the amount of vortex pinning, the sample geometry, the frequency and the sensitivity of measurement, information related to the lower and upper critical fields, the reversible magnetization, the flux flow resistivity and the flux pinning can be obtained. These properties can also be studied by transport and d.c. magnetization, however, a.c. magnetic measurements do not require the attachment of leads to the sample, and the experimental setup, based on phase detection, is simpler than the magnetometers required for d.c. magnetic studies. The price paid for these experimental advantages is a more complex interpretation of the results, which are function of a considerable number of variables.

Magnetic Susceptibility of Superconductors and Other Spin Systems
Edited by R.A. Hein *et al.*, Plenum Press, New York, 1991

Data analysis is even more complex in the case of high temperature [1] and other high κ superconductors [2], due to the existence of an unpinned vortex regime over a wide range of temperatures and fields. The boundary between the pinned and the unpinned regime is often called the irreversibility line [1, 3], and constitutes one of the most interesting features of these materials. Whether this line is a true phase transition (from a vortex liquid to a vortex solid, either crystalline or glassy), or just a dynamic crossover (i.e., the thermal relaxation becomes so fast that the system reaches equilibrium in the time scale of the experiment), is a question of much controversy [4-10]. It is experimentally clear, however, that the magnetic response of the vortex system undergoes a qualitative change in behavior in a narrow range of the field-temperature (H-T) plane [11]. Most of the specific predictions of the various theoretical models for the irreversibility line have, for the most part, focused on the behavior of the current-voltage (I-V) curves, (more precisely, electric field vs current density, E-J). For a given d.c. magnetic field, the E-J curves measured at high enough temperature (but already below the mean field transition temperature, $T_0(H)$) are linear, reflecting the flux-flow response of the unpinned vortices. At low enough temperature, on the other hand, the behavior is similar to that usually observed in low T_c superconductors, the electric field remains very small until a critical value of the current density, J_c, is reached, and then E increases rapidly with J. Near the irreversibility boundary the E-J curves are highly nonlinear, and neither of these two limiting situations give a good description of the data [11].

The goal of this work is to show how, by carefully choosing the various parameters, a.c. susceptibility measurements can be used to study the magnetic response of the vortex array in different regimes. We present here measurements of χ' and χ'' on $YBa_2Cu_3O_7$ single crystals as a function of temperature, d.c. magnetic field, frequency and amplitude of the a.c. field. We correlate these results with the simpler to analyze, but experimentally difficult to obtain, I-V data.

A.C. RESPONSE OF NORMAL METALS AND TYPE II SUPERCONDUCTORS

In this section we will summarize some very well known results [12-15] about the measurement of a.c. susceptibility in normal metals and conventional type II superconductors. Unless otherwise specified, we will use Gaussian units. Suppose that we put a long cylinder of radius R of a normal metal inside a coil that produces a uniform a.c. field $h_{ac} = h_0 e^{i\omega t}$. In addition, a d.c. field H_0 may be superimposed. The metal is characterized by a current independent resistivity ρ_n (we will assume that the material has a permeability of unity). The time varying field will induce eddy currents in the metal which, according to Lenz's law, will try to reduce the variation of the magnetic flux inside the cylinder. As a consequence, there will be a spatially nonuniform distribution of oscillating current and field inside the cylinder, both of which will decay from the surface toward the center. Using Maxwell's equations and Ohm's law, $E = \rho_n J$, a diffusion equation for any of the variables (B, E or J) can be written (see article by A.F. Khoder this volume). The solutions of those equations depend on the sample geometry and can be very complicated for realistic sample shapes. The simplest case is that of a semi-infinite sample with a planar boundary, where the decay is exponential in the coordinate perpendicular to the surface. For a cylinder, the current and field can be expressed in terms of Bessel functions of the radial coordinate. In any case, the decay will occur in a characteristic length known as the "skin depth" of the normal metal, $\delta_n = (\rho_n c^2/2\pi\omega)^{1/2}$. This can be expressed in more conventional *laboratory* units as $\delta_n = 5030\sqrt{\rho_n/f}$, where δ_n is in centimeters, ρ_n in Ohm-cm, and the frequency f in Hertz.

If a second coil is tightly wound around the sample, an alternate voltage proportional to $d\Phi/dt$, where Φ is the magnetic flux through the sample, will be induced. We can define an average internal field $B_{av}(t) = \Phi(t)/\pi R^2$, where R is the radius of the cylinder. This space averaged field can be decomposed into a time independent part (which in this case equals H_0), and a time dependent part, $b_{av}(t)$. *Due to the linearity of the equations governing the behavior (which is a consequence of the current independence of the resistivity, i.e. Ohms law),* $b_{av}(t)$ *will be harmonic,* i.e., it will not have any Fourier components other than at the drive frequency, ω. As a result, a complex a.c. susceptibility, χ can be defined: $b_{av} = (1 + 4\pi\chi)h_{ac}$, where $\chi = \chi' + i\chi''$. Both components of χ are functions only of the ratio of δ_n to the radius of the cylinder. Calculated values of χ' and χ'' are show in Fig. 1 (a), adapted from ref. 15, as a function of $x = \delta/R$. The interpretation of these results is well known. The inductive component χ' measures the amount of screening of the a.c. field, and the resistive component χ'' measures the loss. When $x >> 1$ ($\delta_n >> R$), the a.c. field largely penetrates the sample, so both χ' and $\chi'' \simeq 0$. When $x << 1$ ($\delta_n << R$), the a.c. field is almost completely screened, thus $\chi' \simeq -1/4\pi$. As the screening currents are confined to a narrow region close to the surface, the losses are low, and $\chi'' \simeq 0$. At some intermediate value of x, when $\delta_n \simeq R$, the screening is approximately one half of the maximum and the losses reach a maximum value.

For fixed ω and R, the measurement of χ is equivalent to a measurement of ρ_n. The variable x can be explored by sweeping the temperature, thus allowing a measurement of $\rho_n(T)$, or, if the metal has a magnetoresistance, sweeping of H_0 will allow a measurement of $\rho_n(H)$. The range of ρ_n that can be experimentally accessed can be adjusted by changing ω and R. A very important assumption here is that δ_n does not depend on h_0, i.e. the material has linear I-V behavior. Hence, χ' and χ'' are independent of the amplitude of the a.c. field used in the experiment.

In an ideal type II superconductor without pinning, a viscous flow of the vortices in the mixed state can be induced by an arbitrarily small current. That movement is dissipative and generates an electric field $E = \rho_{ff} J$, where ρ_{ff} is called the flux flow resistivity [16]. The magnitude of this dissipation can be large, $\rho_{ff} = \rho_n \frac{H}{H_{c2}}$, approaching the normal state resistivity at $H = H_{c2}$. Although the microscopic origin is very different than in the case of a normal metal, the flux flow regime is also ohmic (i.e. ρ_{ff} is independent of J), and so the a.c. response in both cases will be very similar. It is only necessary to replace ρ_n with ρ_{ff} in the expression for the skin depth to calculate a flux flow skin depth, δ_{ff}. (We will ignore the effect of the reversible susceptibility which is very small near T_c, see article by J.R. Clem this volume). Here again χ' and χ'' are independent of the amplitude of the a.c. field. This behavior has been observed, for instance, in almost pinning-free samples of Nb alloys [17].

In most cases, conventional type II superconductors contain pinning sites that trap the vortices and prevent their movement until the critical current has been reached. When the external magnetic field is changed, screening currents are established, as in the case of a normal metal or a pinning-free superconductor. However, in contrast to those cases, if pinning occurs and $J \leq J_c$ there will be no dissipation, and thus the current will not decay with time. (We are neglecting thermal activation effects, which will be discussed later). Due to the persistent nature of the supercurrents, d.c. magnetic fields will also be screened. If the variation of the external field is not too small, the resulting distribution of B and J inside the superconductor is well described by the critical state model (CSM)[18], which asserts that the current density in any macroscopic portion of the sample is either J_c or zero, depending on the magnetic history. In the presence of an a.c. field, the current will alternate sign to screen the field variations. Whenever the direction of the current reverses there will be a hysteretic energy loss due to the work done by the a.c. source to reverse the current direction in the sample. The envelope of B will now decay linearly with distance from the surface, in contrast to the exponential decay in the ohmic case. The a.c. current and field will penetrate to a depth $L_p = \frac{c}{4\pi} h_0/J_c$, beyond which the a.c. field is totally screened. A time dependent average internal field can also be defined for this case, but there is an important difference. *Since* L_p *depends on* h_0, $b_{av}(t)$ *also depends on* h_0. This is a consequence of the nonlinear relation between E and J, and implies that $b_{av}(t)$ will be nonharmonic, i.e. the relation between

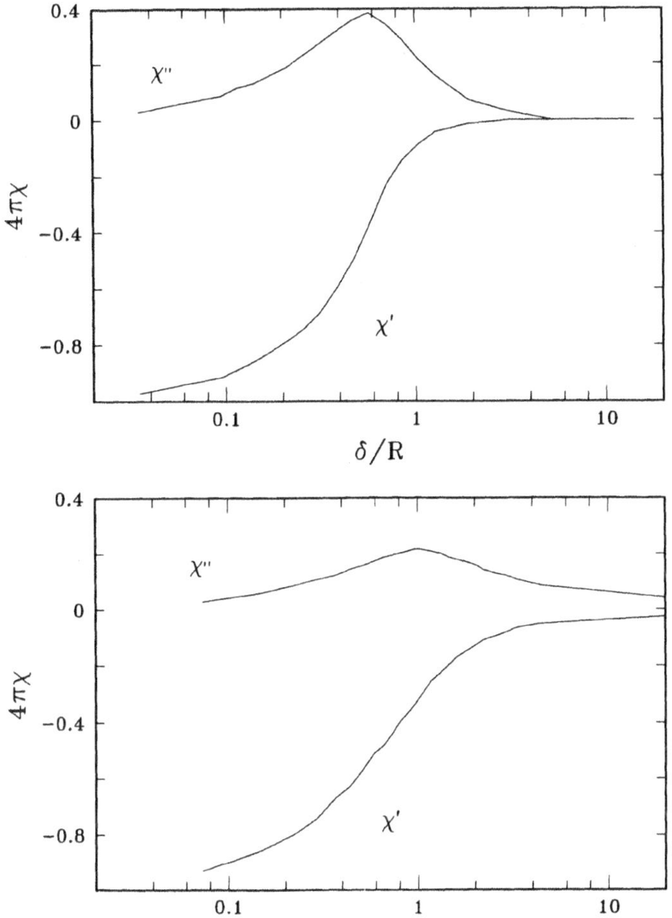

Fig. 1. (a) The calculated a.c. susceptibility response for a cylinder of radius R with a linear resistivity. $4\pi\chi$ is plotted vs. the ratio of the skin depth to the radius. (b) The calculated a.c. susceptibility response for a cylinder of Type II superconductor with radius R. $4\pi\chi$ is plotted vs. the ratio of the Bean penetration depth to the radius. (Adapted from Ref. 15).

$b_{av}(t)$ and h_0 now involves higher Fourier components [19]. Using a lock-in detection techniques we can measure the fundamental components χ' and χ'', which are functions of the ratio of L_p to the characteristic dimension of the sample. The predicted behavior of χ' and χ'' for a cylinder of radius R are shown as a function of L_p/R in Fig. 1 (b), adapted from ref. 15.

The behavior of both systems as a function of the characteristic length, x, is very similar. Both show steps in χ' and peaks in χ'', which in both cases occur when $x \simeq 1$, i.e., when the distance of penetration of the a.c. field is approximately the radius of the cylinder. The difference is that in the case of the normal metal or an unpinned superconductor, δ is proportional to $\omega^{-1/2}$ and is independent of h_0, while for the superconductor with pinning, L_p depends linearly on h_0 and is independent of ω (apart from thermal relaxation effects). For samples of different geometries, the details of the shape of the χ' and χ'' curves will differ slightly, however, the peak in χ'' and the step in χ' will always occur when δ_n, δ_{ff}, or L_p approximately equal a characteristic sample length. Moreover, the dependence of δ on ω and L_p on h_0 holds for any geometry.

THE IRREVERSIBILITY LINE AND THE I-V CURVES

In reality, the behavior of high T_c and other high κ superconductors is more complicated due to thermal activation. As the superconductor is cooled through the mean field transition temperature, $T_0(H)$, fluctuation effects round the sharp change from normal state behavior to flux flow behavior. At some lower temperature vortex pinning develops and the I-V curves develop nonlinear character. Figure 2 shows the resistive transition of a $YBa_2Cu_3O_7$ single crystal in a 7 Tesla field [11]. Also shown in this figure is the extrapolated normal state resistivity and the estimated value of the Bardeen-Stephen flux flow resistivity. There is a region where the resistivity suddenly drops below the Bardeen-Stephen value. The cause of this behavior is not agreed upon, but the I-V character in this region remains linear. Only at lower temperature does nonlinear character develop. Figure 3 shows 19 E-J curves at 7 Tesla for this crystal taken 0.1 K apart from 80.94 K to 78.91 K (this temperature range is indicated in Fig. 2) [11]. The onset in nonlinear behavior is quite sudden, taking place at 80.3 $\pm$0.1 K. For the purposes of this discussion we will call the field dependence of the temperature where this nonlinearity develops the irreversibility line, $H_{irr}(T)$. The rationale for this definition is that nonlinearities in the I-V curves will necessarily lead to irreversible magnetic behavior. This definition is not universally accepted by the High-T_c community.

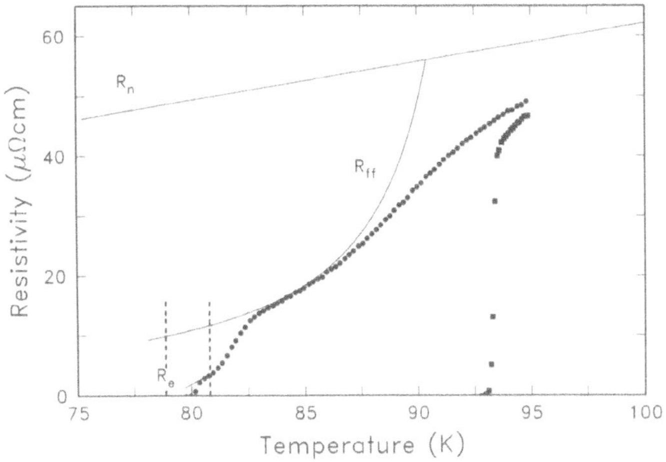

Fig. 2. The resistivity vs. temperature at 0 Tesla (squares) and 7 Tesla (circles). The extrapolation of the normal state resistance, $\mathbf{R_n}$, and an estimate of the Bardeen-Stephen flux flow resistivity at 7 Tesla, $\mathbf{R_{ff}}$, are also shown. Vertical dotted lines indicate the temperature range for the E-J curves shown in Fig. 3. (From Ref. 11).

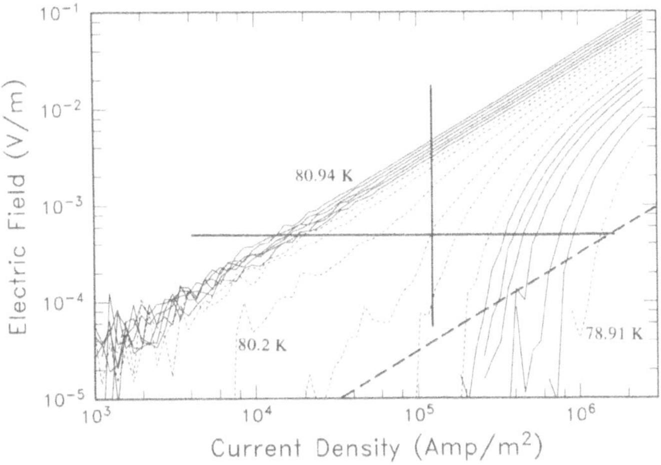

Fig. 3. Nineteen E-J curves measured at 7 T at 0.1 K intervals from 80.94 to 78.91 K. Solid horizontal and vertical lines indicate values of E and J respectively for a typical crystal at half screening in a.c. susceptibility measurements at f = 0.5 MHz and h_0 = 0.05 Oe. Dashed inclined line corresponds to a ρ_{ff} of $3 \times 10^{-10} \Omega/m$ (From Ref. 11).

However our purpose is to describe in detail a particular experimental technique and discuss what the results of this technique imply about the E-J data, not to argue about the particular models and definitions. There is some evidence in Fig. 3 of an additional linear region in the E-J curves at low current in the data at 80.3 and 80.2 K. There is considerable controversy as to whether this linear region at low current continues to zero temperature or vanishes at a nonzero temperature, T_g[4, 5]. However, for the experiments we will discuss here this question is moot. Our experiments are carried out in a frequency-amplitude regime where the existence of a linear resistivity at low current and temperature is not detectable.

The development of nonlinear character in the E-J curves is a clear fingerprint of the presence of vortex pinning, but it does not mean that a critical current can be defined. In fact, it is apparent from Fig. 3 that no persistent current can be sustained at, for example, 80.2 K, because there is a significant E at any finite J. Ideally, the critical state model implies E = 0 for $J < J_c$, which should be reflected in Fig. 3 as *vertical* lines located at $J = J_c(T)$. A tendency towards that behavior is in fact observed in the lower temperature curves in Fig. 3. Of course, the existence of thermal activation implies that in a finite sample there is *always* a finite E at finite J. However, far enough below the irreversibility line E is extremely small for $J < J_c$, and for all practical purposes the screening currents are persistent. The fact that E is not strictly zero will be reflected, for instance, in a slow relaxation in d.c. measurements, and in a small frequency dependence in a.c. measurements.

EXPERIMENTAL SETUP

We have developed a very simple single coil technique[20] to measure the a.c. susceptibility. A small free standing copper coil, ~ 1mm internal diameter is constructed by machining a slot in a plexiglass rod. The coil, ~ 50 turns of 25 μm copper wire is wound in this slot and potted with epoxy. The plexiglass is then dissolved with methylene chloride. The coil typically has an inductance L ~ 2.5 μH and a series resistance R_s ~ 1 ohm at 90 K. The coil is mounted on a silicon chip and attached to a copper heat sink containing a thermometer and heater. Rigid stainless steel coax is used to connect the coil to the room temperature end of the cryostat. The coax adds a series resistance of ~ 6 ohms and has a small temperature (and thus helium level) dependence. The complex impedance of the coil is measured with an HP4194 impedance analyzer. This instrument probes the system with a small a.c. current and deconvolves the measured voltage into L and R_s. This current produces an a.c field of ~0.5 Oe/mAmp. The inductance is a measure of the total flux threading the coil and the series resistance is a measure of the the losses due to the d.c. resistance of the coil and any eddy current losses in the coil or any conductors near the coil. This technique works very well in the frequency range of 20 kHz to 4 MHz with measuring currents of 0.1 to 20 mAmp (h_0 = 0.05 - 10 Oe). The earliest crystals were small enough to fit inside the coil and thus were immersed in a fairly uniform a.c. field. The size of available crystals have grown with time, now occasionally exceeding 3 x 3 mm, and are placed against one surface of the coil. The idealized geometry which most closely resembles our experimental situation is the infinite sheet with the a.c. field applied perpendicular to one surface. The induced current flows in the plane of the surface, which for the samples described here is the a-b plane of a $YBa_2Cu_3O_7$ crystal. The peak in χ'' and the midpoint of the step in χ' represent the temperature where the characteristic length, either δ_{ff} or L_p, is approximately equal to the thickness of the crystal. This technique also works very well for thin films where the geometry even more closely resembles that of a infinite sheet [21]. Above 4 MHz, the response of the system is complicated by a resonances in the cable system, although it is still useable at certain frequencies, and below 20 kHz the signal to noise ratio is low.

The copper heat sink is mounted in a continuous He gas flow insert in a superconducting magnet. The temperature of the gas is controlled at typically 60 K and the temperature of the sample is controlled separately at any desired value above the gas temperature. Data is typically taken by setting the d.c. field while the sample is above the transition temperature and then slowly cooling the sample block. The inductance and series resistance are monitored as a function of the temperature. Figure 4 shows a set of data taken on a single crystal at 1 Tesla d.c. field applied parallel to the c-axis at a frequency of 1 MHz with an applied a.c. field of 0.05 Oe. The data in Fig. 4 is qualitatively similar to that in Figs. 1 (a) and (b). There is a peak in the series resistance and a drop in the inductance. The inductance of the empty coil is quite constant over a wide range of temperature. The series resistance shows a temperature dependence, decreasing more or less linearly due to the temperature dependence of the resistivity of the copper in the coil and the heat sink. The signal due to the empty coil can be measured and subtracted to obtain the signal due to the sample, however for the limited temperature range of interest, 77 K - 95 K, it is usually adequate to use the data above the transition to construct a base line. After the background has been subtracted the inductance is a measure of χ' and the series resistance a measure of χ''.

RESULTS

Flux Flow Regime

As we described above, Figs. 2 and 3 show that at high enough temperature the E-J curves of $YBa_2Cu_3O_7$ are linear, indicating that the vortices move in a flux flow regime. We

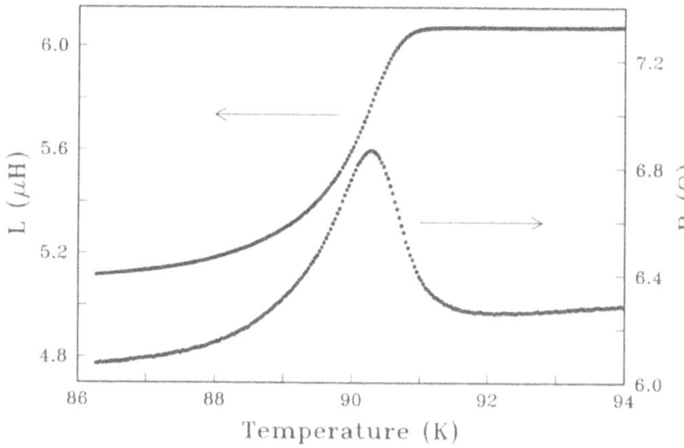

Fig. 4. The inductance and series resistance of the coil containing the $YBa_2Cu_3O_7$ crystal measured at 1 MHz in a d.c. field of 0.5 Tesla using an a.c. field of 0.05 Oe.

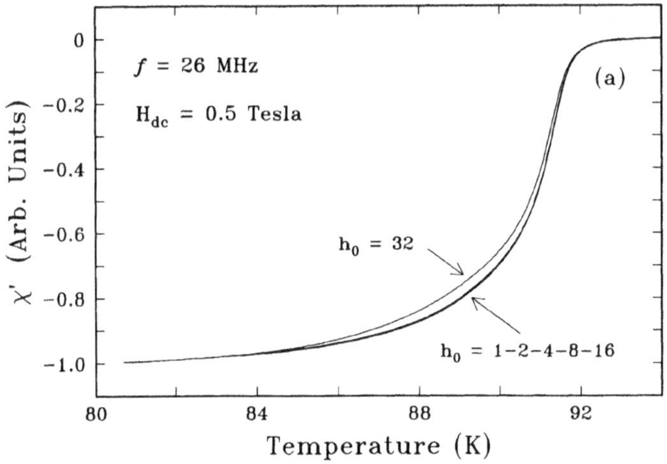

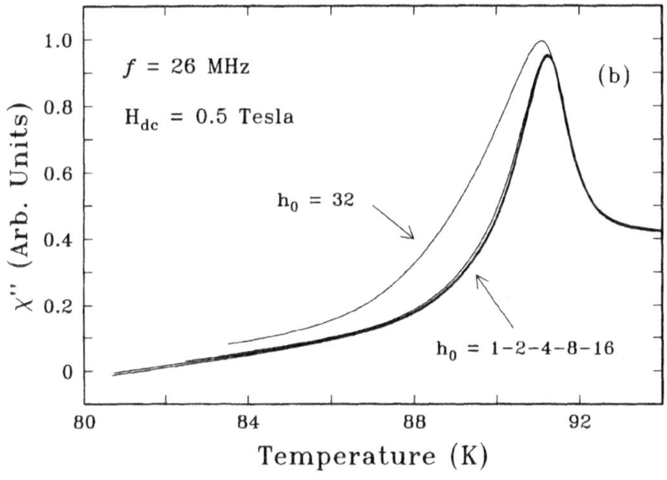

Fig. 5. The a.c. susceptibility response of a $YBa_2Cu_3O_7$ crystal measured in a d.c. field of 0.5 Tesla using a 26 MHz a.c. field at 6 different a.c. field amplitudes h_0. Units for h_0 are arbitrary (see text).

thus expect that the a.c. response should be characterized by δ_{ff}, and therefore χ' and χ'' should be independent of h_0. To observe this behavior, we must select the experimental parameters in such a way that the relevant features of χ' and χ'' occur in this temperature region. This requires that we use a frequency such that, δ_{ff} equals the thickness of the crystal, d = 30 μm, within this temperature region. As the temperature is decreased, ρ_{ff} diminish from $\rho_n \simeq 50$ $\mu\Omega$cm at the transition to about 1 $\mu\Omega$cm at the temperature where the nonlinearities start to develop. The corresponding frequencies are the order of 100 MHz to 3 MHz. Figure 5 (a) shows χ' as a function of temperature as obtained using a frequency of 26 MHz with a d.c. field of 0.5 Tesla applied parallel to the c axis. For our geometry it is not easy to relate quantitatively the change in the inductance of the coil to χ' as it was in the long-cylinder case, so the values of $\chi'(T)$ are shown in arbitrary units. Six curves are actually plotted, each one obtained using an a.c. field of twice the amplitude of the previous one (i.e., h_0 for the various curves follows the relation 1:2:4:8:16:32). The absolute values of h_0 are not known, because at this high frequency a significant fraction of the drive current is diverted by the parasitic capacitance in the system, and does not pass through the coil. However, h_0 is still proportional to the drive current. We estimate that h_0 is less than 1 Oe for the maximum amplitude (i.e., h_0 = 32). It is seen in the figure that the curves taken at amplitudes 1 to 16 coincide perfectly within the experimental resolution, thus proving that the a.c. response is consistent with the linear behavior observed in the flux flow regime. The curve obtained at the maximum amplitude, 32, shows a slight deviation, that may be a consequence of self heating of the crystal. Figure 5 (b) shows χ'' for the same frequency and the same set of h_0, again in arbitrary units. The linearity of the response is also apparent here. Assuming that the χ'' peak occurs at the temperature where δ_{ff} = 30 μm (the sample thickness), the position of the peak corresponds to ρ_{ff} = 9 $\mu\Omega$cm; i.e. $\simeq 0.18\ \rho_n\ (T_0)$.

Critical State Regime

As the frequency is reduced the nonlinear E-J behavior should be reflected in the a.c. response. For f = 100 KHz, this crystal should exhibit the maximum in χ'' when $\rho_{ff} \simeq 0.03$ $\mu\Omega$cm. Inspection of Fig. 3 shows that there is no *linear* E-J curve having such a small resistivity. In that figure we have drawn a straight line corresponding to that value of ρ_{ff}; it intercepts a number of highly nonlinear E-J curves. The effect on the a.c. response is shown in Figure 6 (a) and (b), where $\chi'(T)$ and $\chi''(T)$ respectively are plotted for a d.c. field of 0.5 Tesla, f = 100 KHz, and 6 values of h_0, which again follow the sequence 1:2:4:8:16:32. At this frequency cable effects are negligible so that the absolute values of h_0 are known: h_0 = 1 corresponds to 0.1 Oe. The difference between Figs. 5 and 6 is apparent, the latter having a clear amplitude dependence which is a consequence of nonlinearity in the electrodynamic response.

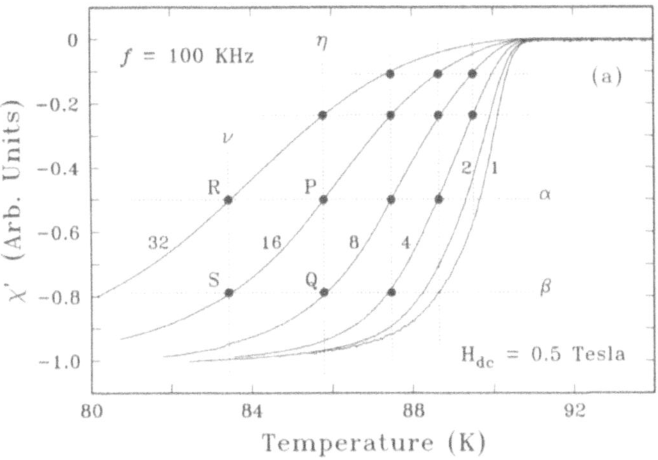

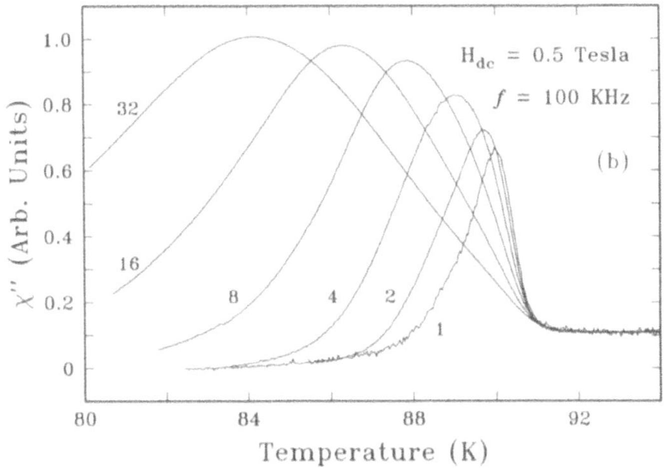

Fig. 6. The a.c. susceptibility response of a $YBa_2Cu_3O_7$ crystal measured in a d.c. field of 0.5 Tesla using a 100 kHz a.c. field. The numbers on each curve indicate the amplitude of the a.c. field h_0 (in arbitrary units). The geometrical construction is used to extract J_c (see text).

As we pointed out before, the observation of nonlinear behavior is not sufficient to prove that the critical state model applies. We will now present a procedure that allows us to test whether the χ' data can be accurately described by a critical state model and, if it is, to derive the temperature dependence of J_c independent of the particular field distribution or sample geometry. We start by noticing that, *provided that the critical state model applies*, for a given sample and coil geometry χ' is only function of L_p. (This is also true for χ'', although we will not analyze it here). Thus in Fig. 6 (a) the horizontal lines such as those marked α and β are lines of constant L_p. On the other hand, J_c is only function of T, so the vertical lines, denoted as η and ν, are lines of constant J_c. This last observation implies that $J_c(P) = J_c(Q)$, where P and Q are points indicated in the figure. The point P belongs to the curve of $h_0 = 16$. The point R has the same L_p as P, but belongs to the curve of $h_0 = 32$. Since the CSM establishes that $J_c \propto h_0 / L_p$, then $J_c(R) = 2\, J_c(P)$. Similarly, the point S of the curve of $h_0 = 16$ has the same L_p as the point Q of the curve of $h_0 = 8$, and so $J_c(S) = 2\, J_c(Q)$. By combining the preceding equalities we find that $J_c(R) = J_c(S)$, implying that R and S should be at the same temperature, as is indeed the case. The property that the four points P,Q,R and S form a rectangle is a consequence of the fact that χ' is completely determined by L_p (and that h_0 for the 3 curves be in the ratio 1:2:4), and thus is a proof of the applicability of the critical state model. In Fig. 6 (a) some other examples of successfully inscribed rectangles are shown, together with some unsuccessful attempts. Notice that this rule is well satisfied by the curves taken at higher amplitudes, and fails with the curves of lower amplitudes. Indeed, the critical state model *must* fail if h_0 is small enough. Since the total displacement of a vortex in each cycle is proportional to the local amplitude of the a.c. field, there is a threshold value of h_0 below which the vortices will not be able to move far enough to leave the pinning site [22]. In that situation, each vortex will oscillate around the minimum of its pinning site rather than jumping from one to another. The response to an a.c. field in this limit will be *linear*, in the sense that it will decay approximately exponentially from the surface, with an *amplitude independent* characteristic distance known as Campbell's penetration depth λ_C. This characteristic length is analogous to the penetration depth in the Meissner state, although is usually much larger. At temperatures where the pinning energy is large enough, the screening length is the larger of L_p and λ_C. A linear response is thus expected for $h_0 < \frac{4\pi}{c} J_c \lambda_C$. At higher temperatures the situation is more complicated [23]; as J_c tends to zero, λ_C diverges and the appropriate characteristic length to describe the linear response at low amplitude is a combination of λ_C and δ_{ff}.

The procedure described above allows us to determine the temperature dependence of J_c. Suppose that in arbitrary units we define that $J_c(P) = 1$. Then moving along the line α we obtain $J_c(R) = 2$. This process can be continued (in both directions) to extract tem-

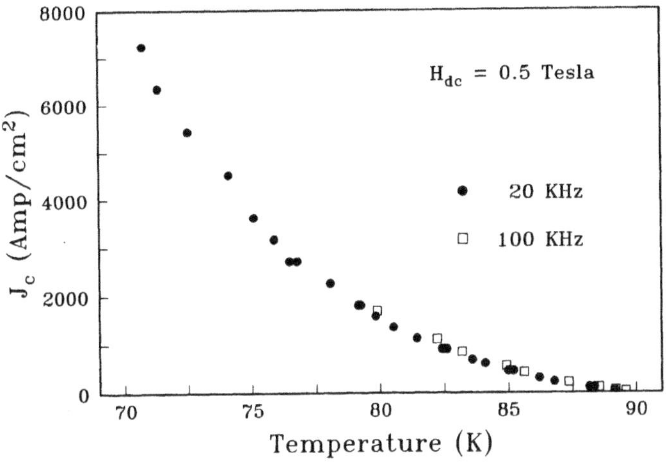

Fig. 7. The critical current density as a function of temperature derived from the a.c. susceptibility data of Fig. 6 and similar data taken at 20 kHz. See text for the determination of the absolute values of J_c.

perature where $J_c = 0.5, 0.25, 0.125$, etc. Along the line β we find $J_c(Q) = 1$, $J_c(S) = 2$, and so forth. The J_c vs temperature so obtained is shown in Fig. 7. The same figure shows the J_c obtained by the same procedure at a frequency of 20 KHz. In that case the geometrical rule described previously is also satisfied. The shape of $J_c(T)$ can be obtained in this way regardless of the geometry of the sample or the coil. The absolute values of J_c are not obtained, but the results at different frequencies can be compared using the fact that data taken with the same h_0 have the same J_c for the same value of χ'. The agreement between the two sets of data is a another clear indication that the critical state model describes the a.c. response in this regime. We can estimate the absolute value of J_c in Fig. 7 by assuming that for our geometry, $L_p = d$ at half screening. The scale of the vertical axis in Fig. 7 has been determined in that way. For instance, at $T = 90$ K we obtain $J_c \simeq 25$ Amp/cm^2. We note that for a crystal of this size, the sensitivity limit of a commercial SQUID magnetometer is ~1000 Amp/cm^2.

Intermediate Regime

We can now analyze the behavior at some intermediate frequency. Data obtained with $f = 9$ MHz are shown in Figure 8. These data are similar to those of Figs. 5 and 6. Figure 8 (a) shows that χ' is not amplitude independent, nor is the critical state condition satisfied.

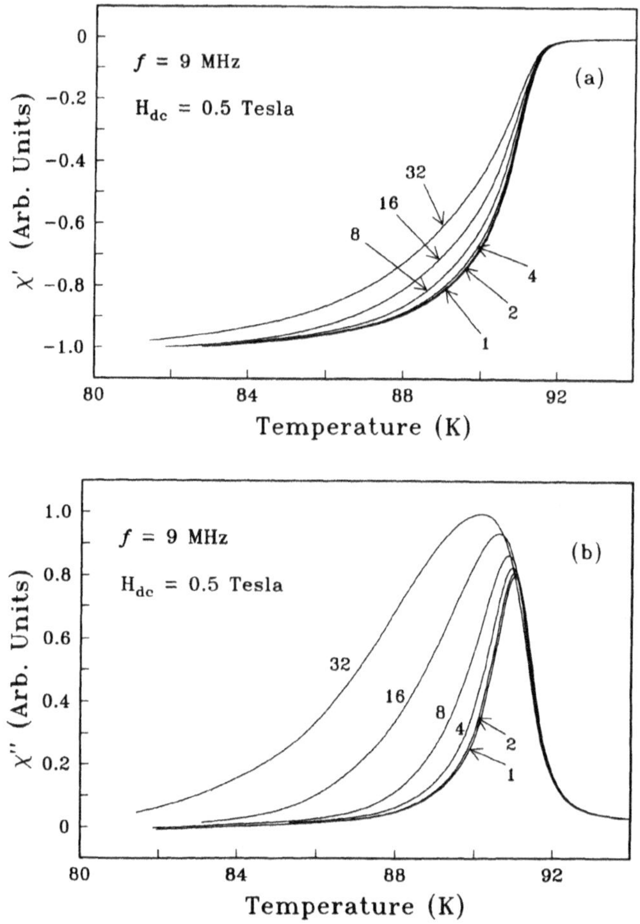

Fig. 8. The a.c. susceptibility response of a $YBa_2Cu_3O_7$ crystal measured in a d.c. field of 0.5 Tesla using a 9 MHz a.c. field. The numbers on each curve indicate the amplitude of the a.c. field h_0 (in arbitrary units).

The curves of $h_0 = 1$ and 2 are superimposed, for $h_0 = 4$ there is a small deviation, then for $h_0 = 8$, 16 and 32 there is a clear amplitude dependence. (Here again parasitic capacitance precludes us from knowing the absolute values of h_0. We estimate that $h_0 = 32$ corresponds to less than 1 Oe). Examination of χ'' in Fig. 8 (b) gives an additional information: the amplitude dependence develops around T = 91 K. This behavior can be qualitatively understood as follows: For T > 91 K the sample is in the flux flow regime. This can be confirmed by examination of Figs. 6 and 7. At lower temperatures the E-J curves are nonlinear. However, at low amplitude we are only exploring the temperature dependence of the low J portion of the E-J curves. The fact that χ' is amplitude independent implies that the E-J curves are linear in that region. With higher h_0 we probe the behavior at larger J, thus exploring a different portion of the E-J curves where E/J is current dependent and therefore, χ' is amplitude dependent. At fixed temperature, the larger h_0 the lower the screening, which implies that E/J is larger than in the initial linear portion. This description is in qualitative agreement with the E-J curves shown in Fig. 3 for temperatures near the onset of nonlinearity. A quantitative description of this intermediate regime is very difficult; the combination of Maxwell's equations with the constitutive relation E(J) produces nonlinear diffusion equations for B, J or E. Moreover, the effect of the Campbell penetration depth [22], λ_C, which is an a.c. effect and therefore not present in the d.c. E-J behavior must be taken into account.

ON THE DETERMINATION OF THE IRREVERSIBILITY LINE

Now with our understanding of the a.c. susceptibility response we can compare the a.c. susceptibility determination of the irreversibility line with that of transport data. Figure 3 shows that the E-J curves develop nonlinear character when $\rho_{ff} \simeq 1 \mu\Omega$cm. (Note: the E and J scales in the transport data at other d.c. fields are very similar). As we calculated above, this value of the resistivity corresponds to $\delta_{ff} = 30$ μm., the typical crystal thickness, when $f \simeq 3$ MHz. If we constrain ourselves to frequencies lower than that, the maximum in χ'' will occur in the irreversible regime. We can estimate the point in the E-J plane where the maximum in χ'' will be observed for given values of ω and h_0. We first notice that if h_0 is small, the current density through the crystal at half screening (which roughly coincides with the peak in χ'') will also be small. For instance, in a 30 μm thick crystal and for $h_0 = 0.05$ Oe, that current density (estimated using $\nabla \times \vec{H} = \frac{4\pi}{c}\vec{J}$), is ~ 12 Amp/cm^2. A vertical line corresponding to this current is shown in Fig. 3. On the other hand, using the Maxwell equation $\nabla \times \vec{E} = \frac{1}{c}\, \partial\vec{B}/\partial t$, the electric field at half screening can be estimated as $E \simeq \frac{1}{c} h_0 \omega d$. For f = 0.5 MHz, h_0 = 0.05 Oe and d = 30 μm, $E \simeq 5 \times 10^{-6}$ V/cm. A horizontal line corresponding to this field is also shown in Fig. 3. The intersection of these two lines roughly indicates the location where the maximum in χ'' and the half screening in χ' would be expected. We see that, *as an experimental fact*, for typical $YBa_2Cu_3O_7$ crystals and for this selection of parameters the maximum in χ'' will occur almost precisely at the temperature where the onset in nonlinear E-J behavior develops. Figure 9 shows the field dependence of the temperature of the peak in χ'', measured a frequency of 0.5 MHz and an a.c. field of 0.05 Oe [11] for the same crystal used for the E-J measurements shown in Fig. 3. Also plotted is the field dependence of the onset of nonlinear E-J behavior derived from sets of data similar to that shown in Fig. 3. The similarity of both determinations is apparent. If h_0 is kept constant and ω is reduced, E will diminish and the intersection will occur at lower temperature, as experimentally observed. If h_0 is increased at fixed ω, the intersection will also shift to lower temperature.

It is not necessary to measure the transport data in order to select an appropriate frequency and amplitude for an accurate determination of the irreversibility line from a.c. susceptibility. By an examination of the frequency and amplitude response it is possible to select parameters such that the nonlinear response develops very near the peak in χ''.

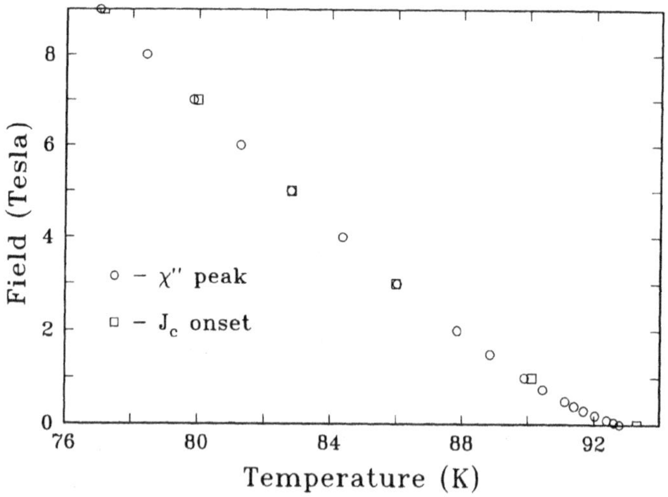

Fig. 9. The field dependence of the peak in χ'' at 0.5 MHz (circles), along with the field dependence of the onset in nonlinear E-J behavior (squares) taken from Fig. 3 and similar sets of curves at other fields. (From Ref. 11).

SUMMARY

We have reviewed the expected a.c. susceptibility response for normal metals and idealized superconductors. We have presented experimental data on single crystal $YBa_2Cu_3O_7$ which show that at high frequency the response is consistent with flux flow behavior and that it is possible to determine the temperature dependence of the flux flow resistivity. By using lower frequency a.c. field and varying the amplitude of the field, we have shown that the behavior is in good agreement with the critical state model and have been able to extract the temperature dependence of the critical current density with high sensitivity. We have also shown that by judicious selection of experimental parameters the onset temperature for nonlinear response can be easily detected.

ACKNOWLEDGEMENT

We are grateful for several useful conversations with J.R. Clem and A.M. Campbell.

REFERENCES

1 K. A. Müller, M. Takashige, and J. G. Bednorz, Phys. Rev. Lett. **58**, 1143 (1987).

2 M. Suenaga, A. Ghosh, D. O. Xu, and D. O. Welch, Phys. Rev. Lett. **66**, 1777 (1991).

3 Y. Yeshurn and A. P. Malozemoff, Phys. Rev. Lett. **60**, 2202 (1988).

4 M. P. Fisher, Phys. Rev. Lett. **62**, 1415 (1989).

5 D. S. Fisher, M. P. Fisher, and D. A. Huse, Phys. Rev. B **43**, 130 (1991).

6 A. Houghton, R. A. Pelcovitz, and A. Sudbo, Phys. Rev. B **40**, 6763 (1989).

7 A. P. Malozemoff, T. K. Worthington, Y. Yeshurun, and F. Holtzberg, Phys. Rev. B **38**, 7203 (1988).

8 P. L. Gammel, L. F. Schneemeyer, and D. J. Bishop, Phys. Rev. Lett. **66**, 953 (1991).

9 C. W. Hagen and R. Griessen, Phys. Rev. Lett. **62**, 2857 (1989).

[10] S. P. Obukhov and M. Rubenstein, Phys. Rev. Lett. **65**, 1279 (1990).

[11] T. K. Worthington, F. Holtzberg, and C. A. Feild, Cryogenics **30**, 417 (1990).

[12] A. Russell, *Treatise on the Theory of Alternating Currents* (Univ. Press, Cambridge, 1914).

[13] R. M. Bozortt, *Ferromagnetism* (D. Van Nostrand, New York, 1951).

[14] J. D. Jackson, *Classical Electrodynamics* (John Wiley & Sons, New York, 1962).

[15] J. R. Clem, in *this volume*.

[16] J. Bardeen and M. J. Stephen, Phys. Rev. **140**, A1197 (1965).

[17] J. R. Clem, H. R. Kerchner, and S. T. Sekula, Phys. Rev. B **14**, 1893 (1976).

[18] C. P. Bean, Phys. Rev. Lett. **8**, 250 (1962).

[19] A. Shaulov and D. Dorman, Appl. Phys. Lett. **53**, 2680 (1988).

[20] T. K. Worthington, W. J. Gallagher, D. L. Kaiser, F. Holtzberg, and T. R. Dinger, Physica **153-155C**, 32 (1987).

[21] L. Civale, T. K. Worthington, and A. Gupta, Phys. Rev. B **43**, 5425 (1991).

[22] A. M. Campbell and J. E. Evetts, Adv. Phys. **21**, 199 (1972).

[23] M. W. Coffey and J. R. Clem, Preprint.

HIGHER HARMONICS OF SINGLE-CRYSTAL $YBa_2Cu_3O_y$ THIN FILMS AND BULK SAMPLES

Hiromasa Mazaki, Kazunuki Yamamoto,* and Hiroshi Yasuoka

Department of Mathematics and Physics
The National Defense Academy
Yokosuka 239, Japan
*Ube Laboratory
Ube Industries Ltd.
Ube 755, Japan

INTRODUCTION

The measurement of complex susceptibility $\chi = \chi' - i\chi''$ in a weak ac magnetic field, is one of the more useful methods for the study of magnetic response of superconductors.[1-4] For sintered high-T_c superconductors, χ generally involves two components, the intragrain bulk component and the intergrain coupling component. These two components can typically be specified by changing the amplitude of the ac measuring field. The intragrain bulk component is essentially insensitive to the change in the field amplitude H_{ac}, while the coupling component is sensitive to changes in $|H_{ac}|$ as small as 10^{-4} mT.[5]

The time-dependent magnetization $M(\omega t)$ of a specimen for magnetic field $H(\omega t) = H_{dc} + H_{ac}\mathrm{Re}[\exp(i\omega t)]$, can be expressed in the form of the Fourier expansion:

$$M(\omega t) = \chi_0 H_{dc} + H_{ac} \sum_{n=1}^{\infty} \mathrm{Re}[\,\chi_n \exp(in\omega t)] \ , \qquad (1)$$

where Re[ ] denotes the real part of the complex variable and ω (= $2\pi f_1$) is the fundamental frequency. The first term $\chi_0 H_{dc}$ is a time independent contribution or "offset" which is due to the presence of the dc magnetic field H_{dc}. The second term is the time dependent component associated with the ac magnetic field $H_{ac}\cos\omega t$. Here we define the harmonic ac susceptibility $\chi_n = \chi_n' - i\chi_n''$ (n = 1, 2, 3...) as the Fourier coefficient of the magnetization, where χ_1 is the fundamental susceptibility and the others are the higher-harmonic susceptibilities.

To elucidate the generation of higher-harmonic components, several models have been proposed. These are roughly divided into two categories. One is constructed on the basis of the critical-state model, the notion of which was first introduced by Bean on the assumption that the critical current density is independent of field.[6] In his model, the existence of odd-harmonic susceptibilities can be derived. Bean's approach was extended by Kim et al.[7] and Anderson,[8] who assumed that every region of the sample can support a critical current density determined only by the local magnetic field. In this model, one can derive only odd-n harmonics when a pure ac field is applied on a superconductor. But if a dc field is superimposed on an ac field, the generation of even-n harmonics takes place.

The other category by which a higher-harmonic susceptibility is predicted is the so-called weak-link model which was proposed by Ishida and Mazaki[9] for a multiconnected Josephson network of low-temperature superconductors. The expressions derived by this model are equivalent to those of Rollins and Silcox.[10] The weak-link model proved the existence of odd-harmonic susceptibilities and qualitatively explained the profiles of χ_n versus temperature T observed for weakly coupled networks.

Recently, Ishida and Goldfarb[11] carried out detailed measurements of χ_n (n = 1, 2,...10) for a sintered $YBa_2Cu_3O_y$ (YBCO) superconductor, where they superimposed a dc field H_{dc} on an ac field. The observed results were analyzed by a theory based on a simplified Kim model for critical current density,[12] and were found to be in good agreement with the temperature- and field-dependent features of χ_n.

In this paper, we will first describe the details of the measuring system used for our ac susceptibility studies. Then we will present the observed χ_n of an Fe-doped YBCO prepared by sintering (10 mm in diameter and 2 mm in thickness), of single-crystal YBCO thin films (1000-6000 Å), and of a single-crystal YBCO bulk specimen (1 x 1 x 0.5 mm^3). Analyses of the data were carried out within the framework of the Kim-Anderson model,[7,8] the modified Bean model,[13] and the Ishida-Mazaki model.[9] Some discussion of the magnetic response of these samples will also be given.

EXPERIMENTAL

Measuring System

The system used to measure the superconducting transition in terms of the fundamental susceptibility $\chi_1 = \chi_1' - i\chi_1''$, is the Hartshorn bridge, of which the schematic diagram is shown in Fig. 1. This bridge essentially consists of two coils M_1 and M_2, and the phase-shift potentiometer R. The cryostat coil M_2, in which a sample is located, consists of two coaxial cylindrical coils. The inner coil (primary) is 7735 turns (in 8 layers and 155-mm long, 923 Ω at room temperature) of Cu wire (0.14 mm in diameter) coated by polyvinyl formal (PVF). The inner diameter of the Lucite (polymethyl methacrylate) bobbin is 29 mm. The outer coil (secondary) is divided into two sections. Each section has 4880 turns of the PVF wire (in 9 layers and 44-mm long) in opposite direction to each other. The inner diameter of the secondary-coil bobbin is 35 mm and the total resistance of the secondary

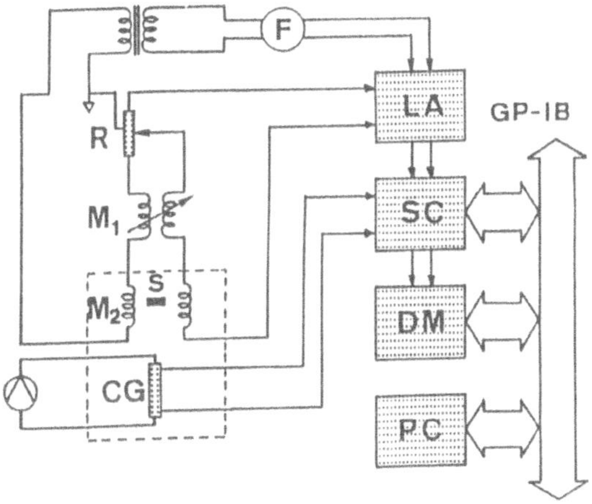

Fig. 1. Schematic diagram of the measuring system of the fundamental susceptibility. M_1: variable standard mutual inductance, M_2: cryostat coil, R: phase-shift potentiometer, S: sample, CG: carbon-glass thermometer, F: function generator, LA: two-phase lock-in amplifier, SC: universal scanner, DM: computing digital multimeter, PC: personal computer.

coil is 747 Ω at room temperature. As a variable standard mutual inductance M_1, Tinsley type 4229 is used, of which the working range is 0.1 μH-1.111 mH with the sensitivity of 0.01 μH. The phase-shift potentiometer is composed of a manganin wire (1 mm in diameter and 1 m in length) and of a slide contact. The two-phase lock-in amplifier is Ithaco 393. The correction factor for calibration of the ac field from the supplied ac current was calculated assuming that the length of the inner cryostat coil is infinite.

The measuring system for the higher-harmonic susceptibility is schematically shown in Fig. 2, where two function generators are used to extract

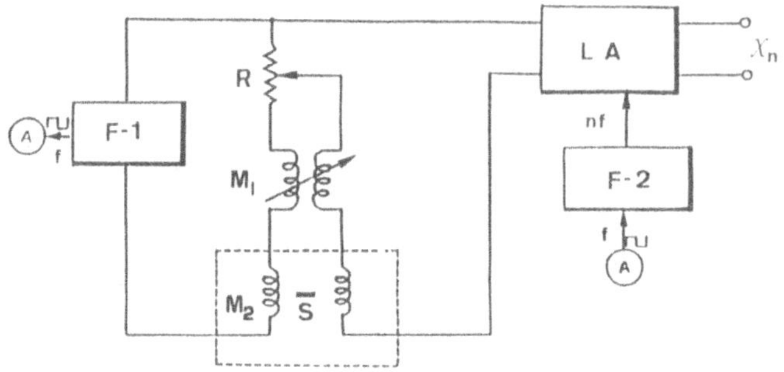

Fig. 2. Schematic diagram of the measuring system of higher-harmonic susceptibility. M_1: variable standard mutual inductance, M_2: cryostat coil, R: phase-shift potentiometer, S: sample, F-1: function generator 1, F-2: function generator 2, LA: two-phase lock-in amplifier.

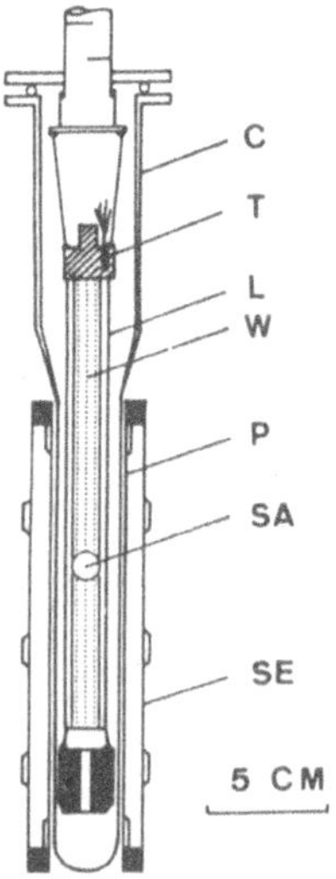

Fig. 3. Details of the sample holder and the adiabatic cell. C: adiabatic cell, T: carbon-glass thermometer mounted in a Cu block, L: lucite pipe, W: bundle of Cu wires, P: primary coil, SA: sample, SE: secondary coil.

the higher-harmonic components. The ac-magnetic field was applied to the sample using the function generator F-1, and the second generator F-2 was used to generate nth-harmonic sinusoidal wave synchronized with F-1. By means of the Lissajous figure on an oscilloscope, we monitored the phase slippage throughout one run of the measurement.

Using the above systems, we can measure the real and imaginary components of χ_n continuously and simultaneously as a function of temperature. The sample is placed inside a vacuum tight pyrex glass chamber which is then inserted into the cryostat coil (see Fig. 3). It is fixed into position so that the sample to be measured is located at the center of the upper section of the secondary coil. After evacuation of the sample chamber, the sample temperature is lowered from room temperature. The temperature variation was controlled by the pressure of He exchange gas (10^{-2}-10^3 Torr), and no heater was used. Since the thermal conductivity of oxide superconductors is rather poor compared to that of usual metallic superconductors, one has to be very careful in the temperature control. In order to minimize the temperature gradient in the sample (or between the sample and the thermometer), the temperature was changed quite slowly. Empirically, the temperature variation of 0.2 K/min was sufficient to eliminate the temperature gradient and the reproducibility of data was satisfactory. A null adjustment of the Hartshorn bridge was made at a sample temperature just above the onset of the superconducting transition. With the bridge in the balanced condition, the phase of the lock-in amplifier was adjusted so that a large change of the bridge inductance produced only an in-phase, or χ', signal with no discernable out-of-phase, or χ'', signal. Temperature was measured with a carbon-glass thermometer. In this experiment, no attempt was made to shield the sample from the Earth's magnetic field, so the applied ac field was superimposed on the Earth's field.

Since the bridge was balanced at the sample temperature just above T_c, the off-balance signal $v(\omega t)$, induced in the secondary circuit by decreasing temperature, is directly proportional to the time derivative of the sample magnetization $M(\omega t)$. Using Eq. (1), one gets

$$v(\omega t) = -\gamma\frac{dM(\omega t)}{dt} = -\gamma H_{ac}\sum_{n=1}^{\infty}\mathrm{Re}[in\omega\chi_n\exp(in\omega t)]$$

$$= -\gamma H_{ac}\sum_{n=1}^{\infty} n\omega\,[-\chi_n'\sin(n\omega t) + \chi_n''\cos(n\omega t)]\,, \tag{2}$$

where γ is a constant which depends on the turn number of coils, on the volume and shape of the sample, as well as on its position in the coil.

In general, $v(\omega t)$ can be expressed by the form of the Fourier expansion

$$v(\omega t) = \sum_{n=1}^{\infty} \mathrm{Re}[a_n \exp(in\omega t)], \tag{3}$$

where $a_n = a_n' - ia_n''$ are the Fourier coefficients which are determined experimentally. From Eqs. (2) and (3), we get the real and imaginary components of nth-harmonic susceptibilities as $\chi_n' = a_n''/\gamma n\omega H_{ac}$ and $\chi_n'' = - a_n'/\gamma n\omega H_{ac}$. To obtain the absolute value of χ_n, γ should be calibrated using a standard sample which has the same volume and shape as the specimen in question. However, since we studied here only the relative values of χ_n normalized by $|\chi_1'|$, calibration of γ was not made.

Sample Preparation

Since the details of the sample preparation and characterization were previously reported,[14-17] we give only brief outlines here.

The Fe-doped YBCO compounds were prepared by the solid-state reaction of 4N-Y_2O_3, $BaCO_3$, CuO, and α-Fe_2O_3.[14] The mixtures were pelletized, heated at 900°C in air for 24 h, ground and reheated. This process was repeated several times and finally, the products were sintered at 930°C in air for 24 h. These sintered samples were fabricated by a conventional oxidization process (slow cooling from 850°C in flowing O_2 gas). The phase identification of the samples was made by the powder x-ray diffraction.

The single-crystal YBCO thin films were prepared by metal evaporation together with a special incorporation of oxygen.[15] The films epitaxially grew with their [001] axis normal to the (100) surface of $SrTiO_3$ single crystal. The phase identification was made by x-ray diffraction patterns using Cu-Kα radiation and reflection high-energy electron diffraction (RHEED) patterns. Scanning of the electron-beam probe over the film surface produced no appreciable changes in the RHEED patterns, suggesting that the films are single crystal.

A single-crystal bulk YBCO was prepared by the flux method,[16,17] where the starting materials were Y_2O_3, $BaCO_3$, and CuO. The resulting bulk sample was then annealed in 200-atm O_2 gas for 120 h at 400°C. X-ray analysis showed that the wide surface (1 x 1 mm^2) was parallel to the c planes (perpendicular to the c axis) and the lattice constants were a = 3.830 Å, b = 3.887 Å, and c = 11.699 Å.

MODEL CALCULATIONS

The basic premise of the critical-state model is that, when a magnetic field is applied to a type-II superconductor, a macroscopic supercurrent circulates in the specimen with a critical current density $J_c(B_i)$, where B_i is the local flux density inside the specimen. Bean[6] derived the full hysteresis loop assuming that J_c is a constant, independent of B_i, and predicted a generation of odd-harmonic susceptibilities. By taking into consideration the effective magnetic field at the sample surface, the Bean model can be improved.[13] This modified Bean model, therefore, involves the lower critical field H_{c1} and the surface barrier ΔH as the parameters.

In the Kim-Anderson model,[7,8] the critical current density is assumed as $J_c(B_i) = k/(B_0 + |B_i|)$, where k and B_0 are constants. Using this relation, Chen and Goldfarb[18] derived the M-H hysteresis loop for the case in which $H_{dc} = 0$, i.e., no superimposed dc magnetic field. In the present work, we extend their derivation to the case where an alternating magnetic field is superimposed on a dc magnetic field, and calculate the fundamental and higher-harmonic susceptibilities as a function of temperature. We emphasize three remarkable points to apply this model. The first is that, if a dc magnetic field is present, generation of even-harmonic susceptibilities is predicted for the case where B_0 does not diverge to infinity. The second is that, when the maximum value of applied field $H_{max} = |H_{dc}| + |H_{ac}|$, i.e., the maximum value of $|B_i|/\mu_0$ ($\mu_0 = 4\pi \times 10^{-7}$ H/m), is comparable with B_0/μ_0, the property of χ_n versus T curves is expected to vary drastically with H_{max}, because the distribution of J_c in the specimen becomes sensitive to the change in H_{max}. The third is that, in an extreme case of $B_0 = 0$ with a fixed H_{dc} (the simplified Kim model[12]) or $B_0 \to \infty$ (the Bean model[6]), the relation between χ_n' and χ_n'' is uniquely determined and consequently no H_{max} dependence appears.

Ishida and Mazaki[9] proposed a phenomenological model to explain the behavior of χ_n of a granular superconductor connected by intergrain weak-link junctions. They considered a superconducting loop which contains the weak-link junctions. When the external magnetic field exceeds a critical value, a magnetic flux quantum penetrates into the loop. They assumed that the intergrain multiconnected network behaves like a single loop as a whole, due to the coherent nature of the specimen. Using this model, they have shown that the main features of the magnetic response of a multiconnected network can be reproduced and that odd-n harmonic susceptibilities are predicted.

When the sample magnetization $M(\omega t)$ is expressed by Eq. (1), the real and imaginary components of χ_n can be calculated by

$$\chi_n' = \frac{1}{\pi H_{ac}} \int_0^{2\pi} M(\omega t)\cos(n\omega t)\,d(\omega t) \ , \quad (4)$$

$$\chi_n'' = \frac{1}{\pi H_{ac}} \int_0^{2\pi} M(\omega t)\sin(n\omega t)\,d(\omega t) \ . \tag{5}$$

For each model described above, $M(\omega t)$ was derived for a period of the applied external field $H(\omega t)$. To carry out an analytic evaluation, a period of the alternating magnetic field ($0 \leq \omega t \leq 2\pi$) was divided into 360 equally spaced discrete points and $M(\omega t)$ was numerically obtained at each point.

The temperature dependence of χ_n comes from the parameters which involve T. The parameter k in the Kim-Anderson model is temperature dependent, but no clear-cut expression has been proposed. Therefore, in the present calculation, we tentatively assumed $k(T/T_c) = k(0)(1 - T/T_c)^{3/2}$, where $k(0)$ was chosen so as to give a transition width similar to that of the experimental curves. This assumption is based on the fact that in the case of $|B_i| = 0$, k is proportional to the critical current density at temperatures near T_c, and for a tunnel junction, $J_c \propto (1 - T/T_c)^{3/2}$.

In the modified Bean model, the temperature dependence of χ_n is introduced through H_{c1}, ΔH, and H^*, where H^* is the full penetration field. Here we take $H_{c1} = 0$, the physical reason for this assumption is discussed below. Since H^* is proportional to the critical current density, we again assume $H^*(T/T_c) = H^*(0)(1 - T/T_c)^{3/2}$. Although the temperature dependence of ΔH still remains to be solved, we found the calculated curves with $\Delta H = \Delta H(0)$ $(1 - T/T_c)$ fit well to the observed curves.

In the Ishida-Mazaki model, the critical field H_m at which flux begins to penetrate into the weak-link loop, is a temperature-dependent parameter. Taking into consideration that H_m is proportional to the critical current density, we assume $H_m(T/T_c) = H_m(0)(1 - T/T_c)^{3/2}$.

RESULTS AND DISCUSSION

Fe-doped YBCO Prepared by Sintering

In the present measurement, we used $YBa_2(Cu_{0.94}Fe_{0.06})_3O_y$ fabricated by the ordinary oxidization treatment. The essential behavior of this sample is the same as that of a pure YBCO, except for T_c. In Fig. 4, we show χ_1' and χ_1'' versus T observed for two different values of $\mu_0 H_{ac}$, 0.01 and 0.1 mT, where f_1 = 132 Hz and no dc field was applied. (The Earth's field was not shielded). As the temperature decreases, the bulk phase first appears, and then the coupling phase grows at lower temperatures. As seen in the figure, the coupling phase sensitively depends on H_{ac} and the temperature where χ_1' saturates, becomes lower for a larger value of H_{ac}. In addition, χ_1'' forms a single peak in the temperature region where the coupling phase grows. This type of profile of χ_1 is usually observed for sintered oxide superconductors and has been qualitatively explained by the weak-link model.[9]

The result of the 2nd-harmonic susceptibility measured for $\mu_0 H_{ac}$ = 0.01 mT, H_{dc} = 0, and f_1 = 132 Hz is shown in fig. 5(A). The absolute value of χ_2 is much smaller than those of the odd-harmonic terms (see below). In a zero dc magnetic field, even-harmonic susceptibilities cannot be generated by any model so far reported. Therefore, we believe the observed χ_2 is caused by the Earth's field. As mentioned before, neither the Bean model nor the Ishida-Mazaki model can account for even-n susceptibilities even for $H_{dc} \neq 0$. In Fig. 5(B), we show the calculated result by the Kim-Anderson model for B_0 = 1 mT, $\mu_0 H_{ac}$ = 0.01 mT, $\mu_0 H_{dc}$ = -0.01 mT, and $k(0)$ = 6 N/m^3. The minus sign on H_{dc} means that the component of the Earth's field parallel to the coil axis is in the opposite direction to the applied ac field. We see the Kim-Anderson model as quite satisfactory for explaining the 2nd-harmonic susceptibility observed here.

In Fig. 6, we show χ_n (n = 3, 5, 7) versus T curves observed for $\mu_0 H_{ac}$ = 0.01 mT, H_{dc} = 0, and f_1 = 132 Hz. Again the Earth's field is present. All of these three higher-harmonic components exhibit quite small, but nonzero values in the temperature region where the coupling phase appears. The nonzero values of χ_n'' indicate that the origin of the higher-harmonic susceptibilities is certainly correlated to the nonlinear response of the intergrain supercurrent (or the sample magnetization) to the applied field.

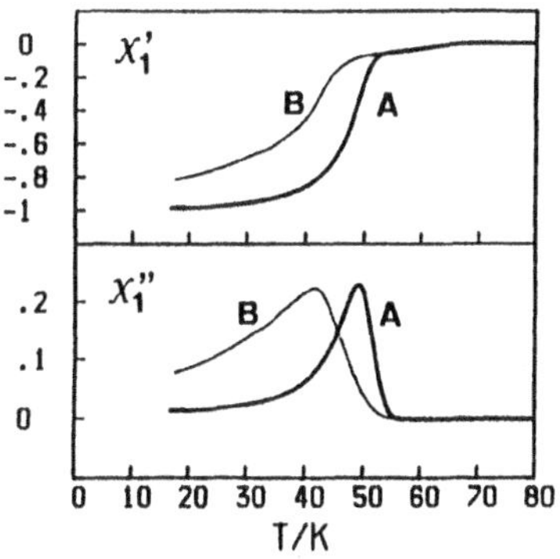

Fig. 4. Real and imaginary parts of the fundamental susceptibility of the Fe-doped YBCO prepared by sintering. A: 0.01 mT, B: 0.1 mT. f_1 = 132 Hz.

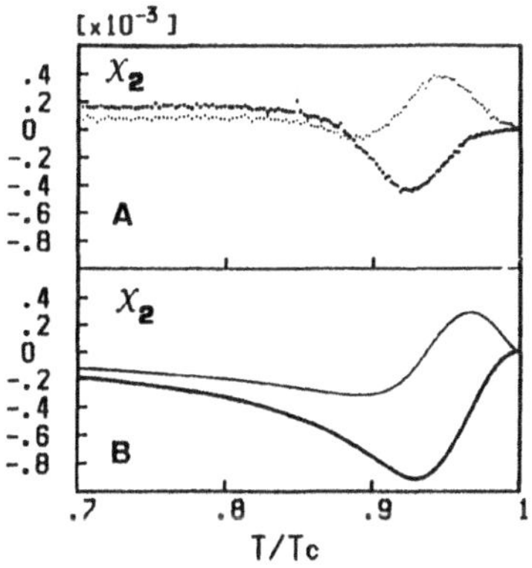

Fig. 5. Measured (A) and calculated (B) χ_2 vs. reduced temperature of the Fe-doped YBCO. Large dots are the real part and small dots are the imaginary part.

Analyses of the data were made in the framework of the Kim-Anderson model[7,8] and of the Ishida-Mazaki model.[9] As a typical result, we show in Fig. 7 the experimental ($\mu_0 H_{ac}$ = 0.01 mT) and calculated χ_7 versus T/T_c curves. The Kim-Anderson model, where B_0 = 1 mT, seems to reproduced the data pretty well. In Fig. 8, we show the observed χ_5 versus T/T_c curves together with calculated curves based on the Kim-Anderson model for three different values of B_0, with $\mu_0 H_{ac}$ = 0.1 mT and f_1 = 132 Hz. The results have also revealed that the value of B_0 must be much larger than 0.1 mT to obtain agreement with experiment.

It is recognized that the generation of the higher-harmonic susceptibilities arises from the supercurrent ruled by the Kim-Anderson model with a *finite* value of B_0, but not by the Ishida-Mazaki model. The main reason for the failure of this latter model probably arises from the assumption that the specimen behaves like a single loop as a whole due to the coherent nature. Assuming that the specimen consists of a number of coaxial weak-link loops, we consider, phenomenologically, the magnetic response of the

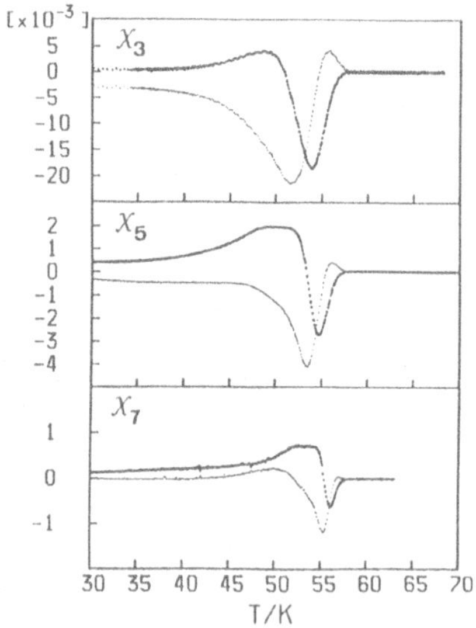

Fig. 6. χ_n (n = 3, 5, 7) vs. T of the Fe-doped YBCO prepared by sintering. $\mu_0 H_{ac}$ = 0.01 mT and f_1 = 132 Hz. Large dots represent the real part and small dots are the imaginary part.

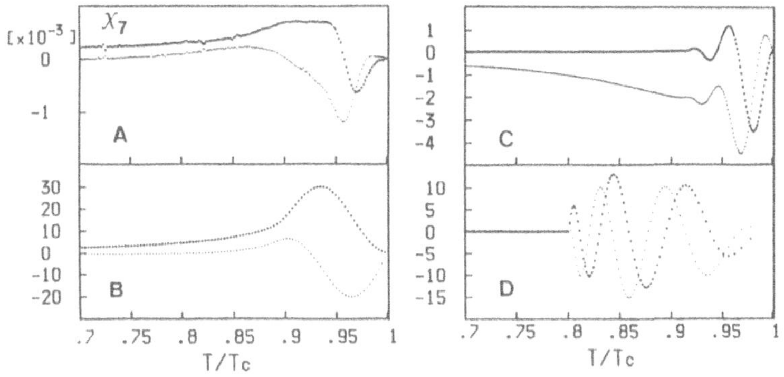

Fig. 7. χ_7 vs. reduced temperature of the Fe-doped YBCO prepared by sintering. A: Measured for $\mu_0 H_{ac}$ = 0.01 mT and f_1 = 132 Hz. B: Calculated by the Kim-Anderson model, where B_0 = 0, $k(0)$ = 0.2 N/m^3. C: Calculated by the same model, where B_0 = 1 mT, $k(0)$ = 60 N/m^3. D: Calculated by the Ishida-Mazaki model, where $\mu_0 H_m$ = 25 x (1 - T/T_c)$^{3/2}$ mT. Large dots represent the real part and small dots are the imaginary part.

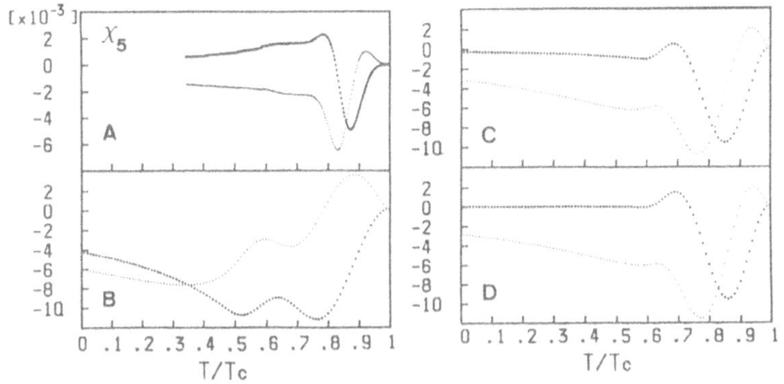

Fig. 8. χ_5 vs. reduced temperature of the Fe-doped YBCO prepared by sintering. A: Measured for $\mu_0 H_{ac}$ = 0.1 mT and f_1 = 132 Hz. B: Calculated by the Kim-Anderson model, where B_0 = 0.1 mT, $k(0)$ = 6 N/m^3. C: Calculated by the same model, where B_0 = 1 mT, $k(0)$ = 60 N/m^3. D: Calculated by the same model, where B_0 = 10 mT, $k(0)$ = 600 N/m^3. Large dots represent the real part and small dots are the imaginary part.

specimen to an applied ac field. When the applied field starts to increase from zero, the flux is, at the initial stage, completely excluded by the shielding supercurrent circulating through the outermost loop. When the field amplitude exceeds the critical value which induces the critical current of the outermost loop I_j, the flux penetrates into this loop. The flux penetration occurs sequentially as the field amplitude increases, and thus the distribution of magnetic flux density in the specimen becomes step like, discretely jumping by ΔB at each loop. Suppose ΔB is small enough and the number of coaxial weak-link loops surrounding the specimen is large; then the distribution of the flux density inside the specimen can be approximated by a continuous line. In addition, if ΔB is a constant for all the coaxial loops, the distribution becomes a linear line, being equivalent to the Bean model.[6] Meanwhile, if I_j is inversely proportional to $(B_0 + |B_i|)$, the distribution of the flux density may obey the Kim-Anderson model.[7,8] In this manner, assuming a large number of loops, the extended Ishida-Mazaki model may be treated as being algebraically equivalent to the conventional critical-state model.

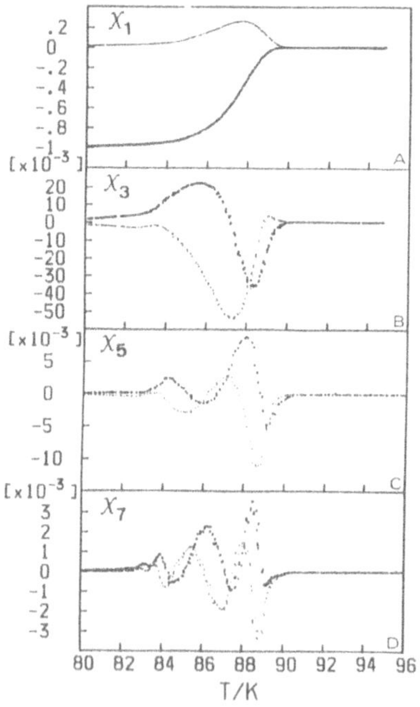

Fig. 9. χ_n (n = 1, 3, 5, 7) vs. T of the single-crystal YBCO thin film, where $\mu_0 H_{ac}$ = 0.1 mT and f_1 = 132 Hz. Large dots are the real part and small dots denote the imaginary part.

Single-Crystal YBCO Thin Films

In Figs. 9(A)-(D), we give the χ_1, χ_3, χ_5, χ_7 versus T curves observed for a 2000-Å single-crystal YBCO thin film, where $\mu_0 H_{ac}$ = 0.1 mT and f_1 = 132 Hz. The absolute values of χ_n decrease rapidly and the profiles become more and more complicated as n increases. Assuming $\mu_0 H_{ac}$ = 0.1 mT, $\mu_0 H^*$ = 10 x $(1 - T/T_c)^{3/2}$ mT, H_{c1} = 0, $\mu_0 \Delta H$ = 1 x $(1 - T/T_c)$, we attempted to reproduce the experimental results by applying the modified Bean model.[13] As demonstrated in Figs. 10(A)-(D), the calculated curves are in good agreement with the observed ones, and the complicated profiles are well reproduced.

On the contrary, the Kim-Anderson model[7,8] and the Ishida-Mazaki model[9] were not successful in reproducing the present experimental results.

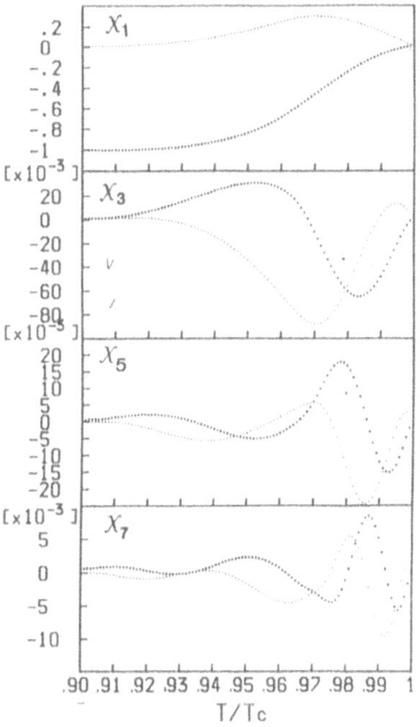

Fig. 10. χ_n (n = 1, 3, 5, 7) vs. reduced temperature curves calculated on the basis of the modified Bean model for $\mu_0 H_{ac}$ = 0.1 mT, H_{c1} = 0, $\mu_0 \Delta H$ = 1 x $(1 - T/T_c)$ mT, and $\mu_0 H^*$ = 10 x $(1 - T/T_c)^{3/2}$ mT. Large dots are the real part and small dots are the imaginary part.

As an example, in Fig. 11, we show χ_7 versus T/T_c calculated by the Kim-Anderson model with $B_0 = 0$ (Fig. 11(B)) and 20 mT (Fig. 11(C)) together with the experimental result (Fig. 11(A)).

To understand why the modified Bean model works well, we discuss the physical reason for choosing $H_{c1} = 0$ and the need of a surface barrier ΔH. When a magnetic field H_a is applied to an ellipsoid of revolution, the internal magnetic field H_i is uniform in the specimen and is written as[19]

$$H_i = (H_a - \frac{NB_i}{\mu_0})/(1 - N) , \tag{6}$$

where N is a demagnetization coefficient and B_i is a magnetic flux density in the specimen. A film is approximated as a flat ellipsoid of revolution whose length of the short axis reaches a limit value of zero, and its demagnetization coefficient is approximated by $N = 1 - (\pi/2)(d/R)$,[19] where $2d$ is the film thickness and R is the radius. Then the magnetization of the specimen M is written as

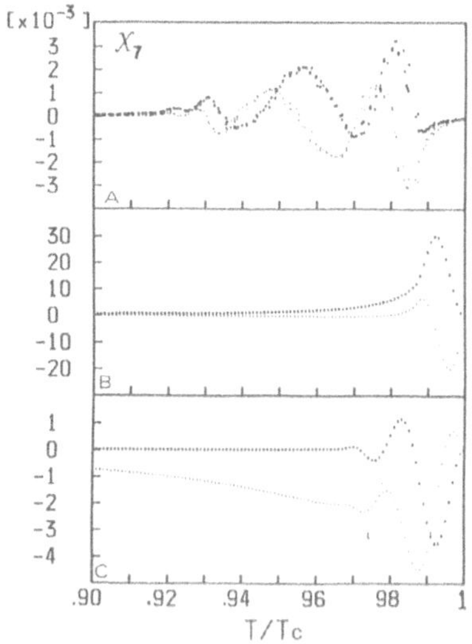

Fig. 11. χ_7 vs. reduced temperature of the single-crystal YBCO thin film. A: Measured for $\mu_0 H_{ac}$ = 0.1 mT, f_1 = 132 Hz. B: Calculated on the basis of the Kim-Anderson model, where $B_0 = 0$, $k(0)$ = 500 N/m^3. C: Calculated by the same model, where B_0 = 20 mT, $k(0) = 5 \times 10^4$ N/m^3. Large dots are the real part and small dots are the imaginary part.

$$M = \frac{B_i}{\mu_0} - H_i = \frac{2R}{\pi d}(-H_a + \frac{B_i}{\mu_0}) . \qquad (7)$$

This equation means that M is proportional to the radius and is inversely proportional to the film thickness. Confirmation of the above discussion can be made by changing R (or the sample area S). For this purpose, we drew lines on a film surface of R = 5 mm, and divided the sample area S (1/4, 1/8, 1/16). Then we compared the saturated values of χ_1' at extremely low temperatures. We found that the saturated values of $|\chi_1'|$ lie on a line corresponding to $M \propto \sqrt{S}$, i.e., $M \propto R$.

When the sample is perfectly diamagnetic (B_i = 0), H_i is enhanced up to ($2R/\pi d$) times of H_a. Therefore, if we choose $\mu_0 H_{c1}$ = 0.01 mT, R = 5 mm, and d = 1000 Å, the effective lower critical field becomes as low as 0.003 mT, which is much smaller than the applied field H_a in the present measurement, suggesting that the assumption H_{c1} = 0 is reasonable.

From the above discussion of the lower critical field, we may conclude that the magnetic flux fully penetrates into the film even at very low temperatures. However, we should note that the higher-harmonic susceptibilities asymptotically go to zero as T decreases. This means that no flux penetration into the film occurs at the very low temperature region. From this fact, we believe that the surface barrier ΔH plays an essential role in preventing flux penetration in the present case, and that this reasonably explains the propriety of adopting the modified Bean model.

Single-Crystal Bulk Sample

The χ_n (n = 1, 2, 3) versus T curves observed for a single-crystal bulk sample are shown in Fig. 12, where the applied ac field is perpendicular to the c planes. The field amplitude $\mu_0 H_{ac}$ is 0.15 mT and f_1 = 132 Hz. The onset temperatures of χ_1' and χ_1'' are the same. The higher-harmonic susceptibilities χ_2 and χ_3 have the same onset temperature as χ_1, and have non-zero values in the temperature region where χ_1'' forms a peak.

Model calculations were carried out within the framework of the Kim-Anderson model,[7,8] where we assumed $k(T/T_c) = k(0)(1 - T/T_c)^{3/2}$, and $k(0)$ was used as a parameter to fit the transition width to the observed results. Three values of B_0 were chosen, B_0 = 0, 0.05, 20 mT. The B_0 = 0 case is equivalent to the simplified Kim model and the case of B_0 = 20 mT nearly corresponds to the Bean model. As a typical result, in Fig. 13 we present the calculated curves of χ_3 versus T/T_c (Figs. 13(B)-(D)), together with the experimental one (Fig. 13(A)). We find that the Kim-Anderson model with B_0 = 0.05 mT fits very well to the observed curves.

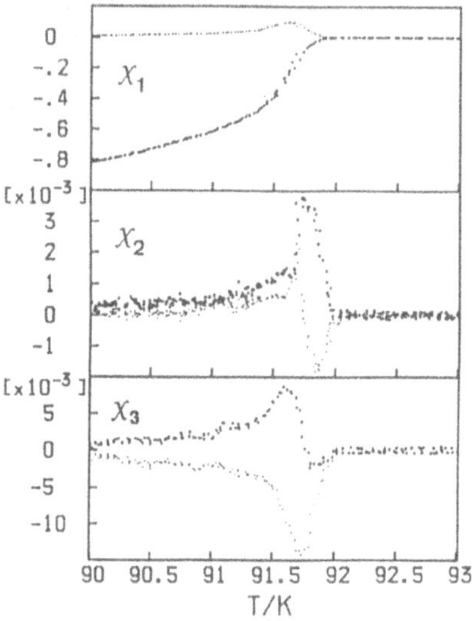

Fig. 12. χ_n (n = 1, 2, 3) vs. T of the single-crystal YBCO bulk specimen observed for $\mu_0 H_{ac}$ = 0.15 mT and f_1 = 132 Hz. Large dots denote the real part and small dots are the imaginary part.

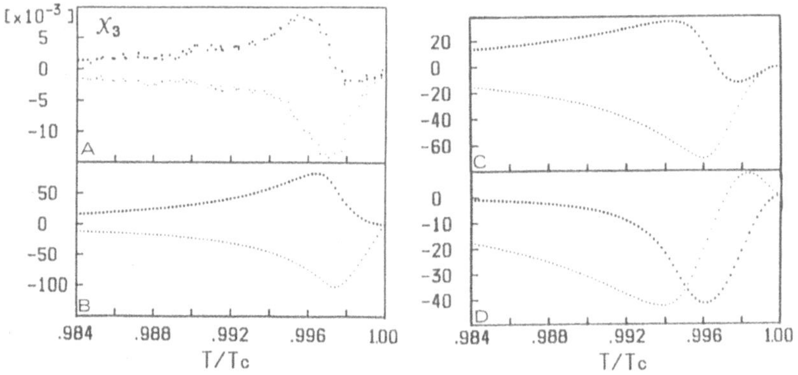

Fig. 13. χ_3 vs. reduced temperature of the single-crystal YBCO bulk specimen. A: Measured for $\mu_0 H_{ac}$ = 0.15 mT, f_1 = 132 Hz. B: Calculated on the basis of the Kim-Anderson model, where B_0 = 0, $k(0)$ = 1 x 10^5 N/m^3. C: Calculated by the same model, where B_0 = 0.05 mT, $k(0)$ = 1 x 10^5 N/m^3. D: Calculated by the same model, where B_0 = 20 mT, $k(0)$ = 1 x 10^7 N/m^3. Large dots are the real part and small dots are the imaginary part.

SUMMARY

We measured the fundamental and higher-harmonic susceptibilities χ_n = $\chi_n = \chi_n' - i\chi_n''$ (n = 1, 2, 3, 5, 7) of high-T_c superconductors. The details of the measuring system which consists of the Hartshorn bridge are described. Using this system, we measured χ_n for three types of samples, an Fe-doped YBCO disk prepared by sintering, single-crystal YBCO thin films (1000-6000 Å) fabricated by activated reactive evaporation, and a single-crystal YBCO bulk specimen prepared by the flux method.

The observed χ_n versus T curves were analyzed within the framework of the Kim-Anderson model,[7,8] the modified Bean model,[13] and the Ishida-Mazaki model.[9] It has been found that for the sintered sample, the Kim-Anderson model with a finite value of B_0 seems to reproduce the experimental results pretty well. For example, the observed temperature dependence χ_5 for an applied ac field of $\mu_0 H_{ac}$ = 0.1 mT and f_1 = 132 Hz, agrees very well with the calculated curve by this model, where B_0 = 1 mT.

The failure of the Ishida-Mazaki model to reproduce the experimental results probably arises from the assumption that the specimen behaves like a single loop as a whole due to the coherent nature of the specimen. However, assuming that the specimen consists of a number of coaxial weak-link loops, the model can be improved, and this extended Ishida-Mazaki model may be treated as being algebraically equivalent to the conventional critical-state model.

For the single-crystal YBCO thin films, we found that the modified Bean model was appropriate to reproduce the temperature dependence of χ_n, but both the Kim-Anderson model and the Ishida-Mazaki model were not successful. In the case of thin films located in an applied field perpendicular to the film surface, the demagnetization coefficient should be extremely large. Taking into consideration the lower critical field and the surface barrier, we discussed the magnetic response of the film against an applied ac field. Due to the large value of the demagnetization coefficient, it is reasonable to expect that the effective lower critical field can be approximated by zero. Nevertheless, from the fact that χ_n asymptotically goes to zero as T decreases, we think no flux penetration into the film occurs at the very low temperature region. This implies that the surface barrier plays an essential role, and this reasonably explain the propriety of adopting the modified Bean model.

The χ_n versus T curves observed for a single-crystal YBCO bulk sample were analyzed within the framework of the Kim-Anderson model, where we assumed $k(T/T_c) = k(0)(1 - T/T_c)^{3/2}$, and $k(0)$ was used as a parameter to fit the transition width to the observed results. These results again show that the model with a finite value of B_0 gives satisfactory profiles of χ_n as a function of temperature.

ACKNOWLEDGMENTS

The authors wish to express their thanks to Professor Y. Bando and Dr. T. Terashima of Kyoto University for their preparation of the single-crystal YBCO thin films and bulk specimen. Thanks are also due to Professor K. Kosuge of Kyoto University and S. Katsuyama of Osaka University for their kind offer of the Fe-doped YBCO sample.

References

1. E. Maxwell and M. Strongin, Filamentary structure in superconductors, Phys. Rev. Lett. 10: 212 (1963).
2. T. Ishida and H. Mazaki, Superconducting transition of electrodeposited technetium, Phys. Rev. B 20: 131 (1979).
3. A. F. Khoder, The superconducting transition and the behavior of the ac susceptibility, Phys. Lett. 94A: 378 (1983).
4. R. A. Hein, ac magnetic susceptibility, Meissner effect, and bulk superconductivity, Phys. Rev. B 33: 7539 (1986).
5. H. Mazaki, M. Takano, R. Kanno, and Y. Takeda, Complex susceptibility of $La_{1.9}Sr_{0.1}CuO_{4-\delta}$, Jpn. J. Appl. Phys. 26: L780 (1987).
6. C. P. Bean, Magnetization of hard superconductors, Phys. Rev. Lett. 8: 250 (1962).
7. Y. B. Kim, C. F. Hempstead, and A. R. Strnad, Magnetization and critical supercurrents, Phys. Rev. 129; 528 (1963).
8. P. W. Anderson, Theory of flux creep in hard superconductors, Phys. Rev. Lett. 9: 309 (1962).
9. T. Ishida and H. Mazaki, Superconducting transition of multiconnected Josephson network, J. Appl. Phys. 52: 6798 (1981).
10. R. W. Rollins and J. Silcox, Nature of the ac transition in the superconducting sheath in Pb-2%In, Phys. Rev. 155: 404 (1967).
11. T. Ishida and R. B. Goldfarb, Fundamental and harmonic susceptibilities of $YBa_2Cu_3O_{7-\delta}$, Phys. Rev. B 41: 8937 (1990).
12. L. Ji, R. H. Sohn, G. C. Spalding, C. J. Lobb, and M. Tinkham, Critical state model for harmonic generation in high-temperature superconductors, Phys. Rev. B 40: 10936 (1989).
13. W. I. Dunn and P. Hlawiczka, Generalized critical-state model of type-II superconductors, Brit. J. Appl. Phys. SER.2 1; 1469 (1968).

14. H. Mazaki, S. Katsuyama, H. Yasuoka, Y. Ueda, and K. Kosuge, Complex susceptibility of $YBa_2(Cu_{1-x}Fe_x)_3O_y$, Jpn. J. Appl. Phys. 28: L1909 (1989).
15. T. Terashima, K. Iijima, K. Yamamoto, Y. Bando, and H. Mazaki, Single-crystal $YBa_2Cu_3O_{7-x}$ thin films by activated reactive evaporation, Jpn. J. Appl. Phys. 27: L91 (1988).
16. S. Shamoto, S. Hosoya, and M. Sato, Single-crystal growth of high-T_c superconductors, Solid State Comm. 66: 195 (1988).
17. K. Yamamoto, K. Hirata, K. Iijima, K. Hayashi, H. Mazaki, H. Yasuoka, T. Terashima, and Y. Bando, Complex susceptibility of Y-system oxide superconductor (II), J. Jpn. Soc. Powder and Powder Met. 37: 799 (1990).
18. D.-X. Chen and R. B. Goldfarb, Kim model for magnetization of type-II superconductors, J. Appl. Phys. 66: 2489 (1989).
19. A. L. Fetter and P. C. Hohenberg, The mixed state of thin superconducting films in perpendicular fields, Phys. Rev. 159: 330 (1967).

THE FREQUENCY DEPENDENCE OF THE AC IRREVERSIBILITY LINE OF SOME HIGH TEMPERATURE SUPERCONDUCTORS

R. B. Flippen

Central Research and Development
E. I. du Pont de Nemours and Co., Inc.
P.O.Box 80228, Wilmington, DE 19880-0228

ABSTRACT

Ac inductance techniques are commonly used to determine transition temperatures and magnetization versus temperature for superconductors by measurement of the real part of the sample susceptibility χ'. The imaginary or absorptive component of the susceptibility χ'' can also be measured by these techniques. The latter typically shows a peak versus temperature for high temperature superconductors (HTSC) which decreases in temperature as the strength of an applied dc magnetic field increases. The curve of the temperature of the peak in χ'' versus dc magnetic field is called the irreversibility line IL. The IL is sometimes interpreted as dividing magnetically reversible from nonreversible regions in the H-T phase diagram of a superconductor. The temperature dependence of the IL is dependent upon the frequency of the measuring ac field, sometimes strongly so, and is different for different HTSC materials as well as for different forms of the same compound. In this work data is presented for the IL of a variety of samples of YBaCuO, BiSrCaCuO, and TlBaCaCuO measured over a frequency range of 0.1 to 20kHz in dc magnetic fields up to 400 Oe. The dependence of the IL on measurement frequency, ac field strength, demagnetizing factors, and sample diameter is discussed in terms of superconductive glass/flux lattice melting/thermally activated flux flow theories, with particular emphasis upon the IL low frequency behavior.

INTRODUCTION

Ac inductance techniques have been used for many years to determine properties of superconductors, particularly transition temperatures and weak magnetic field magnetization versus temperature relations[1]. These techniques generally produce a sample susceptibility consisting of a real or in phase with the ac driving magnetic field component χ' and a component χ'' in quadrature to the driving field[2]. χ' exhibits the magnetic (flux exclusion) state of a superconductor, and χ'' is considered variously as the viscous, lossy, or absorptive part of the susceptibility related to absorption of the magnetic field energy in the sample[3]. In the earlier low temperature superconductors, χ'' has been viewed as resulting from energy dissipation produced by motion of quantized magnetic vortex lines through

Magnetic Susceptibility of Superconductors and Other Spin Systems
Edited by R.A. Hein *et al.*, Plenum Press, New York, 1991

the sample[4]. χ" in these materials typically shows a weakly temperature dependent curve with magnetic field, sometimes with a broad peak as a function of temperature[5]. A different phenomenon is found in the new high temperature superconductors in that a well defined, sometimes sharp χ" peak with temperature is often seen whose temperature dependence varies widely with composition[6]. In ceramic HTSC materials two peaks in χ" with temperature are sometimes seen[7]. The χ" peak-temperature curve in crystalline samples lies between the H_{C1} and H_{C2} temperature curves[8]. Malozemoff et al[8], noting that the χ" curve coincides with a curve determined by dc magnetometer measurements of field-cooled and zero field-cooled magnetization data which separates regions of magnetically reversible from nonreversible behavior, call the χ" peak -T curve the irreversibility line (IL). This definition will be used in the present work.

The position of the IL is observed to depend upon many factors and can vary for different samples of the same material[6]. One of these variables is the frequency of measurement[9], and this factor will be the main subject of the present work; some preliminary studies of this nature have been presented earlier[6]. In the current work extensive results will be given for a ceramic form of $YBa_2Cu_3O_{7-x}O$ which allows for the testing of a number of variables and their effect upon the frequency and other dependences; additional results will be given for other crystalline HTSC materials and compared to those for the ceramic. Current theories of high temperature superconductivity will be examined to see how and if the experimental results fit the theories.

EXPERIMENTAL

Susceptibility measurements were made using conventional ac inductance techniques described previously[10]. A small opposed secondary mutual inductance coil set containing a 1 mm diameter, 5 mm long sample volume and a temperature sensor is surrounded by a heater coil and a magnetic field coil and enclosed in a 0.5 in diameter thin wall stainless steel tube which can be inserted into a standard liquid helium dewar. The secondary coil are each 10 cm long and consist of 5 layers of #38 copper wire. The two sections are separated by 5 mm and are wound in opposition on a 1.5 mm i.d. thin paper form. A primary coil consisting of 3 layers of #38 wire is wound on each secondary, and the number of primary turns on one section are adjusted to give essentially zero output voltage from the secondaries for current in the primaries. An opening is made between the coil pairs for insertion of a Lakeshore diode temperature sensor. A single layer non-inductively wound heater coil is wound of #30 manganin wire on top of the primary coils, insulated from the latter with glass tape. A 4 layer solenoidal field coil 3" long is wound of #28 copper wire on top of the heater coil and also insulated from the heater with glass tape. The whole assembly fits closely within a 0.5" o.d., 10 mil wall stainless steel tube positioned so that a sample in a glass capillary tube can be inserted into the core of one of the coils. The primary and secondary leads (thin coax cable), thermometer leads, heater, and solenoid leads are brought out through the other end of the 4' long tube for the necessary equipment connections. The mass of the assembly is small enough that the temperature of the sample can be adequately controlled by a combination of depth of the tube in the helium dewar and the heater current over the temperature range 4.2°K - 300K. A small temperature dependent background signal is measured in the absence of a sample and subtracted from the signal when a sample is present. An ac current of frequency 10 - 50,000 Hz induces an ac field of 0.01 - 15 Oe in the sample, and a dc magnetic field up to 400 Oe can be applied. The secondary coil signal is detected with a lock-in amplifier for the in phase with the primary current and the signal quadrature component and either plotted directly versus temperature on an X-Y recorder or collected and stored in a computer for later data analysis. A rough calibration of sensitivity was made with a sample of pure lead filling the sample volume by measuring its signal at 4.2K. A detection level of 0.05% superconductor in the sample volume at 1000 Hz was estimated for this apparatus. In practice this calibration could rarely be used due to the well known problems of determining bulk superconductivity with ac techniques[11]. The present work will only be concerned with the quadrature signal from the sample. The phase of the lock-in amplifier was set by measuring a strong X HTSC sample at low temperature and adjusting the in-phase signal for maximum and then using the quadrature phase to determine χ''. This technique was found to be more accurate than trying to adjust for zero phase, and this procedure was found to give consistent results over a range of sample χ strengths and temperatures and gave results in agreement with literature values where comparisons could be made.

SAMPLES

Most HTSC materials cannot be shaped readily to explore the effects of changes of size, demagnetization factors, and surface condition upon the IL behavior. Most ceramic samples which can be shaped or modified have such weak and broad χ'' peaks that the IL dependence cannot be obtained for them. However, a ceramic $YBa_2Cu_3O_{7-x}$ material made in the form of approximately 1 mm diameter rods several cm in length has a relatively sharp, clean χ'' peak which can be followed as a function of a number of variables. This material is made by ICI and described in the literature[12]. Two χ'' peaks are not seen in this substance, apparently because the individual grains are too small to produce a measurable intragranular signal; thus the χ'' peak will be the result of the intergranular properties of the material.

The ICI material allows the demagnetizing factor for the sample to be varied by adjusting the length of the sample with the measuring fields along the axis of the rod; the effect of change in sample diameter can be observed by mechanically removing material from the surface, keeping the same field orientation.

Results for the frequency and magnetic field behavior of the IL of the ceramic material will be compared with results obtained for single crystal and thin film $YBa_2Cu_3O_{7-x}$ and, for comparison with other compositions, with single crystal $Bi_2Sr_2Ca_1Cu_2O_y$ and $Tl_2Ba_2Ca_1Cu_2O_y$. Preparation of these materials has been reported previously[13].

GENERAL RESULTS

Typical plots of the IL for the ICI $YBa_2Cu_3O_{7-x}$, single crystal $YBa_2Cu_3O_{7-x}$, $Bi_2Sr_2Ca_1Cu_2O_y$, $Tl_2Ba_2Ca_1Cu_2O_y$ and two $YBa_2Cu_3O_{7-x}$ thin films are shown in Figure 1 for dc magnetic fields up to 400 Oe. at 20kHz with H = 0.2 Oe.

These data have not been corrected for demagnetizing factors which, as will be seen later, can be important. On this scale of magnetic field, the IL of the single crystal $YBa_2Cu_3O_{7-x}$ is essentially vertical, so that the intergranular IL for ceramic forms of this material is much more field dependent than the IL. However, the IL for single crystal $Bi_2Sr_2Ca_1Cu_2Oy$ and $Tl_2Ba_2Ca_1Cu_2Oy$ is even more field dependent than the ceramic $YBa_2Cu_3O_{7-x}$.

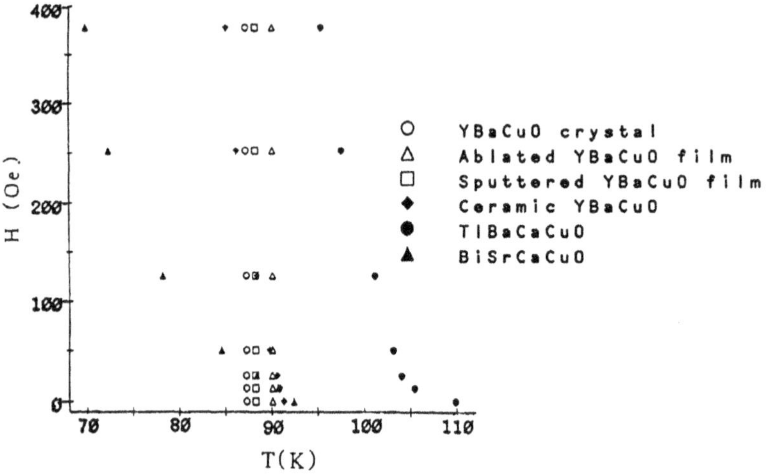

Fig. 1. The irreversibility line at 20 kHz for single crystal and thin films $YBa_2Cu_3O_{7-x}$, $Bi_2Sr_2Ca_1Cu_2O_y$, and $Tl_2Ba_2Ca_1Cu_2O_y$, and ceramic $YBa_2Cu_3O_{7-x}$.

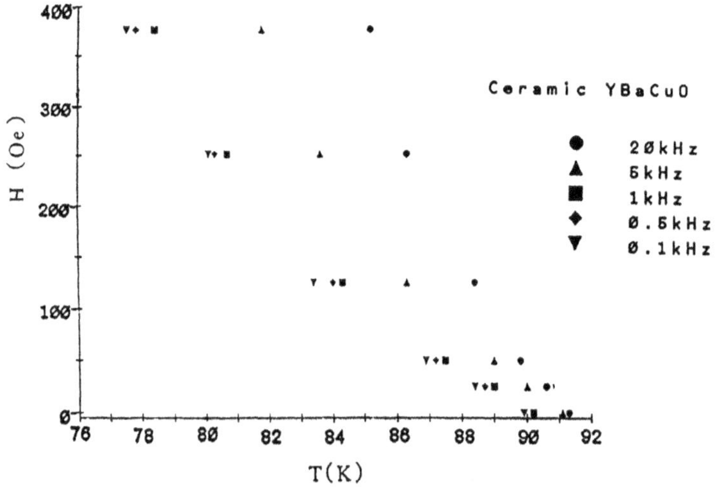

Fig. 2. The irreversibility line of ceramic $YBa_2Cu_3O_{7-x}$ versus frequency.

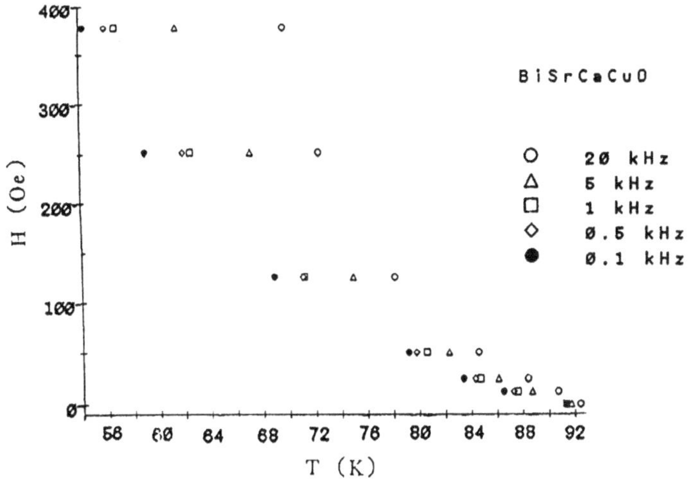

Fig. 3 The irreversibility line of crystalline $Bi_2Sr_2Ca_1Cu_2O_y$ versus frequency.

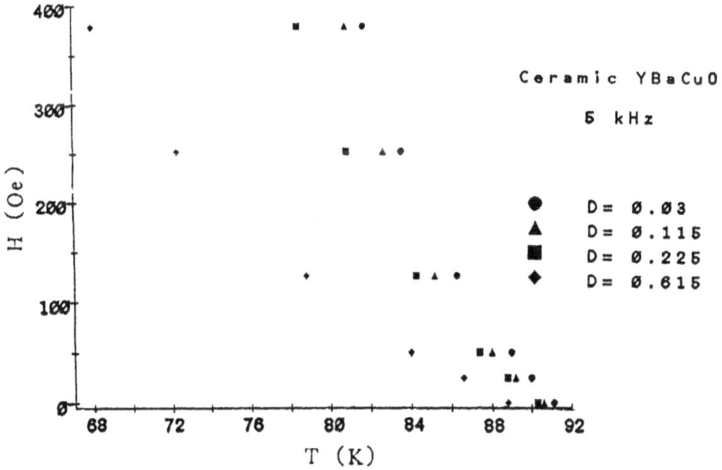

Fig. 4 The irreversibility line at 5 kHz for ceramic $YBa_2Cu_3O_{7-x}$ samples with different demagnetizing factors.

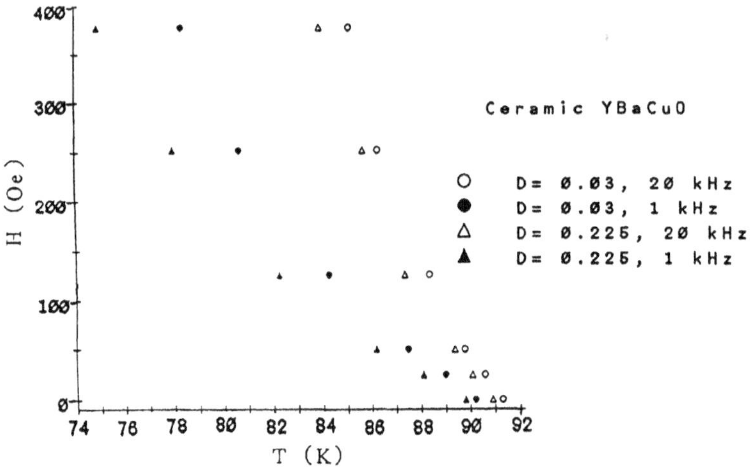

Fig. 5. The irreversibility line of different demagnetizing factor samples of ceramic $YBa_2Cu_3O_{7-x}$ for different frequencies.

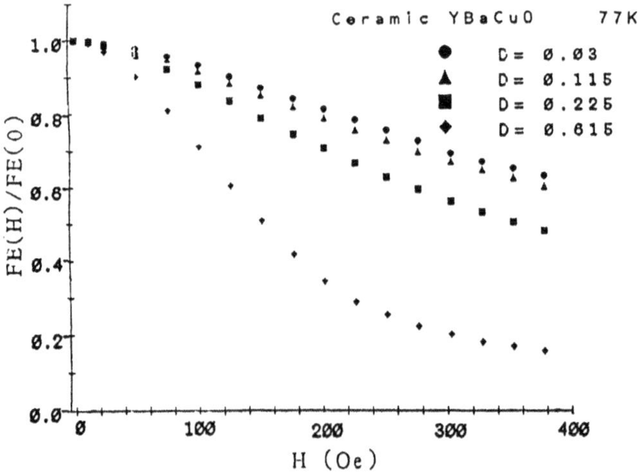

Fig. 6. Flux exclusion versus dc magnetic field at 77K for different demagnetizing factor samples of ceramic $YBa_2Cu_3O_{7-x}$.

The frequency dependence of the ICI material IL from 0.1 kHz to 20 kHz is shown in Figure 2; on this scale the IL of the bulk $YBa_2Cu_3O_{7-x}$ is essentially frequency independent. The frequency dependence of the IL for $Bi_2Sr_2Ca_1Cu_2O_y$ is shown in Figure 3; the dependence for $Tl_2Ba_2CaCu_2O_y$ is similar to that for $Bi_2Sr_2CaCu_2O_y$ but displaced to higher temperatures. The dependence for these materials is even stronger than that for the ceramic $YBa_2Cu_3O_{7-x}$.

The sample demagnetizing factor for the ceramic $YBa_2Cu_3O_{7-x}$ was progressively increased by shortening the sample. Tables of demagnetizing factors for short cylinders were used to determine the factors for the various length/diameter ratios[14]. These results are shown in Figure 4 for 5 kHz, and it is seen that increasing the demagnetizing factor has a pronounced effect of lowering the slope of the IL. Figure 5 compares the IL for two different demagnetizing factors for different frequencies.

Increasing the demagnetizing factor lowers the IL slope, and the lower the measuring frequency, the lower the slope becomes down to about 1 kHz; below this frequency little change occurs. Further data which will be useful in calculating demagnetizing field effects is shown in Figure 6, where relative flux exclusion is plotted versus magnetic field at 77K for the various ceramic sample lengths at 20kHz. Similar plots as the frequency is lowered show increasing dependence of flux exclusion upon magnetic field.

The effect of reducing the sample diameter of the ceramic sample on the IL at 1 kHz is shown in Figure 7. A length of rod was gently scraped around the periphery with a sharp blade to progressively reduce its diameter by stages, with the measurements done for each stage. The IL slope is seen to lower in a nonlinear way as the diameter is reduced. The lower the measuring frequency, the lower the slope, again down to about 1 kHz, below which there is little change. The effect of the strength of the ac measuring field on the IL for 1 kHz is shown in Figure 8. The slope of the IL line is lower the higher the ac field, and the lower the measuring frequency, the greater the effect down to about 1 kHz.

DEMAGNETIZING FIELD EFFECTS

Before discussing the physical phenomena underlying the IL, the effect of changes in the sample internal magnetic field upon the IL should be considered. The internal field $H_{in} = H_{ex} - DM$, where H_{ex} is the external field, M the sample magnetization, and D the sample demagnetizing

Table 1. Calculated internal magnetic field H_i at constant flux exclusion

D	f,kHz	M,a.u.	H_{ex}	H_i
0.030	20	233	215	222
0.115			190	216
0.225			145	215
0.015			78	221

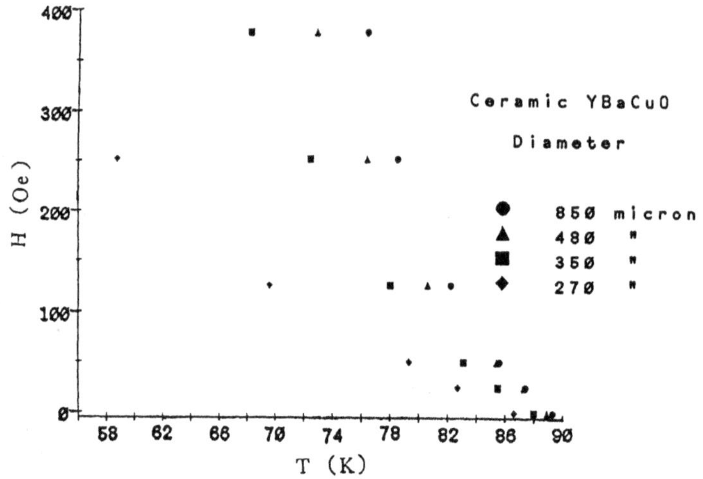

Fig. 7. The irreversibility line of ceramic $YBa_2Cu_3O_{7-x}$ for different diameter samples.

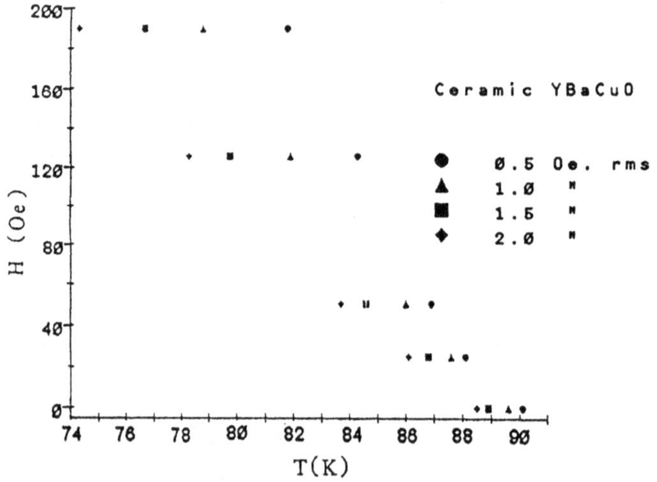

Fig. 8 The irreversibility line of ceramic $YBa_2Cu_3O_{7-x}$ for different ac magnetic fields.

factor[15]. Since for superconductors M is negative and often large, any appreciable demagnetizing factor will cause a demagnetizing field to add to the external field; consequently as the demagnetizing factor increases for a given material, there will be a shift downwards in the slope of the IL. This effect is illustrated in Figure 4 for the ceramic sample and can be quantified by using the data of Figure 6 at constant flux exclusion to calculate the internal field for the various demagnetizing coefficients. The results are shown in Table 1, where a constant internal field results with the demagnetizing field correction. This constant internal field then is the appropriate field to use for the IL. Thus quantitative slopes of the IL must take into consideration demagnetizing fields, particularly samples with large demagnetizing factors such as thin films and flat crystals noted above for Figures 1 and 2. However, the necessity for a demagnetizing field correction says nothing about the observed IL frequency dependence.

THEORIES

The peak in χ'' versus temperature for HTSC materials has been variously defined by Palstra et al[16] as indicating the temperature at which the measuring frequency and magnetic vortex system relaxation time is the same, by Worthington et al[9] as when the ac field just penetrates to the center of the sample, by Malozemoff et al[8] as where irreversibility sets in as the temperature is lowered, and by Nikolo and Goldfarb[7] as arising from flux creep at grain boundaries (for ceramic materials). These interpretations depend upon the model chosen to view the superconductor: vortex glass[17], flux creep[18], thermally assisted flux flow[19], flux lattice melting[20], etc. All these models deal in some way with flux motion dynamics. Gammel[21] has recently critiqued these theories in terms of experimental data for χ'', finding a thermally activated model providing the best fit. However, the crystal of $Bi_2Sr_2Ca_1Cu_2O_y$ used had a large demagnetizing factor, and no mention was made of corrections to the data, which, as shown above, are very important for quantitative discussion. More recently Gammel et al[22] have concluded from high field I-V measurements that data for $YBa_2Cu_3O_{7-x}$ better fits a vortex glass model, while Malozemoff claims magnetic data[23] is better explained by flux lattice melting. No one theory seems to explain in general all or even much of the data for HTSC materials.

FREQUENCY DEPENDENCE

Most thermally activated theories of HTSC materials start with the Campbell and Evetts formulation for flux vortex motion[24] which requires a logarithmic time dependence of the IL. Our previous work has shown that data obtained with the above techniques for crystalline HTSC material can be fitted well to the thermally assisted flux flow (TAFF) theory of Kes and coworkers[19]. This theory predicts that $(1 - T_p/T_c) \propto H^{2/3}$, where Tp is the χ'' peak temperature. A plot of this kind for the ceramic ICI material is shown in Figure 9 and the fit is at least as good as for the crystalline material. The slope varies approximately logarithmically with frequency at higher frequencies as the theory predicts, and the fit is equally good over the range of frequencies used. This result suggests that interparticle as well as intraparticle flux motion in $YBa_2Cu_3O_{7-x}$ is thermally activated. A logarithmic dependence upon the frequency of measurement for χ'' peak temperatures has generally been reported[8], but these measurements are typically well above 1 kHz; as shown here and in other reports[3,7], at least for $YBa_2Cu_3O_{7-x}$, the IL is independent of frequency below about 1 kHz.

DISCUSSION

Some aspects of the results reported here are reasonably interpreted by theories like TAFF; others are less clear. It should be born in mind what is being measured: a change in complex inductance of a coil produced by changes in the permeability of material making up the core of the coil as the temperature/magnetic field/frequency is changed. This permeability is a measure of the magnetic flux state of the superconductor, and dynamic changes in flux produce the results measured. The relatively weak ac field provides a "tickler" field added on to the usually much stronger dc field; this tickler field in essence interrogates the magnetic state of the material.

The TAFF theory as formulated by Malozemoff et al[8] predicts in addition to the H_{dc} dependence, a dependence of the IL upon the thickness of the sample and upon the strength of the ac magnetic field. Figures 7 and 8 both show the type of dependence the theory predicts. Thus this theory is able to predict a number of the magnetic properties of the HTSC materials, at least for higher frequencies and moderate fields.

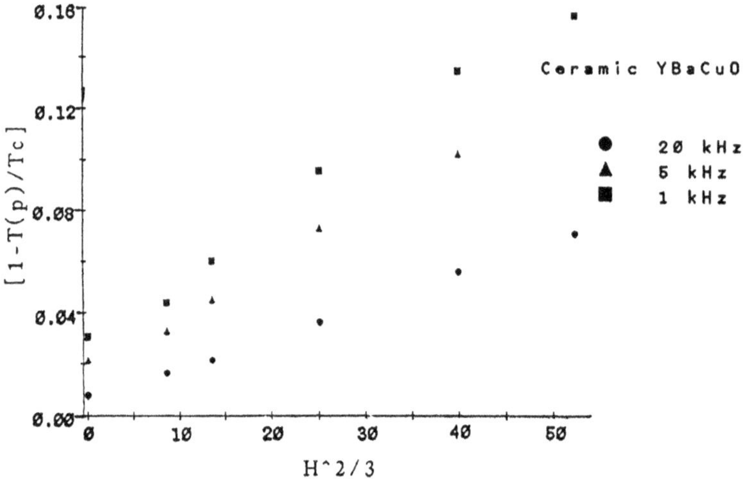

Fig. 9 (1 - Tp/T_C) versus $H^{2/3}$ for ceramic $YBa_2Cu_3O_{7-x}$.

The frequency independence of the results below about 1 kHz suggest that the flux vortices can move in time with the ac field and thus represent a stable state; above this frequency the vortices cannot move in synchronism with the ac field and a different dynamic state results. The vortex-glass picture of superconductivity given by Fisher[25] predicts this kind of behavior in that as the frequency of measurement approaches zero, a finite IL results where the X" peak temperature now represents a "glass transition" temperature for the vortex system. The TAFF theory in contrast predicts an IL approaching zero as the frequency goes to zero. At higher temperatures the glass-vortex theory predicts a power law dependence; over the range of frequencies used in the present work, such a fit is not accurate enough to draw any conclusions. Other work, where such fittings are done[21], find some agreement with experiment, but not as good as with a thermal activation theory.

The present results provide some insight into the nature of the intergranular material in the ICI ceramic $YBa_2Cu_3O_{7-x}$. The flux exclusion versus magnetic field results shown in Figure 6 do not show the rapid falloff of flux exclusion in weak fields characteristic of weak-link connected grains[26], and since the ICI material fits the TAFF theory well, it appears that, other than the stronger magnetic field dependence, the ICI $YBa_2Cu_3O_{7-x}$ has the intrinsic properties of the single crystal material compound. The increased field dependence can be explained by the fact that the material has a porous structure[12], so that there are many regions of small cross-section where the magnetic field lines are concentrated by the repulsion of the surrounding bulk grains. This concentration increases the strength of the local field in the small cross-section areas to above that of the external field, producing an apparent shift downward of the IL. Such local field concentration has been invoked by Askew et al[27] to explain irreversible critical current behavior in $YBa_2Cu_3O_{7-x}$.

CONCLUSIONS

The results reported in this work show a complex behavior of dynamic magnetic properties of some high temperature superconductors over a range of temperatures, magnetic fields, and frequencies of measurement. An irreversibility line IL is defined by the temperature dependence of the peak in X" versus magnetic field measured at a particular frequency. The IL properties of a ceramic and some single crystal HTSC materials are compared and shown to be widely different. The effect of demagnetizing factors upon the IL of a ceramic $YBa_2Cu_3O_{7-x}$ material has been qualitatively determined and shown to have a strong effect upon the position of the IL. The observed much stronger magnetic field dependence of the IL of ceramic $YBa_2Cu_3O_{7-x}$ compared to bulk material has been explained by local magnetic field enhancement due to flux compression in this porous structure. Experimental results have been compared to the predictions of various theories. Thermally activated flux flow provides a good fit to the data for moderate magnetic fields and higher frequencies; however, this theory does not predict the frequency independent IL observed below 1 kHz. Instead the data fits the predictions of a glass-vortex theory of superconductivity in this frequency range; this theory in turn does not fit well higher frequency results. To date no one theory of high temperature superconductivity seems capable of explaining all of the dynamical properties of HTSC materials.

ACKNOWLEDGEMENTS

Helpful discussions were held with T. R. Askew and Andris Suna. The ICI $YBa_2Cu_3O_{7-x}$ was provided by N. McN. Alford, and the thin film samples were provided by D. W. Face.

REFERENCES

1. D. Schoenberg, Pro. Camb. Phil. Soc. 33, 559 (1937).
2. P. Dubots and J. Cave, Cryogenics 28, 661 (1988).
3. D. X. Chen, J. Nogues, and K. V. Rao, Cryogenics 29, 800 (1989).
4. S. T. Sekula, J. Appl. Phy. 42, 16 (1971).
5. J. R. Clem, H. R. Kerchner, and S. T. Sekula, Phy. Rev. B14, 1893 (1976).
6. R. B. Flippen and T. R. Askew, J. Appl. Phy. 67, 4515 (1990).
7. M. Nikolo and R. B. Goldfarb, Phy. Rev. B39, 6615 (1989).

8. A. P. Malozemoff, T. K. Worthington, Y. Yeshurun, F. Holtzberg, and P. H. Kes, Phy. Rev. B38, 7203 (1988).
9. T. W. Worthington, Y. Yeshurun, A. P. Malozemoff, R. M. Yandrofski, F. H. Holtzberg, and J. R. Dinger, J. de Phys. 49, 2093 (1988).
10. R. B. Flippen and T. R. Askew, J. Appl. Phy. 64, 5908 (1988).
11. R. A. Hein, Phy. Rev. B33, 7539 (1986).
12. N. McN. Alford, J. D. Birchall, W. J. Clegg, M. A. Harmer, K. Kendall, and D. H. Jones, J. Materials Sci. 23, 761 (1988).
13. M. Subramanian, J. Calabrese, C. Torardi, J.Gopalakrishnan, T. Askew, R. Flippen, K. Morrissey, U. Chowdhry, and A. Sleight, Science 240, 495 (1988).
14. R. B. Goldfarb and J. V. Minervini, Rev. Sci. Inst. 55, 761 (1984).
15. C. C. Stoner, Phil. Mag. 36, 803 (1945).
16. T. T. M. Palstra, B. Batlogg, R. B. van Dover, L. F. Schneemeyer, and J. V. Waszczak, Phy. Rev. B41, 6621 (1990).
17. K. A. Muller, M. Takashige, and J. G. Bednorz, Phy. Rev. Lett. 58, 1143 (1987).
18. M. R. Beasley, R. Labusch, and W. W. Webb, Phy. Rev. 181, 682 (1969).
19. P. H. Kes, J. Aarts, J. van den Berg, C. J. van den Beek, and J. Mydosh, Supercond. Sci. and Technol. 1, 242 (1989).
20. P. L. Gammel, L.F. Schneemeyer, J. V. Waszczak, and D. J. Bishop, Phy. Rev. Lett. 61, 1666 (1988).
21. P. L. Gammel, J. Appl. Phy. 67, 4676 (1990).
22. P. L. Gammel, L. F. Schneemeyer, and D. L. Bishop, Phy. Rev. Lett. 66, 953 (1991).
23. A. P. Malozemoff, Bull. M.R.S. 15, 53 (1990).
24. A. M. Campbell and J. E. Evetts, Adv. in Phy. 21, 298 (1972).
25. M. P. A. Fisher, Phy. Rev. Lett. 62, 1415 (1989).
26. R. B. Flippen and T. R. Askew, Solid State Comm. 72, 337 (1989).
27. T. R. Askew, R. B. Flippen, K. J. Leary, and M. N. Kunchur, J. Mat. Research , June, 1991.

UTILITY OF THE χ_{ac} RESPONSE IN THE LOW FIELD LIMIT FOR CHARACTERIZING INHOMOGENEOUS SUPERCONDUCTORS

B. Loegel, D. Bolmont and A. Mehdaoui

Laboratoire de Physique et de Structure Electronique des Solides
Faculté des Sciences, Université de Haute Alsace
4, rue des frères Lumière, 68093 Mulhouse Cedex, France

INTRODUCTION

Characterizing a superconducting transition by initial susceptibility leads to highly variable results in inhomogeneous compounds and the shape of the transitions (both for χ' and χ'') depends of numerous parameters such as: fabrication processes / thermal treatments, morphologies / shapes (bulk, wires, thin films, single crystals, powders), stoichiometry, coexistence of several phases, field orientation, etc... One can thus easily understand the difficulty encountered in comparing results obtained in different labs, using various experimental techniques applied on samples of different shapes!

The HTSC are very typical of such problems and we present here our analysis of initial susceptibility measurements in the low field limit which, in spite of numerous parameters governing the transition shapes, allows a comparison between various superconducting systems obtained under different fabrication conditions and morphologies. It is not our aim to give here some theoretical discussion from an expert point of view, but we rather give here a user's point of view, presenting a phenomenological treatment of experimental data. Our analysis leads to a critical dynamic field and estimates the "quality" of an inhomogeneous superconductor in relevance to the strongness of flux pinning in ac fields

I - THE WIDTH OF A MAGNETIC TRANSITION vs APPLIED FIELD AMPLITUDE

The origin of the superconducting transition broadening under ac or dc magnetic fields for a typical hysteretic type II superconductor lies in the magnetic phase diagram as sketched in Fig. 1 which gives the static susceptibility χ=M/H, where M is the magnetic moment of a sample and H is the dc magnetic field whereas the ac susceptibility $\chi_{ac}=(dM/dh)_{H=0}$ is given as $\chi_{ac}=\chi'-i\chi''$ (figure 1(left) for dc fields H_{appl} on the order of one or two Teslas and figure 1(right) for ac fields on the order of one or two Gauss). The transition begins for decreasing temperature, in a given applied field, when $H_{appl} = H_{c2}$ (or $h = H_{c2}$) and ends when $H_{appl} = H_{c1}$ (or $h = H_{c1}$). Diamagnetic shielding is complete at lower temperatures ($\chi = -1$). Onset in ac magnetic fields is less field dependent owing to the steeper slope of $H_{c2}(T)$ in the low field limit. The high temperature superconductors (HTSC) H-T diagram exhibit moreover two peculiar points[1]: (i) very high H_{c2} values, and thus a nearly vertical line for the $H_{c2}(T)$ curve with the scales commonly used for ac fields (ii) low H_{c1} which give rise to incomplete diamagnetic shielding, even at low temperatures and fields.

In characterizing a superconducting transition one must take into account the existence of the above described H - T phase diagram, complicated by the granular structure encountered especially in HTSC. The superconducting ceramics are formed by an assembly of grains (intragrain material with critical temperature T_c^g and critical fields H_{c1}^g and H_{c2}^g) embedded in the intergrain material (with critical temperature T_c^J and critical fields H_{c1}^J and H_{c2}^J).

The grains are superconducting as soon as $T_c^J < T < T_c^g$ but the intergrain material is only superconducting for $T < T_c^J$. Combining now two H - T diagrams for both granular and intergranular phases, we obtain the qualitative result presented on figure 2: χ' presents a

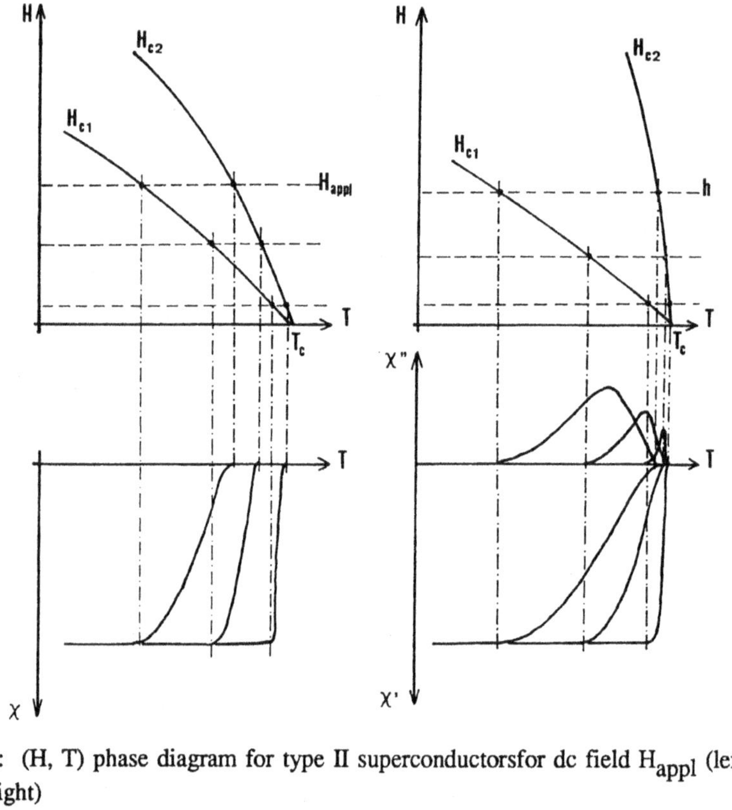

Figure 1: (H, T) phase diagram for type II superconductorsfor dc field H_{appl} (left) and ac field h (right)

plateau and χ'' exhibits two peaks. The detailed behaviours for $\chi'(T)$ and $\chi''(T)$ take account of the characteristic parameters (T_c's and H_c's) and also of the volume fractions of each phase, explaining therefore the great spread of behaviours experimentally observed both for the diamagnetic shielding and the energy losses. Although the field dependences for both $\chi'(T)$ and $\chi''(T)$ are very different in each case we will see below that a careful analysis of such curves in the low field limit ($h_{ac} << 1oe$) is very useful for comparative studies.

THE REAL PART OF THE AC INITIAL SUSCEPTIBILITY AND ITS FIELD DEPENDENCE

Our measurements were done between R.T. and 50K (pumped nitrogen) for ac fields h_{ac} ranging between 10^{-4} Oe and 10^2 Oe ($8\ 10^{-3}$ A/m < h_{ac} < $8\ 10^3$ A/m) and in some cases with superimposed dc fields H_{dc} (1 Oe < H_{dc} < 20 Oe). Results given on figure 3 are obtained for sintered YBaCuO polycrystals.

The response to the driving field h_{ac} is also monitored by an oscilloscope, and is sinusoidal as long as there is a linear relationship between B and H, i.e.àç as long as the

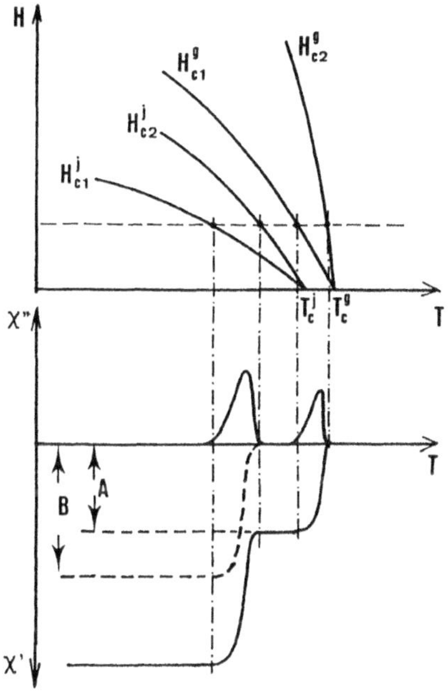

Figure 2: (H, T) phase diagram for type II superconductors with two phases and relevant behaviour of the susceptibility

susceptibility is field independent [2,3,4]. B - H loops take place when vortices are depinned, leading to distorted oscillograms as already observed some years ago in various inhomogeneous superconductors[2,4,5,6,7,8]. The structure of the oscillograms was more recently discussed in details by Müller et al. in the framework of the critical field model[9,10,11].

We observe distorted oscillograms as soon as the vortices are depinned (figure 7), in other words as soon as the applied field exceeds a critical value (depending generally on temperature) which will be called critical dynamic field $h_c^*(T)$ in the case of ac fields. Distortions are thus observed in a strong energy dissipation regime corresponding also to the vicinity of thc χ'' vs T maxima (this point will be discussed in § III).

The shapes observed experimentally for $\chi'(T)$, can be classified into two categories, in relevance to the critical value $h_c^*(T)$ for the applied field:

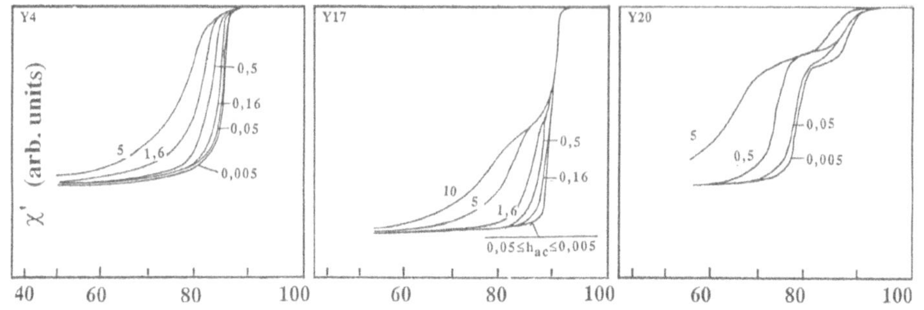

Figure 3: Real part of the susceptibility for some YBCO polycrystals. Numbers give the applied field h_{ac} (Oe).

a) steps disappearing below some field limit: the appearance of field independent $\chi'(T)$ curves signifies that we have reached ac fields h_{ac} below the critical dynamic field $h_c^*(T)$.

b) steps remaining down to the lowest fields: the persistence of steps corresponds otherwise to the existence of phases with significant critical temperature differences (e.g. Y20, figure 3)

In order to determine the value of the critical dynamic field $h_c^*(T)$, we plot the real part $\chi'(T)$ of the susceptibility in reduced coordinates $\chi'(h_{ac})/\chi'(h_{ac\ min})$ vs h_{ac}, where $h_{ac\ min}$ is the lowest driving field h_{ac} used in measurements and with the temperature as a parameter[12,13,14]. Such a plot leads, at constant temperature to a straight horizontal line at ordinate +1 until the critical dynamic field is reached. Departure from the value +1 at a given temperature gives the critical dynamic field for that temperature (figure 4).

We extend our analysis - in order to check the utility of such a test - to other compounds (BiSrCaCuO, BiPbSrCaCuO), other fabrication techniques (zone melting, flux method) and other morphologies (single crystal, wire). The critical dynamic field deduced from $\chi'(T)$ was also estimated from results published by other authors on various inhomogeneous superconducting compounds[15 through 23]. Our analysis depends obviously of $h_{ac\ min}$ and a

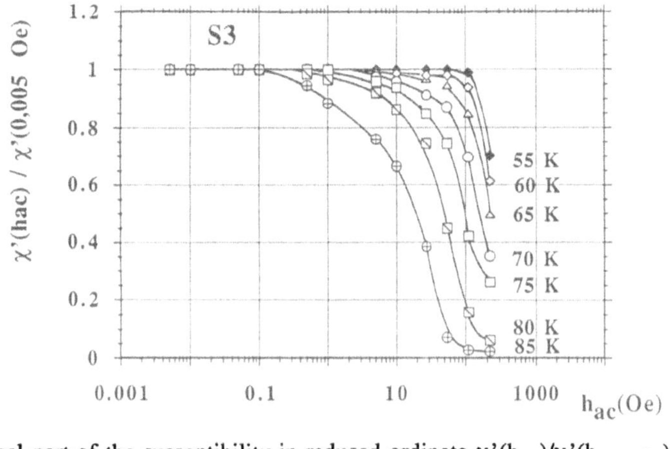

Figure 4: Real part of the susceptibility in reduced ordinate $\chi'(h_{ac})/\chi'(h_{ac\ min})$, $h_{ac\ min}$ = 0.005 Oe, for single crystal S3.

comparative study is only possible if the lowest measuring field lies in the range where $\chi'(T)$ becomes field independent, i.e. $h_{ac\ min} < h_c^*(T)$. We therefore make the assumption -when necessary - that the lowest stated ac field satisfies such a condition. However, an accurate value for $h_c^*(T)$ is sometimes difficult to obtain from the published data owing to the small number of ac fields used.

Geometrical factors and frequency effects are not taken into account by the phenomenological methods we describe below, but most of our conclusion are based on qualitative comparisons and it is noteworthy that similar geometries yield also very different

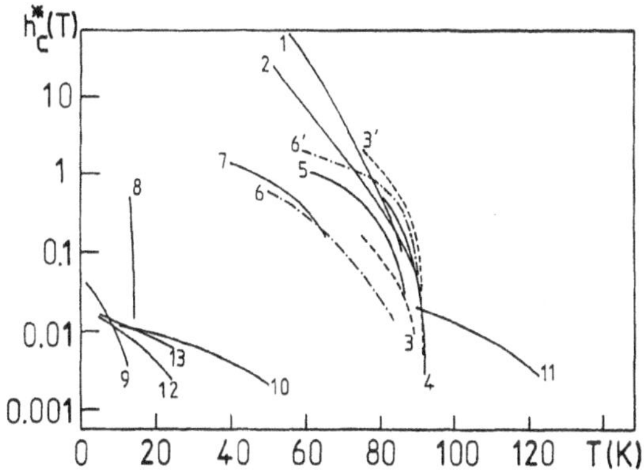

Figure 5: Phase diagram in the ac low field limit. 1: YBCO single crystal S3. 2: zone melted poly-BSCCO. 3: sint. poly-YBCO air-treated[16]. 3': sint. poly-YBCO O_2-treated[16]. 4: sint. poly-YBCO[15]. 5: YBCO wire. 6: sint. poly-YBCO[17]. 6': sol-gel poly-YBCO[17]. 7: YBCO thin film[18]. 8: sint. poly-$PbMo_6S_8$[23].9: poly-NCCO[22]. 10: sint. poly-BSCC 2212[19]. 11: sint. poly-BSCC 2223. 12: sint. poly-LSCO[20]. 13: sint. poly-LCO[21].(sint.: sintered. poly: polycrystal).

behaviours for the initial susceptibility.

All the results we obtained by applying the above analysis are summarized on figure 5 and will be discussed below. But, before discussing these results (obtained from an analysis of the low field limit of the real part of the susceptibility $\chi'(T, h_{ac})$), we will present an alternative method, based on the analysis of the low field limit of the imaginary part of the susceptibility $\chi''(T, h_{ac})$.

THE IMAGINARY PART OF THE ac INITIAL SUSCEPTIBILITY AND ITS FIELD DEPENDENCE.

The peak in the χ'' vs T curves broadens and T_M (temperature of the maxima) decreases with increasing applied ac field h_{ac} (figure 6). A second peak is also frequently mentionned in the literature[16,18,19,22,24,25] and we observed it also in some of our alloys (see Y17 and Y20 on figure 6 and S1 on figure 9).

A detailed study allows one to distinguish two different behaviours by studying - as for the real part of the susceptibility - the field dependences in the low field limit.

On the basis of the H-T diagram (figure 2) we observe distinct maxima T_{M1} and T_{M2}, even in the low field limit, when two phases with significant different critical temperatures T_{c1} and T_{c2} coexist, and the difference $T_{M1} - T_{M2}$ is of the same order of magnitude than $T_{c1} - T_{c2}$. The persistence of two peaks in the low field limit for $\chi''(T, h_{ac})$ corresponds to the persistence of steps in the low field limit for $\chi'(T, h_{ac})$.

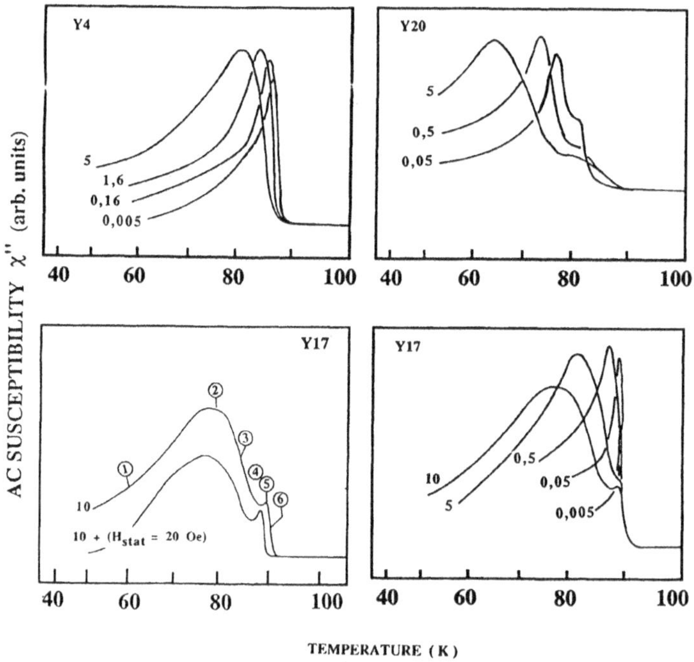

Figure 6: χ'' for some YBCO sintered polycrystals, applied field h_{ac} in Oersted. Number in circles correspond to the oscillograms of figure 7.

The simplest behaviour expected for the T_M vs field dependence is a nearly constant value as long as the applied ac or dc magnetic field is lower than the corresponding critical fields (e.g. as long as energy dissipations are negligible), and a sharp breakdown as soon as the critical value is reached. The situation is somewhat more complicated in the HTSC where Clem's model[26] based on weakly coupled grains describes the granular systems quite well. Vortices move through the intergrain material when e.g. $h_{ac} > H_{c1J}$ (Here H_{c1J} is the intergranular critical field and has a value of approximately 1 0e). Above H_{c1J}, T_M decreases linearly [16,17,26,27]. T_M lies in all cases between T_c and T_{cJ}, where T_c is the critical temperature corresponding to the appearance of shielding effects (χ') and energy losses (χ''), and T_{cJ} is the temperature at which thermal fluctuations destroy phase coherence between grains: Clem's model gives $T_c - T_{cJ} = 1.4$K for $T_c = 95$K. Experimental results for ac magnetic fields show that T_M vs h_{ac} extrapolates not toward T_c for $h_{ac} = 0$, but to a lower value which we denote as T_M°, and $T_c - T_M^\circ$ is of the same order of magnitude as $T_c - T_{cJ}$ (Table I).

A kink in the T_M vs h_{ac} curves appears clearly in some cases (figure 8) and gives directly the critical field which we called critical dynamic field h_c* in the case of ac fields. But many bulk samples exhibit smoothed kinks arising from e.g. a spread of critical fields among the

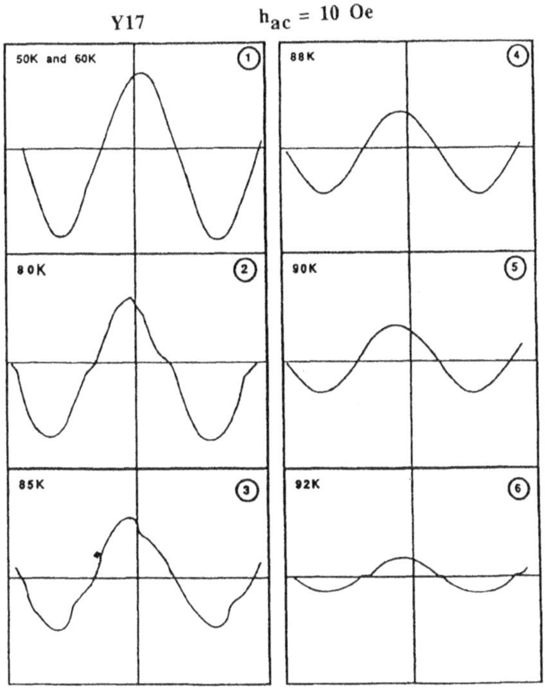

Figure 7: Oscillograms taken at the secondary of the Hartshorn bridge (sample Y17, f=330 Hz).

microcrystals. For this case and for the sake of comparison among superconductors in a wide critical temperature range, we use a semilogarithmic plot of $(T_c - T_M) / T_c$ versus a reduced variable h_{ac}/h_c^* leading to a "universal curve"[28] where the critical dynamic field h_c^* is the single fitting parameter.

We discuss next the results obtained by our analysis both on the imaginary part χ'' and the real part χ' which give the same trends.

PHENOMENOLOGICAL ANALYSIS OF THE ÈXPERIMENTAL RESULTS.

Polycrystalline and bulk high Tc samples.

Table I gives the critical dynamic fields deduced from our phenomenological analysis for some experimental results we obtained on YBaCuO and BiSrCaCuO samples and other data available in the literature including LaBaCuO compounds and Chevrel phase[28,29].

A comparison between the critical field h_c^* - obtained from the above analysis of $\chi''(T, h_{ac})$ - and the critical field $h_c^*(T)$ - obtained in the previous section from the analysis of $\chi'(T, h_{ac})$ - shows that the values obtained in both ways are of the same order of magnitude and the temperature independent critical field h_c^* corresponds roughly to the low temperature limit of $h_c^*(T)$: $h_c^* \approx h_c^*(T)$ for $T \rightarrow 0$.

On the basis of these results we can now draw some general trends for both critical dynamic fields:

1. The highest critical fields are obtained for the most homogeneous cases like the

Table I: Critical dynamic fields h_c^* (Oe) obtained from $\chi''(T, h_{ac})$ for YBaCuO samples. Hc1: first static critical field; T_c : critical temperature. $T_M°$: extrapolated value of T_M for $h_{ac} = 0$)

Réf.	Type	h_c^*(Oe)	H_{c1}(Oe)	$T_c-T_M°$(K)
YBaCuO polycrystals, sintered				
[17]	oxyde powder	0.65	40	4.9
[16]	air treated	0.8		
[11]	intergranular, calculated		2.1	
[16]	oxygen treated	2.6	41	1
	wires	2.0-5.8		
[30]		5.0		
[31]		8.0		
[15]		16.0		
	Y17	3.5	82.5	1.9
	Y4	2.7		
	Y29 (sieved, <40 μm)	0.2		
	Y31 (sieved, 50 μm -60 μm)	0.75		
	Y33 (sieved,106 μm -212 μm)	1.25		
YBaCuO polycrystals, other fabrication techniques				
[18]	thin film	0.5	23	5.8
[17]	sol-gel	8.5	141	0.5
[30]	partial melted, intergranular peak	200		
[30]	partial melted, intragranular peak	350		

Table II: critical dynamic fields h_c^*(Oe) deduced from $\chi''(T, h_{ac})$ for various polycrystalline compounds and for YBCO single crystals (*: with superimposed dc field H_{stat} =8 Oe)

Other polycrystals			YBaCuO single crystals			
Ref.	Type	h_c^*(Oe)	Ref	Type		h_c^*(Oe)
	B2 (phase (2212),Tc=80 K)	0.27	(S1)	h_{ac} // c	T_{M1}	50
	BSCC (phase (2223),Tc=110 K)		(S1)	h_{ac} // c	T_{M2}	50
	(estimated from $\chi'(T)$)	10 x (h_c^*	(S1)	$h_{ac} \perp$ c	T_{MH}	85
		of phase (2212))	(S1)	$h_{ac} \perp$ c	T_{ML}	1.1
[22]	(Nd1.85Ce0.15)CuO4	10^{-2}-10^{-3}	(S1)	$h_{ac} \perp$ c	T_{MH}	34*
[19]	BiSrCaCuO poly, sintered	<0.1 ?	(S1)	$h_{ac} \perp$ c	T_{ML}	4*
[25]	PbMo6S8	1.7	(S3)	$h_{ac} \perp$ c		50
[32]	LaBaCuO	3.2	[34]	YBaCuO single crystal		35
[33]	BiSrCaCuO, zone melted	0.35 ?	[11]	intragranular, calculated		
	BiSrCaCuO, zone melted	5-15?			from $\chi''(T)$	10^2-10^3

intragrain material of Costa et al[30], or single crystals (see § IV - 2) and are of the same order of magnitude as are the intragrain values calculated by Müller et al.[9,10,11].

2. We observe a large spread for the critical fields among a given type of compounds which reflects the high sensitivity of these critical fields to fabrication processes and thermal treatments. In a system like YBCO fabrication processes themselves influence the granularity of the samples:

- Thermal treatment under flowing oxygen results in better critical fields than under air[16].
- Sintered samples exhibit lower h_c*'s than sol-gel fabricated samples[17].
- Partially melted samples show always very high critical fields[30].

3. Independently of the classes of compounds, we observe that, on one hand, the lowest critical fields are always characteristic of sintered samples and, on the other hand, the highest critical fields are obtained for partially melted[30] or zone melted samples[19].

4. It is usually difficult to fit the results for BSCCO and BPbSCCO compounds, nevertheless the h_c*'s order of magnitude for sintered compounds is much lower than for YBCO compounds. This is the case for both the 2212 (Tc = 80K) and 2223 (Tc = 110K) phases.

Single crystals.

The transition shows a very weak dependence on applied ac field. In the low field limit we observe furthermore[14,29] (i) field orientation effects (as already observed[35] the peak of $\chi''(T)$ is broader for h_{ac} applied perpendicular to c axis) (figure 9) (ii) fine structures (figure 9)

The analysis of the field dependence with the above described method leads generally to very high critical dynamic fields h_c* (table II). When applying the analysis of the χ' data

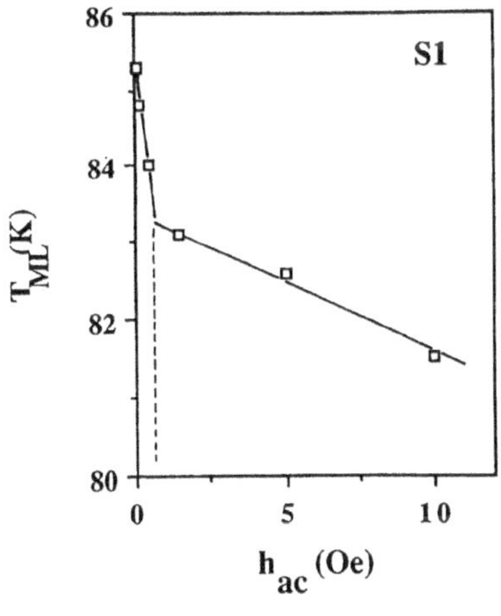

Figure 8: T_M (T_M: Temperature of the maxima of $\chi''(T)$) vs applied field h_{ac} for one of the maxima of the single crystal S1 ($h_{ac} \perp$ c-axis).

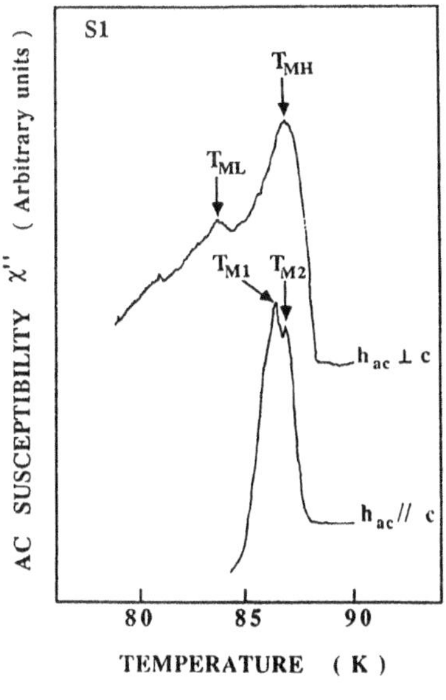

Figure 9: χ'' for YBCO single crystal S1. h_{ac} = 1.6 Oe (f=330 Hz).

described in § II - 2 to single crystals we obtain again the highest values for $h_c^*(T)$. Single crystals have therefore the highest critical fields and only intragrain material and some partially melted polycrystals can compete with them.

V - DISCUSSION

There is clearly a link between the critical dynamic field as defined by us and pinning energy and critical current density (a roughly linear dependence of h_c^* vs j_c can be deduced from some data[17,19]). However single crystals and thin films support current densities well above 10^6 A/cm^2 at 77K while for polycrystalline HTSC the critical current densities are lower by about three orders of magnitude. On the other hand, the critical dynamic fields of single crystals, intragrain material and partially melted or zone melted polycrystalline HTSC lie in the range $5X10^2$ Oe< h_c^*, $h_c^*(T)$ < $5X10^3$ Oe whereas other polycrystalline HTSC and thin films are often lower by more than an order of magnitude.

The high critical fields obtained for single crystals indicate strong pinning, although they are virtually free of inhomogeneities. Pinning can be very efficient when the pinning sites are of atomic sizes and of the same order of magnitude as the correlation length (the presence of "junctions" results from the short intrinsic coherence length of the oxides, combined with the presence of extended crystallographic defects of atomic size , e.g. staking faults and twin boundaries in single crystals[44]). This can also be the case for the CuO planes[45,46] which lead to different pinning forces or activation energies[39] for ac fields // and ⊥ c-axis, in agreement with the h_c^*'s measured in both direction (table II). The fine structure mainly observed in the χ'' vs T curves for h_{ac} ⊥ c-axis are in agreement with the staircase model of Couach et al.[35].

High critical dynamic fields and activation energies are observed when fabrication processes involve melting (zone melting, partial melting, QMG (quench and melt growth), melt processing). We observed high h_c^* (or $h_c^*(T)$) for zone melted BSCCO and the h_c^*'s for partially melted YBaCuO[30] are even higher than those obtained for single crystals. Similarly QMG processed YBaCuO crystals have much higher pinning potentials than single crystals[42] and although the size of inhomogeneities like 211 inclusions present in melt processed YBaCuO is relatively large compared with the coherence length, it was shown[43] that while large flux creep is observed in thin films exhibiting high j_c values, flux creep is very small in melt processed YBaCuO -having a magnitude lower j_c values- as the pinning potential of large normal precipitates is much larger than that of small pinning centers. This is in agreement with the critical dynamic fields we obtain for melted samples and thin films (table I).

Flux creep effects are very small in conventional type II superconductors, but large in HTSC like YBCO and giant in Bismuth and Thallium compounds[36]. Core pinning becomes inefficient at preventing flux motion in the latter compounds.

The critical dynamic fields for BSCCO are also about one order of magnitude lower than for YBCO for similar fabrication techniques, as expected if the pinning force is very weak[14,28,29,37]. Pinning in 2212 phase is even weaker than in 2223 phase[37], in agreement with the results of Reissner et al.[38]. Our results are also in agreement with the drastic difference of the pinning or activation energies between YBCO and BSCCO observed by several groups[38,39,40,41,47].

Very low critical dynamic fields and thus pinning energies in Bi-based polycrystals means that even the small currents used for electrical resistivity measurements lead to magnetic fields which are high enough -in the neighborhood of the critical temperature- for making the electrical resistivity current dependent[37].

CONCLUSION

Initial susceptibility measurements provides a very simple and useful tool for comparing activation energies and/or pinning forces when analyzing the data in the low field limit. The analysis is applied to various superconducting high T_c compounds and also low T_c compounds like Chevrel phases, obtained under different morphologies (Bulk, powders, thin films, wires, single crystals,...) and with different fabrication techniques (Sintering, sol-gel, zone melting, partial melting,...).

Such an analysis should be very profitable for controlling fabrication processes involving e.g. shock processing, melt quenching, melt casting, field oriented precursors and addition or subtitution of elements.

We are also currently studying the frequency dependences of the critical dynamic fields.

REFERENCES

[1] J. E. Drumheller , G. V. Rubenacker, W. K. Ford, J. Anderson, M. Hong, S. H. Liou, J. Kwo and C. T. Chen, Solid State Comm., 64, 509, (1987)

[2] J. R. Clem, H. R. Kerschner and S. T. Sekula, Phys. Rev. B. 14, 1893, (1976)

[3] R. W. Rollins and J. Silcox, Phys. Rev. 155, 404, (1967)

[4] T. Ishida and H. Mazaki, J. Appl. Phys., 52, 6798, (1981)

[5] E. Maxwell and M. Strongin, Phys. Rev. Lett., 10, 212, (1963)

[6] T. Ishida and H. Mazaki, Phys. Rev. B. 20, 131, (1979)

[7] A. Perrin, O. Pena, C. Perrin, Z. Li and M. J. Sergent, J. Phys. France, 49, 301, (1988)

[8] Y. Oda, H. Takenaka, H. Nagano and I. Nakada, Solid State Comm., 35, 887, (1980)

[9] K. H. Müller, J C Mac Farlane and R Driver, Physica C, 158, 69, (1989)

[10] K. H. Müller, J C Mac Farlane and R Driver, Physica C, 158, 366, (1989)

[11] K. H. Müller, Physica C, 159, 717, (1989)

[12] B. Loegel, D. Bolmont and A Mehdaoui, Physica C, 816, (1989)

[13] B. Loegel, A. Mehdaoui and D. Bolmont, J. Less-Common Met., 150, 193, (1989)

[14] B. Loegel, D. Bolmont, H. Dalderop and A. Mehdaoui, Sol. State Comm., to appear

[15] H. Mazaki, M. Takano, Y. Ikeda, Y. Bando, R. Kanno, Y. Takeda and O. Yamamoto Jpn. J. Appl. Phys., 26, L1749, (1987)

[16] R. B. Goldfarb, A. F. Clark, A. I. Braginski and A. J. Panson, Cryogenics, 27, 475, (1987)

[17] F. Gömöry and P. Lobotka, Solid State Commun., 66, 645, (1988)

[18] Ch Neumann, P. Ziemann, J. Geerk and H. C. Li, J. Less-Common Met., 151, 363, (1989)

[19] J. H. P. M. Emmen, V. A. M. Brabers, W. J. M. de Jonge, C. van der Steen, J. H. J. Dalderap, P. M. A. Geppaart and K. Kopinga, J. Less-Common Met., 151, 63, (1989)

[20] P Butaud, P Ségransan, C Berthier and Y Berthier, C Paulsen, J L Tholence and P Lejay, Phys. Rev. B 36, 5702, (1987)

[21] J Beille, R Cabanel, C Chaillout, B Chevalier, G Demazeau, F Deslandes, J Etourneau, P Lejay, C Michel, J Provost, B Raveau, A Sulpice, J L Tholence, et R Tournier, C. R. Acad. Sc. (Paris), 304, 1097, (1987)

[22] G. Hilscher, S. Pöllinger, M. Forsthuber and N. Pillmayr, K. Remschnig and P. Rogl, M. Reissner and W. Steiner , P. Knoll, Physica C., 167, 472, (1990)

[23] M Decroux, D Cattani, J Cors, S Ritter and O Fisher, Supercond. Sci. Technol., to appear

[24] D. X. Chen, R. B. Goldfarb, J. Nogues and K. V. Rao, J. Appl. Phys., 63, 980, (1988)

[25] P. Dubots and R. J. Cave, Cryogenics 28, 661, (1988)

[26] J. R. Clem, Physica C., 153-155, 50, (1988)

[27] H. Küpfer, I. Apfelstedt, R. Flükiger, C. Keller, R. Meier-Hirmer, B. Runtsch, A. Turowski, U. Wiech and T. Wolf, Cryogenics, 28, 650, (1988)

[28] A. Mehdaoui, B. Loegel and D. Bolmont, J. Appl. Phys., 66, 1497, (1989)

[29] B. Loegel, A. Mehdaoui, D. Bolmont, A. Eckhardt and F. Muller, Cryogenics, 30, 623, (1990)

[30] G A Costa, M Ferretti and G L Olcese, V Calzona, M R Cimberle, C Ferdeghini, M Putti and A S Siri, Sol. State Commun., 68, 923, (1988)
[31] E. Babic, D. Drobac, J. Horvat, Z. Marohnic and M. Prester, J. Less-Common Met., 151, 89, (1989)
[32] A Maeda, T Noda, H Matsumoto, T Wada, M Izumi, T Yabe, K Uchinokura and S Tanaka, J. Appl. Phys., 64, 4095, (1988)
[33] J. H P M. Emmen, W. J. M. de Jonge, C. van der Steen, J. H. J. Dalderap, P. M. A. Geppaart and V. A. M. Brabers, Solid State Commun., 66, 1089, (1989)
[34] O. Laborde, J. L. Tholence, P. Monceau, H. Noel, P. Gougeon, J. Padiou, J. C. Levet and M. Potel, Solid State Commun., 69, 513, (1989)
[35] M. Couach, A. F. Khoder, B. Barbara, J. Y. Henry and F. Monnier,Phys. Rev. B. 38, 78, (1988)
[36] G. Deutscher in *"Earlier and Recent Aspects of Superconductivity"*, Springer Series in Sold-State Sciences, Vol. 90, p. 174-200, J.G. Bednorz, K.A. Müller Editors, Springer-Verlag Berlin, Heidelberg (1990)
[37] B. Loegel, A. Mehdaoui and D. Bolmont, Supercond. Sci. Technol., 3, 504, (1990)
[38] M. Reissner, R. Ambrosch and W. Steiner, Proc. LT-19 Satellite Conference on High Temperature Superconductivity, Cambridge (UK), august 13th-15th, 1990, p 436-438,J. Evetts, Editor (Adam Hilger, 1991)
[39] J. Z. Sun, K. Char, M. R. Hahn, T. H. Geballe and A. Kapitulnik, Appl. Phys. Lett., 54, 663, (1989)
[40] K. Kadowaki, N. J. Li, R. de Boer, P. H. Frings and J. J. M Franse, Supercond. Sci. Technol., 4, S88, (1991)
[41] R. Tournier, A. Sulpice, P. Lejay, O. Laborde and J. Beille, J. Magn. and Magn. Mat., 76-77, 552, (1988)
[42] M. Murakami, M. Morita and N. Koyama, Jpn. J. Appl. Phys., 28, L1754, (1989)
[43] M. Murakami, S. Gotoh, H. Fujimoto, K. Yamaguchi, N. Koshizuka and S. Tanaka, Supercond. Sci. Technol. 4, S43, (1991)
[44] G. Deutscher and K.. Müller, Phys. Rev. Lett., 59, 1745, (1987)
[45] M. Tachiki and S. Takahashi , Solid State Commun., 70, 291, (1989)
[46] B. Roas and L. Schultz, G. Saemann-Ischenko, Phys. Rev. Letters, 64, 479, (1990)
[47] M. J. Ferrari, M. Johnson, F. C. Wellstood, J. Clarke, D. Mitzi, P. A. Rosenthal, C. B. Eom, T. H. Geballe, A. Kapitulnik and M. R. Beasley, Phys. Rev. Lett., 64, 72, (1990)

A.C. SUSCEPTIBILITY STUDIES OF TYPE-II SUPERCONDUCTORS: VORTEX DYNAMICS

Xinsheng Ling, and Joseph I. Budnick

Physics Department, University of Connecticut, Storrs, CT 06269

INTRODUCTION

The discovery of high-T_c superconducting oxides has revived interest in studying the vortex dynamics of type-II superconductivity (see two recent review papers of Brandt[1,2]). A very powerful method of investigating the vortex dynamics is the alternating-current (ac) magnetic measurement. Experimentally one usually superimposes a small ac field on a large dc field and determines the real part χ' and the imaginary part χ'' of the susceptibility by measuring the change of inductance and effective resistance of a pickup coil surrounding the sample. Corresponding to the superconducting-to-normal transition, one usually finds a step-like change in χ' and a peak in χ''. Even though there has been a large number of experimental investigations on low temperature superconductors[3,4] as well as on high-T_c superconductors,[5,6] the theoretical interpretations[7,8] of ac susceptibility of a type-II superconductor have been controversial. The controversy over the interpretation of the dissipative χ'' peak is particularly noteworthy in studies of the vortex dynamics of the high-T_c superconductors. In this paper, we present experimental ac susceptibility data on a conventional hard superconductor Nb_3Al and on crystals of $YBa_2Cu_3O_{7-\delta}$. The data will be compared with several theoretical models. Comparison between ac susceptibility and dc magnetization will be made with respect to the issue of the so-called "irreversibility line".[9]

OUTLINE OF MODELS FOR AC SUSCEPTIBILITY

There have been mainly three types of models suggested in literature in various microscopic physical pictures for the ac susceptibility of non-ideal type-II superconductors. The first type of model assumes a temperature (and field) dependent relaxation time τ, which measures how fast the system approaches equilibrium after a disturbance.[10] The resulting expression for the complex susceptibility is usually of the Debye form

$$\chi = \chi_o/(1+i\omega\tau) = \chi' - i\chi'' \qquad (1)$$

where χ_o is the static susceptibility and ω is the frequency of the perturbing ac field, $\chi'/\chi_o = 1/(1+(\omega\tau)^2)$ and $\chi''/\chi_o = \omega\tau/(1+(\omega\tau)^2)$. Fig. 1 (a) shows a plot of χ' and χ'' as a function of $\omega\tau$ for this type of model. In this model, a χ''-peak results at the temperature at which $1/\tau$ reaches the measurement frequency ω. We will see later this Debye form is not appropriate to describe superconductors.

The second type of model emphasizes the diffusive motion of flux lines.[8,11] When a

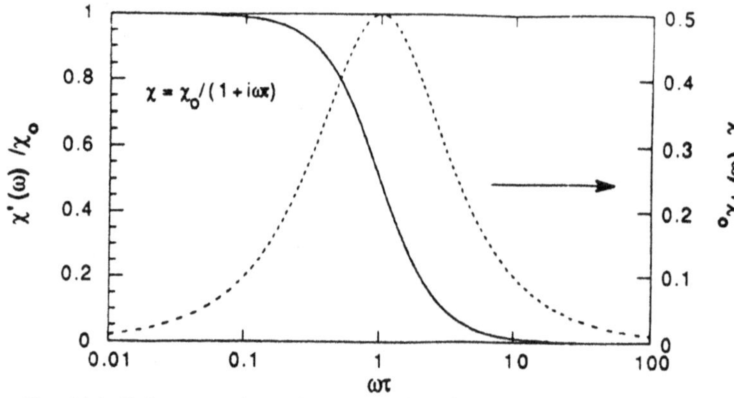

Fig. 1(a): Relaxational model: Normalized real and imaginary parts of AC susceptibility as a function of $\omega\tau$.

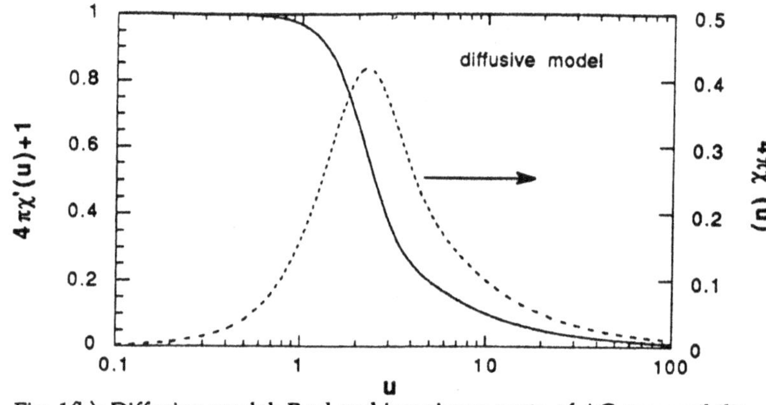

Fig. 1(b): Diffusive model: Real and imaginary parts of AC susceptibility as a function of parameter u.

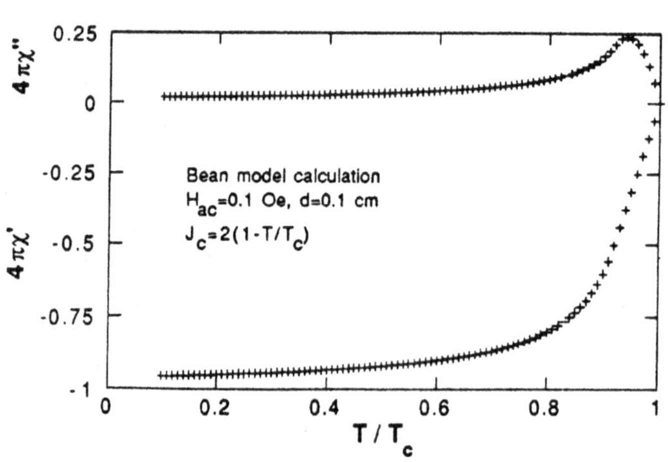

Fig. 1(c): Bean critical-state model: calculated χ' and χ'' as a function of reduced temperature $t=T/T_c$ for assumed $J_c = 2(1\text{-}t)$.

superconductor shows a linear resistivity ρ due to diffusive flux motion, regardless of its origins, the penetration depth of the ac field is a skin-depth $\delta_s = (c^2\rho/2\pi\omega)^{1/2}$. The real and imaginary parts of ac susceptibility have the following forms

$$4\pi\chi' = -1 + (\sinh u+\sin u)/(u(\cosh u+\cos u)) \quad (2a)$$
$$4\pi\chi'' = (\sinh u-\sin u)/(u(\cosh u+\cos u)) \quad (2b)$$

where $u = d/\delta_s$, d is the sample dimension. χ'' reaches a maximum at $u_{max} = 2.25$ corresponding to $\omega_{peak} \sim 0.8c^2\rho(T,H)/d^2$, where c is the speed of light. A plot of χ' and χ'' as a function of u is shown in Fig. 1 (b). A systematic linear response theory was given by Brandt[11] recently. With an ac field modulating a large dc field, Brandt showed that the penetration depth λ_{ac} of the ac field is

$$\lambda_{ac}^2 = \lambda^2 + \lambda_c^2(1-i/\omega\tau)(1+i\omega\tau_o)^{-1}$$

where λ is the London penetration depth, $\lambda_c = (c^2B^2/\alpha_L)^{1/2}$ is the Campbell[12] pinning penetration length, B = flux density, α_L is the elastic restoring force density on flux-lattice (the Labusch parameter); τ is a characteristic time for flux-lattice to relax plastically, defined by $\alpha_L(t) = \alpha_L \exp(-t/\tau)$; $\tau_o = \eta/\alpha_L$ is the relaxation time of the elastically pinned vortex lattice and η is the usual flux-flow viscosity. The magnetic permeability of a thin superconductor is given by

$$\mu(\omega) = \tanh v/v, \qquad v = d/2\lambda_{ac}.$$

For $1/\tau << \omega << 1/\tau_o$, $v = d/2(\lambda^2+\lambda_c^2)^{1/2}$ is real and the superconductor is purely inductive, otherwise $v = (1+i)(\omega d^2/8D)^{1/2}$ where $D = c^2\rho_{fm} = \lambda_c^2/\tau$ for $\omega\tau << 1$ and $D = c^2\rho_{ff} = \lambda_c^2/\tau_o$ for $\omega\tau_o >> 1$, ρ_{fm} is a linear resistivity associated with the plastic relaxation of flux-lattice, and ρ_{ff} is the usual flux-flow resistivity which is associated with the elastic relaxation of flux-lattice. Thus for the cases of $\omega\tau << 1$ or $\omega\tau_o >> 1$, the expressions for the ac susceptibility have the form of eqs. 2 (a) and 2 (b) with $u = 2^{1/2}|v|$.

The third model is the nonlinear response critical-state model,[13] in which the ac susceptibility results from hysteretic penetration of magnetic fluxoids. The expressions for χ' and χ'' (in cgs units) are the following

$$4\pi\chi' = -1+H_{ac}/(4\pi J_c d), \qquad H_{ac}<H^*, \qquad (3a)$$
$$4\pi\chi'' = (1/3\pi^2)H_{ac}/(J_c d); \qquad (3b)$$

$$4\pi\chi' = -\pi J_c d/H_{ac}, \qquad H_{ac}>H^*, \qquad (3c)$$
$$4\pi\chi'' = 4J_c d/H_{ac}-(16\pi/3)J_c^2d^2/H_{ac}^2. \qquad (3d)$$

Eqs. 3(c) and 3(d) are only numerical approximations. A plot of χ' and χ'' as a function of reduced temperature $t = T/T_c$ is shown in Fig. 1 (c), using parameters $H_{ac} = 0.1$ Oe, d = 0.1 cm, and $J_c = 2(1-t)$.

The major differences among the three models are that the first two types of models are independent (linear-response) of the ac field but dependent on the frequency of ac field. The critical-state model does depend (nonlinear-response) on the ac field and could be frequency dependent if the critical current density is time-dependent. Though all the models above can

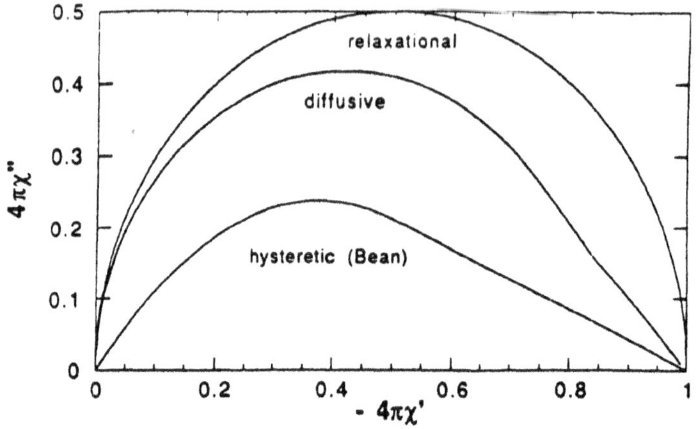

Fig. 2: χ'' vs. χ' for relaxational, diffusive and hysteretic models.

give rise to a peak for χ'', the difference in their predictions for the relationship between χ' and χ'' is very distinct. A plot of $4\pi\chi''$ as a function of $-4\pi\chi'$ is shown in Fig. 2, in which the relaxation model shows symmetry with the χ'' maximum at $4\pi\chi' = -0.5$. The diffusive model shows a slight asymmetry with χ'' at $4\pi\chi' = -0.435$, and the critical-state model shows a distinct asymmetry with a maximum at $4\pi\chi' = -0.375$. We will show that the critical-state model can give best account for the data in zero dc field or dc field perpendicular to ac field, while the low amplitude linear-response data with $H_{dc}//H_{ac}$ can be described by the diffusive Brandt theory.[11]

EXPERIMENTAL METHODS

ac magnetic susceptibility: Two techniques were used for ac magnetic measurements: mutual inductance and self-inductance. A classical Hartshorn type mutual inductance bridge was used for low frequency measurements. The circuit (see Fig. 3) can be divided into two parts, the primary and the secondary. The primary circuit consists of a primary coil (7.5 cm in length, 3500 turns of gauge 38 copper wire), the primary of a fixed transformer M which provides an in-phase voltage for balancing an inductive signal, a current limiting resistor R_1 (1 kΩ) which is also a potentiometer providing an out-of-phase voltage for balancing resistive signal, a shunt resistor R_s (1.2 Ω) for monitoring the current in the primary coil, and a Hewlett-Packard 204C oscillator. The secondary circuit includes two coaxial oppositely wound pickup coils (each has length of 2.5 cm and inner diameter of 0.6 cm, each contains 2500 turns of gauge 42 copper wires) and one of them encloses the sample, a Gertsch ratio-tran along with the secondary of the fixed transformer M, two resistors R_2 (0-0.1 MΩ) and R_3 (10 Ω) forming a voltage divider. The sample is glued with thermal grease on one end of a sapphire rod inserted into the pickup coil. The temperature sensor (silicon diode) is mounted on the other end of the sapphire rod outside the coil. A PAR 129A two-phase lock-in amplifier, with reference frequency at the fundamental, served as a null detector at the differential mode. The operational frequency of this mutual inductance bridge was limited to below 2 kHz by the self-resonance of the coils. The data presented in this paper with this technique was at 107 Hz. The d.c. magnetic field is provided along the coil axis by a copper wire solenoid ($H < 300$ Oe) or normal to the detecting coils axis by an electromagnet ($H < 3500$ Oe). The procedure of balancing the bridge is as follows. The first step is to properly set R_2. R_2 should be small enough to have high sensitivity but large enough to suppress the signal to maintain the bridge balanced in a whole run. This is done by first cooling the sample to about 5 K, then adjust the phase setting on the lock-in such that a full scale adjustment on the potentiometer does not cause one unit of deflection on the quadrature channel. At the same time a full scale adjustment on the ratio-tran should also not cause a deflection in the in-phase channel. Then we can reduce the sensitivity and warm up the sample to above T_c while recording the largest deflection in the in-phase channel (out-of-phase signal χ''). Then we cool the sample back down to 5 K, and adjust R_1 to check if the full scale adjustment of R_1 can cause the same order of deflection in the in-phase channel or reset R_2 so that R_1 can do so. Once R_2 is set and the phase is properly adjusted the circuit is ready for collecting data.

The data were collected during the warm up, due to the better control and sensing of temperature. For small single crystal samples, the filling factor is the main limit of sensitivity. As a compromise, the coils sit in the sample space where temperature is changing in the measurement. The temperature change has two effects. The first is that the coil resistance changes and so does the ac field. The second effect is the phase of the reference signal also changes with temperature. The first effect may be neglected at low ac amplitude, since the total change in the coil current at the temperature range of interest 40 K - 100 K is less than 0.01%, due to the large current-limiting resistance and relatively small coil resistance change. The second effect, which invalidates the previous phase setting on the lock-in, is dealt with by readjusting the phase setting at every 5 K increment in temperature. The phase readjustment is done by holding temperature constant and resetting the phase so that no mixing occurs between the signals of the in-phase and quadrature channels. The total readjustment of phase is about

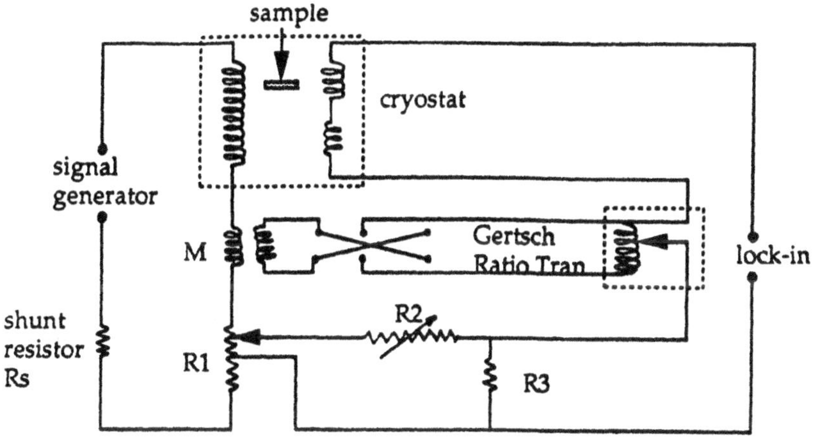

Fig. 3 : A circuit diagram of a modified Hartshorn bridge for AC susceptibility measurements. The lock-in is used as a null detector.

5° from 40 K to 100 K. The accuracy in phase setting is still about 1 part out of 10^5, which is good enough for almost all the measurements. Instead of measuring the unbalance signal by the lock-in directly, the unbalance signal caused by the sample was balanced out by adjusting the ratio-tran and the potentiometer R_1. This procedure takes some extra effort, but the trade back yields much higher sensitivity. After subtracting the background readings, the change on the ratio-tran is the signal in the out-of-phase channel which is the in-phase response ($\sim\chi'$) of the superconductor to the ac field, the change on the potentiometer is the out-of-phase response ($\sim\chi''$).

It was pointed out by A. M. Campbell,[12] that because the lock-in effectively reverses the sign of the signal every half a cycle and then averages over a period of time much longer than the period of the signal, the in-phase signal is the total flux put into the superconductor at the peak of the driving field. The out-of-phase signal is the total flux remaining in the superconductor when the driving field crosses zero. Since the low-pass filter before the output of the lock-in filtered out almost all the higher harmonics, the in-phase signal should indeed be proportional to $4\pi\chi'\eta$, and the out-of-phase signal should be proportional to $4\pi\chi''\eta$, where χ' and χ'' are the fundamental real and imaginary susceptibility and η is the sample-coil coupling constant.

The second type of technique for ac magnetic measurements we used was by directly analyzing the impedance of a coil containing a superconductor. This technique was first used by T. K. Worthington[6] in studies of high-T_c materials. The sample probe consists of a Lakeshore cryogenic miniature radio-frequency cable (type "c1") and a pair of teflon coated fine copper wires put through a Quantum Design SQUID magnetometer sample holder (a stainless steel tube). A small coil (50-150 turns) (wound directly on the sample with fine copper wire (gauge 46, 48, and 50 were used) and held by GE varnish) is connected to the lower end of the Lakeshore cable. A platinum resistor was connected to the other pair of wires for temperature sensing. To avoid mechanical vibration of the coil in the presence of strong magnetic field or large driving current, the coil and loose connecting wires were anchored on a sapphire rod by thermal grease. The temperature sensor was anchored on the same sapphire rod. The impedance and phase angle of the cable-coil-sample system were analyzed by a Hewlett-Packard 4192A LF Impedance Analyzer (5 Hz-13 MHz). The temperature and field (up to 5.5 Tesla) were controlled by the original Quantum Design system computer, while the impedance, phase angle, coil current (~ ac field) and sample temperature were recorded with an IBM PC. It is found that the power dissipation in the coil and the thermometer had to be below 1 mW to avoid heating effects. The ac field amplitude was basically limited by this requirement.

dc magnetization (SQUID technique): For comparison with the ac measurements, the zero-field-cooled and field-cooled dc magnetization were made by using a Quantum Design SQUID magnetometer. The field-cooled measurements were done at both cooling and warming sequence. It is found there is some small hysteresis for the two procedures at low fields (<1 kOe). This hysteresis disappears at higher fields. The travel length of sample was 2 cm with which the field inhomogeneity in the superconducting magnet can be minimized (the field inhomogeneity is about 0.25 Oe at 5 kOe for 2 cm scan).

ZERO DC FIELD OR DC FIELD PERPENDICULAR TO AC FIELD

Fig. 4 (a) and 4 (b) are data collected on the mutual inductance bridge for a bulk melt-cast ingot Nb_3Al (~2x2x10 mm^3) and a single crystal $YBa_2Cu_3O_{6.60}$ (~1x1x0.05 mm^3) grown by flux method. The dip in χ' vs. T of the 3.5 kOe data in Fig. 4 (b) is a real feature which is due to an enhancement of pinning as approaching H_{c2} when vortex lattice becomes soft and can be pinned more effectively (called the peak effect[14,15]). In Fig. 5, a plot of χ'' vs. χ' of the data shown above is contrasted with the Bean model.[13] Because the resistive component of the bridge was not calibrated, all χ'' data in this plot were renormalized so that the χ'' maximum is equal to 0.25, for comparison with models. The unknown filling factor and

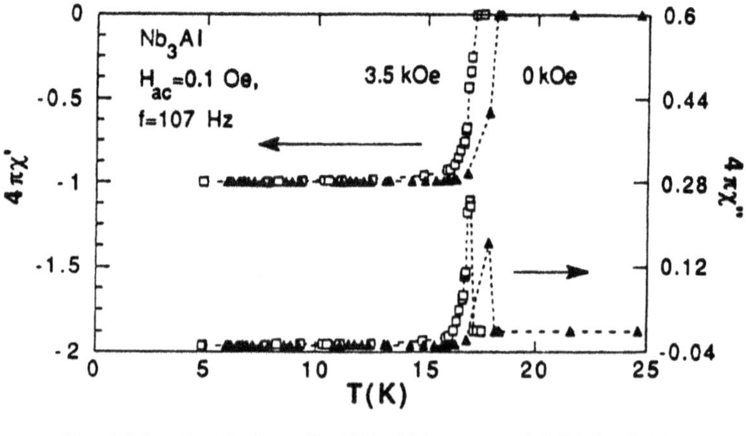

Fig. 4 (a): χ' and χ'' vs. T of Nb_3Al in zero and 3.5 kOe fields. dc field is normal to ac field. f = 107 Hz, H_{ac} = 0.1 Oe.

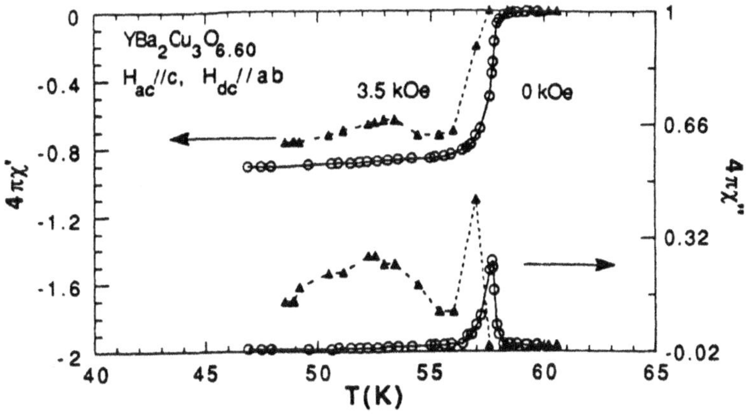

Fig. 4 (b): χ' and χ'' vs. T of $YBa_2Cu_3O_{6.60}$ in zero and 3.5 kOe fields. dc field is normal to ac field. f = 107 Hz, H_{ac} = 0.1 Oe.

demagnetization factor in χ' is dealt with by normalizing χ' data to $4\pi\chi'(10K) = -1$. The negative χ'' in the nearly full shielding range is due to the subtraction of normal state loss from χ''. The data agree with the Bean model quite well in the full penetration range while disagreement appears when approaching full shielding. The disagreement is likely due to the fact that a bulk critical state can not be established if the ac field only penetrates the surface layer.

It is not surprising that the critical-state is relevant even at very low ac field with zero dc field or dc field applied normal to the ac field. With zero dc field, the vortices can travel from the surface to the center of a sample at low ac fields (due to demagnetization effect the local ac field can be larger than H_{c1} even the applied ac field is small), and in the case of dc field applied normal to ac field, the bending of flux lines can build up a critical-state along the direction of the total field.

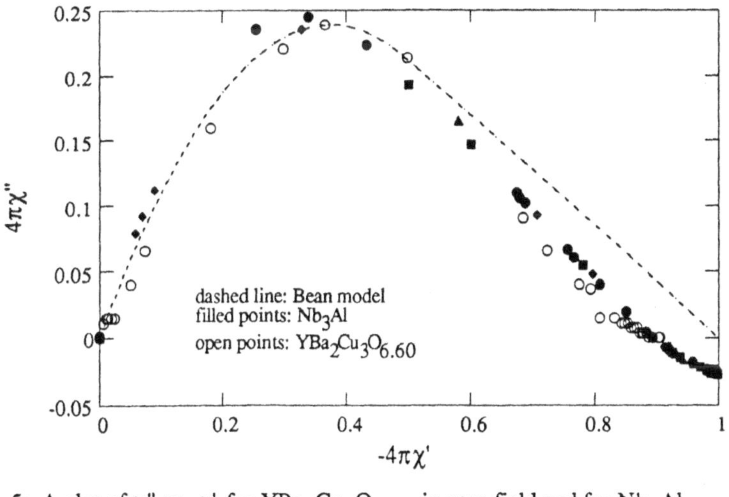

Fig. 5: A plot of χ'' vs. χ' for $YBa_2Cu_3O_{6.60}$ in zero field and for Nb_3Al in zero, 1.1, 2.5 and 3.5 kOe fields, H_{dc} normal to H_{ac}. f = 107 Hz, H_{ac} = 0.1 Oe.

AC FIELD // DC FIELD

For $H_{dc}//H_{ac}//c$, in the range of 5 kOe to 5.5 tesla we find that a typical YBCO crystal shows linear response when the ac field is below 0.5 Oe, while for $H_{dc}//H_{ac}//ab$ the superconductor shows nonlinear response down to minimum accessible field 0.13 Oe when fields are along the ab plane of the crystal. It appears the ac response is dependent on the vortex displacement relative to the inter-pin distances. When ac field and dc field are parallel, for weak pinning, the vortex displacement at the sample surface is about $H_{ac}d/H_{dc}$, where d is the transverse sample dimension. For H_{dc} = 1 Tesla, H_{ac} = 0.2 Oe, d = 0.1 cm, the vortex displacement at the surface is about 20 nm, while the vortex spacing of the flux-lattice at 1 Tesla is $2(1/3)^{1/4}(\phi_o/B)^{1/2} \sim 68$ nm. The onset of nonlinear response at lower ac field for H//ab suggests a shorter inter-pin distance in this field orientation and possibly the ab plane itself acts as a barrier for vortex movement. In Fig. 6 (a) and 6 (b), the linear response ac susceptibilities of a $YBa_2Cu_3O_{6.93}$ crystal (~1x0.78x0.02 mm^3) for $H_{dc}//H_{ac}//c$ in dc fields of 1, 2, 3, 4 and 5 T are shown. The zero-field χ' vs. T is also shown in Fig. 6 (a) for reference. The filling

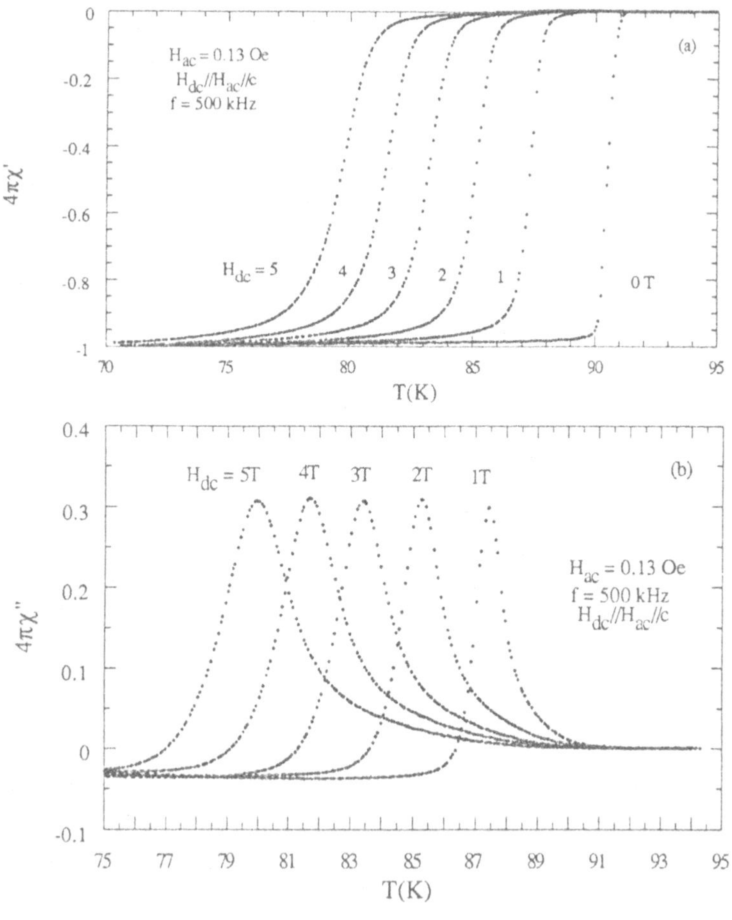

Fig. 6: (a) Linear-response χ' vs. T of a $YBa_2Cu_3O_{6.93}$. H_{dc} = 5, 4, 3, 2 and 1 Tesla. zero-field data is shown for reference.
(b) Corresponding χ'' vs. T. H_{ac} = 0.13 Oe $//H_{dc}//c$, f = 500 kHz

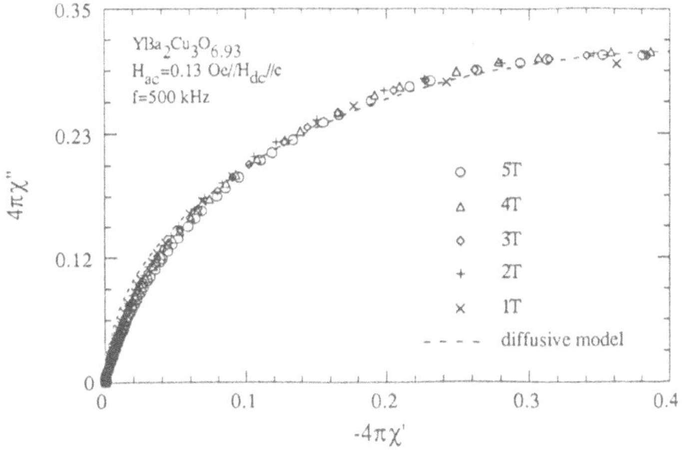

Fig. 7: A plot of χ'' vs. χ' of $YBa_2Cu_3O_{6.93}$ in 1, 2, 3, 4 and 5 T fields. The dashed curve is a fit to eq. 2.

factor and demagnetization factor are dealt with by normalizing the data so that $4\pi\chi'(60K, 0T) = -1$. The normal state contributions to χ' and χ'' were subtracted from the raw data. The high temperature part (above χ'' peaks) of the data in a χ'' vs. χ' plot (Fig. 7) agree with the diffusive model quite well. The dashed line in Fig. 7 is a best fit by eq. 2. The data below χ'' peaks do not agree with eq. 2. The frequency dependence of χ' at various temperatures in a 3 Tesla field is shown in Fig. 8. The data show that the superconductor is nearly pure inductive below 500 kHz at low temperatures and becomes strongly frequency dependent when approaching $T_c(H)$, as predicted by the Brandt theory.[11] The failure of eq. (2) to describe the data below the χ'' peaks is thus understood, because the frequency dependent data (Fig. 8) indicated that at 500 kHz the condition $\omega\tau << 1$ or $\omega\tau_o >> 1$ of the eq. 2 in the Brandt theory is not satisfied below the χ'' peaks.

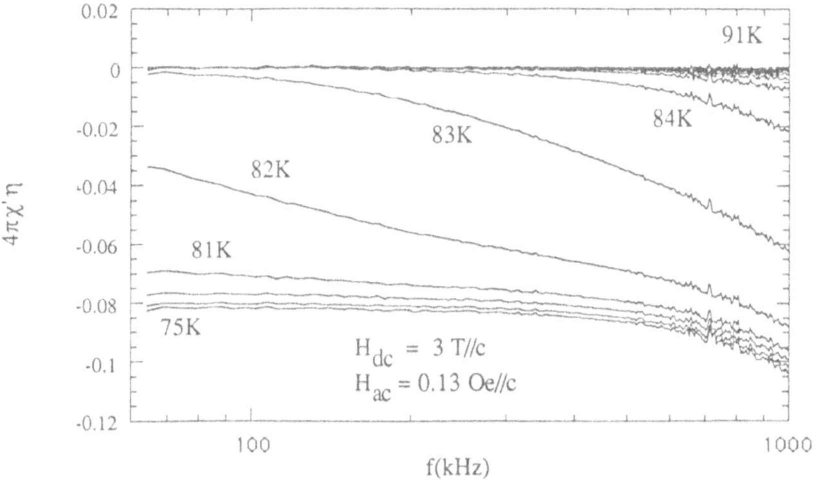

Fig. 8: Linear-response χ' vs. frequency of $YBa_2Cu_3O_{6.93}$ in 3 T//c, at (from below) T = 75, 78, 80, 81, 82, 83, 84, 85, 86, 87, 88, 89, 90 and 91 K .

We have verified the χ'-dip (peak-effect) in several YBCO crystals including the $YBa_2Cu_3O_{6.93}$ studied above in the field orientation of H//ab by the impedance analysis technique. Fig. 9 (a) shows χ' and χ'' vs. T for H_{dc} = 5 kOe and H_{ac} = 10 Oe and H_{dc}//H_{ac}//ab for a $YBa_2Cu_3O_{6.55}$ crystal (~1x1x0.1 mm^3). To check that the χ'-dip is not caused by the mixing of in-phase and out-of-phase signal, $(\chi'^2+\chi''^2)^{1/2}$ vs. T is shown in Fig. 9 (b). A comparison between the above ac data and dc magnetization data is shown in Fig. 10. The χ'-dip is clearly in the regime where the dc magnetization is reversible. This shows that the "irreversibility line" defined by the converging point of zero-field-cooled and field-cooled dc magnetization is not valid. For dc magnetization, at temperatures near T_c the equilibrium magnetization dominates the signal and the "irreversibility line" measured by dc magnetization is a line where the critical current density drops below the resolution of the instrument. For the

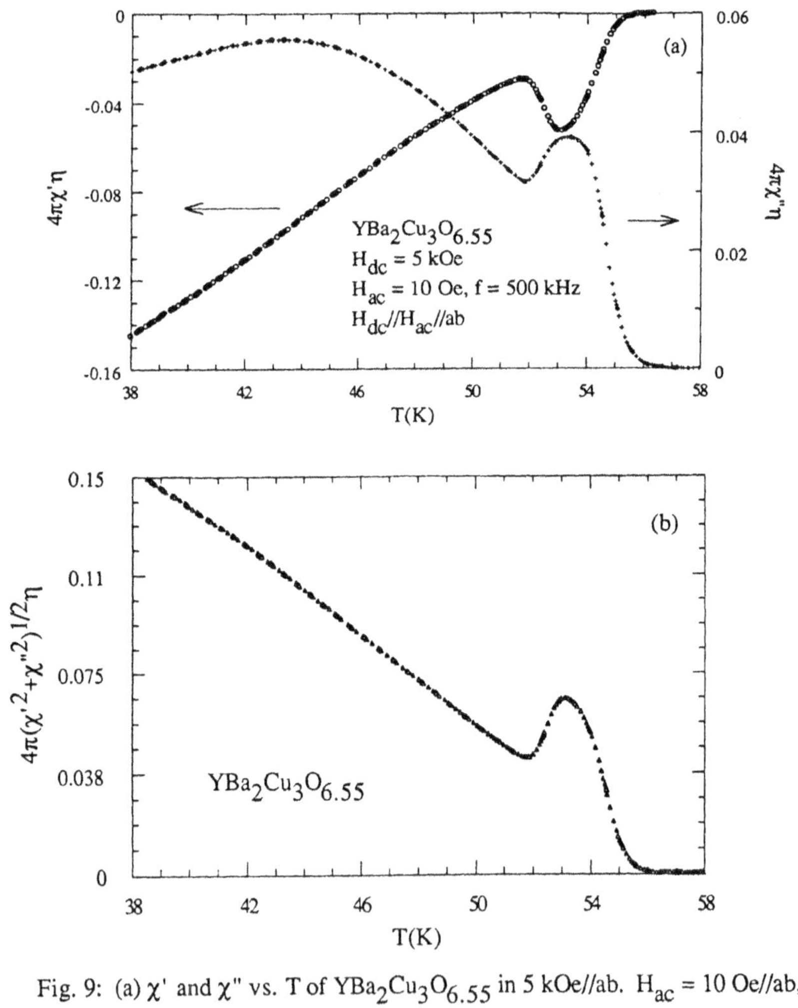

Fig. 9: (a) χ' and χ'' vs. T of $YBa_2Cu_3O_{6.55}$ in 5 kOe//ab. H_{ac} = 10 Oe//ab, f = 500 kHz. (b) $(\chi'^2+\chi''^2)^{1/2}$ vs. T of the data in (a).

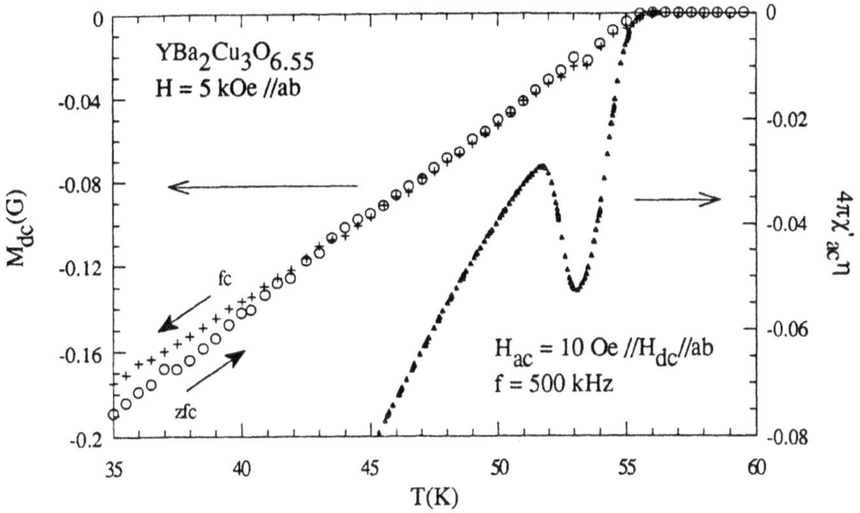

Fig.10: zfc/fc dc magnetization and real part χ' of ac susceptibility vs. temperature of $YBa_2Cu_3O_{6.55}$ in 5 kOe //ab.

Quantum Design SQUID magnetometer, for a YBCO crystal with typical dimensions of $1 \times 1 \times 0.1$ mm^3, the resolution is ~ 10^{-2} G which corresponds to about $30 \times 10^{-2}/0.1 = 3$ A/cm^2 in critical current density.

One can show in the collective pinning theory,[15] on approaching $H_{c2}(T)$, $J_c \sim 1.66B^{-2/3}(1-t)(n^2f_o^4\lambda_o^2/\phi_o^3r_f)^{1/3}$, where B is flux density, $t = T/T_c$, n is the concentration of pins, f_o is the temperature-independent part of the elementary pinning force, λ_o is the zero-temperature penetration length, ϕ_o is the flux quantum and r_f is the effective range of individual pins. With a finite criterion J_c^* (at the dc "irreversibility line"), the scaling form $1-t \sim B^{2/3}$ results. This explains the scaling-law of the "irreversibility line" determined by dc magnetization method in many systems of superconductors.

SUMMARY

1. For zero dc field or dc field applied normal to the ac field, the critical-state is relevant even at low ac field amplitudes, due to large vortex displacement or bending.
2. For $H_{ac}//H_{dc}$, a critical-state can be reached by using large ac fields. At low ac fields a superconductor with pinning can give linear-response signal and can be described by the diffusive Brandt theory.
3. At low ac amplitude with linear-response, the χ'' peak occurs when the plastic relaxation of the vortex lattice takes place in the same time scale as the elastic relaxation, or $\tau \sim \tau_o$. Thus if the plastic relaxation was via thermal activation, that is $\tau = \tau_o\exp(U/kT)$, this may suggest a vanishing barrier U at χ'' peak. Measurements at higher ac amplitudes showed (data not shown due to space limit), however, that the ac susceptibility is amplitude dependent up to T_c, which indicates a persisting barrier. A possible explanation of the fast plastic relaxation in the vortex lattice at and above the χ'' peak is that the plastic relaxation was via fast avalanches,[16] due to strong interactions between the strongly overlapping vortex cores.
4. For H//ab, a peak effect can be observed in the YBCO crystals. This is evidence that the vortex lattice softening enhances pinning, instead of weakening pinning as suggested by the vortex liquid theory.[17]
5. The peak effect occurs above the "irreversibility line" measured by dc method. This proves that the dc "irreversibility line" is a finite constant J_c line which can be explained by the collective pinning theory.[15]

This work was partly supported by a grant from the Department of Higher Education of the State of Connecticut. We are grateful to Dr. B. R. Weinberger of the United Technologies Research Center for lending a Hewlett-Packard Impedance Analyzer. Dr. B. W. Veal and Dr. Donglu Shi of Argonne National Lab have generously provided the YBCO single crystals. We would like to thank Sheng Luo and Court Rossman for their help in experiments.

REFERENCES

1. E. H. Brandt, Int. J. Mod. Phys. B 5, 751 (1991).
2. E. H. Brandt, J. Less Common Metals: Proceedings of the 19th Rare Earth Research Conference, Lexington, Kentucky, July 14-19, 1991.
3. R. W. Rollins, J. Silcox, Phys. Rev. 155, 404 (1967).
4. R. A. Hein, Phys. Rev. B 33, 7539 (1986).
5. T. Ishida, R. B. Goldfarb, Phys. Rev. B 41, 8937 (1990).
6. T. K. Worthington, this volume.
7. H. J. Fink, Phys. Rev. Lett., 16, 447 (1966).
8. V. B. Geshkenbein, V. M. Vinokur, R. Fehrenbacher, Phys. Rev. B 43, 3748 (1991).
9. K. A. Muller, M. Takashige, J. Bednorz, Phys. Rev. Lett. 58, 1143 (1988); Y. Yeshurun, A. P. Malozemoff, Phys. Rev. Lett. 60, 2202 (1988).
10. A. F. Hebard et al., Phys. Rev. B 40, 5243 (1989);M. Nikolo, R. B. Goldfarb, Phys. Rev. B 39, 6615 (1988); L. T. Sagdahl et al., Phys. Rev. B 42, 6787 (1990).
11. E. H. Brandt, submitted to Phys. Rev. Lett. (1991).
12. A. M. Campbell, J. Phys. C (Solid St. Phys.) 2, 1492 (1969).
13. C. P. Bean, Phys. Rev. Lett. 8, 250 (1962); Rev. Mod. Phys. 34, 31 (1964).
14. T. G. Berlincourt, R. Hake, D. H. Leslie, Phys. Rev. Lett. 6, 671 (1961); P. H. Kes, C. C. Tsuei, Phys. Rev. B 28, 5126 (1983).
15. A. I. Larkin, Yu. N. Ovchinnikov, J. Low Temp. Phys. 34, 409 (1979).
16. X. S. Ling, D. Shi, J. I. Budnick, to be published in Physica C: Proc. of M^2S-HTSC III, Kanazawa, Japan, July 22-26, 1991, edited by M. Tachiki et al.; C. Tang, P. Bak, Phys. Rev. Lett. 60, 2347 (1988).
17. D. R. Nelson, Phys. Rev. Lett. 60, 1973 (1988).

A STUDY OF REVERSIBLE AND IRREVERSIBLE MAGNETIZATION BEHAVIOUR IN CONVENTIONAL SUPERCONDUCTORS

S. Ramakrishnan, Ravi Kumar, C.V. Tomy, A.K. Grover, S.K. Malik and P. Chaddah***

Tata Institute of Fundamental Research, Bombay 400 005, INDIA
*Department of Physics, Punjab University, Chandigarh, INDIA
**Bhabha Atomic Research Centre, Bombay 400 085, INDIA

1. INTRODUCTION

The complete expulsion of magnetic flux on cooling a superconductor in a dc magnetic field (Meissner-Oschenfeld effect) ensures that the magnetic response of a superconductor is reversible and that it is a thermodynamic property. Incomplete flux expulsion is, however, a common occurrence in both type I and type II superconductors and such materials display a path-dependent or hysteretic magnetic response. Detailed studies in the high temperature superconductors (HTSC) have resulted in the observation[1] that there exists a thermodynamic reversible region in (H, T) space just below the normal-superconductor phase boundary ($T_c(H)$ line). Here H is the applied field, T is the temperature of the sample. Hysteretic behaviour sets in only below the irreversibility line, (T_r, H_r), and different techniques have been used to determine this (T_r, H_r) line in the HTSC. We have searched for such an irreversibility line in specimens of conventional type I and type II superconductors by both dc and ac magnetization techniques. The following four different procedures were employed to determine the (H_r, T_r) line in HTSC:

(a) Field-cooled (FC) and zero field-cooled (ZFC) magnetization curves in a constant dc field H were measured as a function of temperature. The temperature at which these merge $[T_r(H)]$ gives a point on the irreversibility line.
(b) Hysteretic isothermal magnetizaton, i.e, M *vs* H, curves were measured where M is the sample's magnetization per unit volume and H is the applied field. The field, $[H_r(T)]$, at which the hysteresis vanishes gives a point on the (T_r, H_r) line.
(c) The complex ac magnetic susceptibility ($\chi = \chi'_H + i\chi''_H$) was measured in constant dc field H as a function of temperature. A thermodynamic response requires that χ'_H(T) be positive just below T_c and this is called the *Differential Paramagnetic Effect* (DPE).[2] The temperature at which χ'_H(T) switches from positive to negative is identified with $T_r(H)$.
(d) From the simultaneous measurements of χ''_H(T) as a function of temperature in a constant dc field, the temperature at which it peaks is identified as $T_p(H)$.

In addition to presenting the results of the above series of measurements, the effect which the measuring frequency, used in ac measurements, has on the derived quantities will be briefly addressed. It will be argued that the merger of magnetizations in FC and ZFC gives only a lower limit of $T_r(H)$ and does not necessarily indicate the thermodynamic region near $T_c(H)$. The second method gives $T_r(H)$ values greater than the first one since the $\Delta M(H)$ (from isothermal hysteresis) values are greater than $(M_{FC} - M_{ZFC})$ values. The peak in the $\chi''_H(T)$ curve can arise from the temperature variation of *hysteresis loss* as well as from changes in the *normal state electrodynamics* as one approaches $T_c(H)$. These two need to be separated from each other before one can identify the peak temperature $T_p(H)$ with $T_r(H)$. The DPE is a qualitative feature whose presence is sufficient to imply genuine thermodynamic reversibility.

2. EXPERIMENTAL DETAILS

2.1 Samples

The type I and type II materials chosen for the present study are lead (T_c=7.2 K) and niobium (T_c=9.2 K), respectively. The lead specimen is in the form of a disc (dia. = 4.15 mm, thickness = 2.15 mm). The niobium specimen is taken in the form of a disc (dia. = 3.59 mm, thickness = 0.09 mm) and powder (sieved through 65μ mesh). The Pb specimen is known to display[3] more hysteretic behaviour when the field is applied perpendicular to the plane of the disc compared to the case when H is applied in the parallel direction. The two forms of the niobium specimen also differ in the degree of irreversible behaviour.[4]

2.2 Magnetic Measurements

The dc magnetic measurements have been made using a Quantum Design SQUID magnetometer. The ac susceptibility measurements have been performed using a home-built ac susceptometer.[5] For disc shaped specimens, the dc field (coaxial to the ac field) was either parallel or perpendicular to the plane of the disc. The dc and ac magnetization measurements made include the following:

(i) The isothermal $M - H$ curves at 4.7 K for H parallel and perpendicular to the plane of the Pb and Nb discs. The $M - H$ curves for the Nb powder at 4.7 K and 6.2 K.

(ii) The ZFC and FC magnetizations of the Pb disc at H = 50, 100, 200 and 300 Oe for both orientations. Similar measurements at several dc fields ranging from 20 Oe to 3000 Oe on Nb powder specimens.

(iii) The χ'_H and χ''_H at 21 Hz of the Pb disc at 4.7 K as a function of dc field H values in both orientations.

(iv) The χ'_H and χ''_H at 21 Hz of the Pb disc as a function of T at several H in both orientations. At H = 100 Oe in the perpendicular direction, χ'_H and χ''_H data were also recorded by changing the amplitude (h_{ac}) as well as the frequency of the ac field.

(v) χ'_H and χ''_H at 21 Hz and 210 Hz, of the Nb disc as a function of T for various values of H in the parallel orientation. χ'_H and χ''_H of the Nb powder as a function of T at several frequencies for chosen values of h_{ac} in H = 0 and 100 Oe, respectively.

Most of the results are shown in the figures 1 to 11. The relevant parameters have been specified in the figures and wherever considered appropriate, arrows have been included to indicate the way in which T and/or H were varied.

3. RESULTS AND ANALYSIS

3.1 Lead

Fig. 1 shows $M - H$ curves of the Pb disc at 4.7 K for the parallel and perpendicular orientations. It appears that the field interval over which the forward and reverse magnetization curves overlap is larger in the parallel orientation of the given specimen. Figures 2(a) to 2(d) show χ'_H and χ''_H *vs* H at 21 Hz for the same specimen at 4.7 K. The data in Figs. 2(a) and 2(c) show the appearance of DPE just before

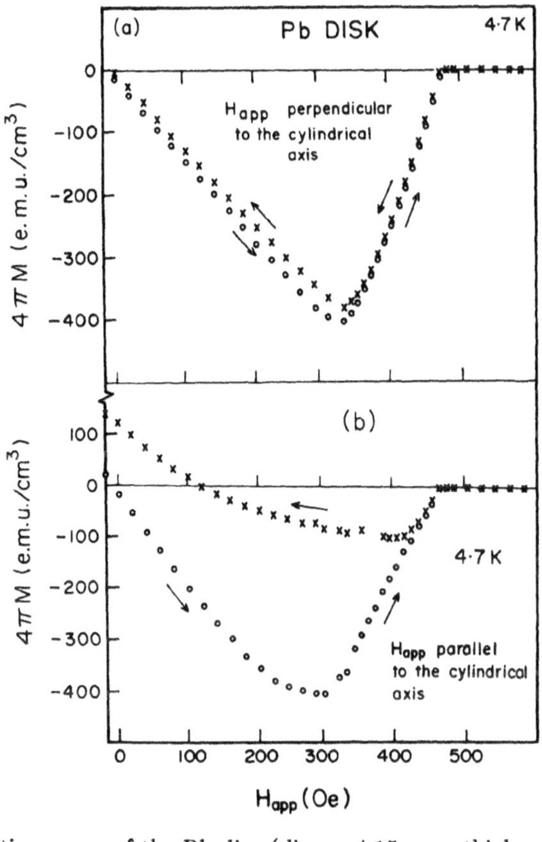

Fig.1. Magnetization curve of the Pb disc (dia. = 4.15 mm, thickness = 2.15 mm) at 4.7 K for H applied nominally parallel and perpendicular to the disc plane.

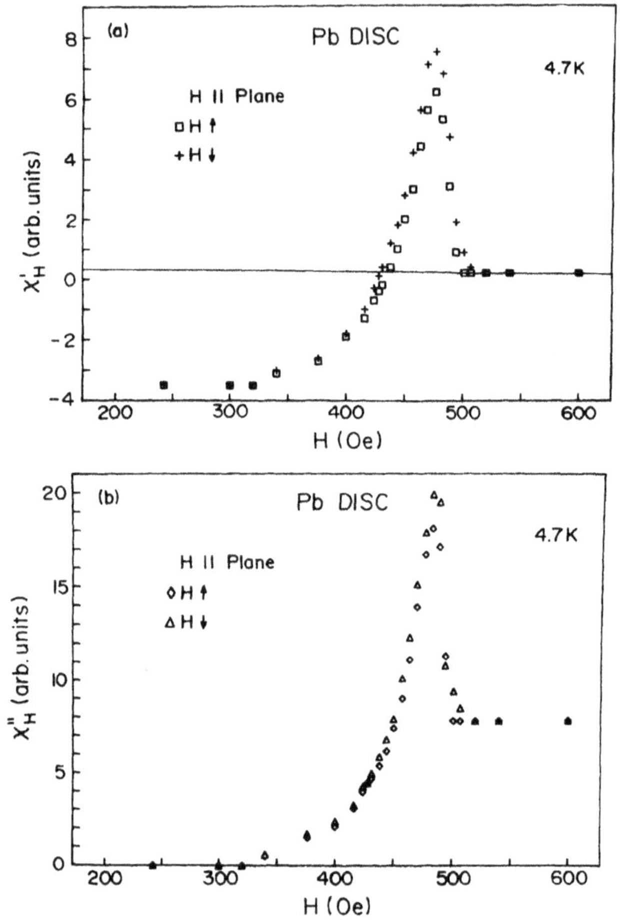

Fig.2. Isothermal χ'_H and χ''_H (measured at 21 Hz with a h_{ac} of 1 Oe rms) *vs* H for the Pb disc at 4.7 K for H applied parallel [(a) and (b)] and perpendicular [(c) and(d)].

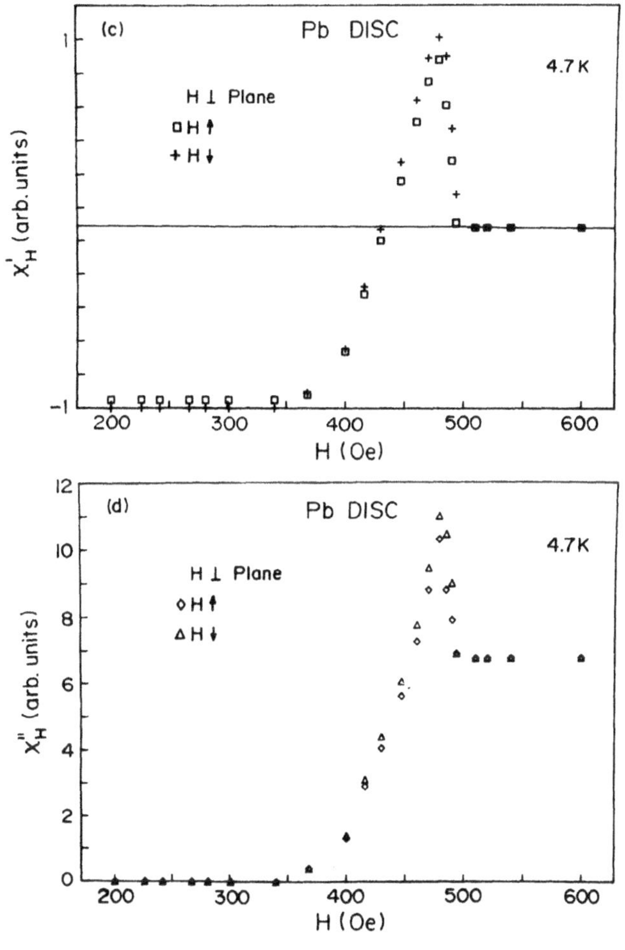

Figure 2. (continued)

the material turns normal. The field interval over which the DPE is present is only marginally greater in the parallel orientation as compared to that in the perpendicular orientation. A comparison of the data in Figs. 2(a) and 2(c) with those in Figs. 1(a) and 1(b), respectively, indicate that χ'_H data do not amount to the tracing of the derivative of the dc magnetization hysteresis curve. It may be specifically noted that the slopes of the forward and reverse magnetization curves between 0 and 300 Oe are very different, however, the measured χ'_H values in the same interval along the forward and reverse field directions are about the same. It had been pointed out by Hein and Falge[2] that the ac susceptibility measurements amount to tracing a minor hysteresis loop over the field interval $\Delta H (= \pm h_{ac})$ around a given $M(H)$ value of the hysteresis curve. The observation of DPE necessitates the existence of nearly overlapping forward and reverse magnetization curves. Thus, the field interval over which DPE is present is a more reliable indicator of quasi-reversible response as compared to the dc $M - H$ curves. No magnetic hysteresis loss is anticipated in the field region of DPE in χ'_H data (cf. Figs. 2(a) and 2(c). However, χ''_H data of Figs. 2(b) and 2(d) show large increase in the said region. We believe that this is a consequence of changes in the normal state electrodynamics as H approaches the critical field value of the superconductor, somewhat akin to the occurrence of peak-like structure in $\chi''_0(T)$ data.[6-9]

Figure 3 shows the plots of $\Delta M(H)$ ($\simeq [M_{FC}(H) - M_{ZFC}(H)]$) *vs* T for the Pb disc specimen in both orientations at the field values indicated. For a given H, $\Delta M(H)$ in the perpendicular orientation is much larger. However, the $T_r(H)$ values that may be identified from $\Delta M \to 0$ criterion appear to be nearly the same in the two orientations. Figures 4(a) to 4(d) show $\chi'_H(T)$ and $\chi''_H(T)$ data recorded at 21 Hz in an ac field of 1 Oe rms at a few fixed dc fields for the Pb disc specimen for parallel and perpendicular orientations. Data recorded at several other fields have not been shown for brevity. Figure 5 shows $\chi'_H(T)$ data at 21 Hz for $H = 100$ Oe in the perpendicular orientation for two fixed values (0.3 and 2 Oe rms) of the ac energizing field. Figure 6 shows $\chi'_H(T)$ data in the same situation at 21 Hz and 210 Hz, obtained with an h_{ac} of 1 Oe rms. In Figs. 4(a) and 4(c), it may first be noted that no paramagnetic effect response is evident in $\chi'_0(T)$ data. However, a dc field of 5 Oe is adequate to elicit a DPE peak (data not shown). The $T_r(H)$ values that may be identified with temperature at which $\chi'_H(T)$ changes from positive to negative values are found to be nearly the same for the two orientations from the data recorded at 21 Hz (cf. Figs. 4(a) and 4(c)). The two sets of data in Fig. 5 show that the $T_r(H)$ values at 21 Hz are not dependent on the amplitude of h_{ac}. However, the two sets of data in Fig. 6 show that $T_r(H)$ value at a given H, estimated by the given criterion, is a sensitive function of the frequency of h_{ac}. $T_r(H)$ value appears to decrease as frequency increases. It is important to note that the $T_r(H)$ values evident from DPE data at 21 Hz are consistent with the corresponding values estimated from dc magnetization data by $\Delta M(H) \to 0$ criterion. This fact appears to establish the efficacy of DPE data at low enough frequency for giving a reasonable estimate for $T_r(H)$ values.

The $\chi''_0(T)$ data in both the orientations show peak-like structures at $T_c(0)$ (see Figs. 4(b) and 4(d)). The peak in $\chi''_H(T)$ shifts to lower temperatures and also broadens as H increases. The peak-like structure in $\chi''_H(T)$ is believed to have contributions from both the changes in normal state electrodynamics on approaching normal-superconducting phase boundary as well as the hysteresis loss in the irreversible region of the superconducting state. The observation, that the temperature intervals over which the DPE exists in $\chi'_H(T)$ data nearly coincide with the corresponding widths of the peaks in $\chi''_H(T)$ data, seems to imply that the latter structures mainly originate from the changes in the normal state electrodynamics. Thus, it may not be appropriate to identify the peak temperature in $\chi''_H(T)$ with the $T_r(H)$ value. The *characteristic temperatures* which can be noted from the experiments conducted on Pb disc specimen are $T_c(H)$, $T_r(H)$ (from DPE at 21 Hz) and $T_p(H)$ (peak temperature in $\chi''_H(T)$).

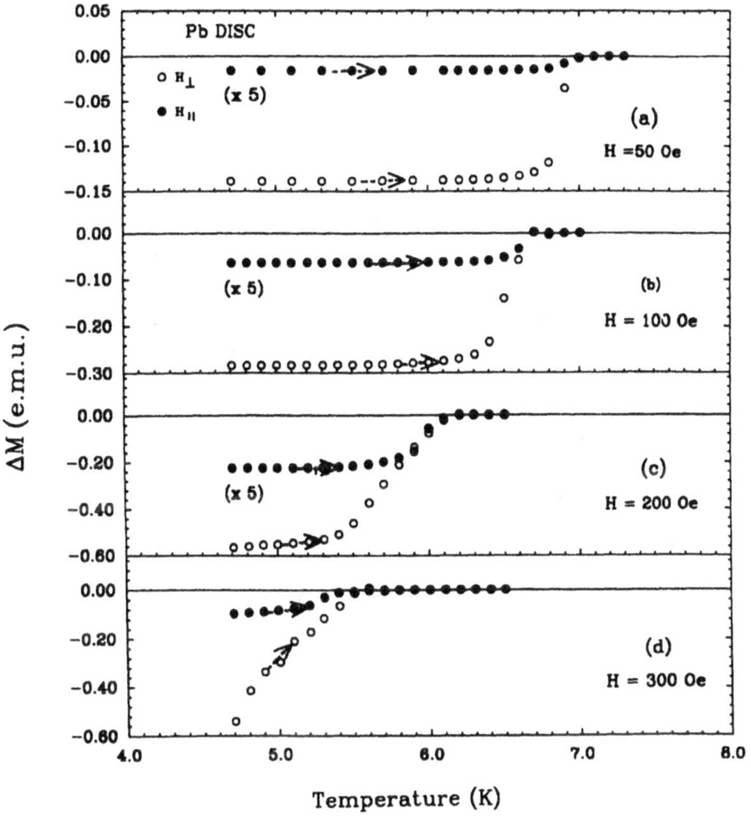

Fig.3. Plots of $\Delta M(H)(= M_{FC}(H) - M_{ZFC}(H))$ *vs* T for the Pb disc for field applied parallel (open circle) and perpendicular (close circle) to disc plane. Arrows mark $T_r(H)$ values by $\Delta M(H) \rightarrow 0$ criterion.

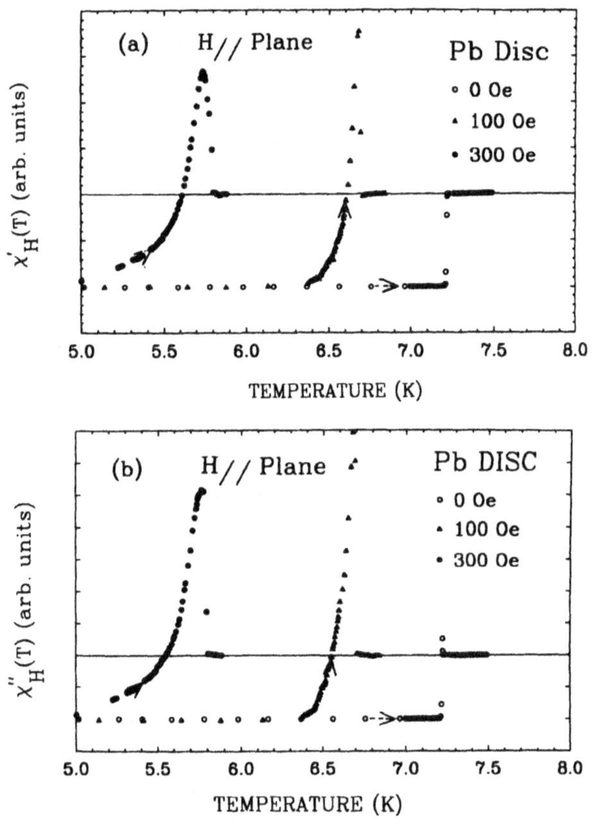

Fig.4. The temperature variation of χ'_H and χ''_H (measured at 21 Hz with a h_{ac} of 1 Oe rms) for the Pb disc specimen in the parallel [(a) and (b)] and perpendicular [(c) and (d)] orientations. The data at different H values have not been scaled to one-another. Wherever considered appropriate, the horizontal line have been drawn through susceptibility values in the normal state for reference purpose only.

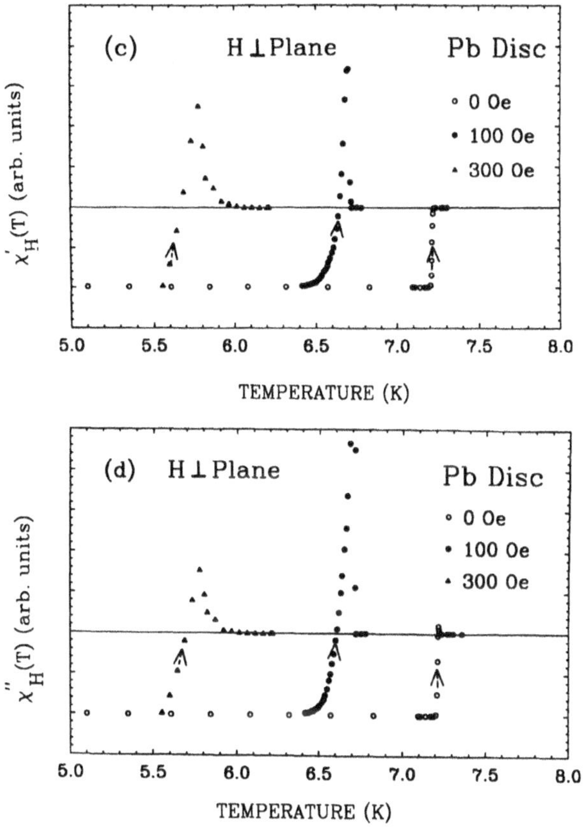

Figure 4. (continued)

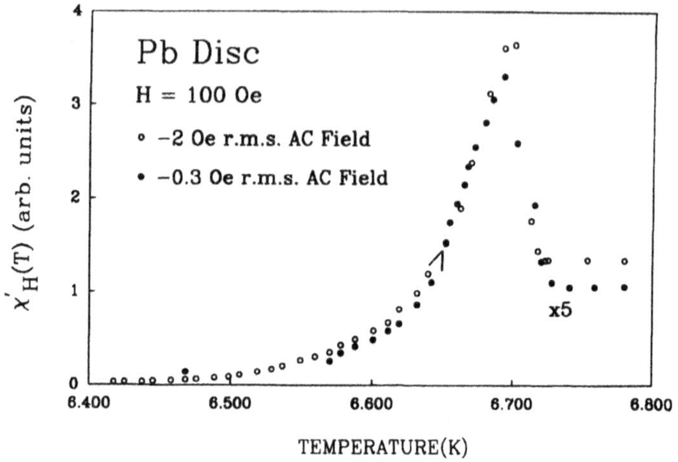

Fig.5. The temperature variation of χ'_H measured at 21 Hz for two values of h_{ac} (0.3 and 2.0 Oe rms, respectively) of the Pb disc specimen for $H = 100$ Oe applied perpendicular to the plane of the disc.

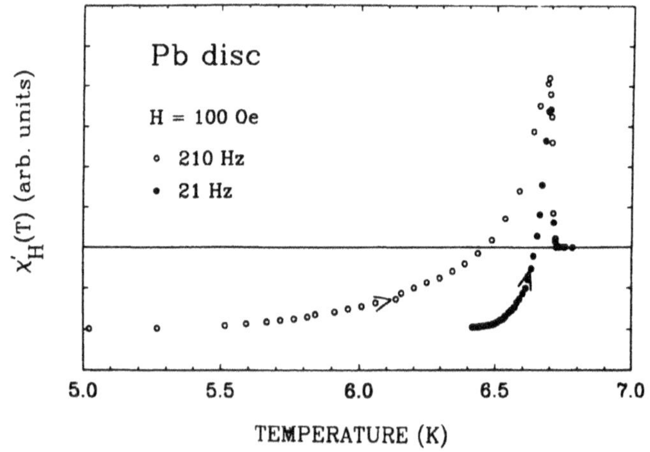

Fig.6. The temperature variation of χ'_H measured with h_{ac} of 1 Oe rms at 21 Hz and 210 Hz, respectively of the Pb disc specimen with $H = 100$ Oe applied perpendicular to the disc plane.

3.2 Niobium

Figure 7 shows a dc magnetization curve of the Nb powder specimen at 6.2 K; the inset shows the forward and reverse magnetization curves near H_{c2} on an expanded scale. Figure 8 shows temperature variation of ZFC and FC susceptibility values in the same specimen at a few H values. The data at several other H values have not been shown for brevity. The $H = 3$ kOe data of Fig. 8 show that χ_{ZFC} and χ_{FC} curves nearly merge into each other $\Delta\chi \rightarrow 0$ by 6.2 K, however, the inset of Fig. 7 indicates that the difference between forward and reverse M-H curves persists for H

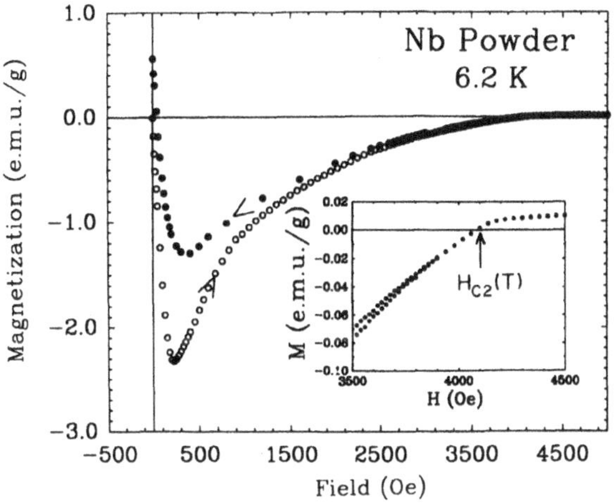

Fig.7. Magnetization curve of the Nb Powder specimen at 6.2 K. The inset shows the forward and reverse $M - H$ curves near H_{c2}.

>3.5 kOe at 6.2 K. Figure 9 shows the temperature dependence of the in-phase, χ', and out-of-phase, χ'', complex ac magnetic susceptibility values of the Nb powder at 21 Hz in h_{ac} of 1 Oe rms and $H = 0$ and 100 Oe, respectively. No DPE or any other peak-like structure can be seen in the χ'_H and χ''_H data of Fig. 9 as well as in other runs made by varying the frequency and amplitude of the ac field. We believe that the failure to observe DPE is a consequence of the absence of genuinely reversible region near the $T_c(H)$ line in the given Nb powder specimen. The situation in this case is that at a given H, though $M_{ZFC} \rightarrow M_{FC}$ as T approaches a quasi-irreversibility temperature (determined *via* $\Delta\chi(H) \rightarrow 0$, marked by arrows in Fig. 8), the width of isothermal hysteresis remains significantly larger. The latter width approaches zero as $H \rightarrow H_{c2}$ (or $T \rightarrow T_c(H)$).

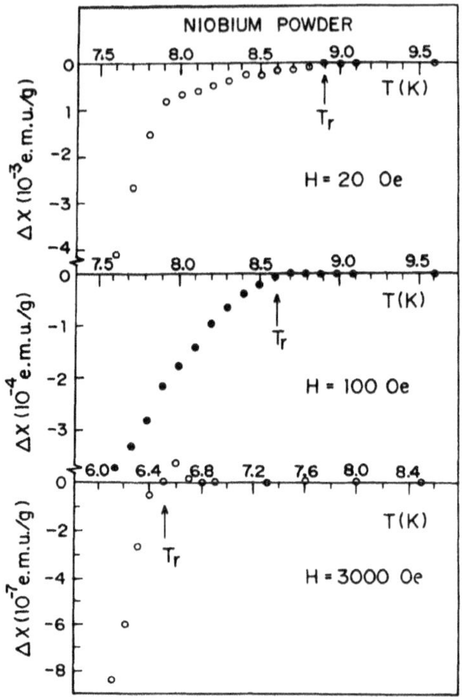

Fig.8. The temperature variation of $\chi(FC)$ - $\chi(ZFC) = 0$ of the Nb powder specimen at H = 20, 100, and 3000 Oe, respectively. The arrows identify the $T_r(H)$ values by $\Delta\chi(H) \to 0$ criterion.

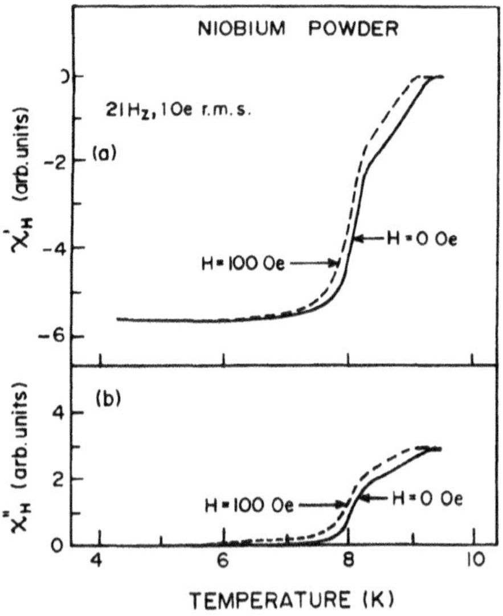

Fig.9. The temperature variation of χ'_H and χ''_H of the Nb powder specimen measured at 21 Hz with h_{ac} of 1 Oe and H = 0 and 100 Oe respectively.

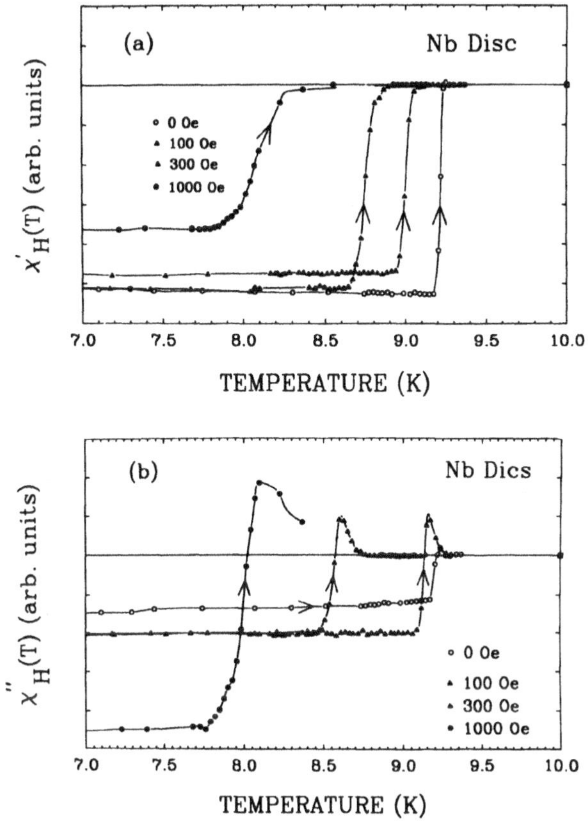

Fig.10. The temperature variation of χ'_H and χ''_H measured at 21 Hz with h_{ac} of 1 Oe rms of the Nb disc at field values indicted, in the parallel orientation. The continuous line joining the data points are just an aid to guide the eye. The data sets at different H are not normalized to one-another.

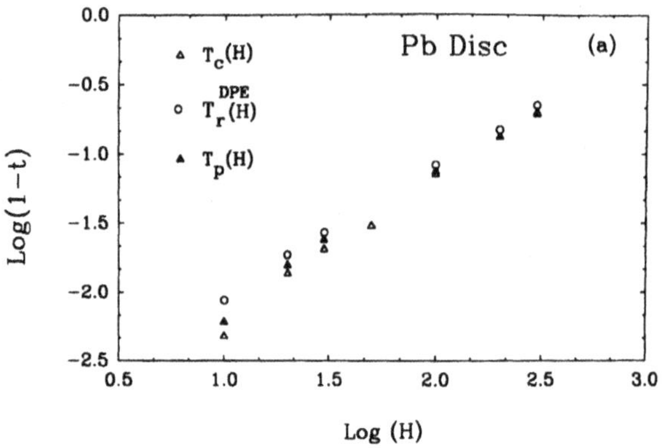

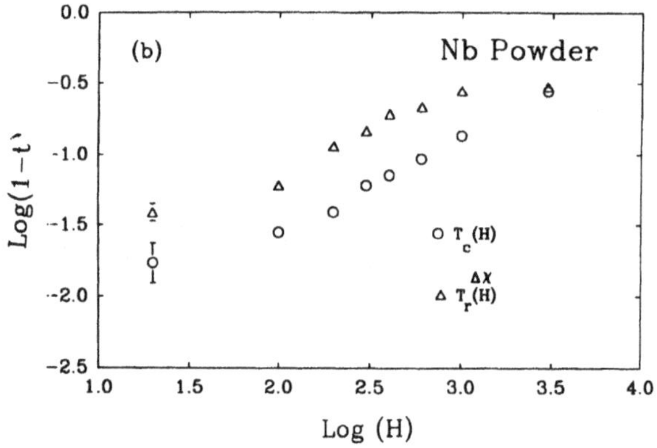

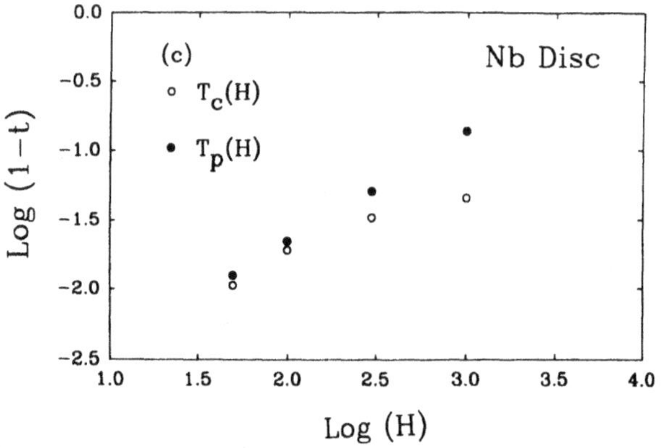

Fig.11. Log-log plots of $(1-t)$ *vs* H for different sets of t values for the Pb disc (a), Nb powder (b) and Nb disc (c) specimens. The three sets of t values for Pb correspond to $T_c(H)/T_c(0)$, $T_r^{DPE}(H)/T_c(0)$ (DPE at 21 Hz) and $T_p(H)/T_c(0)$, respectively. The two sets each for the Nb powder and the disc specimens correspond to $T_c(H)/T_c(0)$ and $T_r^{\Delta\chi}(H)/T_c(0)$, and $T_c(H)/T_c(0)$ and $T_p(H)/T_c(0)$, respectively.

In the case of the Nb disc, field-cooling, for the range of H employed in this study[4] showed that the sample expells no magnetic flux as it enters the superconducting state, i.e. $M_{FC}(H) = 0$. The M_{ZFC} value approaches $M_{FC} = 0$ value only at $T = T_c(H)$. The isothermal forward and reverse $M - H$ curves are seen to meet at (data not shown) $H = H_{c2}(T)$. Figure 10 shows the $\chi'_H(T)$ and $\chi''_H(T)$ data of the Nb disc specimen at several values of H. As anticipated, no DPE-like structure is present in $\chi'_H(T)$ data. However, the peak-like structures are present in $\chi''_H(T)$ data and the structure broadens as H increases. The peak temperature in $\chi''_H(T)$ does not identify the $T_r(H)$ values. The *characteristic temperatures* that can be noted from the experiments conducted on the Nb powder and disc specimens are $T_c(H)$ and $T_r(H)$ (from $\Delta\chi \to 0$ criterion) in the former and $T_c(H)$ and $T_p(H)$ (peak temperature in $\chi''_H(T)$) in the latter, respectively.

3.3 Power law relationship

Müller *et al*[10] had introduced the fitting of $T_r(H)$ values, determined from the merger of M_{ZFC} and M_{FC} curves, to a power law relation, $([1 - T_r(H)/T_c(0)] \propto H^q)$, and found a value of 2/3 for the exponent q in a specimen of cuprate variety of HTSC. We display in Fig. 11 our attempt to fit different characteristic temperatures in Pb and Nb specimens to a power law behaviour. For the Pb disc, $T_c(H)$, $T_r(H)$ and $T_p(H)$ appear to fit to the power law with $q \approx 1.0$. For the Nb powder, $T_c(H)$ and $T_r(H)$ values in the field range of 0.1 to 1.0 kOe can be fitted to the power law with $q \approx 2/3$ and 11/20, respectively. For the Nb disc, $T_c(H)$ data do not appear to fit to the power law, whereas $T_p(H)$ values do with $q \approx 4/5$.

4. SUMMARY AND CONCLUSIONS

To summarize, we have presented experimental data obtained by dc and ac techniques pertaining to the irreversibility phenomenon in superconducting specimens of Pb (type I) and Nb (type II) elements. The physical basis for irreversible behaviour in the two types of superconductors are entirely different. Four different procedures have been followed to ascertain the values of irreversibility temperatures $T_r(H)$. The $T_r(H)$ values determined from dc magnetization data probably only specify a lower limit. The identification of peak temperature in $\chi''_H(T)$ data with $T_r(H)$ value is appropriate only if the contribution from the normal state electrodynamics can be isolated from the peak in $\chi''_H(T)$. The observation of the differential paramagnetic effect (DPE) in $\chi'_H(T)$ is a qualitative feature whose presence is adequate to imply reversibility in magnetization response, however, its efficacy to locate very precisely the $T_r(H)$ and $H_r(T)$ values remains to be established.

It may be argued that before measuring the irreversibility line in HTSC, one must first qualitatively establish the very existence of a "reversible" region by a DPE in $\chi'_H(T)$ data. The recent experiment of Khoder *et al*[11] on Bi- and Tl-based cuprates do show the presence of a DPE. However, prior to them, Hein *et al*[9] did not succeed in their specific search for a DPE in a specimen of $YBa_2Cu_3O_7$ and had surmised that there is no genuine thermodynamic region in that HTSC system. The present data on conventional superconductors appear to strengthen the belief that may stem from the remarks of Hein *et al.*[9] Many interesting theoretical ideas,[12] such as flux lattice melting, flux depinning, vortex liquid to glass transition, *etc.*, which pertain to the nature of irreversibility line, are under active consideration in connection with the physics of HTSC. Some of these may not be relevant for conventional superconductors. However, it is hoped that more experiments akin to the present work would eventually lead to find a reliable way to determine accurate value of $T_r(H)$.

REFERENCES

1. A.P. Malozemoff, MRS Bulletin 15:50 (1990) and references therein.

2. R.A. Hein and R.L. Falge Jr., Phys. Rev. 123:407 (1961).
3. A.K. Grover, Ravi Kumar, P. Chaddah, C.K. Subramanian and V. Sankaranarayanan, Physica C 170:431 (1990).
4. A.K. Grover, P.L. Paulose, P. Chaddah and G. Ravikumar, Physica C 162-164:335 (1989); Pramana-J. Phys. 33:297 (1989).
5. S. Ramakrishnan, S. Sundaram, R.S. Pandit and Girish Chandra, J. Phys. E 18:650 (1985).
6. E. Maxwell and M. Strogin, Phys. Rev. Lett. 10:212 (1963).
7. V.B. Geshkenbein, V.M. Vinokur and R. Fehrenbacher, Phys. Rev. B 43:3748 (1991).
8. R.A. Hein, Phys. Rev. B 33:7539 (1986).
9. R.A. Hein, H. Hojaji, A. Barkatt, H. Shafii, K.A. Michael, A.N. Thorpe, M.F. Ware and S. Alterescu, J. Superconductivity 2:427 (1989).
10. K.A. Müller, M. Takashige and J.G. Bednorz, Phys. Rev. Lett. 58:1143 (1987).
11. A.F. Khoder, M. Cauach and J.L. Jorda, Phys. Rev. B 42:8714 (1990).
12. E.H. Brandt, Themal depinning and melting of flux lattice in high temperature superconductors - a review, to appear in Int. J. Mod. Phys. B (1991).

AC SUSCEPTIBILITY TECHNIQUES APPLIED TO THIN FILM SUPERCONDUCTORS

J. H. Claassen

Naval Research Lab.
Washington, D. C. 20375

INTRODUCTION

The techniques of ac susceptibility have been used to evaluate several properties of thin films: critical temperature, critical current density, and penetration depth. Two basic approaches have been used. The first I would call "conventional", in that the sample is mounted in a conventional ac susceptibility apparatus that is normally used to measure bulk samples. Here the drive and receive coils are both large compared to the lateral dimensions of the sample. A second approach is unique to thin films and involves drive and receive coils that are small compared to the lateral dimensions of the sample. These coils are generally designed to couple only to a region near the center of the sample, thus being insensitive to the exact size or details of the outer edge of the film.

Another distinction could be made between ac and dc measurements. One should recognize that most ac measurements of the sort considered here are done at a fairly low frequency (<1 KHz). To analyze the results of these measurements it is usually adequate to use a quasistatic approximation; ie., to ask what flux would be induced in the receive coil due to sample magnetization if an applied field was slowly swept in an oscillating manner. The reason for using an ac drive for the field is not to reveal any unique physics, but rather to increase the signal-to-noise ratio of the measurement (since the voltage in the receive coil is proportional to the rate of change of the induced flux). With the availability of SQUID sensors or vibrating sample magnetometers which are sensitive to the "dc" magnetic moment of the sample, there is no intrinsic reason to use an ac drive for the field. In practice, dc sensors are difficult to use in comparison to a low-noise amplifier connected to a receive coil, and thus the ac drive technique often is the practical solution.

A major area of current interest is the relaxation of a pinned flux-line lattice towards equilibrium, whether via "flux creep" or some other mechanism. This does introduce a logarithmic frequency dependence into a susceptibility measurement, although the effect tends to be very

Magnetic Susceptibility of Superconductors and Other Spin Systems
Edited by R.A. Hein *et al.*, Plenum Press, New York, 1991

small except close to the transition. In this paper such effects will not be considered, and the quasistatic approximation will be used.

"CONVENTIONAL" AC SUSCEPTIBILITY MEASUREMENTS

I first consider the "conventional" geometry, where the sample is positioned inside coaxial drive and receive coils. In the simplest approximation the receive coil voltage is proportional to the rate of change of the total magnetic moment of the sample. The interpretation of this sort of measurement is quite sensitive to details of the sample orientation and drive level, and it is unfortunate that experimental descriptions often are limited to statements of the form "....... was measured by ac susceptibility".

Transition Temperature Measurements

The most common orientation for transition temperature measurements is with the film surface perpendicular to the axis of the drive and receive coils. An alternating applied field H_e is generated with the drive coil, and the magnetic moment of the sample is detected with the receive coil. In general the internal field B_{in} for any sample that is an ellipsoid of revolution is given by $\mu_o B_{in}=H_e+(1-D)\mathbf{M}$, where $\mathbf{M}$ is its magnetization and D is the so-called demagnetizing factor. (Throughout this paper SI units will be used, where $\mu_o=4\pi x 10^{-7}$ weber/ampere-meter). When it is in the Meissner state, (defined for our purposes as $B_{in}=0$) its magnetization is given by $\mathbf{M}=-H_e/(1-D)$. The film can be approximated as an extremely oblate ellipsoid of revolution with the principle axes a and b identified as the film thickness d and radius R, respectively. (From here on we consider for simplicity only films in the shape of a circle). The demagnetizing factor in this case[1] is $D \approx 1-(d/R)(\pi/4)$. We thus find

$$\mathbf{M}=-H_e/(1-D) \approx -H_e(4/\pi)(R/d), \qquad (1)$$

which is much greater than would occur for a cylindrical sample of the same radius. The apparent enhancement of sensitivity due to the large demagnetizing factor, however, is illusory: the field H next to the sample must be less than H_{c1} everywhere to ensure that it remains in the Meissner state. It is known[2] that at the edge of an ellipsoid the field H increases to $H_e/(1-D)$. Thus the drive field H_e must be reduced by the factor (1-D) to ensure that the field does not exceed H_{c1}. In general, the maximum magnetization of a superconductor in the Meissner state is independent of its shape.

There are some practical disadvantages to measurements of T_c with the applied field perpendicular to the sample. Very often material at the periphery of a thin film sample is thinner or of lower quality than near the center. Yet this configuration is most sensitive to precisely those regions of the sample. Moreover the sample as fabricated rarely is in the shape of a circle; presumably angled corners can only make the H_{c1} threshold even lower.

One consequence of an excitation level that is too high is to broaden the apparent width of the transition. As the temperature is raised toward the transition a point will be reached where $H=H_{c1}$ at the sample edge and flux will start to enter the film. This could be well below the actual transition temperature, but will register in the experiment as the beginning of a low-temperature tail in the measurement. Thus the apparent width of the transition could be a geometric effect, rather than a measure of intrinsic sample properties. In general, with measurements in this geometry it is always necessary to verify that the results are independent of the excitation level.

To avoid these problems one could turn the sample on its side so that its surface is parallel to the applied field. The demagnetizing factor is then ≈0, and the measurement will give a correct indication of the transition width. Because fields penetrate a distance λ into the superconductor, the moment in the Meissner state is given by[3]

$$m=-A\{d-2\lambda\tanh(d/2\lambda)\}, \qquad (2)$$

where A is the film area. When $d<2\lambda$ this becomes very small ($m\approx-Ad^3/12\lambda^2$) and virtually impossible to detect. For thicker films the effective thickness is reduced by 2λ.

The main practical difficulty with the parallel geometry seems to be that it is difficult to avoid having a small component H_n of the applied field perpendicular to the film surface. This will result in a very large diamagnetic moment in this direction, some of which could couple into the receive coil. A special receive coil has been described[4] that discriminates against the film response to normal fields; all the same, it was found necessary to use a film with high H_{c1} (Pb) to calibrate the residual signal due to H_n. In other measurements of superconductor/normal metal multilayers[5] no difficulties related to H_n were reported. Although the signal level was very small, it was possible to obtain data for the temperature dependence of λ using (2) provided extreme care was taken to subtract background terms in the susceptibility data.

Generally the signal corresponding to the superconducting transition is quite small in an unmodified ac susceptibility apparatus, since it is proportional to the volume of the sample no matter what its orientation[6]. The filling factor (volume of sample/volume of receive coil) will necessarily be small in a general purpose apparatus designed for bulk samples. When the sample thickness drops below ~2λ an appreciable signal will only be observed for the perpendicular orientation of the drive field.

Critical Current Measurements

The "conventional" ac susceptibility configuration may also be used to measure the critical current density J_c of a film sample. This is quite appealing, since the usual transport measurement of J_c involves a great deal of processing of the sample (defining a narrow strip, attaching current and voltage leads). Moreover the inductive measurement does not consume the sample, which may be used in further experiments after it has been characterized for critical current.

The essence of this measurement is to apply a sufficiently large excursion of the field H to drive the sample completely into the critical state, and determine the resulting magnetic moment. This basic procedure can be implemented in several ways, the most common of which are "dc" measurements, that is involving a slow sweep of the field H. The moment of the sample is then determined with a SQUID or vibrating sample magnetometer. An ac excitation of the field H may also be used, as long as sufficient amplitude is available to completely magnetize the sample in both directions. The time dependence of the voltage induced in the receive coil will be rather complicated, but the peak magnetic moment of the sample can be extracted. Analysis of these data is identical to the case of a long cylindrical sample and is covered in other papers in this proceeding; see Sections I and II.

The critical state[7] is defined as one where the current density is everywhere equal in magnitude to the critical current density J_c. At a peak excursion of the applied field all the currents will be in the same direction, and the magnetic moment can be related to J_c. The total moment is given by[8]

$$\mathbf{m}=(1/2)\int \mathbf{r}\times\mathbf{J}\ dv \tag{3}$$

where the integral is over the sample volume. Considering first the simplest case where J_c is constant and the currents in a disk-shaped sample are flowing in concentric circles, we find

$$m=(\pi/3)R^3J_c\,d, \tag{4}$$

where R is the radius of the disk and d is its thickness. Note that the average magnetization is the same as for a long cylinder[7].

While the total moment per length of a disk-shaped sample is the same as for a long cylinder in the critical state, the magnetic field profile due to the screening currents (the self-field) is quite different for the two. In the latter case $B_z=\mu_o J_c(R-r)$, where r is the distance from the cylinder axis (which is along the z direction)[9]. For the disk there is no analytic form for B; numerical calculations have been performed[10], showing that B_z is scaled by $\mu_o J_c\,d$ and has a weak divergence at r=0. The magnitude of the self-field is significant for two reasons: first, it establishes the excursion in applied field that is required to place all (or at least most[11]) of the sample in the critical state. This is of the order of $J_c\,d$ in the case of the disk, which is much smaller than the long cylinder where a field excursion $\approx J_c R$ is required. Secondly, the field dependence of J_c will result in a current that is <u>not</u> constant through the sample, since B is not constant. The useful approximation of constant $|J|$ may only be used if $J_c(B)$ varies slowly on a scale of the self field. In the case of a thin film, this is almost always true; for example if d=1 μm, $J_c=10^7$ A/cm^2, the self field is of order 0.1 T.

Whenever a value for J_c is quoted, an electric field criterion is implicitly assumed. In the case of transport measurements this usually is set by the noise level of the voltage amplifier and the distance between probe points. Taking 1 μV and 1 cm for these parameters, a criterion of 1 μV/cm applies. To estimate the equivalent criterion in a magnetic moment measurement, we note that when the sample is in the fully magnetized critical state it is transparent to further increases in the applied field (assuming J_c is independent of field). Thus $\mu_o dB/dt$ is equal to the rate of change of the applied field and is uniform across the sample during some parts of the ac cycle. It is in just these parts of the cycle that the sample's magnetic moment is determined. The azimuthal electric field is given by $E=(r/2)(dB/dt)$[12], so is not uniform across the sample. Since the measurement is heavily weighted by currents at the periphery of the sample, the electric field criterion is characterized by its value at r=R. If we assume that a triangle waveform with maximum field $H_{max}=3J_c d$ is used as the drive (to ensure that the film is in the critical state for at least half the time), we have $dB/dt=4f\mu_o H_{max}$, where f is the drive frequency. Then using the above expression for electric field evaluated at r=R, we find

$$E_c=6\mu_o J_c dRf, \qquad (5)$$

Taking R=5 mm, $J_c=10^6$ A/cm^2, d=3000 Å, f=10 Hz, we find $E_c\approx 10$ μV/cm. Thus the order of magnitude of the electric field criterion is similar to a transport measurement. Note that in the "dc" limit, using a SQUID or VSM, the equivalent frequency (and thus electric field criterion) can be lower by at least 4 orders of magnitude.

It should be stressed that the formalism usually used to discuss the response of magnetic materials is <u>not</u> appropriate for the case of superconductors in the critical state, and tends to confuse rather than illuminate. Thus it is not useful to think in terms of a local magnetization M that is determined by the local value of B, nor do concepts such as the demagnetizing factor of samples, etc., have any utility. It can be seen from (3) that the magnetization of a disk in the critical state is formally given by $-(1/2)rJ_c$, a position dependence that bears no resemblance to the spatial dependence of B. The total magnetic moment is the relevant parameter, and usually is what is measured directly in the experiment.

There are several pitfalls in using (4) to relate the measured moment to the critical current density. An obvious one is that most samples are not circular in shape, and the effective radius R is uncertain. Also the critical current near the edges, which strongly weights the results, is often degraded. If the sample makes a close fit in the receive coil, the signal induced may not be simply proportional to the total moment. One approach to reducing these problems is to sense the self-field B near the center of the film rather than the total moment. Talvacchio has shown[13] that the field at a point spaced z above the center of the film is given by

$$B\approx(\mu_o J_c d/2)\{\ln(2R/z)-1+0.75(z/R)^2+o(z/R)^4\} \qquad (6)$$

This depends only weakly on R, and is weighted most strongly by regions near the center rather than the edges. The field just above the center can be sensed with a small coil if the drive field is oscillated at a reasonably high frequency. Another approach that has been used by the RSRE group[14] is to drive the film into the critical state with a rapid pulse in the drive coil, then sense B with a small Hall effect probe. This system allows a measurement of the flux creep rate as well.

There is a further ambiguity in interpreting the magnetic moment of a film sample in the critical state: In the case of high T_c superconductors, there is a widely held suspicion that the material is subdivided on a fine scale by grain, or twin, boundaries. The supercurrents thus have two components: those flowing between grains and those flowing entirely within grains. Let us consider the case where only the latter exist, ie., the "transport" critical current is zero. If the grain size is a, each grain will have a magnetic moment $m \sim J_{cg}\, d a^3$, where J_{cg} is the critical current within a grain Here it is assumed that the grain size is larger than the film thickness, so Eqn. 4 can be used with $a=R$. One then has $\sim(R/a)^2$ grains in the sample, and the total magnetic moment will be given by $\sim J_{cg} d a R^2$. There is thus a small sensitivity to the critical currents in the grains that cannot easily be distinguished from a moment due to a macroscopic critical current density since the value of a is usually not known.

Having recognized that the conventional ac susceptibility configuration has a small sensitivity to inclusions in a sample, it must be emphasized that the received signal will usually be due to macroscopic shielding currents which reflect an average of the properties of the sample over a large scale. The author is very dubious of a common tendency to ascribe structure in the susceptibility vs. temperature to multiple phases in a sample.

"SCREENING" MEASUREMENTS

A second general approach to measuring properties of superconducting films involves the use of small coils in the drive and receive roles whose diameters are significantly less than the size R of the film. This configuration has been discussed for two cases: with the coils located on opposite sides of the film[15], and on the same side[16]. In both cases the coils are aligned along a common axis which is perpendicular to the film surface. When such a coil is located near the center of a film the inductive response should be only weakly sensitive to the exact radius of the film or its properties near the edges. To analyze the fields and currents generated, it is appropriate to approximate the situation with an infinitely extended film. The finite size of an actual sample is treated as a perturbation that has second order effects on the results. The screening geometry has an obvious advantage when large area films are to be evaluated. It is very compact and lends itself to simple cryostats designed to operate in storage dewars. The fields and currents are localized near the coil and may be calculated quite accurately. Interpretation of results is not sensitive to the shape and properties of the film at its edges.

The basic measurement in the screening configuration is identical to the conventional ac susceptibility system: an ac current is applied to the drive coil and the induced voltage in the receive coil is measured. If the latter is positioned on the reverse side of a superconducting film, one finds that the mutual coupling is very much smaller than when the film is normal. This is true even if the film thickness d is less than its penetration depth λ, a result that is not immediately apparent when one considers that magnetic fields fall off as $\exp(-x/\lambda)$ from the free surface in a bulk sample. To understand this it is useful to consider a related geometry, a cylindrical film of radius R with solenoid-shaped drive and and receive coils positioned outside and inside, respectively. The mutual inductance should be proportional to the ratio B_{int}/B_{ext} of fields inside and outside the cylinder, which has been analyzed by de Gennes[17]. The general dependence of the field in the film is of the form $B=C_1 \exp(-r/\lambda)+C_2 \exp(r/\lambda)$. Two boundary conditions must then be met: at the outer surface the field must equal the applied field B_{ext}. On the inner surface, we must have $\mu_o J=-A/\lambda^2$, from the London equation, where $A=B_{int} R/2$. Substituting $\mu_o J=dB/dx$ (via Maxwell's equation) gives a second condition on the constants C_1 and C_2. The result is[18]

$$B_{int}/B_{ext}=\{\cosh(d/\lambda)+(R/2\lambda)\sinh(d/\lambda)\}^{-1}. \tag{7}$$

The second term dominates, and in the limit $d<\lambda$ we find

$$B_{int}/B_{ext}\approx 2\lambda^2/Rd. \tag{8}$$

Thus the field penetration is very small even when $d<\lambda$ since R is usually much greater than λ^2/d.

The case of a planar film placed between small coils is topologically similar to the cylindrical geometry but is not amenable to an analytical solution. In Refs. 15 and 16 methods are outlined for calculating the dependence of the mutual inductance M on film impedance. The mutual inductance has the same general dependence on film parameters as B_{int}/B_{ext} for the cylinder:

$$M/M_o \sim 2\lambda^2/ad, \tag{9}$$

where M_o is the mutual inductance with no film present and a is the radius of the coil. Eq. 9 was shown[15] to be the general form for the mutual inductance as long as the screening by the film is large; that is, if M/M_o is small. The exact proportionality between mutual inductance and λ^2/d must be determined by a rather forbidding calculation based on the dimensions of the coils and their spacing from the film. Much useful information can be obtained from this configuration, however, without knowledge of the proportionality factor.

Transition Temperature Measurements

To fully characterize the transition of a film it is necessary to determine both the dissipative and reactive parts of its impedance in the transition region. This has been done with the two-coil arrangement by carefully recording the real and imaginary components of the mutual inductance and using the full-fledged formalism of Refs. 15 and 16 . Often a less comprehensive characterization of a film transition is desired, for the purpose of screening samples, etc. The mutual inductance measurement remains quite useful for this sort of sample characterization application. It is found that a high quality sample will exhibit a very abrupt drop in the mutual coupling at the transition temperature. This drop can be analyzed from two perspectives: from the superconducting side, where it is related to the vanishing of the density of superconducting electrons, or from the normal side, where it is related to the resistivity going to zero. Using the simplified analysis given above, we consider both perspectives.

Approached from the superconducting side, the transition corresponds to the divergence of the penetration depth. Consider the following example, for a film of thickness 500 Å, zero temperature penetration depth 2000 Å, in a system where a=1 mm. The midpoint of the transition (M/M_o=0.5) would occur at T= .999T_c, using (9)[19]. Thus any broadening of the large drop in M at the transition would probably be due to sample inhomogeneities rather than the temperature dependence of λ.

Considered as a resistivity transition (approached from the normal side), we use the following expression for the field attenuation of a normal film in the cylindrical geometry:[20]

$$B_{int}/B_{ext}=M/M_o=(1+i\omega\mu_o Rd/2\rho)^{-1} \qquad (10)$$

where ρ is the normal-state resistivity. The second term is negligibly small for typical resistivities in the normal state, so $M/M_o \approx 1$ above T_c. As the resistivity drops to zero so does the mutual coupling[21].

It is of interest to compare the inductive measurement of the transition to a transport measurement. According to (10) the mutual coupling is approximately linear in resistivity only when $\rho<\rho_o=\omega\mu_o ad/2$. For larger values of the resistivity $|M|$ rapidly approaches its maximum value. Thus ρ_o characterizes the maximum resistivity range that is sensed in an inductive measurement. The magnitude of this characteristic resistivity is quite small; for the typical parameters: $\omega/2\pi$=10 KHz, a=1 mm, d=3000 Å, we find $\rho_o \approx 10^{-9}$ Ω-cm. In a system used by the author a drive current was used that resulted in an induced voltage of ~10 μV in the receive coil above the sample T_c, thus giving an overall signal/noise ratio ≈100 for the full transition. At this drive current the current density induced in a 3000 Å film when it is fully superconducting was estimated to be ≈50 A/cm^2. (It is shown below how to calculate this). Let us consider a dc transport measurement using the same current density with voltage probes spaced 1 cm apart and having a normal state resistivity of 1 μΩ-cm. The full transition will correspond to a voltage change of 50 μV, again representing an overall signal/noise ratio ~100. However at resistivity ρ_o (where the inductive transition is just beginning) the voltage across the sample will be only 50 nV, well below the noise level. Thus the inductive technique has much higher sensitivity at the lowest resistivities, but poor sensitivity at the intermediate resistivities which are most usefully probed by a transport measurement. These comments apply equally to measurements in a "conventional" ac susceptibility apparatus. In general the transition "onset" observed with an inductive measurement corresponds to the "zero resistance" point in a transport measurement. The former thus entails a more stringent criterion for T_c.

The transition observed by the inductive technique should be very sharp according to the discussion above. In fact a broadened transition is often observed. Two mechanisms can be invoked to account for a large transition width: On the one hand, the sample could well be inhomogeneous on the scale of the diameter of the coil. Thus the screening currents could pass through regions of varying T_c, resulting in an apparently broad transition. Another source of apparent broadening comes from the fact that the critical current of the film goes to zero as the transition is approached. Thus the screening current induced in the film (50 A/cm^2 in the above example) will exceed the critical current density at some temperature close to T_c, at which point a rise in the mutual inductance will commence. The practical conclusion is that it is necessary to keep the applied field as small as possible, and verify that results are independent of drive current.

One difference between the screening and conventional ac susceptibility configurations is that the former has no sensitivity to purely granular superconductivity. Thus a transition will not be seen unless screening currents can flow on a scale of the coil diameter. This was verified experimentally[22] using a Nb film which was randomly scribed into small islands. After this operation the transition could no longer be observed with the screening technique.

Film Impedance Measurements

Below T_c the mutual coupling measurement can be used to deduce both the dissipative and reactive components of the film impedance. This is discussed in detail in another paper in this proceedings[23]. Well below T_c the reactive part dominates, and is given by (9). In principle an absolute measurement of λ can be obtained, since the coefficient relating λ^2/d and mutual inductance can be calculated for a given coil configuration[15,16]. In practice there is a contribution to the mutual inductance that is independent of λ, arising from two sources: (1) stray coupling in the cryostat wiring, and (2) a small residual coupling due to the finite size of the sample. With sufficient experimental care this offset can be determined and precise measurements of $\lambda(T)$ can be obtained[23]. In practice, however, it is often best to consider the offset in M as an adjustable parameter. This makes it difficult to distinguish between different models of the temperature dependence of λ[24].

One recent application of the mutual inductance measurement is to study the proximity effect in a superconducting/normal metal bilayer[25]. Below T_c we have (see Eqn. 9) $M^{-1} \sim d/\lambda^2$ for a thin superconducting film. Since $1/\lambda^2$ is proportional to the density of superconducting electrons n_s, we can write this as $M^{-1} \sim n_s d$, the total number of electrons per square cm. If a proximity layer is added, n_s is not constant through the superconducting film thickness and assumes a finite value in the normal layer, decaying over a length $\xi_N = (hD/2\pi kT)^{1/2}$ where D is the diffusion constant. In this case it can be shown[25] that M^{-1} is still proportional to the total number of superconducting electrons per square cm: $M^{-1} \sim \int n_s\, dx$. The temperature dependence of this quantity has been shown[25] to be quite different in the case of a normal/superconducting bilayer than for a superconducting film alone.

Fig. 1 shows an example of the mutual inductance data for a superconducting film with and without a normal metal proximity layer. This measurement offers a new technique to study the proximity effect at low temperatures.

Critical Current Measurements

All the applications discussed above involve measurements in the small signal limit, where the coupling is independent of drive amplitude. It is also possible to determine the critical current density of a sample using the screening configuration. The idea is to observe the onset of a nonlinear response as the drive coil current amplitude is raised. In Fig. 2 we show a measurement of the ac voltage induced in the receive

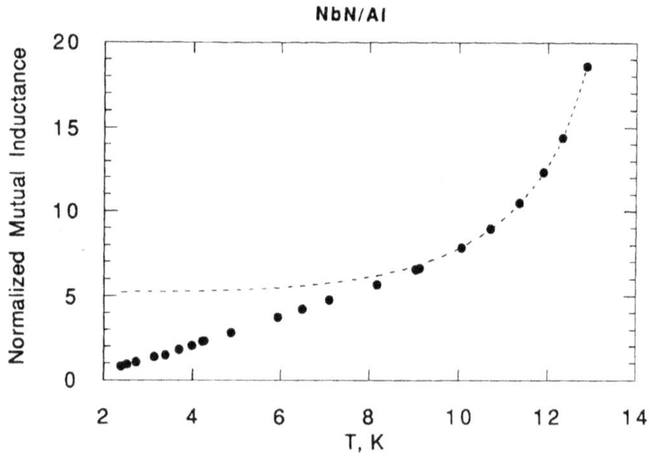

Fig. 1 The mutual inductance of a NbN/Al bilayer. The dashed line represents the data that is obtained in the case of a superconducting film alone. The low temperature drop in mutual inductance is ascribed to the effect of the proximity effect in the normal metal.

coil as a function of the drive coil amplitude for a thin NbN film. Initially the induced voltage is very small, corresponding to nearly perfect screening. Above a threshold drive current a voltage appears, corresponding to reaching the critical current in some regions of the film. At high currents the response approaches a straight line whose slope is ωM_o, ie., the incremental screening (with respect to changes in ac amplitude) has vanished. An obvious interpretation is that during most of the ac cycle the induced currents in the film have saturated at the critical value and flux is able to penetrate to the receive coil. It is reasonable to identify the intercept of the straight line portion of the plot with the drive current axis as a measure of $J_cd=K_c$. (Note that K_c is what is directly measured in any type of critical current measurement on a film sample.)

In order to obtain quantitative information about K_c from a screening measurement it is necessary to calculate the proportionality factor between the coil current and the induced sheet current K=Jd in the film. The approximation of complete screening by the film allows this to be done relatively simply[22]. Briefly, we note that the boundary condition on the field B at the film surface (no normal component) is met if the film is replaced by an image coil carrying the same current. (Here we assume a film of infinite extent). The screening current flowing in the film is then related to the parallel component of B: $K=B_{par}/\mu_o$. From the symmetry of the problem, B_{par} is in the radial direction and K flows azimuthally and in the opposite sense from the coil current. This approximation for K differs from the true value by only a

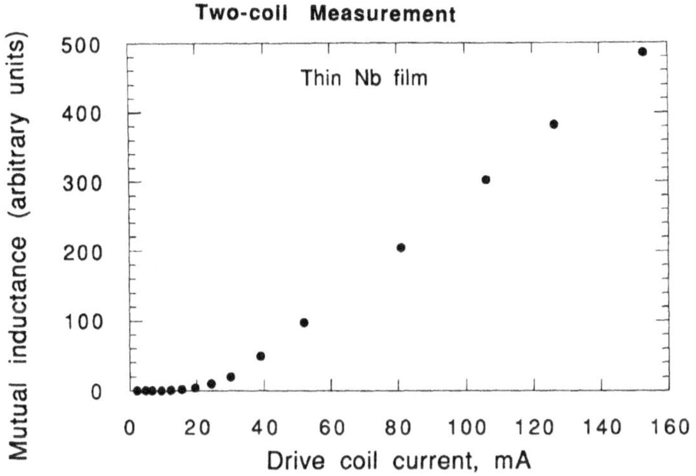

Fig. 2 The voltage induced in the receive coil as a function of ac current amplitude in the drive coil, for a thin (~500 Å) NbN film. The two coils are positioned on opposite sides of the film. The break point is proportional to $J_c d$ in the film.

small factor $\sim\lambda^2/ad$. B_{par} is the sum of the radial fields from the drive coil and its image. The contribution to B_{par} from each turn of the coil can be expressed in terms of elliptic integrals of the first and second kind.[26]

Fig. 3 shows an example of the calculated dependence of K on distance r from the center for a typical coil used in our laboratory. If the coil is pancake-shaped and closely spaced to the sample the screening currents are primarily confined to a region just under the coil. This is the basis for concluding that the results would not be much different for a film of finite extent as long as its radius R is at least,say,twice the coil radius a.

A practical limitation to the maximum K that can be induced in the sample comes from the self-heating in the drive coil. This could cause an unknown rise in the temperature of the sample. A coil used in our laboratory shows evidence of self-heating at 4.2 K when the peak induced film current is K≈1500 A/cm. This obviously sets a limit to the current density that can be measured by this technique. In terms of the nominal 3000 Å thickness, the maximum induced current density is $\approx 5\times10^{7}$ A/cm^2. To measure higher values, some method of thermally isolating the coil from the film while maintaining a sufficiently close spacing for good coupling (<0.5 mm) must be devised.

The screening geometry may also be used to measure the magnetic field dependence of the critical current, as long as the field is applied perpendicular to the film surface. The most important requirement is that the sample be cooled through its transition in the field, thus freezing in a uniform B_z across the film[27]. If instead the field is applied after the sample is cooled, large static shielding currents are established. The static current would then be superimposed on the ac current induced by the measurement, making the result ambiguous.

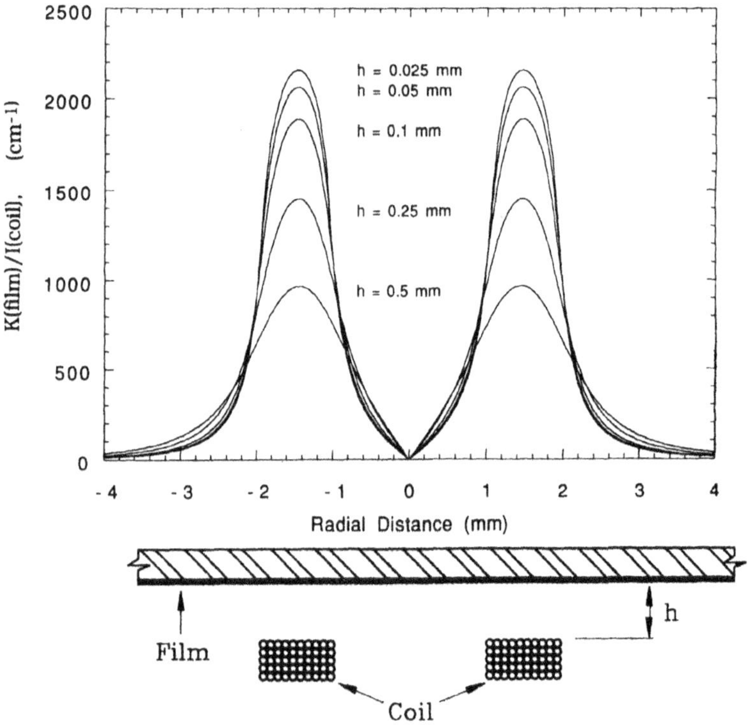

Fig. 3 A calculation the sheet current $K=\int J dz$ induced in a superconducting film when a current I is passed through the adjacent coil, for various spacings h from the film. The film is assumed to extend to infinity and to perfectly screen fields from the reverse side. The coil dimensions are: inner radius=1.05 mm, outer radius=1.95 mm, thickness=0.46 mm, number of turns=300.

In the case of the total moment measurement, it was mentioned that there is a "self-field" due to supercurrents in the sample that adds to the applied field. The analog in the screening geometry is the field due to the coil and the screening currents in the film, referred to as B_{par} above. As was true in the previous case, this is of the order of $\mu_o K_c$, which is usually much less than the scale of fields over which J_c varies significantly. The main difference is that the self-field in the screening geometry is entirely in the radial direction.

One does not expect the development of dissipation in the superconductor to be an abrupt function of J, as is implied in our simple picture above. Rather, the electric field E is a smooth function of J; for example, a flux creep picture gives $E \sim \exp(J/J_o)$[28] when J is well below J_c. Thus one expects rounding in a plot such as Fig. 1, as is always in fact observed. In the low current region one can attempt to relate the voltage induced in the receive coil due to the electric field in the film, by modeling the film as a ring.[29] It is shown that the receive coil voltage is given by $\sim 2\pi aAE$, where A is the area under the current response curve shown in Fig. 3 and a is the average radius of the coil. In addition to this voltage due to dissipation in the superconductor one has the usual linear coupling to the drive coil that was discussed earlier due to the finite value of the penetration depth. The voltage component due to dissipation in the film can only be distinguished from the other terms to the extent that it is nonlinear in the drive current. The sensitivity to the nonlinear electric field component in the response is quite impressive: considering the coil in Fig. 3, with $A \approx 160$ and $a \approx 1.5$ mm, we find 0.1 μV across the coil (easily detected with averaging techniques) corresponds to $E \approx 10^{-9}$ V/cm. A component of voltage that is approximately exponential in drive current over many decades has in fact been observed[29], suggesting an E(J) dependence similar to the flux creep model.

When a superconducting pickup coil coupled to a SQUID magnetometer is used in the role of a receive coil, one has a version of the screening geometry that has very high sensitivity at low sweep rates. This is most useful for observing the initial stages of dissipation in a film. It has been used[30] to detect the onset of irreversible motion of trapped flux vortices in response to a transport current. If this is done at sufficiently low vortex density the effect of interactions between vortices is unimportant and one has a measurement of the so-called elementary pinning force. This may have an important practical use in evaluating materials for use as thin film input transformers for advanced SQUIDs. At low vortex density a conventional transport measurement would be dominated by edge effects and the results would be ambiguous at best. In the screening geometry all fields and currents are confined to regions away from any film edges, and the results are easy to interpret.

Variations of the Screening Measurement

To have an accurate indication of the temperature of a thin film sample it is best to attach the substrate firmly (using pressure or grease) to a high thermal conductivity block whose temperature is recorded. This makes it very difficult to use a second coil on the reverse side. In our lab we have found that measurements of T_c and K_c can in fact be made with only one coil. A simple cryostat has been built that allows a measurement of the transition temperature and critical current density with total turnaround time less than 1 hour, using a He or LN_2 storage dewar. The sample is sandwiched with a spring clip between a coil and a copper thermometer block, and suspended above a helium bath. For more details the reader is referred to Ref. 22.

The single coil T_c measurement is made by monitoring the coil inductance as the temperature is swept. It is found that this drops by more than half if the coil is quite thin in the axial dimension and closely spaced to the sample. We use a phase sensitive amplifier to monitor the coil inductance with more than adequate sensitivity. Pulturak et. al.[31] have described a variation of the single-coil measurement which uses a second identical coil in a bridge configuration. When the sample is normal the bridge is balanced and one has a null signal. Other measurement systems have been described[32] with very high sensitivity which may also be used with very small bulk samples.

The measurement of J_c with a single coil is done by monitoring the drive coil for the onset of a nonlinear response as the ac current is increased. This is most easily done by using a very pure sine wave to drive the coil, and monitor the voltage across the coil for harmonic content. We use a circuit that detects the voltage at the third harmonic of the drive frequency, although a measurement of total harmonic distortion would probably also suffice. Fig. 4 shows a typical result of the harmonic measurement. Note that the shape of the curve is very similar to that obtained with a two-coil system, with a distinctive break point on the drive current axis that is proportional to K_c.

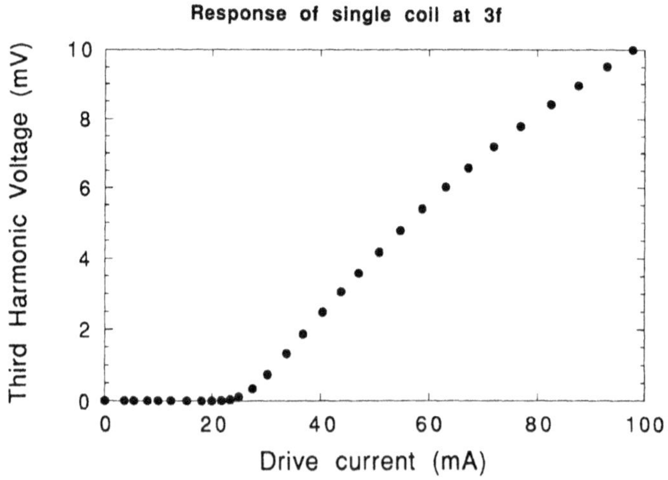

Fig. 4 The third harmonic component of the voltage induced across a drive coil when it is driven with a very pure sine wave, as a function of drive amplitude. The break point corresponds to induced currents in the adjacent film exceeding the critical value.

Another approach to the screening measurement uses a permeable magnetic material to form a core that channels the applied field across a narrow circular gap[33]. There are several potential advantages to this approach; an important one is that the field strength that may be produced at the sample surface is not limited by self-heating of the drive coil but rather by the saturation magnetization of the core. This means that higher values of K_c may be measured. Ref. 33 cites K values up to 4000 A/cm. Another potential advantage is that the screening currents are confined to the region just under the gap. This may provide an unambiguous evaluation of the E(J) dependence in the superconductor. A rather serious difficulty arises from the skin depth in the core, which ideally should be greater than its overall dimensions. Not only do suitable materials have a fairly high conductivity, but the skin depth is proportional to $(\text{permeability})^{-1/2}$. Materials with suitable permeability, ≈ 1000, may not be easily available. It turns out that a frequency of around a few Hz must be used and the voltage induced in the receive coil is quite small.

SUMMARY

Traditional ac susceptibility apparatus, normally used with bulk samples, has been used to evaluate properties of thin films as well. The film is usually oriented with its surface perpendicular to the applied field. For transition temperature measurements it is nécessary to use a much lower amplitude of the applied field than would be appropriate for a bulk sample, due to the extreme demagnetizing factor of a thin film. Otherwise flux will enter the sample at its edges, leading to an apparently broadened transition. Due to the small volume of a film sample, the signal is quite small. Measurements are also possible with the sample surface parallel to the field. In this case an appreciable signal will be seen only if the film thickness is greater than twice its penetration depth. With the field perpendicular, determination of the critical current density may be made, with some uncertainties arising from details of the shape of the film and the possibility of granular effects. The magnitude of field that must be applied to drive the sample entirely into the critical state is modest, $\sim J_c d$.

The author's own experience and bias favor the "screening" geometry for measurements of thin film properties. This configuration is uniquely suited to thin film measurements. Advantages include: ability to measure large area samples, independence of edge effects, reduced size allowing use of a storage dewar for quick measurements. In addition to determinations of T_c and J_c, the screening arrangement allows measurements of the film impedance. To obtain quantitative information a numerical calculation specific to a given experimental configuration is necessary. However, analytical results exist for the related cylindrical geometry and are useful for revealing parameter dependencies and orders of magnitude. T_c and J_c measurements are possible using a single coil. Heating in the drive coil limits the maximum sheet current density ($J_c d$) that may be measured to ~1000 A/cm. Use of a magnetic core can increase this limit significantly.

It is shown that inductive measurements of the superconducting transition are sensitive to a much lower range of sample resistivity than transport measurements; thus care must used when comparing results from the two methods. To compare J_c measurements with transport results, an estimate of the equivalent electric field criterion that applies in the inductive measurement must be made. This depends on sample dimensions and excitation frequency, but often will be higher than the criterion in a careful transport experiment. Thus the apparent J_c will be higher, as has been observed[10]. However the basic sensitivity of the measurement when currents induced in the superconductor are well below the critical value corresponds to an electric field sensitivity much greater than would be available in a transport measurement[29].

Acknowledgement

The author wishes to acknowledge many useful conversations with M. E. Reeves.

REFERENCES

1 J. A. Osborne, Demagnetizing Factors of the General Ellipsoid, Phys. Rev. 67:351 (1945).

2 Michael Tinkham, "Introduction to Superconductivity", McGraw-Hill, New York, (1975), p. 91; D. Shoenberg, "Superconductivity", Cambridge University Press, London, (1952), p. 23.

3 This expression can be obtained by integrating M=B-H across the film thickness. An expression for B(x) is given, for example, in Tinkham's book (ref. 2), Eq. 4-54a.

4 Ch. Neuman, P. Ziemann, J. Geerk, and H. C. Li, Characterization of YBaCuO Films by ac-Susceptibility Measurements, Journal of the Less-Common Metals 151:363 (1989).

5 K Kanoda, H. Mazaki, N. Hosoito, and T. Shinjo, Magnetic-field penetration depth and material parameters of V-Ag multilayered superconductors, Phys. Rev. B 35:8413 (1987).

6 This of course is only true for simply connected samples.

7 Charles P. Bean, Magnetization of High-Field Superconductors, Rev. Mod. Phys. 36:31 (1964).

8 John R. Reitz, Frederick J. Milford, and Robert W. Christy, "Foundations of Electromagnetic Theory", Addison-Wesley, Reading, Mass., (1979), p.166.

9 We have $\mu_0 J_c=(\partial B_r/dz-\partial B_z/dr)$ from Maxwell's equations, and the cylindrical geometry ensures that the first term is zero.

10 M. Däumling and D. C. Larbalestier, Critical state in disk-shaped superconductors, Phys. Rev. B 40:9350 (1989).

[11] The calculated weak divergence in B at the center of a disk really means that the critical state cannot be reached in some small region near the center.

[12] This can be obtained from Maxwell's equation curl(E)=-dB/dt, which in cylindrical coordinates becomes dE/dr+E/r=-dB/dt. Integrating over r yields the given expression.

[13] John Talvacchio, Critical Currents in A-15 Superconductors, Dissertation, Stanford University, 1982 (unpublished). The expression given in (6) is obtained by integrating the Biot-Savart law over the sample assuming a constant $|J|$.

[14] R. G. Humphreys, private communication (Royal Signals and Radar Establishment, St. Andrews Road, Malvern, Worch. WR14 3PS, UK)

[15] A. T. Fiory, A. F. Hebard, P. M. Mankiewich, and R. E. Howard, Penetration depths of high T_c films measured by two-coil mutual inductances, Appl. Phys. Lett. 52:2165 (1988).

[16] B. Jeanneret, J. L. Gavilano, G. A. Racine, Ch. Leemann, and P. Martinoli, Inductive conductance measurements in two-dimensional superconducting systems, Appl. Phys. Lett. 55:2336 (1989).

[17] P. G. de Gennes, "Superconductivity of Metals and Alloys", W. A. Benjamin, Inc., New York, Amsterdam, (1966), p. 194.

[18] Note that this is the the result if the external field is applied after the sample has gone superconducting; ie., there is no trapped flux.

[19] For this calculation we use the two-fluid expression for the penetration depth, $\lambda^2(T)=\lambda^2(0)/[1-(T/T_c)^4]$.

[20] S. Fahey, C. Kittel, and S. G. Louie, Electromagnetic screening by metals, Am. J. Phys. 56:989 (1988).

[21] At very low values of ρ account must be taken of the reactance of the superfluid component. In the simplest two-fluid approximation, this would add a term $Rd/2\lambda^2$ in the denominator of (10).

[22] J. H. Claassen, M. E . Reeves, and R. J. Soulen, A contactless method for measurement of the critical current density and critical temperature of superconducting films, Rev. Sci. Instrum. 62:996 (1991).

[23] A. T. Fiory, this conference.

[24] A. F. Hebard, A. T. Fiory, and D. R. Harshman, Magnetic Penetration Depth of $YBa_2Cu_3O_7$, Phys. Rev. Lett. 62:2885 (1989).

[25] J. H. Claassen, J. E. Evetts, R. E. Somekh, and Z. H. Barber, to be published in Phys. Rev. B1.

[26] J. D. Jackson, "Classical Electrodynamics", Wiley, New York, (1975) p. 177.

[27] If the pinning is strong, the field profile evidently is unchanged when the sample goes superconducting. Even in the opposite limit the field profile will be nearly uniform: because of demagnetizing effects there will be complete field exclusion out to a radius r, defined by $H_{c1} \approx H(r/d)$. Thus the region of zero field is very small, of the order of the film thickness. The same sort of argument suggests that the region of partial shielding only extends a short distance (compared to the coil radius) from the center.

[28] J. D. Hettinger, A. G. Swanson, W. J. Skocpol, J. S. Brooks, J. M. Graybeal, P. M. Mankiewich, R. E. Howard, B. L. Straughn, and E. G. Burkhardt, Flux Creep and High-Field Critical Currents in Epitaxial Thin Films of $YBa_2Cu_3O_7$, Phys. Rev. Lett. 62:2044 (1989).

[29] J. H. Claassen, Inductive Measurements of Critical Current Density in Superconducting Thin Films, IEEE Transactions on Magnetics 25:2233 (1989).

[30] L. H. Allen and J. H. Claassen, Technique for measuring the elementary pinning force in thin films, Phys. Rev. B 39:2054 (1989).

[31] Emil Polturak, Daniel Cohen, David Cohen, and Gad Koren, A Self Contained Inductance Bridge for Rapid NDT of Superconducting Thin Films, this proceedings.

[32] Dimitrios G. Xenikos and Thomas R. Lemberger, ac susceptibility apparatus for measuring the transition temperature of high-T_c crystals, sintered samples, and films, Rev. Sci. Instrum. 60:831 (1989).

[33] T. L. Hylton, M. R. Beasley, and R. C. Taber, An iron-core magnetic inductance probe to measure critical current densities in superconducting thin films, preprint.

A SELF CONTAINED INDUCTANCE BRIDGE FOR RAPID NDT OF SUPERCONDUCTING THIN FILMS

Emil Polturak, Daniel Cohen, David Cohen and Gad Koren

Crown Center for Superconductivity, Physics Department
Technion-Israel Institute of Technology
Haifa 32000, Israel

ABSTRACT

We describe a compact, self contained inductance bridge for rapid testing of superconductive thin films. The bridge requires no adjustments, has a high sensitivity and small offset and can be used for both routine screening of thin films and for basic studies of high temperature superconductors, HTSC, such as the proximity effect.

AC susceptibility is, beside resistivity, the most popular technique for thin film characterization of superconductors. It is especially important when Non Destructive Testing, NDT, is required. The need for NDT testing of films can come about when one has prepared a batch of films for optical investigation of reflectivity, or for patterning of a microwave passive device, etc. Even when one wishes to test the homogeneity of a film deposited on a large wafer without actually cutting it into little pieces, this technique is indispensable. We describe a very simple self contained inductance bridge which can do all of this, and present some results obtained with it.

Most published designs for thin film characterization use a mutual inductance method[1-3]. This usually means the use of two coils, one on each side of the film, and perhaps a third coil to be connected in opposition with the secondary. In some cases, miniature coils are wound especially to fit each sample individually. In order to achieve a large signal to noise, the coils are attached to the film using glue or grease. Although these methods work well in a specialized situation, they are not suitable for rapid processing of many films and lack the ability to compare the results without elaborate calibration procedures. To eliminate these difficulties, we decided to use a self inductance method[4], which compares two flat coils in a bridge configuration shown in Fig. 1. A thin film sample is laid flat on top of one of the coils. Below its transition temperature, the film acts as a ground plane, reducing the self inductance of the coil and throwing the bridge out of balance. Thus, the need to have two coils on both sides of the film is eliminated.

In order to make the device simple to operate, we decided to make the inductance bridge self contained thereby eliminating any need for external adjustments. This required building the bridge in such a way as to minimize offsets caused by thermal drift of the components. The actual layout of the bridge, shown in Fig. 1b, is on a PVC slab. Because PVC is a poor heat conductor, when the slab hangs in the cold gas inside a Liquid Helium dewar, there may be a large temperature difference between its top

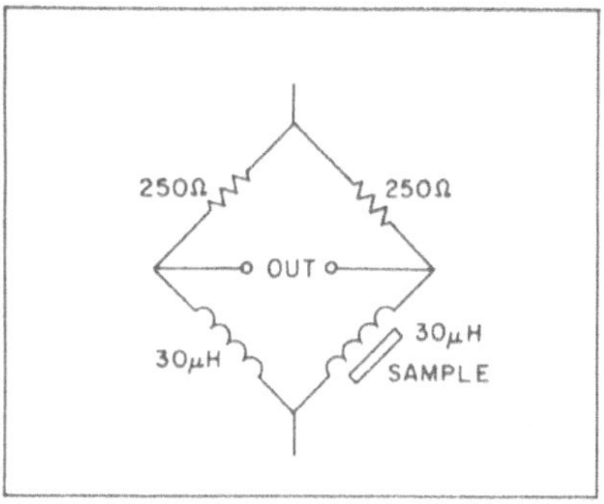

(a)

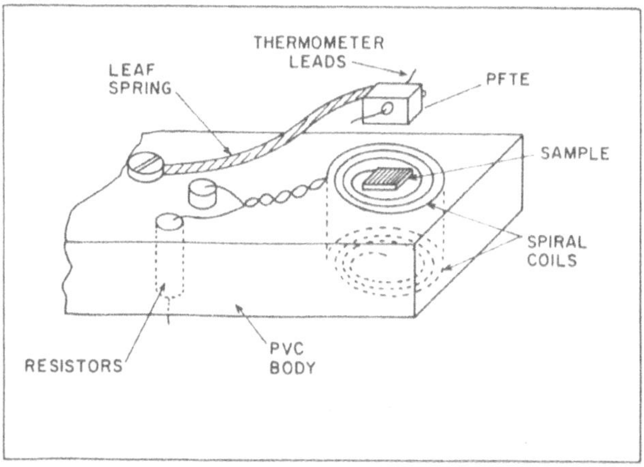

(b)

Fig. 1. (a) - Schematic layout of the self inductance bridge.
(b) - Physical layout of the bridge components on the PVC slab.

and bottom. Thus, the two coils and the two resistors should hang at the same height. The inductors are single layer spiral coils of 100 turns each, wound in a tight groove cut into a cylindrical former cast from 1266 Stycast epoxy loaded with chalk powder. The i.d. and the o.d. are 2mm and 9mm. The inductance is 30 μH per coil. After potting the coils in epoxy, the ends of the cylinder are machined off to within 0.5 mm from the coils, in order to achieve good coupling between the coil and the thin film sample next to it. The coil assembly is glued into the PVC block so that the end of the cylinder near to the "signal" coil is flush with the surface of the PVC slab. Films are laid on top of it and held down by a Teflon tipped leaf spring, as shown in Fig. 1b. Thus, we use no grease, and interchanging samples is trivial. For the bridge resistors, we chose a pair of 500 ohm metal film off the shelf. They seem to be well enough matched for our purpose, although one could do better with non-inductive precision wirewound resistors. At the operating frequency of 50 kHz, the coil impedance is less than 10 Ω. Thus, the bridge works in a constant current mode. The bridge is located in a 30mm diameter vacuum can, which uses a low angle conical grease seal[5]. A single layer mu metal shield is wrapped around the vacuum can. Cooling is achieved by heat conduction of He gas in the vacuum space. All the wiring to the probe is fed through a 3/8" stainless steel tube which can be moved up or down through an O-ring seal made in a flange which fits a standard storage dewar with a 1.5" neck. The temperature of the sample is varied by adjusting the height of the probe over the Liquid Helium. The electronics consist of a two phase lock-in amplifer which is also the excitation source for the bridge, and a DVM which monitors the diode thermometer. All wiring in and out the probe are individually shielded twisted pairs, and the lock-in works in a differential mode. This mode of operation is important to eliminate offsets caused by ground loops.

As an example of the use of the bridge, in Figs. 2 and 3 we show data obtained with two samples: one a 1 μm thick film of YBCO on MgO substrate, and one of an identical film with a 500 Å thick layer of silver deposited on top of it. The films were deposited using a laser ablation technique described previously.[6] In the case of the silver/YBCO film, the deposition of silver was also done in situ, at a substrate temperature of $100^{o}C$. The measurement was done with a 400 mG field (peak to peak) at 40 kHz. First, we point out the flat baseline away from T_c. This shows that the cancellation scheme works quite well. A measurement of the empty probe shows that the total baseline drift between 200K and 4K is about 1% of the signal level obtained with a superconducting film, i.e. negligible. The other thing we point out is the large effect a thin non superconducting metallic film has on the measured screening. Thus, the method is extremely sensitive to the presence of a very thin normally conducting metallic layer on top of the superconductor. Finally, the residual voltage at low T in the real component of the signal is due to imperfectly adjusted phase of the lock in amplifier.

The analysis of the bridge output is done along similar lines to Refs. 2 and 3. The analysis is valid only for excitation fields small enough so that the supercurrents induced in the superconducting film do not exceed the critical current density. We shall remark only briefly on the case of larger fields at the end of this discussion. Without a sample, the voltage across the coil is $L\frac{di_D}{dt}$, with i_D is the constant drive current of the bridge. Assuming a perfect balance, this voltage is exactly cancelled by that of the other coil and the output of the bridge is zero. With a superconducting film near one of the bridge coils, the screening currents in the film create a flux which couples back into the bridge coil with an opposite sign, reducing the seld inductance. These screening currents are related to the magnetic field, or rather the vector potential $\vec{A}(\vec{r})$ by the London equation:

$$\vec{j}(\vec{r}) = -\frac{1}{\mu_0\lambda^2}\vec{A}(\vec{r}) \tag{1}$$

here λ is the penetration depth. The vector potential has to be obtained in a self consistent way, namely it includes the part coming from the driving coil and also from the screening currents in other parts of the film.[2] It is convenient to work with an areal current density $\vec{K}$, defined as $\vec{j} \cdot d$, where d is the film thickness, assumed smaller

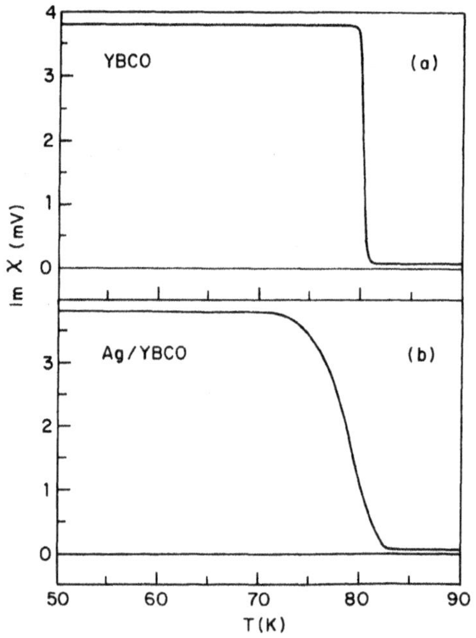

Fig. 2. (a) - Out of phase components of the output of the bridge with an YBCO film. The frequency is 40 kHz and the excitation amplitude 0.2 Gauss.
(b) - The same for an YBCO film with an overlaying 500 Å thick silver film.

than λ. We remark on the thick film case later on. Using Stokes theorem, we rewrite Faraday's law in its integral form as

$$\oint \vec{E} \cdot d\vec{\ell} = \int\int \vec{\nabla} \times \vec{E} \cdot d\vec{s} = -\frac{\partial}{\partial t}\int\int \vec{B} \cdot d\vec{s} = -\frac{\partial}{\partial t}\oint \vec{A} \cdot d\vec{\ell} \tag{2}$$

from which we can express the electric field $\vec{E}$ as $\vec{E} = -\frac{\partial}{\partial t}\vec{A}$. Substituting back into the London equation, we get

$$\vec{E} = \frac{\mu_0 \lambda^2}{d}\frac{\partial}{\partial t}\vec{K} \tag{3}$$

by multiplying Eq. (3) by a length segment $d\ell$, it assumes the familiar form of "potential = inductance $\times \frac{\partial}{\partial t}$ current". The coefficient $\mu_0\lambda^2/d$, which plays the form the inductance, is called L_K, the kinetic inductance. Furthermore, writing $\frac{\partial}{\partial t}\vec{K} = i\omega\vec{K}$, we obtain the equivalent of Ohm's law,

$$\frac{\vec{E}}{i\omega L_K} = \vec{K} \tag{4}$$

which shows that $i\omega L_K$ is the areal impedance of the superconducting film. Finally, we need to incorporate the normal currents into this picture. This is done by adding an areal resistance R in series with the inductance, namely $Z = i\omega L_K + R$, and $\vec{E}/Z = \vec{K}$, where $\vec{K}$ now includes the normal currents as well. Having clarified the role of the kinetic inductance, we now calculate the flux induced in the bridge coil L by the currents in the film. The induced flux density $\vec{B}'(r')$ at a point $\vec{r}'$ in the bridge coil is given by the Biot-Savart law:

$$\vec{B}'(\vec{r}') = \frac{\mu_0}{4\pi}\int \frac{\vec{K}(\vec{r}) \times (\vec{r} - \vec{r}') \cdot d\vec{s}}{|\vec{r} - \vec{r}|^3} \tag{5}$$

with the integration extending over the area of the film. The total induced flux is obtained by integrating $\vec{B}'$ again, this time over the area of the bridge coil. To get absolute values, one has to solve self consistently for the current distribution $\vec{K}$ in the film[2]

$$\vec{K}(\vec{r}) = \frac{-i\omega}{Z}[A_{ext}(\vec{r}) + \frac{\mu_0}{4\pi}\int_{film} \frac{\vec{K}(\vec{r}')}{|\vec{r}' - \vec{r}|} \cdot d\vec{s}] \tag{6}$$

This is done, for example, in ref. 2. This procedure has the advantage that one can also use it for the case where $\vec{K}(\vec{r})$ exceeds the critical current density, K_c. However, one has to know rather well the external potential $\vec{A}_{ext}$. Furthermore, once K_c has been reached, one has to consider flux flow, flux pinning, etc. which is outside the scope of this paper, and is discussed extensively by other authors in this book.

Since we are interested only in comparison of many films having the same geometry, we can assume that the geometrical distribution $\vec{K}(\vec{r})$ is the same for all of them for $|\vec{K}| < |\vec{K}_c|$. As the first approximation we drop the non local term in Eq. (5), thus assuming $\vec{K}(\vec{r}) = -\frac{i\omega}{Z}\vec{A}_{ext}$. Since $\vec{A}_{ext}$ is generated by the bridge, it is proportional in Li_D. Denoting the value of the integral in Eq. (5) by M, we write the induced flux ϕ' through L as

$$\phi' = \frac{i\omega}{Z} L i_D M \tag{7}$$

The induced voltage V is obtained by taking the time derivative of the current i_D in Eq. (7). Separating it into real and imaginary parts, we finally obtain

$$ReV = \frac{\omega^2 RLM}{R^2 + \omega^2 L_K^2} i_D$$
$$ImV = \frac{-\omega^3 LL_K M}{R^2 + \omega^2 L_K^2} i_D \quad (8)$$

Naturally, ReV is proportional to χ'' and ImV to χ'. The constant M can be obtained from the Im V in the low temperature limit. There, $R \to 0$ and Im V then contains a product of M and quantities which are directly measurable. One can do a better approximation by substituting $\vec{A}_{ext}(\vec{r'})$ for $\vec{K}(\vec{r'})$ into the non-local part of Eq. (6). One then gets expressions containing two adjustable constants from the two integrals, with the second one being proportional to i_D^2. The algebra, although messy, is straightforward and will not be done here. In the case of thicker film, one should replace d in the expression for L_K by λ. Looking at Eq. (8), the difference between different films can be labelled as arising from different values of R. Poor quality films will have a higher R, and hence a broader transition. In the case of the proximity sandwich, one effectively has a larger λ as well[7], since the shielding properties of the Ag layer are much weaker than those of YBCO. Since the Ag layer is only weakly superconducting by the proximity effect, its surface region will in effect remain normal and its resistance will therefore be larger. A more quantitative description of this case requires a more complex analysis than presented here, which allows us only to extract an effective value of λ for the Ag/YBCO sandwich and compare it with the value of λ for a homogeneous sueprconductor.

ACKNOWLEDGEMENT

We greatly appreciate the help of L. Wilen and M. Ayalon in the design and construction of the system.

REFERENCES

1. F. Lobotka and G. Gomory, Physica Status Solidi **A109**, 205 (1988).
2. A.T. Fiory, A.F. Hebard, P.M. Mankiewich and R.E. Howard, App. Phys. Lett. **52**, 2165 (1988).
3. B. Jeanneret, J.L. Gavilano, G.A. Racine, Ch. Leemann, and P. Martinoli, App. Phys. Lett. **55**, 2336 (1989).
4. E. Polturak, L. Wilen, D. Cohen and G. Koren, Rev. Sci. Instrum. **61**, 1759 (1990).
5. E.T. Swartz, Rev. Sci. Instrum. **57**, 2848 (1986).
6. G. Koren, E. Polturak, B. Fisher, D. Cohen and G. Kimel, App. Phys. Lett. **53**, 2330 (1988).
7. G. Deutscher and P.G. de Gennes, in *Superconductivity* Vol. 2, R. Parks, ed., (Marcel Dekker, New York), C. 1966.

CHARACTERIZATION OF $Tl_2Ba_2CaCu_2O_8$-FILMS AND SINGLE CRYSTALS BY AN AC-SUSCEPTIBILITY METHOD

Ch. Neumann, Ch. Heinzel, P. Ziemann, W. Y. Lee*

Fakultät für Physik, Universität Konstanz; P.O. Box 5560,
7750 Konstanz, Germany.
*IBM Almaden Research Center, San Jose CA 95120-6099, USA.

ABSTRACT

A specially designed ac-susceptometer is described, which allows one to determine the anisotropy of the magnetic response of small (0.08x0.5x0.6mm) HTSC-crystals or thin films. The technique is illustrated by two examples: 1. Determination of the penetration length $\lambda(0)$ for $Tl_2Ba_2CaCu_2O_8$ films with the external field parallel to the film surface. 2. Anisotropic behavior of the magnetic losses, as given by the imaginary part $\chi''(T)$, in $Tl_2Ba_2CaCu_2O_8$ single crystals with fields perpendicular and parallel to the c-axis. Additional results on a tilted crystal (20° tilt of the c-axis relative to the external field) are discussed in terms of a possible flux line entanglement transition.

INTRODUCTION

Among the high-temperature superconductors, the highly anisotropic systems like BiSrCaCuO and TlBaCaCuO appear to be very attractive from a physics point of view, since here newly predicted phenomena like flux line entanglement[1,2,3], flux line lattice (FLL) melting[4,5], glass-like flux line phases and their transitions[6] or the expected anomalous behavior of magnetic flux parallel to the CuO_2-planes[7,8] should be most pronounced. In the present work, we therefore report new results on $Tl_2Ba_2CaCu_2O_8$ films and single crystals applying an ac-susceptibility technique to characterize the behavior of magnetic flux. For this purpose, a specially designed ac-susceptometer is used, which allows one to distinguish between the magnetization parallel and perpendicular to the c-axes (the films had a c-axis orientation perpendicular to the film plane). By changing the position of a sample relative to the pick-up coils as well as its orientation relative to the direction of the ac-field and monitoring the real and imaginary part of the magnetic susceptibility one is able to determine the anisotropy of the magnetic dissipation. In this way, two qualitatively different loss peaks, which are centered at different temperatures, could be observed and assigned to the alignments of the external ac-field parallel and perpendicular to the c-axis. By tilting the sample by approximately 20° relative to the field direction, a

Magnetic Susceptibility of Superconductors and Other Spin Systems
Edited by R.A. Hein *et al.*, Plenum Press, New York, 1991

superposition of both peaks is found. This leads to an apparent paradox that the component of a flux line perpendicular to the c-axis can move at a lower temperature than the component parallel to the c-axis. Such a behavior suggests the possibility of flux line entanglement.

Additional information can be obtained from the temperature dependence of the real part of χ determined on films. If the film thickness is properly chosen, the penetration behavior of the magnetic field can be observed and models for $\lambda(T)$ can be tested.

EXPERIMENTAL

The samples are investigated in a specially designed vector-lock-in a.c.-susceptometer[9]. The real and imaginary parts of the susceptibility χ are determined as a function of the temperature. Magnetic ac-fields, H_{ac}, in the range of 0.1 Oe to 50 Oe are applied with small superposed dc-fields up to 180 Oe. The frequency can be varied

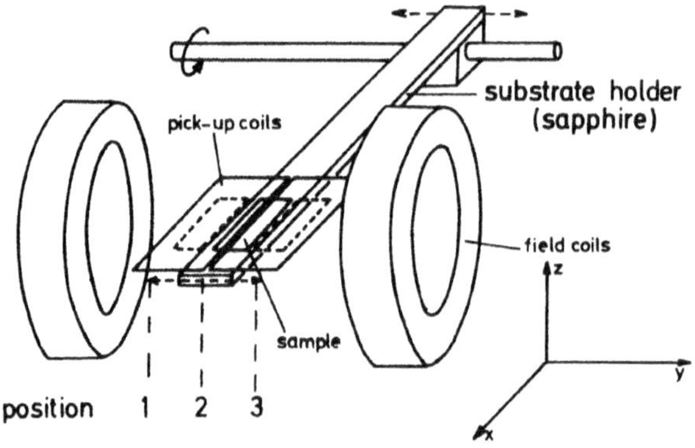

Fig. 1 Schematical view of the essential parts of the ac-susceptometer (not scaled). Diameter of the field coils: 23 mm.

from 31 Hz to 3125 Hz. The magnetic field is produced by a pair of coils (23mm diameter) wound out of Cu-wire (0.1mm diameter) with 1500 turns.

The experimental set-up of the susceptometer is schematically shown in fig. 1. The field and the pick-up coils are mounted on sapphire holders, which are directly connected to the bottom of a ^{4}He-cryostat. Thus, both coil systems are always at temperature of the cooling liquid (liquid ^{4}He or liquid N_2). Therefore temperature drifts and drifts of the phase setting, which is determined with a lead sample at 4.5 K are strongly suppressed. The samples are fixed on the sapphire substrate holder with vacuum grease. This sample holder can be shifted along and rotated around a sapphire axis. At the beginning of each measurement, the sample is moved out of the magnetic field and the residual signal produced by inhomogenieties of the ac-field is compensated by a small coil above one of the counterwise wound pick-up coils. The measurement starts by rotating the sample into its position right in front of the pick-up coils (distance ≃ 0.1 mm). This position is mechanically fixed and thus highly reproducable. The temperature of the sample can be changed by a heater clamped onto the sapphire axis (temperature range $5K<T<150K$) and is determined by a

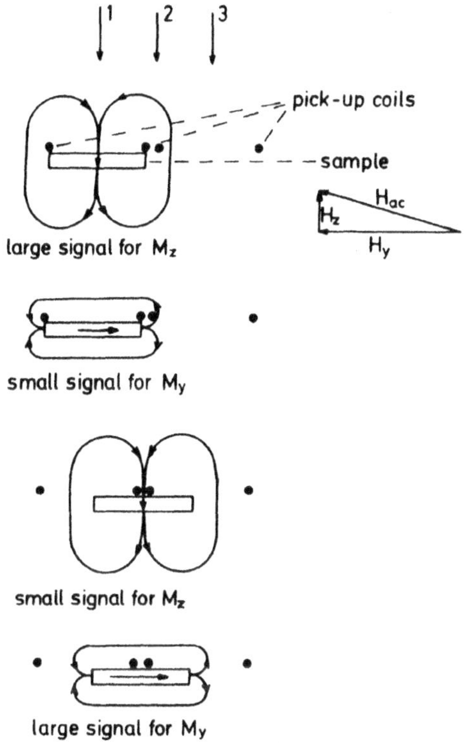

Fig. 2 Separation of the components of the magnetization for a superconducting film by moving the sample under the counterwise wound pick-up coils. The positions 1,2,3 are defined in fig. 1.

diode-thermometer also located on the axis. Remaining small drifts of the system are experimentally determined by moving the sample out of the field coils several times during the measurement. The temperature dependence of the drift is obtained by interpolation between the experimentally determined values and is subtracted from the susceptibility signal. This procedure is performed by a computer, which also controls the rotation of the sample into and out of the field coils. A very detailed description of the apparatus is given by Zibold and Korn[9].

Fig. 2 serves to explain how by shifting the samples to different positions under the pick-up coils information can be obtained on the anisotropy of the magnetization. For this purpose, one exploits the fact that the external ac-field is not perfectly homogeneous along the y-direction (see fig. 1), but exhibits also a small z-component. In case of a thin film sample, this component is perpendicular to the film plane and thereby largely enhanced by demagnetization effects. More specifically, $M = \chi \cdot H_{int}$ with $H_{int} = H_{ac} \cdot (1+N \cdot \chi)^{-1}$. For a superconducting film one has $\chi = -1$ and $N \simeq 1$ leading to large enhancements in the z-direction. But by our experimental arrangement M_z can only be detected if the sample is moved under one of the pick-up coils (positions 1,3). At position 2 (middle) this component delivers no signal due to compensation by the counterwise wound pick-up coils. In the middle-position 2 the magnetization parallel to the film plane is preferentially detected as indicated in fig. 2. By measuring the susceptibility of superconducting Pb-films of thicknesses similar to those of the HTSC-films and using the known $H_c(T)$-curve for Pb, one obtains a calibration factor for each position, i.e. position 1, 2 and 3 of figure 2. This calibration includes a factor depending on the geometry and sensitivity of the apparatus as well as on the demagnetization factor. Thus, no calculated demagnetization factor has to be used[10].

This separation of the different components of the magnetization is only possible for superconducting films due to their large demagnetization effect in z-direction, while other samples like single crystals and sintered material are measured at position 2. In this case different components of the magnetization are obtained by tilting the sample with respect to the field. This tilt is performed by mounting precisely worked copper wedges of different angles between the plate at the sapphire axis and the substrate holder. This copper wedge doesn't significantly change the thermal connection of the sample to the sapphire holder, because we observe no shift of T_c within the resolution of our apparatus.

RESULTS

In fig. 3 the temperature dependence of the real and the imaginary part of the susceptibility of a $Tl_2Ba_2CaCu_2O_8$-film is shown obtained at different positions of the film. The crosses represent measurements at position 2 (middle). No imaginary part can be observed within our resolution and the real part is field independent for ac-fields from 1 Oe to 50 Oe. Such a behavior provides strong evidence, that the temperature dependence of the real part is dominated by the penetration effect of the field, i.e. by the temperature dependence of the penetration length $\lambda(T)$. The dashed line in fig. 3 is a fit to the data assuming a slab-geometry representing the film of thickness d and the field parallel to the surface, i.e.:

$$\chi'(T)=\chi'(0)\cdot(1-(2\cdot\lambda(T)/d)\cdot\tanh(d/(2\cdot\lambda(T)))$$

For the temperature dependence of λ the two-fluid-formula $\lambda(T)=\lambda(0)(1-(T/T_c)^4)^{-1/2}$ was used. This model gives an exellent description of the data within the experimentally observed temperature range $65K < T < 120K$. With a film thickness of 12 μm, a value of $\lambda(0) = 4500$ nm is deduced from the fit, which is significantly larger than typical values reported in the literature for $Tl_2Ba_2CaCu_2O_8$[11]. A possible explanation could be that the measured value is not an intrinsic value but a Josephson penetration depth due to weak links within the film.

The solid line in fig. 3 represents the results found at position 1. Clearly, the component of the magnetization perpendicular to the film, which is preferentially detected at this position, produces magnetic losses as indicated by the peak of the imaginary part (the unphysical sign of this peak is a technical artifact and stems from a 180°-phase shift of the signal at position 1). This is reflected also at the corresponding temperature in the real part behavior, which is a superposition of the magnetic response of a broad transition due to effects of the penetration length in z-direction and a sharp transition to shielding in y-directions.

In Fig. 4a the measurements for the field parallel to the CuO_2-planes are shown. The real part exhibits a broad transition and does not reach to perfect shielding at 70 K. Additionally a peak in the imaginary part is observed. The maximum of this peak shows a strong field dependence which could be described by a power law:

$$H^*(T_{max}) = H^*(0)\cdot(1-(T/T_c))^a$$

with a power a = 1 for this direction[13]. By assuming that for a given amplitude H_{ac},

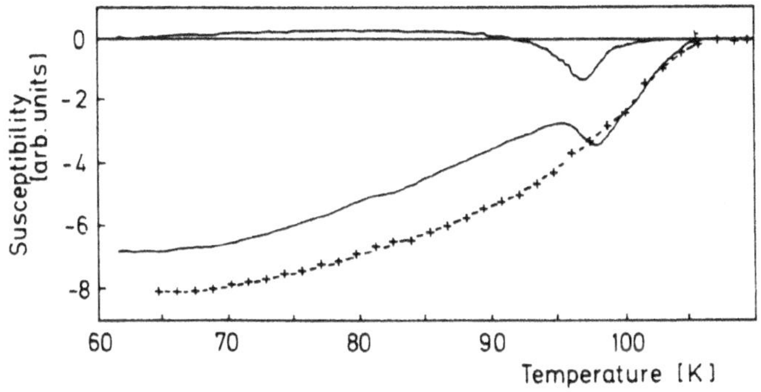

Fig. 3 Ac-susceptibility as a function of temperature for a $Tl_2Ba_2CaCu_2O_8$ film (thickness d = 12μm) at two different positions. H_{ac} = 10 Oe, f = 1049Hz, c-axis perpendicular to the film plane.
crosses: Measurement at position 2 (c. fig. 2): only the magnetization parallel to the film plane is measured. No imaginary part of the susceptibility is detected.
dashed line: Fit to the data assuming a slab-geometry and a temperature dependence of the penetration depth according to the the two fluid model.
solid line: Measurement at position 1 (c. fig. 2): an additional contribution from the component perpendicular to the film plane is observed leading to a peak in the imaginary part. The phase shift of 180° for this component can be explained by the 'wrong' winding sense of the pick-up coils. On position 3 this contribution shows the correct positive sign.

losses start to become observable within our resolution at a temperature $T_1 << T_c$ with $H_{c1}(T_1) = H_{ac}$, an estimate can be obtained for $H_{c1}(0)$. An extrapolation based on $H_{c1}(T) = H_{c1}(0)\cdot(1-(T/T_c)^2)$ leads to an upper limit $H_{c1}(0) < 10$ Oe for field parallel to the CuO_2-planes.

In the following, the results for a $Tl_2Ba_2CaCu_2O_8$ single crystal will be reported. Details about the crystal preparation can be found in [12]. The temperature dependence of the real and imaginary parts of the susceptibility are determined for different ac-fields and different orientations between H_{ac} and the c-axis of the crystal.

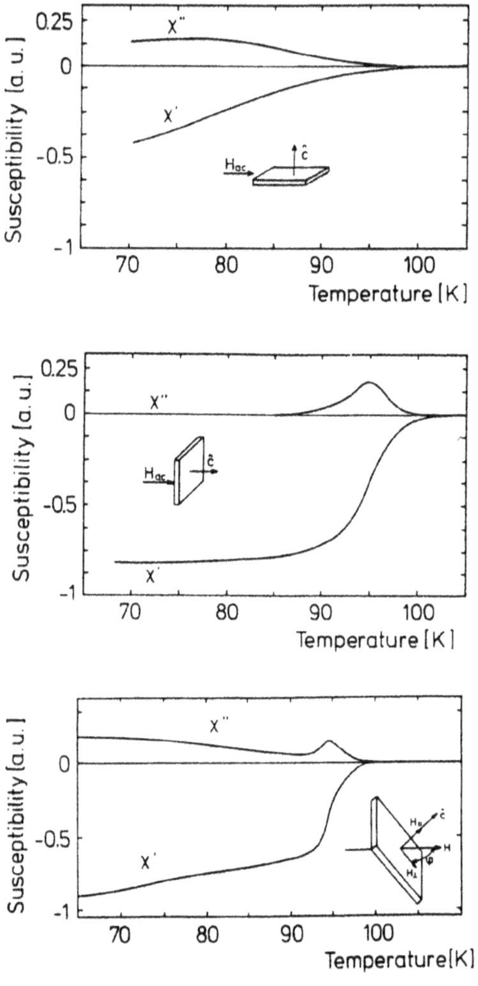

Fig. 4 Susceptibilities of a $Tl_2Ba_2CaCu_2O_8$ single crystal (0.08x0.5x0.6mm) obtained with different orientations between the external ac-field and the CuO_2-planes of the sample. H_{ac} = 10 Oe, f = 1049 Hz.

a) Field orientation parallel to the CuO_2-planes.

b) Field orientation perpendicular to the CuO_2-planes.

c) Field tilted by $\phi = 20°$ relative to the CuO_2-planes.

Rotating the sample by an angle of 90°, i.e. H_{ac} parallel to the c-axis of the crystal, leads to a different behavior (fig. 4b). The shielding transition due to superconductivity is relatively sharp and its field dependence is significantly smaller than for the direction H_{ac} perpendicular to the c-axis. The shift of the loss peaks by increasing fields H_{ac} can again be described by a power law, but with a power $a=1.5$ for this direction. An estimate for the lower critical field obtained in the same way as above leads to a value $H_{c1}(0) = 350$ Oe.

If the observed loss peaks shown in fig. 4a,b are interpreted in terms of a Bean-model (hysteretic losses), the different temperatures of the maxima have to be attributed to the different sample geometries for the z- and y-directions. On the other hand, a clear frequency dependence of the χ''-peak has been observed. A shift to higher temperatures of the order 2K per frequency decade is found in the present case (H_{ac} = 7 Oe, f = 31 Hz - 3.1 kHz). A more detailed study of the frequency dependence of $Tl_2Ba_2CaCu_2O_8$ is given by Nikolo[14]. Such a frequency-dependence can not be explained within a Bean-model, it rather suggests that flux creep/flow is the dominating loss process at least near the χ''-maximum. Within this interpretation, the results of fig. 4a,b indicate that the different components of a flux line become mobile at different temperatures. As a further test of this idea, the sample was tilted by 20° relative to the direction of the ac-field. Within the Bean-model one χ''-peak is expected somewhere between the positions shown in fig. 4a,b. The actual result is given in fig. 4c. Clearly, two peaks can be distinguished and the susceptibility data for the tilted case can be obtained by a superposition of the results found for H_{ac} parallel to the c-axis and H_{ac} perpendicular to the c-axis (quantitatively, a small shift of the χ''-peak to higher temperatures has to be taken into account due to the smaller field component parallel to the c-axis $H_{ac}(20°) = H_{ac}(90°)\cdot(\sin 20°)$. Thus, the results of fig. 4c are taken as further evidence that a Bean-model alone is not appropriate to describe the experimental ac-losses. If these losses are attributed to a viscous movement of the flux lines, one is led to the conclusion that in case of tilted lines their component parallel to the CuO_2-planes becomes mobile at significant lower temperatures than the component parallel to the c-axis. Within this picture, the χ''-peak at lower temperatures could be interpreted as a cutting of the flux lines or a possible entanglement transition, while the peak at higher temperatures corresponds to an overall movement of flux lines, i.e. to flux flow.

ACKNOWLEDGEMENT

We thank Dr. T. Zetterer for supplying us with $Tl_2Ba_2CaCu_2O_8$ crystals. This work was supported by Bundesministerium für Forschung und Technologie (BMFT) and Land Baden-Württemberg (MWK).

REFERENCES

[1] D.R. Nelson, Vortex Entanglement in High-T_c Superconductors, Phys. Rev. Lett. 60, 1973 (1988).
[2] D.R. Nelson and H.S. Seung, Theory of melted flux liquids, Phys. Rev. B 39, 9153 (1989).
[3] S.P. Obukhof and M. Rubinstein, Topological Glass Transition in Entangled Flux State, Phys. Rev. Lett. 65, 1279 (1990).
[4] E.H. Brandt, Thermal Fluctuation and Melting of the Vortex Lattice in Oxide Superconductors, Phys. Rev. Lett. 63, 1106 (1989).

[5] M.V. Feigel'man, V.B. Geshkenbein and A.I. Larkin, Pinning and Creep in Layered Superconductors, Physica C 167, 177 (1990).
[6] M.P.A. Fisher, Quantum Phase Transitions in Disordered Two-Dimensional Superconductors, Phys. Rev. Lett. 65, 923 (1990).
[7] D. Feinberg and C. Villard, Intrinsic Pinning and Lock-in Transition of Flux Lines in Layered Type-II Superconductors, Phys. Rev. Lett. 65, 919 (1990).
[8] P.H. Kes, J. Aarts, V.M. Vinokur and C.J. van der Beek, Dissipation in Highly Anisotropic Superconductors, Phys. Rev. Lett. 64, 1063 (1990).
[9] G. Ziebold and D. Korn, Apparatus for the measurement of the initial susceptibility of quench-condensed spin glass films, J. Phys. E: Sci. Instrum. 12, 490 (1979).
[10] Ch. Neumann, P. Ziemann, J. Geerk and H.C. Li, Characterization of YBaCuO Films by a.c.-Susceptibility Measurements, J. Less-Common Met. 151, 363 (1989).
[11] L. Lichti, K.C.B. Chan, D.W. Cooke and C. Boekema, Coupling strengths and flux pinning in oxide superconductors, Appl. Phys. Lett. 54, 2361 (1989).
[12] H.H. Otto, T. Zetterer and K.F. Renk, Preparation and structural investigation of normalconducting and high-T_c superconducting $(Tl,Pb)Sr_2(Ca,Tl)Cu_2O_{7-y}$, Z. Phys. B: Condensed Matter 75, 433 (1989).
[13] Ch. Heinzel, Ch. Neumann and P. Ziemann, Anisotropy of the Irreversible Magnetic Behaviour of YBaCuO and TlBaCaCuO Single Crystals - A Comparative a.c.-Susceptibility Study, Europhys. Lett. 13, 531 (1990).
[14] M. Nikolo, Design of a high-field susceptometer and measurements of thallium based single crystals, this volume

TEMPERATURE DEPENDENT PENETRATION DEPTHS FROM THE AC MAGNETIC RESPONSE OF THIN FILMS

A. T. Fiory and A. F. Hebard

AT&T Bell Laboratories, Murray Hill, NJ 07974

A method for determining superconducting penetration depths from measurements of a.c. impedances of superconducting films, using a two-coil mutual inductance technique, is reviewed. Results are presented for superconducting states of cuprate films, artificial multilayers, and organic crystals.

I. INTRODUCTION

A two-coil mutual inductance technique was initially used to study Kosterlitz-Thouless superfluid transitions and a.c. dynamics of vortices created by external d.c. magnetic fields in low-T_c superconductive-oxide films.[1] This present paper describes the experimental method and the analysis which is used to obtain complex sheet impedances from measurements of mutual inductances. Selected results for the penetration depth and kinetic inductance of high-T_c and low-T_c superconductors are presented. Also mentioned are some results obtained with a field-effect probe on a superconducting film, which is used to modulate the charge density in the superconducting state at the surface.

II. EXPERIMENTAL PROCEDURE

The experimental apparatus is illustrated in Fig. 1, which shows a film under test placed transverse to the axis of symmetry and located between the drive and receive coils. Not shown in the figure is the sample holder, which is a macor ceramic assembly in which the film and coils are mounted. An a.c. current of constant amplitude I_d (frequency $\omega/2\pi$ in the range 1 to 100 kHz) is applied to the drive coil and the receive-coil voltage amplitude V_r is

Magnetic Susceptibility of Superconductors and Other Spin Systems
Edited by R.A. Hein *et al.*, Plenum Press, New York, 1991

phase-sensitively-detected by a lock-in detector. The in-phase and quadrature components of V_r are used to compute the complex mutual inductance $M = V_r/i\omega I_d$. The a.c. magnetic field produced by the drive coil induces screening-current flow in the plane of the film. These currents attenuate the a.c. pick-up signal at the receive coil. The drive coil is wound astatically, as a quadrupole magnetic source, i.e., the two sections of the coil are connected in opposite polarity. This configuration localizes the a.c. current near the center of the film. A similar configura-

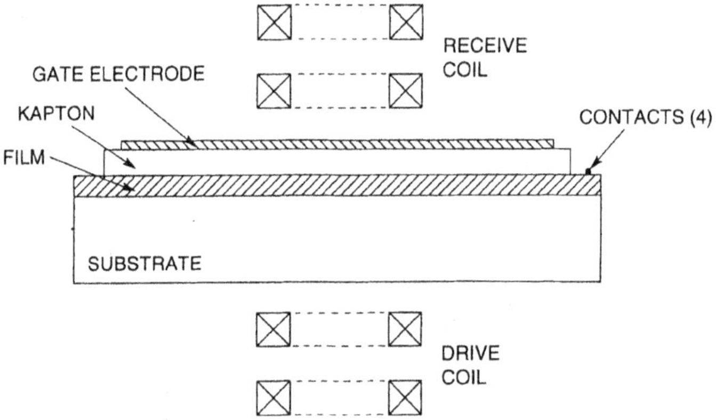

FIG. 1. Cross section of the apparatus for measuring the a.c. impedance of a film from the two-coil mutual inductance. Typical dimensions are: 1.5 mm coil diameter and 1 cm film width. Mounted on the film in the illustration is a gate electrode, separated by a Kapton dielectric, which is used to apply a static charge on the surface of the film in the field-effect measurements discussed in Section VI.

tion of the receive coil localizes the pickup sensitivity to the center of film and also serves to reject extraneous noise. The use of quadrupoles instead of dipole coils reduces sensitivity to the shape of the film, if its mean radius r_F is at least several times larger than the radius of the drive coil, r_d, or the receive coil, r_r. The sensitivity is maximized by placing the receive coil as close as possible to the film. The drive coil, shown facing the substrate, is further away from the film.

Calibration of the system, discussed below, assumes the film is a disk of radius r_F, thickness d_F, and complex resistivity $\rho(\omega)$. Films thinner than the penetration depth λ may be characterized in the superconducting state by a sheet kinetic inductance,

$$L_K = 4\pi\lambda^2/c^2 d_F \tag{1}$$

and a sheet impedance

$$Z = i\omega L_K \ . \tag{2}$$

Since all currents flow transverse to the axis of symmetry, the results do not depend on the axial component of the impedance, provided Z is uniform throughout the film, and thus λ pertains to the in-plane component of the penetration depth.

Calibration Analysis

A special case for the electromagnetic response of an infinite superconducting sheet was calculated by Clem.[2] In practice, his result is applicable for a large-r_F film in the quadrupole geometry of Fig. 1 and may be refined with a correction term for finite r_F. The mutual inductance for single-turn drive and receive coil loops and an infinitely-wide film is

$$M = 4\pi^2 r_d r_r \int_0^\infty dq\, e^{-q(D_d+D_r)} J_1(qr_d) J_1(qr_r) f_1 \ , \tag{3}$$

where D_d the distance from the drive loop to the near (lower) surface of the film, D_r the distance from the receive loop to the near (upper) surface of the film, J_1 the Bessel function, and

$$f_1 = \frac{\operatorname{sech} Qd_F}{[1+\frac{Q^2+q^2}{2qQ}\tanh Qd_F]} \ , \tag{4}$$

and where $Q^2 = q^2+\lambda^{-2}$.

For multiple-turn coils, Eq. (3) is readily extended as a double sum over integrals for the individual windings of the drive and receive coils. The result is independent of the measurement frequency for a film in the pure superconducting Meissner state, or for H close to H_{c1}, since $\omega \ll \Delta/\hbar$, where Δ is the superconducting gap. Near T_c, or when vortices are present, the effect of finite dissipation may be included by replacing λ^2 by $c^2\rho(\omega)/4\pi i\omega$ in the expression for Q, which makes M a complex quantity.[2,3] The effect of finite film width can be approximated by smoothly truncating the integral in Eq. (3) at small wavevector, q, with a Gaussian cut-off function, $f_c = \exp(-q^2/q_c^2)$, and by substituting the following f_1' for f_1 in Eq. (3):

$$f_1' = f_c + (1-f_c) f_1 \ . \tag{5}$$

The value of the cut-off parameter $q_c = 2/\pi r_F$ was determined from a numerical calculation.

A second, numerical method was used to compute the sheet current distribution for two-dimensional thin films.[4] This sheet-impedance model is accurate for film thicknesses satisfying $d_F < \delta(\omega)$, the a.c. skin depth of a resistive medium, or, when the electrodynamics is that of a superconducting medium, satisfying $d_F < 2^{1/2}\lambda$. Representing the film by a circle of radius r_F, which is divided into annular rings of width ε, the azimuthal components k_i of the total sheet current density in each ring are obtained as the solution to the following matrix equation:

$$k_i = \frac{i\omega}{cZ}(a_i + c^{-1}\sum_j G_{ij}k_j) \quad , \tag{6}$$

which expresses the constitutive relation between current density and magnetic-vector potential.[4] Here a_i is the azimuthal component of the external vector potential at r_i, produced by the drive coil, and

$$G_{ij} = c\varepsilon\hat{A}_{loop}(0, r_i, r_j) \ , \qquad i \neq j \ ,$$
$$G_{ii} = 2\varepsilon[(\varepsilon/4r_i)^2 + \ln(2\pi r_i/\varepsilon)] \ . \tag{7}$$

The function $\hat{A}_{loop}(z, r, \rho)$ is the azimuthal component of the vector potential produced by a unit current loop of radius ρ at a transverse distance r from the axis and at a longitudinal distance z along the axis.

$$\hat{A}_{loop}(z, r, \rho) = \frac{4\rho}{ck^2(\rho^2+r^2+z^2+2\rho r)^{1/2}}[(2-k^2)K(k) - 2E(k)] \ , \tag{8}$$

where $k^2 = 4\rho r(\rho^2+r^2+z^2+2\rho r)^{-1}$, and $K(k)$ and $E(k)$ are the complete elliptic integrals. The azimuthal component of the vector potential produced by the drive coil is given by a sum over the turns of the drive coil:

$$a_i = I_d\sum_d p_d\hat{A}_{loop}(z_{dF}, r_i, r_d) \quad , \tag{9}$$

where z_{dF} is the axial distance between the plane of the film and a turn in the drive-coil of polarity p_d and radius r_d. For a given complex impedance Z, Eq. (6) is solved for the complex current distribution, k_i, which is then used to compute the complex mutual inductance:

$$M = M_0 + 2\pi c^{-1} I_d^{-1}\varepsilon\sum_i p_r k_i \hat{A}_{loop}(z_{rF}, r_r, r_i) \quad , \tag{10}$$

where p_r is the polarity of receive-coil turn of radius r_r located an axial distance z_{rF} from the plane of the film. The term for the empty-cell mutual inductance, M_0, is given by

$$M_0 = 2\pi c^{-1} \sum_r \sum_d p_r p_d r_r \hat{A}_{loop}(z_{dr}, r_r, r_d) \quad , \tag{11}$$

where z_{dr} are axial distances between drive and receive coil turns.

Either an analytic or numerical method, as appropriate, can be used to compute the mutual inductance $M = M_I + iM_Q$ for a given complex film impedance $Z = R + i\omega L$. However, for data analysis, one needs to adopt a procedure for inverting the calculated functional dependence $M(Z)$ to obtain Z as a function of M. This is accomplished numeri-

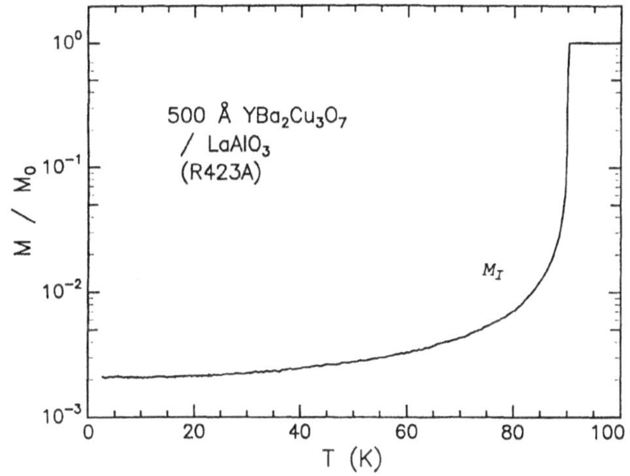

FIG. 2. Temperature dependence of the ratio of the in-phase component of a.c. mutual inductance and the empty-cell mutual inductance (M_0 = 250 nH).

cally, by first computing a table of M values for a suitable mesh of Z values spanning the expected data range. For superconductors, where the results for M fall in the first quadrant of the complex plane, the inversion can be made single-valued by (1) introducing dimensionless reduced mutual inductance parameters,

$$m_1 = (M_0 - M_I)/(M_0 - M_s) \quad ,$$
$$m_2 = M_Q/(M_0 - M_s) \quad , \tag{12}$$

where M_s is the mutual inductance computed in the limit $Z = 0$, and (2) making the transformation

$$u_1 = m_1(m_1^2 + m_2^2)^{-1} - 1 \quad ,$$
$$u_2 = m_2(m_1^2 + m_2^2)^{-1} \quad . \tag{13}$$

Once tables for Z and (u_1, u_2) have been computed for a given film and coil geometry, the mutual inductance data for the film can be analyzed by first computing (u_1, u_2) from the measured M and then interpolating the two tables with a lookup algorithm on a computer to find Z.

III. MUTUAL INDUCTANCE MEASUREMENT

A normalized in-phase component of the mutual inductance for a 500Å $YBa_2Cu_3O_7$ film is shown on a logarithmic scale as a function of temperature in Fig. 2. The film was prepared by vacuum evaporation of a BaF_2-Y-Cu-O precursor onto a polished (001)-$LaAlO_3$ substrate and then oxygen

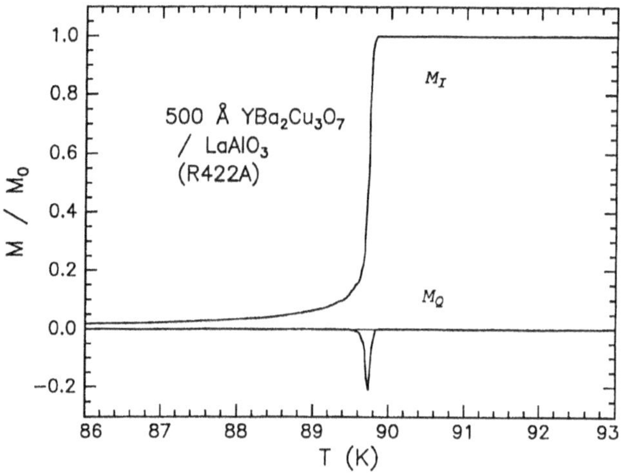

FIG. 3. Normalized mutual inductance components near T_c. The upper curve is the in-phase component and the lower curve is the quadrature component. The external d.c. magnetic field perpendicular to the plane of the film was nulled to less 5 mG; the data were taken at a frequency of 13 kHz with an excitation amplitude in the linear-response region, about 0.1 mG.

annealed using the wet/dry-O_2 process of Mankiewich et al.,[5] which produces an epitaxial film with c-axis oriented perpendicular to the plane of the film. One observes that the mutual inductance drops by a factor of 500 at low temperature. Figure 3 shows the temperature dependence of the in-phase and quadrature components of the mutual inductance in the vicinity of the transition temperature for another 500Å film. The transition width, given as the range over which dissipation is observable in the quadrature component of M, is about 0.2 K. The dissipation peak is caused by vortex-pair excitations, which appear in films of finite thickness near the Kosterlitz-Thouless transition at T_{KT}, a temperature about 1 K below the mean-field or bulk T_c of $YBa_2Cu_3O_7$.[6] Although the transition appears to be very sharp, inhomogeneity in the film also makes a significant contribution to the temperature width of the observed dissipation peak.[7]

Mutual Inductance Calibration

The normalization parameter M_0 used in Figs. 2 and 3 was measured separately as a function of temperature, with the film removed from the sample holder. This procedure corrects for the influences of thermal expansion and temperature-dependent coil resistance on the amplitude and phase gain of the detection system. Since M/M_0 drops to less than 0.01 at low temperature, it is also necessary to accurately determine the experimental baseline, i.e., the signal which corresponds to $Z \to 0$. This was accomplished in the present case by replacing the film with a sheet of Nb foil, 0.025-cm thick, for which the sheet impedance at 4 K is negligible compared to that of the film under test. The baseline measured with the Nb foil in place is small, but not zero, owing to extraneous coupling between the drive and receive circuits. Such an independent measurement of the baseline signal improves the accuracy by which $L(T \to 0)$ and thus $\lambda(0)$ are obtained.

The mutual inductance calibration methods described in the previous section were checked by using test squares of normal conductors – a Cu foil 127 μm thick and an Al foil 7.6 μm thick – where the thicknesses are less than the a.c. skin depths. The resistivities were determined from the 2-coil method as analyzed with the sheet-impedance model for a radius $r_F = (A/\pi)^{1/2}$, where A is the area of the foil. Results agreed from 1 to 5% with 4-probe d.c. resistivity measurements. Effects of finite thickness and electrical anisotropy were also tested using stacks of n electrically-isolated 7.6-μm Al sheets, for n = 1, 2, 3, 5, 8, or 80. These latter tests confirmed that the impedance sheet model is accurate to better than 10% for $d_F < \delta(\omega)$, and that the results are independent of the effective axial component of the resistivity.

V. PENETRATION DEPTH OF $YBa_2Cu_3O_7$

If the temperature-dependent penetration depth of high-T_c superconductors can be measured sufficiently accurately, then it should be possible to distinguish among competing models for the pairing state.[8,9] The superconductor $YBa_2Cu_3O_7$ is probably best suited for such a study since it has modest penetration-depth anisotropy, $\lambda_a \approx 1.2\lambda_b \approx \lambda_c/5$, and films prepared by a variety of methods seem to be of good epitaxial quality. Strongly anisotropic superconductors exhibit pronounced fluctuation effects over a broad temperature range, owing to weaker coupling between cuprate layers. While this is interesting in itself, it complicates separating out the intrinsic penetration depth. However, one caveat to keep in mind is that epitaxial films generally contain a large number of structural defects, mainly stacking faults and microtwins, which force some local flow of supercurrent along the c axis. Since λ_c is determined by interlayer Josephson coupling, and the c-axis conductivity varies approximately linear in T in the normal state, it is possible that a nominal measurement of an average over λ_a and λ_b may also include some temperature-dependent admixture of λ_c.

Figure 4 shows results for the temperature dependence of the inverse kinetic inductance obtained with the two-coil method on a 500Å $YBa_2Cu_3O_7$ film. Plotting the reciprocal in this manner emphasizes the low-temperature region. Similar

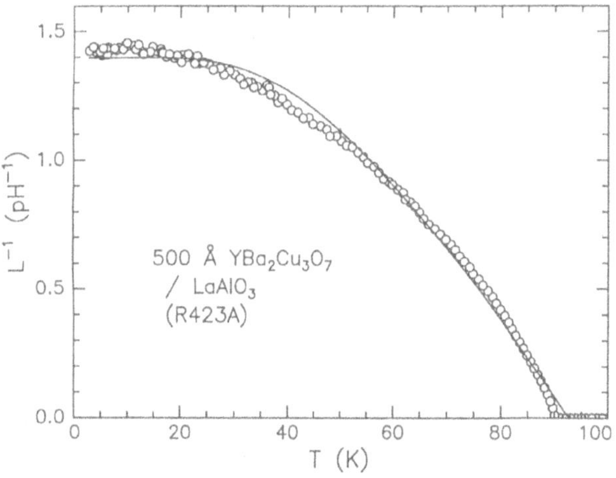

FIG. 4. Reciprocal of the sheet inductance of a film, computed from the mutual inductance data of Fig. 2, using the numerical calibration method described in Sect. II. The curve is a three-parameter fit combining weak-coupling BCS and two-fluid models.

results have been obtained for a number of films of about the same thickness grown on $SrTiO_3$ and $LaAlO_3$ substrates. The solid curve was obtained by a least-squares fit to $L^{-1} = c^2 d_F / 4\pi\lambda^2$, using two terms for λ^{-2}: (1) the weak-coupling BCS theory, and (2) the Gorter-Casimir two-fluid expression, which approximates strong-coupling BCS. The transition temperature was also taken to be an adjustable parameter. The resulting fit apportioned weak-coupling BCS at 94%, two-fluid at 6%, and yielded T_c = 92.6K. This exercise is meant to illustrate that the actual temperature dependence departs significantly from either simple model. The inadequacy of either model to accurately fit the data is given further emphasis in Fig. 5, which shows inverse-square penetration depth data and separate curves for the two models computed for $\lambda(0) = 0.167\,\mu m$ and $T_c = 90\,K$. Generally, what is found is that the onset of temperature dependence occurs at a lower temperature than calculated by weak-coupling BCS, and the curvature near T_c is closer to the behavior of the

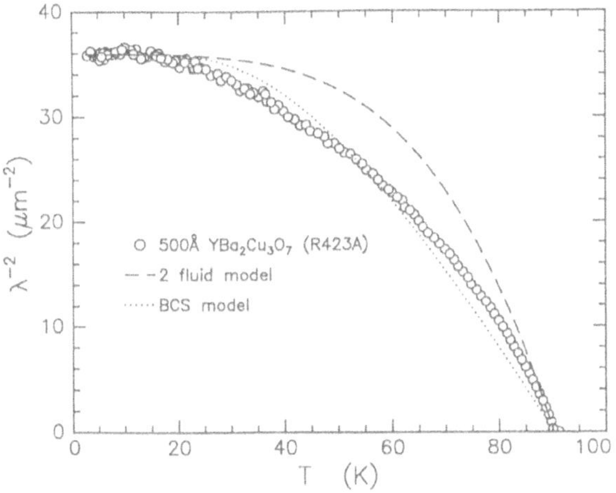

FIG. 5. Inverse-square of the in-plane penetration depth (λ_{ab}^{-2}) computed using the relationship to the kinetic inductance, Eq. (1), and the data of Fig. 4. Dotted and dashed curves illustrate the temperature dependence of the weak-coupling BCS and two-fluid models, respectively, fitted to the data at the lowest temperature ($\lambda(0) = 0.167\mu m$) and taking T_c = 90 K.

two-fluid model. These data show that λ^{-2} approaches a constant at low temperature faster than T^2, which is consistent with s-wave paring. Some films, particularly if they were annealed at high temperature, show a T^2 dependence, which probably indicates defects.[10] This issue is discussed in more detail in Section VIII and Ref. 10.

Figure 6 shows the components of $i\omega Z^{-1}$ for a 150Å film grown on $NdGaO_3$. Although the resistance transition onset is near 90 K for this film, the magnetic superconducting transition is close to 60 K. The intersection of the dotted line with the curve for $Re\{i\omega Z^{-1}\} = L^{-1}$ indicates the location of the Kosterlitz-Thouless temperature[11]

$$T_{KT} = \frac{\Phi_0^2 d_F}{32\pi^2 k_B \lambda^2 (T_{KT})} \quad , \tag{14}$$

which in terms of the kinetic inductance is given by $T_{KT} = 0.98 \cdot 4\pi L_K^{-1}(T_{KT})$. A percolation model has also been used to characterize films with depressed transitions measured magnetically.[12] Intersection with the dashed line indicates the weak-field vortex-lattice melting temperature in two

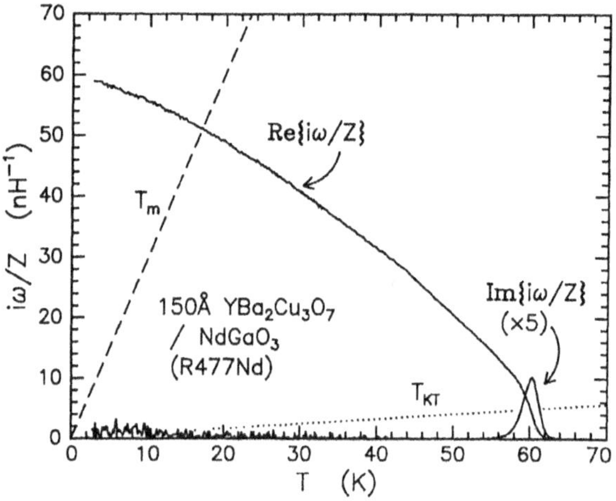

FIG. 6. Temperature dependence of inverse impedance components for a thin film with a depressed transition. Intersection of the real-component curve with the dotted line gives the Kosterlitz-Thouless transition temperature, Eq. (14), and with the dashed curve the weak-field vortex-lattice melting temperature, Eq. (15).

dimensions, as determined from a Brownian dynamics simulation,[13]

$$T_m \approx 0.33 L_K^{-1}(T_m) \quad . \tag{15}$$

Note that the temperature dependence of L^{-1} at low temperatures is significantly more pronounced than for films in which T_c is close to 90 K.

VI. ELECTROSTATIC MODULATION EFFECT

The essential parts of the field-effect probe are shown in Fig. 1. A capacitor is made with the film under test as one electrode and a 100-Å Au film, itself producing negligible screening effect, acting as a gate or counter electrode. The gate is insulated from the superconducting film by a 7-μm Kapton foil. Contacts on the film are used to measure its resistance and also to connect the film to the external charging circuit. The static charge applied to the film is measured with an electrometer.

Electrostatically charging the surface of the film changes the measured mutual inductance, from which the change in the film sheet impedance can be computed. The sign of the modulation, with the impedance decreasing for a positive charge on the film, is that expected for a super-

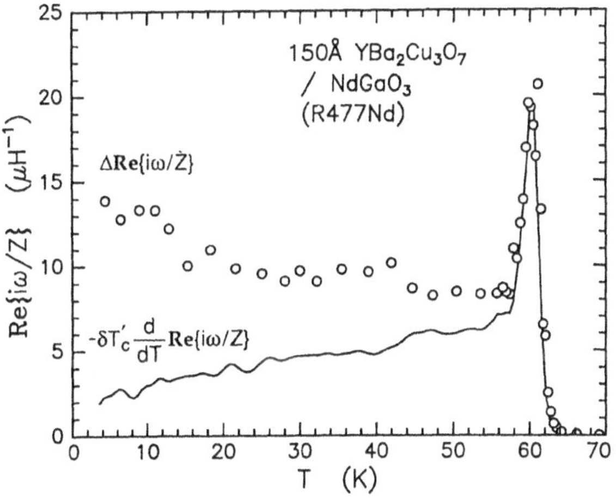

FIG. 7. Electrostatic field-effect modulation of the real component of inverse sheet impedance (points) and the scaled temperature derivative (curve, computed from the data of Fig. 6). The modulation in transition temperature, $\delta T_c' = 5.2$ mK, was chosen to provide overlap in the peaks observed near the transition temperature.

conducting condensate of paired holes.[14] Figure 7 shows the effect on the real component of $i\omega Z^{-1}$ for an applied charge density of 1 mC/m^2. In the region near T_c the modulation in both the in-phase and quadrature components of the mutual inductance scale by a factor $\delta T_c' = 5.2$ mK with their respective temperature derivatives, indicating a modulation of the magnetically-determined transition temperature. This is illustrated in Fig. 7, where the scaled temperature derivative, $-\delta T_c' \cdot d\mathrm{Re}\{i\omega Z^{-1}\}/dT$, is plotted along with $\Delta\mathrm{Re}\{i\omega Z^{-1}\}$. In the limit of zero temperature, one may consider the relation $i\omega Z^{-1} = L_K^{-1} = Ne^2/m^*$, where N is areal carrier density, and m^* the effective carrier mass. The charge modu-

TABLE I. Basal-plane transport parameters for $YBa_2Cu_3O_7$ films. Headings give film identification number, nominal thickness d_F, resistivity ρ, inductive-transition onset T_c, surface mobility μ_s (±5%), carrier concentration n (±5%), effective mass m* (±25%), and penetration depth λ (±10%). μ_s and n are from resistance field effect at 295 K, m* from kinetic inductance field effect at 4 K, λ from kinetic inductance at 4 K.

Film[a]	d_F (Å)	T_c (K)	ρ(295) (μΩcm)	μ_s(295K) ($cm^2V^{-1}s^{-1}$)	n ($10^{21}cm^{-3}$)	m^*/m_e	λ(4K) (μm)
R449A	150	78					0.40
R477Nd	150	65	780	3.0	2.8	12	0.38
R443A	300	89.9	330	4.6	4.1	5.0	0.20
R443Aa	300	89.8	415	3.5	4.3		
R422A	500	90.1	270	4.3	5.4	4.0	0.18
R351A	500	91.4	194	4.6	7.0	4.7	0.18
R423A	500	90.2	270	4.6	5.1	4.6	0.17
R476Nd	500	90.2					0.21
R209S	1000	90.2	418	4.1	3.6	5.5	0.25
R442A	1000	90.1	264	4.4	6.7	~6	0.25
R474A	1000	90.1	274				0.24

[a] Suffixes denote substrate: A = $LaAlO_3$, S = $SrTiO_3$, Nd = $NdGaO_3$.

lation effect on the impedance can then be expressed as $\Delta\mathrm{Re}\{i\omega Z^{-1}\}/\Delta Ne = e/m^*$, with ΔNe being the experimentally applied areal charge density. The result for the 150 Å film is an effective mass $12m_e$ at the surface of the film. Films with T_c near 90 K yield effective masses of $5m_e$.[14] This comparison shows that disorder leads to an enhancement in the effective mass.

In the normal state, the field-effect probe is used to measure the modulation in the sheet resistance of the film and to determine the transport mobility. Table I summarizes the results on various films. By comparing the modulation in the kinetic inductance below T_c with that of the resistance above T_c, one infers that all of the hole carriers in the normal state participate in the superconducting condensate at low temperature.[14]

VII. SUPERLATTICES

Artificial superlattices, prepared by successive deposition of superconducting and insulating layers, provide a means for studying purely magnetic interlayer coupling, where Josephson coupling is effectively eliminated. In zero external magnetic field, the relevant region to examine is near the Kosterlitz-Thouless transition temperature. Figure 8 shows the temperature dependence of the inverse impedance, $i\omega/Z$, measured in a 16-period MoGe multilayer separated by insulating Ge layers.[15] If this system behaves as independently fluctuating two-dimensional sheets, then T_{KT} would be near the intersection of the curve for the real component, $\mathrm{Re}\{i\omega/Z\}$, and the dotted line shown in the figure, which

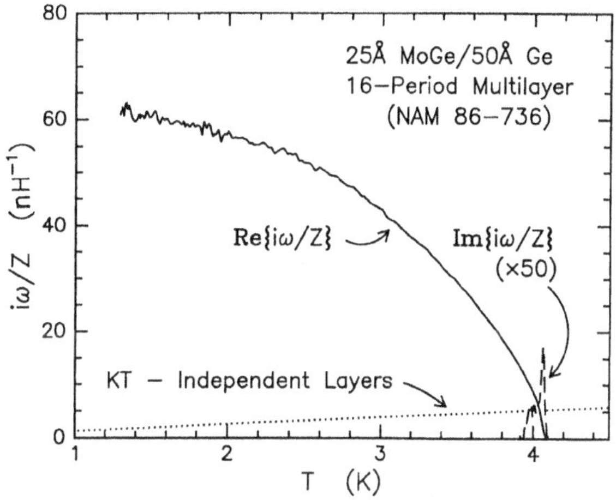

FIG. 8. Temperature dependence of inverse impedance components for a 16-period superlattice film. Intersection of the dotted line with the real component is the point expected for a Kosterlitz-Thouless transition in an individual layer, independent of interlayer interaction.

corresponds to $L^{-1} = 5.3\mathrm{nH}^{-1}$. However, if the entire superlattice fluctuates as a single entity, then the intersection point would be at a factor 16 lower, at $0.33\mathrm{nH}^{-1}$. The transition region indicated by the data, which is the interval spanned by the dissipation peak, overlaps the point predicted for independently fluctuating layers. This is in agreement with theory, in that the thermally-excited vortex pairs giving rise to the observed dissipation near T_{KT} have negligible magnetic interaction because their separation is small compared to the penetration depth.

VIII. MAGNETIC FIELD EFFECTS

The dissipation produced by small stray magnetic fields and the a.c. applied field is generally too small to be observable over most of the temperature range below T_c. On the other hand, if a sufficiently large d.c. magnetic field is applied, the vibration of the vortices in the a.c. field gives rise to dissipation and an increase in inductance. The response depends upon the nature of the intervortex interaction and the distribution of pinning sites.[3] However, for fluxon densities which are low compared to the density of strong pinning sites, the response at low frequency is the sum of a term proportional to the square of the Campbell pinning penetration depth λ_p and the intrinsic kinetic inductance.[16] This model was verified for $YBa_2Cu_3O_7$ films,[16] which exhibit strong core pinning and a pinning penetration depth given by

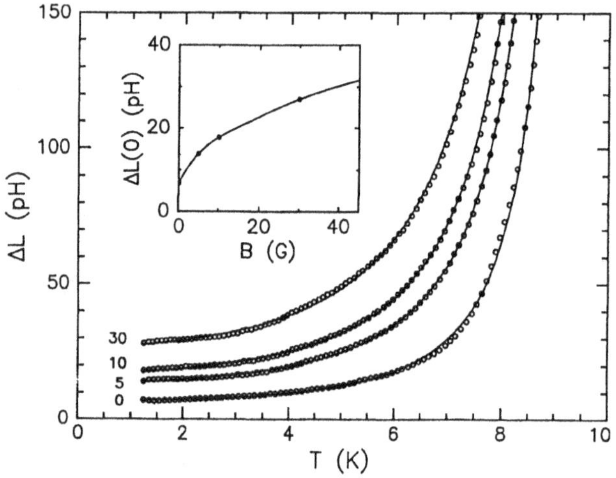

FIG. 9. Temperature and magnetic field dependence of the excess sheet inductance in the high-conductance b-c plane of a $(BEDT\text{-}TTF)_2Cu(SCN)_2$ crystal. The curves are fits to the pinning-defect model, Eq. (17). The coefficients $\Delta L(0)$, plotted in the inset against magnetic field, were determined as fitting parameters.

$$\lambda_p = (N_\phi/\pi)^{1/2}\phi_0/H_c \quad , \tag{16}$$

where N_ϕ is the vortex density and H_c is the critical field. When λ_p^2 becomes larger than λ^2 at high flux density, the temperature dependence of the kinetic inductance crosses over to that of H_c^{-2}, which is $[1-(T/T_c)^2]^{-2}$. Interestingly, this dependence, which is T^2 in leading order at low temperature, has also been observed at *zero* magnetic field in a number of materials.[10] It suggests the existence of numerous internal defects sites where vortex cores may spontaneously nucleate, e.g., in pairs at the ends of cracks within the material.

Examples of both effects are shown in Fig. 9, which plots the change in the inductance of a $(BEDT{-}TTF)_2Cu(SCN)_2$ crystal, an organic superconductor in the form of a platelet 50 μm thick and 1 to 2 mm across, with $T_c = 9.5\,K$. The data were taken after initially cooling the crystal in the constant d.c. B fields shown. The intrinsic penetration depth is less than 1 μm, so the kinetic inductance is much smaller than the observed inductance. However, the inductance increases rapidly with temperature and magnetic field. The curves through the points in Fig.9 are the function,

$$\Delta L = \Delta L(0)[1-(T/T_c)^2]^{-2} \quad , \tag{17}$$

which fits the temperature dependence of the pinning penetration-depth model, $\Delta L = 4\pi\lambda_p^2/c^2 d_F$. The variation with B suggests a weak-link type of internal structure with defects approximately 2 μm apart.

CONCLUSIONS

The two-coil a.c. mutual inductance method described in this paper provides sensitive measurements of sheet impedances in the superconducting state of thin films. In zero or very weak d.c. magnetic fields, the impedance is inductive, apart from a dissipative region near the phase transition, allowing the in-plane penetration depth to be computed from the kinetic inductance to within about 1 K of T_c. The temperature dependence of the penetration depth in the best films can be fit moderately well to weak-coupling BCS, but significant systematic deviations are observed which are suspected to be due to extrinsic defects. Electrostatic charge modulation of the surface of a film causes a shift in the transition temperature and can be used to determine the Cooper-pair effective mass in the surface layer. In electrically-isolated superlattices, the phase transition is found to be dominated by vortex-pair fluctuations within the individual layers. Oscillation of pinned vortices and defects in the material are several mechanisms which enhance the effective penetration depth over its intrinsic value, and yield a T^2 leading term in the low-T temperature dependence.

Acknowledgments

The authors wish to express their thanks to numerous colleagues for invaluable contributions to experiments reviewed in this paper and to J. R. Clem, for providing

Ref.2. R. P. Minnich, W. I. Glaberson, R. H. Eick collaborated on the experimental apparatus; P. M. Mankiewich, M. L. O'Malley, R. E. Howard, J. R. Kwo, M. Hong, C. E. Rice, A. F. J. Levi, M. Anzlowar, K. I. Baldwin, P. L. Gammel, T. Siegrist, E. Coleman, J. M. Phillips, M. P. Siegal, L. F. Schneemeyer, and J. V. Waszczak collaborated in the work on high-T_c cuprates and provided numerous films and crystals; N. A. Missert and M. R. Beasley collaborated in the work on superlattice films; R. C. Haddon collaborated in the work on organic superconductors.

REFERENCES

1. A. T. Fiory, A. F. Hebard, and W. I. Glaberson, Phys. Rev. B 28:5075 (1983).
2. J. R. Clem, "Mutual Inductance between a Driver Coil and a Pickup Coil with a Thin-Film Type-II Superconductor Containing Vortices of Flux Density B_o in Between", *unpublished memorandum, part VIII* (2 April 1981).
3. Theoretical forms taken by the complex resistivity $\rho(\omega)$ or film impedance $Z(\omega)$ for given models of vortex dynamics have been presented by J. R. Clem and by E. H. Brandt (to be published).
4. A. T. Fiory, A. F. Hebard, P. M. Mankiewich, and R. E. Howard, Appl. Phys. Lett. 52:2165 (1988). The authors wish to thank M. Karkut for noting the misprints in this reference, which in the present text have been corrected. An alternate two-coil method, with both drive and receive coils on the same side of the film, was presented by B. Jeanneret, J. L. Gavilano, G. A. Racine, Ch. Leemann, and P. Martinoli, Appl. Phys. Lett. 55:2336 (1989).
5. P. M. Mankiewich, J. H. Scofield, W. J. Skocpol, R. E. Howard, A. H. Dayem, and E. Good, Appl. Phys. Lett. 51:1753 (1987).
6. A. T. Fiory, A. F. Hebard, P. M. Mankiewich, and R. E. Howard, Phys. Rev. Lett. 61:1419 (1988).
7. L. C. Davis, M. R. Beasley, and D. J. Scalopino, Phys. Rev. B 42:99 (1990).
8. J. Rammer, Europhys. Lett. 5:77 (1988).
9. J. Annett, N. Goldenfeld, and S. R. Renn, Phys. Rev. 43:2278 (1991).
10. A. F. Hebard, A. T. Fiory, M. P. Siegal, J. M. Phillips, and R. C. Haddon, Phys. Rev. Lett. (*submitted*).
11. B. I. Halperin and D. R. Nelson, J. Low Temp. Phys. 36:599 (1979).

12. Ch. Leemann, Ph. Flückiger, V. Marsico, J. L. Gavilano, P. K. Srivastava, Ph. Lerch, and P. Martinoli, Phys. Rev. Lett. 64:3082 (1990).

13. A. T. Fiory, Phys. Rev. B 28:236 (1983).

14. A. T. Fiory, A. F. Hebard, R. H. Eick, P. M. Mankiewich, R. E. Howard, and M. L. O'Malley, Phys. Rev. Lett. 65:3441 (1990); *Erratum: ibid.* 66:845 (1991).

15. N. A. Missert, PhD. Thesis, Stanford University (1989).

16. A. F. Hebard, P. L. Gammel, C. E. Rice, and A. F. J. Levi, Phys. Rev. B 40:5243 (1989).

MAGNETIC DETECTION OF OPTICAL EXCITATIONS IN HTSC THIN FILMS BY a.c. SUSCEPTI BILITY MEASUREMENTS

C. Giovannella, A. Fontana, P. Cikmach*

Dip. di Fisica - Universita' di Tor Vergata
Via E. Carnevale, s.n.
I - 00173 Roma

INTRODUCTION

The recent interest in the optical response of High T_c superconductor thin films (HTSC) has been motivated by the hope of obtaining broad band detectors working at liquid nitrogen temperatures. A summary of previous research on this subject can be found in ref. 1, 2.

Beside the technological interest, experimental findings reported up to now, leave open fundamental questions as to the nature of the response of the HTCS thin films to an optical excitation. Is it possible to observe a non-bolometric response? If it is, what is the physical mechanism causing the non-bolometric response?

Non-bolometric effects on the transport characteristics of low T_c superconducting thin films (LTCS) Josephson Junctions (or LTCS thin films) have been observed[3] and ascribed to the breaking of Cooper pairs caused by the incident photons, and to the consequent creation of a non-equilibrium density of quasiparticles; that is quasiparticles that relax with a characteristic time of the order of one nanosecond. As far as the HTCS materials are concerned, possible non-bolometric responses have been suggested in order to explain the following three experimental observations:

a) a fast component of the optical response whose relaxation time is of the order of one nanosecond, shown by time resolved transport measurements under a pulsed optical excitation;[4]

b) an increase of the photoresponse intensity with decreasing temperature, T, for T lower than 50 °K;[5,6]

c) an increasing value of the ratio between the photoresponse intensity (PR) and the derivative of the resistive curve, dR/dT, in the proximity of T_{coff}, the offset temperature of the resistive transition, i.e. that temperature for which the detected resistivity becomes lower than the instrumental resolution (a bolometric response would give a constant value down to T_{coff}).[7]

---

*Permanent Address: Solid State Physics Institute, University of Latvia, 8 Kengaraga Str., 226063 Riga, Latvia

The former two have been ascribed by some authors to the possible generation of a non-equilibrium density of quasiparticles; an interpretation that has been questioned by some other authors.[8]

On the other hand, the observation of a non costant value of the PR/(dR/dT) ratio has been attributed to a photoinduced phase slippage,[9] or to a photoenhanced flux creep.[10] Since in samples containing a large number of natural junctions (like the HTCS thin films) there cannot exist flux creep without phase slippages, in our opinion one has to consider the two models in some respects equivalent, and distinguish between phase slippages occuring in the presence of a magnetic field gradient (creep), and those occuring at thermal equilibrium (fluctuactions).

Very recently the problem of the fast component of the optical response seems to have provided a convincing, although partial, solution. Indeed, by means of time resolved reflectivity measurements,[11] and absorption measurements,[12] it has been shown that non-bolometric effects due to a non-equilibrium quasiparticle density can be observed only for times shorter than 2-5 ps. These observations imply:

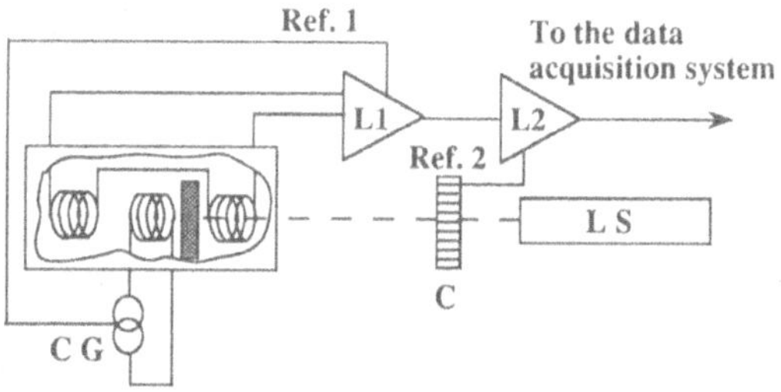

Fig.1 Experimental setup.CG: Current Generator, LS: light source, C: chopper, L1 and L2: Lock-in amplifier;

a) that the same effects can hardly be detected by transport measurements.

b) That the transient response studied by resistive measurements is likely to be bolometric. As an alternative explanation, relaxation effects due to a rearrangement of a non-equilibrium distribution of magnetic field could be invoked. Their characteristic time scales however, are closer to the micro and milliseconds, rather than to nanoseconds.

The purpose of this paper is to show how magnetic measurements can contribute to the above discussion. In particular, we will show that, by using an ac susceptibility probe, it is possible to perform magnetic detection of optical excitations (MDOE).

EXPERIMENTAL

The experimental set up as well as the samples used in our experiments have been described in detail elsewhere.[13,14,15,17] However, we will recall briefly their main characteristics. The experimental set up was put in a flux-flow cryostat and is schematically shown in fig.1. The ac susceptibility signal is detected by a flux exclusion probe utilizing a lock in amplifier, L1. The reference signal to L1 (usually at 5 kHz) is supplied by the current generator, CG. The light source, a 200 W halogen lamp was mechanically chopped at 33 Hz, supplied, at the sample surface, a power density of about $1W/cm^2$ The variation induced by the light beam on the χ' and χ'' signals, namely $\Delta\chi'$ and $\Delta\chi''$, are detected by means of the the lock-in amplifier L2. In the configuration of fig. 1 the apparatus allows for the simultaneous recording of χ', χ'' and $\Delta\chi'$, or $\Delta\chi''$.

The YBCO samples employed in the experiments are 3000-4000 Å thick films with the a-axis oriented ⊥ to the substrate. All the films show an initial T_{coff} ranging between 86 and 89 °K that degrades with the time, in some cases in few days in others

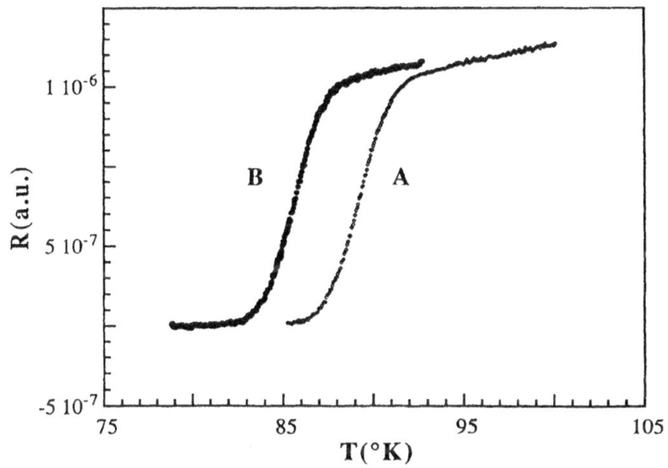

Fig.2 A: Resistivity curve of sample 50.1. B: Resistivity curve of sample 50.1 under illumination

after several months, depending on the preparation protocol and the rate of moisture-exposure. The stability of T_{coff} during the present experiments has been carefully checked. More information on the structural and electrical characteristics of the samples employed can be found in the references 14,15.

The resistivity curve of the sample #50.1 taken before (and checked to be stable after) the MDOE experiments is shown in fig. 2 as curve A. Curve B was recorded with the light on to estimate the thermal shift due to the optical excitation, which is about 3-4 °K. This thermal shift has to be compared with that induced by a pulsed laser beam, ~ 20-50 °K,[8] during the time resolved transport experiments, or the 10-100 mK induced by the mechanically chopped HeNe laser beam, usually used during steady state or low frequency transport measurements.[16]

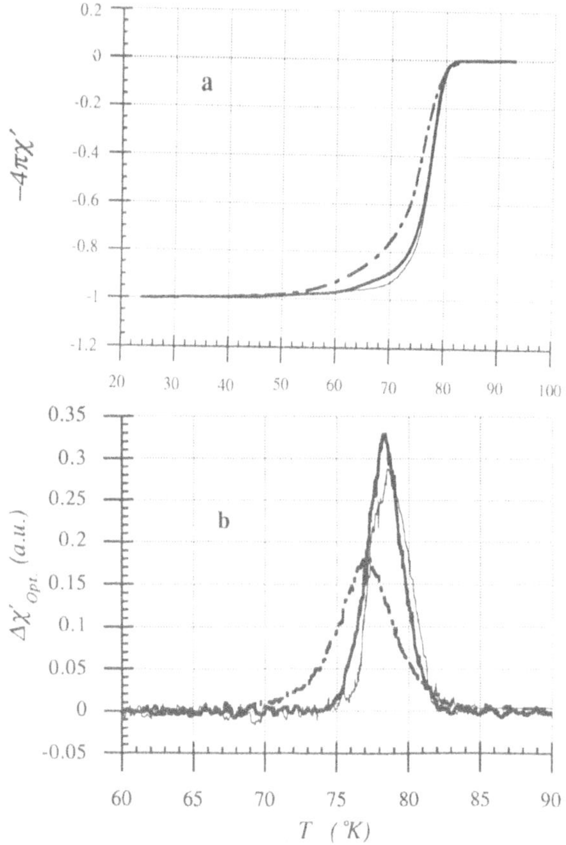

Fig 3 a) χ' vs T; b) $\Delta\chi'$ vs T; for three different values of the a.c. magnetic probing field:
0.4 Oersted (line)
2.2 Oersted (heavy line)
11.2 Oersted (dot-dash line)

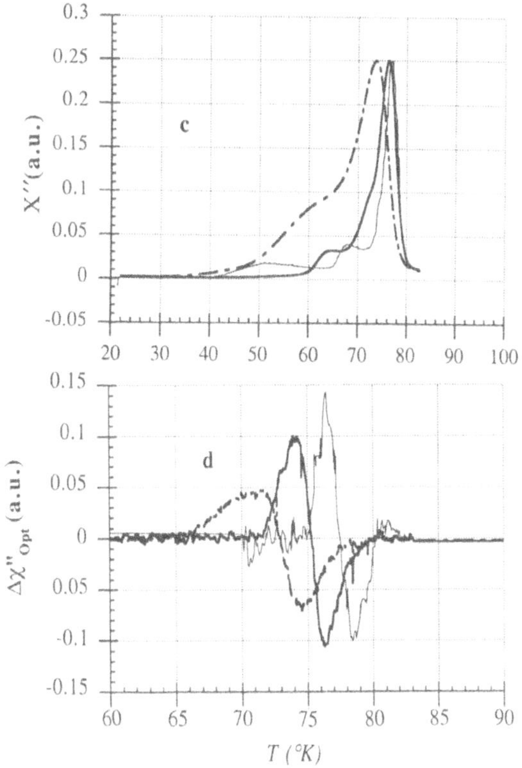

Fig 3 c) χ" vs T; d) Δχ" vs. T for three different values of the a.c. magnetic probing field:
0.4 Oersted (line)
2.2 Oersted (heavy line)
11.2 Oersted (dot-dash line)

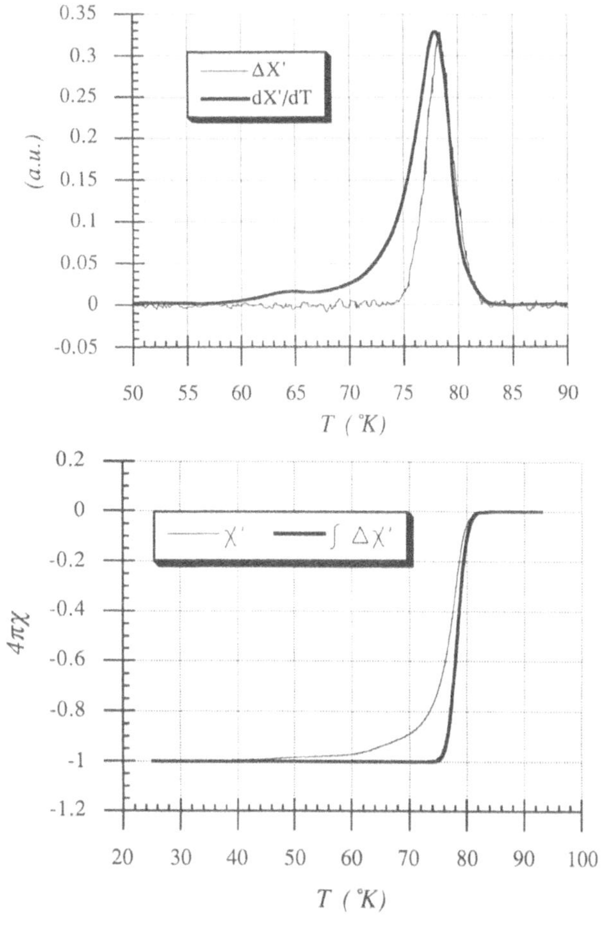

Fig. 4 Upper figure: $\Delta\chi'$ and $d\chi'/dT$ vs. T;
Lower figure: $\int\Delta\chi'$ dT and χ' vs. T.

RESULTS AND DISCUSSION

MDOE signals in the proximity of T_{co}

In fig. 3 we show a series of typical χ', χ'', $\Delta\chi'$ and $\Delta\chi''$ curves recorded for three different values of the ac magnetic probing field.

We point out that $\Delta\chi'$ and $\Delta\chi''$ become detectable only for temperatures lower than T_{coff}. This illustrates one of the main advantages of using magnetic techniques. Transport measurements are sensitive only to the percolative path with the higher critical current (all the other being shunted) while susceptibility measurements test the

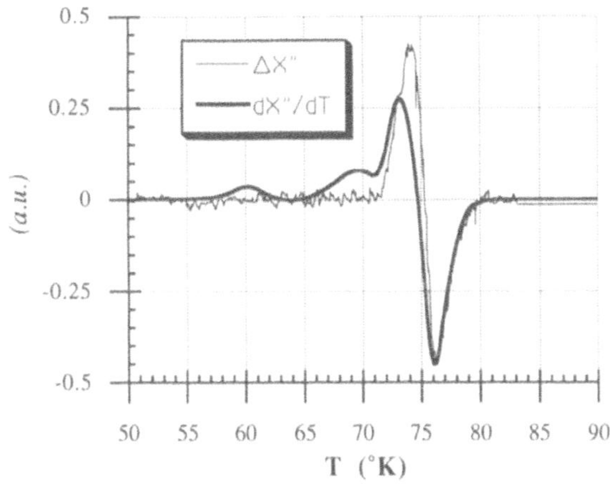

Fig. 5 $\Delta\chi''$ and $d\chi''/dT$ vs. T

average properties of the whole sample. Moreover the magnetic susceptibility does not depends on the effect of contact resistance.

Figures 4 (upper figure) and 5 show respectively the comparison between $\Delta\chi'$ and the derivative of χ', $d\chi'/dT$; and between $\Delta\chi''$ and the derivative of χ'', $d\chi''/dT$. The lack of superposition between the two pair of curves is clearly demonstrated. In the following we will focus the discussion only on the comparison between $d\chi'/dT$ and $\Delta\chi'$, since the other component behaves very similarly in many respects. Since the sharpness of the $\Delta\chi'$ curve (see also the comparison between the integral of $\Delta\chi'$ and χ', fig. 4 (lower figure).

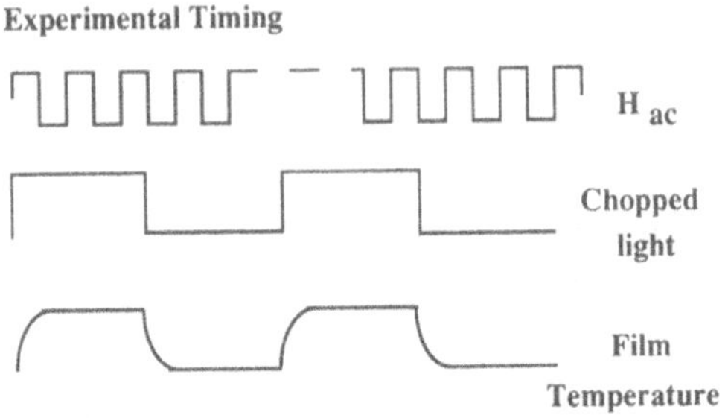

Fig. 6 Timing of the experiment.

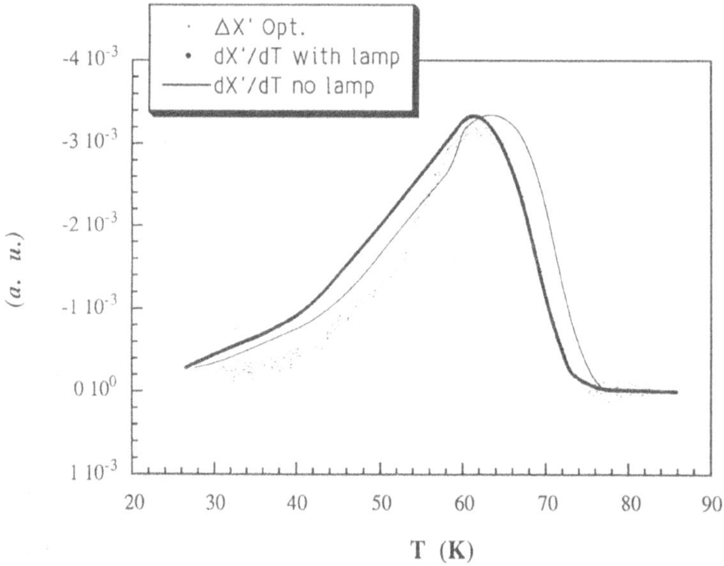

Fig.7 $\Delta\chi'$ vs. T (dot);
$d\chi'/dT$ with lamp on vs T (solid line);
$d\chi'/dT$ with no lamp (line);

was almost unexpected, the observed difference between $d\chi'/dT$ and $\Delta\chi'$ deserves a careful analysis to show that it does not arise from an experimental artifact. Problems, indeed, might arise from several experimental facts like the following: a)the two curves, $\Delta\chi''$ and $d\chi''/dT$, are recorded at two different frequencies; b) the thermal shift induced by the light is such (3-4 K) that probably we have to consider the ratio between finite increments, $D\chi'/DT$ rather than the derivative of χ'. Minor problems may also arise: c) from the different timing of the two curves (see fig.6), that causes them to be recorded at slightly different temperatures, and from d) possible variations of the thermal conductance of the film, and/or of the substrate, with decreasing T.

Let us discuss these facts one by one.

It is well known that an increase of the measuring frequency shifts the transition of the susceptibility curves toward higher temperatures, and sharpens the transition somewhat (see ref. 17 and references quoted therein). Since $\Delta\chi'$ was recorded at a lower frequency than χ', frequency effects per se can not explain the lack of superposition between the two curves.

Since the timing of the measurement is different for $\chi'$and $\Delta\chi'$, one has to check that this fact does not substantially affect the shape of the susceptibility curve. In fact χ' is recorded partially in a darkened condition and partially in an illuminated condition. This has been done for the sample #22, a sample with a degraded $T_{coff} \sim 75$ K; in fig.7. Here we show the dark susceptibility, the "dark-light" susceptibility (i.e. the susceptibility recorded in an illuminated condition using the experimental timing shown in fig. 6) of #22 and for comparison $\Delta\chi'$.

Apart from a rigid shift, no significant changes in the shape are visible in the proximity of the $\Delta\chi'$ tail that could explain the difference between the two curves. Using these same curves, we have studied the possible effects of finite increments; in fig.8a $\Delta\chi'$ is compared with $D\chi'/DT$ derived from measured susceptibility curves. Again, we were not able to explain the differences between χ' and $\Delta\chi'$. The same operation has been done for #50.1 using the value of the thermal shift measured from transport measurements; the result is shown in fig.8b.

Although different experimental timings of χ' and $\Delta\chi'$ have no evident effect on the shape of the susceptibility curve, they result in a difference in the average measuring temperatures of χ' and $\Delta\chi'$. In fact this is the explanation we give for the initial undershoot of the plot $\Delta\chi'$-$d\chi'/dT$ versus T, fig.9 .

It remains to consider the possible changes in the thermal conductivity of the film-substrate "system" with decreasing T. These are expected to give a contribution proportional to T^3. Fig.10 clearly shows that even this effect cannot be invoked to explain the differences between χ' and $\Delta\chi'$.

After the above consideration of the experimental data we have to conclude that in the proximity of, and somewhat below, T_{coff} HTCSTFs respond to light excitation in a non-bolometric way.

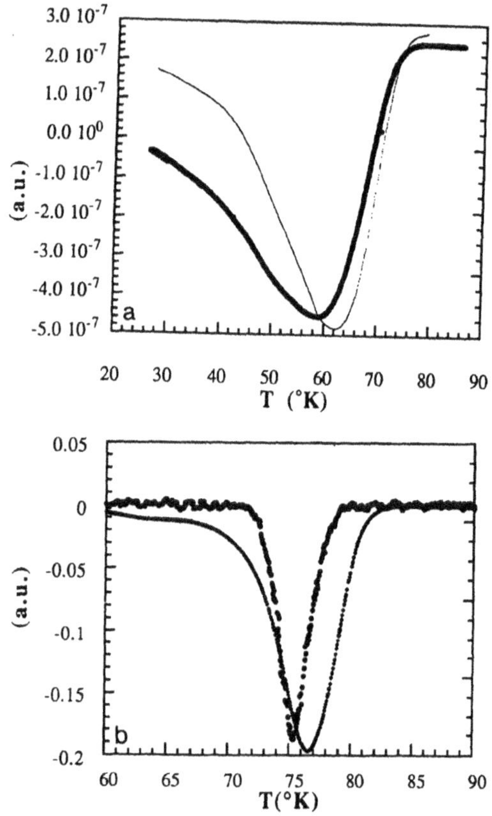

Fig.8 a)$\Delta\chi'$ vs. T (dot)
$D\chi'/DT$ vs. T (solid line); sample #22 (up)
b)$\Delta\chi'$ vs. T (dot)
$D\chi'/DT$ vs. T (circles); sample #50.1 (down)

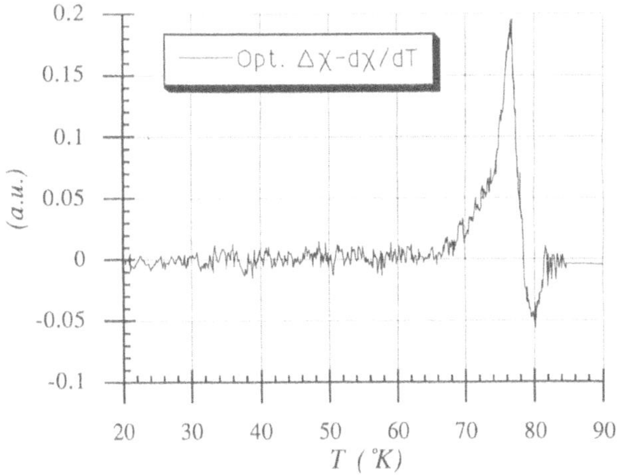

Fig. 9 $\Delta\chi'$-$d\chi'/dT$ difference plotted versus the temperature T.

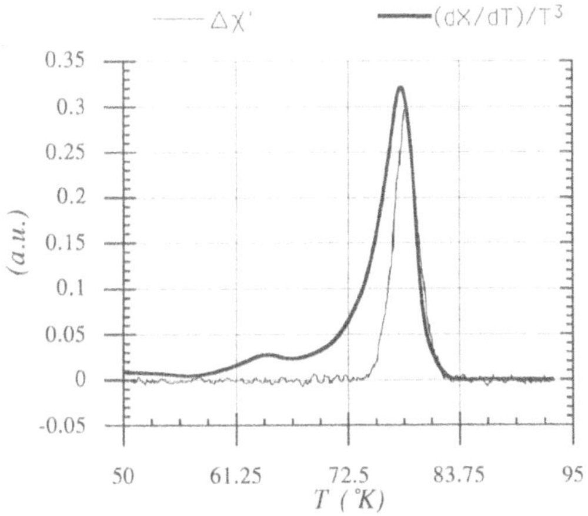

Fig.10 $\Delta\chi'/T^3$ ratio plotted versus the temperature T.

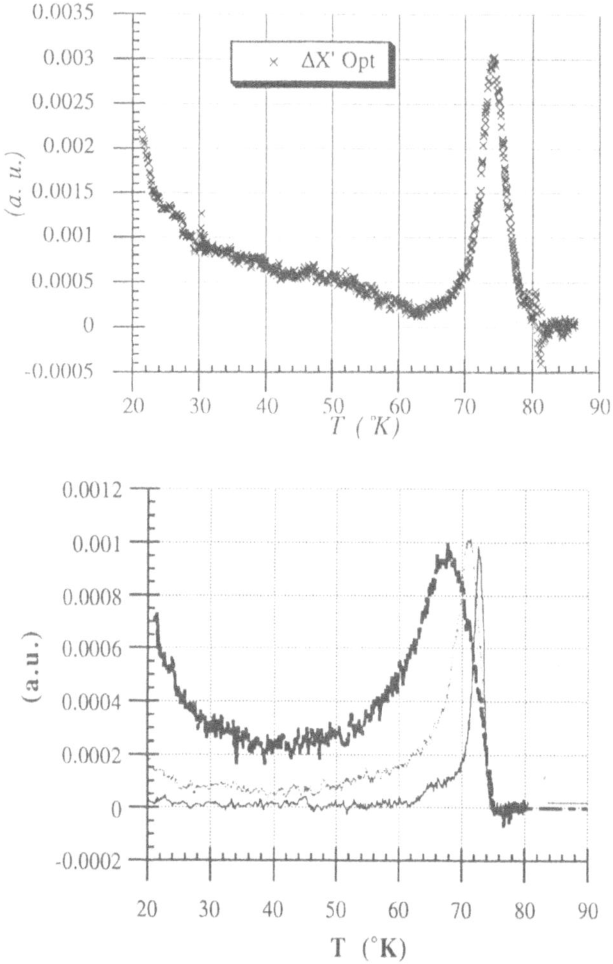

Fig.11 Upper figure: $\Delta\chi'$ vs T for an inhomogeneous film

Lower figure: $\Delta\chi'$ vs T for three different intensities of the a.c. magnetic probing field:
0.4 Oersted (line)
2.2 Oersted (heavy line)
11.2 Oersted (dotted line); sample #45.

MDOE signals at low temperature

Before discussing a possible mechanism for the the non-bolometric response, we would like to consider and discuss the low temperature MDOE regime. In some of the samples examined (less homogeneous, or homogeneous but having a lower T_{coff}), we found evidence for a rise of $\Delta\chi'$ with decreasing T, which is similar to what has been observed by resistivity measurements[8] and by means of a microwave detection of the optical response (MDOR).[6] The MDOR signal has been considered by the authors as possible evidence for non-bolometric response. They ascribe the low temperature increase of the intensity of the photoresponse to the exponential increase of the recombination time of the quasiparticle with decreasing T. However after an inspection of the figures reported in ref. 6, we have to conclude that their experimental findings do not justify their conclusion.

Our MDOE data are reported in figures 11. In particular the lower one shows how the increase of the MDOE signal with decreasing T depends on the applied probing

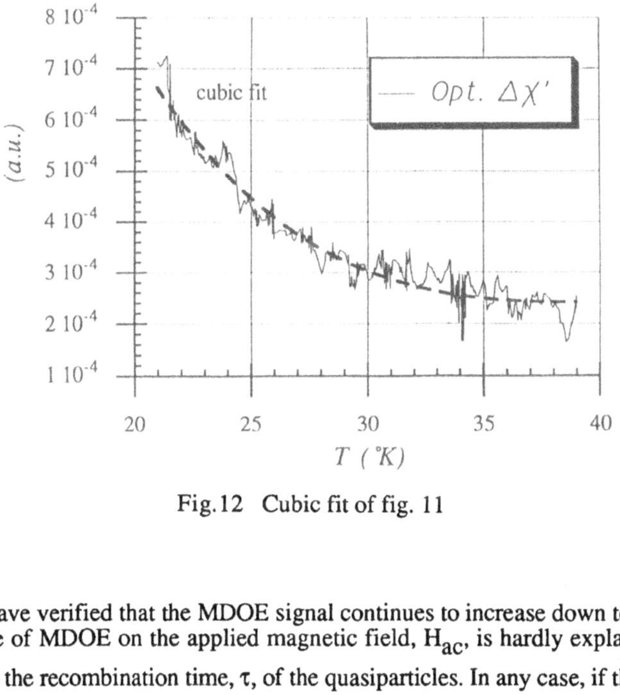

Fig.12 Cubic fit of fig. 11

field. We have verified that the MDOE signal continues to increase down to 5 °K. The dependence of MDOE on the applied magnetic field, H_{ac}, is hardly explainable by a variation in the recombination time, τ, of the quasiparticles. In any case, if the variation of τ were responsible, one should expect the signal to increase exponentially. Instead the fit of fig.12, shows that it increases as T^3, as one would expect from the variation of the thermal conductance of the film-substrate system. The dependence on H_{ac}, as well as the observation of the same signal in less homogeneous films, suggest that the variation in the thermal conductance occurs in the film, and this because of the weakening of the superconducting properties induced by H_{ac} in a portion of the sample volume. Our conclusion agrees with that reported in ref 8, where the thermal conductance of the film-substrate system has been worked out from resistivity measurements.

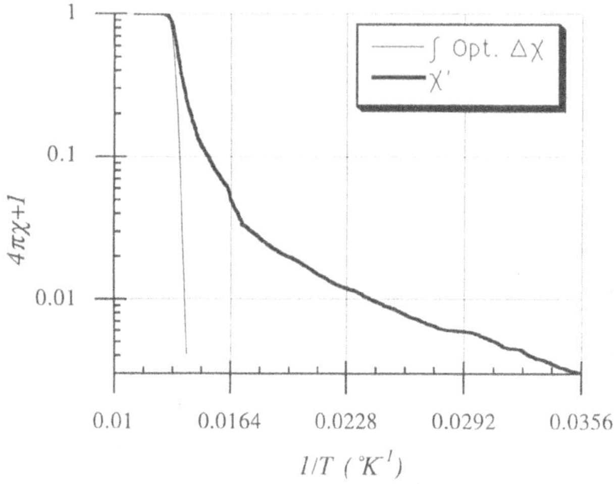

Fig.13 $\int \Delta\chi' \, dT$ and χ' vs. 1/T.

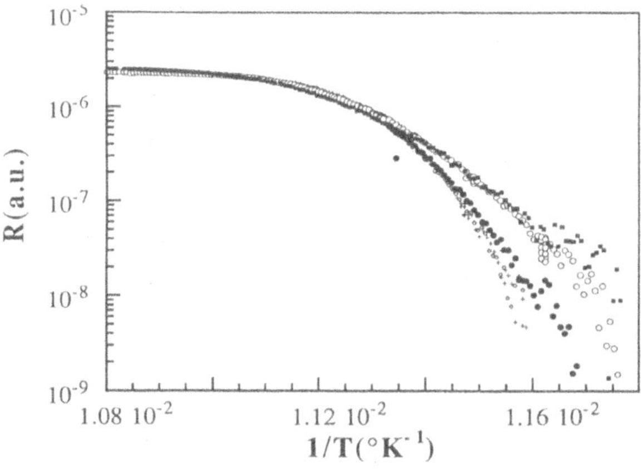

Fig.14 R vs 1/T with 5 different bias current: 1mA (closed squares); 500 μA (open circles); 100 μA (closed circles); 10 μA (open squares); 1 μA (crosses).

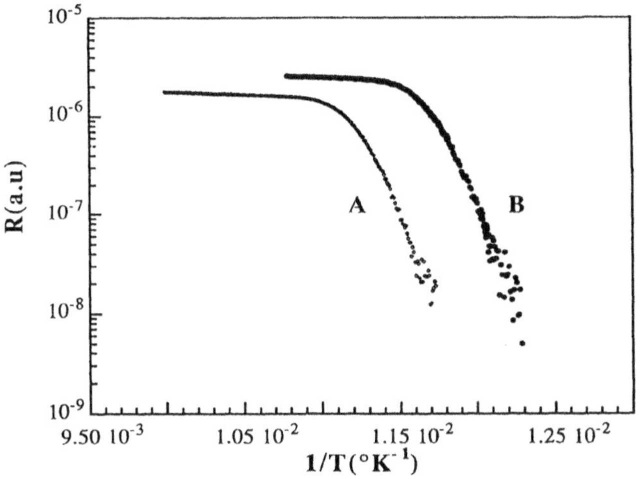

Fig.15 R vs 1/T A: no illumination, B: with illumination.

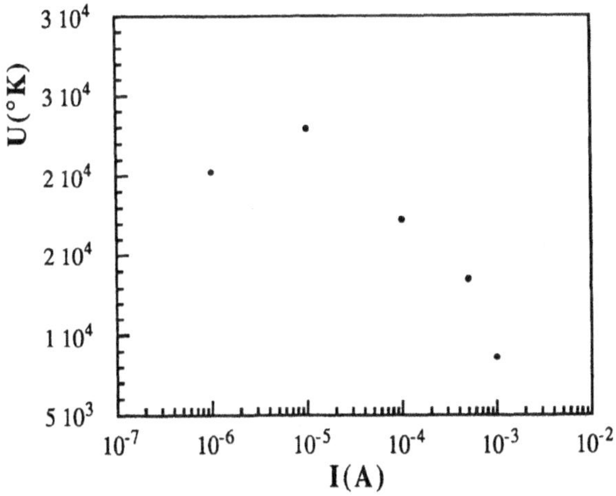

Fig.16 Activation energy U vs I_b

Comparison between MDOE and resistivity measurements

Replotting fig. 4b on a logaritmic scale vs 1/T, fig.13 , one realizes that while the susceptibility curve does not indicate any activated mechanism the $\Delta\chi'$ curve does. Evidence of activated processes have also been observed in resistivity measurements in the proximity of T_{coff} for $\rho < 0.15\rho_n$ (the resistivity of the normal state).[10] In the same temperature region it has been possible to observe a photoresponse, and a dependence of the activation energy on the biasing current, I_b. The physical mechanism suggested has been a photoenhanced flux creep.

The samples we investigated by MDOE do not show any non bolometric photoresponse when studied by resistivity, at least in the limit of our experimental

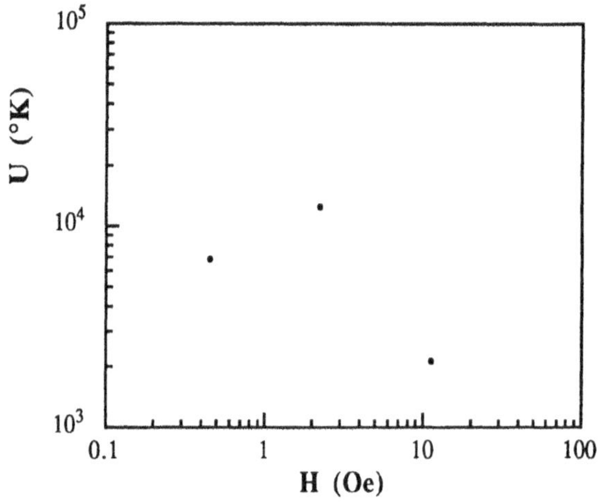

Fig.17 Activation energy U vs H_{ac}

resolution.[18] However, they show a clearly discernable dependence of the activation energy, U, with I_b, even for zero d.c. magnetic field, see fig.14 . We checked that the values of the activation energy derived from the resistive curves were not affected by the incident light beam. As an example we show in fig.15 the same data of fig.2 replotted on a log scale versus 1/T. The fits of the data do not show any effect due to the light. U has been extracted from the semilog plots of figures 14 and 15 using the region in which R decreases linearly and sharply, and neglecting small corrections due to a $(1-T/T_{off})^{\beta}$ term; for a discussion of this term see for example ref. 19 and references quoted therein. The behaviour of U as a function I_b is shown in fig.16. There is a central region where U seems to vary logaritmically with I_b as reported in ref. 10, but we observed clear deviations from this trend for low and high I_b; we think that these deviations are real and actually they are also visible in the data reported in ref. 10 for some values of the current density.

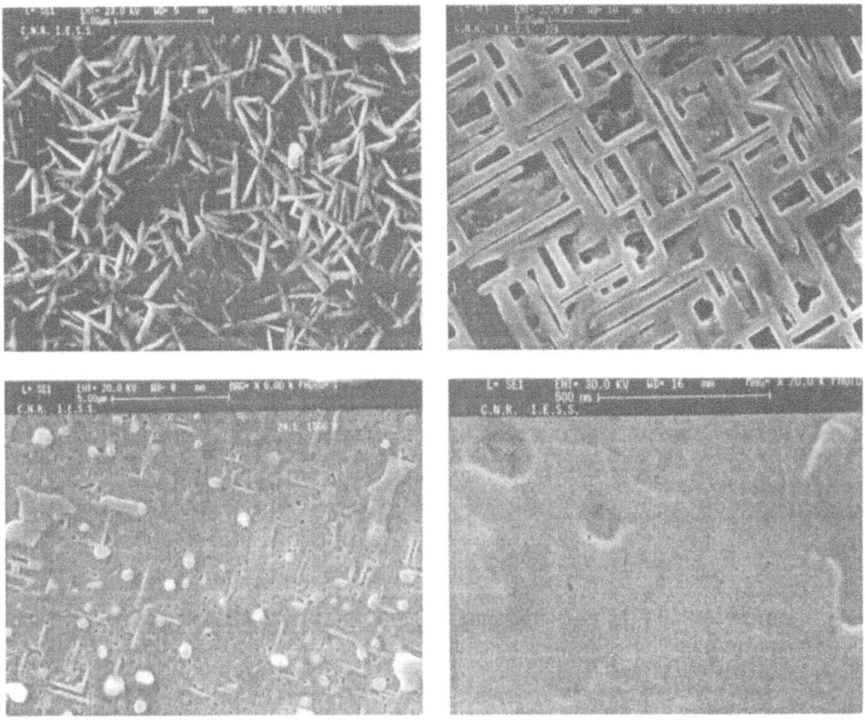

Fig.18 SEM pictures of 4 samples having different thickness d:
a)d=7500 A b)d=3500 A
c)d=1500 A d)d=600 A
The cross-linked structure is formed by crystallites with the a-axis perpendicular to the plane of the substrate; the c-axis lies along the shortest dimension of the crystallites; the different scale of the plots have been chosen in order to make some of the stuctures more visible.

In fig.17 we show the variation of U with H_{ac} , derived from the MDOE experiments. The observed behaviour is very similar to that of fig.16 and suggests a similar physical mechanism underlying both. The lower values of the activation energy deduced from the MDOE curves may be justified by the following observations. Transport measurements are sensitive to the properties of just a part of the sample, the paths that support the highest critical current. MDOE on the other hand tests the sample as a whole. If the physical mechanism behind the photoresponse is related to the motion of magnetic flux, it seems plausible that the part of the sample having the highest critical current is also characterized by a higher value of the activation energy.

We thus suggest that resistivity and MDOE are probing the same physical mechanism. Even if we believe that the motion of magnetic flux is involved, we are not yet in the position to say if we are dealing with creep or fluctuactions. The MDOE region that can be fitted by a thermal activation law is the one delineated by 0 and 60-70% of the intensity of the MDOE peak; a region where the normalized χ' varies between -0.6 and -0.2. From previous studies we know that this range of χ' values is out of the region in which the flux creep in a gradient of field is the predominant mechanism.[20] Thus we tentatively suggest as a possible mechanism the motion of flux through potential barriers in an essentially homogeneous magnetic field. However small local field imbalance may always be present.

As far as the nature of the barriers is concerned, we may speculate that these occur at the connective points of the cross-linked structure of our samples, see fig.18. These barriers weaken with increasing T and may explain the rounding and the decrease of the MDOE signal. The role played by junctions is also revealed by very recent resistive measurments under microwave excitation.[21]

CONCLUSIONS

In conclusion we have shown that the optical response of a superconductor can be conveniently studied by the MDOE technique; a technique that is sensitive to the average properties of the sample as a whole, and not only to a part of it as is the resistivity. We have given evidence for the existence of a non-bolometric response in the proximity of T_{coff}, and have shown that the low temperature signal can be well explained by a variation of the thermal conductance of the film. From a comparison with transport measurements, we conclude that MDOE and resistivity are sensitive to the same physical mechanism. Finally we have suggested as a possible mechanism the motion of magnetic flux through barriers constituted by the junctions connecting the a-oriented crystallites of our films; such a motion occurs in an essentially constant magnetic field.

REFERENCES

1. J. Talvacchio, M. G. Forrester and A. I. Braginski, in "Science and Technology of Thin-Film Superconductors", ed. by R. Mc Connell and S.A. Wolf, Plenum Press, New York, 1989
2. A. Frenkel, in the Proceedings of the Workshop on "HTCS Thin Films: Properties and Applications", ed. by M. De crescenzi, C. Giovannella, M. Iannuzzi, D. Fiorani and R. Vaglio, Physica C, in print.
3. C. C. Chi, M. M. T. Loy and D.C. Cronemeyer, Phys Rev. B 23: 124 (1980) and references therein
4. A. Frenkel, M.A. Saifi, T. Venkatesan, Chinion Lin, X.D. Wu and A. Inam, Appl Phys. Lett. 54:1594 (1989) and references therein
5. J.C. Culbertson, U. Strom, S.A. Wolf, P. Skeath, E.J. West, and W.K. Burns, Phys. Rev. B 39:12359 (1989)

6. R. Kaplan, W.E. Carlos, E.J. Cukauskas and J. Ryu, J. Appl. Phys. 67:4212 (1990)
7. E. Zeldov, N.M. Amer, G. Korena and A. Gupta, Phys. Rev. B 39:9712 (1989)
8. A.I. Braginski, M.G. Forrester and J. Talvacchio, Proceedings International Superconductivity Electronics Conference, Tokyo, June 1989
9. M. Leung, P.R. Broussard, J.H. Claassen, M. Osofsky, S.A. Wolf and U. Strom Appl. Phys. Lett. 51:2046 (1987)
10. E. Zeldov, N.M. Amer, G. Koren, A. Gupta, R.J. Gambino and M.W. McElfresh, Phys. Rev. Lett. 62:3093 (1989)
11. S.G. Han, Z.V. Vardeny, K.S. Wong, O.G. Symko and G. Koren, Phys. Rev. Lett. 65:2708 (1990)
12. J.M. Chwalek, C. Uher, J.F. Whitaker, G.A. Mourou, J. Agostinelli and M. Lelental, Appl. Phys. Lett. 57:1696 (1990)
13. P. Cickmach, A. Fontana, C. Giovannella, M. Iannuzzi and R. Messi in the Proceedings of the Workshop on "HTCS Thin Films: Properties and Applications", ed. by M. De crescenzi, C. Giovannella, M. Iannuzzi, D. Fiorani and R. Vaglio, Physica C, in print.
14. P. Cikmach, M. Diociaiuti, A. Fontana, F. Gauzzi, C. Giovannella, M. Iannuzzi, C. Lucchini, V. Merlo, R. Messi, L. Paoluzi, ang G. Petrocco, in "Modern Aspects of Superconductivity", ed. R. Suryanarayanan, I.I.T.T. International, Paris, 1989
15. P. Cocciolo, M. Gentili, C. Giovannella, M. Iannuzzi, L. Luciani, V. Merlo, R. Messi and L. Paoluzi, J. of Less-Comm. Met. 164-165:374 (1990)
16. M.G. Forrester, J. Talvacchio and A.I. Braginski, Proceedings of GACIAC Workshop on High Tc Superconductivity, Huntsville, Alabama, May 1989
17. C. Giovannella, C. Lucchini, B. Lecuyer, L. Fruchter, IEEE Transation of Magnetism 25:3521 (1989); M. Nikolo and R.B. Goldfarb, Phys. Rev. B 39:6615 (1989) and R.A. Hein, Phys. Rev. B 33:7539 (1986)
18. P. Cocciolo, R. Messi and M. Iannuzzi, private communication
19. O Brunner, L. Antognazza, J.M. Triscone, L. Mieville, O. Fisher, submitted to Phys. Rev. Lett
19. C. Lucchini, C. Giovannella, R. Messi, B. Lecuyer, L. Fruchter and M. Iannuzzi, Phys. Stat. Sol. (b) 157:K123 (1990)
20. B.G. Boone, R.M. Sova, K. Moorjani, W.J. Green and B.E. Grabow, J. Appl. Phys. 69:2676 (1991)

AC SUSCEPTIBILITY OF DILUTE MAGNETIC SYSTEMS

Gwyn Williams

Department of Physics
University of Manitoba
Winnipeg, Canada, R3T 2N2

Much of the current interest in the ac magnetic susceptibility of dilute magnetic systems, and its variation with static applied magnetic field, was initiated by the series of measurements performed nearly two decades ago by Cannella and Mydosh[1] on the AuFe system. Such measurements, carried out at fairly low frequencies (~150 Hz) and driving fields (~5 Oe), revealed the presence of a very sharp cusp in a plot of the ac susceptibility against temperature (near 10 K for a 1 at.% Fe sample) which was rapidly smeared into a rather uninteresting broad maximum by quite modest dc magnetic fields ($H_a \geq$ 2-300 Oe) applied in a direction parallel to the ac driving field. The principal reason that such a result generated wide interest was because it revealed, for the first time in this type of system, a sharp anomaly in a response function (magnetic, thermal, electrical, etc.) invariably regarded by experimentalists as a signal of a potential phase transition. Indeed, the sharpness of this cusp in the zero field ac magnetic susceptibility ($\chi(H=0,T)$) coupled with the relationship between the latter and the derivatives of the Gibb's function (G):

$$\chi = -\left(\frac{\partial^2 G}{\partial H^2}\right) \tag{1}$$

suggested the possibility of a second order phase transition (in the Ehrenfest scheme). Furthermore, if such a transition actually did take place, it was to a ground state of somewhat different character from those previously studied (ferromagnet, antiferromagnet, spiral, etc.); the lack of a spontaneous magnetisation at low temperature in systems such as AuFe taken together with the assumed random distribution of moment bearing impurities over various lattice sites led to suggestions that the rapid suppression in the magnetic response below the cusp temperature (T_{sg}) resulted from the freezing of these moments in uncorrelated directions with no long-range order, as depicted in figure 1; the so-called spin-glass ground state.[2]

With hindsight it has become clear why the transition to the spin-glass state was so difficult to quantify - its unprecedented non- linearity; fields of just a few hundred Oersted, while representing a very small fraction of T_{sg} (typically $\mu_B H_a/k_B T_{sg} \simeq 10^{-3}$ where μ_B is the Bohr magneton and k_B is the Boltzmann constant), influence the response near T_{sg} quite significantly. As a consequence of the many problems raised by this and other complications in the behaviour of spin-glasses (hysteretic response, waiting time effects, etc.) it seemed expedient to first examine the ac magnetic response - and its variation with superimposed static fields - of dilute magnetic systems with a

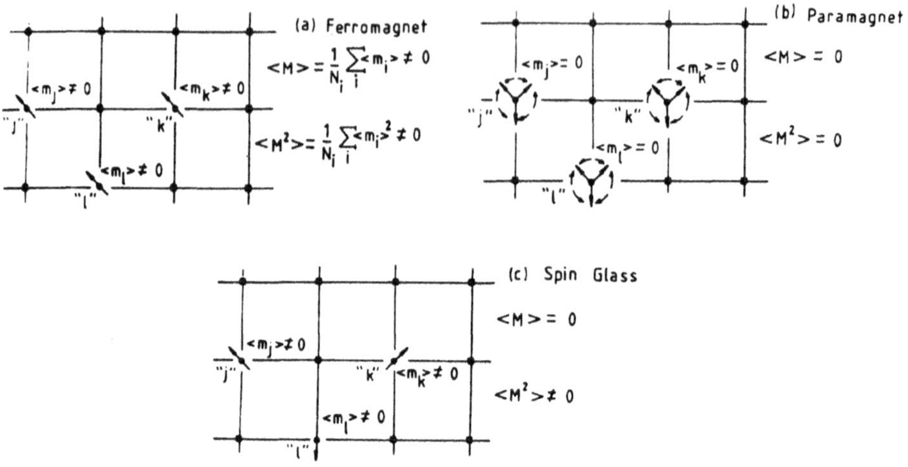

Fig. 1. An illustration of the difference between (a) a ferromagnet below T_c (b) a paramagnet and (c) a spin-glass below T_{sg}; the choice of the order parameter $q = \langle M^2 \rangle$ for the latter is made apparent.

ferromagnetic ground state; the characteristics of the latter are well established and many of the critical exponents associated with the phase transition accompanying entry into this ground state had been measured by a variety of conventional (and well understood) techniques. Furthermore, sinc many theoretical approaches[2] had predicted that both ferromagnetic and spin-glass ground states could occur in the same system by varying the concentration of the magnetic species, the possibility of first investigating the ferromagnetic transition and then (through the relatively uncomplicated adjustment in composition alone) the spin-glass response, held considerable experimental appeal.

As discussed below, measurements of the ac susceptibility of ferromagnets have revealed a new technique for investigating critical exponents which, in several ways, is superior to other methods; the technique is, however, limited by several constraints (principally the occurrence of "technical" magnetic hardness). It can also be very useful in elucidating many features of spin-glasses, although here its application is substantiall more constrained, mainly by effects associated with the abbreviated time scale for the measurement.

Before discussing the results of specific experiments, a review of the theoretical background is appropriate as it provides some rationale for the experimental procedures adopted. The fundamental question that both theory and experiment in this area are attempting to answer is whether a system of mutually interacting impurity spins coupled by either an Ising or Heisenberg like Hamiltonian[3]

$$\mathscr{H} = - \sum_{i<j} J_{ij} \vec{S}_i . \vec{S}_j \qquad (2)$$

in which the effective exchange constant J_{ij} varies in both magnitude and sign, exhibits a phase transition at finite temperature for an arbitrary distribution of the J_{ij}'s. That the J_{ij}'s in dilute systems are non-uniform results primarily from the fact that direct exchange is precluded by the relatively large separation between magnetic sites. The interimpurity coupling is thus indirect, being mediated by the well known RKKY conduction electron polarisation.[4] Briefly the conventional exchange coupling between an impurity spin ($\vec{S}_i$) at site "i" and a conduction electron spin ($\vec{\sigma}$)

$$\mathcal{H}^{ce}_{ex} = -J\vec{S}_i.\vec{\sigma} \qquad (3)$$

results in a long-ranged oscillatory conduction electron spin polarisation which admits both ferromagnetic (parallel) and antiferromagnetic (anti-parallel) coupling between the central site "i" and more distant impurity sites, with an effective interimpurity coupling of the form shown in (2), in which

$$J_{ij} \sim \frac{J^2\cos(2k_F R_{ij})}{(2k_F R_{ij})^3} \qquad (4)$$

(here k_F is the Fermi wavevector of the conduction electron gas and R_{ij} the interimpurity separation). A number of theoretical approaches to this problem model it by restoring translational invariance to the spin system, replacing the effects of spatial disorder (entering via the R_{ij}'s) directly by an exchange bond disorder. The particular model discussed below is a generalisation[5] to arbitrary spin S of an effective-field approach[6,7] based on an Ising-model Hamiltonian incorporating a uniform applied field H:

$$\mathcal{H} = -\sum_{i<j} J_{ij}\, S_i S_j - H\sum_i S_i \quad ; \quad -S \leq S_i \leq S \quad . \qquad (5)$$

Various approximations to treat the effects of disorder can be introduced through assumptions about the moments of the exchange bond distribution. If the latter is assumed to be a Gaussian (as in the much discussed Sherrington-Kirkpatrick (SK) model[8]), then this approach yields a set of coupled equations for the thermal and exchange averaged magnetisation $m = \langle\langle S_i\rangle_T\rangle_J$ (the conventional order parameter for a ferromagnet) and the quantity $q = \langle\langle S_i\rangle_T^2\rangle_J$ (the spin-glass order parameter - figure 1):

$$m = \frac{1}{\sqrt{2\pi}} \int_{-\infty}^{\infty} SB_S[\beta S(\bar{J}_0 m + \bar{J}q^{1/2}\alpha + H)]e^{-\alpha^2/2}d\alpha \qquad (6)$$

$$q = \frac{1}{\sqrt{2\pi}} \int_{-\infty}^{\infty} S^2B_S^2[\beta S(\bar{J}_0 m + \bar{J}q^{1/2}\alpha + H)]e^{-\alpha^2/2}d\alpha \qquad (7)$$

$\bar{J}_0$ and $\bar{J}$ measure the first and second moments of this distribution while B_S is the Brillouin function for spin S. Despite the fact that for S=½ these equations become identical to those of the SK model, this effective field approach (although neglecting the Onsager reaction term) avoids the spurious behaviour associated with replica symmetry breaking. The low temperature thermodynamic properties are well behaved (the expressions for both the entropy and free energy are different from the SK model) and the third law is not violated.

The coupled eqs.(6) and (7) can be solved numerically for m and q using Newton's method, and the phase diagram - figure 2 - results. As expected for a bond disorder $\eta = \bar{J}_0/\bar{J} \gg 1$, each spin experiences a positive exchange coupling of essentially constant magnitude and a ferromagnetic ground state evolves. In the opposite extreme, when $\bar{J}_0 = 0$, as many spins experience a positive exchange interaction of a given magnitude as experience a negative interaction of the same magnitude and they freeze with as many pointing up as down - the spin-glass state. As is typical of mean field theory, a change in the ground state structure occurs at $\bar{J}_0/\bar{J} = 1$.

From the outset it was expected that the mean-field nature of such a model would preclude it from predicting reliably various critical exponents near any of the transition lines. Furthermore, the one dimensional nature of

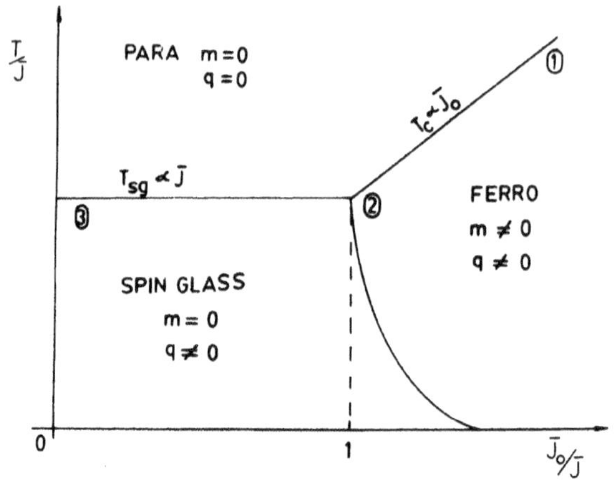

Fig. 2. The effective mean-field model phase diagram.

this Ising model - which nevertheless predicts a paramagnetic to ferromagnetic to spin-glass sequence of transitions on cooling near η=1 - does no reveal the richness in structure associated Heisenberg models in this intermediate regime,[9] as discussed later. The strength of the model, however, arises from its ability to yield values for the susceptibility $\chi = \partial m/\partial H$ everywhere, which can be compared directly with experiment. It accounts surprising well for the systematics displayed by such data over the entire phase diagram.

The experimental approach that seemed most reasonable to follow was to first examine the ac response at a well defined (and supposedly well understood) ferromagnetic transition (near point 1 in figure 2), and once this ha been established, to move along the transition line by varying the composition of the system towards point 2 (the so-called multicritical point)

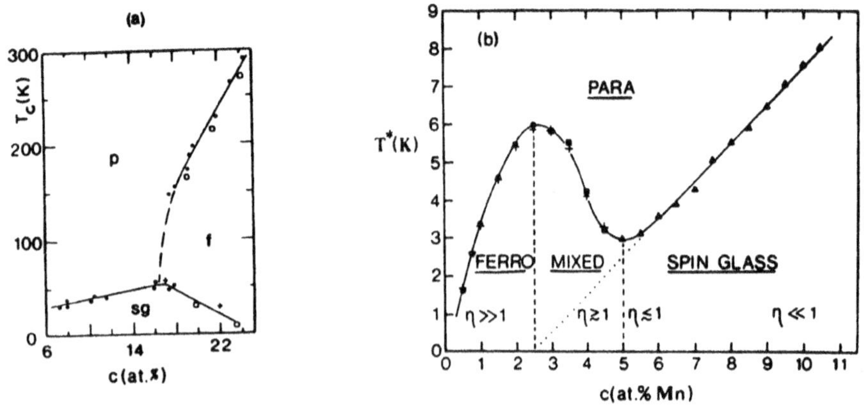

Fig. 3. The experimentally determined phase diagram for (a) AuFe, (b) PdMn.

investigating the effects of increasing bond disorder on the ferromagnetic response. Finally, the spin-glass susceptibility could be measured by moving between points 2 and 3. In real systems of course, as opposed to model systems, the relationship between bond disorder (the abscissa of figure 2) and concentration c (the experimental variable) can be complicated. In systems like AuFe there appears to be an almost direct relationship between c and η so that the phase diagram for this system (figure 3a) bears a strong superficial resemblance to model predictions (figure 2). By contrast the phase diagram in other systems such as PdMn (figure 3b) looks "flipped-around", with a ferromagnetic phase at a lower composition than that of the spin-glass. This occurs because of exchange enhancement effects in the Pd band structure[10] which results in a suppression of the oscillations in the conduction electron spin polarisation to large distances from the polarising site (leading to the giant magnetic moment phenomenon).[11] At concentrations as low as a few tenths$^+$ of an atomic %Mn this results in the average inter-impurity separation lying inside the first zero in this modified polarisation (i.e. direct overlap of exchange enhanced polarisation clouds from adjacent impurities) leading to a ferromagnetic ground state. As the Mn concentration is increased, however, so does the probability of near neighbour Mn-Mn direct exchange. Since the latter is expected to be antiferromagnetic,[14] it provides the bond competition characteristic of spin-glasses. In this system therefore, alloys containing around 0.5 at.% Mn represent good ferromagnets, and the decrease in the ratio η from large, positive values near 0.5-1 at.% can be accomplished by increasing the Mn concentration. Comparing the data in figure 3b with model predictions (figure 2), η is expected to be near unity around 5 at.% Mn decreasing below 1 above this concentration when the spin-glass phase is entered (the possibility of re- entrant behaviour between 2.5 and 5 at.% Mn is discussed later).

Measurements of the in-phase (real part) of the ac (differential) susceptibility of numerous samples were carried out in a modified linear phase-locked susceptometer[15] operating at 2.4 kHz with driving fields down to 25 mOe r.m.s. The detection coil system consisted of two liquid nitrogen cooled, balanced coils connected in series opposition, as shown in figure 4. Static biasing fields up to 1 kOe were applied through a second liquid nitrogen cooled solenoid mounted coaxially with respect to the detection system. The sample being investigated was suspended in a bundle of thirty or so fine copper wires extending into both detection coils, and located in the tail section of a glass Dewar system. The overall length of these wires was around 15 cms, and they were soldered at their upper end into a cone shaped copper block onto which a heater had been wound. The latter was used to provided heating rates of between 0.01 and 0.2 K/min between various refrigerant fixed points (the lower rates being used at lower temperature). The sample temperature was measured by either a calibrated Ge resistor (1.5-20 K) or a Au-0.03 at.% Fe vs Chromel thermocouple (10-300 K) located just below the cone block and in good thermal contact with the sample. While the absolute accuracy of these sensors lies in the range $\pm(0.1\text{-}1)\%$, relative temperatures can be determined to somewhat greater precision.

The samples themselves generally consisted of one (for zero or low field measurements) or more (at higher fields) well annealed strips secured in a sandwich form with adjacent faces electrically insulated from each other. Typical specimen dimensions were $(1.5\times0.2\times0.01)$ cm^3 (with rounded corners). The associated demagnetising factors - found by treating these samples as ellipsoids with principal axes equal to the dimensions given, and evaluating

---

$^+$At much lower Mn concentration the impurities, on average, reside in regions in emu/cc.Oe of oscillatory polarisation, and a spin-glass state similar to that in, say, Au+1 at.% Fe is expected to evolve at low temperature[12]; such a situation has been reported for the giant moment PdFe system.[13]

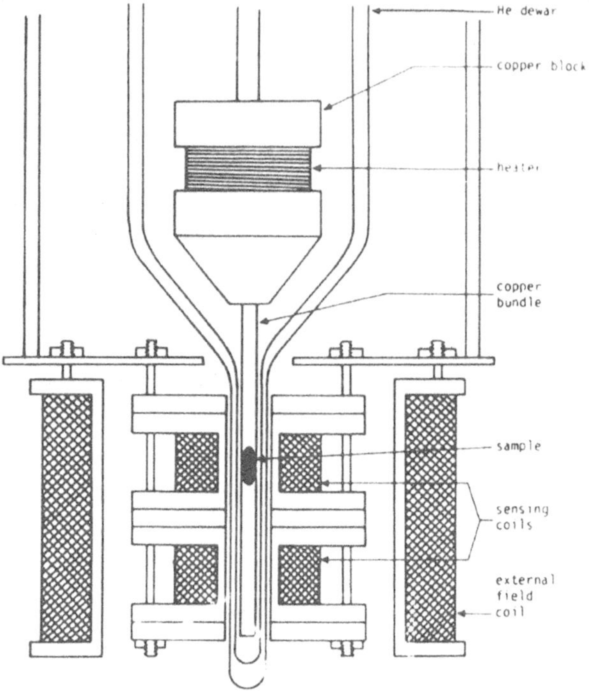

Fig. 4. A schematic illustration of the susceptometer cryostat.

the corresponding elliptic integrals[16] - were $N \simeq 0.05$-0.1, nearly two order of magnitude smaller than that for spherical samples. This geometry was specifically adopted to minimize both the corrections necessary to evaluate the internal field (H_i) from the applied field (H_a) and the volume magnetisation (M, in emu/cc), and those made to find the true differential volume susceptibility ($\chi_t = \partial M/\partial H_i$ in emu/cc-Oe) from the measured response ($\chi_m = \partial M/\partial H_a$). This can be seen from the usual relationship

$$H_i = H_a - NM \tag{8}$$

which leads directly to

$$\chi_t = \frac{\chi_m}{(1-N\chi_m)} \tag{9}$$

In addition, in a ferromagnet below its Curie temperature T_c, <u>if</u> the measure susceptibility is to <u>remain</u> at the demagnetising constraint limit of $\chi_m = N^-$ set by eq.(9) (corresponding to H_i=0 in eq.(8)), then the use of an oscillating driving field H_a means that M must similarly oscillate, with the demagnetising factor N being the proportionality constant between them. Thus for an H_a of set magnitude, smaller N values demand larger amplitude oscillations in M to maintain H_i=0. Correspondingly more coherent rotation and/or domain wall motion is necessary in specimens with small N if the above condition is to be met. Consequently anisotropy effects are far more likely to b observed in small N samples, and this is the principal reason why the measured response does <u>not</u> reach the demag limit of N^{-1} in many systems we have studied.

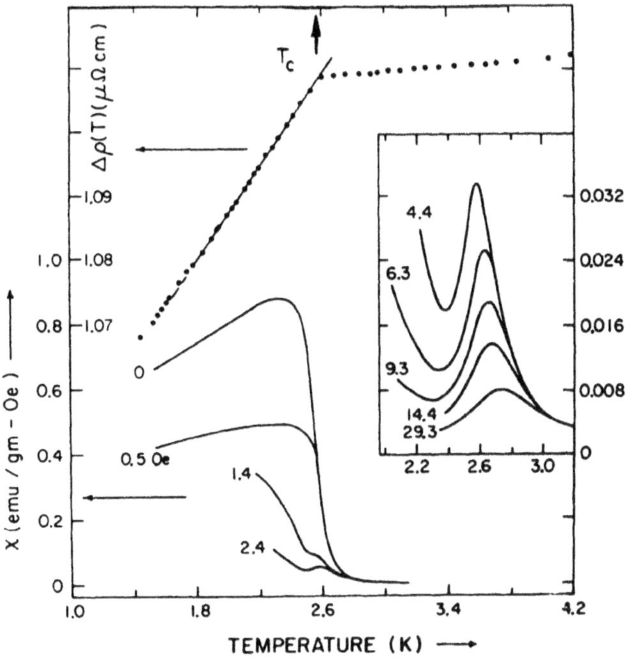

Fig. 5. The incremental resistivity $\Delta\rho(T)$ (upper portion) and the ac susceptibility $\chi(H,T)$ of Pd-0.75 at.% Mn as a function of temperature (with an ac field amplitude of 25 mOe rms at 2.4 kHz). The numbers shown against the susceptibility curves represent the static biasing fields (in Oe). The insert shows the secondary peak structure near 2.6K on an enlarged scale.

With the use of a specific sample geometry it is necessary to calibrate the susceptometer with standard samples with comparable filling factors; we have used the compound Gd_2O_3. Despite these precautions absolute susceptibilities are still estimated to be uncertain to possibly ±5%, although relative values can be measured with a precision exceeding 1 in 10^3.

Figure 5 shows the results of measurements on a typical ferromagnetic sample, Pd- 0.75 at.% Mn. At the top of this figure one can see the incremental resistivity $\Delta\rho(T)$ (= $\rho_{Alloy}(T) - \rho_{Pd}(T)$); in the paramagnetic phase $\Delta\rho(T)$ is essentially temperature independent, and on entering the ferromagnetic ground state near $T_c \simeq 2.6$ K the incremental resistivity falls abruptly (a sharp anomaly in the electrical response function). This fall is readily understandable in terms of the local moment-conduction electron coupling of eq.(3):

$$\mathscr{H}^{ce}_{ex} = -J\vec{S}_i \cdot \vec{\sigma} = -J\{S_i^z\sigma^z + \frac{1}{2}(S_i^+\sigma^- + S_i^-\sigma^+)\} \qquad (10)$$

The presence of an internal field below T_c renders the transverse scattering channels associated with the raising (S_i^+) and lowering (S_i^-) operators[17] inelastic, hence they progressively freeze out with decreasing temperature.

The linear decrease in $\Delta\rho(T)$ with temperature immediately below T_c is, incidentally, a prediction of mean-field theory.[18] Immediately below the resisitvity data, and on the same temperature scale, are the ac susceptibilities. In zero static biasing field the ac susceptibility $\chi(0,T)$ climbs rapidly as T_c is approached from above, passing through a maximum (the Hopkinson peak⁺) at a temperature somewhat below T_c. The application of an external static biasing field to systems like PdMn, which have a relatively low net moment which approaches (technical) saturation in low applied fields ("soft" materials), results in a rapid suppression in both amplitude and temperature of this principal maximum. This facilitates the observation of smaller secondary peaks. The detailed behaviour of these secondary peaks ar shown in the insert in figure 5; they decrease in amplitude and move upward in temperature as the applied field increases. Such peaks are a direct manifestation of critical fluctuation in a system approaching a second order paramagnetic to ferromagnetic transition and they are uniquely revealed by susceptibility measurements, as the following argument shows.

The conventional static scaling law equation of state[20] relates the reduced magnetisation m to the linear scaling fields $h \sim H_i/T_c$ and $t = |T-T_c|/T_c$ via the equation:

$$m(h,t) = t^{\beta}F\left(h/t^{\gamma+\beta}\right) \tag{11}$$

Such a theory, of course, specifies only the argument of the scaling functio F not its general form; nevertheless, it leads to the following asymptotic power law dependences

$$m(0,t) \propto t^{\beta}: \qquad T < T_c \tag{12}$$

$$\chi(0,t) \propto t^{-\gamma}: \qquad T > T_c \tag{13}$$

$$m(h,0) \propto h^{1/\delta}: \qquad T = T_c \tag{14}$$

(provided the Widom equality[21] holds, viz: $\gamma = \beta(\delta-1)$ (15))

which have been widely exploited experimentally to estimate the critical exponents γ,β and δ. By contrast, scaling relationships involving the susceptibility $\chi(h,t)$ are not nearly so well established; equation (11) yields a susceptibility[22]

$$\chi(h,t) = \frac{\partial m}{\partial h} = t^{-\gamma}\dot{F}\left(h/t^{\gamma+\beta}\right)$$

(where $\dot{F}$ indicates the derivative of F with respect to its argument); on re-arranging

$$\begin{aligned}\chi(h,t) &= \left(\frac{1}{t^{\gamma+\beta}}\right)^{\frac{\gamma}{\gamma+\beta}} \dot{F}\left(\frac{h}{t^{\gamma+\beta}}\right) = h^{-\gamma/\gamma+\beta}\left(\frac{h}{t^{\gamma+\beta}}\right)^{\frac{\gamma}{\gamma+\beta}} \dot{F}\left(\frac{h}{t^{\gamma+\beta}}\right) \\ &= h^{(1/\delta)-1}G\left(\frac{h}{t^{\gamma+\beta}}\right)\end{aligned} \tag{16}$$

where $G(x) = x^{\gamma/\gamma+\beta}\dot{F}(x)$ and (15) has been assumed to hold. A maximum in this susceptibility, measured in fixed field h, as a function of temperature occurs when

---

⁺The Hopkinson maximum is not a result of critical effects, but arises from "technical" processes[19]; specifically the rapid increase in anisotropy (mos probably arising from spin-orbit coupling) with decreasing temperature[3] below T_c, particularly when its value begins to exceed the ac field.

$$\left(\frac{\partial\chi}{\partial t}\right)_h = 0 \qquad \text{i.e. if } G = 0$$

This latter condition is satisfied if the argument of this function is a constant (when the function itself is a constant) viz. the temperature t_m of the maxima in $\chi(h,t)$ occur at field h such that the argument of G satisfies

$$\frac{h}{t_m^{\gamma+\beta}} = C \tag{17}$$

As a corollary, since the scaling function G is a constant at these maxima, then their amplitude depend on field alone, being given by

$$\chi(h,t_m) \propto h^{(1/\delta)-1} \tag{18}$$

Thus the static scaling law predicts the occurrence of maxima in $\chi(h,t)$ which move upward in temperature with increasing field along a "cross-over" line (discussed below) in the (h,t) plane, given by

$$t_m \propto h^{1/\gamma+\beta} \tag{19}$$

($(\gamma+\beta)$ being termed the "cross-over" exponent), while the susceptibility at these maxima $\chi(h,t_m)$ - eq.(18) - displays the same dependence on field as does the susceptibility at T_c- eq.(14)*, decreasing in amplitude with increasing field (as $\delta>1$). The predictions of eqs.(18) and (19) are precisely the behaviour observed experimentally. Furthermore, measurements along this cross-over line have the distinct advantage of being independent of the choice of T_c and are thus potentially superior experimentally to measurements along the critical isotherm ($t=0$, $T=T_c$, eq.(14)) which specifically require T_c to be identified.

The term cross-over line appears particularly appropriate when the physical origin of these susceptibility peaks is discussed. The line of maxima given by eq.(19) (along which $(\partial\chi/\partial t)_h=0$) delineates a temperature dominated regime ($t>>h$) in which $(\partial\chi/\partial t)<0$ from a field dominated regime ($h>>t$) where $(\partial\chi/\partial t)>0$. This can be seen in a semi-quantitative way from the fluctuation-dissipation theorm[20]:

$$\chi(H,T) \sim \frac{1}{T}\left\{\langle S_z^2\rangle - \langle S_z\rangle^2\right\} \tag{20}$$

In the high temperature regime $\langle S_z^2\rangle \to S(S+1)/3$ and $\langle S_z\rangle \to 0$ when $\chi(H,T)$ varies as T^{-1}, so $(\partial\chi/\partial t)_h < 0$; in the opposite limit, if T_c is approached from above in non-zero field the magnetisation is driven towards saturation, an effect which becomes more pronounced the closer T_c is approached. Thus, fluctuations in the magnetisation decrease in finite field as the temperature is lowered towards T_c leading, via eq.(20), to a decrease in the differential susceptibility. This susceptibility does not vanish, of course, at T_c, as

*In ferromagnets there appear to be three solutions to the condition G = constant summarised in eq.(17). A non-trivial solution occurs along the cross-over line, and one trivial solution occurs at $t=0$ ($T=T_c$), both of which are discussed above. A further trivial solution appears as $t\to\infty$ in finite field, when the argument of the scaling function, the scaling function itself and $\chi(h,t\to\infty)$ become zero. A similar situation has been reported recently for the electric susceptibility near a ferroelectric transition.[23] By contrast, the scaling function in spin-glasses seems to admit the two trivial solutions alone.[24] Physically, this difference may result from a direct strengthening of the scaling field h in ferromagnets by the uniform applied field H_a, in contrast to the effects of the latter on the conjugate field (a random staggered field) of spin-glasses.[25]

eq.(14) shows, although this argument indicates that $(\partial\chi/\partial t)_h > 0$ as $t \to 0_+$. The maximum in $\chi(h,t)$ discussed above must accompany the behaviour outlined in the limits $t \to 0_+$ and $t \to \infty$. This general argument is confirmed by calculations of the differential susceptibility $\chi(h,t)$ in the specific case of the mean effective-field model,[5] as shown in figure 6. An interesting corollary in the behaviour of paramagnetic systems near T=0 can also be found.[26]

To return to the experimental data of figure 5, a more stringent test o the applicability of eqs.(18) and (19) is carried out in figures 7 and 8. Figure 7b shows the peak susceptibility data $\chi(H_i,T_m)$ (corrected for back-

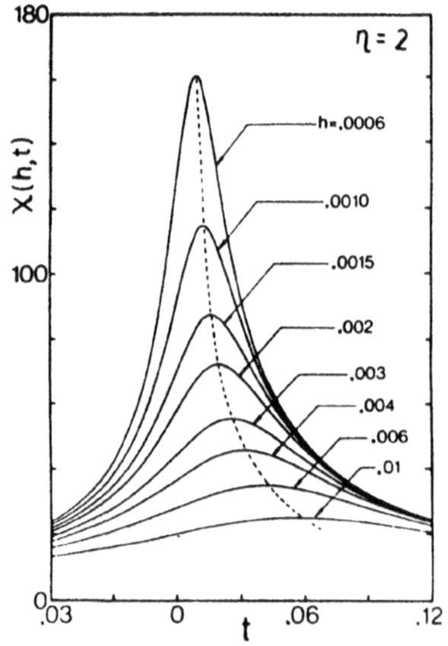

Fig.6. The susceptibility $\chi(h,t)$ in various fixed fields h predicted by the effective mean-field model for $\eta=2$; the dotted line represents the cross-over line.

ground and demagnetising effects) from figure 5 plotted against the internal field⁺ H_i on a double-logarithmic scale. The straight line nature of the resulting plot over the entire field range examined ($2.5 \leq H_i \leq \sim 1000$ Oe) confirms the power law prediction of eq.(18); the slope of the line drawn yields

$$\delta = 4.0 \pm 0.1$$

⁺The internal field is found from eq.(8), in which the magnetisation is estimated by numerical (trapezoidal rule) integration of experimental susceptibility data, viz:

$$M = \int_0^{H_i} \chi_t dH_i = \int_0^{H_a} \chi_m dH_a$$

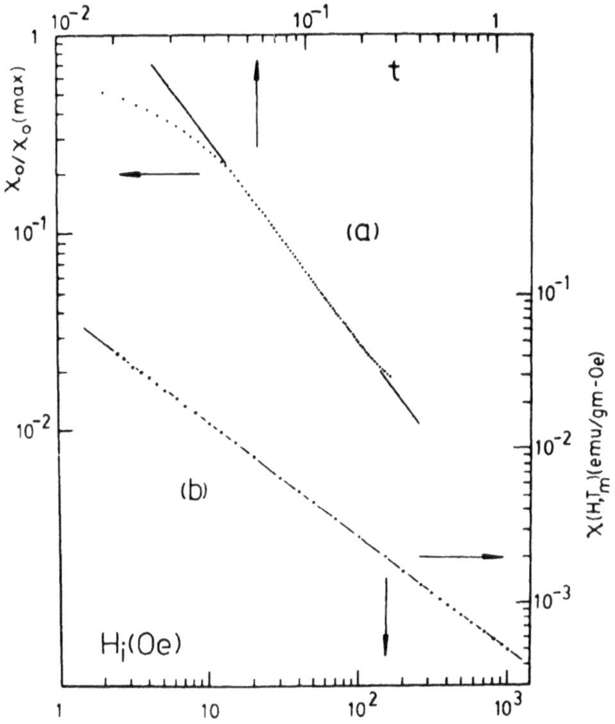

Fig. 7. Double logarithmic plots for the 0.75 at.% Mn sample of (a) the zero-field susceptibility against the reduced temperature t(with T_c=2.53(2)K) and (b) the peak susceptibility plotted against the internal field. In both plots the susceptibility has been corrected for background and demagnetising effects.

close to that found in elemental ferromagnets such as Ni(δ=4.2),[27] but smaller than both that found in a number of amorphous systems using the same technique (δ = 4.85±0.15 in $Fe_{91}Zr_9$)[28] and the value of δ=4.80 predicted by renormalisation group methods for the isotropic 3-dimensional Heisenberg model.[29] Further, values of δ estimated from conventional measurements along the critical isotherm and from data acquired along the cross-over line in Metglas 2826A yield the same value for this exponent,[30] confirming the experimental equivalence of the two techniques.

Figure 8 shows the susceptibility peak temperature t_m plotted against the internal field. Again, the double-logarithmic nature of this plot combined with the resulting straight line at all but the lowest fields confirm the power law prediction of eq.(19); the straight line drawn corresponds to the cross-over exponent

$$(\gamma+\beta)^{-1} = 0.544$$

While these data are consistent with cross-over exponents in the range 0.54-0.56 (the isotropic, 3-dimensional Heisenberg model yields γ=1.386, β=0.365 so $(\gamma+\beta)^{-1}$ = 0.571),[29] we favour the lower values as these allow the Widom relationship, eq.(15), to be satisfied. Obviously the ordinate in this

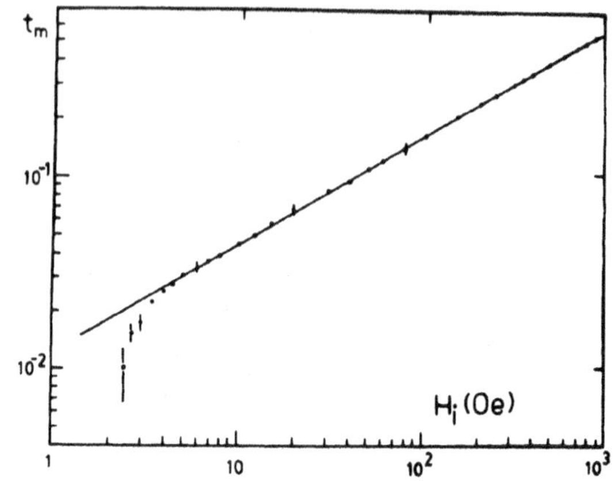

Fig. 8. A double logarithmic plot of the reduced susceptibility peak temperature $t_m(= (T_m - T_c)/T_c)$ against the internal field. The line drawn corresponds to the cross-over exponent $(\gamma+\beta)^{-1}=0.544$.

latter figure requires a specific choice be made for T_c. This is done by plotting the measured susceptibility peak temperatures T_m against $\sqrt{H_i}$ (a reasonable first approximation for eq.(19)), and extrapolating to $H_i=0$ to ge a first estimate for T_c. This is subsequently adjusted by a small amount (1-10 mK) until a consistent set of plots - as in figures 7a and 8 - result. Figure 7a shows a double-log plot of the zero-field susceptibility (normalised to its peak value which, in the ferromagnetic PdMn samples we have studied, represents some 60-80% of the calculated demag. limit N^{-1}) against the reduced temperature t. The straight line drawn through these data between $4\times10^{-2} \leq t \leq 3\times10^{-1}$ yields $\gamma=1.36$ (±0.05).

The deviations from a power law behaviour evident at low field/temperature in figures 7a and 8 are believed to arise from small residual anisotrop effects (even in the case of a supposedly Mn^{2+} $3d^5$ S-state configuration). This residual anisotropy eventually blocks the rapid increase in $\chi(0,t)$ as $t\to0$ and its presence also leads to a regular contribution in $\chi(h,t)$ which is not saturated in low field ($\leq$10 Oe). The latter causes the peak temperature in low field to be underestimated. The presence of this anisotropy can be inferred indirectly from a comparison of the measured $\chi(h,t)$ (figure 5) and the calculated response (figure 6) below the peak. The related giant moment system PdFe appears to be a "technically softer" system, as these effects ar suppressed to lower fields/temperatures[31].

The use of a scaling function also means that the susceptibility can be written as an (even) power series in the field, as dictated by symmetry arguments, viz:

$$\chi(h,t) = t^{-\gamma}\dot{F}\left(\frac{h}{t^{\gamma+\beta}}\right) = \frac{1}{t^{\gamma}} - \frac{h^2}{t^{3\gamma+2\beta}} + \frac{h^4}{t^{5\gamma+4\beta}} - \ldots\ldots \qquad (21)$$

For reasons which will become clear in the following section on spin-glasse: the leading non-linear term in this response $(h^2/t^{3\gamma+2\beta})$ can be estimated b: plotting the measured response against H_i^2 at a number of fixed temperatures,[32] as shown in figure 9. The coefficient a(T) of this H_i^2 term is plotted against the reduced temperature t on a double logarithmic scale in

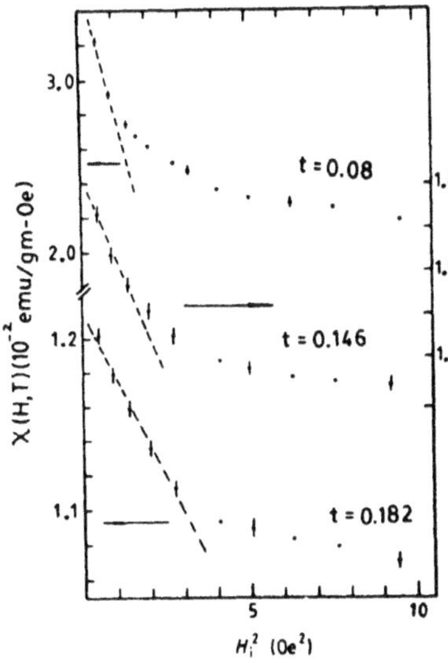

Fig. 9. The field-dependent susceptibility plotted against the square of the internal field at the reduced temperatures indicated.

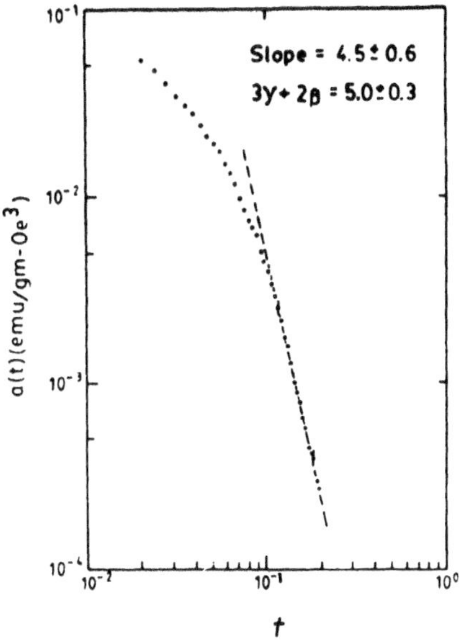

Fig. 10. The coefficient of the h^2 term plotted against the reduced temperature on a double logarithmic scale.

figure 10, which shows that for $t \geq 9\times10^{-2}$ the dependence of a(t) predicted from eq.(21), $a(t) \propto t^{-(3\gamma+2\beta)}$, is confirmed; the slope of this plot is 4.5 ± 0.6 while the value of $3\gamma+2\beta = 5.0\pm0.3$ from the data summarised in figures 7 and 8 above. Nevertheless, even in a good ferromagnet, deviations away from this power law dependence are evident below $t \sim 9\times10^{-2}$. The probable causes of this are evident from an inspection of figure 9. This figure shows that the field range over which the H_i^2 term dominates the field dependent response shrinks markedly as $t \to 0$, when H_i^4, H_i^6 and higher terms become increasingly important. In fact it appears that the dominance of the H_i^2 term covers a field range comparable with the smallest biasing field used near $t \simeq 10^{-1}$. The influence of the next most dominant term in H_i^4 can be seen from eq.(21) to lead to an underestimate for a(t) as is indeed observed experimentally. This type of experiment indicates the difficulties inherent in attempting to establish critical behaviour by observing particular terms in a power series expansion of $\chi(h,t)$ and the rather restricted range of the critical region that they access.

In figure 11 the entire field and temperature dependence of the field-dependent response is summarised in a single scaling plot. From eqs.(16), (17) and (18), the ratio

$$\frac{\chi(h,t)}{\chi(h,t_m)} = \frac{G(h/t^{\gamma+\beta})}{G(h/t_m^{\gamma+\beta})} \sim G(h/t^{\gamma+\beta}) \qquad (22)$$

(since the denominator is a constant); equation (22) indicates that a plot of the normalised susceptibilities against the argument of the scaling function, or its inverse $t/h^{1/(\gamma+\beta)}$, should yield a universal curve, the shape of which specifies the scaling function G for arbitrary values of its argument.

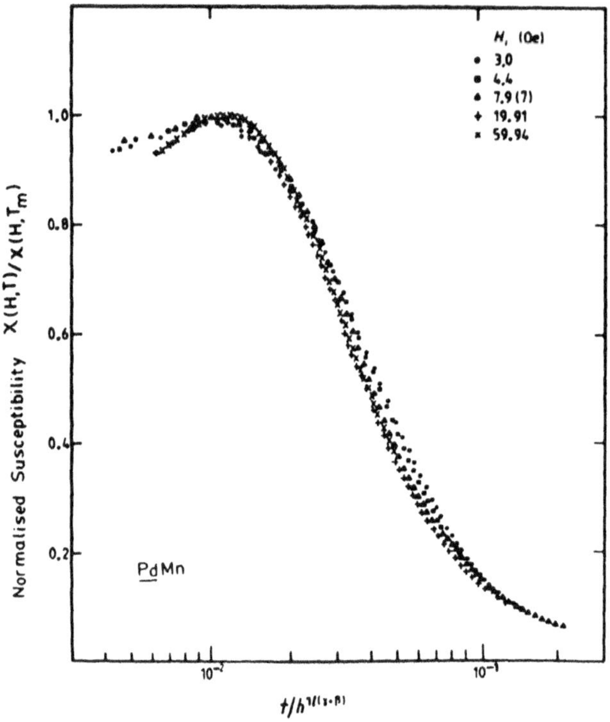

Fig. 11. The normalised field and temperature dependence susceptibility plotted against the inverse of the argument of the scaling function, using T_c=2.53 and $(\gamma+\beta)$=1.84, for various fields.

Having established the behaviour of a good ferromagnet the procedure originally outlined can be followed. The Mn concentration can be increased from near 0.75 at.% to values of 4-4.5 at.% which, according to the phase diagram (figure 3b), corresponds to the bond distribution changing from one with $\eta \gg 1$ to one with $\eta \geqq 1$. Qualitatively the behaviour of $\chi(H,T)$ is substantially unaltered; the field-dependent susceptibility exhibits a peak above T_c, the amplitude of which decreases while the peak temperature T_m increases with increasing field, as before.[22,33] Between 3 and 3.5 at.% Mn, however, these critical peaks begin to require a larger field to initially resolve them,[33] and once resolved, they appear broader. These changes becom more apparent when quantitative analysis is attempted. Tests of the power law dependence of the peak heights - as in figure 7b - begin to exhibit curvature, leading to an effective exponent δ^* that decreases with increasir field. Initially such curvature barely exceeds the experimental uncertainty,[22] (so that between 1 and 2.5 at.% Mn a single effective exponent appears to characterise the data), and this effective exponent decreases monotonicaly with increasing composition. Between 3 and 4.5 at.% Mn this curvature becomes quite marked,[33] as shown in figure 12, enabling different straight lines to be drawn through the data at low and high field; the corresponding effective exponents are $\delta^{LF} \simeq 4.8 \pm 0.2$ (close to the 3-dimensional Heisenberg model prediction) while $\delta^{HF} \simeq 3 \pm 0.2$ (near the mean field value). The conclusions drawn from these results were that while asymptotic exponent values $(h \to 0)$ are essentially unchanged (indeed the δ value for the 0.75 at.% sample <u>may</u> be underestimated), effective exponent values decrease with increasing field, an effect that becomes more pronounced as $\eta \to 1$.

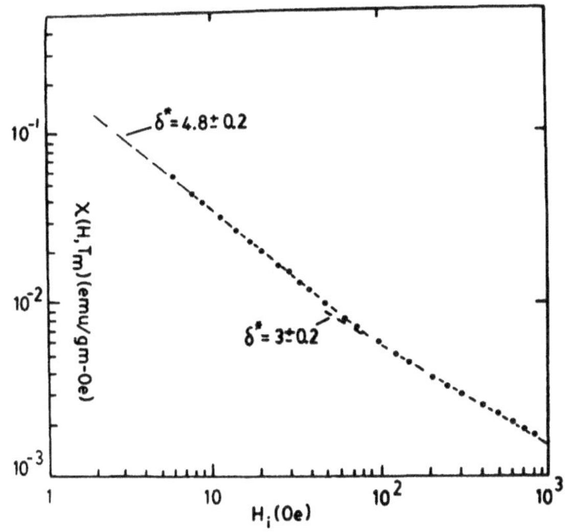

Fig. 12 The susceptibility peak heights plotted against the internal field for Pd+4.5 at.% Mn; the lines drawn to low and high field yield the effective exponent values shown.

Figure 13(a) reproduces values of the effective exponent δ^* as a function of field h for various bond distributions η generated by the effective mean-field model[34] (i.e. by following the field dependence of the peak amplitude in the numerically generated $\chi(h,t)$). While the asymptotic exponent value (δ=3.0) does not agree with experiment (this is a mean-field model predicting mean-field exponents γ=1.0, β=0.5, δ=3.0), the systematics displayed by this exponent as a function of h and η are well reproduced. Figure 13(b) presents similar model generated data for β^* (the temperature dependence of the susceptibility peak yields the cross-over exponent, but since γ=1.0 everywhere in this model β^* may be inferred directly). Such behaviour is difficult to observe in PdMn due to uncertainties associated with the broadening of the susceptibility peaks above about 3 at.% Mn; however, the systematics of this variation of β^* with t are consistent with the behaviour of the cross-over line[35] in NiMn and with the variation of β^* deduced from the spontaneous magnetisation in a number of crystalline and amorphous ferromagnets.[36]

The effective susceptibility exponent γ^* also shows the influence of bond disorder quite dramatically - figure 14; its asymptotic value ($\gamma^* \simeq 1.35$ for $t \leq 2\times10^{-2}$) is consistent$^+$ with Heisenberg model predictions,[29] but at higher temperatures it displays a hump characteristic of many disordered systems.[36] The latter cannot be explained by the effective mean-field model (in which γ=1.0 everywhere), but it can be accounted for by a correlated molecular field theory.[37]

In summary, for the ferromagnetic alloys, the presence of bond disorder

$^+$The fall in γ^* to yet lower values as $t\to0$ is a direct consequence of the failure of $\chi(0,t)$ to increase without limit as a result of residual anisotropy effects. This behaviour is particularly visible in samples with small demag. factor and is not, we believe, evidence of cross-over to other exponent/model values.

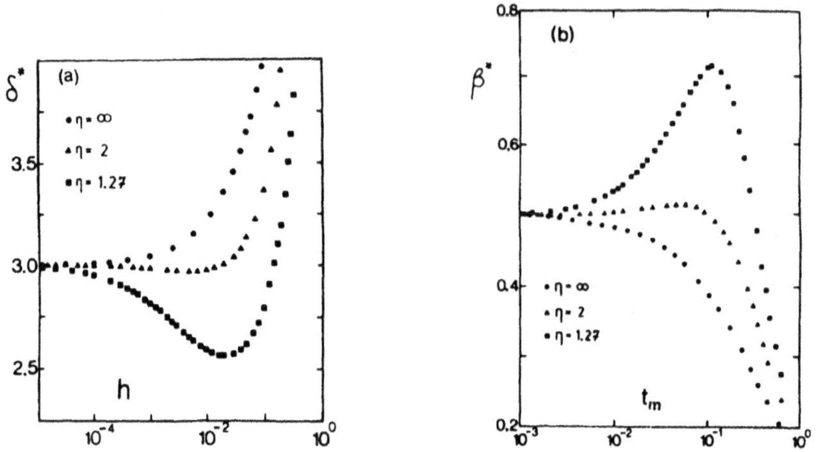

Fig. 13. Model generated effective exponent values as a function of field/temperature for various values of the ratio η; (a) δ* from the susceptibility peak amplitude along the cross-over line and (b) β* deduced from the cross-over exponent.

does not appear to effect the asymptotic exponent values, although the region of applicability of the scaling law with asymptotic exponent values shrinks progressively in the (h,t) plane as the bond disorder increases and η→1. The effective mean-field model reproduces the systematics of this variation quite well. Such variations are not confined to the ac differential susceptibility alone, but are also predicted to appear in conventional Arrott-type scaling plots.[38]

By contrast the behaviour of χ(H,T) in the spin-glass phase of this

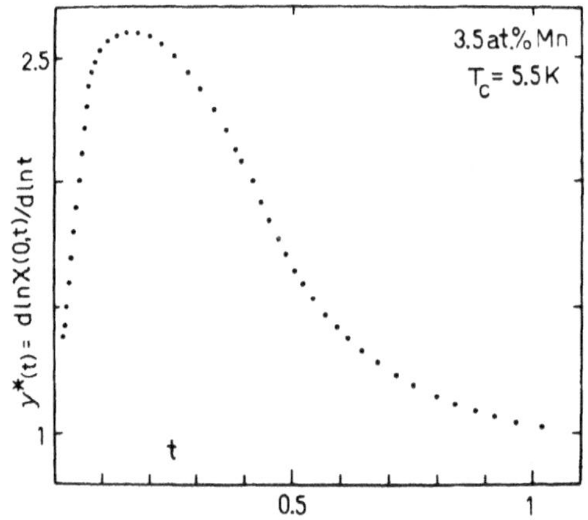

Fig. 14. The effective susceptibility exponent plotted against reduced temperature for the 3.5 at.% Mn sample.

system ($c \geq 5$ at.% Mn) is quite different. As figure 15 shows the field dependent response now exhibits but a single peak, with both the amplitude and temperature of this maximum$^+$ decreasing with increasing field. Despite the contrast in behaviour with the ferromagnetic regime, experimental pragmatism again leads one to plot the (normalised) peak heights against the (reduced) field ($h = g\mu_B H_a / k_B T_{sg}$) in analogy with the procedure adopted in figure 7b. Clearly no power law relationship exists here (figure 16); in fact this figure demonstrates that initially the peak susceptibility $\chi(h,T_p)$ is independent of field, a trivial, non-critical result in complete contrast with the rapid (initial) field dependence exhibited by a ferromagnet at T_c or

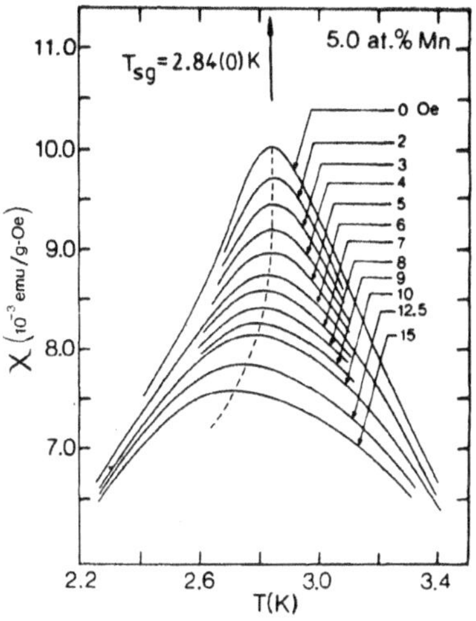

Fig. 15. The field dependent susceptibility plotted against temperature for the 5 at.% Mn sample in various static biasing fields (marked in Oe).

along the cross-over line. The solid lines in this figure represent the results of the predictions of the effective mean-field model using the η values shown. These best fit η values (being less than unity) are consistent with the phase diagram - figure 3b, in particular the appearance of a spin-glass phase at $c \geq 5$ at.% Mn. Furthermore, these η values decrease with increasing concentration beyond 5 at.%, as expected.[39]

The success of this model in fitting the peak variation shown in figure 16 prompted a detailed study of its predictions for the susceptibility; asymptotic expansions of the coupled equations yield[40]:

$^+$The maximum in the zero-field susceptibility decreases dramatically with increasing composition between 3 and 5 at.% Mn, from 45% of the calculated demagnetisation limit, to 4%.

$$\chi(h,t) = S\frac{(S+1)}{3}\left[(t+1-\eta)^{-1} - \frac{h^2}{t}\{t + \frac{3}{(t+2)}\}\frac{(t+1)}{(t+1-\eta)^4}f(S) + O(\frac{h^4}{t^3})\right] \quad (23)$$

in which $t = |T-T_{sg}|/T_{sg}$, $3k_BT_{sg} = S(S+1)\bar{J}$ and $f(S) = (2S^2+2S+1)/120$. This result is consistent with the critical component in the non-linear susceptibility $\chi_{NL}(h,t)$ in the spin-glass phase being expressed in the form of a scaling function[41]

$$\chi_{NL}(h,t) = \chi(0,t) - t^{\beta'}F(\frac{h^2}{t^{\gamma'+\beta'}}) \approx \chi(0,t) - \frac{h^2}{t^{\gamma'}} + \frac{h^4}{t^{2\gamma'+\beta'}} \quad (24)$$

although eq.(23) yields mean-field values for the exponents (primes are used to distinguish spin-glass from ferromagnetic exponents), $\gamma'=1$, $\beta'=1$, $\delta'=2$, which are not expected to be correct (h^2 is the conjugate field). Further-

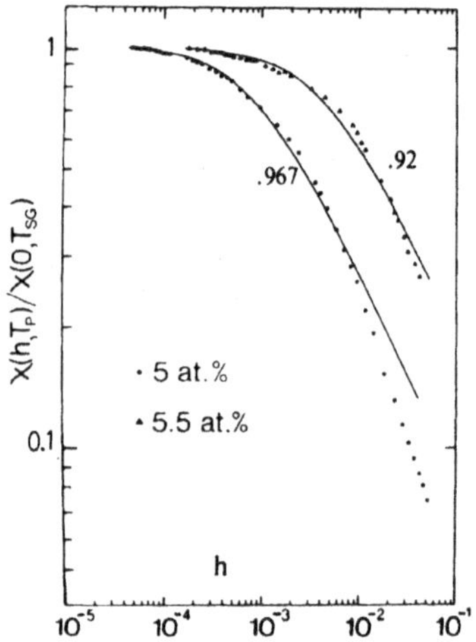

Fig. 16. Double logarithmic plots of the normalised peak susceptibility against the reduced field h ($g\mu_BH_a/k_BT_{sg}$) for two samples. The solid curves represent model calculations using the η values shown.

more, while $\chi(0,t)$ is not singular, it varies as $(T-\theta)^{-1}$, $\theta \propto J_0 < T_{sg}$, the field dependent susceptibility does contain a divergent contribution (viz. the spin-glass order parameter q couples to the non-linear response). Specifically eq.(24) predicts (i) that $\chi(h,t)$ should be dominated initially by an H^2 contribution the amplitude of which increases as $t \to 0$ ($T \to T_{sg}$); and (ii) at a given reduced temperature t the influence of the bond distribution is reflected in the $(t+1-\eta)^{-4}$ term which indicates a progressively larger H^2 amplitude as η increases towards unity. Detailed numerical solutions of the coupled eqs.(6) and (7) confirm the above, and also indicate[42] that the range of dominance of χ_{NL} by the H^2 term decreases as $T \to T_{sg}$ when H^4, H^6 and higher order terms of alternating sign play an increasingly important role. Experimental measurements[43] on PdMn spin-glasses confirm the systematics of these model predictions. Figure 17 demonstrates the presence of an H^2 term in $\chi(H,T)$ the amplitude a(t) of which increases while the range of dominance decreases as T approaches T_{sg}=2.84(0)K (from figure 15). Figure 18 confirms the influence of the bond distribution - via η, although this dependence appears to be stronger than the model result. Figure 19 summarises the measured temperature dependence of the H^2 coefficient a(t) in two samples with different bond distributions. The dramatic increase in this coefficient as T_{sg} is approached is clearly evident from this figure, as is the dependence on the ratio η (at any reduced temperature this coefficient is smaller in the 5.5 at.% sample (η=0.92) than in the 5 at.% specimen (η=0.97)).

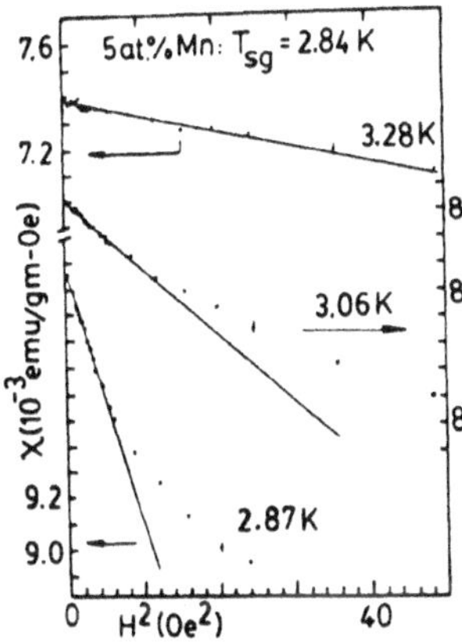

Fig. 17. The susceptibility plotted against the square of the field at three temperatures above T_{sg}

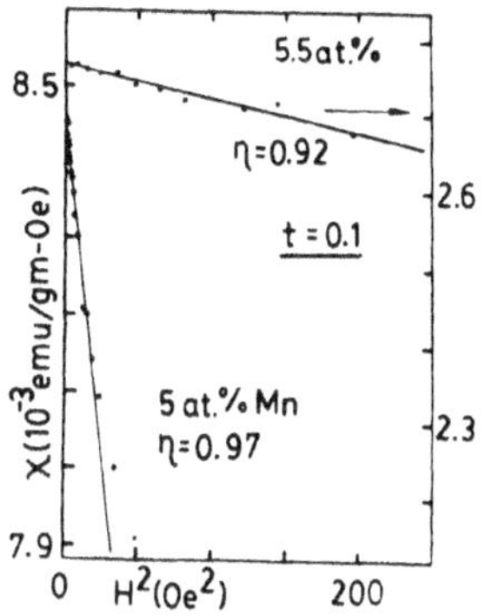

Fig. 18. A plot of the susceptibility against the square of the field at the same reduced temperature in two samples.

Figure 19 displays data acquired both above and below T_{sg}, indeed these data indicate that this coefficient exhibits an essentially symmetric behaviour about T_{sg}. This symmetry would again be in agreement with the predictions of the effective mean-field model[42] which, due to its avoidance of the replica trick, does not suffer from the instabilities[2],[44] associated with replica symmetry breaking below T_{sg}. From an experimental point of view however, the use of an ac technique to investigate the magnetic response below T_{sg} raises several questions resulting from the well documented time dependent effects (viz. irreversibilities, cooling rate dependences and waiting time effects) that occur in this regime.[45] While it is currently unclear whether these latter effects should be considered part of the spectrum of critical fluctuations (the contribution from coherent rotation and/or domain wall motion in ferromagnets - often the principal source of hyseresis below T_c in them - certainly should not), the effects of critical slowing down near T_{sg} are well established. The relaxation time τ associated with the latter exhibits a form similar to that appropriate for a second order transition, viz:

$$\tau = \tau_0 t^{-z\nu} F(h^2/t^{\gamma'+\beta'}) \tag{25}$$

although the value for the dynamical exponent $z\nu$ ($\lesssim 10$) is very large in spin-glasses.[46] The influence of such slowing down becomes quite apparent when the ac data of figure 19 are replotted on a double logarithmic scale - figure 20. A power law dependence is observed only for $t \gtrsim 10^{-1}$, with quite severe flattening being evident in these plots at lower reduced temperature (i.e. as $T \rightarrow T_{sg}$ from above and below). There is a superficial similarity between figure 20 and the corresponding plot in the ferromagnetic regime (figure 10) and while the source of the "rounding" discussed there - increasingly important H^4, H^6 and higher order contributions - may also be playing some role here, critical slowing down is believed to be the primary source of the

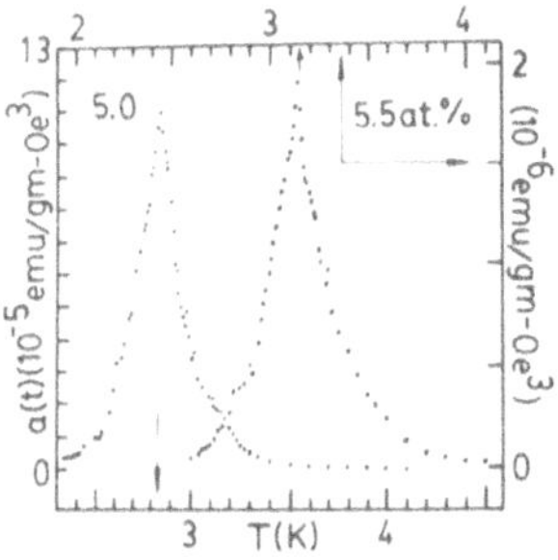

Fig. 19. The variation of the coefficient of the H^2-term with temperature in the 5 and 5.5 at.% Mn samples.

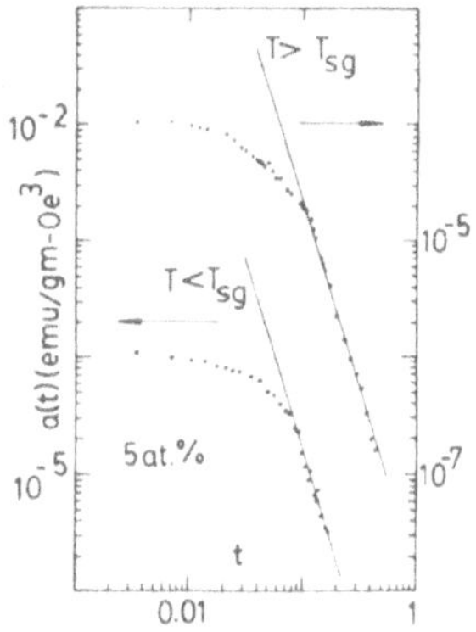

Fig. 20. The data of fig. 19 replotted on a double logarithmic scale against the reduced temperature for the 5 at.% Mn specimen.

flattening evident in spin-glasses. Nevertheless, the width of the accessible critical region for the leading non-linear coefficient appears comparable in ferromagnets and spin-glasses (figures 10 and 20). Unfortunately the exponent values deduced from the straight line portions of such plots (yielding $\gamma'_+ = 3.2 \pm 0.35$ and $\gamma'_- = 3.2 \pm 0.6$ in the 5 at.% Mn sample) are subject to far greater uncertainty in the spin-glass regime. The principal source of this uncertainty is evident in eq.(23). The presence of a substantial regular contribution (i.e. a non-divergent contribution) in the non-linear response, which lead to an expression for a(t) of the form:[40]

$$a(t) \propto \frac{1}{t} g(t) \quad ; \quad g(t) = (1+t)(t^2+2t+3)/(t+2)(t+1-\eta)^4 \tag{26}$$

Here g(t) represents a non-universal correction to scaling; while $g(t) \rightarrow 3[2(1-\eta)^4]$ (a constant) as $t \rightarrow 0$, so that the asymptotic exponent value ($\gamma'=1$ in this model) is unaffected as expected, away from t=0 this correction term modifies the temperature dependence of a(t). This leads to an apparent increase in the exponent. Such corrections are much weaker in the analogous ferromagnetic case, as can be seen from figure 10 and its associated discussion, in agreement with model predictions as g(t) in the ferromagnetic regime is given by:[40]

$$g(t) = (1+t)[1+3\eta^2/\{(1+t)^2-\eta^2\}] \qquad (27)$$

The suggested influence of corrections to scaling in the spin-glass regime is confirmed by a number of quasi static magnetisation studies and very low frequency ($\leq 10^{-3}$ Hz) ac susceptibility measurements. In Pd+6 at.% Mn these yield[47] smaller exponent values ($\gamma' = 2.0\pm0.2$, $\beta' = 0.9\pm0.15$, $\delta' \simeq 3$) at lower reduced temperatures ($2\times10^{-3} \leq t < 10^{-1}$). These lower exponent values, while agreeing with corresponding estimates for the case[48] of Mn in Ag (from which evidence supporting a symmetric transition was also obtained), are nevertheless dependent on the choice[46,49] of the critical temperature T_{sg}.

Despite the constraints noted above on the use of the ac susceptibility to probe the magnetic response of spin-glasses such as PdMn, it is clear that

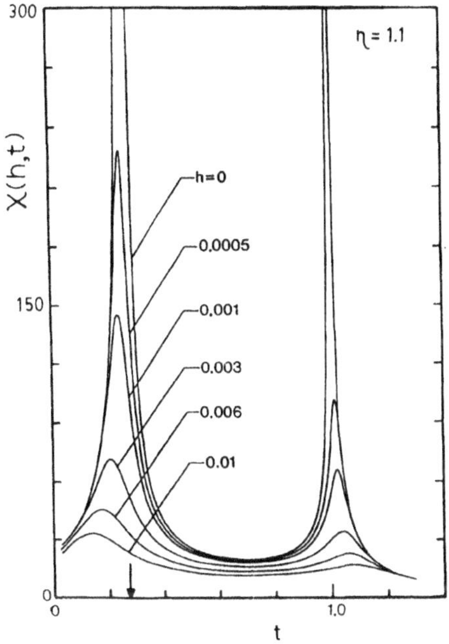

Fig. 21. The field and temperature dependent susceptibility calculated from the effective mean-field model in the re-entrant regime.

this technique can still reveal a clear anomaly in the non-linear susceptibility close to T_{sg} (uncertainties in exponent values notwithstanding), an anomaly that is observable as T_{sg} is approached from both above and below. It is this latter feature which is particularly useful in the investigation of systems exhibiting supposedly re-entrant behaviour, as discussed below.

Inspection of the PdMn phase diagram - figure 3b - suggests that on cooling samples containing between 2.5 and 5 at.% Mn, the system should first exhibit a paramagnetic to ferromagnetic transition, and on further cooling a second transition to a spin-glass state*. This sequence of transitions is predicted by both the effective mean-field and the SK model. In the latter however the ferromagnet-spin-glass boundary occurs at a temperature below the de Almeida-Thouless (AT) line,[51] so that the replica-symmetric solution of the SK model (eqs.(6) and (7) with $S=\frac{1}{2}$), being unstable, should be replaced by the Parisi solution[52] with an order function in place of the order parameter q. Here a modified ferromagnetic or "mixed" phase evolves instead of the spin-glass phase. In vector models this mixed phase appears below the Gabay-Toulouse (GT) line[53] and is characterized by the coexistence of ferromagnetic ordering amongst the longitudinal spin components and transverse spin-glass ordering. The theoretical appeal of models with a spontaneously broken symmetry has resulted in some attempts[54] to justify qualitatively the use of the replica symmetric solution - due to its equivalence with the coupled equations of the effective mean-field model (at least for $S=\frac{1}{2}$) - below T_{sg} in the presence of a field, at least for systems with anisotropy (for example,[55,56] a Dzyaloshinski-Moriya type interaction). Irrespective of whether this justification is finally judged acceptable, comparison between experiment and theory is always appropriate.

One of the characteristic features displayed by the field dependent susceptibility $\chi(H,T)$ in the supposedly re-entrant or mixed regime of systems like PdMn, is the appearance of a double peaked structure.[33] These peaks exhibit the individual features of the separate transitions discussed above, viz. while both peaks are suppressed in amplitude, the upper peak moves upwards and the lower peak moves downwards in temperature as the applied field increases. Current broken symmetry models are incapable of reproducing this structure whereas the effective mean-field model does[57] - figure 21. This latter model also suggests that conventional ferromagnetic scaling should occur as T_c is approached from above while the non-linear response should scale in the manner summarised by eq.(24) as T_{sg} is approached from below. (In particular, the coefficient a(t) should diverge,[58] although the associated exponent is expected to be different from the model predicted value of $\gamma'_- = 4.0$ obtained from figure 22. This last exponent value is obtained from numerically generated data, unlike those values quoted after eq.(24) which were obtained from asymptotic expansions of the coupled equations. This procedure cannot be used near the re-entrant boundary since q is not small there[58]). The effective mean-field model also suggests the presence of an anomaly in the non-linear response as the re-entrant boundary is approached from the high temperature side. Here the presence of a symmetry breaking internal field (associated with a spontaneous magnetisation that vanishes at the re-entrant boundary in Ising models[8,58]) leads to an initial linear dependence of $\chi(H,T)$ on field. The coefficient of this linear term behaves as $(T-T_{sg})^{-5/2}$ for $T>T_{sg}$. Not only is such a dependence believed to be inconsistent with Heisenberg model predictions, due to the differences outlined above, but in real systems the presence of ferromagnetic domain structure above T_{sg} complicates any analysis of data acquired between T_c and T_{sg}.

Experimentally, both the phase diagram and the qualitative behaviour[33] of $\chi(H,T)$ in samples containing 4 and 4.5 at.% Mn are consistent with re-entrant behaviour. A detailed comparison, however, between experiment and

*Note that the paramagnetic-spin-glass boundary extrapolates to $T_{sg}=0$ near 2.5 at.% Mn. The fcc structure has 42 sites in the shell containing first, second and third near neighbours, so that there is a 100% probabililty of finding two Mn atoms in this shell at 2.38 at.%. With antiferromagnetic Mn-Mn coupling out to third neighbour distances,[50] the composition at which a spin-glass ground state first emerges appears to coincide remarkably well with occurrence of an antiferromagnetic component (and hence a competition) in the total exchange coupling.

model prediction does not support such a conclusion. An investigation of the lower peak position as a function of field,[59] reproduced in figure 23 in the case of the 4.5 at.% Mn sample, shows that it is inconsistent with the location of the second transition line shown in figure 3b. More convincingly measurements of the non-linear response in the 4 at.% sample reveal that it decreases monotonically with decreasing temperature below T_c with no indication of any anomaly in the vicinity of the supposedly re-entrant boundary, as indicated in figure 24. The lower peak in these intermediate concentration PdMn samples actually appears to eminate from the principal (Hopkinson) maximum in $\chi(0,T)$, and while no specific anisotropy mechanisms has been identified as the source of this maximum, no clear link between it and those interactions presumably responsible for the re-entrant phase boundary exists.

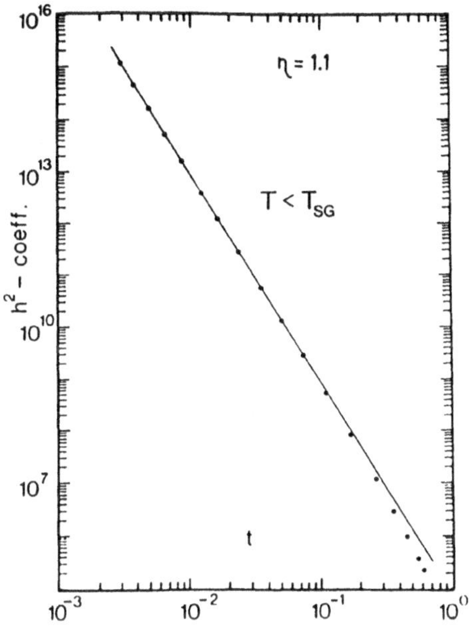

Fig. 22. A double logarithmic plot of the model generated coefficient of the h^2 term against reduced temperature as T_{sg} is approached from below in the re-entrant regime.

The above behaviour is substantially different from that found[60] in (Pd+0.35 at.% Fe) + 5 at.% Mn. A two-peaked structure is again evident in $\chi(H,T)$, but here the lower one emerges near 4K well below the Hopkinson maximum at 9.2K and the peaks associated with the ferromagnetic cross-over line which first appear some 0.2K higher. Comparisons of the peak amplitudes in various fixed fields confirm the model results shown in figure 21, the lower peak being stronger. A complete analysis of the response along the cross-over line can be carried out; this yields a unique value for $\delta=4.1\pm0.1$ over the entire field range examined ($3\leq H_i<1$ kOe) with no hint of curvature in the associated power-law plot as might be expected near $\eta\gtrsim1$ in this re-entrant regime. The finite width of the bond distribution does show up in both the

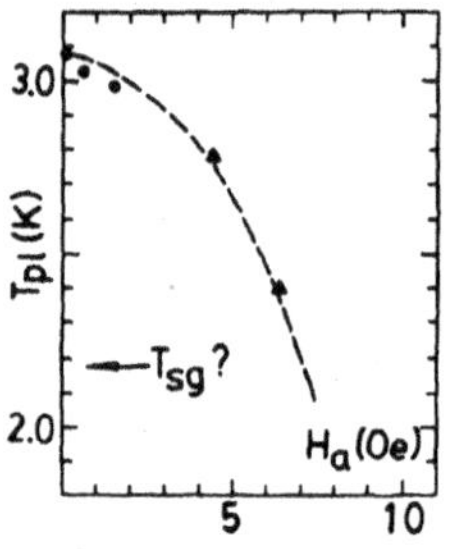

Fig. 23. The lower peak temperature plotted against the applied field for the 4.5 at.% Mn sample.

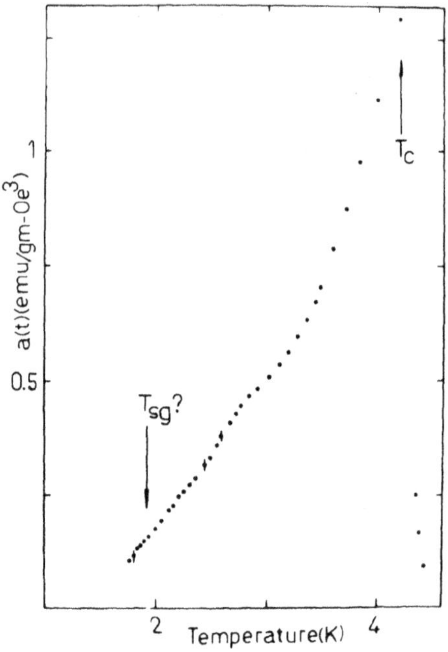

Fig. 24. The coefficient of the H^2 term (estimated from a two field measurement) plotted against temperature for the 4 at.% Mn specimen.

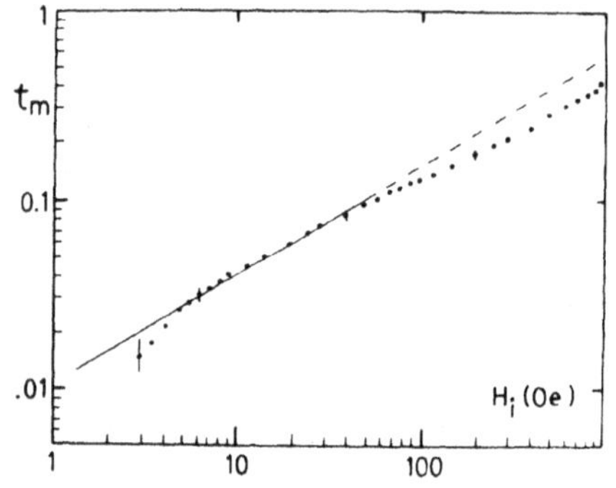

Fig. 26. The behaviour of the cross-over line in Pd(Fe,Mn).

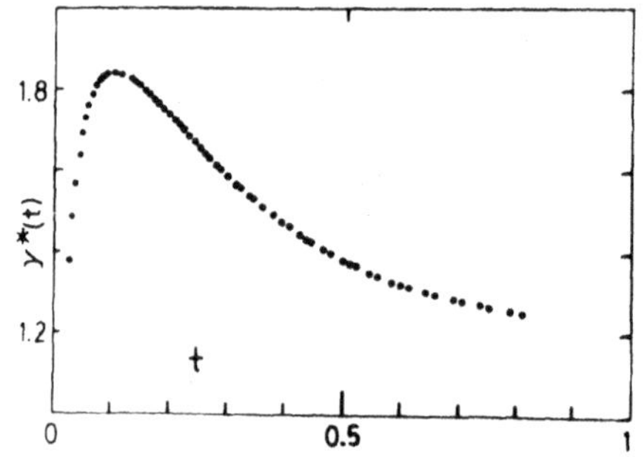

Fig. 25. The effective susceptibility exponent plotted against temperature for the Pd(Fe,Mn) sample.

temperature dependence of the effective susceptibility exponent γ^* (figure 25) and the cross-over line (figure 26). The asymptotic behaviour of the former is consistent with 3-d Heisenberg model values as does the line drawn in the latter at lower fields which corresponds to $(\gamma+\beta)^{-1}=0.57$. These, and other tests involving the temperature dependence of the H^2 coefficient and the scaling of the entire field-dependent susceptibility, confirm the character of this upper transition.

The behaviour of the non-linear response - in particular of the H^2 term a(T) - in the vicinity of the supposedly re-entrant boundary is established in a manner similar to that shown in figures 9 and 17. The results[60] are similar to those shown in these figures, the magnitude of a(T) increases while its range of dominance of $\chi(H,T)$ decreases as the re-entrant boundary is approached from below, in qualitative agreement with model results*. Figure 27 indicates the temperature dependence of the coefficient a(T) over the temperature range examined, and reveals a sharp maximum in this coefficient at 4.07K. A double logarithmic plot of these data acquired below T_{sg} against the reduced temperature $t = |T-T_{sg}|/T_{sg}$, based on the choice of $T_{sg} = 4.07K$ (from figure 27) is reproduced in figure 28. Comparisons between these data and those shown below T_{sg} in figures 19 and 20 for the direct paramagnetic-spin-glass transition in Pd+5 at.% Mn indicate that the increase in a(t) as $t \to 0_-$ is much weaker in this supposedly re-entrant system. Further, the exponent value of $\gamma'_- = 3.6\ (\pm 0.6)$ deduced from the line drawn in figure 28, while being fairly close to model values, should be regarded with even more reservation that those estimated from figure 20; it is obtained from a very limited temperature range and at large reduced temperature values where corrections to scaling are expected to be considerable.

There are, of course, many objections to this type of analysis in supposedly re-entrant systems. In real (3-dimensional) spin systems the nature of the coupling between the measured, longitudinal response and the conjectured spin-glass freezing of the transverse components of the spins is not clear.[61] For Pd(Fe,Mn) in particular, (i) no additional experimental data accurately determining/confirming T_{sg} currently exist; (ii) the measured anomaly in a(t) is clearly not divergent, and is thus much weaker the Ising model predictions; in fact (iii) this anomaly is weaker than that measured near the direct para-spin-glass transition in PdMn. Points (ii) and (iii) may indicate that the critical dynamics are even more constrained near the re-entrant boundary. Some of these uncertainties have been removed, at least in part, by very recent measurements[28] of $\chi(H,T)$ in amorphous FeZr. Here the low temperature anomaly in a(T) has been shown to correspond with the onset of transverse spin freezing deduced from in-field Mössbauer studies.[62] The temperature dependence of a(T) remains comparably "sloppy" to that in Pd(Fe,Mn) so that the critical nature of the transverse freezing still needs to be established. Further measurements of the ac susceptibility (probably at ultra low frequencies ≲1 Hz) on a range of systems are clearly needed. In addition to studies on potentially re-entrant systems the behaviour of compounds such as $CeFe_2$ should also be investigated. The zero-field susceptibilities of this compound[63] and its pseudobinaries $Ce(Fe_{1-x}Co_x)_2$ and $Ce(Fe_{1-x}Ru_x)_2$ are similar to that of supposedly re-entrant alloys, although

---

*In this system the leading non-linear contribution to $\chi(H,T)$ above T_{sg} is also quadratic in H. This contradicts the model result, possibly because the spontaneous magnetisation existing above T_{sg} is obscured by domain structure which might lead to a strong suppression of the predicted linear term.

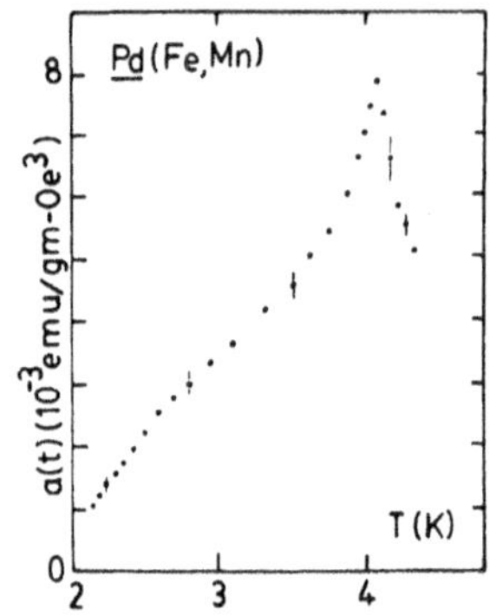

Fig. 27. The variation of the coefficient of the H^2 term in the vicinity of the lower candidate transition.

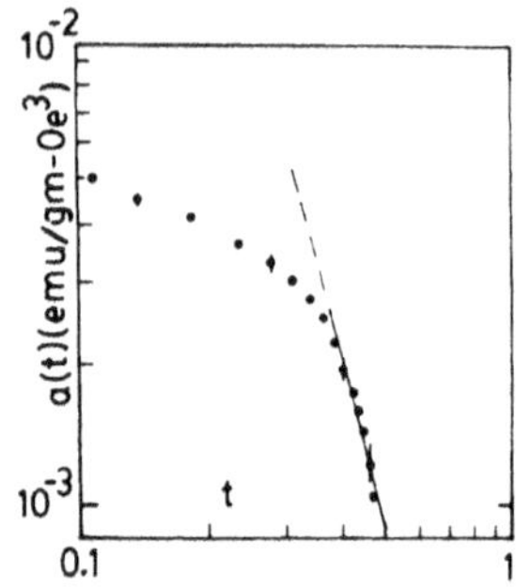

Fig. 28. The data of fig. 27 on a double logarithmic scale against the reduced temperature below T_{sg}.

here neutron scattering measurements[64] have been interpreted as indicating that a lower temperature transition to a canted spin arrangement occurs in the Co and Ru doped systems only. The behaviour of a(T) near this lower transition appears important to establish, and a comparison of the behaviour of a(T) in doped samples (which do exhibit a canting transition) and "pure" $CeFe_2$ all of which display similar zero-field behaviour would be interesting (if only in a negative sense[65]).

In summary, despite the numerous associated problems, measurements of the field-dependent ac susceptibility have been, and will continue to be, a powerful and useful probe of the various ground states that occur in magneti systems.

Acknowledgments: It is a pleasure to thank numerous colleagues, post-docs and former students, in particular Drs. I. Maartense (who designed the present susceptometer and participated in the early experiments), R.M. Roshk (who collaborated on the model calculations) and H.P. Kunkel, along with S.C Ho, E.C. Zastre, H. Ma, and Z. Wang for their many contributions to this work. Financial support from NSERC and the Research Board of the University of Manitoba has also been essential.

References

1. V. Cannella and J. A. Mydosh, Phys. Rev. B6:4220 (1972).
2. K. Binder and A. P. Young, Rev. Mod. Phys. 58:801 (1986).
3. A. H. Morrish, "The Physical Principles of Magnetism", Wiley, New York (1965).
4. M. A. Ruderman and C. Kittel, Phys. Rev. 96:99 (1954); K. Yosida, Phys. Rev. 106:893 (1957); T. Kasuya, Prog. Theor. Phys. (Kyoto) 16:45 (1956).
5. R. M. Roshko and G. Williams, J. Phys. F: Met. Phys. 14:703 (1984).
6. T. Kaneyoshi, J. Phys. C: Solid State Phys. 8:3415 (1975).
7. B. W. Southern, J. Phys. C: Solid State Phys. 9:4011 (1976).
8. D. Sherrington and S. Kirkpatrick, Phys. Rev. Lett. 35:1792 (1975).

9 D. M. Cragg and D. Sherrington, Phys. Rev. Lett. 49:1190 (1982).
10. D. J. Kim and B. B. Schwartz, Phys. Rev. Lett. 20:201 (1968).
11. J. Crangle, Phil. Mag. 5:335 (1960); R. M. Bozorth, P. A. Wolff, D. D. Davis, V.B. Compton, and J.H. Wernick, Phys. Rev. 122:1157 (1961).
12. G. J. Nieuwenhuys, Phys. Lett. 67A:237 (1978).
13. R. P. Peters, Ch. Buchal, M. Kubota, R. M. Mueller, and F. Pobell, Phys. Rev. Lett. 11:1108 (1984).
14. T. Moriya, Proc. Int. School Phys. "Enrico Fermi" (Course XXXVII), Academic Press, New York (1967).
15. I. Maartense, Rev. Sci. Instrum. 41:657 (1970).
16. J. A. Osborn, Phys. Rev. 67:351 (1945).
17. J. D. Patterson, "Introduction to the Theory of Solid State Physics", Addison-Wesley, Reading (1971).
18. P. D. Long and R. E. Turner, J. Phys. C: Solid State Phys. S3:127 (1970).
19. S. Chikazumi, "Physics of Magnetism", Wiley, New York (1964).
20. H. E. Stanley, "Introduction to Phase Transitions and Critical Phenomena", Clarendon Press, Oxford (1971).
21. B. Widom, J. Chem. Phys. 43:3898 (1965).
22. S. C. Ho, I. Maartense, and G. Williams, J. Phys. F: Met. Phys. 11:699 (1981).
23. B. Westwanski and B. Fugiel, Phys. Rev. B43:3637 (1991).
24. A. P. Malozemoff and Y. Imry, J. Mag. Mag. Mat. 31-34:1425 (1983).
25. This suggestion was made to us by B.W. Southern.
26. H. P. Kunkel, R. M. Roshko, and G. Williams, Phys. Rev. B37:5880 (1988).
27. J. S. Kouvel and M. E. Fisher, Phys. Rev. 136A:1626 (1964).
28. H. Ma, H. P. Kunkel, and G. Williams, J. Phys: Condens. Matter 3:5563 (1991).
29. L. C. LeGuillou and J. Zinn-Justin, Phys. Rev. B31:3976 (1981).
30. P. Gaunt, S. C. Ho, G. Williams, and R. W. Cochrane, Phys. Rev. B23:251 (1981).
31. Z. Wang, Ph.D. Thesis, University of Manitoba (1991).
32. H. P. Kunkel and G. Williams, J. Phys. F: Met. Phys. 18:1271 (1988).
33. S. C. Ho, I. Maartense, and G. Williams, J. Phys. F: Met. Phys. 11:1107 (1981).
34. K. Kornik, H. P. Kunkel, R. M. Roshko, and G. Williams, Solid State Commun. 76:993 (1990).
35. H. P. Kunkel, R. M. Roshko, W. Ruan, and G. Williams, Phil. Mag. B64:153 (1991).
36. S. Kaul, J. Mag. Mag. Mat. 53:5 (1985).
37. M. Fähnle, J. Mag. Mag. Mat. 65:1 (1987).
38. I. Yeung, R. M. Roshko, and G. Williams, Phys. Rev. B34:3456 (1986).
39. E. C. Zastre, R. M. Roshko, and G. Williams, J. Appl. Phys. 57:3447 (1985).
40. I. Yeung, R. M. Roshko, and G. Williams, J. Mag. Mag. Mat. 68:39 (1987); R. M. Roshko, W. Ruan, and G. Williams, Phys. Rev. B37:9806 (1988).
41. J. Chalupa, Solid State Commun. 22:315 (1977); M. Suzuki, Prog. Theor. Phys. 58:1151 (1977).
42. R. M. Roshko and G. Williams, J. Mag. Mag. Mat. 50:31 (1985).
43. E. C. Zastre, R. M. Roshko, and G. Williams, Phys. Rev. B32:7597 (1985).
44. D. M. Cragg, D. Sherrington, and M. Gabay, Phys. Rev. Lett. 49:158 (1982); M. A. Moore and A. J. Bray, J. Phys. C: Solid State Phys. 15:L301 (1982).
45. P. Nordblad, P. Svedlindh, L. Lundgren, and L. Sandlund, Phys. Rev. B33:645 (1986); L. Lundgren, P. Nordblad, and P. Svedlindh, Phys. Rev. B34:8164 (1986); M. Alba, M. Ocio, and J. Hamman, Europhys. Lett. 2:45 (1986).

46. S. Geschwind, D. A. Huse, and G. E. Devlin, Phys. Rev. B41:4854 (1990).
47. B. R. Coles and G. Williams, J. Phys. F: Met. Phys. 18:1279 (1988).
48. H. Bouchiat, J. Phys. (Paris), 47:71 (1986); L. P. Levy, Phys. Rev. B38:4963 (1988).
49. J. J. Prejean, E. Carré, P. Beauvillain, and J. P. Renard, J. Phys. (Paris), C8:995 (1988).
50. J. J. Smit, G. J. Nieuwenhuys, and L. J. de Jongh, Solid State Commun. 30:243 (1979).
51. J. R. L. de Almeida and D. J. Thouless, J. Phys. A11:983 (1978).
52. G. Parisi, Phys. Rev. Lett. 43:1754 (1979); J. Phys. A13:1101 (1980); J. Phys. A13:1887 (1980); J. Phys. A13:L115 (1980); Phys. Rev. Lett. 50:1946 (1983).
53. M. Gabay and G. Toulouse, Phys. Rev. Lett. 47:201 (1981).
54. G. Williams, Can. J. Phys. 65:1251 (1987).
55. G. Kotliar and H. Sompolinsky, Phys. Rev. Lett. 53:1751 (1984).
56. A. Fert, N. de Courtenay, and I. A. Campbell, J. Appl. Phys. 57:3398 (1985).
57. K. Kornik, M.Sc. Thesis, University of Manitoba (1989).
58. K. Kornik, R. M. Roshko, and G. Williams, J. Mag. Mag. Mat. 81:323 (1989).
59. S. C. Ho, I. Maartense, and G. Williams, J. Appl. Phys. 53:2235 (1982).
60. H. P. Kunkel and G. Williams, J. Mag. Mag. Mat. 75:98 (1988).
61. M. Katori and M. Suzuki, Prog. Theor. Phys. 74:1175 (1985).
62. D. H. Ryan, J. O. Ström-Olsen, R. Provencher, and M. Townsend, J. Appl. Phys. 64:5787 (1988).
63. S. B. Roy and B. R. Coles, Phys. Rev. B39:9360 (1989).
64. S. J. Kennedy, A. P. Murani, J. K. Cockcroft, S. B. Roy, and B. R. Coles, J. Phys: Condens. Matter 1:629 (1989).
65. Very recent measurements on undoped $CeFe_2$ (H.P. Kunkel, M.S. Westmore, and G. Williams, Phil. Mag. B (in press)) indicate that a(T) displays a single peak around T_c, with no second anomaly at lower temperature.

SPIN-GLASS AND SUPERCONDUCTING PROPERTIES

Jean-Louis Tholence

Centre de Recherches sur les Très Basses Températures
C.N.R.S., BP 166 X, 38042 Grenoble-Cédex, France

INTRODUCTION

Soon after their discovery of high T_c superconductivity in oxides, K.A. Muller *et al*[1] pointed out some similarities between the magnetization of $(La_{1-x}Ba_x)_2CuO_4$ and spin glasses. In particular, an irreversible and time-dependent component of the magnetization is observed below the so-called "irreversibility line" :

$$H_{irr}(T) \propto \left(\frac{T_c-T}{T_c}\right)^{3/2},$$

reminiscent of[2] the irreversibility below the de Almeida-Thouless (AT) line in spin glasses.[3,4] These similarities between high T_c superconductors and spin glasses exist also for various magnetic systems, classical superconductors, assemblies of Josephson junctions, etc ... Their physical origins are very different, but a better knowledge of one of these systems[5] can help one to study and understand the properties of new systems. Also, when the size of the system becomes of the order of the magnetic or superconducting correlation lengths, ξ, one expects strong changes in the properties from 3d to 2d behaviour. After a short review of the main properties of spin glasses and related systems, I'll underline similar and different behaviors observed in high T_c superconductors.

SPIN GLASSES

Spin-glass properties are observed in many magnetic materials, in which, instead of participating in a long range magnetic order, the spins can be considered as frozen in the random direction of the local magnetic field they experience. This spin-glass state is due to the oscillating character of the long-range RKKY interaction in dilute alloys (<u>Cu</u>Mn, <u>Ag</u>Mn, <u>Au</u>Fe, ...), or to the frustration introduced either by the dilution with non-magnetic ions or by the disorder (between spins and/or interactions) in concentrated systems with short range interactions ($Eu_xSr_{1-x}S$, $Eu_xGd_{1-x}S$, ...). Qualitatively, spin-glass like properties are observed in a large variety of magnetic materials which may be summarized as :

1°) A cusp in the thermal dependence of the susceptibility measured in a small alternating field (a.c. susceptibility) of pulsation or angular frequency ω. At high temperature this susceptibility obeys a simple Curie or Curie-Weiss law.[6]

2°) A difference between the magnetization measured during cooling in a small continuous (d.c.) field (it is the field cooled magnetization, f.c.), and the magnetization measured for increasing temperatures, after applying the field at low temperature (or zero field cooled magnetization, z.f.c.). This difference $M_{f.c} - M_{z.f.c}$ is time-dependent and corresponds to a hysteresis, or remanent magnetization, effect.[7-10]

Magnetic Susceptibility of Superconductors and Other Spin Systems
Edited by R.A. Hein *et al.*, Plenum Press, New York, 1991

Only a more careful study of these properties allows one to determine which systems are simple superparamagnets and which are spin glasses.

1. Superparamagnets

In an assembly of superparamagnetic particles, non interacting or weakly interacting, the magnetic moments can relax between equilibrium positions with a relaxation time $\tau = \tau_o \exp(W/kT)$ where $\tau_o \sim 10^{-13}$ s and W is a potential barrier[11-12]. τ increases abruptly when the temperature decreases.[10] In an experiment with characteristic measuring time t, an anomaly will occur when τ becomes larger than t. For t = 1s this will occur at a blocking temperature, T_b, such that $W/kT_b = \ln(1/\tau_o) = 13 \ln 10 \sim 30$ [11-12]. In the case of a distribution of relaxation times, anomalies in the a.c. susceptibility at T_g (for instance the maximum of χ_{ac} or a 5 % deviation from the high temperature Curie-Weiss law) are associated with the maximum or with some average of the largest relaxation times of the distribution. The same

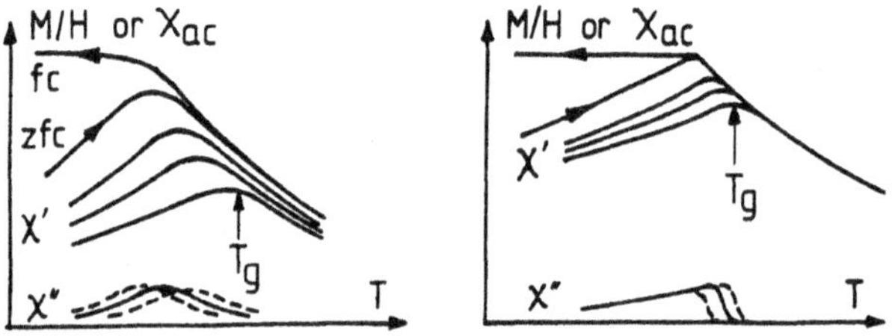

Fig. 1 : Field cooled (f.c.) and zero field cooled (z.f.c.) d.c. susceptibility defined as $(M/H)_{H\to 0}$ for : a) superparamagnetic particles, b) spin glass such as CuMn and compared with the a.c. susceptibility. The irreversibility and frequency dependence of $\chi_{a.c}$ (real part:χ', imaginary part:χ'') extend well above the temperature of $\chi_{max.}$ in the case of a superparamagnet.

Arrhenius law is observed to describe the time dependence, T_g (t), of these anomalies in χ_{ac} .

$$t = \tau_o \exp(W/kT_g),$$

In the case of interacting particles another parameter is needed, and the Vogel-Fulcher law :

$$t = \tau_o \exp(W/k(T_g - T_o)),$$

with $0 < T_o < T_g$, has been found successfull in many cases[13,15]. In superparamagnets T_o is small compared to T_g, which corresponds to large (Arrhenius like) frequency dependences of χ_{ac} and T_g. These large variations of χ_{ac} and T_g with frequency correspond to large $(T_g - T_o)/T_g$ values (> 0.5) which can be taken as a criterion to classify various freezing magnetic systems . Moreover, time dependent effects are still observed above T_g.

On the contrary, in good spin-glasses, T_o is close to T_g, and small variations of χ_{ac} and T_g with frequency are associated with small $(T_g-T_o)/T_g$ values (typically : $\Delta T_g/T_g\Delta\log_{10}\nu \sim \Delta\chi_{ac}/\chi_{ac}\Delta\log_{10}\nu \sim (T_g-T_o)/T_g < 0.2$ for audio frequencies ν)[15,16]. Moreover time effects are only observed around and below T_g.

Finally the magnetization of superparamagnets, which can be developed as a function of H/T, $M = a_1(H/T) + a_3(H/T)^3 + a_5(H/T)^5$ ...., has regular field and temperature dependences. In particular no divergence of any non linear term (b3, b5, ...) is observed if a comparison with the effective temperature dependence of the first term $\chi_o H$ is made in the series expansion[13] :

$$M = \chi_o H - b_3(\chi_o H)^3 + b_5(\chi_o H)^5 \ \ldots .$$

On the contrary, in good spin-glasses, the non linear terms diverge at the transition temperature with critical exponents characteristic of the transition.

At low temperature, logarithmic time-dependent effects are observed on the thermo-remanent magnetization (TRM) obtained after field cooling, and the isothermal remanent magnetization (IRM) obtained after zero field cooling. In superparamagnets these effects are associated with a large distribution P(W) of activation barriers : $W = kT_b \ln(\tau/\tau_o)$ between two level systems These potential barriers can be overcome by thermal activation or after a corresponding waiting time. This results in similar effects for a temperature change (ΔT) or a lnt change (Δlnt) with a possible superposition of the magnetization curves as a function of the composite variable : kT ln t [2,9,14].

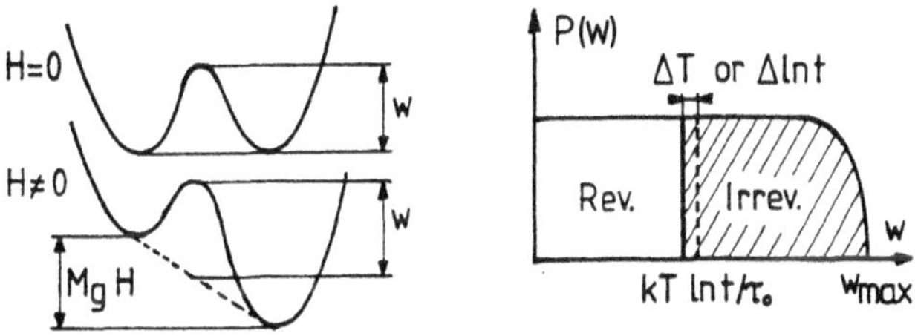

Fig. 2 : Exemple of a two level system. Activation above the potential barrier,W , can be obtained by a temperature change ΔT, a waiting time Δlnt, or ... a magnetic field. The reversible magnetization is due to the low energy side of the P(W) distribution of potential barrier.

With such a large distribution, at a temperature T, and with a measuring time t, all magnetic entities having $T_b > T$ and $\tau > t$ can contribute to the low temperature hysteresis. In particular the saturated remanent magnetization is due to all the particles situated on the right of $kT \ln(t/\tau_o)$. The temperature dependence of the saturated remanent magnetization is found to decrease exponentially with temperature $M \sim M_o \exp(-T/T_o')$ with $T_o' \sim T_g/3$, which depends not only on P(W) but also on the value of the magnetic moments having activation barriers W. The low field TRM is observed to decrease linearly with T.

A more complete description of the field dependence of the magnetization can be obtained if, the effect of a magnetic field H on this distribution of potential barriers is considered[14]. For instance in figure 3, after zero field cooling, all the particles on the left of $kT \ln t/\tau_o$ give a reversible contribution to the magnetization, proportional to the field H and increasing with $kT \ln t/\tau_o$, like the reversible susceptibility. A fraction, H/H_{max}, (where H_{max} allows to return entities of potential barrier W_{max}) of the entities on the right of $kT \ln t/\tau_o$ gives an irreversible contribution to the magnetization proportional to H^2, before becoming proportional to H in fields larger than H_{max}. In particular the shape of the low field hysteresis curve, the general field dependence of M(H), IRM(H) and TRM(H) can be predicted using such Preisach's diagrams[14].

2. Spin Glasses

In these systems the a.c. susceptibility cusp is sharper, its frequency dependence smaller, and all irreversibility effects disappear just above T_g. The frequency dependence of χ_{ac} below the cusp temperature T_g can be characterized by the quantity $(1/T_g)dT_g/d\log_{10}(t)$ which is constant and of the order of a few thousandths for CuMn. Experimentally the frequency dependence of the cusp of the a.c. susceptibility can be described either by a Vogel-Fulcher law[15] :

$$\tau = \tau_o \exp[W/k(T-T_o)],$$

where $\tau_o \sim 10^{-13}$s and $0 << T_o \lesssim T_g$, or by a power law where $T_o < T_c < T_g$ and $z\nu \sim 7\text{-}10$ [16]:

$$\tau = \tau_o[T/(T-T_c)]^{z\nu}.$$

These laws are very difficult to distinguish experimentally, since about the same quality of fits is obtained, and one can be developed as an expansion of the other. One can notice that T_o (of the VF law) is smaller than T_c (power law) which itself corresponds better to the

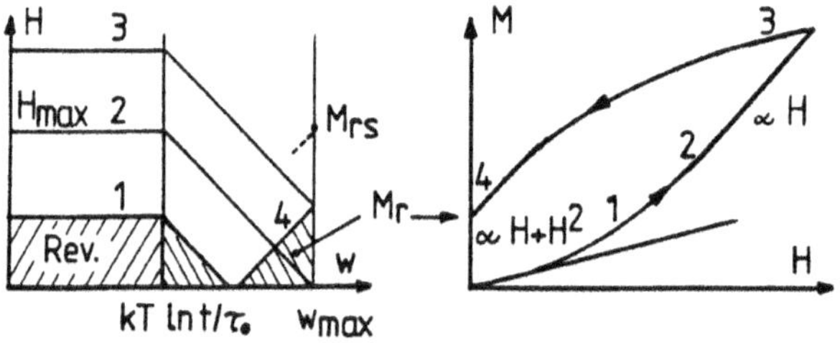

Fig. 3 : The field effect on the activation of two-level systems can describe the complex field dependence of M, using Preisach like diagrams[14]. The magnetization contains a reversible term proportional to H, and an irreversible one, proportional to H^2 for $H < H_{max}$, then proportional to H.

anomaly observed in field cooled magnetization curves. This would be in favor of the power law[16]. Moreover, a small T_o (compared to T_g) corresponds to a large value $z\nu$ (> 10), while, for T_o close to T_g, $z\nu$ tends to a value of 7 or 8. This gradual change in the frequency dependence of T_g from system to system explains the difficulty in characterizing the spin-glass transition.

A transition is usually characterized by a correlation length ξ which diverges when the transition temperature, T_c, is approached from above : $\xi = \xi_o(1-T_c/T)^{-\nu}$. The relaxation time associated to "entities" of size ξ depends on the model. In the critical slowing down model, W/kT is proportional to $\ln\xi$ and[17,32] $\tau \sim \tau_o(1-T_c/T)^{-z\nu}$. Assuming that W is a power of ξ a variation $\tau \sim \tau_o \exp(B/(T-T_o)^\sigma$ is expected.[18] Other frequency dependences are expected, which always are similar to a VF or a power law (the VF law corresponding in fact to the first terms of a series expansion of the power law). The variety of values of critical exponents $z\nu$ between 7 and ∞ remains one problem of spin-glasses. No transition (i.e. $T_c = 0$) is expected theoretically for Heisenberg spin glasses,[19] and $z\nu \sim 7$ is obtained numerically for 3d Ising spin glasses [20]. Experimentally, Heisenberg like systems seem to exhibit the transition ($z\nu \sim 7$) expected for Ising spin glasses.

Another aspect of the spin-glass transition has been pointed out by M. Suzuki.[21] It focuses on the divergence of non-linear terms in the low field magnetization : $M = a_1(H/T) + a_3(H/T)^3 + a_5(H/T)^5 + \dots$. In a ferromagnet the first term diverges at T_c. In spin glasses the first term is regular, only higher order terms diverge giving critical exponents γ and β :

$$a_3 = a_3^{\infty}(1 - T_c/T)^{\gamma}$$

$$\text{and} \quad a_5 = a_5^{\infty}(1 - T_c/T)^{-2\gamma-\beta}.$$

In spin glasses, such a divergence of $a_3(T)$ and $a_5(T)$ at T_c was initially observed in the non-linear a.c. susceptibility by Miyako *et al.*[22], then in the magnetization of several systems with small frequency dependence of T_f. On the contrary, it is not observed for systems for which $T_c \sim 0$ (and if one takes care to compare the non-linear terms to the effective temperature dependence of the linear one)[12].

Experimental values $\gamma \simeq 3$ and $\beta = 1$ are usually found[28] for the critical exponents of Heisenberg spin glasses, while smaller γ values (1-2)[23], characteristic of an Ising behavior, have been found in small fields (i.e. H< 1000 Oe) and closer to T_c[23]. Two opposite interpretations have been given: i, Heisenberg spin glasses present a transition and γ can be underestimated in low field experiments if T_c is overestimated[28], ii, the existence of any weak anisotropy energy gives a Ising character to the low field transition of an Heisenberg spin-glass[23]. Another exponent, δ, gives the field dependence of the magnetization at T_c :

$$M \propto H^{2/\delta} \text{ at } T_c.$$

Large magnetic fields destroy the frozen spin-glass state, up to a paramagnetic state. The effect of a field allows one to define critical lines, the de Almeida-Thouless and the Gabay-Toulouse lines, with characteristic temperature dependences :

$$H_{A.T.} \simeq (1 - T/T_c)^{3/2},$$

$$H_{G.T.} \simeq (1 - T/T_c)^{1/2}$$

The de Almeida-Thouless line separates the spin-glass state of an Ising spin-glass from its paramagnetic state. In Heisenberg spin glasses, the Gabay-Toulouse line separates the paramagnetic state (high field, high temperatures) from the freezing of transverse degrees of freedom, the total freezing arriving only below the A-T line. Although the de Almeida-Thouless line is usually identified experimentally from the onset of magnetic irreversibility, the existence of the G.T. line is more controversial. A cross-over line $[(T-T_c)/T_c] \propto H^{2/\phi}$ marks, at high temperature, the departure of the paramagnetic state from a Curie or Curie-Weiss law.[24,25]

Irreversibility lines (with A.T. like thermal dependence), are also found at the onset of irreversibility, in other systems, like superparamagnetic particles, high T_c superconductors, ... However, since the irreversibility is time dependent, the irreversibility lines are also time dependent. Then, only accurate measurements of this time dependence (for instance from a.c. susceptibility measurements) allow one to demonstrate that some characteristic relaxation time τ diverges for a non zero temperature (in a magnetic field) when the measuring frequency ω tends to zero. Therefore, only in the case where the irreversibility line still exists for $\omega = 0$, can it be identified with the A.T. transition line.[26,27] Obviously, there is no transition line if, already in null field, $T_c(h=0,\omega=0) \simeq 0$.

The low temperature irreversibility of spin glasses gives a saturated remanent magnetization which decreases exponentially with temperature[7,8,9,10,28] for a given measuring time.

$$M_{r_s} = M_{r_o} \exp(-T/T_o').$$

The magnetization is a universal function of the $T\ln(t/\tau_o)$ variable[28], therefore at low temperature one expects to have :

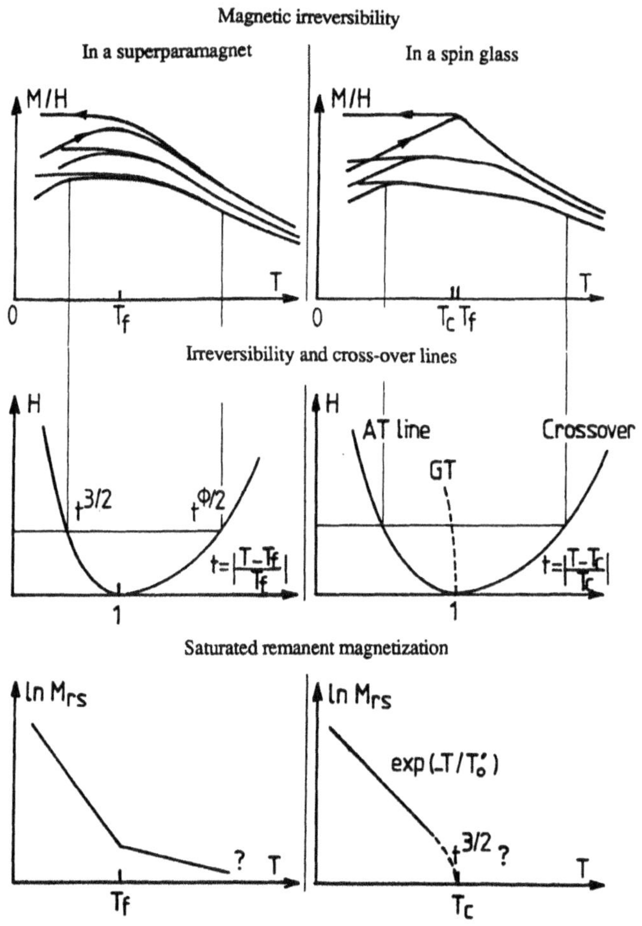

Fig. 4 : Comparison of the irreversibility observed in a superparamagnet (Fe particles in Al_2O_3)[13] or CuMn. Critical lines (AT, GT lines) and exponents are predicted for the spin-glass transition. The magnetic behavior of a superparamagnet (T_c=0) exhibits some similarities with that of a spin-glass.

$$M_{r_s} = M_{r_o} \exp\left[-(T/T_o)\ln t/\tau_o\right] \quad \text{with } T_o \sim 10 \times T_c$$

and $$M_{r_s} \sim M_{r_o} (t/\tau_o)^{T/T_o}.$$

which corresponds to the time dependence observed experimentally.

This time dependence is often approximated (for limited times of measurement and accuracy) by a logarithmic law :

$$M_{r_s} = M_{r_{so}}(1-\ln t),$$

Close to T_c a stretched exponential[29-31] :

$$M_{r_s} \sim M_{r_{so}} \exp\left[-(t/\tau)^{\beta(T)}\right],$$

or a power law for the time dependence:

$$M_{r_s} \sim M_{r_{so}}(t/\tau_o)^{\alpha(T/T_g)},$$

are observed as well as an effect of the waiting time (before a change in field is made to look at time effects).

Both these temperature and time dependences of the remanent magnetization indicate that it results from a large distribution, P(W), extending down T = 0. However the low temperature irreversibility of spin glasses cannot simply be described as an assembly of two-level systems as for superparamagnets, for which the ground state would always be the same. On the contrary, the different possibilities for each spin to freeze in different positions creates other possibilities for further freezing of other spins when the temperature is lowered. This reasoning led to the ideas of the growing of an infinite cluster[32] or of a hierarchical model[33] which give an infinity of possible ground states, coming from the high temperature side.

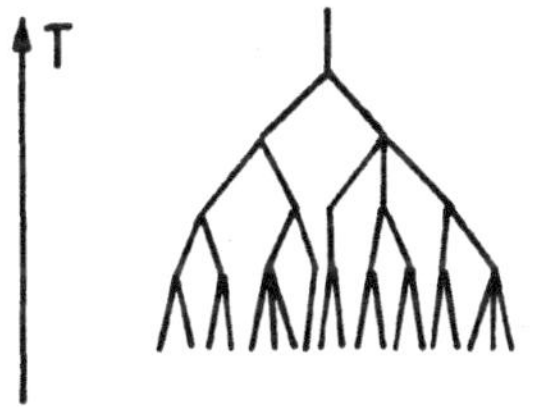

Fig. 5 : Hierarchical model, with multiple possibilities of reaching ground states from high temperature.

In this case it is possible to justify a gaussian distribution[33,34] of $\ln\tau$, which Castaing and Souletie[35] have found to be a consequence of the Kadanoff's renormalisation applied to spin-glasses.There are several ways to determine the distribution $P(\ln\tau)$. The Cole-Cole diagrams close to T_c (χ'' is plotted versus χ' for various frequencies) allow one to reach the distribution of relaxation times.[36,37] Lundgren *et al.*[38] have shown that χ' decreasing linearly with $\ln(\omega)$ also means that there exists a large distribution $g(\ln\tau)$ with a correlation $\chi''(\omega) = -(\pi/2)[d\chi'(\omega T)/d\ln\omega]$ well verified experimentally and $T\chi'' \propto g(\ln\tau)$. Also from the time dependence of the magnetization $S = dM/d\ln t$ one can get $P(W) \sim S/T$ which is found to be constant at low temperature. All these techniques are perturbative since the applied field is changed, and only noise experiments allow one to reach the distribution of relaxation times without changing the system.[23]

3. Summary

Some of the properties associated with the spin-glass transition (for instance the A.T. line), appear to be much more general, since a similar behavior is observed in other magnetic or non-magnetic systems.[1-3,13] It is the case in particular of :

1°) The $t^{3/2}$ behavior of the $H_{irr}(T)$ irreversibility line in low field.
2°) The exponential behavior of the saturated remanent magnetization at low temperature: $M_{r_s} \sim M_{r_{so}} \exp(-T/T_o')$ with $T_o' \sim T_c/3$.
3°) Universal laws to represent irreversible properties as a function of $T\ln t/\tau$ which come from 4°).
4°) A large distribution for the logarithm of the relaxation time $\ln\tau$. The exact form of which depends on the model, and can, for instance, be checked through noise experiments[23]

The study of dilute alloys CuMn, AgMn with RKKY interactions ... or of some more concentrated systems with short-range interactions ($Eu_xSr_{1-x}S$, $Eu_{0.5}Gd_{0.5}S$, Mn alumino-silicates ... has given arguments in favor of the 3D spin-glass transition. Many other classes of systems (random anisotropy systems like DyNi[39], random field systems, Ising or XY alloys, dilute antiferromagnets...) have been studied which qualitatively present similar properties but have "critical exponents" intermediate between those of canonical spin glasses (CuMn, AgMn, ...) and superparamagnets.

A.c. susceptibility measurements have been a powerful tool to characterize the spin-glass transition, starting from the cusp which was the first indication of some kind of transition. The divergence of non-linear terms of the magnetization was first shown from a study of the harmonics in the a.c. susceptibility. The small frequency dependence of anomalies in $\chi_{a.c.}$ allowed one to set some limit for $T_c(\omega=0)$ either in null field, or in a field,to get the de Almeida-Thouless transition line from the irreversibility.

One can expect drastic changes in the spin-glass properties of films, when the size of the film becomes small compared to the coherence length ξ. The effects must be important in the case of a RKKY spin glass where a change from a 3D transition to a 2D one is possible. In that case the properties of a thin film should be compared to those of, for instance, a 2D spin glass like $Fe_{0.3}Mg_{0.7}Cl_2$[40] which does not have a transition at finite temperature (T_c = 0). Several studies on thin films have been reported.[41-43] The comparison between a 10^4 Å and a 30 Å film gives a T_c reduced from 66 K down to ~ 26 K. At the same time the quantity $(1/T_f)dT_f/d\log_{10}t$ increases from 0.005 to 0.025 indicating stronger dynamical effects also shown in the exponent $z\nu$ of $t \sim \tau_o[(T_f-T_c)/T_c]^{-z\nu}$ which increases from 9 to 19. All these properties are in better agreement with calculations[44,45] and theory[46] for a 2D Ising spin glass since $\ln t \propto T_f^{-2.6}$ fits better the frequency dependence of T_f. This behavior has been followed for thicknesses W_{SG} down to 1 or 2 monolayers of CuMn and AgMn with a thickness dependence of $T_f \propto W_{SG}^{0.8}$ as predicted[46] for a 2D Ising spin glasses with T_f still larger than 0 for 1 monolayer film.

HIGH T_c SUPERCONDUCTORS (HTSC) MAGNETIZATION

The magnetic properties of high T_c superconductors have been extensively studied, and now some of the properties of classical type II superconductors are revisited in the light of what have been learnt from HTSC.

The magnetization of HTSC is of quite different origin than in other magnetic systems, since it is due to screening currents and the penetration of vortices inside the superconducting material. However, some similarities can be noted : after zero field cooling, the screening of an external field by surface currents is perfect up to H_{c_1} and the apparent magnetization $M = -(1/4\pi)\,H$ (per cm^3). Then vortices start to penetrate the sample with a gradient of the induction dB/dr equal to the critical current density J_c. The magnetization can be approximated using the Bean model where a characteristic dimension of the section perpendicular to the magnetic field is introduced, as well as the pinning force which fixes dB/dr.

Then, as with Preisach's diagrams, the magnetization above H_{c1} can be derived giving deviations in H^2 from the initial slope. Characteristic fields for the minimum magnetization (1) and to saturate the remanent magnetization (2) are marked in fig 6.

In very small fields ($H \lesssim 100$ Oe) the magnetization of a ceramics sample is due to currents circulating at the surface of the sample, through Josephson junctions between grains (when the field increases, the flux penetrates between the grains). The Bean model applies, using the sample diameter, and spin-glass like properties are observed (see for instance Z. Koziol[47]) : a small frequency dependence of T_g at the maximum of χ'', $\omega = \omega_o(T_g-T_c)^{z\nu}$ with $z\nu = 3$, an irreversibility line in a d.c. field H given by $T_g(0)-T_g(H) \propto H^{2/3}$. Spin-glass like properties can be attributed to the distribution of Josephson junctions, pinning centers, vortices or macroscopic flux lines, etc ... which dominate the low field superconductivity. However different behaviors for the irreversibility are observed depending on the sample (Barbara[48], Giovanella[49]).

In larger fields, the Bean model still applies for ceramics samples, but one must use the grain size (or the crystal size for crystals) since only the intra-grain currents tend to screen the applied field (fig.7). The magnetization above H_{c1}, or the change ΔM in magnetization

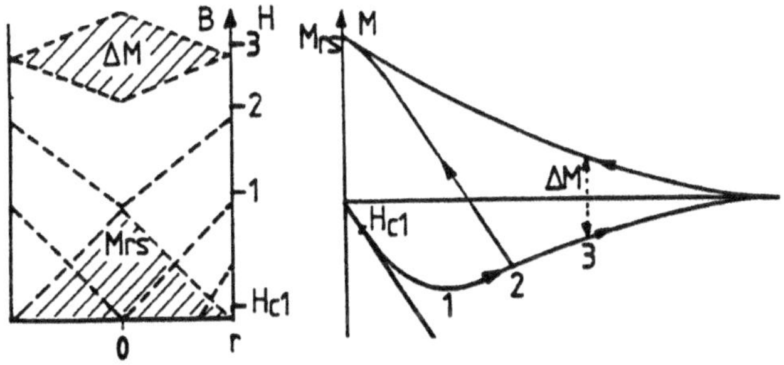

Fig. 6 : The magnetization of HTSC can be constructed from the Bean model. Some improvements consist to take dB/dr varying with r to take into account non linear sample size effects, and DM decreasing to zero when the field increases up to H_{c2}.

for increasing or decreasing fields, or the remanent magnetization M_{rs} can be used to determine the critical current density, Jc.,below Hirr.(T)[50-53] In the case of anisotropic superconductors the various components of Jc can be determined from a study of the magnetization for fields applied along various directions [54-57].

The main features obtained from magnetization measurements (fig.8) are that :

1°) The saturated remanent magnetization decreases exponentially with T, at low temperature (T < 50 K)

$$M_{r_s}(T) = M_{r_s}(0)\exp(-\frac{T}{T_o'}),$$

with $T_o' \sim 20$ K for $YBa_2Cu_3O_7$ and $T_c = 92$ K, for a field applied parallel to the c-axis or perpendicular to the c-axis.[58]

As in the case of spin glasses the time and temperature dependence of M_{r_s} can be written more generally as :

$$M_{r_s}(T,t) \sim M_{r_{so}} \exp(-\frac{kT}{U_o} \ln \frac{t}{\tau_o}),$$

with: $U_o \sim T_o' \ln t/\tau_o \sim 200\text{-}600K.$

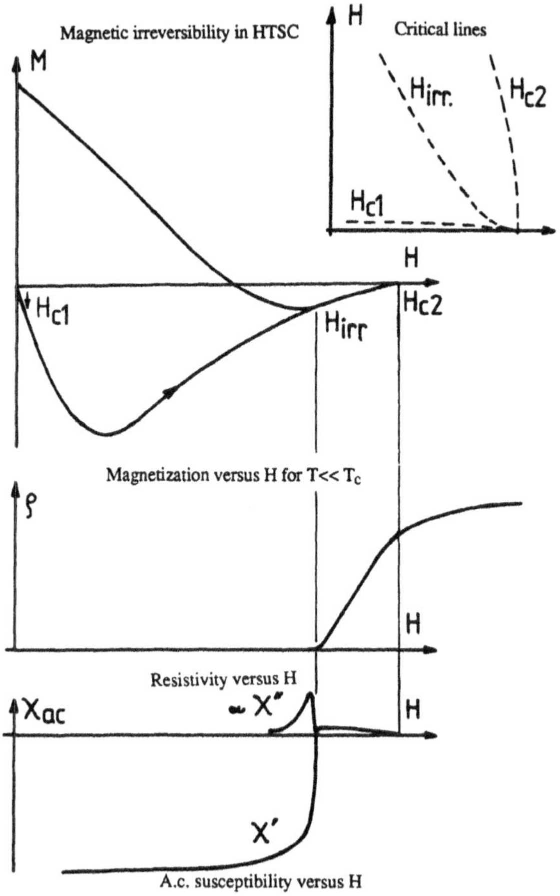

Fig. 7 : Critical fields H_{c1}, H_{c2}, and possible experimental determinations of $H_{irrev.}(T)$ from the magnetization, the resistivity, or the a.c. susceptibility. The detection and study of non linear effects (in resistivity, susceptibility ...measurements) are expected to give the frequency dependence of $H_{irrev.}(T)$ and to characterize a possible transition.

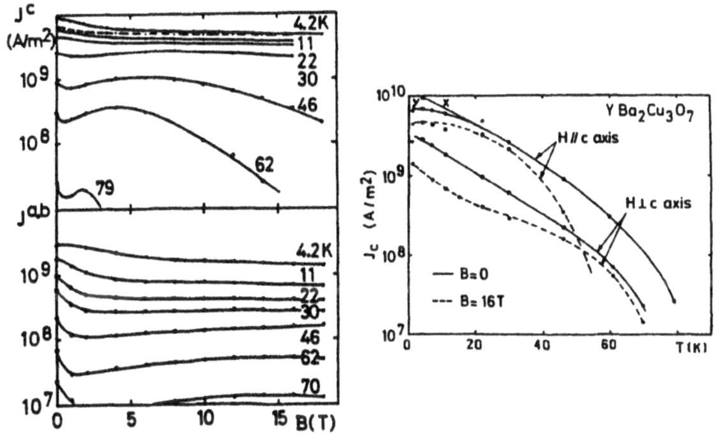

Fig. 8 : Magnetization of a single crystal of YBCO, for two field directions. Note the magnetic jumps observed at low temperature, and the cigar shape of the magnetic hysteresis at high temperature. The field dependence, $J_c(H)$, and exponential behavior of $J_c(T)$ in null field, are shown on the lower part (from ref. 58).

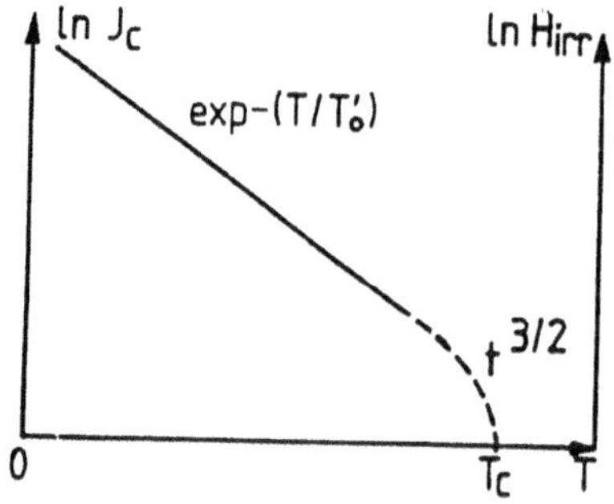

Fig. 9 : Expected variations of $J_c(T)$, $M_{r_s}(T)$ or $H_{irr}(T)$

2°) The anisotropy is conserved in the all temperature range for $J_c(0)$. It is of ~ 4-5 for the average values of J_c, obtained after H has been applied parallel or perpendicular to the c axis. Larger anisotropies between the various components of J_c are found.

3°) In $YBa_2Cu_3O_7$ a maximum in $J_c(H)$ is observed in a large field, which could be due to a pinning by oxygen defects which in a field become normal (on a microscopic scale) and pinning centers.[58]

4°) This magnetic irreversibility disappear on the irreversibility line where $\Delta M = 0$. We have shown that on the same irreversibility line, $\rho(T)$ tends to zero, $\chi'(T)$ becomes diamagnetic, $\chi''(T)$ starts to increase (and presents a maximum at lower temperatures), with small differences being attributable to time effects in the experimental definition of $H_{irr}(T)$ (fig.7).[53] Obviously, $J_c(T)$ as deduced from magnetic measurements vanishes on the irreversibility line. However, resistive measurements have shown the persistence of critical currents above the irreversibility line in thin films.

5°) The irreversibility line has been shown to vary as $[(T_c-T)/T_c]^{3/2}$ close to T_c. Far from T_c (it is the case for BiSCCO[(59)] where the irreversibility fields are smaller) $H_{irr} \propto \exp(-T/T'_o)$.[59] However, this thermal behavior of $H_{irr}(T)$ is not particular to HTSC, it has been observed since then in a Chevrel phase[60] and classical type II superconductors[61].

6°) The critical current J_c decreases as $t^{3/2}$ (where $t = (T_c - T)/T_c$) close to T_c. From these experimental observations one can make a parallel between the general thermal dependence of $J_c(T)$ and $H_{irr}(T)$ which would result from the conservation of the shape of the irreversible magnetization for increasing T : $M_{r_s} \propto H_{irr}$. Both $J_c(T)$ and $H_{irr}(T)$ should decrease as $\exp(-T/T'_o)$ at low temperature, then as $t^{3/2}$ close to T_c (as shown in fig.9). However, this low temperature behavior of $H_{irr}(T)$ can be out of reach experimentally, for instance in the case of $YBa_2Cu_3O_7$ where H_{irr} becomes too high.Otherwise, $H_{irr}(T)$ can be strongly reduced in nonhomogeneous systems, which is often the case with Bi or Tl compounds, and give a variety of thermal behaviors.

The irreversibility line of high T_c superconductors is associated with the melting of the vortex lattice.If this melting is due to a thermally activated process on the pinning of vortices the time dependence of the irreversibility line is expected to be Arrhenius like, without any transition. In other cases (vortex glass, effect of the disorder on the vortex lattice,transiton from a 2D to a 3D superconductivity, etc....) a kind of transition is expected close to the irreversibility line, which, therefore, would tend to a limit for infinite measuring time. The mixed state above the irreversibility line and below $H_{c2}(T)$, still needs to be studied in more detail, in HTSC and classical type II superconductors.[59-61]

Experimentally, the frequency dependence of $H_{irr}(T)$ is often weak [50-52], however, it is not sure that its time dependence can be fitted in general with a power law :

$$\omega \sim \omega_0[(T-T_c)/T_c]^{z\nu}$$

which would be in favor of a real transition line, T_c (H), when ω tends to zero. Like for any transition, non linear behaviors are expected (for instance in the M(H), V(I), a.c.susceptibility, electrical or magnetic noise, electrical or magnetic losses....) at this transition, and sometimes observed[67]. Obviously, the transition at T_c (H), must be characterized by various critical exponents which still need to be accuratly predicted in various models, and determined experimentally on good crystals.

Typical behavior for HTSC which have a marked 2D character (Bi or Tl phases) has been predicted with important effects of 2D fluctuations.[62-63]

HTSC Films

This 2D like behavior can be studied in artificial systems. For instance in $YBa_2Cu_3O_7$, the effect of interactions between successive CuO planes must be responsible for the 3D character of the superconductivity. Very careful studies of sandwiches with layers of (superconducting) $YBa_2Cu_3O_7$[64-65] and (normal) $PrBa_2Cu_3O_7$ have allowed one to observe the effect of the decrease of the interactions when the number of $PrBa_2Cu_3O_7$ layers increase, and to follow the superconducting temperature of a decreasing number of $YBa_2Cu_3O_7$ layers. Limiting T_c values of 50 K and 10 K have been obtained respectively for two (24 Å) or one (12 Å) $YBa_2Cu_3O_7$ layers[64,65], implying the existence of superconductivity in an isolated single cell layer .

SUMMARY

In HTSC the observed $T^{3/2}$ behavior of $J_c(T)$ and $H_{irr}(T)$ is just the Ginzburg-Landau behavior close to T_c. The low temperature exponential behavior of $H_{irr}(T)$ in Bi compounds has been attributed to the critical field at which the "quasi 2D" superconductivity localized in CuO_2 planes at high temperature, extends to the normal part (between superconducting planes) and becomes 3D.[59] However, since then, the same exponential like behavior has been found in Chevrel phases[60] and, possibly, in classical type II superconductors[61] where there is no structural reason for "2D like" superconductivity. It appears then that this is a much more general behavior, which probably appears as soon as a large enough distribution of activation (or pinning) energies exists, possibly due to the disorder on the vortex lattice.

In spite of the similarities in the magnetic behavior of spin glasses (or related magnetic systems) and HTSC, the physical origins of these magnetizations are very different. In HTSC, the magnetization is due to vortices penetrating inside the superconductor. Their number, at the difference of spins in a magnetic system, strongly depends on the magnetic field, on the temperature, on the number and efficiency of pinning centers (critical current), on the defects which can favor their penetration, on the anisotropy and on the geometry of the sample (Bean model). The variety of pinning centers (oxygen defects efficient in high field in the case of $YBa_2Cu_3O_7$, other atomic defects, twins more efficient in lower fields ...) and the possibility of multiple pinning centers on the same vortex might be at the origin of a large distribution of pinning energies in HTSC and ... in other superconductors. The irreversibility line and its time dependence certainly depend on the sample quality. Several experimental criteria for the irreversibility line can be proposed : onset of diamagnetic χ', onset of χ'', deviations from the linearity of the V(I) curves[67], onset of a critical current[67], noise, torque, sound velocity experiments, appearence and divergence of non-linear effects (3rd, 5th .. harmonics in χ'), etc ... . Then, the existence of a "transition" line can be proved by extrapolation to infinite measuring time of $T_c(H_{irr})$ and the transition characterized by the determination of critical exponents above,at,and below the obtained transition temperature... as was done for the spin-glass transition[23].

N.B.:Some evidence in favor of a vortex-glass transition has recently been found in the a.c. impedance of thin films by H.K. Olsen et al.[68].

Remark

Magnetic jumps are observed on the magnetic hysteresis cycle of many magnetic materials (including spin glasses) and of type II superconductors (including HTSC[58]). They appear in the critical regime either spontaneously, or after a small perturbation (field change, T change, waiting time t ...) and are the catastrophic (avalanche effect) consequence of a weak local heating which starts a macroscopic magnetization reversal (in magnetic materials) or flux penetration (HTSC). In HTSC these effects must be stopped, through a better thermalization of the superconductors and improved pinning, for technological applications.

Aknowledgement

I thank J. Souletie for fruitful discussions when preparing this paper.

References

1. K.A. Muller, T. Takashige and J.G. Bednorz, Phys. Rev. Lett. **58**, 1143 (1987).
2. J.J. Préjean, J. Souletie, Phys. Rev. Lett. **60**, 1884 (1988).
3. A detailed discussion of critical lines can be found in: Y. Yeshurun and A.P. Malozemoff, Phys. Rev. Lett. **60**, 2202 (1988).
4. J.R.L. de Almeida and D.J. Thouless, J. Phys. A **11**, 983 (1978).
5. A discussion of reentrant spin glasses is given by Gwyn. Williams, this workshop.
6. V. Cannella, J.A. Mydosh and J.I. Budnick, J. Appl. Phys. **42**, 1689 (1971).
7. J.L. Tholence and R. Tournier, J. Physique **35**, C4-229 (1974) ; see also Physica **86-88B**, 875 (1977).
8. J. Souletie and R. Tournier, J. Low Temp. Phys. **1**, 95 (1969).
9. J.J. Préjean, J. Physique **39**, C6-907 (1978).
10. H. Bouchiat and P. Monod, J. Mag. Mag. Mat. **3**, 175 (1982).
11. L. Néel, Ann. Geophys. **5**, 99 (1949).
12. E.P. Wohlfarth, Phys. Lett. **70A**, 489 (1979).
13. D. Fiorani, J.L. Tholence and J.L. Dormann, J. Phys. C **19**, 5495 (1986).
14. J. Souletie, J. Physique **49**, 1211 (1988) ; J. Physique **51**, 883 (1990) and references therein.
15. J.L. Tholence, Solid State Commun. **35**, 113 (1980) ; Physica B **126**, 157 (1984).
16. J. Souletie and J.L. Tholence, Phys. Rev. B **32**, 516 (1985).
17. C.L. Henley, Phys. Rev. Lett. **43**, 2030 (1985).
18. D.S. Fisher and D.A. Huse, Phys. Rev. Lett. **56** (1986) 1601.
19. A.J. Bray, M.A. Moore and P.A. Young, Phys. Rev. Lett. **56**, 2641 (1986).
20. A. Ogielski and I. Morgenstem, Phys. Rev. Lett. **54**, 924 (1985) ; A. Ogielski, Phys. Rev. B **32**, 7384 (1985)..
21. M. Suzuki, Prog. Theor. Phys. **58**, 1151 (1977).
22. S. Chikazawa, C.J. Sandberg and Y. Miyako, J. Phys. Soc. Jpn **50**, 2884 (1981).
23. H. Bouchiat, N. de Courtenay and P. Monod, Jpn J. Appl. Phys. **26-3**, 1951 (1987).
24. B. Barbara, A.P. Malozemoff, and Y. Imry, J. Appl. Phys. **53**, 7662 (1982) ; A.P. Malozemoff, S.E. Barnes and B. Barbara, Phys. Rev. Lett. **51**, 1704 (1983).
25. For reviews on spin glasses and related systems see : Magnetic glasses, by K. Moorjani and J.M.D. Coey in Methods and Phenomena **6** (Elsevier, 1984, Eds. S.P. Wolsky and A.W. Czanderna).

26. C. Paulsen, S.J. Williamson and H. Maletta, Phys. Rev. Lett. **59**, 128 (1987).
27. N. Bontemps, J. Rajchenbach, R.V. Chamberlin and R. Orbach, Phys. Rev. B **30**, 6514 (1984).
28. R. Omari, J.J. Préjean, J. Souletie, J. Physique **44**, 1069 (1983).
29. R.V. Chamberlin, G. Mozurkevich and R. Orbach, Phys. Rev. Lett. **52**, 867 (1984).
30. M. Alba, M. Ocio and J. Hamman, Europhysics Lett. **2**, 45 (1986).
31. L. Lundgren, J. Physique **49**, C8-1001 (1988) and references therein.
32. R. Rammal et A. Benoit, Phys. Rev. Lett. **43** (1985).
33. R.G. Palmer, D.L. Stein, E. Abrahams and P.W. Anderson, Phys. Rev. Lett. **53**, 958 (1984).
34. K.L. Ngaï, Comments Solid State Phys. **9**, 127 (1979).
35. B. Castaing and J. Souletie, J. Phys. I-**1**, 403 (1991).
36. C. Dekker, A.F.M. Arts, H.W. de Wijn, A.J. van Duyneveld and J.A. Mydosh, Phys. Rev. Lett. B **61**, 1780 (1988).
37. A.J. Dirkmaat, G.J. Nieuwenhuys, J.A. Mydosh, P. Kettler and M. Steiner, Phys. Rev. B **36**, 352 (1988).
38 L. Lundgren, P. Svedlindh and O. Beckman, J. Mag. Mag. Mat. **25**, 33 (1981).
39. B. Dieny and B. Barbara, Phys. Rev. Lett. **57**, 1169 (1986).
40. J.P. Redoules, R. Bendaoud, A.R. Fert, D. Bertrand, J. Vétel, J.C. Lasjaunias, J. Souletie and J. Ferré, J. Physique **51**, 1911 (1990).
41. G.G. Kenning, J.M. Slaughter, and J.A. Cowen, Phys. Rev. Lett. **59**, 2596 (1987).
42. L. Sandlung, P. Granberg, L. Lundgren, P. Nordblad, P. Svedlindh, J.A. Cowen and G.G. Kenning, Phys. Rev. B **40**, 869 (1989).
43. L. Hoines, R. Stubi, R. Loloee, J.A. Cowen and J. Bass, Phys. Rev. Lett. **66**, 1224 (1991).
44. A.T. Ogielski, Phys. Rev. B **32**, 7384 (1985).
45. W. Kinzel and K. Binder, Phys. Rev. B **29**, 1300 (1984).
46. D.S. Fisher and D.A. Huse, Phys. Rev. B **38**, 373 (1988) ; B **38**, 386 (1988) ; B **36**, 8937 (1987).
47. Z. Koziol, Physica C **159**, 281 (1989).
48. B. Barbara, A.F. Khoder, M. Couach and J.Y. Henry, Europhys. Lett. **6**, 621 (1988).
49. C. Giovannella, C. Chappert and P. Beauvillain, Europhys. Lett. **5**, 535 (1988).
50. J.H.P.M. Emmen, G.M. Stollman and W.J.M. de Jonge, Physica C **169**, 418 (1990).
51. T.K. Worthington, W.J. Gallagher, D.L. Kaiser, F.H. Holtzberg and T.R. Dinger, Physica C **153-155**, 32 (1988).
52. A.P. Malozemoff, T.K. Worthington, Y. Yeshurun, F. Holtzberg and P.H. Kes, Phys. Rev. B **38**, 7203 (1988).
53. O. Laborde, J.L. Tholence, P. Monceau, H. Noel, P. Gougeon, J. Padiou, J.C. Levet and M. Potel, Solid State Commun. **69**, 509 (1989).
54. E.M. Gyorgy, R.B. van Dover, K.A. Jackson, L.F. Schneemeyer and J.V. Waszczak, Appl. Phys. Lett. **55**, 283 (1989) ;
55. F.M. Sauerzopf, H.P. Wiesinger and H.W. Weber, Cryogenics **30**, 650 (1990).
56. J.C. Martinez, thesis (Grenoble 1991).
57 J.L. Tholence, M. Saint-Paul, O. Laborde, P. Monceau, M. Guillot, H. Noel, J.C. Levet, M. Potel, J. Padiou and P. Gougeon, in Studies of "High Temperature Superconductors", Vol. 6, p. 37 (Nova Science Publishers Inc., New York, 1990, Ed. Narlikar).
58 J.L. Tholence, H. Noel, J.C. Levet, M. Potel and P. Gougeon, Solid State Comm.**65**, 1131 (1988).
59 P. de Rango, B. Giordanengo, R. Tournier, A. Sulpice, J. Chaussy, G. Deutscher, J.L. Genicon, P. Lejay, R. Retoux and B. Raveau, J. Physique **50**, 2857 (1989).
60 C. Rossel, E. Sandvold, M. Sergent, R. Chevrel and M. Potel, Physica C **165**, 233 (1990).
61 M. Suenaga, A.K. Ghosh, Youwen Xu and D.O. Welch, Phys.Rev.Lett. **66**, (1991).
62 D.A. Huse and H.Sebastian Seung, Phys. Rev. **B42** , 1059 (1990).
63 A.P. Malozemoff and M.P.A.Fisher, Phys. Rev.**B42**, 6784 (1990).
64 J.M. Triscone, Ø. Fischer, O. Brunner, L. Antognazza and A.D. Kent, Phys. Rev. Lett. **64**, 804 (1990).
65 L. Antognazza, J.M. Triscone, O. Brunner, M.G. Karkut and Ø. Fischer, Physica B **165-166**, 471 (1990).

66 Q. Li, X.X.D. Wu, A. Inam, S. Vadlamannati and W.L. McLean, Phys. Rev. Lett. **64** (1990).
67 L. Civale and T.K. Worthington, this workshop.
68 H.K. Olsen, R.H. Koch, W. Eidelloth and R.P. Robertazzi, Phys. Rev. Lett. **66** , 2661 (1991).

NOVEL LOW FIELD AC MAGNETIC SUSCEPTIBILITY TECHNIQUE IN UHV:

MAGNETISM OF *hcp* Gd(0001).

Francisco H. Salas

Departamento de Física, Facultad de Ciencias
Universidad de Oviedo, E-33007 Oviedo, Spain
and Departamento de Física, Instituto Tecnológico y
de Estudios Superiores de Monterrey (ITESM)
64849 Monterrey, N.L., México.

INTRODUCTION

Gadolinium was the fourth ferromagnetic element to be discovered and ever since it has drawn a lot of attention among scientists due to a number of interesting properties, some of which were initially reported as long ago as 1935 by Urbain et al.[1]. Gadolinium is unique for its high magnetic moment and for its special Curie point lying just at room temperature. In addition, gadolinium shows large anomalies in practically all its magnetic, elastic, thermal, and electrical properties.[2] Recently, an UHV-compatible low field ac initial magnetic susceptibility [$\chi^{ac} \equiv \chi(\omega$=constant, $T)$] technique has been developed and applied to the system *hcp* Gd(0001)/ *bcc* W(110),[3] and the so-called "butterfly susceptibility" was recorded as well[4]. In this paper I present a review of the first experiment in which χ^{ac} was measured *in situ* in ultrahigh vacuum (UHV) by the use of an inductive method. This novel technique, having nanometer resolution, has motivated further experimental work on ultrathin films.[5]

EXPERIMENTAL DETAILS

Experimental setup for measuring $\chi^{ac} \equiv \chi(\omega$=constant, $T)$

The experimental setup[3] for measuring χ^{ac} in UHV consists of the modified Hartshorn bridge-type susceptometer shown in Fig. 1. A Wavetek 183 is used as a signal generator containing a quartz-crystal oscillator in order to stabilize the operating frequency, the latter being in the audio frequency range. The primary coil is driven by the signal generator via a transformer (not shown in the figure). The secondary consists of two identical solenoidal sections wound in opposition to each other and connected in series. The output of the secondary coils, which is proportional to the magnetic susceptibility of the sample, is detected by a two-phase lock-in amplifier (PAR 5204) operated in the external modulation mode, so that the commutating frequency of the lock-in will correspond to the stabilized frequency of the signal generator. Actually the coil system is placed *ex situ* by fitting it to the UHV chamber through a coaxial quartz finger. This allows one to *in situ* keep the Gd(0001) thin film within the quartz finger during the measurements as seen in Fig. 2, whereas the ac magnetic field is applied *ex situ* as shown in Fig. 3.

Magnetic Susceptibility of Superconductors and Other Spin Systems
Edited by R.A. Hein *et al.*, Plenum Press, New York, 1991

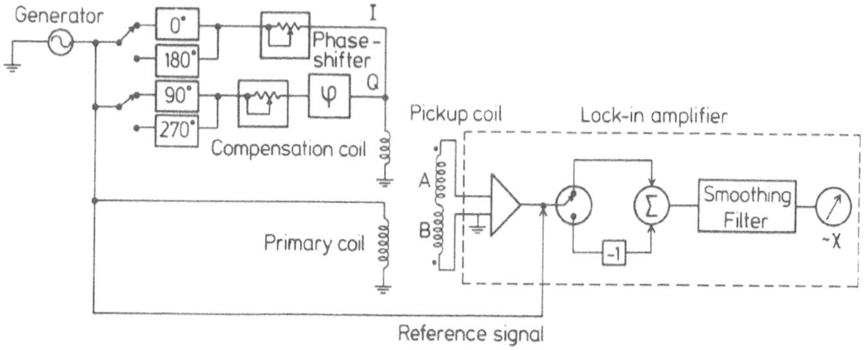

Figure 1. Experimental setup. (as taken from Ref.3)

The electronic balance of the secondary part of the coil system in a given range of sample susceptibilities is done by means of a passive network driving a compensation coil. A small fraction of the signal of the generator is sampled and separated into its components in-phase I and quadrature Q. Then I, Q and the phase are carefully changed until the signal of the secondary is nulled out.

Although the W(110) substrate used has the advantage of presenting a nonmagnetic behaviour, it nevertheless has the problem of the eddy currents arising from its metallic character. It is very important to do the electronic balance of the ac bridge once the Gd(0001)/W(110) film is placed into one of the secondary coils and at a temperature where no paramagnetic Gd signal is detected, e.g., T = 350K . However, due to the problem of the eddy currents, it is also necessary to adjust the phase of the lock-in amplifier for each operating frequency. This should be done

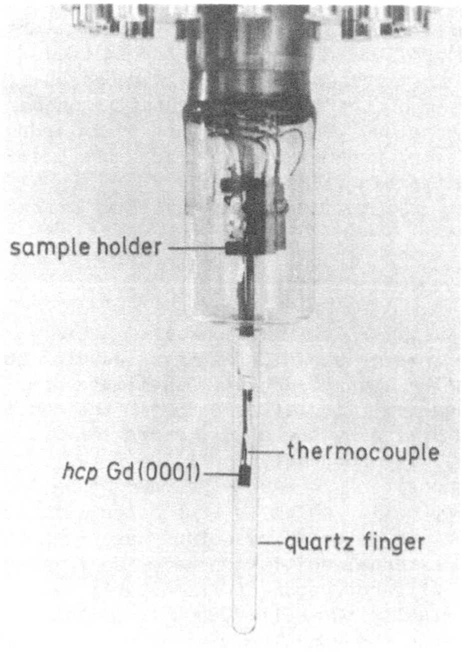

Figure 2. *In situ* Gd(0001) thin film in UHV.

Figure 3. *Ex situ* pickup coil.

prior to the evaporation of the Gd film while the W substrate is kept inside the pickup coil. In particular, it was found that the phase of the lock- in amplifier, at which the signal induced by W vanishes (in both ferromagnetic- and paramagnetic phases of Gd), corresponds approximately to the phase at which a maximum Gd signal (between the background and the Hopkinson maximum[6]) is observed. In other words, the Gd and W signals were held in quadrature to each other with an accuracy of ±2° in the temperature range 250 K < T < 325 K. Therefore, these measurements only refer to the in-phase part of the ac magnetic susceptibility of the Gd sample.

With regard to the sample preparation, the Gd(0001) films are evaporated (in UHV conditions) onto the surface of an Auger clean W(110) single crystal in order to get an epitaxial growth[3,7,8]; the deposition of Gd(0001) onto W(110) in terms of the Stranski-Krastanow growth mode[9] is monitored by taking the peak-to-peak ratio of the $Gd_{138/140}$ and $W_{163/169}$ Auger lines versus deposition time, the deposition rate being equal to 1.2 picometers per second.

Experimental setup for measuring the "butterfly susceptibility" $\chi_{\alpha\beta} \equiv \chi(\omega=\text{constant}, T=\text{constant}, H_{dc})$

By slightly modifying the above experimental setup it is possible to measure the "butterfly susceptibility", i.e., with a dc biasing field, H_{dc}, applied to the sample and measuring χ^{ac} under isothermal- and isofrequency conditions. This quasi-static method has been used by authors for many years.[10] Nevertheless, the present experiment[4] on Gd(0001)/W(110) is the first one in which the butterfly susceptibility is measured *in situ* in UHV by the use of an inductive method. In order to do this, H_{dc} is applied by means of a pair of Helmholtz coils either longitudinally or transversally to the alternating field (H_{ac}). Both H_{dc} and H_{ac} are applied *ex situ*. A quasi-statical sweep of the bias field allows χ^{ac} to be recorded as a function of H_{dc} under isofrequency- and isothermal conditions, i.e., $\chi_{\alpha\beta} \equiv \chi(\omega=\text{constant}, T= \text{constant}, H_{dc})$. Here α refers to the

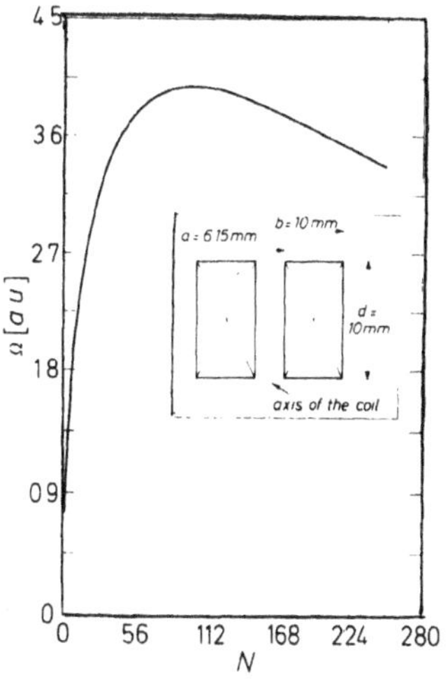

Figure 4. Geometric factor of the signal-to-noise ratio versus the number of turns in the radial direction. The inset shows (not to scale) the cross section of the pickup coil. (as taken from Ref.3)

biasing mode (as explained below) and β refers to the angle between the easy axis of magnetization[11] and H_{dc}. From $\chi_{\alpha\beta}$ one gets valuable information about the reversal process of the magnetization in ferromagnetic thin films. In particular, two cases are studied:

(i) transverse biased initial susceptibility (TBIS): $\alpha=t$ (from "transversal") , i.e., $H_{dc} \perp H_{ac}$ and (ii) longitudinal biased initial susceptibility (LBIS): $\alpha=l$ (from "longitudinal") , i.e., $H_{dc} \parallel H_{ac}$.

Optimization of the pickup coil

In measuring *in situ* the ac magnetic susceptibility of thin films, it is crucial to optimize the geometrical characteristics of the pickup coil.[3]

The cross section of the pickup coil used in the present experiment is shown (not to scale) in the inset of Fig. 4. The signal being induced by the sample in the pickup coil is proportional, among other things, to the operating frequency, the modulation amplitude, and the number of turns of the pickup coil.

The signal-to-noise ratio can be written in a first approximation as

$$\frac{\text{signal}}{\text{noise}} \propto \frac{N_T}{\bar{r}^2} \times \frac{1}{\sqrt{4kTR\Delta f}} . \tag{1}$$

Here $\bar{r} = \sqrt{ab}$ is the geometrical radius of the coil (a and b refer to the inner and outer radius, respectively), N_T is the total number of turns of the coil, and $\sqrt{4kTR\Delta f}$ is the Johnson noise (k is the Boltzmann constant, T is the absolute temperature, R is the coil resistance, and Δf is the bandwidth in which the measurements are made).

The total number of turns is

$$N_T = \left(\frac{d}{\phi}\right) N, \tag{2}$$

where d is the vertical length of the pickup coil, ϕ is the wire diameter, and N is the number of turns in the radial direction.

On the other hand, the coil resistance is

$$R = \rho \frac{l}{s} = \rho \frac{\left(2\pi aN + \pi\phi\ (N-1)N \right) \cdot \dfrac{d}{\phi}}{\pi \dfrac{\phi^2}{4}} , \tag{3}$$

where ρ is the wire electrical resistivity, l is the wire length, and s is the cross-section of the wire.

By substituting the values of $\bar{r}$, N_T, and R into the expression (1), one obtains

$$\sqrt{4kT\Delta f}\ \frac{\text{signal}}{\text{noise}} \propto \frac{1}{2} \sqrt{\frac{\phi d}{\rho}}\ \frac{N/(\phi aN + a^2)}{\sqrt{2aN + \phi N^2 - N\phi}} . \tag{4}$$

Since the area of the W substrate amounts to $8.5\times 5.15\ mm^2$, d was chosen to be 10mm in order to attain the largest vertical filling factor. Thus the problem now consists of maximizing (4) with respect to N. To do that, Eq. (4) is rewritten and a geometrical factor Ω is defined as follows:

$$\Omega \equiv \left[(\phi a N + a^2)\sqrt{\frac{2a-\phi}{N}} + \phi \right]^{-1} . \tag{5}$$

In Fig. 4, Ω is shown as a function of N for ϕ=0.04 mm (diameter of the Cu wire that was wound) and a=6.15 mm (≈ radius of the quartz finger adapted to the UHV chamber). This curve has a maximum for N=95, which corresponds from Eq. (2) (if the pitch factor is equal to 1) to an ideal value of N_T=23,750. The actual pitch factor (for b=10mm and d= 10mm) was equal to 0.421, giving a value of N_T =10,000. If one considers that the pitch factor affects both vertical and radial winding directions in the same way, then the pickup coil has approximately 62 turns in the radial direction. This value of N still is very favourable as is seen in the curve shown in Fig. 4. In fact, one gets $\Omega(N=62)/\Omega(N=95) \approx 0.97$.

The primary coil consisted of two layers (1,347 turns altogether) wound onto the coils **A** and **B** of Fig. 1 using Cu wire with a diameter of 0.1mm. The small compensating coil consisted of 20 turns, which were wound onto the coil **B** also with a Cu wire of 0.1 mm of diameter. In addition, the coils **A** and **B** were separated by a distance of 13mm in order to obtain a good differential signal.

RESULTS AND DISCUSSION

Results for $\chi^{ac} \equiv \chi(\omega=\text{constant}, T)$

After having prepared and characterized the Gd(0001) thin film, the latter is moved *in situ* with a conventional (x, y, z, ϕ) manipulator into the UHV clean quartz finger adapted to the UHV chamber until the sample is placed carefully in the position of the *ex situ* pickup coil as shown in Figs. 2 and 3.

χ^{ac} was measured during the cooling cycle in a driving rms field of ≈ 2 Oe (applied in the plane of the film) when the samples were approximately 20 min old. The sample temperature was monitored by a precalibrated (PtAuPd 52/46/2 - PtRh 95/5) thermocouple. Typical results of these measurements are depicted in Fig. 5; the induced voltage (in μV) is shown as a function of the output voltage of the pallaplat thermocouple (in mV). The arrows indicate some relevant temperatures. In Fig. 5 one sees that the *in situ* UHV-compatible ac magnetic susceptibility technique is sensitive to 10 nm corresponding to 10^{16} Gd atoms. Thus, the sensitivity limit of these measurements is about 10^{16} atoms/Oe_{rms}. For $T \gtrsim 298K$, a rapid fall in sensitivity accompanied by increased experimental scatter was observed. Working with a higher frequency f causes a large instability in the background of the signal due to the fact that the eddy current losses are proportional to f^2.[12]

In the temperature dependence of χ^{ac} (see Fig. 5a), three ranges are distinguished, viz., (i) for 250K < T < 287K , χ^{ac} decreases nearly linearly with increasing temperature, (ii) for 287K < T < 290K, the rotation of Gd spins is thermally assisted[13] and the Hopkinson effect[6] takes place, i.e., χ^{ac} goes through a local maximum[14] centered around the so-called Hopkinson temperature T_H= 289±1K and (iii) above T_H, the proximity of the temperature to the Curie point ($T_C \approx 292K$) produces a rapid decrease in the signal detected. Note that the inflection point (just above T_H) is identified as T_C. It is important to note that T_H is frequency independent at least in the frequency range 180 Hz < f < 480 Hz. The occurrence of a local maximum in χ^{ac} of Gd samples might be explained in terms of the thermally assisted spin rotations. The frequencies of thermal fluctuations of magnetic moments of Gd are about 12 meV, while the energies of d-f exchange excitations and "s-l" excitations among d electrons are equal to 0.7 eV and 0.325 eV, respectively.[15] On the other hand, the product kT_H in Gd is close to 25 meV. Thus this energy range is propitious to activate thermally the spin rotations.

Fig.6 shows the effect of film contamination on χ^{ac} for a Gd(0001) film 80 nm thick (the same sample as in Fig.5a).[14] The quenching of the Hopkinson maximum upon contamination of the film caused by exposing it to atmospheric conditions for 7 hrs is evident. As usual, the gain or loss of voltage expressed in decibels is twenty times the logarithm of the voltage ratio. Thus, the signal attenuation at T_H = 289±1K is found to be[14] $A_V \approx 20 \log_{10}(0.98/5.8) = -15.4$ dB , which is attributed to the high enthalpy of reaction of gadolinium with oxygen: -16.7×10^5 J/mol.[16] Note

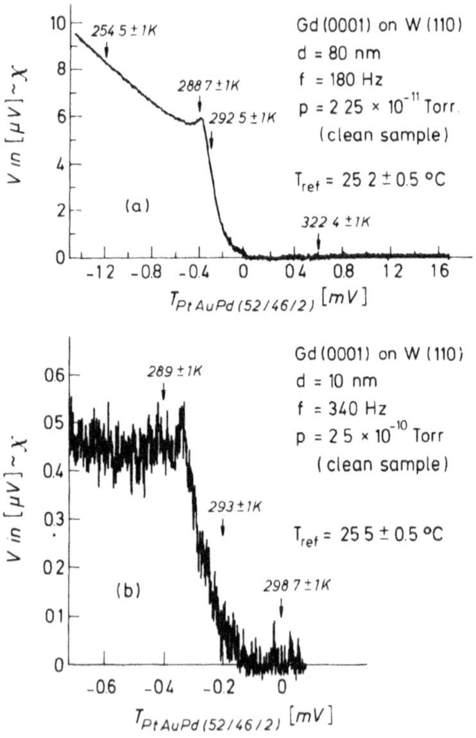

Figure 5. Voltage induced by the gadolinium samples in the pickup coil versus output voltage of the thermocouple used. (as taken from Ref. 3)

that the values of $V_{in}(T=T_H)$, before (5.8μV) and after (0.98μV) contamination of the film, are the values of V_{in} (see Fig.6) normalized with respect to 180 Hz. In both cases of Fig.6 the driving rms field was about 2 Oe.

An additional interesting point refers to the sharpness in the drop of χ^{ac} for quasi-bulk thicknesses. This is illustrated in Fig.7 for a Gd(0001) film 12 nm thick, this film thickness being equal to about 42 atomic layers. The *ex abrupto* change in χ^{ac} within 4K above the Hopkinson maximum indicates[17] that quasi-bulk Gd(0001) clearly retains the first-order magnetic phase transition, which characterizes the surface of a semi-infinite Gd(0001) system[18]. This has been interpreted as the occurrence of tricritical behaviour[19], i.e., the change from a second-order phase transition to a first-order one.

Results for $\chi_{\alpha\beta} \equiv \chi(\omega=\text{constant}, T=\text{constant}, H_{dc})$

In Fig.8, three butterflies are compared with each other and are qualitatively in good agreement. In Fig.8a, an "authentic butterfly" is shown as a function of bias field. The name of the measurement method becomes evident *per se*. In this figure also appear both magneto-optic Kerr-effect (MOKE)[7]- and TBIS[4] hysteresis loops. The magnetic state of the film is determined by the bias field and such spectra indicate how easy, or how difficult, it is for H_{ac} to "agitate" the magnetic domains, i.e., how easy, or how difficult, it is for a butterfly to flutter at the corresponding points on the χ^{ac} versus T curve. In Fig.9 the TBIS loops are shown at various frequencies and temperatures. For the data of Figs.9

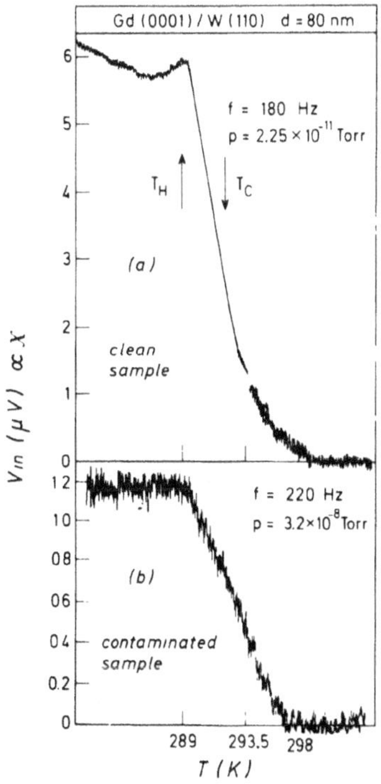

Figure 6. effect of film contamination on χ^{ac}. (as taken from Ref.14)

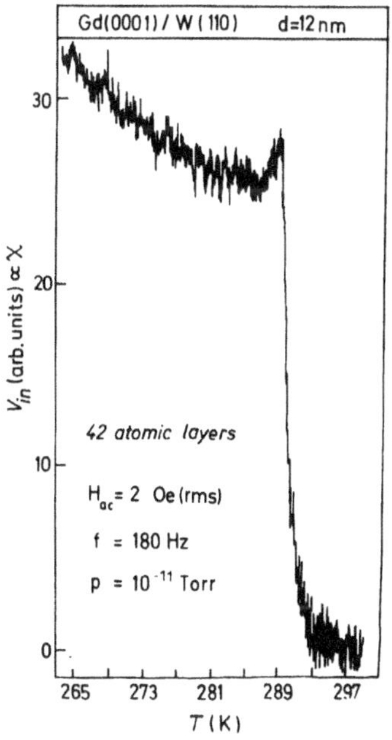

Figure 7. χ^{ac} of quasi-bulk Gd(0001).(as taken from Ref.17)

(a,b,c,d) the bias field was applied in the basal plane. In Fig. 9a, a virgin curve can be seen. The paramagnetic phase (Fig. 9c) shows the hysteresis disappearing and one also observes the intrinsic drift of the pickup coil. In Fig. 9d the typical loop for uniaxial anisotropy is shown, and in Fig. 9e reversible processes are observed in the case of biasing the film perpendicularly to the basal plane. In Figs. 10 and 11 the LBIS and TBIS loops are depicted for a thickness of 40 nm. Both TBIS and LBIS are hysteretic. From the maxima in the curves of Figs. 9a and 9b it can be concluded that the coercive field (H_c) is equal to 26 and 22 Oe for f=480 Hz and f=340 Hz, respectively. At $H_{dc} = H_c$ the macroscopic magnetization vanishes. Similarly, Figs. 10 (a,b) show that H_c= 13.9 Oe. In all cases of Figs. 10 and 11 the film was biased in the basal plane.

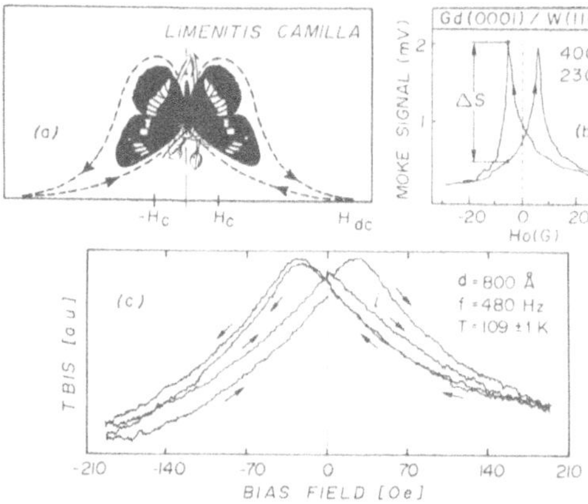

Figure 8. Three butterflies: (a) *Limenitis Camilla* (Ref. 4), (b) MOKE (Ref. 7), and (c) TBIS (Ref. 4).

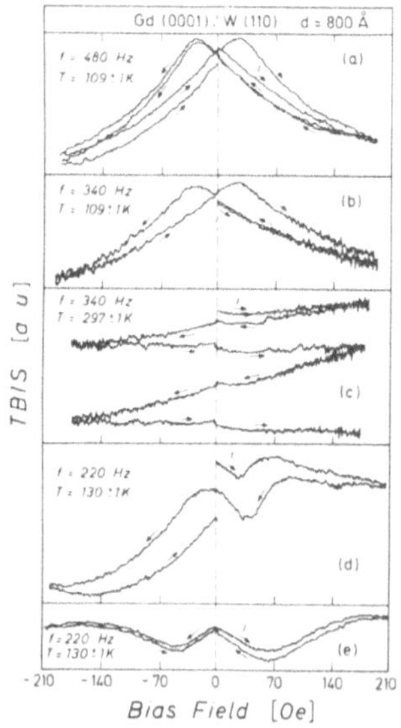

Figure 9. *In situ* TBIS loops. (as taken from Ref. 4)

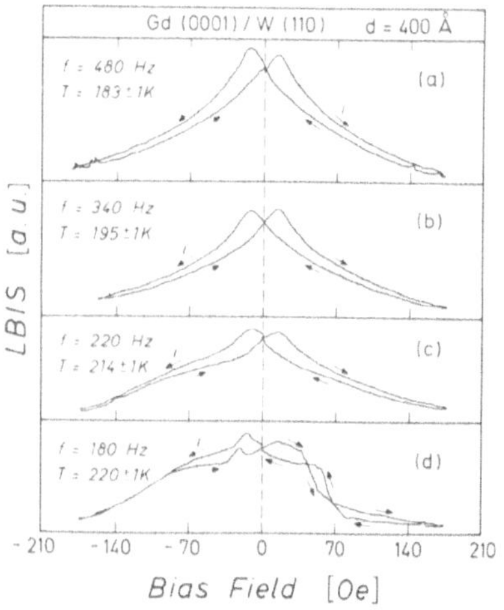

Figure 10. *In situ* LBIS loops. (as taken from Ref. 4)

Finally, in Fig. 12 a few representative Frölich-Kennelly plots are shown. The representation of the inverse susceptibility as a function of the applied field was proposed by Kennelly[20] a century ago. In all cases of Fig. 12, the existence of macroscopic anisotropy becomes clear due to the fact that the linear extrapolation of the data for high fields does not cut the bias field axis at the origin.[21] This means that the effective field acting on the film consists on both an applied and a macroscopic anisotropy field. In this study, the Stoner-Wohlfarth[22] type coherent rotations of magnetization are predominant for bias fields $\gtrsim$ 165 Oe. However, for bias fields $\lesssim$ 165 Oe (shaded areas) the elastic forces may play a similar role as the role of the magnetic forces, and therefore, irreversible processes are observed. Presumably, the plots shown in Fig. 12 suggest that the domain structure in zero applied field has the nature of that discussed by Kittel[23], in which domains are magnetized perpendicular to the plane of the film.

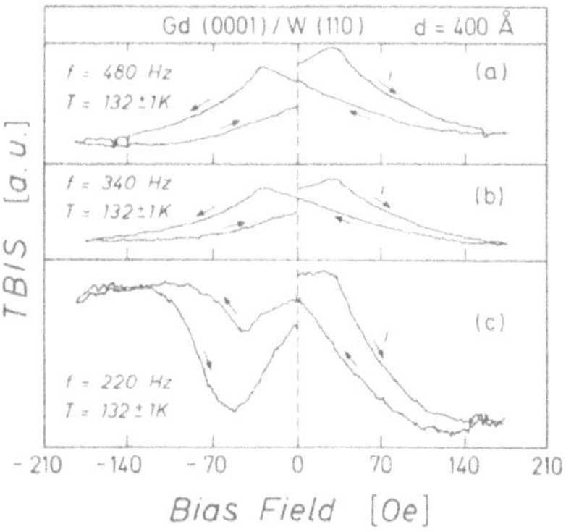

Figure 11. *In situ* TBIS loops.

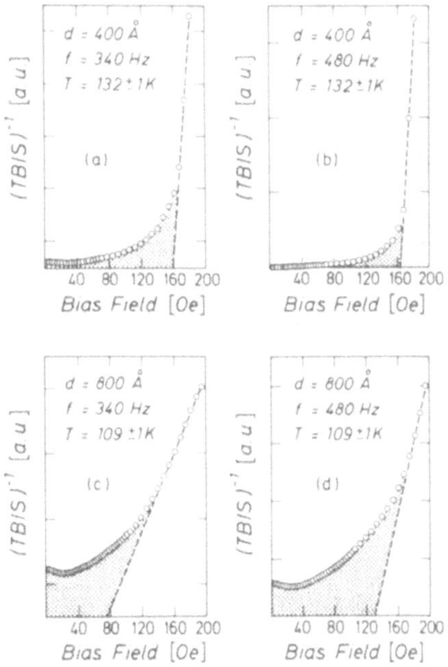

Figure 12. Frölich-Kennelly plots. (as taken from Ref. 4)

SUMMARY

The conventional method of ac magnetic susceptibility has been combined with the ultrahigh-vacuum technique offering new possibilities in studying UHV-clean magnetic thin films in the low-applied-field regime. This technique allows one to scrutinize, for example, the Hopkinson effect, the butterfly susceptibility, and the critical behaviour of magnetic thin films. The sensitivity limit, for films thicker than 10 nm, is equal to 10^{16} atoms/ Oe_{rms}. *Hcp* Gd(0001) exhibits a Hopkinson maximum at T_H=289±1 K and tends to reach the Stoner-Wohlfarth field threshold at about 165 Oe. Gd(0001) films 12 nm thick (42 atomic layers) might present tricritical behaviour, i.e., the change from a second-order magnetic phase transition to a first-order one.

ACKNOWLEDGMENTS

I would like to thank the European Community (EURAM Grant No. MA1E-0081-C) and the División de Ciencias y Humanidades del ITESM (Monterrey, Nuevo León, México) for their partial financial support which made possible my participation in this workshop. I would also like to give special thanks to Professor J. M. Alameda for his support at the University of Oviedo (Spain) and to Dr. M. Mirabal-García for his help during part of the experiments carried out at the Institut für Atom- und Festkörperphysik, Freie Universität Berlin (Germany). I also thank Professor M. A. Pérez (CINVESTAV-IPN, México) and Professor E. Olivares Santana (Aguascalientes, México) for their support. The technical assistance of M. A. Muñiz is also acknowledged. Finally, I gratefully thank Professor R. A. Hein (University of Washington), Dr. T. L. Francavilla (NRL, Washington, D.C.), and Dr. D. H. Liebenberg (ONR, Arlington, VA) for inviting me to participate in the workshop.

REFERENCES

1. G. Urbain, P. Weiss, and F. Trombe, Un nouveau métal ferromagnétique, le gadolinium, C. R. Acad. Sci. (Paris) 200:2132 (1935).
2. B. Coqblin, "The Electronic Structure of Rare-Earth Metals and Alloys: the Magnetic Heavy Rare-Earths," Academic, London (1977).
3. F. H. Salas and M. Mirabal-García, *In situ* ac magnetic susceptibility of gadolinium thin films, J. Appl. Phys. 66:5523 (1989).
4. F. H. Salas and M. Mirabal-García, *In situ* butterfly susceptibility of Gd thin films, J. Magn. Magn. Mater. 83:109 (1990).
5. U. Stetter and K. Baberschke (unpublished).
6. J. Hopkinson, Magnetic properties of alloys of nickel and iron, Proc. R. Soc. Lond. 48:1 (1890) ; S. Chikazumi, "Physics of Magnetism," Wiley, New York (1964), pp. 247 and 273; I. Maartense and G. Williams, Onset of magnetic order in the giant moment system PdCo, J. Phys. F 6:L121 (1976).
7. D. Weller, Ph.D. thesis, Universität zu Köln (1985), and Jülich Report No. Jül-2026 (1985).
8. J. Kolaczkiewicz and E. Bauer, The adsorption of Eu, Gd and Tb on the W(110) surface, Surf. Sci. 175:487 (1986).
9. I. N. Stranski and L. Krastanow, Zur Theorie der orientierten Ausscheidung von Ionenkristallen aufeinander, Sitz.ber., Oesterr. Akad. Wiss. Math.-Nat.wiss. Kl.II, Math. Phys. Tech. Wiss. (Vienna) 146:797 (1938); L. C. Feldman and J. W. Mayer, "Fundamentals of Surface and Thin Film Analysis," Elsevier, New York (1986), p.138.
10. R. M. Bozorth, "Ferromagnetism," Van Nostrand, Princeton (1951), p.542; R. A. Hein and R. L. Falge, Jr., Differential paramagnetic effect in superconductors, Phys. Rev. 123:407 (1961); E. J. Torok, R. A. White, A. J. Hunt, and H. N. Oredson, Measurement of the easy-axis and H_k probability density functions for thin ferromagnetic films using the longitudinal permeability hysteresis loop, J. Appl. Phys. 33:3037 (1962).
11. F. Milstein and L. B. Robinson, Direction of easy magnetization in gadolinium, Phys. Rev. 177:904 (1969) and references therein.
12. R. M. Bozorth, "Ferromagnetism," Van Nostrand, Princeton (1951), p.778.
13. O. Popov and M. Mikhov, Thermal activation and the Hopkinson effect, J. Magn. Magn. Mater. 82:29 (1989).
14. F. H. Salas and M. Mirabal-García, Quenching of the Hopkinson maximum under contamination in the system Gd(0001)/W(110), Phys. Rev. B 41:10859 (1990).
15. A. B. Beznosov, V. V. Eremenko, and V. P. Gnezdilov, d-f Exchange resonance in gadolinium: theory and experiment, J. Magn. Magn. Mater. 43:243 (1984).

16. D. B. Basler and M. F. Berard, Measurement of oxygen diffusion in Dy_2O_3 and Gd_2O_3 by studying high-temperature oxidation of the metals, J. Am. Ceram. Soc. 57:447 (1974).
17. F. H. Salas, First-order magnetic phase transition in quasi-bulk Gd(0001), J. Phys.: Condens. Matter 3:2839 (1991).
18. D. Weller and S. F. Alvarado, Possible evidence for a first-order magnetic phase transition on the Gd(0001) surface, Phys. Rev. B 37:9911 (1988).
19. T. Kaneyoshi, "Introduction to Surface Magnetism," Chemical Rubber Company, Boca Raton (1991), to be published; T. Kaneyoshi, Surface magnetism; magnetization and anisotropy at a surface, J. Phys.: Condens. Matter 3:4497 (1991).
20. A. E. Kennelly, Magnetic reluctance, Trans. Am. Inst. Elec. Engrs. 8:485 (1891).
21. J. M. Alameda, private communication.
22. E. C. Stoner and E. P. Wohlfarth, A mechanism of magnetic hysteresis in heterogeneous alloys, Philos. Trans. R. Soc. Lond. A 240:599 (1948).
23. C. Kittel, Theory of the structure of ferromagnetic domains in films and small particles, Phys. Rev. 70:965 (1946).

MAGNETICALLY MODULATED RESISTANCE TECHNIQUES

B. F. Kim, K. Moorjani, F. J. Adrian and J. Bohandy

The Johns Hopkins University Applied Physics Lab
Johns Hopkins Road
Laurel, Maryland 20723

INTRODUCTION

The method of magnetically modulated resistance (MMR) is a variant of resistance vs. temperature measurements in which the magnetic field dependence of resistance vs. temperature is measured. This general technique is particularly useful for study and characterization of superconductors because the superconducting phase transition and weak link effects, present in granular samples, are magnetic field dependent. Because the field dependence of the intrinsic phase transition is different from that of extrinsic effects due to granularity, it is possible to separately identify MMR responses due to these effects. Thus the species dependent and sample dependent responses can be separately recorded. Another important aspect of this technique is its anisotropic response to macroscopically oriented samples. This allows detection of oriented weak link structures in single crystals[1,2] which would be difficult, if indeed possible, by other methods. In this paper, we present an overview of this method and some examples from recent studies.

MAGNETICALLY MODULATED RESISTANCE

A schematic diagram of an MMR system is shown in Figure 1. The sample is subjected to a magnetic field consisting of a dc component, H_0, and a collinear ac modulation component, H_m. The resistance, R, of the sample is measured at the modulation frequency by phase detection while the temperature of the sample is slowly varied. The phase detected resistance or MMR signal is recorded vs. temperature. In actual practice, the unmodulated resistance vs. temperature is recorded simultaneously to allow comparison between the modulated and unmodulated resistance measurements. As explained later, this allows identification of granular effects in the MMR response.

The physical basis for this type of measurement is straightforward. The resistance of a superconductor is a function of the absolute value of magnetic field. Therefore if the field consists of a dc component, H_0, and a small ac component, $H_m \sin \omega t$, we can write the resistance as

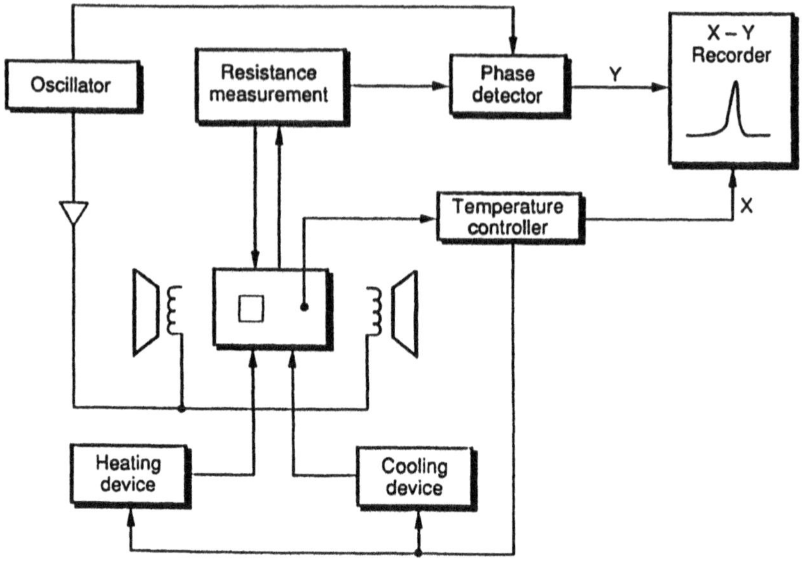

Fig.1. Magnetically modulated resistance system diagram.

$$R(H,T) = R(|H_o + H_m \sin\omega_m t|, T) = R(H_o, T) + \left(\frac{\partial R}{\partial H}\right)_T H_m \sin\omega_m t \qquad (1)$$

which is valid if $H = H_o + H_m \sin\omega_m t$ does not change sign (i.e., $H_o < H_m$). Phase detection at ω_m then yields a signal proportional to $\frac{\partial R}{\partial H}$.

APPARATUS

The MMR technique can be implemented in a number of ways, the principal variation being the frequency of the driving currents. The two most common variants of MMR presently in use are magnetically modulated electrical resistance[3] (MAMER) which uses dc or low frequency (~ 20 Hz) current and magnetically modulated microwave absorption[4] (MAMMA) which uses microwave frequencies. Each has its own advantages for particular types of studies and sample specimens. For example, the MAMER method requires electrically continuous samples while the MAMMA method, being a contactless technique, can be implemented on a variety of shapes and forms of superconducting samples including electrically discontinuous samples, such as powders, or bulk and thin film samples, which are electrically continuous.

For the case of microwave absorption, the sample is typically placed in a microwave cavity in the region of maximum rf magnetic field and minimum rf electric field as shown in Fig. 2 which also shows the current and field directions relative to the sample for a microwave TE_{102} cavity. A microwave bridge used in the reflection mode is employed. A dc bias magnetic field and a small ac magnetic field are simultaneously applied to the sample. The bias field must be larger than the peak ac modulation field so that the polarity of the total magnetic field remains unchanged. The resistance-dependent voltage generated in the sample is phase-detected at

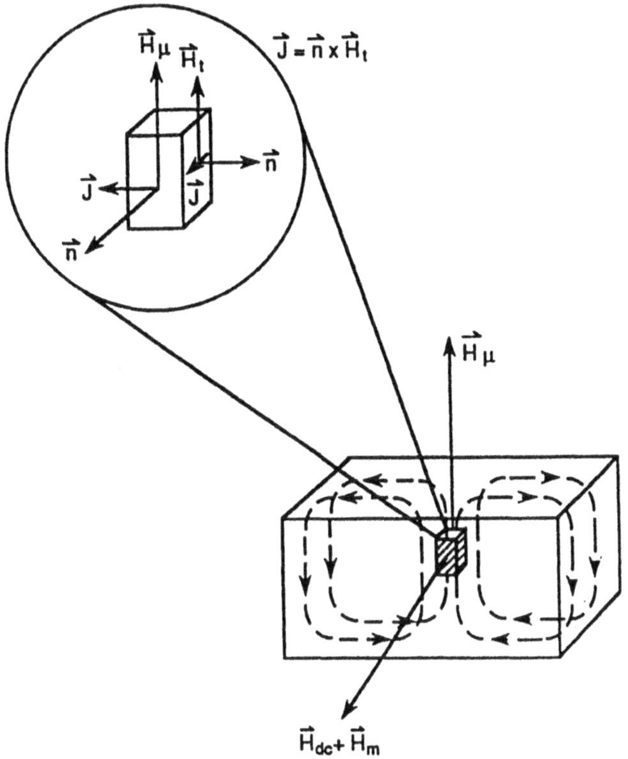

Fig. 2. Sample geometry in the MAMMA experiment. The dashed lines represent the microwave magnetic field in the TE_{102} cavity.

the modulation frequency and recorded as a function of temperature. As we stated previously, it is important to simultaneously record the unmodulated resistance. With a two channel data collection system the unmodulated microwave absorption can be simultaneously recorded by sampling the output of the microwave crystal detector before phase-detection. The result is a plot of unmodulated and magnetically modulated microwave absorption as a function of temperature. Typical values of experimental parameters are 9.3 GHz microwave frequency, 1-200 mW of microwave power, 10-100 KHz field modulation frequency, and 0.1 - 5 G field modulation amplitude.

MAMER is implemented by substituting four-point probe resistance measurement equipment for the microwave bridge. The resistance-dependent voltage in this case is generated by a dc current or a low frequency (~20 Hz) ac current if one so chooses. This is conceptually a simpler experiment than MAMMA because the microwave bridge and cavity are not required. In practice, however, there are additional considerations to take into account. First, if one uses the low frequency ac technique for measuring sample resistance, a second lock-in amplifier is required.

Second, and more important experimentally, a large dc offset voltage frequently appears at the field modulation frequency due to pickup in the leads used in the electrical measurements. This can be minimized by proper geometry, but our approach was to eliminate it using double modulation. A low frequency (23 Hz in our case) current modulation is imposed on the sample in addition to the 10 KHz magnetic field modulation. The output of the 10 KHz lock-in amplifier is fed into a second lock-in and phase-detected at 23 Hz. The output of the second lock-in is a MAMER response which is free from the baseline offset due to the 10 KHz pickup. The dc resistance can be simultaneously recorded in this case by phase-detecting the voltage generated in the sample using a third lock-in also tuned to 23 Hz. Even though the cavity is not needed for the MAMER measurements, it is convenient to do so because both the dc and microwave experiments can be conducted without adjusting the sample. The two methods actually measure responses from different parts of a superconductor; a very important feature when dealing with inhomogeneous samples. In the following section, we discuss in detail the nature of the physical measurements involved in the MAMMA and MAMER methods.

THE MMR SIGNAL

We now discuss, in some detail, the physical characteristics of MMR measurements as implemented by the MAMER and MAMMA techniques. This discussion is offered here to provide a technical basis for interpretation of MMR data obtained under various circumstances and to provide a basis for selection of MAMMA or MAMER for different types of samples and MMR experiments.

The electrical resistance of a sample (at $T \geq T_c$) is the physical quantity which determines the MMR signal, but the precise relation between resistance and MMR signal is different for the MAMER and MAMMA techniques. In the case of MAMER, the signal is a voltage which is derived from the sample resistance through Ohm's law. Therefore, the relevant physical parameter is the bulk resistivity of the sample species. The directions of the currents in the sample are determined by the electric field applied to the sample electrodes and the locus of the path of least electrical resistance. This is, conceptually, the simpler of the two methods. In the MAMMA technique, the signal results from absorption of microwave energy by the sample which is typically located in a resonant cavity. The relationship between the MAMMA signal and sample resistance, and the direction of the sample currents depends upon the disposition of microwave fields at the sample. In the following discussion, we will present this relationship in detail for a common type of MAMMA apparatus.

There are a number of ways to implement the MAMMA technique, particularly with regard to the configuration used to impress microwave energy onto the sample. For many and perhaps most research groups, the most straightforward method is to use an electron spin resonance (ESR) spectrometer where the sample is contained in a resonant cavity. This type of apparatus is particularly useful and convenient for studying anisotropy in the MAMMA response of oriented samples. We will, therefore, confine our subsequent discussion to this type of apparatus.

Because of the similarity between the MAMMA technique and conventional ESR, it is worth noting the essential difference in the physical phenomena observed in these two methods. In ESR, the object of study is microwave absorption due to magnetic dipole transitions between atomic or molecular Zeeman levels in a magnetic field. Because absorption occurs only when the microwave quantum energy ($E = \hbar\omega$) matches the Zeeman energy differences, the absorption is a resonant absorption with respect to magnetic field (the frequency, ω, is nominally constant). The term magnetic resonance refers to this type of absorption even though the transitions are, in some cases, broad. If the sample has a non-zero electrical conductivity, then another form of absorption occurs due to induction of eddy currents in the sample. The latter mechanism, which will be shown to be the relevant mechanism in the MAMMA method, produces a non-resonant absorption. It is the abrupt decrease in this non-resonant absorption at $T = T_c$ which is the signature for the superconducting phase transition in the unmodulated microwave signal.

We now describe the character of the unmodulated microwave signal at temperatures above T_c under the assumption that the MAMMA apparatus is in an ESR type configuration. The characteristics of ESR spectrometers have been described in detail by Feher[5] although his analysis in some respects applies only to magnetic resonance absorption. This analysis, however, is readily extended to encompass non-resonant absorption due to eddy currents. Thus the following description is couched in Feher's model of the response of ESR reflection spectrometers.

It is assumed that the ESR spectrometer crystal detector is biased to operate in the linear region. The signal detected, (due to the wave reflected from the sample cavity) which is denoted V_{REFL} is written

$$V_{REFL} \sim \frac{(VSWR-1)}{(VSWR+1)} V \tag{2}$$

where VSWR is the voltage standing wave ratio in the cavity branch of the microwave bridge and V is the voltage associated with the wave which is incident on the cavity. We are concerned with the change in the output voltage, ΔV_{REFL}, which occurs due to microwave absorption in the sample cavity. This absorption produces a loss in the cavity which directly affects the quality factor of the cavity Q. The maximum value of ΔV_{REFL} can be written[5]

$$\frac{\Delta V_{REFL}}{V} = \pm\frac{\sqrt{2}}{4}\frac{\Delta r}{r} = \pm\frac{\sqrt{2}}{4}\frac{\Delta Q}{Q_o} \tag{3}$$

where r is the cavity resistance and Q_o is the unloaded cavity Q (i.e., the Q of the cavity without a sample). The sign depends upon whether the cavity is overcoupled or undercoupled. Q is expressed as

$$Q = \omega\frac{\text{Energy stored}}{\text{Average power dissipated}} \tag{4}$$

$$= \omega\frac{\frac{1}{8\pi}\int_{V_c} H_1^2 dv}{P_1 + \frac{1}{2}\omega\int_{V_s} H_1^2 \chi'' dv + P_e} \tag{5}$$

where H_1 is the microwave magnetic field intensity, P_1 the average power dissipated in the unloaded cavity, V_c and V_s the cavity and sample volumes respectively, χ'' is the imaginary part of the magnetic susceptibility of the sample owing to the presence of intrinsic magnetic dipole moments, and P_e is the average power dissipated in the sample by eddy currents. Using the binomial theorem, when P_1 is the dominant term in the denominator,

$$Q = Q_o - Q_o^2\left[\frac{4\pi\int_{V_s} H_1^2\chi''dv}{\int_{V_c} H_1^2 dv} + \frac{8\pi}{\omega}\frac{P_e}{\int_{V_c} H_1^2 dv}\right] \tag{6}$$

$$\frac{\Delta Q}{Q_o} = \frac{Q_o - Q}{Q_o} = Q_o\left[4\pi\eta\chi'' + \frac{8\pi}{\omega}\frac{P_e}{\int_{V_c} H_1^2 dv}\right] \tag{7}$$

where

$$\eta = \frac{\int_{V_s} H_1^2 dv}{\int_{V_c} H_1^2 dv} \tag{8}$$

is the filling factor.

This expression differs from that of Feher[5] by the inclusion of the term containing P_e.

The eddy current loss, P_e, is formally written

$$P_e = \frac{1}{2}\int_{V_s} \vec{E}^* \cdot \vec{J}\, dv = \frac{1}{2}\int_{V_s} J^2\rho\, dv \tag{9}$$

where $\vec{J}$ is the current density in the sample and ρ the resistivity. At microwave frequencies, the fields and currents are generally confined to a region near the surface. The distance from the surface encompassed by this region is the skin depth, δ which can be expressed as

$$\delta = \sqrt{\frac{2\rho}{\mu\omega}} \tag{10}$$

when $1 >> \omega\varepsilon r$, where ε is the electric permitivity and μ the magnetic permeability. The integration in (9) can, therefore, be restricted to a region in the neighborhood of the surface encompassing a thickness, Δ, which is small but greater than δ. Then

$$P_e = \frac{1}{2}\int_{V_s} J^2\rho\, dv = \frac{1}{2}\int_{\Delta} d\zeta \int_S J^2\rho\, d\sigma \tag{11}$$

where $d\zeta$ is normal to the surface, $d\sigma$ is a surface element, and S is the surface of the sample. Integration of (11) using $J = J_o e^{-\zeta/\delta}$ and taking $\delta << \Delta$ yields

$$P_e = \frac{1}{4}\int_S J_s^2 \rho_s d\sigma \tag{12}$$

where

$$\vec{J}_s = \delta \vec{J} \tag{13}$$

is the surface current density per unit length and

$$\rho_s = \left(\frac{\rho}{\delta}\right) = \left(\frac{\mu\omega\rho}{2}\right)^{1/2} \tag{14}$$

is the corresponding surface resistivity. The last result follows by using the the expression for δ in (10). $\vec{J}_s$ is related to the tangential component of the magnetic field, H_t, at the surface[6] by

$$\vec{J}_s = \vec{n} \times \vec{H}_t \tag{15}$$

where $\vec{n}$ is a unit vector normal to the surface. Then P_e can be written

$$P_e = \frac{1}{4}\int_S H_t^2 \rho_s d\sigma \tag{16}$$

Thus, the maximum sensitivity of the MAMMA apparatus is

$$\left|\frac{\Delta V_{REFL}}{V}\right| = \frac{\sqrt{2}}{4}\frac{\Delta Q}{Q_o} = \frac{\sqrt{2}}{4} Q_o\left[4\pi\eta\chi'' + \frac{2\pi}{\varpi}\rho_s \eta'\right] \tag{17}$$

where

$$\eta' = \frac{\int_s H_t^2 ds}{\int_{V_c} H_1^2 dv} \tag{18}$$

is a modified filling factor. The (unmodulated) microwave signal, therefore, has two components: that due to the existence of intrinsic magnetic dipoles in the sample, and that due to the microwave induced eddy currents. The latter component is proportional to the surface resistivity of the sample.

The intrinsic MAMMA response, which is the signature for a superconducting phase transition, is due only to the second term in equation (17). At microwave frequencies the contribution of χ" is small compared to the eddy current loss for metals and superconducting materials.. Thus only the second term in equation (17) is relevant to MAMMA for studies of superconductivity.

The relation between MAMMA and MAMER can now be examined in light of equations (15) and (17). It is immediately seen that the surface resistivity is the relevant physical parameter in MAMMA while, as previously stated, that for MAMER is the bulk resistivity. It is also seen that the currents are determined by the tangential component of the magnetic field at the surface of the sample in the MAMMA method. Since both methods exhibit a peak response at T_c, they are equivalent in their ability to ascertain the occurrence of a superconducting transition at a particular critical temperature. For more detailed characterization, however, each technique has merits which, to some extent, complement that provided by the other. It has already been stated that MAMER requires electrically continuous samples while MAMMA does not. Furthermore, since MAMMA depends upon the surface resistivity while MAMER the bulk resistivity, each method probes different sample regions. The MMR method of choice depends to some extent upon the type of sample being studied as well as the nature of the experimental study. If, for example, the sample were a single crystal, then it might be advantageous to study the anisotropy of the MMR response due to both intrinsic phase transitions and to the possible existence of oriented weak links in the crystal. Valid interpretations of such data require knowledge of the directions of both current and field in the sample. The magnetic field, being externally applied in both MAMMA and MAMER, is known in either case, but the currents are another matter. In MAMMA, the current directions are given by equation (15). Since the field directions in the cavity are known, the current directions are easily obtained (assuming that the sample geometry is known). In MAMER, on the other hand, the current orientations are generally known only approximately since these are determined by the electric field and path of least resistance between the external probe leads on the sample. Furthermore, even if the currents were strictly unidirectional between the external current probes, it would be very inconvenient to conduct extensive orientation studies since this would require remounting new probe leads for each orientation. Therefore, in most cases MAMMA is better suited for observing anisotropic effects in oriented samples.

In polycrystalline materials such as bulk sintered ceramic superconductors, the current and field directions are irrelevant for understanding MMR data. Other considerations may be important, however, particularly the magnitudes of the field and currents. We will see later that in certain types of samples, weak link effects have significant current and field dependencies. Studies of these effects by MMR require a reasonably large range of current and field. In studies of this type where a high range of current is required and where precise knowledge of current direction is unimportant, the MAMER technique is the more suitable method. Examples of studies in using both MAMMA and MAMER will be described later.

INTERPRETATION OF MMR RESPONSE

We have previously shown that MMR signals are a measure of the magnetic field derivative of the sample resistance vs. temperature. There are two classes of MMR response which have been identified; one which is associated with the bulk superconductor material, which is the intrinsic response, and those associated with superconductor weak link structures which are extrinsic responses.[3,8] The intrinsic and extrinsic responses generally have different temperature, field, and current dependencies and therefore can be separately identified from one another. Granular samples exhibit extrinsic MMR responses because of the natural occurrence of weak links. We shall see, however, that even single crystal specimens often exhibit extrinsic MMR responses.

The intrinsic response is due to the change in T_c with magnetic field. In the vicinity of the transition temperature, T_c, the measured response at the modulation frequency is proportional to $(\partial R/\partial H) = (\partial R/\partial T)(\partial T_c/\partial H)$ so that for a superconducting transition $[(\partial R/\partial H) \neq 0)]$ a peak is observed at T_c with amplitude proportional to both $\partial R/\partial T$ and $\partial T_c/\partial H$.[4] The factor $(\partial T_c/\partial H)$ appearing in the MMR response assures that a peak is observed only for superconducting transitions and therefore helps differentiate such transitions from transitions that exhibit a precipitous fall in resistance (e.g., insulator-metal transitions) but which are not magnetic field dependent.[7] The factor $\partial R/\partial T$, on the other hand, implies that the shape of the MMR response in the neighborhood of T_c is proportional to the temperature derivative of the unmodulated resistance signal. This is the primary characteristic for identifying the intrinsic peak. The MMR extrinsic features are identified by their location in temperature, which is generally less than T_c, and by the fact that their shapes are not proportional to $\partial R/\partial T$. (These arguments are valid for type I superconductors, and type II superconductors at low fields. At high fields, some type II superconductors may exhibit intrinsic MMR responses with shapes which depart from dR/dT due to flux flow effects.)

The observation that the signature of a superconducting transition in the MMR techniques is a peak located at T_c has also proved to be a convenient way to detect multiple superconducting phases in inhomogeneous samples where a peak corresponding to each superconducting phase will result. In favorable cases, peaks separated by 1 K are easily resolved. (It is, of course, necessary to correctly classify multiple peaks as being intrinsic peaks before one can infer the existence of multiple superconducting phases.) In highly inhomogeneous samples, quantitative information about a broad distribution of T_c's is easily obtained.

We now illustrate these principles with two examples. Figure 3 shows the resistance and MAMER response vs. temperature for an NbN thin film[3] which is non granular. The MAMER response contains a single intrinsic peak at T_c. As expected, the intrinsic MAMER response is proportional to the derivative of the dc resistance with respect to temperature. In Fig. 4, the MAMMA and unmodulated microwave absorption of a granular thin film specimen of Bi-Sr-Ca-Cu-O is shown. In this example a rather pronounced weak link response occurs in addition to a relatively diminished intrinsic peak.

ANISOTROPIC EFFECTS

It is well known that some superconductors exhibit anisotropic properties. The critical current and field dependence of T_c are highly anisotropic in $YBa_2Cu_3O_{7-y}$ for example. Josephson tunneling also occurs in some single crystal superconductors, but their orientation dependence has, in general, not been extensively studied. The MAMMA method is particularly advantageous for detection of oriented Josephson junctions in single crystals. There are two conditions which must be satisfied in order to record the occurrence of these junctions; the currents must traverse the junctions and the magnetic field must penetrate the junctions. These

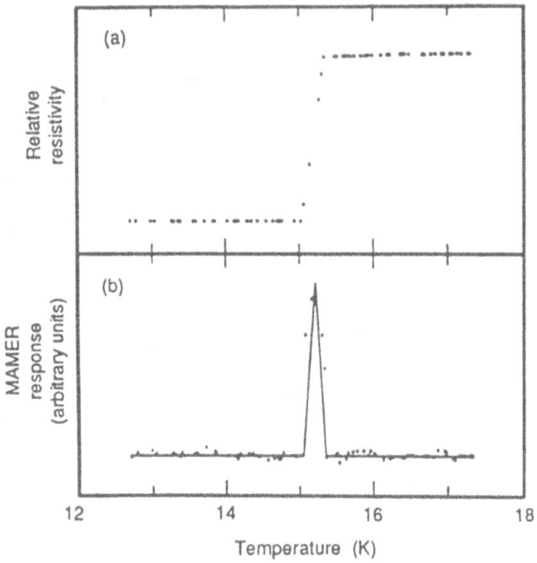

Fig. 3. The dc resistance (a) and the magnetically modulated electrical resistance (b) vs. temperature of a thin film of NbN. H_{dc} = 30 G, H_{mod} = 5 G and I = 100 µA.

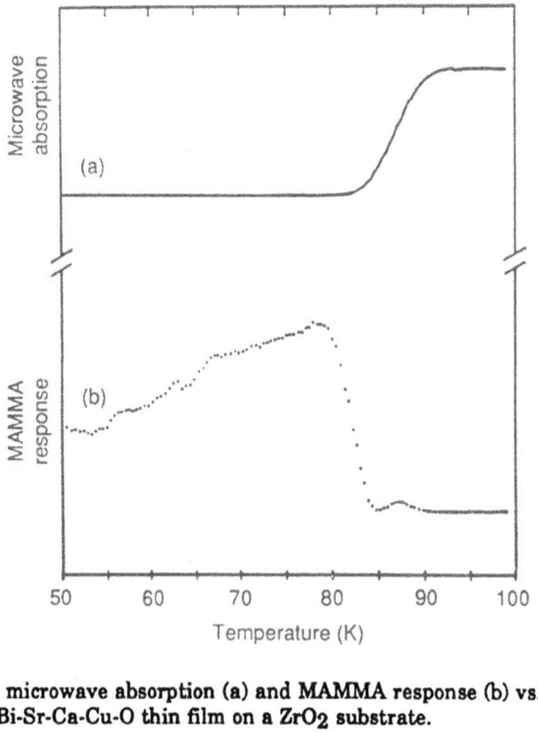

Fig. 4. Direct microwave absorption (a) and MAMMA response (b) vs. temperature for a Bi-Sr-Ca-Cu-O thin film on a ZrO_2 substrate.

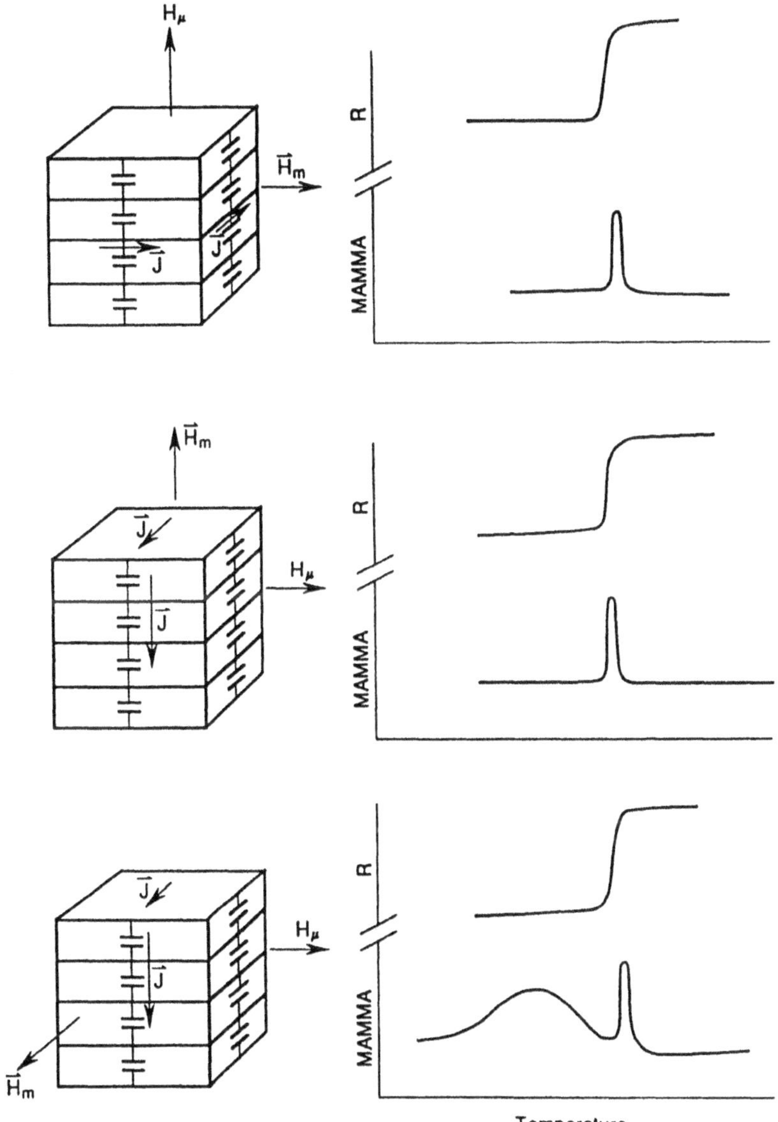

Fig. 5. An illustration of the effect of magnetic modulation field and current orientation on the MAMMA response to oriented weak links.

conditions simplify, to some extent, the determination of the orientation of the junctions responsible for the MAMMA signal. Figure 5 illustrates this by considering the cases of a single crystal with oriented Josephson junctions. In (a) the currents do not traverse the junctions while in (b) the field does not penetrate the junctions. Therefore in both of these cases the MAMMA response contains only an intrinsic peak. In (c) both conditions are satisfied and both a weak link response and an intrinsic response occurs in the MAMMA signal. One could easily infer the orientation of the junctions from this type of experiment.

A recent study was done on the organic superconductor κ-$(BEDT\text{-}TTF)_2Cu(NCS)_2$ using these principles.[1] The MAMMA responses for a single crystal of this material for two orientations of field are shown in Fig. 6. The MAMMA response for the first case contains a weak link

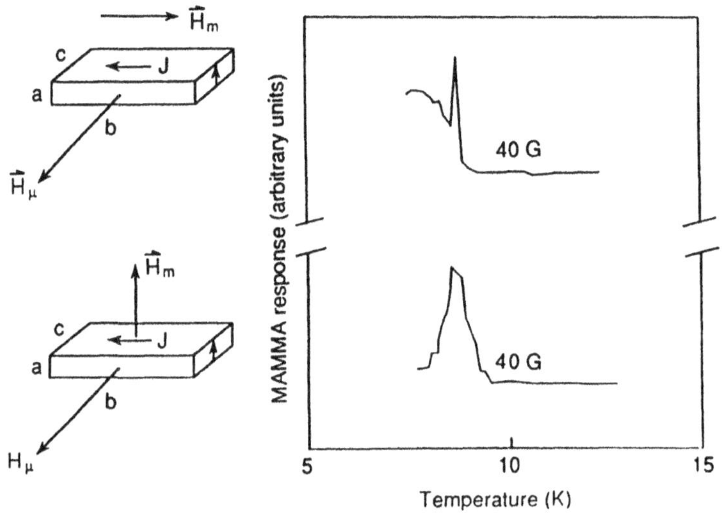

Fig. 6. **MAMMA response vs. temperature of κ-$(BEDT\text{-}TTF)_2Cu(NCS)_2$ for two orientations of magnetic modulation field, indicating the existence of Josephson tunneling between the conduction planes.**

response while the second case contains only the intrinsic peak response at 8.3 K. The currents are the same in each case but the modulation field, H_m, penetrates the bc conduction planes in the first case but does not in the second. This is consistent with the existence of Josephson tunneling between the bc conduction planes.

We now present a number of examples of recent work which should serve to further illustrate and perhaps encourage use of this technique for studies of superconductivity.

The unmodulated microwave absorption and the MAMMA response for two samples of $YBa_2Cu_3O_7$ are shown in Fig. 7a and b. The data in Fig. 7a is for a high quality fully oxygenated detwinned single crystal with H_o = 30 G applied along the c-axis. An extremely sharp transition at T_c = 93.5 K is observed in the MAMMA response whose full width at half maximum is less than 0.3 K. This single crystal sample was specifically prepared under stringent processing conditions, after single crystal samples investigated earlier showed somewhat reduced values of T_c and/or multiple superconducting transitions as revealed in the MAMMA response, due to either slight reduction in the oxygen content or the ubiquitous existence of twin planes in the material. Carefully prepared bulk samples of $YBa_2Cu_3O_y$ can also exhibit sharp superconducting transitions as seen in Fig. 7b for a

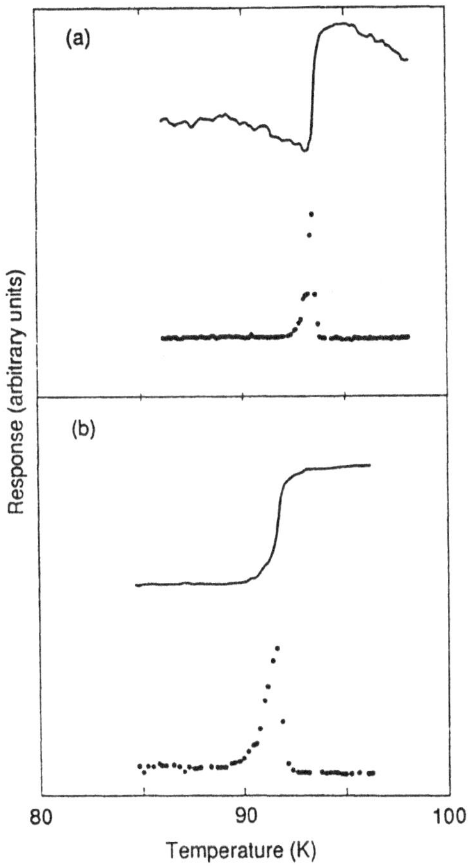

Fig. 7. Direct microwave absorption and MAMMA response vs. temperature for a detwinned YBCO single crystal (a), and a bulk YBCO sample prepared by quench and melt growth (b).

quench and melt growth sample. The value of T_c = 92.2 K and the full width at half maximum of the MAMMA peak are respectively somewhat lower and larger than those observed for the detwinned single crystal discussed above (see Fig. 7a) but one would nevertheless conclude that the sample is essentially fully oxygenated and contains large size single crystal grains that exhibit a single superconducting transition.

As stated earlier, inhomogeneities often exist in cuprate oxide high temperature superconductors. Depending on the nature of inhomogeneities, they can lead to either multiple superconducting phases and/or coexistence of superconducting and nonsuperconducting phases in the same sample. We now consider examples of the former case.

Multiple superconducting phases with closely spaced values of T_c or a more or less continuous distribution of T_c's are not easily resolved in dc resistivity measurements. Even SQUID measurements on inhomogeneous samples generally only show a broad (> 2-3 K wide) diamagnetic transition. The MMR methods, however, depending as they do on the derivative of the resistance, more easily resolve multiple superconducting transitions when they are present in an inhomogeneous sample. Of particular importance is the fact that in single crystal samples, the MMR phase transition signature is generally anisotropic with respect to the current and field directions. Because of this, it is sometimes possible to interpret intrinsic MMR peaks as belonging to particular structures in inhomogeneous single crystals. There are two effects which cause anisotropy in the intrinsic MMR peaks. First, anisotropy in dT_c/dH with respect to the direction of H in a crystal leads to corresponding variations in the peak height which is proportional to this quantity. This is illustrated in Fig. 8 which shows the unmodulated

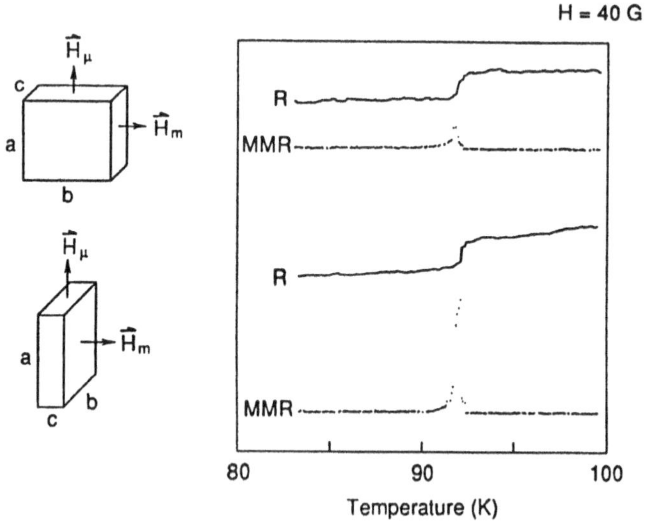

Fig. 8. Direct microwave absorption and MAMMA response vs. temperature for a YBCO single crystal with two different orientations of modulation field. The amplitude of the MAMMA response is anisotropic due to the anisotropy of dT_c/dH.

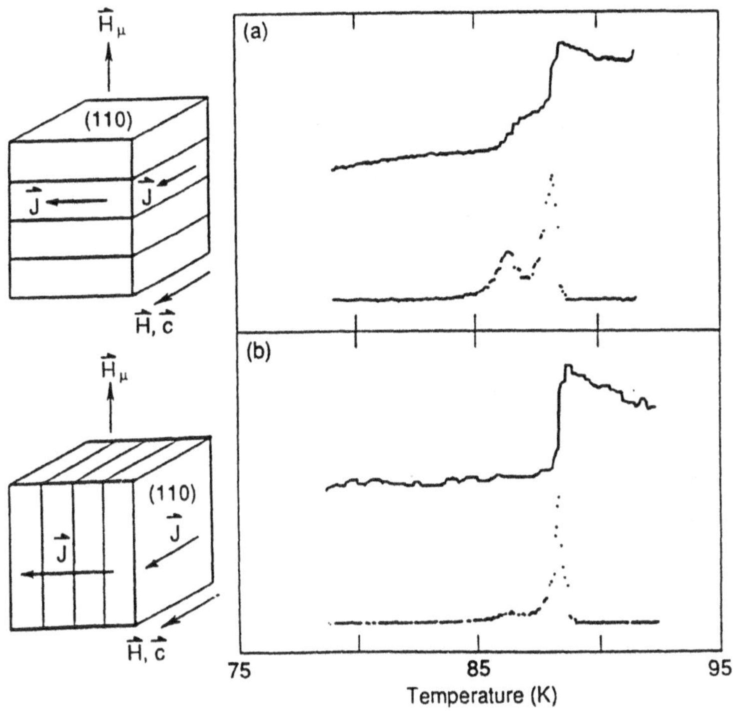

Fig. 9. Unmodulated (solid) and MAMMA response (dotted) of a low purity twinned single crystal of YBCO. (a) Microwave magnetic field perpendicular to (110) and (b) Microwave magnetic field parallel to (110) $\vec{H}_{dc} \parallel \vec{C}$ in both cases.

and MAMMA response of a YBCO crystal for two orientations of the magnetic modulation field. The ratio of the MAMMA peak for the two orientations is equal to the corresponding ratio of dT_c/dH. Second, orientation of a minor phase can cause variations in peak height with respect to the current direction, as exemplified in Fig. 9 using a twinned single crystal of $YBa_2Cu_3O_7$. This sample contains a single twin habit and SQUID measurements show a diamagnetic susceptibility that drops over a temperature range of 3 K. The MAMMA measurements, however, clearly resolve two superconducting phases located at 88 K and 86 K. The double transition is also seen in the unmodulated microwave absorption (top curve in Fig. 9a) which ascertains that the second MAMMA peak arises from a second superconducting phase and is not due to weak link effects. The low value (88 K) of the higher T_c and the existence of two superconducting phases are likely to be related to incomplete oxygenation and the presence of impurities, since fully oxygenated samples of $YBa_2Cu_3O_7$, as discussed earlier, indeed show a single transition in the 92-93 K range (see Fig. 7a). The minor phase with T_c at 86 K is apparently oriented since the corresponding MAMMA response is anisotropic with respect to the orientations of the induced microwave currents. This minor phase manifests itself in the case where currents are parallel to the twin boundaries on all faces (Fig. 9a) and is largely absent when the currents intersect normally the twin planes on two faces (Fig. 9b). Although we have not yet determined the mechanism for this anisotropy in the MAMMA response of the minor phase, we suspect that the minor phase is due to the presence of impurities in the twin planes.

Another example of inhomogeneity and multiple superconducting phases is seen in Fig. 10 for a single crystal of $La_{1.88}Sr_{0.12}CuO_4$. The dc resistivity measurements on another crystal from the same batch showed a sharp drop at ~33 K and in SQUID measurements a broadened magnetic transition was seen. However, dc measurements on the crystal used here indicate a relatively sharp superconducting transition centered at ~30 K with no indication of multiple phases. The multiphasic character of the sample is, however, clearly revealed in the MAMER response (Fig. 10a) with several closely spaced peaks.[2] The temperature dependence of the microwave resistance (Fig. 10b - top) of the same sample shows a considerably broader superconducting transition centered at 28 K while the MAMMA response (Fig. 10b - bottom) clearly shows the presence of several different phases with T_c's spread over approximately 5 K. These measurements provide an excellent example of the complimentary nature of MAMER and MAMMA techniques. The dc resistance and MAMER reflect the best superconducting path through the sample and thus should, and does, yield the narrowest transition and highest T_c. The microwave resistance and MAMMA measurements, on the other hand, provide information about all regions of the sample penetrated by the microwaves. In the case of a single crystal sample, this is primarily the surface regions

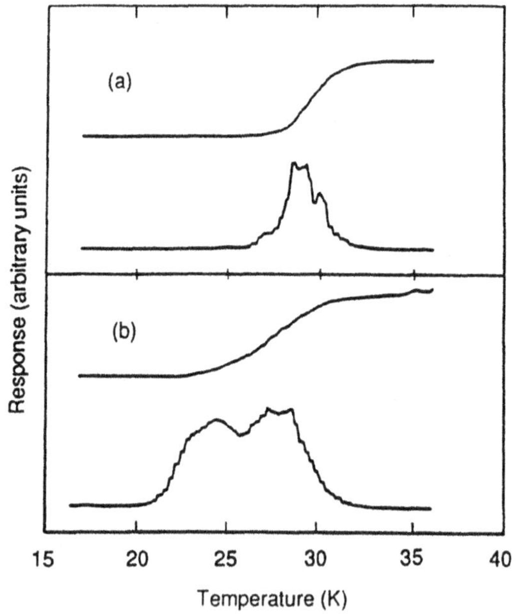

Fig. 10. (a) Electrical resistance (top) and magnetically modulated resistance (bottom) vs. temperature in a $(La_{0.94}Sr_{0.6})_2CuO_4$ single crystal. (b) Microwave absorption (top) and magnetically modulated microwave absorption (bottom) vs. temperature in the same sample.

as the skin depth at microwave frequencies is small compared to the crystal dimensions. In the present case, the superconducting properties of these surface regions are considerably poorer, particularly regarding the presence of multiple phases, than in the best superconducting path through the sample. Accordingly, microwave measurements are specially important for examining specimens that are employed in experiments involving crystal surfaces.

WEAK LINK SUPERCONDUCTIVITY

Thin granular films of conventional metallic superconductors in the presence of rf radiation or current are known to exhibit anomalous effects understood in terms of superconducting tunneling among naturally occurring weak links. The bulk sintered samples of high temperature superconducting oxides normally consist of a myriad of small randomly oriented crystalline grains and exhibit broadened dc resistance and diamagnetic transitions and carry low critical currents. These properties are generally ascribed to the presence of structural disorder (e.g., defects and dislocations) as well as compositional disorder (e.g., impurities and grain boundaries) in the bulk samples, leading to the formation of weak links that affect the critical current and field dependence of the sample. (We should note that these low field effects should be distinguished from high field effects which are attributed to flux flow and flux creep.[10,11]) Given the small, highly anisotropic, values of the coherence lengths in superconducting ceramics, even very small scale disorder of ≅ 10 Å can lead to weak links. The presence of Josephson junction-type weak links in sintered samples of $YBa_2Cu_3O_7$ was conjectured soon after these materials became available. Our conjecture for this possibility was based on the observation of a rising base line and increased noise in the MAMMA response at temperatures below the intrinsic peak.[9] Subsequently, it was argued that the tail on the low temperature side of the resistance vs. temperature curve of granular superconductors is due to weak links[12,13] and that the behavior of critical current could be fitted to a model based on an array of parallel Josephson junctions.[14] Initial MMR studies of weak links in granular high temperature superconductors were done on polycrystalline bulk specimens using MAMER.[8] We now present a description of these results.

Figs. 11a through d show the MAMER spectra and simultaneously recorded resistance curves of a sample of $EuBa_2Cu_3O_{7-y}$ (EBCO) in a static magnetic field of 500 G, 30 G, 10 G, and 2 G, respectively, using a 5 mA excitation current. In each case the MAMER response contains an intrinsic and an extrinsic or weak link peak which are resolved, with the exception of case (d). The effect of magnetic field is to broaden the extrinsic peak and move it to lower temperatures. A similar effect was observed with regard to the current.

We have modeled the results of the MAMER experiment, taking into account the behavior of both intrinsic and weak link responses. (This model also applies to the MAMMA experiment.) The conceptual approach and results will be described briefly here; for more details, the reader is referred to reference 8. It was assumed that the sample was electrically equivalent to a parallel array of inequivalent Josephson junctions, each characterized by a different value of H_0, the field which produces a single quantum of flux in the junction. The distribution of the values of H_0 was assumed to be Gaussian because the method of sample preparation was

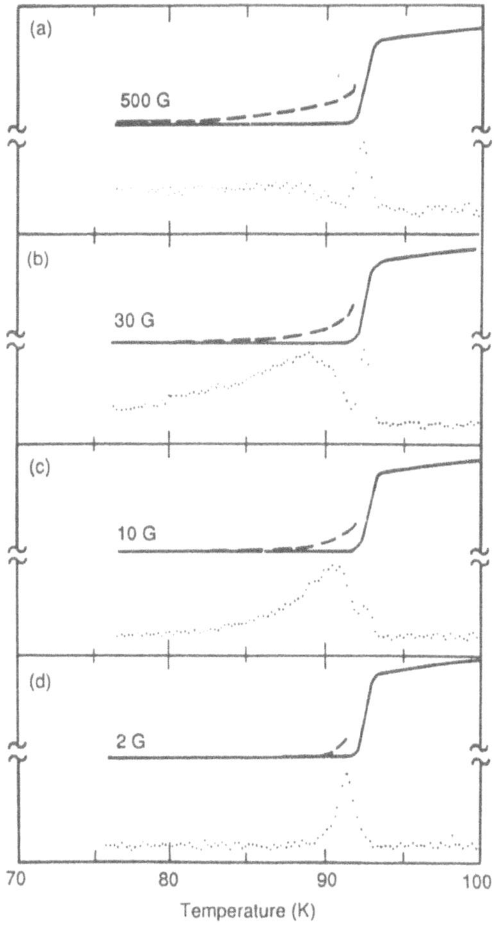

Fig. 11. Magnetic field dependence of the MAMER response (dotted) and dc resistance (x1 solid, x10 dashed) of $EuBa_2Cu_3O_{7-y}$ using a current of 5 mA. (a), (b) and (c) 5 G modulation amplitude, (d) 1 G modulation amplitude.

expected to yield a Gaussian distribution of grain sizes. Since the junctions were presumed to be formed by the interfaces between grains, it followed that the distribution of junction sizes and hence H_0 should also be Gaussian. A quadratic temperature dependence for the critical current was used and the MMR signal was thus calculated using a simulation of the dependence of the intrinsic resistance vs. temperature in the region of T_c. The mean and variance of the H_0 distribution, as well as the critical current at T = 0 (assumed to be the same for all junctions) were adjustable parameters.

A comparison of the data in Fig. 11 and the corresponding model spectra in Fig. 12 indicates that the model simulates the general behavior of the weak link peaks relative to that of the intrinsic peaks for different magnetic fields. In particular, the model correctly predicts that the weak link peak shifts to lower temperatures and broadens with increasing field and current. It also correctly shows that the peak height decreases with increasing field.

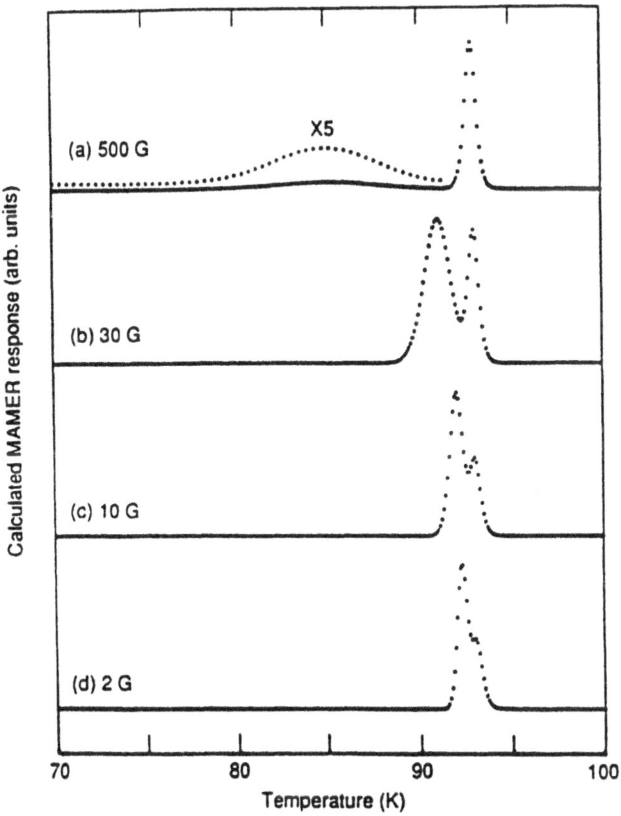

Fig. 12. Calculated field dependence of the MAMER response of $EuBa_2Cu_3O_{7-y}$ using a current of 5 mA.

Other types of granular samples, particularly thin films, can exist in which the distribution of weak links is much different from that used in the preceding study. The corresponding MAMMA and MAMER spectra can be dramatically different.

Previously we discussed the MAMER response of a NbN thin film (Fig. 3). The film is non-granular and the MAMER response is a single sharp peak arising from the intrinsic superconducting transition located at 15.3 K. Note that the MAMER spectrum is a derivative of the resistance curve. We have never observed such a clean response in any of the thin films of cuprate oxide superconductors whether prepared in our laboratory or provided by other researchers. A particularly good example of a poor thin film is shown in Fig. 13 for a Bi-Sr-Ca-Cu-O sample deposited by laser ablation on a MgO substrate. Judging from the microwave resistance curve, an intrinsic MAMER peak should be observed at 87 K. Instead a broad peak located at 78 K is seen. The MAMER response is not proportional to the temperature derivative of the resistance and the entire response arises from weak links in this highly granular film. An example, noted previously, of a sample where both intrinsic and weak link features are present is seen in Fig. 4 which depicts the MAMMA response of a laser ablated film of Bi-Sr-Ca-Cu-O deposited on a ZrO_2 substrate. A small intrinsic peak observed at 87 K corresponds well to the inflection in the non-modulated microwave absorption, while the bulk of the signal at temperatures below 80 K is associated with weak links.

CONCLUSION

MMR techniques offer some unique advantages for characterization of superconductors. Foremost is the ability to separately record the intrinsic superconducting phase transition and effects due to sample granularity. This is important because in some cases (high temperature super

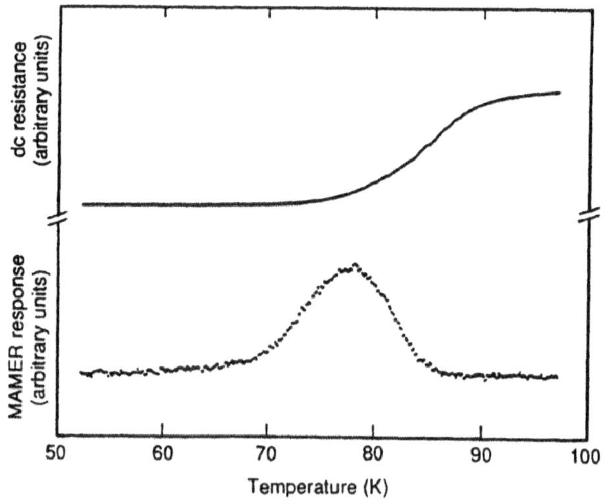

Fig. 13. MAMER response (dotted) and dc resistance (solid) vs. temperature of a Bi-Sr-Ca-Cu-O thin film on MgO substrate.

conductor "epitaxial" thin films), it is difficult to detect granularity by other methods. This same advantage extends to single crystal samples with the additional benefit that the orientations of weak link structures can be inferred.

ACKNOWLEDGMENT

This work was supported by the Department of the Navy, Space and Naval Warfare Systems Command, under Contract No. N00039-89-C-0001.

REFERENCES

1. J. Bohandy, B. F. Kim, F. J. Adrian, K. Moorjani, S. D. D'Arcangelis, and D. O. Cowan, Phys. Rev. B 43:3724 (1991).
2. F. J. Adrian, B. F. Kim, K. Moorjani, and J. Bohandy, Phys. Rev. B (in press).
3. J. Bohandy, T. E. Phillips, F. J. Adrian, K. Moorjani, and B. F. Kim, Mod. Phys. Lett. B. 3:933 (1989).
4. B. F. Kim, J. Bohandy, K. Moorjani, and F. J. Adrian, J. Appl. Phys. 63:2029 (1988).
5. G. Feher, Bell Syst. Tech. J. 36:449 (1957).
6. See, for example, J. A. Stratton, "Electromagnetic Theory," McGraw-Hill (1941).
7. Materials exhibiting magneto-resistance will also elicit a response in this type of measurement. The nature of the response, however, is generally different from that of superconductors for which the response is due to a magnetic field dependent phase transition.
8. B. F. Kim, J. Bohandy, T. E. Phillips, F. J. Adrian, and K. Moorjani, Physica C 161:76 (1989).
9. J. Bohandy, B. F. Kim, F. J. Adrian and K. Moorjani, Phys. Rev. B 39:2733 (1989).
10. M. Tinkham, Phys. Rev. Lett. 61:1658 (1988).
11. Y. Yeshurun and A. P. Malozemoff, Phys. Rev. Lett 60:2202 (1988).
12. P. England, T. Venkatesan, X. D. Wu, A. Inam, M. S. Hegde, T. L. Cheeks and H. G. Craighead, Appl. Phys. Lett. 53:2336 (1988).
13. M. A. Dubson, S. T. Herbert, J. J. Calabrese, D. C. Harris, B. R. Patton and J. C. Garland, Phys. Rev. Lett. 53:1061 (1988).
14. G. Paternó, C. Alvani, S. Casadio, U. Gambardella and L. Maritato, Appl. Phys Lett. 53:609 (1988).

HIGH-FIELD AC SUSCEPTOMETER DESIGN FOR MEASUREMENTS OF SUPERCONDUCTING SINGLE CRYSTALS AND RESULTS

M. Nikolo and A. M. Hermann

Physics Department,
University of Colorado
Boulder, Colorado 80309-0390

INTRODUCTION

The use of ac induction techniques involving ac mutual inductance bridges and variances thereof have been discussed in detail in Section I of this volume. As has been emphasized in Section I and elsewhere[1-3], a wealth of information can be obtained from measurements of the complex ac magnetic susceptibility, $\chi^{ac}(H,T)=\chi'(H,T)-i\chi''(H,T)$ where H is the total magnetic field acting on the sample and T its temperature. Examples of the utility of these measurements involving topics of current and significant interest are: (a) the study of the "loss" peak as demonstrated by a peak in plots of $\chi_H''(T)$ vs. T and its relationship to the onset of magnetic irreversibility[4,5] and (b) the dependence on the frequency of the ac measuring field of the normal to superconducting transition temperature, T_c, as manifested by the change in $\chi_H'(T)$. Studies of this frequency shift allows one to determine pinning energies involved in thermally activated flux motion occuring in single crystals and multigrained bulk samples[1,6].

In both of these types of measurements, one needs to operate with applied fields in excess of the lower critical magnetic field $H_{c1}(T)$ in order to study the effects of flux creep and flux flow upon $\chi^{ac}(H,T)$. In the case of single crystal samples this requires the capability of using relatively large ac measuring fields, $H_{ac}=H_0e^{-i\omega t}$ with $|H_0|\approx 10^2\text{-}10^3$ Oe. Magnetic fields of this order of magnitude exceed the capability (0.01-10 Oe) of currently available commercial ac susceptometers. In this article we discuss design criteria used, and fabrication techniques employed, to construct an ac susceptometer for studies of single crystals with the capability of using measuring fields as high as 500 Oe.

MEASUREMENT PRINCIPLES

Measuring the induced voltage with a lock-in amplifier the real, χ', and the imaginary, χ'', parts of the complex ac magnetic susceptibilty can be separated. A classical susceptibility setup is shown in Figure 1. Here an alternating magnetic field is generated by a solenoid that serves as the primary coil in a transformer circuit. The solenoid is driven with an ac current source with variable amplitude and frequency.

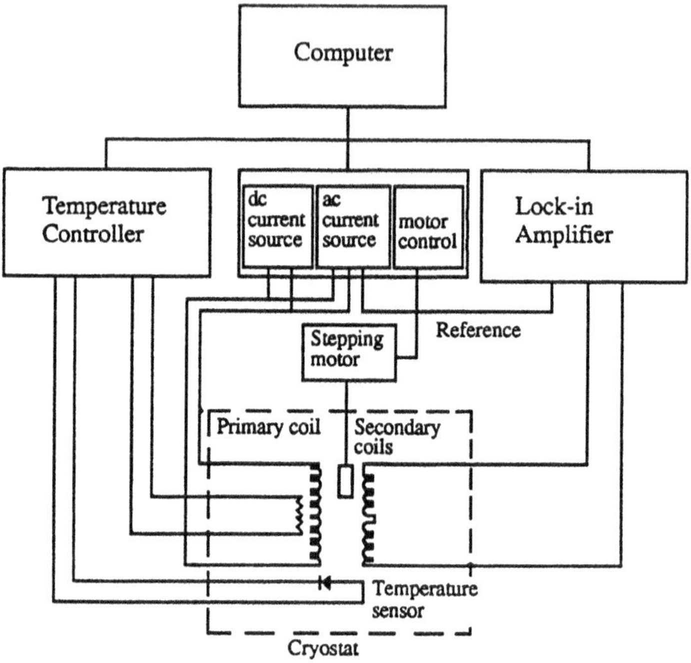

Figure 1. Classical schematic of an ac susceptometer.

Two identical sensing coils are positioned symmetrically inside the primary coil and serve as the secondary coils in the measuring circuit. For clarity the primary and pick up coils are shown separately in Figure 1. The two sensing coils are connected in series opposition in order to cancel the voltages induced by the ac field itself or voltages induced by unwanted external sources. The stepping motor, a feature used mostly in commercial designs, moves the sample holder from the center of one to the center of the other sensing coil.

When a sample is placed within one of the sensing coils, the voltage balance is disturbed and the measured voltage is proportional to the amplitude of χ, the first harmonic of the ac susceptibility of the sample, and several other parameters $v = (1/\alpha)\ VfH_a\chi$, where α is the calibration constant, V is the sample volume, and H_a and f are the magnitude and frequency of the applied magnetic field.

First, one has to decide how large an ac magnetic field is needed to be generated inside the primary coil and what is the maximum desired, driving frequency. The long solenoid approximation for the magnetic field, B, inside the solenoid, $B = \mu_0 NI/l$, where μ_0 is the permeability of free space, N is the number of turns, l is the length of the solenoid, and I is the current, can be applied.

Once the maximum required current is determined, a suitable current source together with the frequency generator have to be selected. The total impedance of the primary coil will be the underlying factor in selecting the proper cuurent source. Assuming that we can neglect the capacitive coupling, Z, the total impedance is $Z = (R^2 + (\omega L)^2)^{1/2}$. Here R is the resistance, L is the self-inductance of the coil, and ω is the angular frequency of the driving ac field. The long coil approximation can be used to estimate the self-inductance $L = \mu_0 N^2 A/l$ where A is the area of a single loop.

It becomes obvious that in the attempt to increase the magnetic field, B, by increasing the number of turns, one also increases the self-inductance, L, by an even faster rate. This, in turn, reduces B by reducing the current in the coil. Therefore in the attempt to increase B, it can actually be reduced. This consideration can be ignored for the design of a low field (≤10 Oe) susceptometer where the current source limitations are not serious. But for ac fields in the region of several hundred oersted, this becomes a formidable obstacle. One way of resolving this problem is to connect a power amplifier to the ac current source.

Knowing the maximum field desired, one gets an approximate idea of the maximum current needed given the approximate dimensions of the coil. The length/diameter ratio should always be at least 2, in order to maintain an acceptible degree of uniformity of the field inside the coil. Another issue is the choice of the wire diameter for the coil. Obviously, the larger the diameter, the higher the current and therefore the possible field, but at the same time, the smaller the number of turns per layer possible and therefore the larger the volume that the coil fills, an important considertation when overall dimensions are a key factor.

In designing the coil configuration, one can quite accurately calculate the voltage signal obtained given the coil geometry, the applied field and frequency, and the sample volume[7]. This can give one an idea in advance of how carefully the coils will need to be balanced to sense the induced voltage due to the sample; in another words, the system's sensitivity can be accurately estimated. Also, see discussions by Salas and Rillo in this volume.

For the sensing coils, one should strive for dimensions such that the sample volume is not negligible compared to that of the coil volume. This is because the pick-up voltage is measured by comparing the flux displaced in one coil against the flux present in the empty, balancing coil. One can have exactly the same number of turns on each coil such that the resistances are the same, but any slight imperfection in the winding will change the self-inductance of each coil and therefore the balance. Winding the coils with a very thin wire, remains an art form, especially if one tries to wind many layers without any imperfections and a small residual voltage will be seen by the lock-in amnplifier.

One can try to compensate the imbalance by inventing a number of methods to improve the balance of the coils. The most effective one is to use a stepping motor and to subtract the upper and lower coil signals to obtain the net sample signal. But this design is not always practical for it is material and labor intensive. A simpler method of balancing the coils is to connect a variable ratio transformer across the two sensing coils with the variable center tap connected where the two sensing coils are connected in series opposition. The balance point can change as either temperature, applied magnetic field, or frequency is varied and one has to be very careful. Another method is to use the nonuniform, decreasing field at the ends of the primary coil. By sliding the primary coil slightly along the two sensing coils, one can obtain a better balance.

A HIGH-FIELD AC SUSCEPTOMETER FOR SINGLE CRYSTAL MEASUREMENTS

The design of a high-field ac susceptometer for the measurement of single crystals is shown here. The average lateral dimension of single crystals to be investigated is 1 mm, and therefore an extreme sensitivity is desired. Figure 2 shows the dimensions of the sample holder, limited by the size (1.6 × 3.2 mm) of the smallest miniature temperature diode found (LakeShore DT-450). To maximize the sensitivity, the smallest possible dimensions are selected and the sample holder is designed around the size of the diode.

A Stanford Research Systems Lock-in amplifier, SR-530, is used to detect the signal from

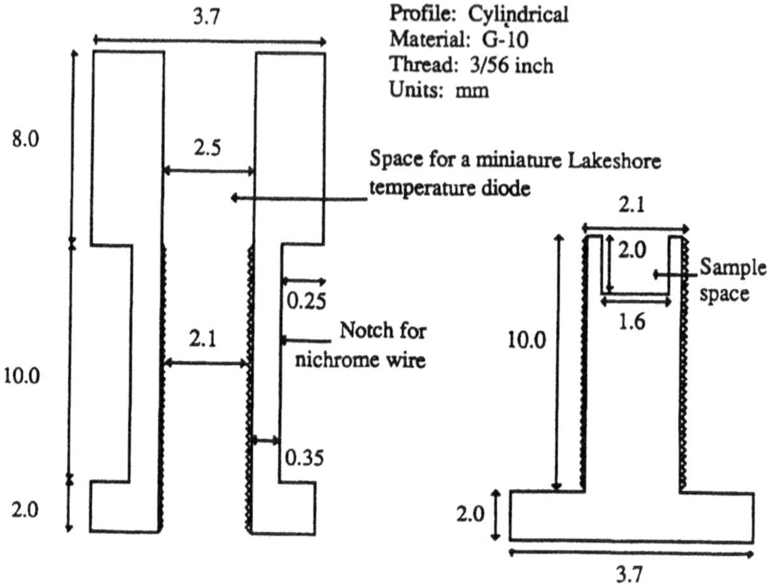

Figure 2. Schematic of a miniature sample holder.

the pick-up coils. The temperature is maintained using an Omega CYC91 controller. A nichrome heater coil (32 AWG, LakeShore) is folded in half and twisted to eliminate it's magnetic effects. Its current is of the order of a few milliamps and less compared to the current driving the primary coil which is several orders of magnitude larger. The primary coil is driven with a California Instruments 3213K high power constant rms current source.

The sample holder is attached to a quartz tube with 4 mm OD. A 9-pin connector was attached at the other end of the quartz tube with 2 connections for the nichrome heater wire and 4 connections for the DT-450 temperature diode. The pick-up coils are wound with a 42 AWG (0.003" diameter) wire in 6 layers and their ID is 5 mm. The total number of turns in each pick-up coil is 670.

Figure 3 shows the dimensions of the primary and pick-up coils. The pick-up coils are wound on a coil form made out of G-10 plastic. The upper end of the form is attached to a quartz tube. The primary coil, wound separately on another coil form, is slid over the secondary coils. The primary and secondary coil configuration, attached to the quartz tube, can then be inserted to a liquid nitrogen bath and the sample holder tube is inserted inside the lower pick-up coil. Since the coils are immersed in liquid nitrogen and separated from the sample heater, they remain at a constant temperature and therefore a voltage drift due to coil expansion is avoided.

Given the dimensional parameters, the calibration constant, α, is calculated to be 1.28 (SI)[7]. Assuming a voltage resolution of 10^{-7} volt and a measuring ac magnetic field of 100 Oe, at 1000 Hz, gives a practical sensitivity of 1.28×10^{-8} emu. This design allows ac magnetic fields as large as 500 Oe (rms).

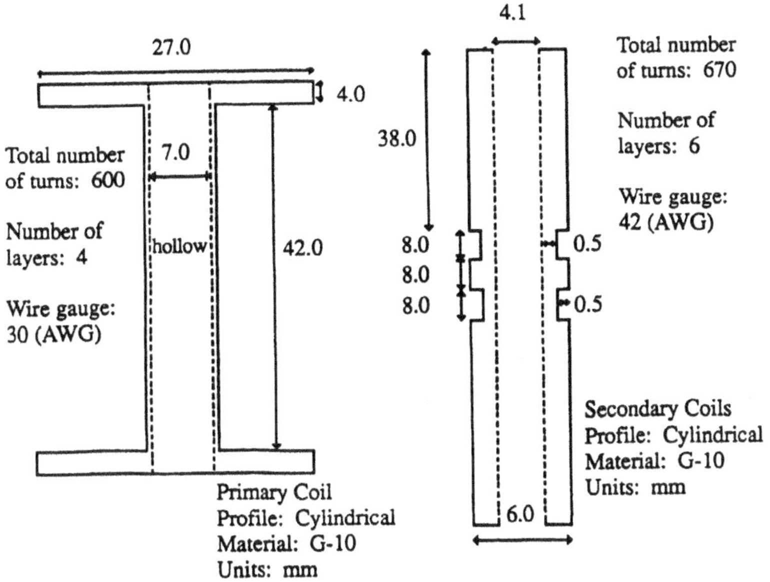

Figure 3. Primary and secondary coil configuration.

SINGLE CRYSTAL RESULTS

Figure 4 shows the data obtained for a $Tl_2Ba_2CaCu_2O_x$ single crystal at an applied ac magnetic field of 100 Oe (rms) and at three separate driving frequencies. The size of the crystal is approximately 1.5 $\times$ 2.0 mm across and about 30 μm thin. Only the real part of the susceptibility, χ', is observed here. The imaginary part, χ'', producing a considerably smaller signal than that of the real part, could not be separated from the background signal.

The real part of the susceptibility shows a sharp drop at the transition temperature that depends on temperature, the field amplitude and frequency. As frequency is increased, the transition drop moves to a higher temperature. Figure 4 shows a typical frequency shift for an applied field of 100 Oe rms. The measured χ' signal is larger for higher frequencies. The figure is normalized such that at low temperature χ' susceptibility is the same for all frequencies. Here, we are only looking for the temperature of the midpoint of the transition drop and this normalization of χ' will not affect it.

As a reference point for specifying the temperature shift, the midpoint of the drop, T_0, is used. The frequency induced shift in transition temperature increases as the applied magnetic field is increased from 15 Oe to 200 Oe. The frequency shift is as large as 15 K per frequency decade in a field of 200 Oe rms in the hetz to kilohertz frequency range at which the experiment is performed. It is assumed that giant flux creep is observed here.

Since the hopping rate is proportional to the time derivative of the measuring, or driving field, the hopping rate will be proportional to the driving frequency of the ac field signal. Therefore the experimental data is modeled by the Arrhenius-like expression $f=f_0\exp(-E_a/kT_0)$. Here E_a is the thermal activation energy, k is the Boltzmann's constant, f is the driving

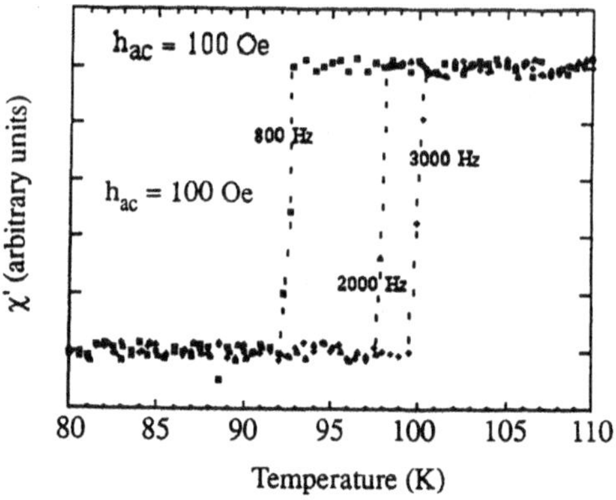

Figure 4. Real part, χ', of ac susceptibility versus temperature at several frequencies for a $Tl_2Ba_2CaCu_2O_x$ single crystal.

frequency, and f_0 is a term proportional to the characteristic hopping frequency. The activation energy E_a is given by E_a=U-FVd where F is the Lorentz force, V is the flux bundle volume, and d and U are the effective width and height of a pinning potential barrier at F=0.

For each measuring magnetic field a linear fit of ln (f) vs. $1/T_0$ is obtained by interpoltation. Typically, the fit is a result of about five data points. The activation energy is the computed from the slope of the various fits as a function of the applied measuring field[6]. The energies vary from 288 meV at 15 Oe to 25 meV at 200 Oe. It is estimated that the accuracy of the energies quoted is ±5% from the fit of ln (f) vs $1/T_0$ for each field. For fields smaller than 1: Oe, the frequency shift is no longer apparent i.e., there is no observable flux creep. This enable: one to estimate H_{c1} at about 100 K to be approximately 10 Oe.

The activation energies are quite small compared to the integranular energies reported by one of the authors (M.N.) previously[1]. Thus it takes very little energy, especially at higher fields, to depin the flux bundles at temperatures of measurement close to T_c. At a given temperature, higher frequencies allow less time for flux to penetrate the crystal. This tends to improve its shielding capacity. In other words, at a higher frequency the critical state field profil has less time to relax and therefore the effective pinning force density increases with increasing frequency. In this regime one is approaching the crossover from flux creep to flux flow. The strong frequency dependence means that a considerable millisecond flux creep is observed. A more detailed analysis of the single crystal data is presented in a separate paper[6].

SUMMARY

We have discussed the principle of measurement of ac susceptibility and applied it to a design of a simple, high-field ac susceptometer with sufficient sensitivity to detect a signal from single crystals. Various aspects of the design geometry were examined and optimal features wer selected. We have demonstrated that single crystal ac susceptometer data can be useful to study flux creep and flux flow effects, and have shown how flux creep activation energies can be measured by the ac susceptibility method.

REFERENCES

1. M. Nikolo and R. B. Goldfarb, Phys. Rev. B **39**, 6615 (1989).
2. K.-H. Müller, M. Nikolo, and R. Driver, Phys. Rev. B **43**, 7976 (1991).
3. K.-H. Müller, M. Nikolo, N. Savvides, and R. Driver, "Ac Susceptibility of High-Temperature Ceramic Superconductors: Critical State and Creep at Grain Boundaries and in Grains," *Proceedings of the 3rd International Symposium on Superconductivity*, Sendai, Japan, 6-9 November, 1990, Advances in Superconductivity III, Springer-Verlag, Tokyo 1991.
4. Y. Yeshurun and A. P. Malozemoff, Phys. Rev. Lett. **60**, 2202 (1988).
5. A. P. Malozemoff, T. K. Worthington, Y. Yeshurun, F. H. Holtzberg, and P. H. Kes, Phys. Rev. B **38**, 7203 (1988).
6. M. Nikolo, W. Kiehl, H. M. Duan, and A. M. Hermann, "Flux Creep Activation Energies in $Tl_2Ba_2CaCu_2O_x$ Single Crystals", submitted to Phys.Rev. B.
7. M. Coach, A.F. Khoder, and F. Monnier, Cryogenics **25**, 695 (1985).

BALANCING COILS FOR 10 MHZ SUSCEPTIBILITY SIGNALS

W. L. Hults and J. L. Smith

Los Alamos National Laboratory
Los Alamos, New Mexico 87545, USA

INTRODUCTION

The de Haas-van Alphen (dHvA) effect is the low temperature oscillatory magnetization of a metal in a high magnetic field. Analysis of the oscillations yields such information as the cross sectional area of the Fermi surface extrema, the effective electron masses for extremal areas, and the electron scattering rates (or mean free paths) for the electron orbits of these areas. In the early fifties, Shoenberg[1] pioneered the use of a pair of balanced pick-up coils, with a sample in one coil, in pulsed magnetic fields to measure the dHvA effect in the highest possible magnetic fields. The oscillations in these approximately 10 Tesla (T) fields were detected at a frequency of order 10^5 Hz. Recent work[2] to measure the Fermi surface parameters of $YBa_2Cu_3O_{6.97}$, a high temperature superconductor, in fields to about 100 T produced oscillations of order 10^7 Hz. At these frequencies, we found that new ways had to be developed to balance coils because the capacitive reactance dominates. In this paper we describe the problems, our solutions, and the experimental results from our coils designed to operate at these higher frequencies in a system where it must work the first time.

$YBa_2Cu_3O_{6.97}$ has a superconducting transition temperature in zero magnetic field of 93 K. If the Ginzburg-Landau relationship is assumed, $dH_{c2}/dT = 1.7$ T K^{-1}, it suggests that H_{c2} (T = 4.2 K) is in excess of 100 T. However, the value of H_{c2} has not been established to our satisfaction. The question of working above or below H_{c2} often arises, but it is clear that dHvA signals can be observed below H_{c2}.[3] Thus there is no need to discuss this very interesting question for these proceedings.

EXPERIMENTAL SITUATION

High temperature superconductors have fairly high normal state electrical resistivities, which means a short mean free path for the electrons at low temperatures. The dHvA

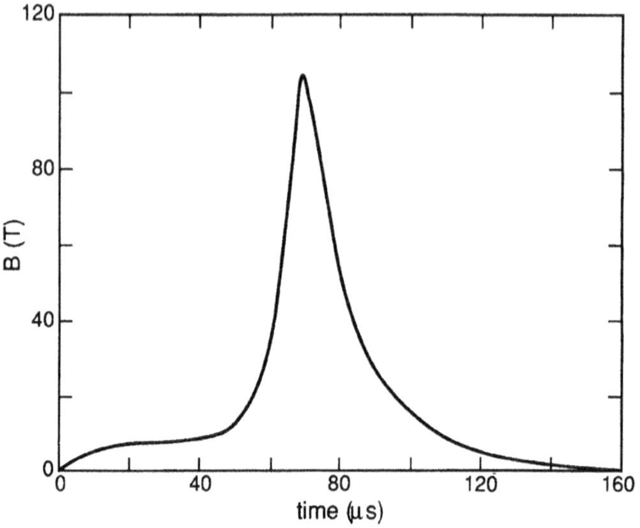

Fig. 1. Typical magnetic field pulse as a function of time. The coil was usually destroyed at ≈80 microseconds and the remainder of this curve was extrapolated.

requirement that an electron in a cyclotron orbit should roughly complete one circle before being scattered suggests the use of very high magnetic fields. Explosively driven pulsed magnetic fields in the 100-200 T range can be produced[4] with half widths of about 20 microseconds as shown in Fig. 1. The fields are produced in a one-turn magnet made of a brass block (5 x 5 x 7.6 cm) with a 1.6 cm hole in it. Temperatures down to about 2 K can be provided by pumping liquid helium through an orifice in vacuum-jacketed glass tubing from a 4 K helium reservoir in a 2-liter glass liquid nitrogen dewar.[5]

The compound $YBa_2Cu_3O_{6.97}$ was ground to an average particle size of less than 10 micrometers and stirred into epoxy that was then cured in a 4.2 T magnetic field at room temperature. This process resulted in a dispersion of particles in cured epoxy which have their c-axis alligned in the direction of the magnetic field and thus allowed the Fermi surface extrema in the a-b plane to be measured. Analysis by x-ray diffraction of the samples showed alignment of the c-axis to better than two degrees. These small particles are insulated from each other which: allowed an estimate of their temperature rise from eddy current heating[2] (less than 0.2 K); permitted the field to penetrate between the particles; and permitted fields changing at the rate of 10^7 T/sec to penetrate into the particles. The epoxy sample was machined into a 1 mm dia. x 6 mm rod. In accordance with Chap. 3 of Ref. 1, the signal was estimated to be in the range of 0.01 to 0.10 V for a sample diameter of 1 mm and a 100 turn coil wound directly onto the sample. The pulsed field induces a voltage of about 1 kV across such a coil; this prohibits a second layer of turns because a potential of 2 kV between layers would result. High voltage insulation, as such, would lack geometric precision. These considerations required that a

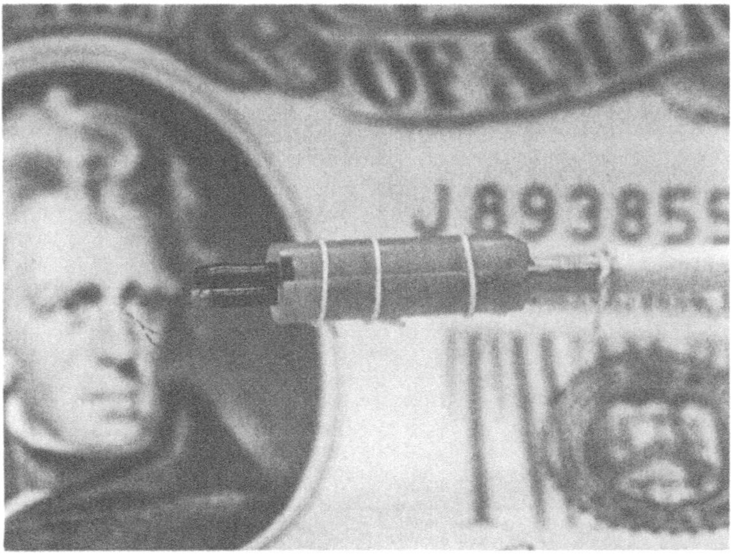

Fig. 2. Pair of detection coils prior to final balancing. The length of the last turn was adjusted (the loose lead is shown) to match the electrical characteristics of the coils. The top, black coil is epoxy filled with oriented superconductor particles.

balance between the sample coil and the compensating (pure epoxy) coil must be at least 1 part in 10^4.

The limited dynamic range of fast analog-to-digital converters made impossible the alternative approach of recording the signal from each coil and subtracting them later. Thus, the coils must be connected in opposition and only their difference recorded. No signal averaging was possible as the experiment destroys sample, coils, and dewar. Thus techniques had to be developed that would reproducibly balance the coils.

EXPERIMENTAL DEVELOPMENT AND RESULTS

Initially, we used a large homogeneous modulation coil driven by a lockin amplifier through a power amplifier with our sample coils positioned in the center. The capacitance in the modulation coil prevented useful fields as we increased frequency. After some wasted time, we realized that it was adequate to match the resistance, capacitance, and inductance of the coils using an HP 4275A LCR Meter operating at 4 MHz. It was then obvious that the pure epoxy and superconductor-filled epoxy must be machined to match with great precision. We have found that the two 1 mm diameter, 4-5 mm long rods could be matched in diameter, taper, and roundness to about 10 micrometers using a large shadowgraph at 50x magnification for physical inspection and sorting through many rods.

About 90 turns of 30 micrometer diameter copper wire with

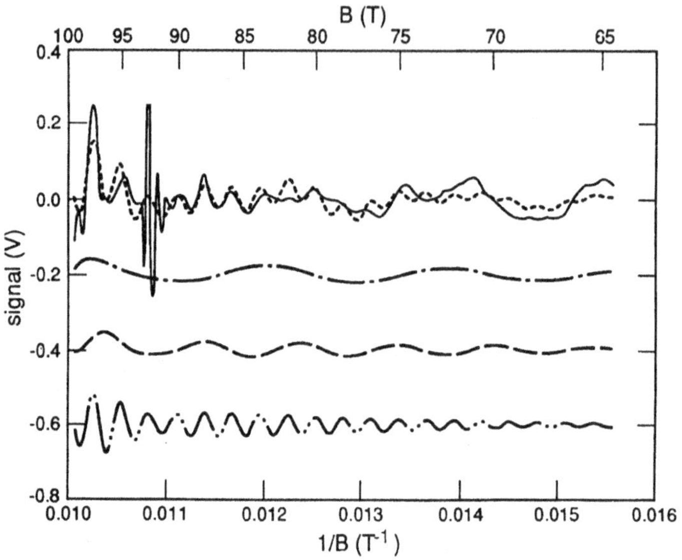

Fig. 3. Solid line is the signal output from the passive filter where the time dependence has been eliminated so that it is plotted as a function of the inverse of the applied field. The dashed line is the sum of the three bottom curves which represents a fit to the signal with frequencies of a 0.53 kT, 0.78 kT, and 3.51 kT (from Ref. 2). Thus the top curves represents the measured (solid) and calculated (dashed) curves.

Beldure insulation were then wound on each rod and fastened with a hint of heavily thinned GE 7031 varnish. The signal leads of 120 micrometer insulated copper wire were twisted, put in woven fiberglass spaghetti, and connected to the twisted coil leads that were tied down. The other ends of the coils were connected together and a light coat of Stycast 1266 epoxy was put over all but the last few coil turns. This stage in the coil assembly is shown in Fig. 2. The final matching was achieved by adjusting the last turn of one coil while measuring L, C, and R of each coil sequentially. Then the end was epoxied to secure all of the loose wire while allowing for the curing and shrinking of the epoxy which predictably shifts the inductance down a few parts in 10^4. Typical values for each finished coil were L = 4 microhenries, C = 400 picofarads, and R = 4 ohms; the coils were identical in all respects to 0.2%. Cooling the matched set to liquid nitrogen temperature resulted in L increasing and C decreasing by 1% each but with no change in their matching. In general, we avoided thermal cycling until cooling for the actual experiment.

Tests of matched coils on pure epoxy in the pulsed field showed an overall compensation to typically within one half of a turn with the coils, twisted leads, high-pass filter, 15 meters of co-axial cable, and a terminator in the system. The use of a 0.5 MHz high-pass filter allowed us to eliminate almost all of the magnetic field generator noise but passed the useful signals. This lowered the balance condition from the 1 part in 10^4 mentioned earlier. A typical filtered signal after the peak field is shown in Fig. 3; the time variable has been eliminated using the field calibration obtained from integrating the output of a small uncompensated coil. The plot was made as a function of the inverse of the applied field because dHvA oscillations are periodic in B^{-1}. The fast pulses just below 0.011 T^{-1} are a high-frequency characteristic of the generator that always appeared just after peak field.

The dashed curves are the results of a Lifshitz-Kosevich fit to three different pieces of the Fermi-surface derived from the T = 2.3 K solid-line experimental trace as discussed in Ref. 2. For our purposes here, it sufficed to sum the three dashed curves (with their arbitrary zero offsets) and to plot the upper dashed curve on top of the data trace for a comparison of experiment and theory. With increasing time, the field decayed through radiation, as well as coil deformation and breakup; it is obvious that the goodness of the fit deteriorates at larger values of B^{-1}. Nonetheless, it is clear that a signal coil was adequately compensated in the 10 MHz range by a matched empty coil for use in a fairly harsh magnetic environment.

ACKNOWLEDGMENTS

We are indebted to N. V. Coppa, C. M. Fowler, B. L. Freeman, J. C. King, A. Migliori, F. M. Mueller, J. A. O'Rourke, T. H. Pierce, and R. F. Rockage for discussions and assistance. This work performed under the auspices of the U. S. Department of Energy, Office of Basic Energy Sciences, Division of Materials Sciences.

REFERENCES

1. D. Shoenberg, "Magnetic Oscillations in Metals," Cambridge University Press, Cambridge (1984).
2. C. M. Fowler, B. L. Freeman, W. L. Hults, J. C. King, F. M. Mueller, and J. L. Smith, submitted to Phys. Rev. Lett.
3. J. E. Graebner and M. Robbins, Phys, Rev. Lett. 36:422 (1976); T. Maniv, R. S. Markiewicz, I. D. Vagner, and P. Wyder, Physica C 153-155:1179 (1988).
4. C. M. Fowler, R. S. Caird, W. B. Garn, and D. J. Erikson, IEEE Trans. on Magnetics MAG 12:1018 (1976).
5. J. L. Smith, C. M. Fowler, B. L. Freeman, W. L. Hults, J. C. King, and F. M. Mueller, to appear Proc. Miami Workshop: Electronic Structure and Mechanisms for High Temperature Superconductivity, Plenum, NY (1991).

REVERSIBILITY LINE MEASUREMENTS IN A SQUID WITHOUT SAMPLE MOVEMENT

A.R. Perry, A.M. Campbell

IRC in Superconductivity
Madingley Road
Cambridge, U.K.

ABSTRACT

The standard method of measuring magnetisation with a SQUID magnetometer involves moving the sample between the measuring coils. The inhomogeneity of the magnet field can cause the hysteresis loop to collapse in hysteretic materials and will upset the flux distribution in flux creep measurements. Two ways are described for determining reversibility lines more accurately than can be done from the standard magnetisation curves. In the first, a standard scan is used but sophisticated signal analysis of the response is used to detect magnetisation changes caused by inhomogeneity in the magnet field during the scan. In the second technique, the sample sits in a small copper coil which is used to impose a small ripple field on the steady field of the magnet. A compensating coil minimises direct interaction of the drive field with the SQUID and the magnetisation of the sample is measured directly from the SQUID output. It was found that very large drift occurs in this output, which is normally removed by the instrument software before the results are presented. This had to be removed to get significant answers. When this was done there was a clear difference between the reversibility line[1] as measured by D.C. magnetisation and that observed by this method.

STANDARD MEASUREMENT USING A SQUID

A SQUID magnetometer consists of three essential elements:

1. A source of a uniform dc magnetic field.
2. A system of 'pickup' coils within which the sample is moved.
3. A SQUID sensor which measures the current in the 'pickup' coils.

The current in a superconducting pickup coil changes in step with flux moving in and out. The flux motion in the coil is reduced when the dipole magnetic source is further away from the coil. When a point magnetic source is a distance x along the axis away from a coil of radius R_C, the signal in the coil is proportional to $S_C(x)$

$$S_C(x) = \frac{R_C^2}{2\left(x^2 + R_C^2\right)^{3/2}} \tag{1}$$

[1] see section *I* of this volume

The coil configuration used in most SQUID systems is called *second differential* because of the shaped response it provides. The configuration has a central coil and an additional coil on each side of the central coil. The central coil has twice as many turns as the outer coils and is wound in the opposite direction. A sketch of the configuration and the signal response along the displacement path is shown in figure 1. With the coils spaced a distance D_C apart, the total signal in the wire of the configuration is

$$S(x) = f_F\left[-S_C(x - D_C) + 2S_C(x) - S_C(x + D_C)\right] \tag{2}$$

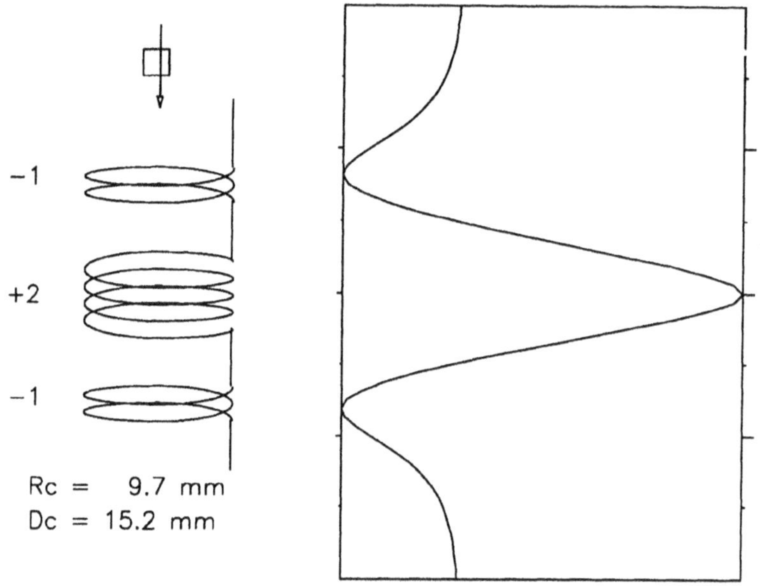

Figure 1

A sketch of the pickup coil configuration and a trace of its sensitivity to a coaxial dipole. The parameters R_C and D_C correspond to those of the magnetometer used.

This configuration is frequently chosen because the sensitivity to magnetic disturbances decreases very rapidly with distance and provides good rejection of external fields.

<u>Short scan length requirement</u>

Superconducting magnets are designed to have a constant field strength in the central region. The MPMS [2] magnet, for example, has variations of the order of 10^{-5} of the ambient field. This is sufficient for variations to be less than $1\,mT$ in the high field regime. This is sufficiently small to ignore in most applications, but not for measurement on hysteretic superconductors.

In order to measure the value of J_c of a sample, the amount of flux that can be pinned is measured. The amount of pinned flux is generally found as the difference between the upper and lower limits of the major hysteresis loop. J_c decreases with

[2] Manufactured by Quantum Design

increasing field, reducing the difference between the two measurements, leading to small hysteresis amplitudes $\ll 1\,mT$ in high fields.

When the hysteresis becomes comparable to the variation in ambient field during a scan of the sample, the pinned flux will be moved in and out of the sample during the scan, ie the displacement of the sample, resulting in an apparent loss of hysteresis. This limits the ability of the magnetometer to detect J_c well before the limits imposed by the SQUID sensor sensitivity.

It is possible that many reports of an unpinned flux liquid in the literature are due to the inhomogeneity effect of the magnet. We have made measurements of the field inhomogeneity of our magnet and the effects on magnetisation measurements have been published[**1**]; subsequently Quantum Design published a Technical Advisory Note[**2**] to cover this point.

The magnet field shows less variation in the central region so that the sensitivity to hysteresis can be improved using shorter scan lengths. From a $60\,mm$ scan length, that can show all three peaks of the response function, a move to $30\,mm$ should give a factor of 10 improvement in sensitivity to hysteresis. For $20\,mm$ and shorter scans there is a more significant problem in setting up the position of the sample to be in the centre of the scan.

Hysteresis information from DC results

The data was analysed by assuming that the response can be split into four functions, S_1 to S_4. The functions were assumed to take the same shape for all fields and temperatures in a given sample, but to vary in amplitude. Initial shapes of the functions were guessed and a linear regression used to find their amplitudes for each response in the run by minimising the residual errors.

The total error from all the responses was found by summing all the individual residual errors. The function shapes were then adjusted by a gradient descent regression, changing each of the points in them by a small amount and repeating the linear regression. The process was repeated until the total error was reduced to the minimum possible. In this way the shapes and amplitudes of the relevant functions which contributed to any of the response curves could be determined independently of the initial assumptions about them. This also results in a major reduction in the noise.

The functions which appear from this analysis have the following physical interpretations.

S_1 A clean response function for a point source sample as in figure 1. The amplitude of this function is proportional to the average magnetisation M during the scan.

S_2 Has the shape of the derivative of the response function with respect to sample position in the scan. The effect of combining some of S_2 with S_1 is to move the apparent position of the sample. The occurrence of S_2 is explained by movement of the sample in the course of the experiment due to continuous thermal contraction of the the sample mount.

S_3 Has the same shape as the response function of the empty sample carrier. The sample carrier moment varies differently with temperature and field from the sample moment. This allows its response to be separated from that of the sample and subtracted from it.

S_4 An unusually shaped response function which is caused by variation in the magnetisation M during the scan as the sample cycles through a minor hysteresis loop during the measuring scan, as described by Blunt[**1**].

Other functions such as those for quadrapole moments have not been detected in the small samples normally used.

Example Analysis

A sample of sintered $YBa_2Cu_3O_{7-\delta}$ was measured using the MPMS system with a short scan length of $34\,mm$ at a temperature of $77\,K$. The scan length is close to the lower limit for reliable operation unless custom software is used. The difference in magnetisation between the increasing and decreasing field $\Delta M(H) = M(H\downarrow) - M(H\uparrow)$, referred to as hysteresis and is shown in figure 2, shows the flux pinning potential of the material decreasing into noise before the field reaches $4\,T$. This conflicts with ac experimental results[**3**], where the hysteresis continues to about $9\,T$ at $77\,K$.

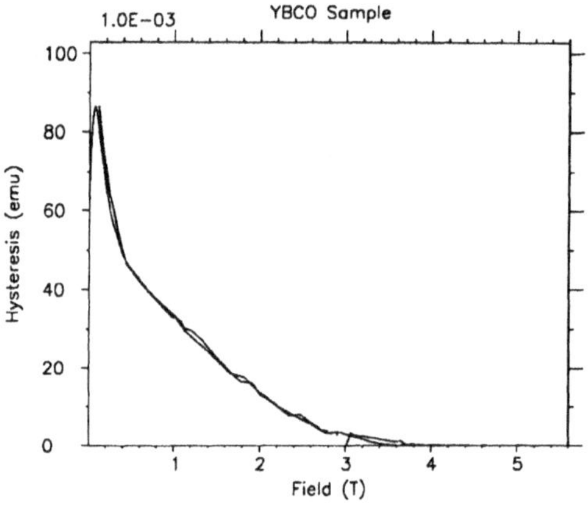

Figure 2
Apparent reduction in the hysteresis of the sample at $77\,K$

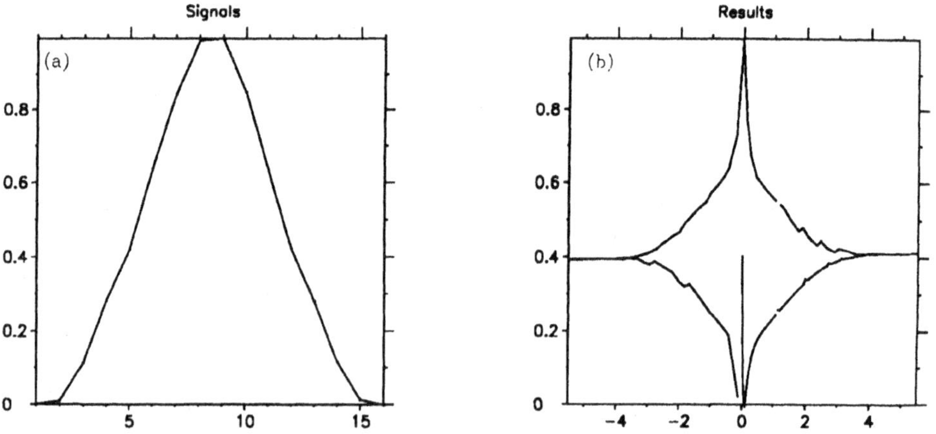

Figure 3
Component of results corresponding to S_1. Left is the signal shape detected during the scan, right is the amplitude, which is proportional to magnetisation. As in figure 2, the value is zero for magnetic fields above $4\,T$.

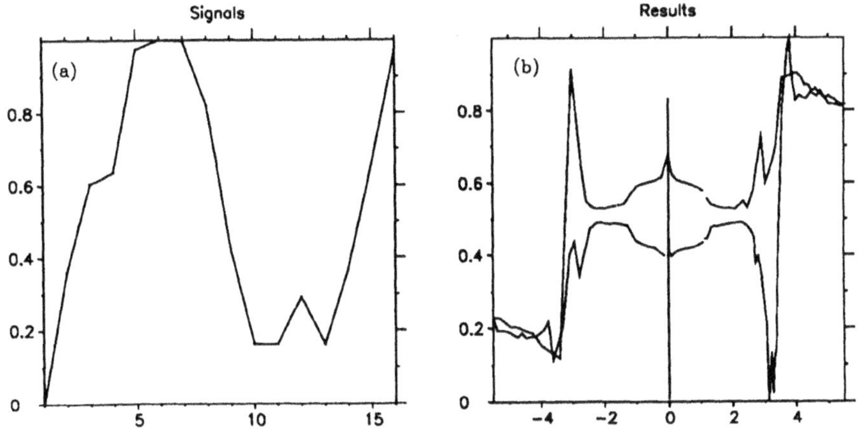

Figure 4

Component of results corresponding to S_4. Left is the signal shape detected during the scan, right is the amplitude, which is proportional to minor hysteresis loop during scan. Unlike figure 2, this has a non-zero value in entire range.

The first function detected, shown in figure 3a, is characteristic of S_1. The associated amplitudes, shown in 3b, are the same as the results used to generate figure 2.

The second function, shown in figure 4a, is S_4 due to the sample magnetisation changing during the displacement. S_4 is caused entirely by hysteresis since the change in reversible magnetisation during the scan is much smaller. The sign of the amplitude in figure 4b is determined by the sign of the mean magnetisation. Up to about $3.5\,T$, we get different values for increasing and decreasing fields corresponding to the hysteresis in the magnetisation curve in figure 3. Above this value, the measured magnetisation is the same for increasing and decreasing fields so only a single curve appears in figure 4b. However, the fact that the signal appears at all shows that the sample is being put around a minor hysteresis loop during the scan and that the reversibility line is therefore well above the fields where the magnetisation hysteresis reaches zero (figure 2) and cannot be reached at $77\,K$ in this apparatus. The amplitude can be seen to be reducing at higher fields (and linear extrapolation gives a reversibility field of $9\,T$ at $77\,K$).

Although this is a sensitive detector of hysteresis, the amplitude of the hysteresis depends on the curvature of the field profile of the magnet. This profile would have to be measured at each field to get quantitative results.

STATIONARY SAMPLE MEASUREMENTS

To avoid the problems involved with moving the sample during measurement, it would obviously be desirable to keep the sample stationary while putting it round a minor hysteresis loop and retrieve the magnetisation directly from the SQUID output.

Direct Oscillation of the Magnet

A small oscillation was imposed on the dc field of the superconducting magnet with the same sample as used earlier in this paper. A complex response was seen.

Minor loop information was observed, but the results did not change when the sample and sample holder were removed from the system. It follows that the response was dominated by hysteresis in the magnet, detected by the very small imbalance in the SQUID coil configuration.

Supplementary Coil Technique

For this type of measurement, the sample was placed in one coil of a probe into which a low frequency current was injected. The magnetisation of the sample was detected by monitoring the SQUID output. The probe and sample are not moved during the experiment.

The probe consists of a pair of coils wound in series opposition. In operation the coils produce a magnetic field with odd symmetry about the centre. An axial space allows the sample to be placed in the centre of one of the coils. A sketch of the configuration is shown in figure 5.

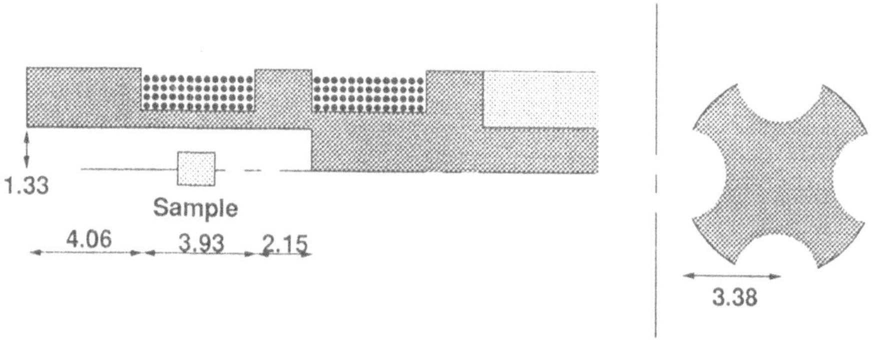

Figure 5
Two cross-sections of the probe, showing fins required for heat dissipation.

The SQUID sensing region has three coils arranged with mirror symmetry. There is a peak of maximum sensitivity at the centre of symmetry which is broader than the coil probes are long. A sketch of this is shown in figure 1.

Balancing the Coils

To minimise the coupling between the drive coils and SQUID, the position of the probe was set so that the point of symmetry is in the middle of the central peak of the sensor response function. The odd symmetry of the probe field cancels against the even symmetry of the sensor response, making the *Drive* field invisible to the sensor, while the sample is still near the middle of the peak and is visible.

The stepper motor responsible for the sample motion during ordinary MPMS measurements is used to achieve the balance position. The positioning accuracy of $\approx 2\mu m$ is sufficient for use with small currents in the coils. In order to operate with larger currents the central tap between the two coils is available to the electronics. A small proportion of the drive current is injected into this tap and alters the magnetic balance slightly. This change, generally $< 0.1\%$, will not affect the field seen by the sample but is sufficient to provide the necessary fine tuning.

After coarse and fine balancing, the sample is centred in the magnetic field of the main magnet and is also subject to the fields generated by the coil in which it sits. Application of current to the coils causes the sample to cycle around minor magnetic hysteresis loops. The instantaneous magnetisation is detected by the SQUID sensor superposed on the constant offset that is due to the residual imbalance of the coils.

Detecting the correct balance position for the coils manually is a time consuming operation as the sensitivity of the SQUID sensor requires a very accurate initial positioning if the most sensitive ranges are to be available. The coil configuration had been chosen for ease of balancing, but it was found that a skilled operator could locate the correct point to the required accuracy in about $1\frac{1}{2}$ hours. A software based search is thus used, completing after 10 minutes, to acquire sufficient information that a regression can determine the required position. The fast search can reduce the amount of data required by relating its measurements to the known relationship between balance signal and position. It chooses the position of the next data point to minimise the uncertainty in the regression.

Probe design

The size of the coils in the probes is determined by the small diameter of the sample space. The coils have an inner diameter of $4\,mm$ and an outer diameter between $7\,mm$ and $8\,mm$. Their length is generally about $4\,mm$ and their spacing $< 2\,mm$. The volume of the coil contains a varying number of turns of copper wire; different probes use different diameters of the wire as the field density to power dissipation ratio is not altered by the wire cross section. The cross section is chosen for matching to the power supply and for ease of coil winding.

Large currents are required to investigate samples with a large J_c in order that they can be forced around a hysteresis loop. These cause localised heating of the sample region of the sample mount. The gas environment in the sample space cannot remove heat rapidly leading to increases in sample temperature during measurement.

In limited temperature ranges, where an inert liquid cryogen is available, the sample space can be filled with the liquid and the probe operated within it. This provides a much greater capacity for heat removal.

The main probe body is made from OFHC copper to improve heat removal from the sample region. The probe body has four longitudinal grooves cut into the surface which increase the surface area for a small reduction in cross section. These fins (see figure 5 right) offer a larger area for heat conduction to the gas or liquid surroundings. This helps to maintain the sample at a constant temperature during measurement.

Results from Coil Probes

Figure 6 shows the magnetisation detected in the standard way using the MPMS dc method at $0.5\,T$ for the sample of $YBa_2Cu_3O_7$.

The apparatus was not designed for this type of measurement, but the results from the standard technique may be compared with the equivalent in phase and out of phase components as measured by the stationary coil technique in figures 7 and 8. The current amplitude was kept constant.

The secondary T_c at $60\,K$ is visible in the loss signal but not apparent in the moments of the dc results. The oscillation period used for these measurements was about $30\,s$.

The problem with acquiring results over large temperature ranges is that the sample holder and probe assembly undergoes thermal contraction during the experiment and changes the balance condition as seen by the SQUID sensor. The balance condition was satisfied near $60\,K$, a linear correction was used that is appropriate in the range up to $150\,K$. The thermal contraction is not linear at low temperatures resulting in the apparent reduction in the real part of the moment below $40\,K$ in figure 7.

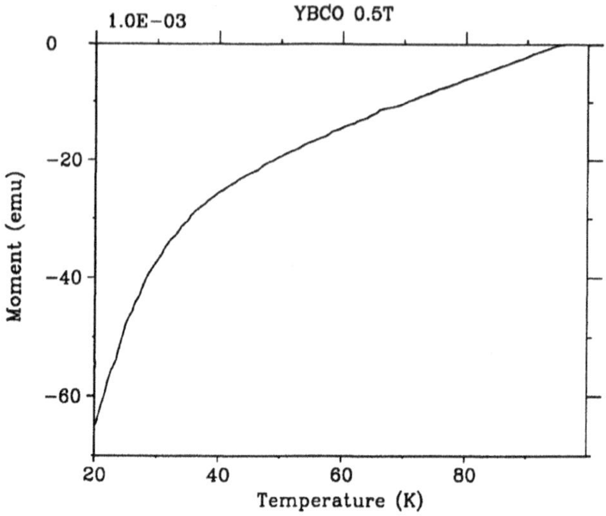

Figure 6
Standard dc measurement of sample, cooled in Zero Field, when warming slowly in $0.5\,T$ applied field.

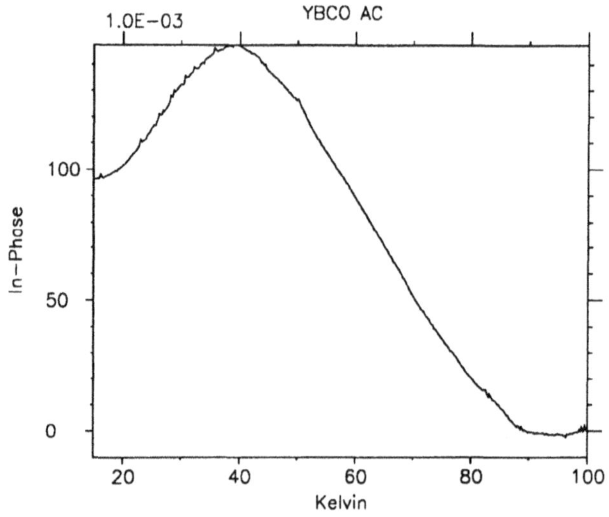

Figure 7
In-Phase component measurement using probe.

The temperature range between $70\,K$ and $100\,K$ was measured, with measurements spanning the range of fields from 0 to $5.5\,T$ to permit the reversibility boundary region to be examined.

The in phase and out of phase signals are shown in figures 9 and 10. These figures show a contour map of the signals (with field on the x axis and temperature on the y axis) together with a perspective view of the surface.

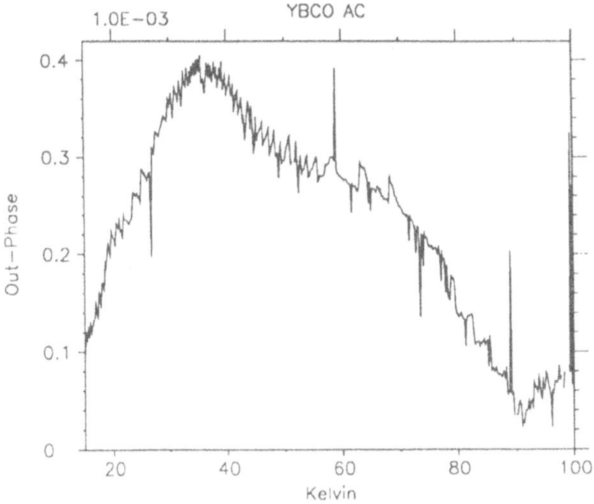

Figure 8

Out-of-Phase conp)onent measurement using probe.

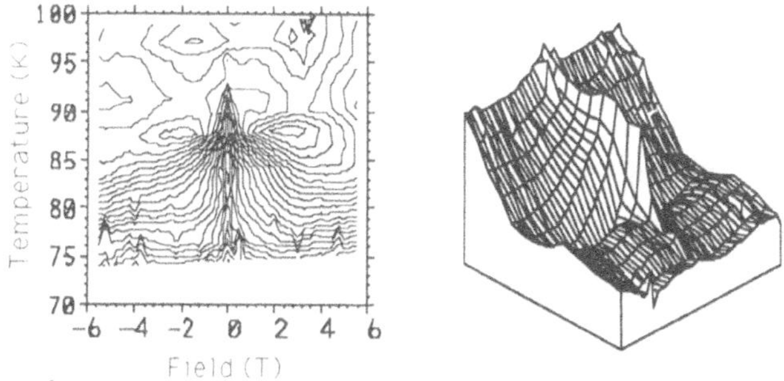

Figure 9

Contour map of the in phase signal from the probe, with a view of the projected surface. The onset of flux exclusion near 90 K is shown in figure 11.

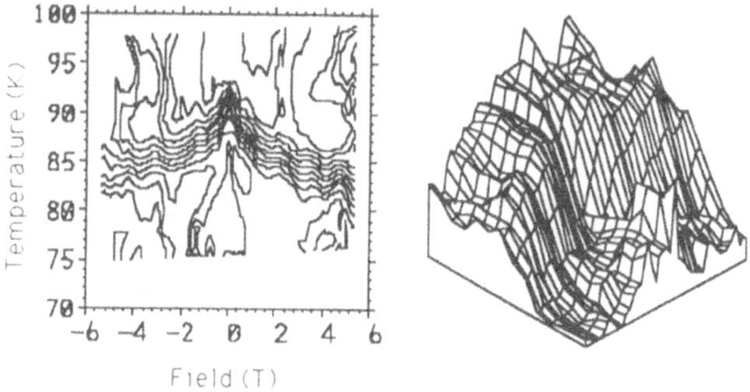

Figure 10

Contour map of the loss signal from the probe, with a view of the projected surface. The onset of loss $B_{Rev}(T)$ is shown in figure 11.

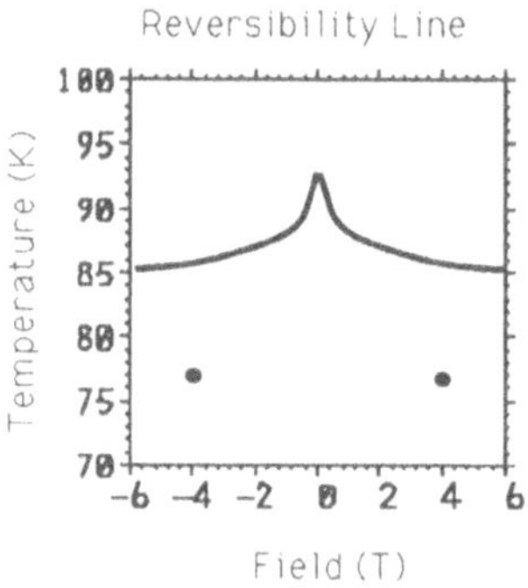

Figure 11

Reversibility line constructed by inspection of figure 10.

Edges are visible in both figures, shown by the upper limit of the close contours section, which are due to the cessation of hysteretic properties of the sample. The line along this edge, shown in figure 11, is the reversibility line for the sample. The two points marked on the figure represent the estimates of reversibility derived the standard dc magnetisation results in figure 2. The reversibility fields found by this technique coincide with those found by the signal analysis technique that resulted in 4b, when repeated similarly at different temperatures.

CONCLUSIONS

The analysis of the response curve from the moving sample measurements shows that hysteresis extends well above the point where the dc magnetisation curves join.

The use of supplementary drive coils to put the sample round a minor hysteresis loop can also be used to derive the true reversibility line and is a sensitive technique for getting absolute values of hysteresis at frequencies down to dc.

REFERENCES

[1] BLUNT J., PERRY A.R., CAMPBELL A.M., LIU R.S. *An investigation of the appearance of positive magnetic moments on field cooling some superconductors* Physica C., 175 (1991) 539-544

[2] QUANTUM DESIGN *Variation of Magnetic Field over the Scan Length* Technical Advisory # 1, Aug 1989

[3] ZHENG D-N. *Pers. Comm.* March 1991

AUTOMATIC TEMPERATURE CALIBRATION DURING SWEPT TEMPERATURE MAGNETISATION MEASUREMENTS

A.R. Perry, A.M. Campbell

IRC in Superconductivity
Madingley Road
Cambridge, U.K.

ABSTRACT

Most temperature controllers overshoot when changing temperature, which causes major problems in measuring the magnetisation of hysteretic superconductors. Continuous sweeping avoids this problem but introduces lags between sample temperature and the thermometer which vary for different sample sizes and materials. A common way to allow for temperature lags is to do the experiment at different sweep rates and extrapolate to zero sweep rate. It is shown that this is an inefficient and time consuming solution and that a better method is to impose a ripple temperature variation on the sweep. The variation of the amplitude and phase of the ripple in the magnetisation give enough information for the thermal time constants to be determined at the same time as the average magnetisation is being measured. The temperature can be corrected for any sample during the course of a magnetisation measurement. In this work, the sample is modelled as a single body connected by a thermal resistance to the cold source.

THE PROBLEM OF SAMPLE TEMPERATURE HISTORY

The system temperature is controlled by actuators such as heaters, valves and pumps. Information from temperature sensors is collected by the controlling system which sets these actuators appropriately. Inappropriate use of the actuators can result in thermal oscillations and overshoots. Many temperature controllers do not consider oscillations and overshoots to be significant and give no indication to the operator that they have occurred.

Estimation of T_c and J_c of a superconductor by magnetic methods is normally performed by estimating the shielding current flowing around the material. If the sample temperature overshoots during a measurement of T_c, magnetisation results show a loss of superconductivity on returning to the set temperature. If the sample is not completely cooled when a measurement of J_c is started, the depressed superconductivity during the calibration at the beginning of the experiment causes the results to show an increase in J_c later in the experiment as the sample cools.

It is thus very important to stabilise the system temperature prior to starting an experiment and to maintain smooth control of it throughout. One way around this problem is to ramp the temperature continuously. However, this introduces a different form of error that is due to thermal lags between the sample and the temperature sensor.

Surfactants in Solution, Volume 11, Edited by K.L. Mittal and D.O. Shah, Plenum Press, New York, 1991

A TECHNIQUE FOR CONTROLLING TEMPERATURE: DRIFT MODE

There are so many problems in accurately controlling the sample temperature that an alternative approach is often taken. The cryostat is set to gradually change temperature by fixed heating and cooling actuator settings. The system moves towards the desired temperature and will continue past it. It does not stop or change direction. This means that stranded magnetisation states and the problems associated with them are avoided.

MPMS, the SQUID magnetometer product of Quantum Design, can perform such Drift Mode measurements. The operator must determine the appropriate actuator settings manually, a mechanism which offers minimal control of ramp rate.

Gross variations in the thermal properties of the cryostat system require that actuator settings vary with temperature region in order to keep the drift rate within reasonable bounds. The effect of changing drift rate during an experiment and the subsequent difficulty in estimating sample temperatures led us to develop a more powerful Drift Mode. In this version, the temperature remains under active control but the temperature setpoint is smoothly ramped through the range of interest. When the ramp is started, stranded magnetisation states are initially encountered. As the ramp proceeds, the temperature will pass that of the stranded state, unstranding it. The controller settles to a moving target as easily as to a stationary one, so after the usual settling time the sensor temperature ramps smoothly at a controlled rate.

This extended drift mode leads to a more even distribution of results through the temperature range and eliminates the inadvertent destruction of the cryomagnetic history of the specimen.

MULTIPLE RATE DRIFTING - AN EASY AND INEFFICIENT SOLUTION.

The thermal response of the equipment changes as the heat capacities and conductivities vary with temperature. If the system behaves linearly then the relevant theory described by Richards[**1**] may be applied. In practice this means that the different components of the cryostat must be closer together in temperature than the difference over which the thermal response changes. The variation of the thermal response can then be considered a background and safely ignored. In this case, the thermal parameters of the system can be expressed as a set of time constants which are independent of thermal history, even though they may still vary strongly with temperature range.

Assuming this independence criterion, results obtained by performing drift mode experiments at different thermal ramping rates r may be combined. Assuming the ramping temperature to be locally linear with time, the transfer function from the sensor to the sample can be replaced by a simple time lag τ.

$$\begin{aligned} T_{sensor} &= T_{system} \\ &= T_o + r(t - t_o) \\ T_{sample}(t) &= T_{sensor}(t - \tau) \\ &= T_{sensor}(t) - r\tau \end{aligned} \tag{1}$$

For two measuring scans with different ramp rates, the value of τ delays the readings differently. For small sections of the data at each temperature the results from different scans can be compared to determine the apparent temperature difference for equivalent features.

Such a technique can successfully operate with, for example, a superconducting transition as shown in figure 1. Close inspection shows that the curves have apparent

T_c's of 92.3 and $98.0 \pm 0.1\,K$ for drift rates of 0.1 and $1.0\,K/min$ respectively and thus the sections near T_c will superimpose if displaced by $5.7\,K$.

$$\begin{aligned} &\text{From (1)} \\ &T_{sample} = T_1 - r_1\tau \\ &\qquad\quad = T_2 - r_2\tau \\ &\text{thus} \\ &\tau = \frac{\Delta T}{r_2 - r_1} \\ &\text{where}\quad \Delta T = T_2 - T_1 \end{aligned} \tag{2}$$

In this case, the two drift rates r_1, r_2 were 0.1 and $1.0\,Kmin^{-1}$, which for $\Delta T = 98.0 - 92.3\,K$ yields a local $\tau = 6.3 \pm 0.2\,min$. The T_c of this test sample of $YBa_2Cu_3O_{7-\delta}$ is thus actually $91.7 \pm 0.2\,K$ in a magnetic field of $1\,T$.

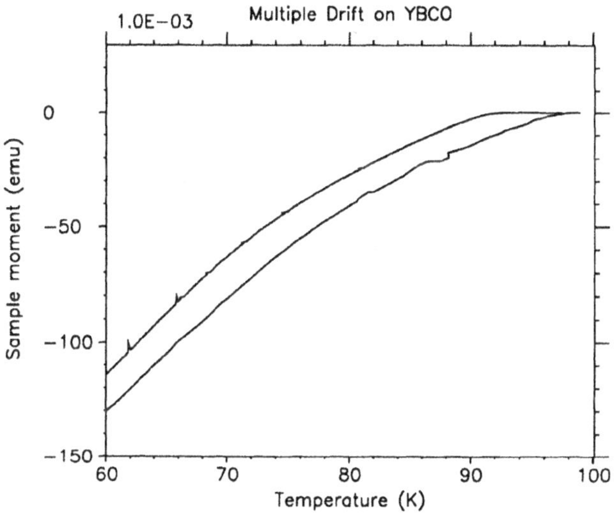

Figure 1

Magnetisation of a sample on warming in a $1\,T$ field, two curves result from different warming rates.

In figure 1 a small sample of a pure superconductor yielded two smooth clean curves. The sample was not removed or in any way changed between the runs in order that shape factors, scan length adjustments and other problems remain unaltered between the two measurements. τ can be estimated graphically over the whole temperature range. Since we have a 1 : 1 mapping between M and T, the results can be reversed to show $T(M)$ and the difference between corresponding temperatures found. Such a plot of $T_{0.1}(M) - T_{1.0}(M) = \Delta T(M)$ is shown in figure 2 where the trend of ΔT

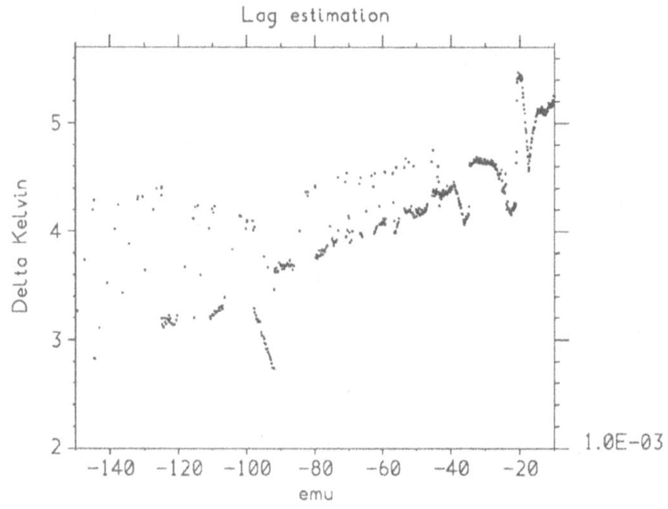

Figure 2
The difference in temperature, for equivalent magnetisations, between the two curves shown in figure 1.

increases from 3 to over 5 K as the magnetisation reduces across the temperature range investigated.

This useful technique is not usually possible on commercially available cryostats due to restrictions imposed by their temperature controllers, although on some magnetometers careful artificial restructuring of the time constants used by the temperature controller can give the desired effect. Reliance on recognisable sharp features within the magnetisation results leads to poor performance where the magnetisation variation is smooth. The operator can then only compare the values of τ for similar samples and other temperature ranges and approximate to the true value.

The multiple scan method is inherently inefficient. In the above example the faster scan was quite sufficient to measure the required magnetisation but the error in sample temperature means that the results are uncalibrated and so are only qualitative. Instead of cooling the sample once and making one measurement scan, it was necessary to cool the sample twice and make an initial scan with a given r and then a second scan taking ten times longer. The factor of ten can be reduced at the expense of greater errors in the resulting τ. The total amount of machine time used was a factor of 12 more than that required if the sample and the sensor were at the same temperature.

USE OF A RAMPED RIPPLE - AN ELEGANT AND TIME EFFICIENT SOLUTION

Given the inefficiency of the multiple rate method, a much more efficient method was required. This was developed as follows:

When the sensor temperature is slowly ramped, the error in the sample temperature can be described by a single parameter L measuring how much time elapses between the sensor and sample passing a given temperature.

When the temperature of the system is forced into oscillation and detected by the sensor, the sample is unable to keep in step and its temperature lags with an oscillation of reduced amplitude. This can be described by the two parameters in the complex gain $G(\omega)$. We define the ratio of the temperature oscillation amplitudes as the magnitude of the complex value $|G|$, generally less than unity. The argument $\angle G$ is the phase angle by which the sample is ahead of the sensor, generally negative.

We may detect small variations in sample temperature by monitoring the changes in magnetisation signal that occur and relating these to the observed changes in sensor temperature.

Theoretical Treatment

- We assume that the sensor is at the same temperature as the bulk thermal mass of the system.
- We model the thermal characteristics connecting the sample to the system, but describe this by the apparent characteristics from the sensor to the sample, since the sensor and the system are at the same temperature.
- The sample is modelled by a lumped thermal mass connected to the sensor by a thermal impedance. This lag is described by a time constant τ.
- We also add a delay of duration d to account for the time taken for gas to flow from the sensor to the sample.

The transfer function generated by standard methods [1] from this model is

$$G(\omega) = \frac{e^{-i\omega d}}{1 + i\omega\tau} \tag{3}$$

In the new technique, instead of causing the sensor temperature T_{sensor} to ramp smoothly, active control is used to superimpose a small ripple $\alpha \cos \omega t$ into the slope. If $\alpha\omega$ is less than the ramp rate g then the sample temperature T_{sample} never reverses direction and the sample temperature cannot exhibit hysteretic behaviour.

A check must be made to ensure that the sensor temperature T_{sensor} is oscillating as expected, since inappropriate control parameters could cause the system temperature not to follow the intended profile.

We linearise the problem by considering a small temperature range around a particular temperature T_o. $T_{sensor}(t)$ will be the ramp with the sinusoidal ripple imposed and the magnetisation M will show a similar ripple. We can find the expected response from the measurement in terms of the sensor behaviour T_{sensor} and the transfer function as follows

$$\begin{aligned} M(T) &= M(T_o) + (T - T_o)\left.\frac{dM(T)}{dT}\right|_{T=T_o} \\ M(T) &= M_o + (T - T_o)M_g \\ &\text{where} \begin{cases} M_o = M(T_o) \\ M_g = \left.\dfrac{dM(T)}{dT}\right|_{T=T_o} \end{cases} \end{aligned} \tag{4}$$

We replace cos () by the real component of the complex exponential, reduce the amplitude by $G(\omega)$, and delay the ramp by L

$$\begin{aligned} \cos \omega t &= \Re e^{i\omega t} \\ T_{sensor}(t) &= T_o + gt + \Re\left(\alpha e^{i\omega t}\right) \\ T_{sample}(t) &= T_o + g(t - L) + \Re\left(G(\omega)\alpha e^{i\omega t}\right) \end{aligned} \tag{5}$$

A standard result [2] permits L to be deduced from $G(\omega)$

$$\begin{aligned} L &= -\lim_{\omega \to 0} \frac{G(\omega) - 1}{i\omega} \\ &= d + \tau \end{aligned} \tag{6}$$

From (5) and (4) above we can deduce the variation in magnetisation of the sample.

$$\begin{aligned} M(t) &= M(T_{sample}) \\ &= M_o + M_g g(t - L) + M_g \Re \left(G(\omega) \alpha e^{i\omega t} \right) \end{aligned} \tag{7}$$

From measured values of the magnetisation variation, we can use this to determine the two parameters of $G(\omega)$. To perform an experiment, a value of ω is chosen in order that the argument of $G(\omega)$, $\angle(G)$, may be expected to be about 45°. A fit is applied to each of $M(t)$ and $T_{sensor}(t)$ with a small window that is just large enough for a few oscillations of the temperature setpoint to be present. This fit finds the arithmetic mean value K, the phasor $C + iS$ of the frequency component and the local gradient R.

$$fit(t) = K + R(t - \bar{t}) + \Re\left[(C + iS)e^{i\omega t}\right] \tag{8}$$

The values of the parameters of (8) found in the fits are used to calculate the physical parameters of the system using (5) and (7) above, as follows. Subscript T refers to the temperature fit and subscript M refers to the magnetisation fit. Three terms define the temperature ramp, two terms define the linear approximation to the magnetisation and one complex term defines the transfer function

$$\begin{aligned} \alpha &= \sqrt{(C_T^2 + S_T^2)} \\ T_o &= K_T \\ g &= R_T \\ M_o &= K_M + LR_M \\ M_g &= R_M / R_T \\ G(\omega) &= \frac{C_M + iS_M}{C_T + iS_T} \left(\frac{R_T}{R_M} \right) \end{aligned} \tag{9}$$

but from equation (3)

$$\begin{aligned} |G(\omega)| &= \frac{1}{\sqrt{1 + \omega^2 \tau^2}} \\ \angle G(\omega) &= -d - \tan^{-1} \omega\tau \end{aligned} \tag{10}$$

thus

$$\tau = \frac{1}{\omega} \sqrt{|G|^{-2} - 1} \qquad d = -\angle G - \tan^{-1} \omega\tau \tag{11}$$

We can thus calculate the two parameters τ and d as a function of temperature. To reduce noise, the parameters are determined from smoothed graphs over the whole temperature range and the smoothed parameters are used in the transfer function to correct every T_{sensor} in the experiment to a T_{sample}.

Example Analysis

A sample of approximate composition $YBa_2Cu_3O_7$ was measured with the parameters listed below and then analysed as described above.

$g =$	$0.07\,K\,min^{-1}$	Ramp rate
$2\pi/\omega =$	$40\ min$	Period
$\alpha\omega/g =$	1	Modulation depth
$T_1 =$	$10\ K$	Temperature start
$T_2 =$	$97\ K$	Temperature finish
$\mu_0 H =$	$1\ Tesla$	Applied field
$l_s =$	$5\ cm$	Sample scan length

The value of $G(\omega)$ resulting from the calculations is shown on the left in figure 3, estimates of the thermal lag τ and thermal delay d are shown on the right. Each fit results in one point on each of the graphs. It is clear from these results that for this particular measurement there was a negligible amount of thermal delay and thermal lag varied between about 3 and 8 min depending on the temperature.

The difference between the sensor and calculated sample temperatures $T_{sensor} - T_{sample}$ is shown in figure 4 and has a mean of $g(\tau + d) = 0.5\,K$. This value is small, because the ramp rate was a factor of 7 smaller than usual. In ordinary use, this technique would have corrected temperature errors of the order of $4\,K$.

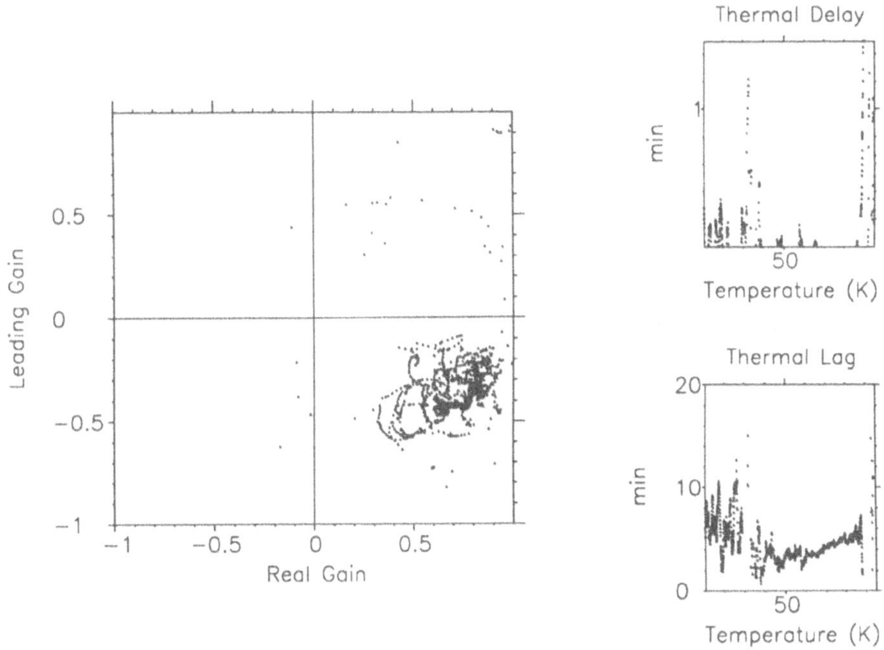

Figure 3

Summary of the complex gain from thermal oscillation of the sensor to magnetisation oscillation in the sample. Variation of the model parameters d and τ against temperature. The value of d is too small to be measured.

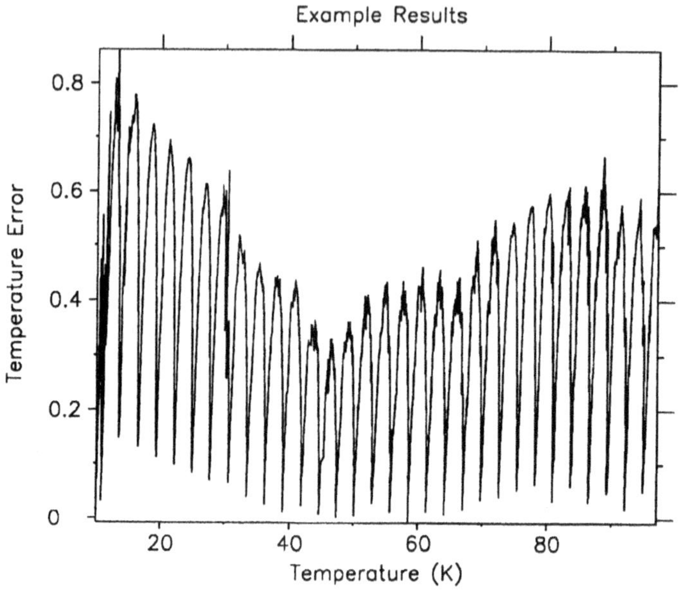

Figure 4

The difference between the sensor and the modelled sample temperature $T_{sensor} - T_{sample}$, varying with temperature. The mean is $g(\tau + d)$.

- There are obvious benefits in these two methods over Drift Mode. the second method is an improvement over the first.
- The new technique allows measurement to be as fast as a single standard drift mode run.
- The parameters of the model are estimated separately for different temperatures, so single experiments that scan through large temperature ranges are possible.
- The model parameters measured of the sample are used. The sample need not be compared with similar standard samples, so that unlike methods that predict the parameters by comparison with test runs, these inaccuracies need not occur.
- The recalculation of the sample temperature from the sensor also allows the model to correct for temporary failures in the temperature control.

There are still problems on systems with a constant temperature difference between the sensor and sample due to a flowing gas environment. While this new method is still an improvement over other methods, the sample temperatures resulting from it do not take the constant difference into account. Cryostats such as that of the MPMS, with a virtually stagnant gas sample chamber, do not suffer this additional problem.

REFERENCES

[1] Richards R.J.; chapter 2 *An Introduction to Dynamics & Control* Longman Group (1979)

[2] Richards R.J.; page 12 *An Introduction to Dynamics & Control* Longman Group (1979)

Appendix

Irreversibility Line Debate

The Irreversibility Line

Q. Y. Chen

Honeywell, Systems Research Center
MN65-2600, 3660 Technology Drive
Minneapolis MN 55418

Measurements of AC susceptibility for type II superconductors, including high temperature superconductors, largely reveal two types of energy losses: (1) hysteresis and (2) flux motion. Using a small ac signal H_{ac} superimposed on a dc field, the points (H_{dc},T) on the H-T plane for the χ" peaks have been used to determine the "irreversibility line" (IL) for a high-temperature superconductor. While the IL has been treated as a thermodynamic boundary between the vortex liquid and vortex solid (based on the notion of vortex melting), which thus should be an intrinsic property, its dependence on the sample sizes, ac amplitudes, and frequencies have been observed, although opposite results also exist. For the former, the line shifts toward higher temperature as the frequency increases and toward lower temperature as the sample size decreases or as the ac amplitude increases.

The hysteresis contribution, quite similar to that due to domain-wall motion in ferromagnetic materials, originates from the orientation change of trapped flux and should be independent of the frequency of applied field. Under such circumstances, superconducting screening is effective and the critical state model gives a good general account of the magnetic behavior. The work by Maxwell and Strongin has laid much of the physical foundation for interpreting the basic feature of χ-T curves of most superconductors. Later works on granular superconductors further take the intergranualar and intragranular critical currents into account, but the fundamental physics is quite the same. For strong-pinning type II superconductors, the Bean critical state model and its modifications have provided a simple picture for a field penetration and the associated current distribution. In these materials the energy loss is mainly due to the magnetic hysteresis and χ is therefore independent of frequency.

I have illustrated in my article that with increasing applied field and frequency the flux pinning would be weakened and flux-motion effects will become more significant. Consequently, the superconducting screening effect will decrease and eventually be overwhelmed by the regular ac screening effects mediated by the flux flow induced resistivity. In the weak-pinning regime, the ac magnetic behavior can be understood in the context of flux-motion-induced resistive dissipation which is governed by the magnetic diffusion equation. In a sense, one may regard such materials as a viscous vortex-dragging oscillatory system.

When using ac susceptibility as a means of determining the irreversibility line, it is important that one establish whether it is the hysteresis or the flux-motion effect that is dominant. The dc irreversibility line on the H-T plane is the boundary above which the induced dc magnetization current in the superconductor cause the vortices to move. The ac irreversibility line, however, delineates a demarcation line whose origin needs to be further determined; in the flux-motion dominated case, regular skin-effect may play a major role.

Generally speaking, it is possible that the width of χ will be much broader in this case as compared with the superconductivity related cases. Discussions on the skin-effect and on the energy dissipations can be found in classical electrodynamics textbooks.

Unless it can be certain that absolutely no ac frequency or amplitude dependence can be found in measuring χ'', the controversy over the physical meaning of the ac irreversibility line remains unresolved. While lack of accuracy, the dc technique in principle is the only magnetic method, in my judgement, that would give least dispute on the validity of "irreversibility line" data interpretation.

Remarks on the Irreversibility Line

A. Campbell

Research Center in Superconductivity
University of Cambridge
West Cambridge Site Madingley Road
Cambridge CB3 OHE, United Cambridge

1) The different results from different experimenters can be explained in terms of the different flux line amplitudes involved.

If we have low displacements compared with the vortex spacing the system is linear and the critical state is not involved. As we cool the sample the resistance decreases and we get a peak in the loss when the skin depth is equal to the particle size. This peak in the loss is independent of amplitude, but depends on frequency and sample size. It occurs at a higher temperature than the zero resistivity point, or the point where the magnetic hysteresis goes to zero.

If we have large displacements we use the critical state model. As the sample is warmed the loss increases as J_c decreases, going through a peak as the critical state reaches the center of the sample. This peak is independent of frequency but depends on the amplitude and sample size. It occurs below the point of zero resistance and dc reversibility.

2) There are two irreversibility lines which occur at very different temperatures. One is the intergrain value when the junction $i_c\phi_0$ reaches kT, the other when the grains become reversible.

3) If it is due to a phase transition in the vortex lattice it must be independent of the direction of the current with respect to the crystal planes. Resistive measurements and recent inductive measurements by Asquinazi do not support this (F. de la Cruz and C Duran, Proc. 6th Int. Workshop on Critical Currents, Cambridge 1991, to be published in Super. Sci. and Tech.).

Comments upon Irreversibility Line Measurements

R. B. Flippen

E. I. Du Pont de Nemours and Company
Central Research and development Department
Experimental Station, PO Box 80228
Wilmington DE 19880-0228

During the special session discussing the irreversibility line (IL), it was evident that there was a misunderstanding concerning the use of the χ'' peak temperature as a function of magnetic field to define the irreversibility line. This peak is seen both as a function of the strength of the ac magnetic field (zero or small dc field present) and for a very weak ac field in the presence of much larger dc fields. Only the latter measurement relates to the IL, for the following reason: the peak in χ'' at a given temperature in the former situation is (in the critical state model) at the point where the magnetic field just penetrates to the center of the sample and is a function of the ac field strength and the geometry of the sample. The peak appears at this point because maximum magnetic energy absorption takes place here, and a peak will occur whether there is an IL or not in the material. On the other hand, for a weak (<1 Oe) ac field in the presence of a larger dc field, the change in magnetization of the sample due to the ac field represents a very minor hysteresis loop on the overall magnetization of the material; an ac measurement of χ in effect samples the magnetic state of the superconductor. However small the ac field, χ still exhibits a response to motion of magnetic flux in the magnetic state of the material, so that, for example, effects of flux pinning will be seen.

In the latter measurement above for a material with an IL, as the temperature is lowered, absorption as seen by the size of χ'' increases in the reversible region as flux motion becomes more sluggish with decreasing temperature. Below the IL flux pinning sets in, and as progressively more flux lines get tied up at pinning centers, less and less magnetic energy is absorbed and χ'' decreases. Thus χ'' should peak at the boundary between reversible and nonreversible (pinning) regions, i.e., at the IL, and the line can be traced by the χ'' peak temperature as a function of dc field. The IL determined in this way agrees closely with the IL determined by FC/ZFC SQUID measurements for a number of materials (to be published). Since the IL from χ'' measurements is frequency dependent, presumably from thermally activated processes, this comparison is made using a measurement frequency low enough to approximate a dc measurement.

Comments on the Irreversibility Line

Carlo Giovannella

Departimento de Fisica
Universita di Roma Tor Vergata
Via Emanuale Carnivale
00173 Rome, Italy

In this meeting a certain amount of time has been devoted to discuss the possible meaning, if there is any, of the irreversibility line measured first by K.A. Muller, M. Takashige and J.G. Bednorz.[1] The observation of irreversibility lines on the H-T plane, is not per se a signature of a phase transition. Very important, in fact, is also the dependence of the position of the irreversibility line on the H-T plane on the measuring frequency. If its position depends strongly on the measuring frequency, i.e. on the time window of the experiment, this line has probably nothing to do with critical phenomena. A phase transition occurs, by definition, because of a sudden change in the state of the system under investigation; and when it occurs, we should be able to define a critical region (in general very narrow in temperature), static and dynamical critical exponents and to observe a critical, i.e. a sharp, slowing down of the dynamics; all the above quantities can be experimentally measured and determined.

However, since the disorder, that is present in a large part of the physical systems, generates a distribution of coupling strengths, one might observe non-critical slowing down also in systems that one expects to undergone a critical transition. In this situation, is it still possible to define a critical transition and a scaling? The answer is yes, if one makes use of the concepts of self-similarity and percolation. The concept of self-similarity apply to the disorder, so that the system has to appear the same on whichever scale one chooses.

A percolation transition can be defined either in the real space or in the space of the configuration. The relation between the two has not yet been clarified. In the first case a critical transition occurs when a connected cluster, for example, of frozen spins (if we are considering a paramagnetic-spin glass, P-SG, transition from above) spans the whole volume of the system; and this in spite of the existence of a distribution of clusters of different sizes not yet frozen. The above scenario is that of the Fractal Cluster Model, FCM.[2] In the case of the configurational space, and considering always as example the P-SG transition, one can say that a percolation transition occurs when the connectivity of the space is broken, and only unconnected areas of accessible states remain.[3]

Granular superconductors, like the high critical temperature oxides, because of the presence of a large number of Josephson Junctions can be considered as an ensemble of X-Y spins characterized (because of the disorder) by a distribution of coupling strengths. Their behavior was examined in the framework of the FCM and the results were encouraging enough to suggest the occurrence of a percolation transition.[4] In the FCM framework, however, the irreversibility line has not any relevant meaning; it just tell to the experimentalist that its apparatus is sensitive enough to detect the first loop of junctions, or the first ensemble of loops of junctions, that, having their phases frozen are in the condition to trap magnetic flux. The definition of the experimental point at which the percolation transition occurs is not completely unambiguous and has been discussed in ref. 4.

The role played by the disorder, either positional or compositional, on the properties of the system can be well understood only coming back to ordered systems (using systems simulated at a computer) and introducing in them the disorder in a controlled way. Indeed, many of the characteristics of granular systems that have been experimentally measured, like for example hysteresis cycles with a butterfly form, can be well reproduced.[5]

The last point I want to stress is the need of performing measurements in a situation in which the system can be considered homogeneous and not disturbed by, for example, the applied field. In fact, we can not pretend to observe a critical transition if, for example, there exist a macroscopic field gradient in the sample. Critical transition, thus, can be studied only in very week magnetic fields and in the proximity of the critical transition (in the critical region), where the sample is in average fully penetrated by the field.

1. K.A. Müller, M. Takashige, J.G. Bednorz, Phys. Rev. Lett. 58:1143 (1987)
2. A.P. Malozemoff and B. Barbara, J. Appl. Phys. 57:3410 (1985) and references therein
3. I.A. Campbell, Phys. Rev. B 37:9800 (1988) and references therein
4. C. Giovannella, Phys. Stat. Sol. (b) 154:273 (1989) and A. Giannelli and C. Giovannella, Physica A 168:277 (1990) and references therein
5. A. Giannelli and C. Giovannella, J. of Less-Comm. Met. 164 & 165:1488 (1990) and references therein

Experimental Determination of the Irreversibility line: Criticism of some techniques and new proposals.

C. Rillo, R. Navarro

Instituto de Ciencia de Materiales de Aragon,
CSIC-Universidad de Zaragoza,
50009 Zaragoza, Spain.

Irreversibility effects appear in classical type II superconductors due to the pinning forces acting on the magnetic flux lines (vortex) and experimentally may be detected by the associated hysteresis effects. In high temperature superconductors the same characteristics holds, but before to reach the $H_{c2}(T)$ line of the H-T phase diagram the so called irreversibility line, $H_{irr}(T)$ is easily detected (Muller et al, 1987; Malozemoff et al, 1988). Above this boundary the vortex start to move quite freely. Without going into the discussion of the nature of $H_{irr}(T)$ it is clear that different experimental techniques and criteria used in its determination, even for the same sample, gives great differences in both shape and value. Consequently, many comparisons performed to elucidate the more appropriated theoretical model as well as its physical origin are lacking of the required unambiguous experimental results for $H_{irr}(T)$. Thus a critic revision of such techniques may give the necessary insight to the problem and this is the aim of this short discussion.

The most direct and easy to interpret measurements of $H_{irr}(T)$ are those associated with the magnetization hysteresis-loop. Other determinations using electric transport properties, to the above difficulties of the magnetic study, add the presence of Lorentz forces giving further complications. Common magnetic derivation of $H_{irr}(T)$ are the following:

i) Magnetization. Field cooled (FC) and zero field cooled (ZFC) dc magnetization, M_{FC} and M_{ZFC} respectively, has been used in the derivation of $H_{irr}(T)$. Upon increasing the temperature, the joining of the $M_{ZFC}(T)$ and $M_{FC}(T)$ curves recorded at a given field H_o takes place at a temperature, T_o, giving a point of the irreversibility line; $H_{irr}(T_o)=H_o$. The obvious experimental limit of this technique is the dispersion of the data (noise), $\Delta M(T)$ (Xu and Suenaga, 1991). Then, T_o is the temperature at which the difference $M_{ZFC}(T)$-$M_{FC}(T)$ becomes greater than $\Delta M(T)$. Indeed as $M_{ZFC}(T)$ and $M_{FC}(T)$ are smoothly joining curves the error bars in T_o are important and careful experiments with all this precautions are difficult and scarce (Suenaga et al., 1991).

ii) Ac susceptibilily. The maximum of the out-of-phase component, $\chi''(T)$ of the ac susceptibility, recorded at low frequency and some ac field amplitude, h_o, which holds at some temperature T_{max} determine $H_{irr}(T)$ by the relation $H_{irr}(T_{max})=h_o$ (Emmen et al,

1990). Moreover, also the onset temperature of $\chi''(T)$, T_{onset}, have been used; $H_{irr}(T_{onset})=h_o$ (Civale et al., 1991). The absorption χ'' is proportional to the ac magnetization hysteresis loop area which is non-zero in the presence of irreversibility. Obviously $\chi''(T)\neq O$ at $T>T_{max}$ and irreversibility may exist above T_{max} thus there is not physical reasons for the χ'' maximum criteria to define $H_{irr}(T)$ and its use give big contradictions as frequency dependent values. Furthermore, although in superconducting slabs and cylinders the simplest critical state model predicts $h^*(T_{max})=h_o^*$ (h^* being the field needed for full magnetic penetration) for other geometries and models this coincidence does not take place (Navarro and Campbell, 1991).

On basis of the above arguments, the use of the $\chi''(T)$ onset criteria will be more consistent. However, skin depth penetration effects may also give non zero $\chi''(T)$ values (Khoder, 1991) even in normal conductors. Thus the analysis of χ'' may be carefully performed and may give erroneous values.

iii) Non Linear effects. The onset of higher order non linear effects in the ac susceptibility may overcome some of the mentioned difficulties of χ'' in the determination of $H_{irr}(T)$ (Shaulov and Dorman, 1988). In particular the modulus of the third harmonic, $\chi_3(T)$, has the greater intensity and do not appear by skin penetration effects. Thus we have proposed (Angurel et al., 1991) its measurements for a better determination of $H_{irr}(T)$ using ac techniques. Preliminary experiments performed on thin films and sintered powder samples (Angurel et al., 1991) proves that this criteria yields field amplitude independent values decreasing the experimental uncertainties.

REFERENCES

Angurel L.A. et al., to be published in Proceedings EMRS-ICAM'91 (1991)
Civale L et al, Phys. Rev B.43 5425 (1991)
Emmen J.H.P.M. et al, Physica C 169 418 (1990)
Khoder A.F. Proceedings of the 54 workshop (1991)
Malozemoff A P et al, Phys. Rev. B 38 7203 (1988)
Muller K.A. et al, Phys. Rev. Lett. 58 408 (1987)
Navarro R. and Campbell L.J. submitted to Phys. Rev. 8(1991)
Shaulov A and Dorman D. Appl. Phys. Lett. 53 2680-(1988)
Suenaga M. et al, to be published in Supercond. Sci. Tech. (1991)
Xu Youwen and Suenaga M. Phys. Rev. B 43 5516 (1991)

PARTICIPANTS

Dr. E. Babic
Faculty of Science University of Zagreb
PO Box 162-Marullcev trg 19
41001 Zagreb Croatia
Yugoslavia

Dr. R. Budhani
Dept. of Appl. Science Building 480
Brookhaven National Laboratory
Upton New York 11973
USA

Dr. Joseph I. Budnick
Physics Department
University of Connecticut
Storrs CT 06269
USA

Dr. A. Campbell
Res. Ctr. in Supercond. Univ. of Cambridge
West Cambridge Site Madingley Road
Cambridge CB3 OHE
United Kingdom

Dr. J. Cave
Materials Technology Hydro-Quebec
1800 Montee Ste Julie
Varennes Province of Quebec
Canada

Dr. D. -X. Chen
Dept. de Fisica Grup d'Electromagnetisme
Edifici C Univ. Autonoma de Barcelona
08193 Bellaterra (Barcelona)
Spain

Dr. Quark. Chen
Honeywell Systems Research Center
1070 Lyndale Ave. South Bloomington
MN 55420
USA

Dr. L. Civale
IBM Thomas J. Watson Research Center
PO Box 218
Yorktown Heights NY 10598
USA

Dr. J. Claassen
Code 6344
Naval Research Laboratory
Washington DC 20375-5000
USA

Dr. J. Clem
A 517 Physics Dept.
Iowa State University
Ames IA 50011
USA

Dr. M. Couach
Centre D'Etudes Nucleaires de Grenoble
Avenue des Martyrs
85X 38041 Grenoble Cedex
France

Dr. J. Diederichs
Physics Department
Washington University
1 Brooking Drive Campus Box 1105
St. Louis Missouri 63130-4899

Dr. A. Fiory
AT&T Bell Laboratories
Research Physics Division ID 460
600 Mountain Ave. Murry Hill NJ 07974
USA

Dr. R. B. Flippen
E. I. Du Pont de Nemours and Company
Central Res. & Dev. Dept. Exp. Station
PO Box 80228
Wilmington DE 19880-0228

Dr. T. L. Francavilla
Code 6344
Naval Research Laboratory
Washington DC 20375-5000
USA

Dr. C. Giovannella
Departimento de Fisica
Universita di Roma Tor Vergata
Via E. Carnivale
00173 Roma Italia

Dr. R. B. Goldfarb
National Inst. of Standards and Tech.
325 Broadway
Boulder CO 80303
USA

Dr. F. Gömöry
Institute of Electrical Engineering
Slovak Academy of Sciences
842 39 Bratislava
Czechoslovakia

Dr. R. A. Hein
Department of Physics FM-15
University of Washington
Seattle WA 98195
USA

Dr. W. E. Henry
Department of Physics
Howard University
PO Box 761
Washington DC 20001

Dr. H. Khan
Fors. Inst. für Edelmetalle und Metallchemie
Katharinenstraza 17.7070
Schwabisch Gemund
Germany

Dr. A. F. Khoder
Centre D'Etudes Nucleaires de Grenoble
Avenue des Martyrs
85X 38041 Grenoble Cedex
France

Dr. B. Kim
Johns Hopkins Univ. - Appl. Phys. Lab.
Johns Hopkins Road
Laurel MD 20723
USA

Dr. J. K. Krause
Lakeshore Cryotronics Inc.
64 East Walnut Street
Westerville Ohio 43081
USA

Dr. S. Kumar
Quantum Magnetics
11578 Sorrento Valley Road Suite 30
SanDiego CA 92121
USA

Dr. D. Liebenberg
Office of Naval Research
800 N. Quincy Street
Arlington VA 22217-5000
USA

Dr. Xinsheng Ling
University of Connecticut
Physics Department
Storrs CT 06269
USA

Dr. C. Loegel
Lab. de Phys. et de Spect. Elect.
Faculte des Sciences et Techniques
4 rue des freres Luminere 68093 Mulhouse
Cedex France

Dr. I. Maartense
Mater. Lab. Wright Dev. Lab.
WRDC/MLPO
Ohio 45433-6533
USA

Dr. Z. Marohnic
Faculty of Science
University of Zagreb
PO Box 162-Marullcev trg 19
41001 Zagreb Croatia Yugoslavia

Dr. P. Martinoli
Inst. de Phys. Univ. de Neuchatel
Rue Breguet 1
CH- 2000 Neuchatel
Switzerland

Dr. H. Mazaki
Dept. of Math. and Phys. Nat. Defense
Academy
Hashirimizu 1-10-20 Yokosuhe 239
Kanagawa Prefecture
Japan

Dr. K. Moorjani
Johns Hopkins University - Appl. Phys.
Lab.
Johns Hopkins Road
Laurel MD 20723
USA

Dr. K. H. Muller
CSRIO Division of Applied Physics
PO Box 218
Lindfield NSW 2070
Australia

Dr. R. Navarro
Condensed Matter and Statistical Physics
MS-B262
Los Alamos National Laboratory
Los Alamos New Mexico 87545

Dr. Ch. Neumann
Universitat Konstanz Fakultat fur Physik
Universitatsstrasse10 Postfach 5560
D-7750 Konstanz 1
Germany

Dr. M. Nikolo
University of Colorado Physics Department
Campus Box 390
Boulder Colorado 80309
USA

Dr. A. R. Perry
Res. Ctr. in Supercond. Univ. of Cambridge
West Cambridge Site Madingly Road
Cambridge CB3 OHE
United Kingdom

Dr. E. Polturak
Department of Physics
Technion-Israel Institute of Technology
32 000 Haifa
Israel

Dr. S. Reich
Weizmann Institute of Science
Department of Materials Research
Rehovot 76100
Israel

Dr. C. Rillo
Inst. de Ciencia de Materiales de Aragon
CSIC-Univ. de Zaragoza Facultad de Ciencias
50009 Zaragoza
Spain

Dr. F. H. Salas
Prol. Colon # 816
Aquascalientes, Ags.
Mexico

Dr. A. Sanchez
Department de Fisica
Universitat Autonoma de Barcelona
08193 Bellaterra (Barcelona)
Spain

Dr. E. -W. Scheidt
Fach Berich Physik Freie Universitat Berlin
Institut fur Experimental Physik
Arnimalle 14 D-1000
Berlin 33 FRG

Dr. J. S. Schilling
Washington University
Campus Box 1105
One Brookings Drive
St. Louis Missouri 63130-4899

Dr. J. L. Smith
Los Alamos National Laboratory
MS K763
PO Box 1663 Los Alamos NM
87545

Dr. M. Suenaga
Boookhaven National Labotatory
Bldg. 480
Upton NY
11973

Dr. J. R. Thompson
Physics Department
University of Tennessee
Knoxville TN 37996-1200
USA

Dr. J. L. Tholence
CRTBT/CNRS 25 Avenue des Martyrs
166X-Centre de Tri
F 38042 Grenoble-Cedex
France

Dr. V. Wang
Lakeshore Cryotronics Inc.
64 East Walnut Street
Westerville Ohio 43081
USA

Dr. G. Williams
Department of Physics
University of Manitoba
Winnepeg MB R3T 2N2
Canada

Dr. T. Worthington
IBM Thomas J. Watson Research Center
PO Box 218
Yorktown Heights NY 10598
USA

Dr. D. G. Xenikos
Department of Physics
Ohio State University
174 West 18th Avenue
Columbus OH 43210

INDEX

* in-depth accounts
** this is the seminal critical state model (CSM) and is referred to throughout the proceedings.

GPSR Compliance

The European Union's (EU) General Product Safety Regulation (GPSR) is a set of rules that requires consumer products to be safe and our obligations to ensure this.

If you have any concerns about our products, you can contact us on ProductSafety@springernature.com

In case Publisher is established outside the EU, the EU authorized representative is:

Springer Nature Customer Service Center GmbH
Europaplatz 3
69115 Heidelberg, Germany

Batch number: 10798362

Printed by Printforce, the Netherlands